T0210790

Lecture Notes in Computer Science 8580

Beniamino Murgante Sanjay Misra
Ana Maria A.C. Rocha Carmelo Torre
Jorge Gustavo Rocha Maria Irene Falcão
David Taniar Bernady O. Apduhan
Osvaldo Gervasi (Eds.)

Computational Science and Its Applications – ICCSA 2014

14th International Conference
Guimarães, Portugal, June 30 – July 3, 2014
Proceedings, Part II

Volume Editors

Beniamino Murgante, University of Basilicata, Potenza, Italy
E-mail: beniamino.murgante@unibas.it

Sanjay Misra, Covenant University, Ota, Nigeria
E-mail: sanjay.misra@covenantuniversity.edu.ng

Ana Maria A.C. Rocha, University of Minho, Braga, Portugal
E-mail: arocha@dps.uminho.pt

Carmelo Torre, Politecnico di Bari, Bari, Italy
E-mail: torre@poliba.it

Jorge Gustavo Rocha, University of Minho, Braga, Portugal
E-mail: jgr@di.uminho.pt

Maria Irene Falcão, University of Minho, Braga, Portugal
E-mail: mif@math.uminho.pt

David Taniar, Monash University, Clayton, VIC, Australia
E-mail: david.taniar@infotech.monash.edu.au

Bernady O. Apduhan, Kyushu Sangyo University, Fukuoka, Japan
E-mail: bob@is.kyusan-u.ac.jp

Osvaldo Gervasi, University of Perugia, Perugia, Italy
E-mail: osvaldo.gervasi@unipg.it

ISSN 0302-9743 e-ISSN 1611-3349
ISBN 978-3-319-09128-0 e-ISBN 978-3-319-09129-7
DOI 10.1007/978-3-319-09129-7
Springer Cham Heidelberg New York Dordrecht London

Library of Congress Control Number: 2014942987

LNCS Sublibrary: SL 1 – Theoretical Computer Science and General Issues

Typesetting: Camera-ready by author, data conversion by Scientific Publishing Services, Chennai, India

Printed on acid-free paper

Springer is part of Springer Science+Business Media (www.springer.com)

Welcome Message

On behalf of the Local Organizing Committee of ICCSA 2014, it is a pleasure to welcome you to the 14th International Conference on Computational Science and Its Applications, held during June 30 – July 3, 2014. We are very proud and grateful to the ICCSA general chairs for having entrusted us with the task of organizing another event of this series of very successful conferences.

ICCSA will take place in the School of Engineering of University of Minho, which is located in close vicinity to the medieval city centre of Guimarães, a UNESCO World Heritage Site, in Northern Portugal. The historical city of Guimarães is recognized for its beauty and historical monuments. The dynamic and colorful Minho Region is famous for its landscape, gastronomy and vineyards where the unique *Vinho Verde* wine is produced.

The University of Minho is currently among the most prestigious institutions of higher education in Portugal and offers an excellent setting for the conference. Founded in 1973, the University has two major poles: the campus of Gualtar in Braga, and the campus of Azurém in Guimarães.

Plenary lectures by leading scientists and several workshops will provide a real opportunity to discuss new issues and find advanced solutions able to shape new trends in computational science.

Apart from the scientific program, a stimulant and diverse social program will be available. There will be a welcome drink at Instituto de Design, located in an old Tannery, that is an open knowledge centre and a privileged communication platform between industry and academia. Guided visits to the city of Guimarães and Porto are planned, both with beautiful and historical monuments. A guided tour and tasting in Porto wine cellars, is also planned. There will be a gala dinner at the Pousada de Santa Marinha, which is an old Augustinian convent of the 12th century refurbished, where ICCSA participants can enjoy delicious dishes and enjoy a wonderful view over the city of Guimarães.

The conference could not have happened without the dedicated work of many volunteers, recognized by the coloured shirts. We would like to thank all the collaborators, who worked hard to produce a successful ICCSA 2014, namely Irene Falcão and Maribel Santos above all, our fellow members of the local organization.

On behalf of the Local Organizing Committee of ICCSA 2014, it is our honor to cordially welcome all of you to the beautiful city of Guimarães for this unique event. Your participation and contribution to this conference will make it much more productive and successful.

We are looking forward to see you in Guimarães.

Sincerely yours,

Ana Maria A.C. Rocha
Jorge Gustavo Rocha

Preface

These 6 volumes (LNCS volumes 8579-8584) consist of the peer-reviewed papers from the 2014 International Conference on Computational Science and Its Applications (ICCSA 2014) held in Guimarães, Portugal during 30 June – 3 July 2014.

ICCSA 2014 was a successful event in the International Conferences on Computational Science and Its Applications (ICCSA) conference series, previously held in Ho Chi Minh City, Vietnam (2013), Salvador da Bahia, Brazil (2012), Santander, Spain (2011), Fukuoka, Japan (2010), Suwon, South Korea (2009), Perugia, Italy (2008), Kuala Lumpur, Malaysia (2007), Glasgow, UK (2006), Singapore (2005), Assisi, Italy (2004), Montreal, Canada (2003), and (as ICCS) Amsterdam, The Netherlands (2002) and San Francisco, USA (2001).

Computational science is a main pillar of most of the present research, industrial and commercial activities and plays a unique role in exploiting ICT innovative technologies, and the ICCSA conference series has been providing a venue for researchers and industry practitioners to discuss new ideas, to share complex problems and their solutions, and to shape new trends in computational science.

Apart from the general track, ICCSA 2014 also included 30 workshops, in various areas of computational sciences, ranging from computational science technologies, to specific areas of computational sciences, such as computational geometry and security. We accepted 58 papers for the general track, and 289 in workshops. We would like to show our appreciation to the workshops chairs and co-chairs.

The success of the ICCSA conference series, in general, and ICCSA 2014, in particular, was due to the support of many people: authors, presenters, participants, keynote speakers, workshop chairs, Organizing Committee members, student volunteers, Program Committee members, Advisory Committee members, international liaison chairs, and people in other various roles. We would like to thank them all.

We also thank our publisher, Springer–Verlag, for their acceptance to publish the proceedings and for their kind assistance and cooperation during the editing process.

We cordially invite you to visit the ICCSA website `http://www.iccsa.org` where you can find all relevant information about this interesting and exciting event.

June 2014

Osvaldo Gervasi
Jorge Gustavo Rocha
Bernady O. Apduhan

Organization

ICCSA 2014 was organized by University of Minho, (Portugal) University of Perugia (Italy), University of Basilicata (Italy), Monash University (Australia), Kyushu Sangyo University (Japan).

Honorary General Chairs

Antonio M. Cunha	Rector of the University of Minho, Portugal
Antonio Laganà	University of Perugia, Italy
Norio Shiratori	Tohoku University, Japan
Kenneth C. J. Tan	Qontix, UK

General Chairs

Beniamino Murgante	University of Basilicata, Italy
Ana Maria A.C. Rocha	University of Minho, Portugal
David Taniar	Monash University, Australia

Program Committee Chairs

Osvaldo Gervasi	University of Perugia, Italy
Bernady O. Apduhan	Kyushu Sangyo University, Japan
Jorge Gustavo Rocha	University of Minho, Portugal

International Advisory Committee

Jemal Abawajy	Daekin University, Australia
Dharma P. Agrawal	University of Cincinnati, USA
Claudia Bauzer Medeiros	University of Campinas, Brazil
Manfred M. Fisher	Vienna University of Economics and Business, Austria
Yee Leung	Chinese University of Hong Kong, China

International Liaison Chairs

Ana Carla P. Bitencourt	Universidade Federal do Reconcavo da Bahia, Brazil
Claudia Bauzer Medeiros	University of Campinas, Brazil
Alfredo Cuzzocrea	ICAR-CNR and University of Calabria, Italy

Marina L. Gavrilova	University of Calgary, Canada
Robert C. H. Hsu	Chung Hua University, Taiwan
Andrés Iglesias	University of Cantabria, Spain
Tai-Hoon Kim	Hannam University, Korea
Sanjay Misra	University of Minna, Nigeria
Takashi Naka	Kyushu Sangyo University, Japan
Rafael D.C. Santos	National Institute for Space Research, Brazil

Workshop and Session Organizing Chairs

Beniamino Murgante	University of Basilicata, Italy

Local Organizing Committee

Ana Maria A.C. Rocha	University of Minho, Portugal (Chair)
Jorge Gustavo Rocha	University of Minho, Portugal
Maria Irene Falcão	University of Minho, Portugal
Maribel Yasmina Santos	University of Minho, Portugal

Workshop Organizers

Advances in Complex Systems: Modeling and Parallel Implementation (ACSModPar 2014)

Georgius Sirakoulis	Democritus University of Thrace, Greece
Wiliam Spataro	University of Calabria, Italy
Giuseppe A. Trunfio	University of Sassari, Italy

Agricultural and Environment Information and Decision Support Systems (AEIDSS 2014)

Sandro Bimonte	IRSTEA France
Florence Le Ber	ENGES, France
André Miralles	IRSTEA France
François Pinet	IRSTEA France

Advances in Web Based Learning (AWBL 2014)

Mustafa Murat Inceoglu	Ege University, Turkey

Bio-inspired Computing and Applications (BIOCA 2014)

Nadia Nedjah | State University of Rio de Janeiro, Brazil
Luiza de Macedo Mourell | State University of Rio de Janeiro, Brazil

Computational and Applied Mathematics (CAM 2014)

Maria Irene Falcao | University of Minho, Portugal
Fernando Miranda | University of Minho, Portugal

Computer Aided Modeling, Simulation, and Analysis (CAMSA 2014)

Jie Shen | University of Michigan, USA

Computational and Applied Statistics (CAS 2014)

Ana Cristina Braga | University of Minho, Portugal
Ana Paula Costa Conceicao Amorim | University of Minho, Portugal

Computational Geometry and Security Applications (CGSA 2014)

Marina L. Gavrilova | University of Calgary, Canada
Han Ming Huang | Guangxi Normal University, China

Computational Algorithms and Sustainable Assessment (CLASS 2014)

Antonino Marvuglia | Public Research Centre Henri Tudor, Luxembourg
Beniamino Murgante | University of Basilicata, Italy

Chemistry and Materials Sciences and Technologies (CMST 2014)

Antonio Laganà | University of Perugia, Italy

Computational Optimization and Applications (COA 2014)

Ana Maria A.C. Rocha — University of Minho, Portugal
Humberto Rocha — University of Coimbra, Portugal

Cities, Technologies and Planning (CTP 2014)

Giuseppe Borruso — University of Trieste, Italy
Beniamino Murgante — University of Basilicata, Italy

Computational Tools and Techniques for Citizen Science and Scientific Outreach (CTTCS 2014)

Rafael Santos — National Institute for Space Research, Brazil
Jordan Raddickand — Johns Hopkins University, USA
Ani Thakar — Johns Hopkins University, USA

Econometrics and Multidimensional Evaluation in the Urban Environment (EMEUE 2014)

Carmelo M. Torre — Polytechnic of Bari, Italy
Maria Cerreta — University of Naples Federico II, Italy
Paola Perchinunno — University of Bari, Italy
Simona Panaro — University of Naples Federico II, Italy
Raffaele Attardi — University of Naples Federico II, Italy

Future Computing Systems, Technologies, and Applications (FISTA 2014)

Bernady O. Apduhan — Kyushu Sangyo University, Japan
Rafael Santos — National Institute for Space Research, Brazil
Jianhua Ma — Hosei University, Japan
Qun Jin — Waseda University, Japan

Formal Methods, Computational Intelligence and Constraint Programming for Software Assurance (FMCICA 2014)

Valdivino Santiago Junior — National Institute for Space Research (INPE), Brazil

Geographical Analysis, Urban Modeling, Spatial Statistics (GEOG-AN-MOD 2014)

Giuseppe Borruso	University of Trieste, Italy
Beniamino Murgante	University of Basilicata, Italy
Hartmut Asche	University of Potsdam, Germany

High Performance Computing in Engineering and Science (HPCES 2014)

Alberto Proenca	University of Minho, Portugal
Pedro Alberto	University of Coimbra, Portugal

Mobile Communications (MC 2014)

Hyunseung Choo	Sungkyunkwan University, Korea

Mobile Computing, Sensing, and Actuation for Cyber Physical Systems (MSA4CPS 2014)

Saad Qaisar	NUST School of Electrical Engineering and Computer Science, Pakistan
Moonseong Kim	Korean Intellectual Property Office, Korea

New Trends on Trust Computational Models (NTTCM 2014)

Rui Costa Cardoso	Universidade da Beira Interior, Portugal
Abel Gomez	Universidade da Beira Interior, Portugal

Quantum Mechanics: Computational Strategies and Applications (QMCSA 2014)

Mirco Ragni	Universidad Federal de Bahia, Brazil
Vincenzo Aquilanti	University of Perugia, Italy
Ana Carla Peixoto Bitencourt	Universidade Estadual de Feira de Santana Brazil
Roger Anderson	University of California, USA
Frederico Vasconcellos Prudente	Universidad Federal de Bahia, Brazil

Remote Sensing Data Analysis, Modeling, Interpretation and Applications: From a Global View to a Local Analysis (RS2014)

Rosa Lasaponara	Institute of Methodologies for Environmental Analysis National Research Council, Italy
Nicola Masini	Archaeological and Monumental Heritage Institute, National Research Council, Italy

Software Engineering Processes and Applications (SEPA 2014)

Sanjay Misra	Covenant University, Nigeria

Software Quality (SQ 2014)

Sanjay Misra	Covenant University, Nigeria

Advances in Spatio-Temporal Analytics (ST-Analytics 2014)

Joao Moura Pires	New University of Lisbon, Portugal
Maribel Yasmina Santos	New University of Lisbon, Portugal

Tools and Techniques in Software Development Processes (TTSDP 2014)

Sanjay Misra	Covenant University, Nigeria

Virtual Reality and its Applications (VRA 2014)

Osvaldo Gervasi	University of Perugia, Italy
Lucio Depaolis	University of Salento, Italy

Workshop of Agile Software Development Techniques (WAGILE 2014)

Eduardo Guerra	National Institute for Space Research, Brazil

Big Data:, Analytics and Management (WBDAM 2014)

Wenny Rahayu	La Trobe University, Australia

Program Committee

Jemal Abawajy	Daekin University, Australia
Kenny Adamson	University of Ulster, UK
Filipe Alvelos	University of Minho, Portugal
Paula Amaral	Universidade Nova de Lisboa, Portugal
Hartmut Asche	University of Potsdam, Germany
Md. Abul Kalam Azad	University of Minho, Portugal
Michela Bertolotto	University College Dublin, Ireland
Sandro Bimonte	CEMAGREF, TSCF, France
Rod Blais	University of Calgary, Canada
Ivan Blecic	University of Sassari, Italy
Giuseppe Borruso	University of Trieste, Italy
Yves Caniou	Lyon University, France
José A. Cardoso e Cunha	Universidade Nova de Lisboa, Portugal
Leocadio G. Casado	University of Almeria, Spain
Carlo Cattani	University of Salerno, Italy
Mete Celik	Erciyes University, Turkey
Alexander Chemeris	National Technical University of Ukraine “KPI”, Ukraine
Min Young Chung	Sungkyunkwan University, Korea
Gilberto Corso Pereira	Federal University of Bahia, Brazil
M. Fernanda Costa	University of Minho, Portugal
Gaspar Cunha	University of Minho, Portugal
Alfredo Cuzzocrea	ICAR-CNR and University of Calabria, Italy
Carla Dal Sasso Freitas	Universidade Federal do Rio Grande do Sul, Brazil
Pradesh Debba	The Council for Scientific and Industrial Research (CSIR), South Africa
Hendrik Decker	Instituto Tecnológico de Informática, Spain
Frank Devai	London South Bank University, UK
Rodolphe Devillers	Memorial University of Newfoundland, Canada
Prabu Dorairaj	NetApp, India/USA
M. Irene Falcao	University of Minho, Portugal
Cherry Liu Fang	U.S. DOE Ames Laboratory, USA
Edite M.G.P. Fernandes	University of Minho, Portugal
Jose-Jesus Fernandez	National Centre for Biotechnology, CSIS, Spain
Maria Antonia Forjaz	University of Minho, Portugal
Maria Celia Furtado Rocha	PRODEB and Universidade Federal da Bahia, Brazil
Akemi Galvez	University of Cantabria, Spain
Paulino Jose Garcia Nieto	University of Oviedo, Spain
Marina Gavrilova	University of Calgary, Canada
Jerome Gensel	LSR-IMAG, France

Maria Giaoutzi	National Technical University, Athens, Greece
Andrzej M. Goscinski	Deakin University, Australia
Alex Hagen-Zanker	University of Cambridge, UK
Malgorzata Hanzl	Technical University of Lodz, Poland
Shanmugasundaram Hariharan	B.S. Abdur Rahman University, India
Eligius M.T. Hendrix	University of Malaga/Wageningen University, Spain/Netherlands
Hisamoto Hiyoshi	Gunma University, Japan
Fermin Huarte	University of Barcelona, Spain
Andres Iglesias	University of Cantabria, Spain
Mustafa Inceoglu	EGE University, Turkey
Peter Jimack	University of Leeds, UK
Qun Jin	Waseda University, Japan
Farid Karimipour	Vienna University of Technology, Austria
Baris Kazar	Oracle Corp., USA
DongSeong Kim	University of Canterbury, New Zealand
Taihoon Kim	Hannam University, Korea
Ivana Kolingerova	University of West Bohemia, Czech Republic
Dieter Kranzlmueller	LMU and LRZ Munich, Germany
Antonio Laganà	University of Perugia, Italy
Rosa Lasaponara	National Research Council, Italy
Maurizio Lazzari	National Research Council, Italy
Cheng Siong Lee	Monash University, Australia
Sangyoun Lee	Yonsei University, Korea
Jongchan Lee	Kunsan National University, Korea
Clement Leung	Hong Kong Baptist University, Hong Kong
Chendong Li	University of Connecticut, USA
Gang Li	Deakin University, Australia
Ming Li	East China Normal University, China
Fang Liu	AMES Laboratories, USA
Xin Liu	University of Calgary, Canada
Savino Longo	University of Bari, Italy
Tinghuai Ma	NanJing University of Information Science and Technology, China
Sergio Maffioletti	University of Zurich, Switzerland
Ernesto Marcheggiani	Katholieke Universiteit Leuven, Belgium
Antonino Marvuglia	Research Centre Henri Tudor, Luxembourg
Nicola Masini	National Research Council, Italy
Nirvana Meratnia	University of Twente, The Netherlands
Alfredo Milani	University of Perugia, Italy
Sanjay Misra	Federal University of Technology Minna, Nigeria
Giuseppe Modica	University of Reggio Calabria, Italy

José Luis Montaña	University of Cantabria, Spain
Beniamino Murgante	University of Basilicata, Italy
Jiri Nedoma	Academy of Sciences of the Czech Republic, Czech Republic
Laszlo Neumann	University of Girona, Spain
Kok-Leong Ong	Daekin University, Australia
Belen Palop	Universidad de Valladolid, Spain
Marcin Paprzycki	Polish Academy of Sciences, Poland
Eric Pardede	La Trobe University, Australia
Kwangjin Park	Wonkwang University, Korea
Ana Isabel Pereira	Polytechnic Institute of Braganca, Portugal
Maurizio Pollino	Italian National Agency for New Technologies, Energy and Sustainable Economic Development, Italy
Alenka Poplin	University of Hamburg, Germany
Vidyasagar Potdar	Curtin University of Technology, Australia
David C. Prosperi	Florida Atlantic University, USA
Wenny Rahayu	La Trobe University, Australia
Jerzy Respondek	Silesian University of Technology, Poland
Ana Maria A.C. Rocha	University of Minho, Portugal
Humberto Rocha	INESC-Coimbra, Portugal
Alexey Rodionov	Institute of Computational Mathematics and Mathematical Geophysics, Russia
Cristina S. Rodrigues	University of Minho, Portugal
Octavio Roncero	CSIC, Spain
Maytham Safar	Kuwait University, Kuwait
Chiara Saracino	A.O. Ospedale Niguarda Ca' Granda - Milano, Italy
Haiduke Sarafian	The Pennsylvania State University, USA
Jie Shen	University of Michigan, USA
Qi Shi	Liverpool John Moores University, UK
Dale Shires	U.S. Army Research Laboratory, USA
Takuo Suganuma	Tohoku University, Japan
Sergio Tasso	University of Perugia, Italy
Ana Paula Teixeira	University of Tras-os-Montes and Alto Douro, Portugal
Senhorinha Teixeira	University of Minho, Portugal
Parimala Thulasiraman	University of Manitoba, Canada
Carmelo Torre	Polytechnic of Bari, Italy
Javier Martinez Torres	Centro Universitario de la Defensa Zaragoza, Spain
Giuseppe A. Trunfio	University of Sassari, Italy
Unal Ufuktepe	Izmir University of Economics, Turkey
Toshihiro Uchibayashi	Kyushu Sangyo University, Japan

Mario Valle	Swiss National Supercomputing Centre, Switzerland
Pablo Vanegas	University of Cuenca, Equador
Piero Giorgio Verdini	INFN Pisa and CERN, Italy
Marco Vizzari	University of Perugia, Italy
Koichi Wada	University of Tsukuba, Japan
Krzysztof Walkowiak	Wroclaw University of Technology, Poland
Robert Weibel	University of Zurich, Switzerland
Roland Wismüller	Universität Siegen, Germany
Mudasser Wyne	SOET National University, USA
Chung-Huang Yang	National Kaohsiung Normal University, Taiwan
Xin-She Yang	National Physical Laboratory, UK
Salim Zabir	France Telecom Japan Co., Japan
Haifeng Zhao	University of California at Davis, USA
Kewen Zhao	University of Qiongzhou, China
Albert Y. Zomaya	University of Sydney, Australia

Reviewers

Abdi Samane	University College Cork, Ireland
Aceto Lidia	University of Pisa, Italy
Afonso Ana Paula	University of Lisbon, Portugal
Afreixo Vera	University of Aveiro, Portugal
Aguilar Antonio	University of Barcelona, Spain
Aguilar José Alfonso	Universidad Autónoma de Sinaloa, Mexico
Ahmad Waseem	Federal University of Technology Minna, Nigeria
Aktas Mehmet	Yildiz Technical University, Turkey
Alarcon Vladimir	Universidad Diego Portales, Chile
Alberti Margarita	University of Barcelona, Spain
Ali Salman	NUST, Pakistan
Alvanides Seraphim	Northumbria University, UK
Álvarez Jacobo de Uña	University of Vigo, Spain
Alvelos Filipe	University of Minho, Portugal
Alves Cláudio	University of Minho, Portugal
Alves José Luis	University of Minho, Portugal
Amorim Ana Paula	University of Minho, Portugal
Amorim Paulo	Federal University of Rio de Janeiro, Brazil
Anderson Roger	University of California, USA
Andrade Wilkerson	Federal University of Campina Grande, Brazil
Andrienko Gennady	Fraunhofer Institute for Intelligent Analysis and Informations Systems, Germany
Apduhan Bernady	Kyushu Sangyo University, Japan
Aquilanti Vincenzo	University of Perugia, Italy
Argiolas Michele	University of Cagliari, Italy

Athayde Maria Emília Feijão Queiroz	University of Minho, Portugal
Attardi Raffaele	University of Napoli Federico II, Italy
Azad Md Abdul	Indian Institute of Technology Kanpur, India
Badard Thierry	Laval University, Canada
Bae Ihn-Han	Catholic University of Daegu, South Korea
Baioletti Marco	University of Perugia, Italy
Balena Pasquale	Polytechnic of Bari, Italy
Balucani Nadia	University of Perugia, Italy
Barbosa Jorge	University of Porto, Portugal
Barrientos Pablo Andres	Universidad Nacional de La Plata, Australia
Bartoli Daniele	University of Perugia, Italy
Bação Fernando	New University of Lisbon, Portugal
Belanzoni Paola	University of Perugia, Italy
Bencardino Massimiliano	University of Salerno, Italy
Benigni Gladys	University of Oriente, Venezuela
Bertolotto Michela	University College Dublin, Ireland
Bimonte Sandro	IRSTEA, France
Blanquer Ignacio	Universitat Politècnica de València, Spain
Bollini Letizia	University of Milano, Italy
Bonifazi Alessandro	Polytechnic of Bari, Italy
Borruso Giuseppe	University of Trieste, Italy
Bostenaru Maria	"Ion Mincu" University of Architecture and Urbanism, Romania
Boucelma Omar	University Marseille, France
Braga Ana Cristina	University of Minho, Portugal
Brás Carmo	Universidade Nova de Lisboa, Portugal
Cacao Isabel	University of Aveiro, Portugal
Cadarso-Suárez Carmen	University of Santiago de Compostela, Spain
Caiaffa Emanuela	ENEA, Italy
Calamita Giuseppe	National Research Council, Italy
Campagna Michele	University of Cagliari, Italy
Campobasso Francesco	University of Bari, Italy
Campos José	University of Minho, Portugal
Cannatella Daniele	University of Napoli Federico II, Italy
Canora Filomena	University of Basilicata, Italy
Cardoso Rui	Institute of Telecommunications, Portugal
Caschili Simone	University College London, UK
Ceppi Claudia	Polytechnic of Bari, Italy
Cerreta Maria	University Federico II of Naples, Italy
Chanet Jean-Pierre	IRSTEA, France
Chao Wang	University of Science and Technology of China, China
Choi Joonsoo	Kookmin University, South Korea

Choo Hyunseung	Sungkyunkwan University, South Korea
Chung Min Young	Sungkyunkwan University, South Korea
Chung Myoungbeom	Sungkyunkwan University, South Korea
Clementini Eliseo	University of L'Aquila, Italy
Coelho Leandro dos Santos	PUC-PR, Brazil
Colado Anibal Zaldivar	Universidad Autónoma de Sinaloa, Mexico
Coletti Cecilia	University of Chieti, Italy
Condori Nelly	VU University Amsterdam, The Netherlands
Correia Elisete	University of Trás-Os-Montes e Alto Douro, Portugal
Correia Filipe	FEUP, Portugal
Correia Florbela Maria da Cruz Domingues	Instituto Politécnico de Viana do Castelo, Portugal
Correia Ramos Carlos	University of Evora, Portugal
Corso Pereira Gilberto	UFPA, Brazil
Cortés Ana	Universitat Autònoma de Barcelona, Spain
Costa Fernanda	University of Minho, Portugal
Costantini Alessandro	INFN, Italy
Crasso Marco	National Scientific and Technical Research Council, Argentina
Crawford Broderick	Universidad Catolica de Valparaiso, Chile
Cristia Maximiliano	CIFASIS and UNR, Argentina
Cunha Gaspar	University of Minho, Portugal
Cunha Jácome	University of Minho, Portugal
Cutini Valerio	University of Pisa, Italy
Danese Maria	IBAM, CNR, Italy
Da Silva B. Carlos	University of Lisboa, Portugal
De Almeida Regina	University of Trás-os-Montes e Alto Douro, Portugal
Debroy Vidroha	Hudson Alley Software Inc., USA
De Fino Mariella	Polytechnic of Bari, Italy
De Lotto Roberto	University of Pavia, Italy
De Paolis Lucio Tommaso	University of Salento, Italy
De Rosa Fortuna	University of Napoli Federico II, Italy
De Toro Pasquale	University of Napoli Federico II, Italy
Decker Hendrik	Instituto Tecnológico de Informática, Spain
Delamé Thomas	CNRS, France
Demyanov Vasily	Heriot-Watt University, UK
Desjardin Eric	University of Reims, France
Dwivedi Sanjay Kumar	Babasaheb Bhimrao Ambedkar University, India
Di Gangi Massimo	University of Messina, Italy
Di Leo Margherita	JRC, European Commission, Belgium

Di Trani Francesco	University of Basilicata, Italy
Dias Joana	University of Coimbra, Portugal
Dias d'Almeida Filomena	University of Porto, Portugal
Dilo Arta	University of Twente, The Netherlands
Dixit Veersain	Delhi University, India
Doan Anh Vu	Université Libre de Bruxelles, Belgium
Dorazio Laurent	ISIMA, France
Dutra Inês	University of Porto, Portugal
Eichelberger Hanno	University of Tuebingen, Germany
El-Zawawy Mohamed A.	Cairo University, Egypt
Escalona Maria-Jose	University of Seville, Spain
Falcão M. Irene	University of Minho, Portugal
Farantos Stavros	University of Crete and FORTH, Greece
Faria Susana	University of Minho, Portugal
Faruq Fatma	Carnegie Melon University,, USA
Fernandes Edite	University of Minho, Portugal
Fernandes Rosário	University of Minho, Portugal
Fernandez Joao P	Universidade da Beira Interior, Portugal
Ferreira Fátima	University of Trás-Os-Montes e Alto Douro, Portugal
Ferrão Maria	University of Beira Interior and CEMAPRE, Portugal
Figueiredo Manuel Carlos	University of Minho, Portugal
Filipe Ana	University of Minho, Portugal
Flouvat Frederic	University New Caledonia, New Caledonia
Forjaz Maria Antónia	University of Minho, Portugal
Formosa Saviour	University of Malta, Malta
Fort Marta	University of Girona, Spain
Franciosa Alfredo	University of Napoli Federico II, Italy
Freitas Adelaide de Fátima Baptista Valente	University of Aveiro, Portugal
Frydman Claudia	Laboratoire des Sciences de l'Information et des Systèmes, France
Fusco Giovanni	CNRS - UMR ESPACE, France
Fussel Donald	University of Texas at Austin, USA
Gao Shang	Zhongnan University of Economics and Law, China
Garcia Ernesto	University of the Basque Country, Spain
Garcia Tobio Javier	Centro de Supercomputación de Galicia (CESGA), Spain
Gavrilova Marina	University of Calgary, Canada
Gensel Jerome	IMAG, France
Geraldi Edoardo	National Research Council, Italy
Gervasi Osvaldo	University of Perugia, Italy

Giaoutzi Maria	National Technical University Athens, Greece
Gizzi Fabrizio	National Research Council, Italy
Gomes Maria Cecilia	Universidade Nova de Lisboa, Portugal
Gomes dos Anjos Eudisley	Federal University of ParaÃba, Brazil
Gomez Andres	Centro de Supercomputación de Galicia, CESGA (Spain)
Gonçalves Arminda Manuela	University of Minho, Portugal
Gravagnuolo Antonia	University of Napoli Federico II, Italy
Gregori M. M. H. Rodrigo	Universidade Tecnológica Federal do Paraná, Brazil
Guerlebeck Klaus	Bauhaus University Weimar, Germany
Guerra Eduardo	National Institute for Space Research, Brazil
Hagen-Zanker Alex	University of Surrey, UK
Hajou Ali	Utrecht University, The Netherlands
Hanzl Malgorzata	University of Lodz, Poland
Heijungs Reinout	VU University Amsterdam, The Netherlands
Henriques Carla	Escola Superior de Tecnologia e Gestão, Portugal
Herawan Tutut	University of Malaya, Malaysia
Iglesias Andres	University of Cantabria, Spain
Jamal Amna	National University of Singapore, Singapore
Jank Gerhard	Aachen University, Germany
Jiang Bin	University of Gävle, Sweden
Kalogirou Stamatis	Harokopio University of Athens, Greece
Kanevski Mikhail	University of Lausanne, Switzerland
Kartsaklis Christos	Oak Ridge National Laboratory, USA
Kavouras Marinos	National Technical University of Athens, Greece
Khan Murtaza	NUST, Pakistan
Khurshid Khawar	NUST, Pakistan
Kim Deok-Soo	Hanyang University, South Korea
Kim Moonseong	KIPO, South Korea
Kolingerova Ivana	University of West Bohemia, Czech Republic
Kotzinos Dimitrios	Université de Cergy-Pontoise, France
Lazzari Maurizio	CNR IBAM, Italy
Laganà Antonio	Department of Chemistry, Biology and Biotechnology, Italy
Lai Sabrina	University of Cagliari, Italy
Lanorte Antonio	CNR-IMAA, Italy
Lanza Viviana	Lombardy Regional Institute for Research, Italy
Le Duc Tai	Sungkyunkwan University, South Korea
Le Duc Thang	Sungkyunkwan University, South Korea
Lee Junghoon	Jeju National University, South Korea

Lee KangWoo	Sungkyunkwan University, South Korea
Legatiuk Dmitrii	Bauhaus University Weimar, Germany
Leonard Kathryn	California State University, USA
Lin Calvin	University of Texas at Austin, USA
Loconte Pierangela	Technical University of Bari, Italy
Lombardi Andrea	University of Perugia, Italy
Lopez Cabido Ignacio	Centro de Supercomputación de Galicia, CESGA
Lourenço Vanda Marisa	University Nova de Lisboa, Portugal
Luaces Miguel	University of A Coruña, Spain
Lucertini Giulia	IUAV, Italy
Luna Esteban Robles	Universidad Nacional de la Plata, Argentina
Machado Gaspar	University of Minho, Portugal
Magni Riccardo	Pragma Engineering SrL, Italy, Italy
Malonek Helmuth	University of Aveiro, Portugal
Manfreda Salvatore	University of Basilicata, Italy
Manso Callejo Miguel Angel	Universidad Politécnica de Madrid, Spain
Marcheggiani Ernesto	KU Lueven, Belgium
Marechal Bernard	Universidade Federal de Rio de Janeiro, Brazil
Margalef Tomas	Universitat Autònoma de Barcelona, Spain
Martellozzo Federico	University of Rome, Italy
Marvuglia Antonino	Public Research Centre Henri Tudor, Luxembourg
Matos Jose	Instituto Politecnico do Porto, Portugal
Mauro Giovanni	University of Trieste, Italy
Mauw Sjouke	University of Luxembourg, Luxembourg
Medeiros Pedro	Universidade Nova de Lisboa, Portugal
Melle Franco Manuel	University of Minho, Portugal
Melo Ana	Universidade de São Paulo, Brazil
Millo Giovanni	Generali Assicurazioni, Italy
Min-Woo Park	Sungkyunkwan University, South Korea
Miranda Fernando	University of Minho, Portugal
Misra Sanjay	Covenant University, Nigeria
Modica Giuseppe	Università Mediterranea di Reggio Calabria, Italy
Morais João	University of Aveiro, Portugal
Moreira Adriano	University of Minho, Portugal
Mota Alexandre	Universidade Federal de Pernambuco, Brazil
Moura Pires João	Universidade Nova de Lisboa - FCT, Portugal
Mourelle Luiza de Macedo	UERJ, Brazil
Mourão Maria	Polytechnic Institute of Viana do Castelo, Portugal
Murgante Beniamino	University of Basilicata, Italy
NM Tuan	Ho Chi Minh City University of Technology, Vietnam

Nagy Csaba	University of Szeged, Hungary
Nash Andrew	Vienna Transport Strategies, Austria
Natário Isabel Cristina Maciel	University Nova de Lisboa, Portugal
Nedjah Nadia	State University of Rio de Janeiro, Brazil
Nogueira Fernando	University of Coimbra, Portugal
Oliveira Irene	University of Trás-Os-Montes e Alto Douro, Portugal
Oliveira José A.	University of Minho, Portugal
Oliveira e Silva Luis	University of Lisboa, Portugal
Osaragi Toshihiro	Tokyo Institute of Technology, Japan
Ottomanelli Michele	Polytechnic of Bari, Italy
Ozturk Savas	TUBITAK, Turkey
Pacifici Leonardo	University of Perugia, Italy
Pages Carmen	Universidad de Alcala, Spain
Painho Marco	New University of Lisbon, Portugal
Pantazis Dimos	Technological Educational Institute of Athens, Greece
Paolotti Luisa	University of Perugia, Italy
Papa Enrica	University of Amsterdam, The Netherlands
Papathanasiou Jason	University of Macedonia, Greece
Pardede Eric	La Trobe University, Australia
Parissis Ioannis	Grenoble INP - LCIS, France
Park Gyung-Leen	Jeju National University, South Korea
Park Sooyeon	Korea Polytechnic University, South Korea
Pascale Stefania	University of Basilicata, Italy
Passaro Pierluigi	University of Bari Aldo Moro, Italy
Peixoto Bitencourt Ana Carla	Universidade Estadual de Feira de Santana, Brazil
Perchinunno Paola	University of Bari, Italy
Pereira Ana	Polytechnic Institute of Bragança, Portugal
Pereira Francisco	Instituto Superior de Engenharia, Portugal
Pereira Paulo	University of Minho, Portugal
Pereira Ricardo	Portugal Telecom Inovacao, Portugal
Pietrantuono Roberto	University of Napoli "Federico II", Italy
Pimentel Carina	University of Aveiro, Portugal
Pina Antonio	University of Minho, Portugal
Pinet Francois	IRSTEA, France
Piscitelli Claudia	Polytechnic University of Bari, Italy
Piñar Miguel	Universidad de Granada, Spain
Pollino Maurizio	ENEA, Italy
Potena Pasqualina	University of Bergamo, Italy
Prata Paula	University of Beira Interior, Portugal
Prosperi David	Florida Atlantic University, USA
Qaisar Saad	NURST, Pakistan

Quan Tho	Ho Chi Minh City University of Technology, Vietnam
Raffaeta Alessandra	University of Venice, Italy
Ragni Mirco	Universidade Estadual de Feira de Santana, Brazil
Rautenberg Carlos	University of Graz, Austria
Ravat Franck	IRIT, France
Raza Syed Muhammad	Sungkyunkwan University, South Korea
Ribeiro Isabel	University of Porto, Portugal
Ribeiro Ligia	University of Porto, Portugal
Rinzivillo Salvatore	University of Pisa, Italy
Rocha Ana Maria	University of Minho, Portugal
Rocha Humberto	University of Coimbra, Portugal
Rocha Jorge	University of Minho, Portugal
Rocha Maria Clara	ESTES Coimbra, Portugal
Rocha Maria	PRODEB, San Salvador, Brazil
Rodrigues Armanda	Universidade Nova de Lisboa, Portugal
Rodrigues Cristina	DPS, University of Minho, Portugal
Rodriguez Daniel	University of Alcala, Spain
Roh Yongwan	Korean IP, South Korea
Roncaratti Luiz	Instituto de Fisica, University of Brasilia, Brazil
Rosi Marzio	University of Perugia, Italy
Rossi Gianfranco	University of Parma, Italy
Rotondo Francesco	Polytechnic of Bari, Italy
Sannicandro Valentina	Polytechnic of Bari, Italy
Santos Maribel Yasmina	University of Minho, Portugal
Santos Rafael	INPE, Brazil
Santos Viviane	Universidade de São Paulo, Brazil
Santucci Valentino	University of Perugia, Italy
Saracino Gloria	University of Milano-Bicocca, Italy
Sarafian Haiduke	Pennsylvania State University, USA
Saraiva João	University of Minho, Portugal
Sarrazin Renaud	Université Libre de Bruxelles, Belgium
Schirone Dario Antonio	University of Bari, Italy
Schneider Michel	ISIMA, France
Schoier Gabriella	University of Trieste, Italy
Schutz Georges	CRP Henri Tudor, Luxembourg
Scorza Francesco	University of Basilicata, Italy
Selmaoui Nazha	University of New Caledonia, New Caledonia
Severino Ricardo Jose	University of Minho, Portugal
Shakhov Vladimir	Russian Academy of Sciences, Russia
Shen Jie	University of Michigan, USA
Shon Minhan	Sungkyunkwan University, South Korea

Shukla Ruchi	University of Johannesburg, South Africa
Silva J.C.	IPCA, Portugal
Silva de Souza Laudson	Federal University of Rio Grande do Norte, Brazil
Silva-Fortes Carina	ESTeSL-IPL, Portugal
Simão Adenilso	Universidade de São Paulo, Brazil
Singh R K	Delhi University, India
Soares Inês	INESC Porto, Portugal
Soares Maria Joana	University of Minho, Portugal
Soares Michel	Federal University of Sergipe, Brazil
Sobral Joao	University of Minho, Portugal
Son Changhwan	Sungkyunkwan University, South Korea
Sproessig Wolfgang	Technical University Bergakademie Freiberg, Germany
Su Le Hoanh	Ho Chi Minh City Technical University, Vietnam
Sá Esteves Jorge	University of Aveiro, Portugal
Tahar Sofiène	Concordia University, Canada
Tanaka Kazuaki	Kyushu Institute of Technology, Japan
Taniar David	Monash University, Australia
Tarantino Eufemia	Polytechnic of Bari, Italy
Tariq Haroon	Connekt Lab, Pakistan
Tasso Sergio	University of Perugia, Italy
Teixeira Ana Paula	University of Trás-Os-Montes e Alto Douro, Portugal
Teixeira Senhorinha	University of Minho, Portugal
Tesseire Maguelonne	IRSTEA, France
Thorat Pankaj	Sungkyunkwan University, South Korea
Tomaz Graça	Polytechnic Institute of Guarda, Portugal
Torre Carmelo Maria	Polytechnic of Bari, Italy
Trunfio Giuseppe A.	University of Sassari, Italy
Urbano Joana	LIACC University of Porto, Portugal
Vasconcelos Paulo	University of Porto, Portugal
Vella Flavio	University of Rome La Sapienza, Italy
Velloso Pedro	Universidade Federal Fluminense, Brazil
Viana Ana	INESC Porto, Portugal
Vidacs Laszlo	MTA-SZTE, Hungary
Vieira Ramadas Gisela	Polytechnic of Porto, Portugal
Vijay NLankalapalli	National Institute for Space Research, Brazil
Villalba Maite	Universidad Europea de Madrid, Spain
Viqueira José R.R.	University of Santiago de Compostela, Spain
Vona Marco	University of Basilicata, Italy

Wachowicz Monica	University of New Brunswick, Canada
Walkowiak Krzysztof	Wroclav University of Technology, Poland
Xin Liu	Ecole Polytechnique Fédérale Lausanne, Switzerland
Yadav Nikita	Delhi Universty, India
Yatskevich Mikalai	Assioma, Italy
Yeoum Sanggil	Sungkyunkwan University, South Korea
Zalyubovskiy Vyacheslav	Russian Academy of Sciences, Russia
Zunino Alejandro	Universidad Nacional del Centro, Argentina

Sponsoring Organizations

ICCSA 2014 would not have been possible without the tremendous support of many organizations and institutions, for which all organizers and participants of ICCSA 2014 express their sincere gratitude:

Universidade do Minho
Escola de Engenharia

Universidade do Minho
(http://www.uminho.pt)

University of Perugia, Italy
(http://www.unipg.it)

University of Basilicata, Italy (http://www.unibas.it)

Monash University, Australia
(http://monash.edu)

Kyushu Sangyo University, Japan
(www.kyusan-u.ac.jp)

Associação Portuguesa de Investigação Operacional
(apdio.pt)

Table of Contents

Workshop on Computational Optimization and Applications (COA 2014)

IMRT Beam Angle Optimization Using DDS with a Cross-Validation Approach for Configuration Selection 1
Joana M. Dias, Humberto Rocha, Brígida Ferreira, and Maria do Carmo Lopes

IMRT Beam Angle Optimization Using Non-descent Pattern Search Methods 17
Humberto Rocha, Joana M. Dias, Brígida Ferreira, and Maria do Carmo Lopes

Maximizing Expectation on Vertex-Disjoint Cycle Packing 32
João Pedro Pedroso

On Modelling Approaches for Planning and Scheduling in Food Processing Industry 47
G.D.H. Claassen and Eligius M.T. Hendrix

A Multiobjective Approach for a Dynamic Simple Plant Location Problem under Uncertainty 60
Joana M. Dias and Maria do Céu Marques

Joint Scheduling and Optimal Charging of Electric Vehicles Problem 76
Ons Sassi and Ammar Oulamara

Automatic Clustering Using a Genetic Algorithm with New Solution Encoding and Operators 92
Carolina Raposo, Carlos Henggeler Antunes, and João Pedro Barreto

On Simplicial Longest Edge Bisection in Lipschitz Global Optimization 104
Juan F.R. Herrera, Leocadio G. Casado, Eligius M.T. Hendrix, and Inmaculada García

Heuristics to Reduce the Number of Simplices in Longest Edge Bisection Refinement of a Regular n-Simplex 115
Guillermo Aparicio, Leocadio G. Casado, Boglárka G-Tóth, Eligius M.T. Hendrix, and Inmaculada García

Multiple Roots of Systems of Equations by Repulsion Merit Functions 126
Gisela C.V. Ramadas, Edite M.G.P. Fernandes, and Ana Maria A.C. Rocha

Branch and Bound Based Coordinate Search Filter Algorithm for Nonsmooth Nonconvex Mixed-Integer Nonlinear Programming Problems 140
Florbela P. Fernandes, M. Fernanda P. Costa, and Edite M.G.P. Fernandes

Solving Multilocal Optimization Problems with a Recursive Parallel Search of the Feasible Region 154
Ana I. Pereira and José Rufino

Stiction Detection and Quantification as an Application of Optimization 169
Ana S.R. Brásio, Andrey Romanenko, and Natércia C.P. Fernandes

On the Properties of General Dual-Feasible Functions 180
Jürgen Rietz, Cláudio Alves, José Manuel Valério de Carvalho, and François Clautiaux

A Global Optimization Approach Applied to Structural Dynamic Updating 195
Marco Dourado, José Meireles, and Ana Maria A.C. Rocha

A Hybrid Heuristic Based on Column Generation for Two- and Three-Stage Bin Packing Problems 211
Filipe Alvelos, Elsa Silva, and José Manuel Valério de Carvalho

Experiments with Firefly Algorithm 227
Rogério B. Francisco, M. Fernanda P. Costa, and Ana Maria A.C. Rocha

A New Branch-and-Price Approach for the Kidney Exchange Problem 237
Xenia Klimentova, Filipe Alvelos, and Ana Viana

A Study of the Complexity of an Infeasible Predictor-Corrector Variant of Mehrotra Algorithm 253
Ana Paula Teixeira and Regina Almeida

A Multi-start Tabu Search Approach for Solving the Information Routing Problem 267
Hela Masri, Saoussen Krichen, and Adel Guitouni

IMRT Beam Angle Optimization Using Electromagnetism-Like Algorithm 278
Humberto Rocha, Ana Maria A.C. Rocha, Joana M. Dias, Brígida Ferreira, and Maria do Carmo Lopes

Improving Branch-and-Price for Parallel Machine Scheduling 290
Manuel Lopes, Filipe Alvelos, and Henrique Lopes

Workshop on Computational Geometry and Security Applications (CGSA 2014)

Fast Parallel Triangulation Algorithm of Large Data Sets in E^2 and E^3 for In-Core and Out-Core Memory Processing 301
Michal Smolik and Vaclav Skala

A Robust Key Management Scheme Based on Node Hierarchy for Wireless Sensor Networks 315
A.S.M. Sanwar Hosen, Gideon, and Gi-hwan Cho

A Method to Triangulate a Set of Points in the Plane 330
Taras Agryzkov, José L. Oliver, Leandro Tortosa, and José F. Vicent

3D Network Traffic Monitoring Based on an Automatic Attack Classifier 342
Diego Roberto Colombo Dias, José Remo Ferreira Brega, Luis Carlos Trevelin, Bruno Barberi Gnecco, João Paulo Papa, and Marcelo de Paiva Guimarães

Topology Preserving Algorithms for Implicit Surfaces Simplifying and Sewing 352
Aruquia Peixoto and Carlos A. de Moura

Closest-Point Queries for Complex Objects 368
Eugene Greene and Asish Mukhopadhyay

How Similar Are Quasi-, Regular, and Delaunay Triangulations in $\mathbb{R}^3$? 381
Donguk Kim, Youngsong Cho, Jae-Kwan Kim, Yuan-Shin Lee, and Deok-Soo Kim

Workshop on Cities, Technologies and Planning (CTP 2014)

WebGIS Solution for Crisis Management Support – Case Study of Olomouc Municipality 394
Rostislav Netek and Marek Balun

Sensing World Heritage: An Exploratory Study of Twitter as a Tool for Assessing Reputation 404
Vasco Monteiro, Roberto Henriques, Marco Painho, and Eric Vaz

Municipal Building Regulations for Energy Efficiency in Southern Italy 420
Eleonora Riva Sanseverino, Raffaella Riva Sanseverino, Gianluca Scaccianoce, and Valentina Vaccaro

On Fractal Complexity of Built and Natural Landscapes 437
Andrei Bourchtein, Ludmila Bourchtein, and Natalia Naoumova

Using Ontologies to Support Land-Use Spatial Data Interoperability.... 453
Falk Würriehausen, Ashish Karmacharya, and Hartmut Müller

Crowdsourced Monitoring, Citizen Empowerment and Data Credibility: The Case of Observations.be 469
Jihad Farah

City as Commons: Study of Shared Visions by Communities on Facebook.. 486
Maria Célia Furtado Rocha, Pablo Vieira Florentino, and Gilberto Corso Pereira

Evaluating Urban Development Plans: A New Method in the Toolbox.. 502
Valerio Cutini

Characteristics of Sprawl in the Naples Metropolitan Area. Indications for Controlling and Monitoring Urban Transformations 520
Rocco Papa and Giuseppe Mazzeo

ClickOnMap: A Framework to Develop Volunteered Geographic Information Systems with Dynamic Metadata 532
Wagner Dias de Souza, Jugurta Lisboa-Filho, Jean Henrique de Sousa Câmara, Jarbas Nunes Vidal Filho, and Alcione de Paiva Oliveira

Government Tools for Urban Regeneration: The Cities Plan in Italy. A Critical Analysis of the Results and the Proposed Alternative 547
Antonio Nesticò and Gianluigi De Mare

A Model for the Economic Evaluation of Energetic Requalification Projects in Buildings. A Real Case Application 563
Antonio Nesticò, Gianluigi De Mare, Pierfrancesco Fiore, and Ornella Pipolo

The Paradigm of the Modern City: *SMART and SENSEable Cities* for Smart, Inclusive and Sustainable Growth 579
Ilaria Greco and Massimiliano Bencardino

The Geographic Turn in Social Media: Opportunities for Spatial Planning and Geodesign 598
Michele Campagna

An Agent-Based Model as a Tool of Planning at a Sub-regional Scale ... 611
Fernando Pereira da Fonseca, Rui A.R. Ramos, and Antônio Nélson Rodrigues da Silva

Some Preliminary Remarks on the Recreational Business District in the City of Sassari: A Social Network Approach 629
Silvia Battino, Giuseppe Borruso, and Carlo Donato

Orienteering and Orienteering Yourself. User Centered Design Methodologies Applied to Geo-referenced Interactive Ecosystems 642
Letizia Bollini

User Experience & Usability for Mobile Geo-referenced Apps. A Case Study Applied to Cultural Heritage Field 652
Letizia Bollini, Rinaldo De Palma, Rossella Nota, and Riccardo Pietra

PALEOBAS: A Geo-application for Mobile Phones – A New Method of Knowledge and Public Protection of the Paleontological Heritage of Basilicata (Southern Italy) 663
Maurizio Lazzari, Agostino Lecci, and Nicola Lecci

GIS Assessment and Planning of Conservation Priorities of Historical Centers through Quantitative Methods of Vulnerability Analysis: An Example from Southern Italy 677
Maurizio Lazzari, Maria Serena Patriziano, and Giovanna Alessia Aliano

A Methodological Approach to Integrate Ontology and Configurational Analysis 693
Antonia Cataldo, Valerio Di Pinto, and Antonio M. Rinaldi

If Appleseed Had an Open Portal: Making Sense of Data, SEIS and Integrated Systems for the Maltese Islands 709
Saviour Formosa

Involving Citizens in Public Space Regeneration: The Experience of "Garden in Motion" 723
Sara Lorusso, Michele Scioscia, Gerardo Sassano, Antonio Graziadei, Pasquale Passannante, Sara Bellarosa, Francesco Scaringi, and Beniamino Murgante

Smart City or Smurfs City 738
Beniamino Murgante and Giuseppe Borruso

Territorial Specialization in Attracting Local Development Funds: An Assessment Procedure Based on Open Data and Open Tools 750
Francesco Scorza and Giuseppe Las Casas

Using Spatiotemporal Analysis in Urban Sprawl Assessment and Prediction 758
Federico Amato, Piergiuseppe Pontrandolfi, and Beniamino Murgante

A New Design Method for Managing Spatial Vagueness in Classical Relational Spatial OLAP Architectures 774
Elodie Edoh-Alove, Sandro Bimonte, and Yvan Bédard

Growing Sustainable Behaviors in Local Communities through Smart Monitoring Systems for Energy Efficiency: RENERGY Outcomes 787
Francesco Scorza, Alessandro Attolico, Vincenzo Moretti, Rosalia Smaldone, Domenico Donofrio, and Giuseppe Laguardia

Author Index 795

IMRT Beam Angle Optimization Using DDS with a Cross-Validation Approach for Configuration Selection

Joana M. Dias[1,2], Humberto Rocha[2], Brígida Ferreira[3,4], and Maria do Carmo Lopes[3,4]

[1] Faculdade de Economia, Universidade de Coimbra, 3004-512 Coimbra, Portugal
[2] Inesc-Coimbra, Rua Antero de Quental, 199, 3000-033 Coimbra, Portugal
[3] I3N, Departamento de Física, Universidade de Aveiro, 3810-193 Aveiro, Portugal
[4] Serviço de Física Médica, IPOC-FG, EPE, 3000-075 Coimbra, Portugal
joana@fe.uc.pt, hrocha@mat.uc.pt, brigida@ua.pt, mclopes@ipocoimbra.min-saude.pt

Abstract. Radiation incidences (angles) that are used in Intensity Modulated Radiation Therapy (IMRT) treatments have a significant influence in the treatment clinical outcome. In clinical practice, the angles are usually chosen after a lengthy trial and error procedure that is significantly dependent on the planner's experience and time availability. The use of optimization models and algorithms can be an important contribution to the treatment planning, improving the quality of the solution reached and decreasing the time spent on the process. This paper describes a Dynamically Dimensioned Search (DDS) approach for IMRT beam angle optimization. Several different sets of parameters and search options were analyzed. Computational tests show that the final outcome is strongly influenced by these choices. This motivated the use of a cross-validation based procedure for choosing the algorithm's configuration, considering a set of ten retrospective treated cases of head-and-neck tumors at the Portuguese Institute of Oncology of Coimbra.

Keywords: Dynamically Dimensioned Search, IMRT, Beam Angle Optimization, Derivative-Free Methods.

1 Introduction

Radiation therapy is one of the treatments used for cancer patients. Its aim is to destroy cancer cells through radiation, but at the same time spare healthy tissue that can also be damaged by radiation. The patient is usually immobilized on a coach that can rotate. The radiation is delivered through the use of a linear accelerator mounted on a gantry that can rotate along a central axis parallel to the couch. The rotation of the couch combined with the rotation of the gantry allows radiation from almost any angle around the tumor. Intensity Modulated Radiation Therapy is one type of radiation therapy where it is possible to modulate the radiation intensities that are delivered to the patient from each radiation incidence. This modulation is achieved through the use of a multileaf collimator. The collimator has left and right leaves that can block radiation. By moving these leaves it is possible to create different intensity profiles

B. Murgante et al. (Eds.): ICCSA 2014, Part II, LNCS 8580, pp. 1–16, 2014.

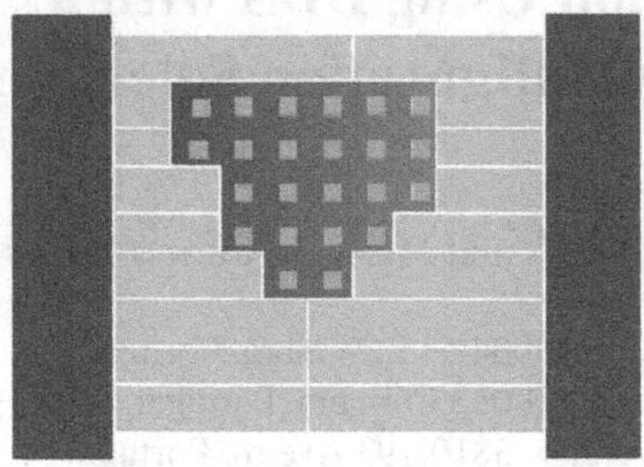

Fig. 1. Illustration of a multileaf collimator (with nine pairs of leaves)

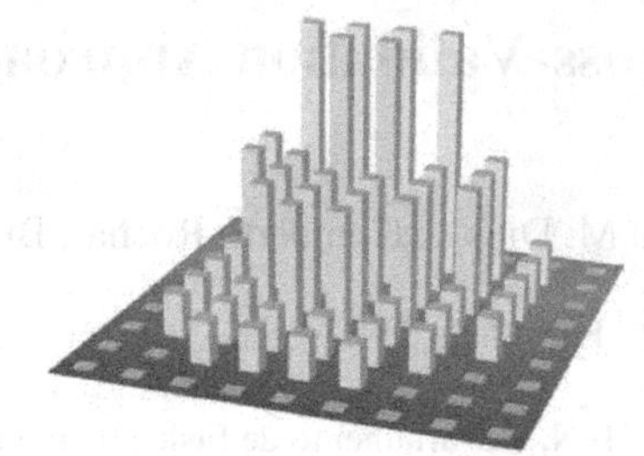

Fig. 2. Illustration of a beamlet intensity map (9×9)

(Fig. 1 and Fig. 2). Conceptually, this is equivalent to consider that, instead of having one single radiation beam from each radiation incidence used in the treatment, we can have a discretization of this beam into beamlets, each one with a given radiation intensity.

The possibility of modulating the radiation intensities increases the precision of the treatment and can be very important in diminishing treatment's side effects as it is possible to better spare cells that we do not want to irradiate. Nevertheless, it requires a complex planning procedure where many different and interconnected decisions have to be made by the planner, beginning by deciding how many and which angles to use in the treatment (the best angles can often be non-intuitive).

Whenever a patient is referred to an IMRT treatment, the medical doctor will delineate in the patient's computed tomographies (CT) the structures of interest: areas that should be treated plus a safety margin (Planning Target Volumes – PTV) and also the organs that should be spared (Organs at Risk - OAR). The medical doctor will also establish the medical prescription (Table 1), defining the desired dose for PTVs and mean or maximum doses to OARs (that sometimes are not possible to achieve). It is then up to the medical physicists to plan the treatment, by interacting with a Treatment Planning System (TPS) that simulates the behavior of the linear accelerator and calculates the radiation dose that will be deposited in the patient. By a trial and error procedure, several different treatment parameters are tried until a treatment plan that is considered admissible by the planner and that the planner thinks it is hard to improve is reached.

Table 1. Prescribed doses for all the structures considered for IMRT optimization

Structure	Mean dose	Maximum Dose	Prescribed Dose
Spinal cord	–	45 Gy	–
Brainstem	–	54 Gy	–
Left parotid	26 Gy	–	–
Right parotid	26 Gy	–	–
PTV1	–	–	70.0 Gy
PTV2	–	–	59.4 Gy
Body	–	80 Gy	–

Most of the times, the quality of the treatment plan is dependent on the planner's experience and time availability.

Another approach is to consider inverse planning, where the trial and error procedure is totally or partially replaced by the use of mathematical models and optimization algorithms. The mathematical models are characterized by being large-scale, non-linear and multi-modal, with objective functions that are computationally expensive to calculate. Global optimization algorithms that are derivative-based can easily be trapped in one of the many existing local minima.

In this paper we are concerned with the definition of the best radiation angles to use, considering the number of angles fixed *a priori* (Beam Angle Optimization – BAO). The BAO problem has been tackled using several different methodologies: scoring methods ([1]); methods based on the concept of beam's eye view ([2, 3, 4, 5]); response surface approaches ([6]); derivative-free approaches ([7]); mixed integer programming approaches ([8]); simulated annealing ([2, 9, 10]); particle swarm optimization ([11]); genetic algorithms ([12, 13, 14]) among others (see, for instance, [15, 16, 17, 18]).

In this paper, we consider an approach based on Dynamically Dimensioned Search (DDS). Several computational tests were done to understand the influence of the algorithm's configuration in the final outcome. The choice of the best configuration to use is not trivial, and we propose the use of a cross-validation procedure. This work follows a first preliminary experiment, where only one set of parameters was tested and that showed encouraging results [19].

The next section describes the mathematical model used. Section 3 describes the DDS algorithm. Computational results are shown and discussed in section 4. Section 5 states the main conclusions and presents future research ideas.

2 BAO Mathematical Model

The treatment planning process consists in determining, for a given patient, the angles that will be used in the treatment (BAO), the radiation intensities (fluences) for each of the angles (fluence map optimization - FMO), and the way that the multileaf collimator leaves should move to produce the desired fluence patterns (segmentation).

In this paper we are concerned with the BAO problem and we consider that the number of angles to use, k, will be fixed *a priori* by the planner. This means that we aim at finding out which is the best set of k angles out of every possible combination. For each beam angle set, we have to find a way of calculating the quality of this set. This can only be done after performing the fluence optimization, where the best radiation intensities for each of the angles considered will be calculated. To solve this FMO, the patient is discretized into voxels (small volume elements) and the radiation dose that is deposited in each of the patient's voxels is computed using the superposition principle, i.e., considering the contribution of each beamlet. Typically, a dose matrix D is constructed from the collection of all beamlet intensities, by indexing the rows of D to each voxel and the columns to each beamlet, i.e., the number of rows of matrix D equals the number of voxels (V) and the number of columns equals the number of beamlets (N) from all beam directions considered. Therefore we can say that the total dose received by the voxel i is given by $\sum_{j=1}^{N} D_{ij} w_j$, with w_j the weight of beamlet j. Usually, the total number of voxels considered reaches the tens of thousands. If we define Θ as the set of all possible angles, then a basic formulation for the BAO problem can be defined as follows:

$$\min f(\theta_1, \theta_2, \cdots, \theta_k) \tag{1}$$

$$\text{subject to } \theta_1, \cdots, \theta_k \in \Theta \tag{2}$$

Many mathematical optimization models and algorithms have been proposed for the FMO problem and it is out of the scope of this paper to discuss the pros and cons of those models. In this paper, a convex penalty function voxel-based nonlinear model is used [20], such that, each voxel is penalized considering the square difference of the amount of dose received by the voxel and the amount of dose desired/allowed for the voxel. This formulation yields a quadratic programming problem with only linear nonegativity constraints on the fluence values [21]. Let T_i be desired dose for voxel i, $\underline{\lambda}_i$ and $\overline{\lambda}_i$ the penalty weights of underdose and overdose of voxel i, respectively, and $(\bullet)_+ = \max\{0, \bullet\}$. Then the model can be defined as follows:

$$Min_w \sum_{i=1}^{V} \left[\underline{\lambda}_i \left(T_i - \sum_{j=1}^{N} D_{ij} w_j \right)_+^2 + \overline{\lambda}_i \left(\sum_{j=1}^{N} D_{ij} w_j - T_i \right)_+^2 \right] \tag{3}$$

$$\text{s.t. } w_j \geq 0, j = 1, \cdots, N \tag{4}$$

Although this formulation allows unique weights for each voxel, weights are assigned by structure only so that every voxel in a given structure has the weight assigned to that structure divided by the number of voxels of the structure [21]. This nonlinear formulation implies that a very small amount of underdose or overdose may be accepted in clinical decision making, but larger deviations from the desired/allowed doses are decreasingly tolerated. Objective function (1) is calculated by (3), considering only beamlets that belong to $(\theta_1, \theta_2, \cdots, \theta_k)$. This objective function is computationally expensive to calculate, taking up to a few minutes, depending on the patient itself and the number of angles considered.

3 DDS Algorithm

Considering the BAO problem, the DDS algorithm has as main advantages the fact that it is derivative-free, being able to escape from local minima, and the fact that it is possible to define *a priori* the number of objective function evaluations that will be performed. This is especially important when dealing with a computationally expensive objective function.

The DDS algorithm begins with any admissible solution of the problem that becomes the current solution. In each iteration the algorithm finds a new solution by randomly perturbing the current one. Whenever a better solution is found, it becomes the current solution that, in turn, will be perturbed. The DDS algorithm can be interpreted as a random search process, considering searchable neighborhoods that will decrease in size as the algorithm iterates. This will promote a more global search at the beginning of the search and a more local search in the final iterations. In each iteration, each variable (angle) will be perturbed with a given probability. This probability decreases with the increase in the number of iterations, so that less and less angles are changed as the algorithm progresses. The magnitudes of the perturbations are randomly sampled from a normal distribution with mean 0. It is not necessary to consider an upper or lower bound for each variable, since an angle of -10º, for instance, is equal to 350º or an angle of 370º is equal to 10º. In our implementation of the algorithm we followed [22], considering some adaptations described in [23]. The algorithm's parameters are as follows:

- *r_init* represents the initial standard deviation considered;
- *r_max* and *r_min* represent the maximum and minimum admissible standard deviations considered;
- *N* represents the maximum number of iterations (an upper limit to the number of objective function evaluations, since in each iteration at most one solution is evaluated);
- *l_success* and *l_failure* determine a change in the current standard deviation due to successive successful or unsuccessful iterations (a success meaning that the objective function value has improved).

The algorithm has as input an admissible solution to the problem (that can be randomly generated) and returns as output an improved admissible solution (it is not possible to guarantee that it is optimal). The algorithm behavior can be described as follows:

1. Set counter $i \leftarrow 1$; Define the initial admissible solution *x_current* and evaluate this solution (*f_current*). *f_best*←*f_current*; *x_best*←*x_current*; *success*←0; *failure*←0; *r*←*r_min*.
2. Calculate the probability of any given variable (angle) be perturbed as $p(i) = 1 - \ln(i)/\ln(N)$. For each decision variable *x_best*(*j*), *j*=1,…,*k*, add the variable to the set *J* with probability *p*(*i*).
3. For every variable *x_best*(*j*), $j \in J$, perturb randomly this variable considering a normal distribution $N(0,r)$. This perturbed solution will constitute the new *x_current*.
4. Evaluate *x_current*. If *f_current*<*f_best*, then *f_best*←*f_current*; *x_best*←*x_current* ; *success*← *success* +1 and *failure*←0. Else *success*←0 and *failure*← *failure*+1.

5. If $failure \geq l\_failure$ then $r \leftarrow \min(r/2, r\_min)$.
6. If $success \geq l\_success$ then $r \leftarrow \max(2r, r\_max)$.
7. $i \leftarrow i+1$. If $i \geq N$ then stop, else go to 2.

Steps 2 and 3 of the algorithm are responsible for calculating a new current solution in a random manner (by randomly deciding which angles to perturb and the magnitude of the perturbation). Given the specificities of the BAO problem, we also guarantee that the current solution does not have two adjacent angles that are too near each other. From a clinical point of view, angles that are less than 4º apart are considered the same. The evaluation of the current solution in step 4 is done by resorting to the optimization of the FMO problem, considering the angles defined by the current solution. Step 5 introduces a dynamic in the DDS algorithm considering that after some failed trials it is time to look for solutions in a narrower neighborhood, and when there are successful trials the searchable neighborhood can be wider.

4 Computational Experiments

The DDS algorithm was tested considering ten clinical examples of retrospective treated cases of head-and-neck tumors at the Portuguese Institute of Oncology of Coimbra (IPOC). A typical head-and-neck treatment plan consists of radiation delivered from 5 to 9 equally spaced coplanar orientations around the patient. The optimization of the angles has an increased importance when fewer angles are used. Being able to deliver a high quality treatment with fewer angles is beneficial for both the patient and the health institution. From the institution point of view, fewer angles mean faster treatment times, so that more patients can be treated. From the patient point of view, the faster the treatment the better because it is more likely that the patient does not change his position significantly during the treatment, which contributes to more accurate treatment results. For these reasons, treatments with 5 coplanar beams were considered.

In order to facilitate convenient access, visualization and analysis of patient treatment planning data, as well as dosimetric data input for treatment plan optimization research, the computational tools developed within MATLAB and CERR – computational environment for radiotherapy research ([24]) are used widely for IMRT treatment planning research. The ORART – operations research applications in radiation therapy ([25]) collaborative working group developed a series of software routines that allow access to influence matrices, which provide the necessary dosimetry data to perform optimization in IMRT. CERR was elected as the main software platform to embody our optimization research. Our tests were performed on a Intel Core i7 CPU 2.8 GHz computer with 4GB RAM and Windows 7. We used CERR 3.2.2 version and MATLAB 7.4.0 (R2007a). The dose was computed using CERR's pencil beam algorithm (QIB). For each of the ten head-and-neck cases, the sample rate used for Body was 32 and for the remaining structures was 4, resulting in 20,874 to 24,158 voxels and 948 to 1,283 beamlets for the 5-beam equispaced coplanar treatment plans. An automatized procedure for dose computation for each given beam angle set was developed, instead of the traditional dose computation available from IMRTP module accessible from CERR's menubar. This automatization of the dose computation was essential for integration in our DDS algorithm. To address the convex nonlinear formulation of the FMO problem we used a trust-region-reflective algorithm (*fmincon*) of MATLAB 7.4.0 (R2007a) Optimization Toolbox. For this set of patients, each instance of the FMO problem can take from 56 seconds to 350 seconds to be calculated, depending on the patient and on the set of beam angles considered.

4.1 Clinical Examples

The selected clinical examples were signalized at IPOC as complex cases where proper target coverage and organ sparing, in particular parotid sparing, proved to be difficult to obtain. The patients' CT sets and delineated structures were exported via Dicom RT to CERR. Since the head-and-neck region is a complex area where, e.g., the parotid glands are usually in close proximity to or even overlapping with the target volume, careful selection of the radiation incidence directions can be determinant to obtain a satisfying treatment plan.

The spinal cord and the brainstem are some of the most critical organs at risk (OARs) in the head-and-neck tumor cases. These are serial organs, i.e., organs such that if only one subunit is damaged, the whole organ functionality is compromised. Therefore, if the tolerance dose is exceeded, it may result in functional damage to the whole organ. Thus, it is extremely important not to exceed the tolerance dose prescribed for these types of organs. Other than the spinal cord and the brainstem, the parotid glands are also important OARs. The parotid gland is the largest of the three salivary glands. A common complication due to parotid glands irradiation is xerostomia (the medical term for dry mouth due to lack of saliva). This decreases the quality of life of patients undergoing radiation therapy of head-and-neck, causing difficulties to swallow. The parotids are parallel organs, i.e., if a small volume of the organ is damaged, the rest of the organ functionality may not be affected. Their tolerance dose depends strongly on the fraction of the volume irradiated. Hence, if only a small fraction of the organ is irradiated the tolerance dose is much higher than if a larger fraction is irradiated. Thus, for these parallel structures, the organ mean dose is generally used instead of the maximum dose as an objective for inverse planning. In general, the head-and-neck region is a complex area to treat with radiotherapy due to the large number of sensitive organs in this region (e.g., eyes, mandible, larynx, oral cavity, etc.). For simplicity, in this study, the OARs used for treatment optimization were limited to the spinal cord, the brainstem and the parotid glands. For the head-and-neck cases in study the PTV was separated in two parts with different prescribed doses: PTV1 and PTV2. The prescription dose for the target volumes and tolerance doses for the OARs considered in the optimization are presented in Table 1. The parotid glands are in close proximity to or even overlapping with the PTV which helps explaining the difficulty of parotid sparing. Adequate beam directions can help on the overall optimization process and in particular in parotid sparing.

4.2 Results

For each BAO problem, the DDS algorithm was executed considering different configurations for the algorithm. The parameters that are expected to have a greater impact in the algorithm's outcome are the initial standard deviation (*r_init*), *l_failure* and *l_success*. The last two are responsible for the evolution of the *r* parameter. However, after some preliminary tests, it was possible to conclude that there are seldom two consecutive successful iterations, so that if *l_success* takes values greater than 1 this is equivalent to never changing *r* according to step 6 of the algorithm. Parameters *r_max* and *r_min* can and should be defined considering the specificities of the problem. In this case it was considered *r_max* equal to 90º and *r_min* equal to 3º. As we are randomly perturbing an angle using a normal distribution of mean 0 and standard deviation *r*, we know that there is 95% of probability of generating a perturbation value that belongs to the interval [-2*r*, 2*r*]. Notice that the greatest perturbation that is interesting to consider is 180º. Table 2 presents the values that were considered for parameters *r_init* and *l_failure*. Regarding *l_success*, it was considered to be fixed to 1 for the reasons exposed. The choice of the *r_init* values was motivated by the number of angles considered and the equidistant solution where all angles are 72º apart.

Table 2. Values of the Parameters

r_init	18	36	72
l_failure	5	20	

It is also interesting to consider the DDS algorithm when *r* stays constant throughout the algorithm's execution. This means that steps 5 and 6 are not considered.

The choice of increasing *r* in successive successful iterations and decreasing it after a sequence of failed iterations is an option that can be as justifiable as doing exactly the opposite. Notice that the algorithm's convergence is being guaranteed by the fact that the probability of perturbing each variable decreases iteration after iteration. So, we chose to also test this different version of the algorithm (Steps 5 and 6 will be replaced by Steps 5a and 6a).

5 a. If *failure* ≥ *l_failure* then *r*= max(2*r*,*r_max*).
6 a. If *success*≥*l_success* then *r*= min(*r*/2,*r_min*).

We have also tested a simpler rule, where *r* is randomly generated after *l_failure* successive failed iterations (Steps 5 and 6 are replaced by Step 5b).

5 b. If *failure* ≥ *l_failure* then *r* is randomly generated using a uniform distribution in [*r_min*, *r_max*].

For each of the ten patients, and for each version of the algorithm, five different runs were considered because of the random nature of the algorithm. A total of *N* equal to 200 iterations was considered. The initial solution considered was always the equidistant solution, as this is most of the times also the solution used in clinical practice. So, we are interested in measuring the improvement of the objective function value of the final solution (*f*DDS) when compared with the equidistant initial one (*fequi*). This improvement is calculated as $(fequi - fDDS)/fequi$. Before showing the global computational results, it is also worth to look at the influence of each parameter in the algorithm's behavior.

When the standard deviation *r* is kept constant, then we should expect a smooth behavior with smaller *r_init* values. Fig. 3 depicts the situation for a run of the algorithm considering patient 5. This patient was randomly selected, and similar behaviors are observed in the other patients.

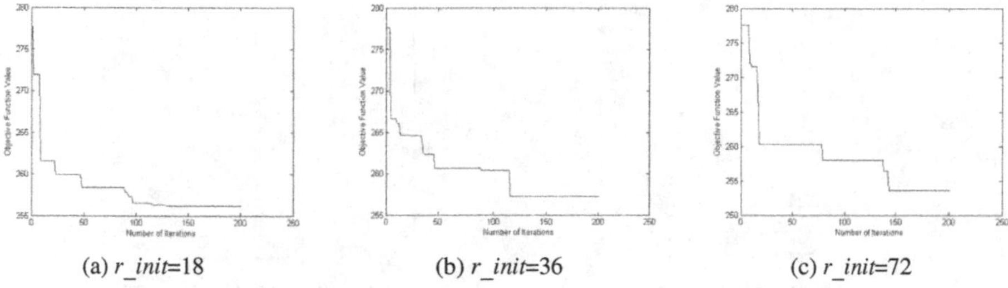

(a) *r_init*=18 (b) *r_init*=36 (c) *r_init*=72

Fig. 3. Algorithm's behavior for different *r_init* values

A similar behavior can be seen even when *r_init* is indeed only an initialization parameter. Smaller initial values are associated with smoother objective function values transitions. As the change in *r* is considered as dividing or multiplying its value by 2, the influence of *r_init* is present in all iterations. This can be seen in Fig. 4 and Fig. 5.

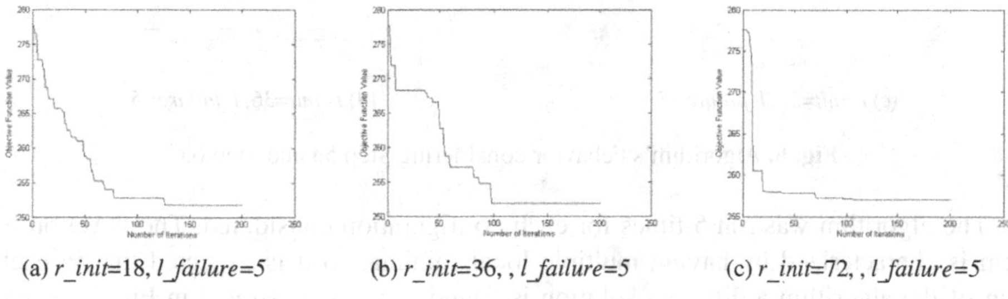

(a) *r_init*=18, *l_failure=5* (b) *r_init*=36, , *l_failure=5* (c) *r_init*=72, , *l_failure=5*

Fig. 4. Algorithm's behavior for different *r_init* values, *l_failure*=5

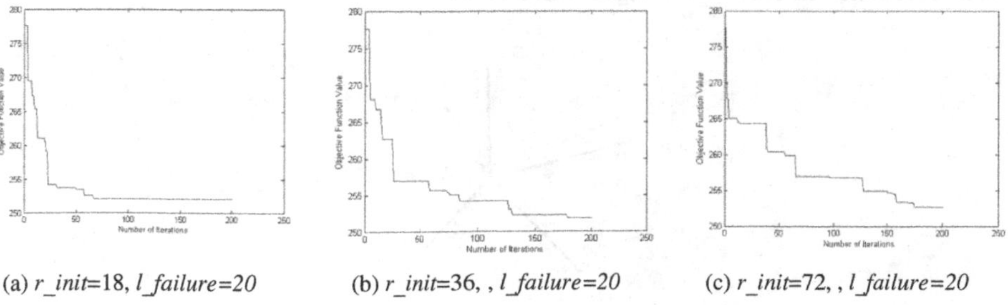

(a) *r_init*=18, *l_failure=20* (b) *r_init*=36, , *l_failure=20* (c) *r_init*=72, , *l_failure=20*

Fig. 5. Algorithm's behavior for different *r_init* values, *l_failure*=20

The impact of the *l_failure* parameter is more visible with greater values of *r_init*. By inspection of Fig. 4 and Fig. 5, we can see that smaller values of *l_failure* promote a faster convergence of the algorithm when *r_init* is equal to 36 or 72.

The option of using Step 5a and Step 6a allows a steepest descent in early iterations, but diminishes the successful search iterations as the algorithm progresses. This behavior is more pronounced for greater values of *r_init*.

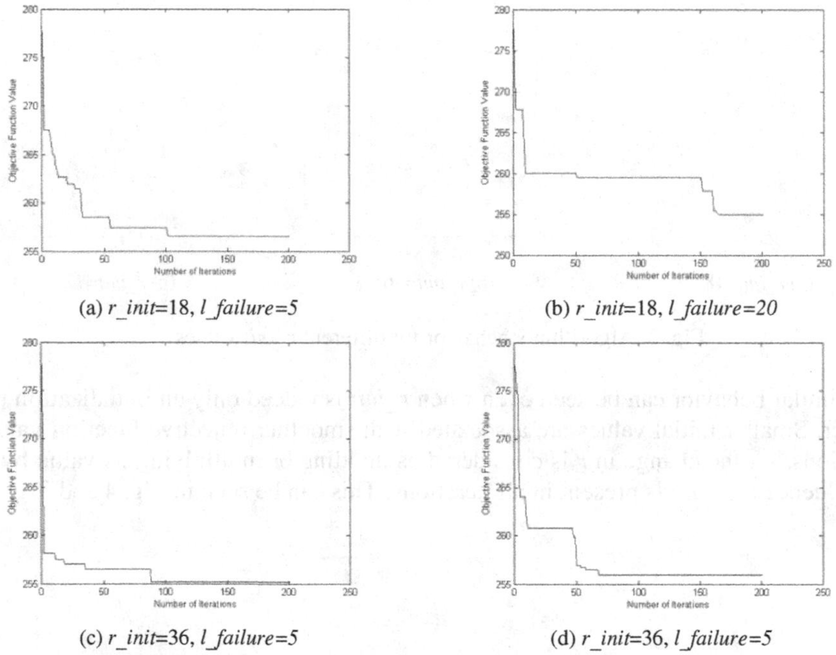

(a) *r_init*=18, *l_failure*=5 (b) *r_init*=18, *l_failure*=20

(c) *r_init*=36, *l_failure*=5 (d) *r_init*=36, *l_failure*=5

Fig. 6. Algorithm's behavior considering Step 5a and Step 6a

The algorithm was run 5 times for each configuration considered. The BAO problem is characterized by having multiple local minima, so it is expected that in each run of the algorithm a different solution is found. This is illustrated in Fig. 7, where the equidistant solution is shown [black solid line] together with 5 other solutions that were calculated in each of the algorithm's runs.

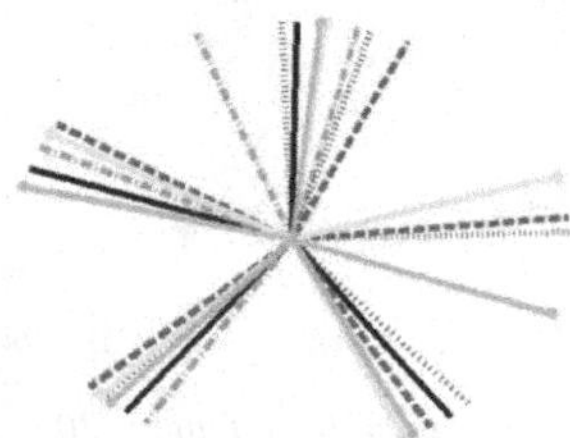

Fig. 7. Different runs of the algorithm usually end up with different solutions

Table 3. Improvement in the objective function value (Mean values)

r_init	l_failure	Steps	Patients 1	2	3	4	5	6	7	8	9	10	Average improvement
18	-	-	2,75%	6,74%	7,49%	5,12%	7,28%	5,88%	14,65%	7,22%	6,82%	2,93%	6,69%
36	-	-	3,14%	7,05%	8,03%	4,38%	7,56%	5,59%	15,48%	8,33%	5,58%	2,42%	6,76%
72	-	-	3,24%	6,17%	8,67%	3,78%	7,88%	6,83%	14,16%	8,36%	5,43%	2,04%	6,66%
18	5	5, 6	3,00%	5,58%	7,22%	4,52%	7,77%	4,75%	14,94%	6,67%	5,54%	2,46%	6,25%
18	20	5, 6	3,24%	5,76%	7,71%	4,88%	8,15%	6,41%	14,40%	9,29%	6,63%	1,96%	6,84%
36	5	5, 6	3,22%	5,87%	8,74%	4,41%	7,85%	5,28%	14,43%	7,12%	5,92%	3,30%	6,61%
36	20	5, 6	3,25%	5,95%	8,82%	4,72%	7,64%	6,43%	14,57%	7,95%	5,95%	3,65%	**6,89%**
72	5	5, 6	2,71%	6,59%	7,32%	4,56%	6,94%	5,70%	15,31%	6,62%	5,10%	1,73%	6,26%
72	20	5, 6	2,93%	7,00%	7,03%	4,35%	7,99%	6,04%	13,29%	7,91%	6,17%	2,47%	6,52%
18	5	5a, 6a	3,05%	6,22%	7,17%	4,76%	7,32%	6,25%	16,81%	7,94%	5,39%	3,57%	6,85%
18	20	5a, 6a	3,19%	6,19%	8,92%	4,56%	7,46%	5,04%	16,44%	6,98%	5,20%	2,28%	6,63%
36	5	5a, 6a	3,06%	6,97%	7,39%	4,29%	7,56%	5,66%	15,71%	9,18%	5,25%	3,19%	6,82%
36	20	5a, 6a	3,27%	6,74%	6,90%	5,00%	7,20%	5,63%	14,77%	7,25%	6,51%	2,97%	6,62%
36	-	5b	3,22%	6,40%	7,23%	4,70%	8,37%	5,94%	14,52%	7,08%	5,74%	2,39%	6,56%
	Average		3,09%	6,38%	7,76%	4,57%	7,64%	5,82%	14,96%	7,71%	5,80%	2,67%	
	maximum		3,27%	7,05%	8,92%	5,12%	8,37%	6,83%	16,81%	9,29%	6,82%	3,65%	
	minimum		2,71%	5,58%	6,90%	3,78%	6,94%	4,75%	13,29%	6,62%	5,10%	1,73%	

Table 3 shows the average improvement achieved in the objective function value. For each patient, the highest mean improvement obtained is highlighted. We see that the choice of the algorithm's configuration can have an important impact in the quality of the solution reached. There is no single configuration that appears as being the best one for a significant part of the patients: most algorithms are the best for one or two patients at the most. In clinical practice, due to time constraints, it is not possible to run the algorithm with different configurations and then choose the best solution reached. So, how should we decide which configuration to consider? One trivial choice would be to consider the one that would, on average, be the best one over all the patients tested. This approach can, however, be misleading.

The approach proposed in this paper is to consider cross-validation. This means that we select a set of patients, and with this set of patients all versions of the algorithm are ran. The best configuration, on average, for this set of patients, is then applied to the rest of the patients not belonging to this "cross-validation set". We have chosen *leave-one-out* cross-validation:

1. Select one patient *j* at a time. Consider a set constituted by all patients but *j*.
2. Run all versions of the algorithm, 5 times each. Calculate the mean improvement over all patients for each version of the algorithm.
3. Choose the version of the algorithm that presents the best objective value improvement. Apply this version of the algorithm to patient *j*, running the algorithm 5 times and recording the results.
4. Repeat the process for every available patient.

This *leave-one-out* cross-validation procedure can be implemented in a clinical setting, since the time constraints that exist are mainly concerned with guaranteeing that new patients are treated as soon as possible. So, it would be feasible to run several times the algorithm for each already treated patient, keeping a database with these results, and resorting to this database whenever it is necessary to choose a given version of the algorithm to apply to a new patient. The set of patients to include in this set could even consider some measures of similarity between patients.

Applying this procedure with our set of 10 patients, the results are as depicted in Table 4.

Table 4. Computational results when parameters are chosen by cross-validation

patient	*r_init*	*l_failure*	algorithm	*fequi*	mean *fDDS*	% improvement	Standard deviation
1	36	20	Steps 5,6	387,28	374,70	3,25%	1,07
2	36	20	Steps 5,7	72,93	68,59	5,95%	1,28
3	18	5	Step 5a, 6a	187,65	174,20	7,17%	3,61
4	36	20	Steps 5,6	156,37	148,99	4,72%	1,30
5	36	20	Steps 5,7	277,60	256,40	7,64%	2,25
6	36	5	Step 5a, 6a	165,58	156,21	5,66%	1,36
7	36	20	Steps 5,6	40,35	34,48	14,57%	0,72
8	36	20	Steps 5,7	166,08	152,87	7,95%	2,10
9	18	5	Step 5a, 6a	124,25	117,55	5,39%	1,65
10	18	20	Steps 5,7	186,44	182,77	1,96%	2,06

On average we are able to improve the objective function value 6,43%. For many other optimization problems, this would seem as a modest improvement. However, IMRT optimization problems have specificities that make the improvement in the objective function only one amongst several other criteria that can be used to assess the quality of the proposed optimization algorithm. More than the value of an objective function, the impact on the quality of the treatment for each patient is what really matters. The objective function is just a way of guiding the search for a better solution, but it cannot represent the whole set of complex features that have to be taken into account when assessing and considering admissible a given treatment plan.

A metric usually used for plan evaluation is the volume of PTV that receives 95% of the prescribed dose. Typically, 95% of the PTV volume is required as a minimum. These metrics are displayed for the ten cases in Fig. 8, considering the equidistant solution and the best and worst solutions out of the 5 solutions generated for each patient. The horizontal lines represent 95% of the prescribed dose. Satisfactory treatment plans should obtain results above these lines. By simple inspection we can verify the advantage of DDS treatment plans that have an improved tumor irradiation metric for most cases compared to equidistant treatment plans.

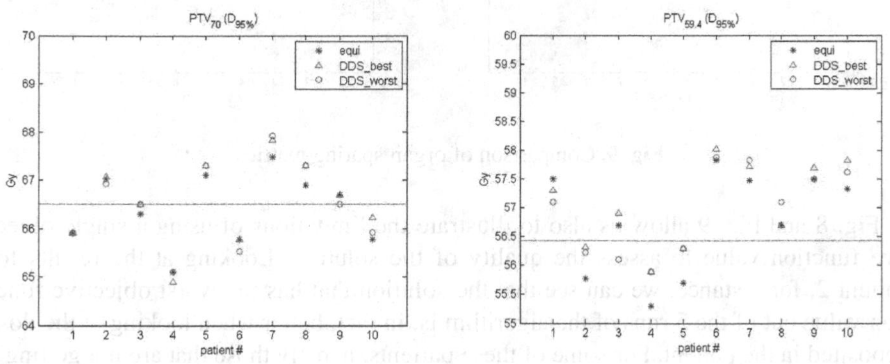

Fig. 8. Comparison of target irradiation metrics using DDS and equidistant treatment plans

In order to verify organ sparing, mean and/or maximum doses of OARs are usually displayed. These metrics are displayed for the ten cases in Fig. 9. The horizontal lines represent the tolerance mean or maximum dose for the corresponding structures. Satisfactory treatment plans should obtain results under these lines. For spinal cord, all treatment plans satisfy the maximum dose tolerance. For brainstem, treatment plans fulfill the maximum dose tolerance in almost all tested cases. Considering the mean dose limit for parotids, it was achieved less times. Looking at the right parotid, about half the patients receive an amount of radiation above what is desired. For the left parotid, the DDS optimized solutions guarantee a desirable level of radiation for 8 of the patients. Observing Fig. 9, it is perceivable that DDS treatment plans outperform equidistant treatment plans in terms of mean dose obtained.

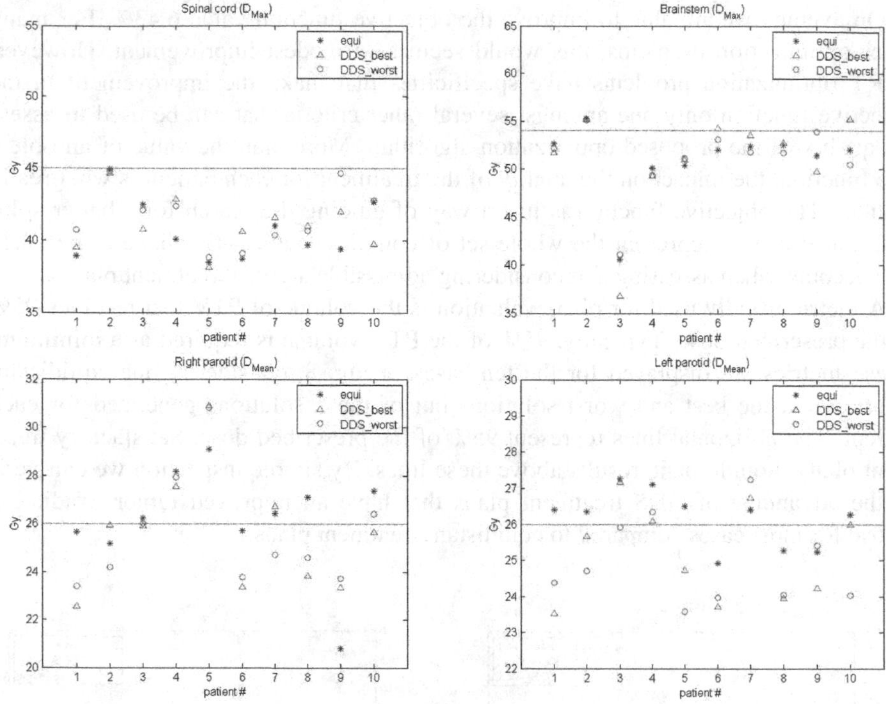

Fig. 9. Comparison of organ sparing metrics

Fig. 8 and Fig. 9 allow us also to illustrate the limitations of using a single objective function value to assess the quality of the solution. Looking at the results for patient 2, for instance, we can see that the solution that has the worst objective function value out of the 5 runs of the algorithm is, in fact, better when looking at the dose deposited in the patient. For some of these patients, namely those that are not getting a sufficient coverage of the PTV even with the optimized solutions, the next step would be to plan treatments with an increased number of radiation angles.

5 Conclusions

BAO problem is a very difficult global optimization problem, characterized by being a large, non-linear and multi-modal problem with a computationally expensive objective function. The DDS approach presented in this paper has as major advantages the fact that it is easily implemented, it is possible to determine the number of function evaluations that are performed and is a derivative-free search strategy that will not get easily trapped in a local minimum. Computational results show that the approach is capable of improving the equidistant solution. The calculation of optimized solutions are important not only contributing to the improvement of the treatment delivered to the patient considering the number of radiation incidences usually determined *a priori*, but also allowing the planner to conclude that it will be necessary to increase the number of angles in order to reach an admissible treatment plan.

Further work will consider some changes in the proposed algorithm, namely embedding the DDS concept of neighborhood into a Simulated Annealing approach. It will also be necessary to consider the calculation of sets of solutions, instead of one single solution, that can illustrate the multiobjective inherent nature of this problem.

Acknowledgements. This work was supported by FEDER, COMPETE, iCIS (CENTRO-07-ST24-FEDER-002003), Portuguese Foundation for Science and Technology under project grants PEst-OE/EEI/UI308/2014FCT, PTDC/EIA-CCO/121450/2010.The work of H. Rocha was supported by the European Social Fund and Portuguese Funds.

References

1. D'Souza, W., Meyer, R.R., Shi, L.: Selection of beam orientations in intensity-modulated radiation therapy using single-beam indices and integer programming. Physics in Medicine and Biology 49, 3465–3481 (2004)
2. Lu, H.M., Kooy, H.M., Leber, Z.H., Ledoux, R.J.: Optimized beam planning for linear accelerator-based stereotactic radiosurgery. International Journal of Radiation Oncology, Biology, Physics 39, 1183–1189 (1997)
3. Goitein, M., Abrams, M., Rowell, D., Pollari, H., Wiles, J.: Multi-dimensional treatment planning: II. Beam's eye-view, back projection, and projection through CT sections. International Journal of Radiation Oncology* Biology* Physics 9, 789–797 (1983)
4. Pugachev, A., Xing, L.: Pseudo beam's eye-view as applied to beam orientation selection in intensity-modulated radiation therapy. International Journal of Radiation Oncology* Biology* Physics 51, 1361–1370 (2001)
5. Pugachev, A., Xing, L.: Computer-assisted selection of coplanar beam orientations in intensity-modulated radiation therapy. Physics in Medicine and Biology 46, 2467–2476 (2001)
6. Aleman, D., Romeijn, H., Dempsey, J.: A Response Surface-Based Approach to Beam Orientation Optimization in IMRT Treatment Planning. In: IIE Annual Conf. Exposition, Orlando, FL (2008)
7. Rocha, H., Dias, J., Ferreira, B.C., Lopes, M.C.: Selection of intensity modulated radiation therapy treatment beam directions using radial basis functions within a pattern search methods framework. Journal of Global Optimization 57, 1065–1089 (2013)
8. Lee, E.K., Fox, T., Crocker, I.: Simultaneous beam geometry and intensity map optimization in intensity-modulated radiation therapy. International Journal of Radiation Oncology, Biology, Physics 64, 301–320 (2006)
9. Djajaputra, D., Wu, Q., Wu, Y., Mohan, R.: Algorithm and performance of a clinical IMRT beam-angle optimization system. Physics in Medicine and Biology 48, 3191–3212 (2003)
10. Bortfeld, T., Schlegel, W.: Optimization of beam orientations in radiation therapy: some theoretical considerations. Physics in Medicine and Biology 38, 291–304 (1993)
11. Li, Y., Yao, D., Yao, J., Chen, W.: A particle swarm optimization algorithm for beam angle selection in intensity-modulated radiotherapy planning. Physics in Medicine and Biology 50, 3491 (2005)

12. Dias, J., Rocha, H., Ferreira, B., Lopes, M.C.: A genetic algorithm with neural network fitness function evaluation for IMRT beam angle optimization. Central European Journal of Operations Research (2013) (accepted for publication)
13. Wu, X., Zhu, Y., Dai, J., Wang, Z.: Selection and determination of beam weights based on genetic algorithms for conformal radiotherapy treatment planning. Physics in Medicine and Biology 45, 2547–2558 (2000)
14. Li, Y., Yao, J., Yao, D.: Automatic beam angle selection in IMRT planning using genetic algorithm. Physics in Medicine and Biology 49, 1915 (2004)
15. Ehrgott, M., Johnston, R.: Optimisation of beam directions in intensity modulated radiation therapy planning. OR Spectrum 25, 251–264 (2003)
16. Lim, G.J., Cao, W.: A two-phase method for selecting IMRT treatment beam angles: Branch-and-Prune and local neighborhood search. European Journal of Operational Research 217, 609–618 (2012)
17. Das, S.K., Marks, L.B.: Selection of coplanar or noncoplanar beams using three-dimensional optimization based on maximum beam separation and minimized nontarget irradiation. International Journal of Radiation Oncology, Biology, Physics 38, 643 (1997)
18. Craft, D.: Local beam angle optimization with linear programming and gradient search. Physics in Medicine and Biology 52, N127–N135 (2007)
19. Dias, J., Rocha, H., Ferreira, B., Lopes, M.C.: IMRT Beam Angle Optimization Using Dynamically Dimensioned Search. In: Zhang, Y.-T. (ed.) The International Conference on Health Informatics. IFMBE Proceedings, vol. 42, pp. 1–4. Springer, Heidelberg (2013)
20. Aleman, D.M., Kumar, A., Ahuja, R.K., Romeijn, H.E., Dempsey, J.F.: Neighborhood search approaches to beam orientation optimization in intensity modulated radiation therapy treatment planning. Journal of Global Optimization 42, 587–607 (2008)
21. Romeijn, H.E., Ahuja, R.K., Dempsey, J.F., Kumar, A., Li, J.G.: A novel linear programming approach to fluence map optimization for intensity modulated radiation therapy treatment planning. Physics in Medicine and Biology 48, 3521–3542 (2003)
22. Tolson, B., Shoemaker, C.: Dynamically dimensioned search algorithm for computationally efficient watershed model calibration. Water Resources Research 43, W01413 (2007)
23. Regis, R.G., Shoemaker, C.A.: Combining radial basis function surrogates and dynamic coordinate search in high-dimensional expensive black-box optimization. Engineering Optimization 45, 529–555 (2013)
24. Deasy, J.O., Blanco, A.I., Clark, V.H.: CERR: A computational environment for radiotherapy research. Medical Physics 30, 979–985 (2003)
25. Deasy, J., Lee, E.K., Bortfeld, T., Langer, M., Zakarian, K., Alaly, J., Zhang, Y., Liu, H., Mohan, R., Ahuja, R.: A collaboratory for radiation therapy treatment planning optimization research. Annals of Operations Research 148, 55–63 (2006)

IMRT Beam Angle Optimization Using Non-descent Pattern Search Methods

Humberto Rocha[1], Joana M. Dias[1,2], Brígida Ferreira[3,4], and Maria do Carmo Lopes[3,4]

[1] INESC-Coimbra, Rua Antero de Quental, 199 3000-033 Coimbra, Portugal

[2] Faculdade de Economia, Universidade de Coimbra, 3004-512 Coimbra, Portugal

[3] I3N, Departamento de Física, Universidade de Aveiro, 3810-193 Aveiro, Portugal

[4] Serviço de Física Médica, IPOC-FG, EPE, 3000-075 Coimbra, Portugal

hrocha@mat.uc.pt, joana@fe.uc.pt, brigida@ua.pt, mclopes@ipocoimbra.min-saude.pt

Abstract. The intensity-modulated radiation therapy (IMRT) treatment planning is usually a sequential process where initially a given number of beam directions are selected followed by the fluence map optimization (FMO) considering those beam directions. The beam angle optimization (BAO) problem consists on the selection of appropriate radiation incidence directions in radiation therapy treatment planning and may be decisive for the quality of the treatment plan, both for appropriate tumor coverage and for enhancement of better organs sparing. This selection must be based on the optimal value of the FMO problem otherwise the resulting beam angle set has no guarantee of optimality and has questionable reliability. Pattern search methods (PSM) have been used successfully to address the BAO problem driven by the optimal fluence value of the FMO problem. PSM are iterative methods generating a sequence of non-increasing iterates such that iterate progression is solely based on a finite number of function evaluations in each iteration, without explicit or implicit use of derivatives. Typically, in IMRT optimization, the quality of the solutions obtained is not simply related to the final value of an objective function but rather judged by dose-volume histograms or considering a set of physical dose metrics. These dose metrics can be simply described as obtaining a minimum prescribed dose for the target volumes (the regions that have to be irradiated) and a maximum or mean tolerance dose values for the remaining surrounding structures (the regions that should be spared). The goal of this paper is to present a non-descent PSM that can be guided both by an objective function formulation of the FMO problem and by physical dose metrics. Four retrospective treated cases of head-and-neck tumors at the Portuguese Institute of Oncology of Coimbra are used to discuss the benefits of non-descent PSM for the optimization of the BAO problem.

Keywords: Pattern Search Methods, IMRT, Beam Angle Optimization.

B. Murgante et al. (Eds.): ICCSA 2014, Part II, LNCS 8580, pp. 17–31, 2014.

1 Introduction

Radiation therapy is with surgery and chemotherapy one of the three main treatment approaches for cancer, used for around 50% of all patients. With this modality, patients are irradiated with beams of ionizing radiation attempting to sterilize all cancer cells while minimizing the collateral effects on the surrounding healthy organs and tissues. An important type of radiation therapy is intensity-modulated radiation therapy (IMRT), a modern technique where the radiation beam is modulated by a multileaf collimator allowing the irradiation of the patient using non-uniform radiation fields from selected angles. The ionizing radiation is generated by a linear accelerator mounted on a gantry that can rotate along a central axis and is delivered with the patient immobilized on a couch that can rotate. The rotation of the couch combined with the rotation of the gantry allows radiation from almost any angle around the tumor. Despite that fact, the use of angles that lay in the plane of rotation of the gantry, i.e. coplanar angles, is predominant. The selection of appropriate radiation incidence directions in radiation therapy treatment planning – beam angle optimization (BAO) problem – is important for the quality of the treatment plan [5,13], both for appropriate tumor coverage and for better organ sparing. However, in clinical practice, the beam angle number and directions are typically selected in a time-consuming trial-and-error procedure by a dosimetrist. The fact that the BAO problem is a highly non-convex optimization problem with many local minima [4], yet to be solved in a satisfactory way within a clinically acceptable time frame, helps explaining the current clinical practice.

The BAO problem is the first problem that arises in treatment planning, but its optimal solution is highly dependent on the optimal solution of the fluence map optimization (FMO) problem – the problem of deciding what are the optimal radiation intensities associated with each set of beam angles. When the BAO problem is not based on the optimal FMO solutions, the resulting beam angle set has no guarantee of optimality and has questionable reliability since it has been extensively reported that optimal beam angles for IMRT are often non-intuitive [21]. Obtaining the optimal solution for a beam angle set is time costly and even if only one beam angle is changed in that set, it is necessary to calculate the radiation dose that is being deposited in the patient's tissues – organs and tumors. Therefore, methods that avoid being easily trapped in local minima and that require few function value evaluations to progress and converge are advantageous. The pattern search methods (PSM) framework has been used by us to address the BAO problem successfully due to its ability to avoid local entrapment and its need for few function value evaluations to converge [16,17,18,19].

PSM are iterative methods generating a sequence of non-increasing iterates such that iterate progression is solely based on a finite number of function evaluations in each iteration, without explicit or implicit use of derivatives. In IMRT, the final value of an objective function is not a complete and unique measure of the quality of the solution obtained. In clinical practice, the quality of the solution obtained is rather judged by dose-volume histograms or considering a set of physical dose metrics. These dose metrics can be simply described as obtaining a minimum prescribed dose for the target volumes (the regions that have to be irradiated) and a maximum or mean tolerance dose values for the remaining surrounding structures (the regions that should be spared). The goal of this paper is to present a non-descent PSM that can be guided both by the objective function of the FMO problem and by the dose metrics. Typically, the progression of PSM is determined by the decrease in the objective function value. In this paper, small increases of the objective function value are allowed whenever some physical feature of the problem is improved at the cost of other feature(s) that still remain within the limit(s) prescribed. Four retrospective treated cases of head-and-neck tumors at the Portuguese Institute of Oncology of Coimbra are used to discuss the benefits of non-descent PSM for the optimization of the BAO problem. The paper is organized as follows. In the next section we describe the BAO problem. Non-descent PSM framework is presented in section 3. Computational tests using clinical examples of head-and-neck cases are presented in section 4. In the last Section we have the conclusions.

2 Beam Angle Optimization in IMRT Treatment Planning

The BAO problem is a quite difficult problem to solve since it is a highly non-convex optimization problem with many local minima – see Fig. 1. In most of the previous works on BAO, the entire range $[0°, 360°]$ of gantry angles is discretized into equally spaced beam directions with a given angle increment, such as 5 or 10 degrees, where exhaustive searches are performed directly or guided by a variety of different heuristics including simulated annealing [3], genetic algorithms [10], particle swarm optimization [12] or other heuristics incorporating a priori knowledge of the problem [15]. Although those global heuristics can theoretically avoid local optima, globally optimal or even clinically better solutions can not be obtained without a large number of objective function evaluations. On the other hand, the use of single-beam metrics has been a popular approach to address the BAO problem as well, e.g., the concept of beam's-eye-view [14]. Despite the computational time efficiency of these approaches, the quality of the solutions proposed cannot be guaranteed since the interplay between the selected beam directions is ignored.

In order to model the BAO problem as a mathematical programming problem, a quantitative measure to compare the quality of different sets of beam angles is required. For the reasons presented in Section 1, our approach for modeling the BAO problem uses the optimal solution value of the FMO problem as the measure of the quality for a given beam angle set. Thus, we will present the formulation of the BAO problem followed by the formulation of the FMO problem we used. Here, we will assume that the number of beam angles is defined a priori by the treatment planner and that all the radiation directions lie on the same plane.

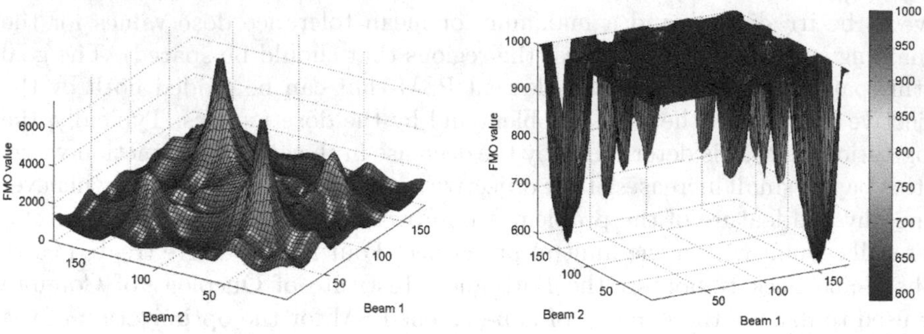

Fig. 1. 2-beam BAO surface (left) and truncated surface (right) to highlight the many local minima

2.1 BAO Model

Let us consider n to be the fixed number of (coplanar) beam directions, i.e., n beam angles are chosen on a circle around the CT-slice of the body that contains the isocenter (usually the center of mass of the tumor). In our formulation, instead of a discretized sample, all continuous $[0°, 360°]$ gantry angles will be considered. Since the angle $-1°$ is equivalent to the angle $359°$ and the angle $361°$ is the same as the angle $1°$, we can avoid a bounded formulation. A simple formulation for the BAO problem is obtained by selecting an objective function such that the best set of beam angles is obtained for the function's minimum:

$$\begin{aligned} &\min f(\theta_1, \ldots, \theta_n) \\ &s.t. \ (\theta_1, \ldots, \theta_n) \in \mathbb{R}^n. \end{aligned} \tag{1}$$

Here, for the reasons stated before, the objective $f(\theta_1, \ldots, \theta_n)$ that measures the quality of the set of beam directions $\theta_1, \ldots, \theta_n$ is the optimal value of the FMO problem for each fixed set of beam directions. The FMO model used is presented next.

2.2 FMO Model

In order to solve the FMO problem, i.e., to determine optimal fluence maps, the radiation dose distribution deposited in the patient needs to be assessed accurately. Each structure's volume is discretized into small volume elements (voxels) and the dose is computed for each voxel considering the contribution of each beamlet. Typically, a dose matrix D is constructed from the collection of all beamlet weights, by indexing the rows of D to each voxel and the columns to each beamlet, i.e., the number of rows of matrix D equals the number of voxels (N_v) and the number of columns equals the number of beamlets (N_b) from all beam directions considered. Therefore, using matrix format, we can say that the total dose received by the voxel i is given by $\sum_{j=1}^{N_b} D_{ij}w_j$, with w_j the weight of beamlet j. Usually, the total number of voxels is large, reaching the tens of thousands, which originates large-scale problems. This is one of the main reasons for the difficulty of solving the FMO problem.

For a given beam angle set, an optimal IMRT plan is obtained by solving the FMO problem - the problem of determining the optimal beamlet weights for the fixed beam angles. Many mathematical optimization models and algorithms have been proposed for the FMO problem, including linear models [20], mixed integer linear models [11] and nonlinear models [2]. Here, we will use this later approach that penalizes each voxel according to the square difference of the amount of dose received by the voxel and the amount of dose desired/allowed for the voxel. This formulation yields a quadratic programming problem with only linear non-negativity constraints on the fluence values [20]:

$$\min_w \sum_{i=1}^{N_v} \frac{1}{v_S} \left[\underline{\lambda}_i \left(T_i - \sum_{j=1}^{N_b} D_{ij}w_j \right)_+^2 + \overline{\lambda}_i \left(\sum_{j=1}^{N_b} D_{ij}w_j - T_i \right)_+^2 \right]$$

$$s.t. \quad w_j \geq 0,\ j = 1, \ldots, N_b,$$

where T_i is the desired dose for voxel i of the structure v_S, $\underline{\lambda}_i$ and $\overline{\lambda}_i$ are the penalty weights of underdose and overdose of voxel i, and $(\cdot)_+ = \max\{0, \cdot\}$. This nonlinear formulation implies that a very small amount of underdose or overdose may be accepted in clinical decision making, but larger deviations from the desired/allowed doses are decreasingly tolerated [2].

The FMO model is used as a black-box function and the conclusions drawn regarding BAO coupled with this nonlinear model are valid also if different FMO formulations are considered.

3 Non-descent Pattern Search Methods

PSM framework will be briefly described followed by the presentation of the proposed non-descent PSM algorithm tailored for the BAO problem.

3.1 Pattern Search Methods Framework

PSM use the concept of positive bases (or positive spanning sets) to move towards a direction that would produce a function decrease. A positive basis for $\mathbb{R}^n$ can be defined as a set of nonzero vectors of $\mathbb{R}^n$ whose positive combinations span $\mathbb{R}^n$ (positive spanning set), but no proper set does. A positive spanning set contains at least one positive basis. It can be shown that a positive basis for $\mathbb{R}^n$ contains at least $n+1$ vectors and cannot contain more than $2n$ [8]. Positive bases with $n+1$ and $2n$ elements are referred to as minimal and maximal positive basis, respectively. Commonly used minimal and maximal positive bases are $[I \ \ -e]$, with I being the identity matrix of dimension n and $e = [1 \ldots 1]^\top$, and $[I \ -I]$, respectively. The motivation for directional direct search methods such as PSM is given by one of the main features of positive basis (or positive spanning sets) [8]: there is always a vector $\mathbf{v}^i$ in a positive basis (or positive spanning set) that is a descent direction unless the current iterate is at a stationary point, i.e., there is an $\alpha > 0$ such that $f(x^k + \alpha \mathbf{v}^i) < f(x^k)$. This is the core of directional direct search methods and in particular of PSM. The notions and motivations for the use of positive bases, its properties and examples can be found in [1,8].

PSM are iterative methods generating a sequence of non-increasing iterates $\{x_k\}$. Given the current iterate x^k, at each iteration k, the next point x^{k+1}, aiming to provide a decrease of the objective function, is chosen from a finite number of candidates on a given mesh M_k defined as

$$M_k \; = \; \{x^k + \alpha_k \mathbf{V}\mathbf{z} : \; \mathbf{z} \in \mathbb{Z}_+^p\},$$

where α_k is the mesh-size (or step-size) parameter, $\mathbb{Z}_+$ is the set of nonnegative integers and $\mathbf{V}$ denote the $n \times p$ matrix whose columns correspond to the p ($\geq n+1$) vectors forming a positive spanning set.

PSM are organized around two steps at every iteration. The first step consists of a finite search on the mesh, free of rules, with the goal of finding a new iterate that decreases the value of the objective function at the current iterate. This step, called the search step, has the flexibility to use any strategy, method or heuristic, or take advantage of a priori knowledge of the problem at hand, as long as it searches only a finite number of points in the mesh. The search step provides the flexibility for a global search since it allows searches away from the neighborhood of the current iterate, and influences the quality of the local minimizer or stationary point found by the method.

If the search step fails to produce a decrease in the objective function, a second step, called the poll step, is performed around the current iterate. The poll step follows stricter rules and, using the concepts of positive bases, attempts to perform a local search in a mesh neighborhood around $\mathbf{x}^k$, $\mathcal{N}(\mathbf{x}^k) \; = \; \{\mathbf{x}^k + \alpha_k \mathbf{v} : \; \text{for all } \mathbf{v} \in P_k\} \subset M_k$, where P_k is a positive basis chosen from the finite positive spanning set $\mathbf{V}$. For a sufficiently small mesh-size parameter α_k, the poll step is guaranteed to provide a function reduction, unless the current iterate is at a stationary point [1]. So, if the poll step also fails to produce a function reduction, the mesh-size parameter α_k must be decreased. On the other hand, if both the search and poll steps fail to obtain an improved value for the objective function, the mesh-size parameter is increased or held constant. The most common choice for the mesh-size parameter update is to halve the mesh-size parameter at unsuccessful iterations and to keep it or double it at successful ones. The PSM framework is summarized in Algorithm 1.

3.2 Non-descent Pattern Search Methods for BAO

PSM are derivative-free optimization algorithms widely used for the minimization of non-convex functions such that iterate progression is solely based on a

Algorithm 1. Pattern search methods framework

Initialization:

- Set $k = 0$.
- Choose $\mathbf{x}^0 \in \mathbb{R}^n$, $\alpha_0 > 0$, and a positive spanning set $\mathbf{V}$.

Iteration:

1. Search step: evaluate f at a finite number of points in M_k with the goal of decreasing the objective function value at $\mathbf{x}^k$. If $\mathbf{x}^{k+1} \in M_k$ is found satisfying $f(\mathbf{x}^{k+1}) < f(\mathbf{x}^k)$, go to step 4 and expand M_k. Both search step and iteration are declared successful. Otherwise, go to step 2 and search step is declared unsuccessful.
2. Poll step: this step is only performed if the search step is unsuccessful. If $f(\mathbf{x}^k) \leq f(\mathbf{x})$ for every $\mathbf{x}$ in the mesh neighborhood $\mathcal{N}(\mathbf{x}^k)$, go to step 3 and shrink M_k. Both poll step and iteration are declared unsuccessful. Otherwise, choose a point $\mathbf{x}^{k+1} \in \mathcal{N}(\mathbf{x}^k)$ such that $f(\mathbf{x}^{k+1}) < f(\mathbf{x}^k)$, go to step 4 and expand M_k. Both poll step and iteration are declared successful.
3. Mesh reduction: let $\alpha_{k+1} = \frac{1}{2} \times \alpha_k$. Set $k \leftarrow k + 1$ and return to step 1 for a new iteration.
4. Mesh expansion: let $\alpha_{k+1} = \alpha_k$ (or $\alpha_{k+1} = 2 \times \alpha_k$). Set $k \leftarrow k + 1$ and return to step 1 for a new iteration.

finite number of function evaluations in each iteration, without explicit or implicit use of derivatives. In IMRT, the objective function value is not a unique measure of the quality of a given solution. For the assessment of the clinical expected outcome, dose-volume histograms and a set of physical dose metrics should be considered. The tumor to be treated plus some safety margins is called planning target volume (PTV). For PTV, an important dose metric is the volume of PTV that receives 95% of the prescribed dose. Typically, 95% of the PTV volume is required. Mean and/or maximum doses are usually the most important dose metrics for the organ's at risk (OARs). There are many different ways to incorporate these dose metrics in a flexible PSM framework. They can be used within the poll step to decide which point to choose when more than one point improves the best objective function value. They can be used to accept a trial point if it improves the best objective function value and also improves the dose metrics. Our strategy, based on extensive numerical experiments, is to accept a trial point if it improves the best objective function value or if it improves the dose metrics and the objective function value is within a radius of the best objective function value. This proposal considers that the choice of beam angle sets with similar objective function value of the current beam angle set, i.e. $f(x^{k+1}) < f(x^k) + \epsilon$, should be made using dose metrics, directing the algorithm to regions of the search space where better dose metrics are obtained.

Deciding which set of beam angles have the best dose metrics is not straightforward. For the PTV we consider the dose metric fulfilled if 95% of the PTV volume receives more than 95% of the prescribed dose. For an OAR the dose metric is fulfilled if the maximum dose (or the mean dose depending on the type of organ) is under the prescribed values. In the context of the BAO process, we consider that a beam angle set improves the dose metrics of the current beam angle set when all the dose metrics already fulfilled remain fulfilled and one or more dose metrics yet to be fulfilled are improved. For example, for a current beam angle set that satisfies dose metrics for all structures except one OAR, a beam angle set is considered as improving the dose metrics if it improves the dose metric of the OAR yet to be fulfilled and all the other structure's dose metrics remain fulfilled. Necessarily some of the already fulfilled dose metrics will be deteriorated but the prescribed dose limits will continue to be assured. When and while all structures have dose metrics fulfilled, the BAO process progression is only determined by decreases on the objective function value, meaning that structures considered more important in the objective function value will be better spared/irradiated. The strategy adopted here attempts to maximize the number of structures with the dose metrics fulfilled which roughly correspond to fulfill the prescribed doses by the medical doctor. It should be highlighted that, in our tests, most of the times, we could not obtain a treatment plan that fulfill the prescribed doses for all structures. The strategy sketched for non-descent PSM to address the BAO problem is presented in Algorithm 2.

Algorithm 2. Non-descent PSM for BAO

Initialization:

- Set $k = 0$.
- Choose $\mathbf{x}^0 \in \mathbb{R}^n$, $\alpha_0 > 0$, $\epsilon > 0$ and a positive spanning set $\mathbf{V}$.
- Compute $f(x^0)$ and the dose metrics for the beam angle set $\mathbf{x}^0$.

Iteration:

1. Search step: evaluate f at a finite number of points in M_k with the goal of improving the current beam angle set. If $x^{k+1} \in M_k$ is found satisfying $f(x^{k+1}) < f(x^k)$ or $f(x^{k+1}) < f(x^k) + \epsilon$ and the dose metrics of x^{k+1} improve the dose metrics of x^k, go to step 4, expand M_k and compute the dose metrics of the new best beam angle set. Both search step and iteration are declared successful. Otherwise, go to step 2 and search step is declared unsuccessful.
2. Poll step: this step is only performed if the search step is unsuccessful. If there is a point $x^{k+1} \in \mathcal{N}(x^k)$ such that $f(x^{k+1}) < f(x^k)$ or $f(x^{k+1}) < f(x^k) + \epsilon$ and the metrics of x^{k+1} improve the metrics of x^k, both poll step and iteration are declared successful, go to step 4 and compute the dose metrics of the new best beam angle set. Otherwise, if every x in the mesh neighborhood $\mathcal{N}(x^k)$, fail to improve the current beam angle set then go to step 3 and shrink M_k. Both poll step and iteration are declared unsuccessful.
3. Mesh reduction: let $\alpha_{k+1} = \frac{1}{2} \times \alpha_k$. Set $k \leftarrow k + 1$ and return to step 1 for a new iteration.
4. Mesh expansion: let $\alpha_{k+1} = \alpha_k$ (or $\alpha_{k+1} = 2 \times \alpha_k$). Set $k \leftarrow k + 1$ and return to step 1 for a new iteration.

The efficiency of PSM improved significantly by reordering the poll directions according to descent indicators built from simplex gradients [7]. Adding to the efficiency provided by an insightful reordering of the poll directions, the search step was recently provided with the use of minimum Frobenius norm quadratic models to be minimized within a trust region, which can lead to a significant improvement of direct search for smooth, piecewise smooth, and noisy problems [6]. For implementation and comparison of the non-descent PSM algorithm for the BAO problem, we use as basis the last version of SID-PSM [6,7] which is a MATLAB implementation of the PSM.

4 Computational Results for Head-and-Neck Clinical Examples

The non-descent PSM algorithm was tested using four clinical examples of retrospective treated cases of head-and-neck tumors at the Portuguese Institute of Oncology of Coimbra (IPOC). In general, the head-and-neck region is a complex area to treat with radiotherapy due to the large number of sensitive organs in this region (e.g., eyes, mandible, larynx, oral cavity, etc.). For simplicity, in this study, the OARs used for treatment optimization were limited to the spinal cord, the brainstem and the parotid glands. The spinal cord and the brainstem are some of the most critical OARs in the head-and-neck tumor cases. These are serial organs, i.e., organs such that if only one subunit is damaged, the whole organ functionality is compromised. Therefore, if the tolerance dose is exceeded, it may result in functional damage to the whole organ. Thus, it is extremely important not to exceed the tolerance dose prescribed for these type of organs. Other than the spinal cord and the brainstem, the parotid glands are also important OARs. The parotid gland is the largest of the three salivary glands. A common complication due to parotid glands irradiation is xerostomia (the medical term for dry mouth due to lack of saliva). This decreases the quality of life of patients undergoing radiation therapy of head-and-neck, causing difficulties to swallow. The parotids are parallel organs, i.e., if a small volume of the organ is damaged, the rest of the organ functionality may not be affected. Their tolerance dose depends strongly on the fraction of the volume irradiated. Hence, if only a small fraction of the organ is irradiated the tolerance dose is much higher than if a larger fraction is irradiated. Thus, for these parallel structures, the organ mean dose is generally used instead of the maximum dose as an objective for inverse planning optimization. For the head-and-neck cases in study, PTV was separated in two parts with different prescribed doses: PTV1 and PTV2. The prescription dose for the target volumes and tolerance doses for the OARs considered in the optimization are presented in Table 1.

Our tests were performed on a 2.66Ghz Intel Core Duo PC with 3 GB RAM. The patients' CT sets and delineated structures are exported via Dicom RT to a freeware computational environment for radiotherapy research – CERR [9]. We used CERR 3.2.2 version and MATLAB 7.4.0 (R2007a). An automatized

Table 1. Prescribed doses for all the structures considered for IMRT optimization

Structure	Mean dose	Max dose	Prescribed dose
Spinal cord	–	45 Gy	–
Brainstem	–	54 Gy	–
Left parotid	26 Gy	–	–
Right parotid	26 Gy	–	–
PTV1	–	–	70.0 Gy
PTV2	–	–	59.4 Gy
Body	–	80 Gy	–

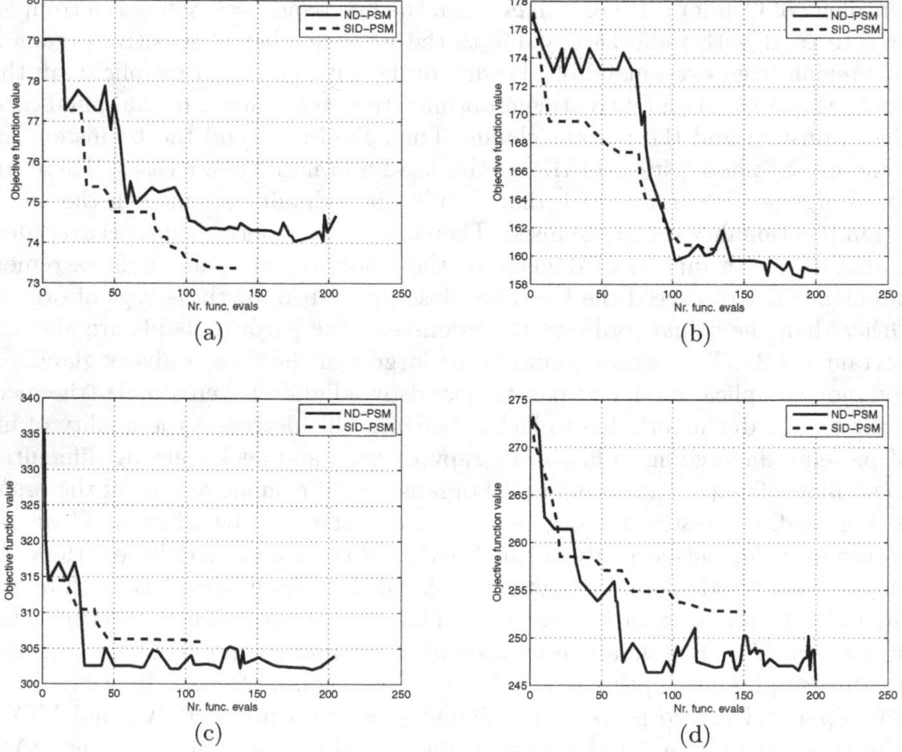

Fig. 2. History of the 5-beam angle optimization process using *SID-PSM* and *ND-PSM*, considering the equispaced configuration (equi) as starting point, for cases 1 to 4, 2(a) to 2(d) respectively

procedure for dose computation for each given beam angle set was developed, instead of the traditional dose computation available from IMRTP module accessible from CERR's menubar. This automatization of the dose computation was essential for integration in our BAO algorithm. To address the convex nonlinear formulation of the FMO problem we used a trust-region-reflective algorithm (*fmincon*) of MATLAB 7.4.0 (R2007a) Optimization Toolbox.

We choose to implement the non-descent PSM algorithm taking advantage of the availability of an existing PSM framework implementation used successfully by us to tackle the BAO problem [16,17,18,19] – the last version of SID-PSM [6,7]. The spanning set used was the positive spanning set ($[e \ \ -e \ \ I \ \ -I]$. Each of these directions corresponds to, respectively, the rotation of all incidence directions clockwise, the rotation of all incidence directions counter-clockwise, the rotation of each individual incidence direction clockwise, and the rotation of each individual incidence direction counter-clockwise.

Treatment plans with five to nine equispaced coplanar beams are used at IPOC and are commonly used in practice to treat head-and-neck cases [2]. We considered plans with five coplanar beams because the importance of BAO increases when a lower number of beam directions is considered. Therefore, treatment plans of five coplanar orientations were obtained using *SID-PSM* and using non-descent PSM algorithm denoted *ND-PSM*. These plans were compared with the typical 5-beam equispaced coplanar treatment plans denoted *equi*. Since we want to improve the quality of the typical equispaced treatment plans, the starting point considered is the equispaced coplanar beam angle set.

The history of the BAO comparing the objective function value decrease versus the number of function evaluations for the four clinical cases of head-and-neck tumors using *SID-PSM* and *ND-PSM* are displayed in Fig. 2. Due to its non-descent nature, ND-PSM required a larger number of function evaluations to converge. However, most of the times, it lead to better results in terms of objective function value obtained. The main purpose of the strategy delineated for the non-descent PSM algorithm was to obtain solutions with better dose metrics results regardless of the final objective function obtained. Nevertheless, the strategy of directing the search to neighborhoods with better dose metrics apparently also benefits the obtention of improvements in the final objective function value.

Despite the improvement in FMO value, as referred previously, the quality of the results can be perceived considering a variety of metrics. A metric usually used for plan evaluation is the volume of PTV that receives 95% of the prescribed dose. Typically, 95% of the PTV volume is required. The occurrence of coldspots, less than 93% of PTV volume receives the prescribed dose, and the existence of hotspots, the percentage of the PTV volume that receives more than 110% of the prescribed dose, are other measures usually used to evaluate target coverage. Mean and/or maximum doses of OARs are usually displayed to verify organ sparing.

The results regarding targets coverage are presented in Table 2. Using only 5 beam directions makes harder to obtain a satisfactory target coverage. We can

Table 2. Target coverage obtained by treatment plans

Case	Target coverage	*ND-PSM*	*SID-PSM*	*equi*
1	PTV1 at 95 % volume	67.07 Gy	67.32 Gy	67.22 Gy
	PTV1 % > 93% of Rx (%)	99.32	99.52	99.39
	PTV1 % > 110% of Rx (%)	0.00	0.00	0.00
	PTV2 at 95 % volume	56.82 Gy	56.12 Gy	55.47 Gy
	PTV2 % > 93% of Rx (%)	96.93	96.07	95.26
	PTV2 % > 110% of Rx (%)	6.07	5.96	6.24
2	PTV1 at 95 % volume	65.77 Gy	65.07 Gy	64.57 Gy
	PTV1 % > 93% of Rx (%)	95.18	94.88	93.64
	PTV1 % > 110% of Rx (%)	0.00	0.00	0.02
	PTV2 at 95 % volume	56.57 Gy	56.72 Gy	56.32 Gy
	PTV2 % > 93% of Rx (%)	96.59	96.54	96.16
	PTV2 % > 110% of Rx (%)	24.88	25.19	25.03
3	PTV1 at 95 % volume	67.27 Gy	66.97 Gy	66.97 Gy
	PTV1 % > 93% of Rx (%)	99.21	99.08	99.21
	PTV1 % > 110% of Rx (%)	0.00	0.00	0.00
	PTV2 at 95 % volume	55.02 Gy	55.07 Gy	54.17 Gy
	PTV2 % > 93% of Rx (%)	94.88	94.85	94.19
	PTV2 % > 110% of Rx (%)	12.43	12.30	12.16
4	PTV1 at 95 % volume	65.27 Gy	64.97 Gy	64.42Gy
	PTV1 % > 93% of Rx (%)	95.23	94.73	93.93
	PTV1 % > 110% of Rx (%)	0.00	0.00	0.00
	PTV2 at 95 % volume	58.07 Gy	58.22 Gy	58.22 Gy
	PTV2 % > 93% of Rx (%)	98.61	98.52	98.20
	PTV2 % > 110% of Rx (%)	37.15	37.37	37.64

verify that optimized treatment plans consistently obtained slightly better target coverage numbers compared to *equi* treatment plans. On the other hand, target coverage numbers are favorable to *ND-PSM* treatment plans compared to *SID-PSM* treatment plans, particularly when 95% of the prescribed dose of PTV1 (66.5 Gy) or PTV2 (56.43 Gy) is not fulfilled. Organ sparing results are shown in Table 3. All the treatment plans fulfill the maximum dose requirements for the spinal cord and the brainstem. However, as expected, the main differences reside in parotid sparing. The optimized treatment plans clearly improve the usually clinically used equispaced treatment plans. The *equi* treatment plans could never fulfill parotid sparing while *ND-PSM* treatment plans always fulfill the parotid's mean dose requirements except for the right parotid in case 3. Although *SID-PSM* treatment plans reduced the parotid's mean dose as well, it does not manage to fulfill the dose limits prescribed as many times as *ND-PSM* treatment plans. Curiously, despite the final objective function value being worse for *ND-PSM*, case 1 illustrates the benefits of the non-descent strategy with respect to parotid sparing.

Table 3. OARs sparing obtained by treatment plans.

Case	OAR	Mean Dose (Gy)			Max Dose (Gy)		
		ND-PSM	*SID-PSM*	*equi*	*ND-PSM*	*SID-PSM*	*equi*
1	Spinal cord	–	–	–	44.10	40.05	41.32
	Brainstem	–	–	–	53.18	51.90	51.86
	Left parotid	25.85	25.81	26.63	–	–	–
	Right parotid	24.68	26.60	26.32	–	–	–
2	Spinal cord	–	–	–	39.59	41.26	40.30
	Brainstem	–	–	–	43.78	40.34	39.49
	Left parotid	25.72	26.31	26.58	–	–	–
	Right parotid	25.11	24.34	26.45	–	–	–
3	Spinal cord	–	–	–	37.96	37.94	38.10
	Brainstem	–	–	–	50.43	50.68	50.20
	Left parotid	24.43	24.37	27.02	–	–	–
	Right parotid	26.74	29.00	29.44	–	–	–
4	Spinal cord	–	–	–	39.37	40.14	38.68
	Brainstem	–	–	–	52.94	52.12	52.26
	Left parotid	22.29	23.32	26.86	–	–	–
	Right parotid	21.39	24.07	26.96	–	–	–

5 Conclusions

The ultimate goal of treatment planning is to be able to obtain a treatment plan that is in accordance with the medical prescription in terms of radiation dose distribution. Usually, the medical prescription will define prescribed doses to the target volumes, and mean or maximum tolerance doses to the organs at risk. This paper proposes an alternative approach to the BAO problem whose optimization process is simultaneously guided by dose metric goals and by an objective fucntion value. The PSM framework had already proved to be a suitable approach for the resolution of the non-convex BAO problem. For the clinical cases retrospectively tested, the use of dose metrics as decision criteria in our tailored approach showed a positive influence on the quality of the local minimizer found. The improvement of the solutions for the head-and-neck cases tested lead to high quality treatment plans with better target coverage and with improved organ sparing.

Acknowledgements. This work was supported by QREN under Mais Centro (CENTRO-07-0224-FEDER-002003) and FEDER funds through the COMPETE program and Portuguese funds through FCT under project grant PTDC/EIA-CCO/121450/2010. This work has also been partially supported by FCT under project grant PEst-OE/EEI/UI308/2014. The work of H. Rocha was supported by the European social fund and Portuguese funds from MCTES.

References

1. Alberto, P., Nogueira, F., Rocha, H., Vicente, L.N.: Pattern search methods for user-provided points: Application to molecular geometry problems. SIAM J. Optim. 14, 1216–1236 (2004)
2. Aleman, D.M., Kumar, A., Ahuja, R.K., Romeijn, H.E., Dempsey, J.F.: Neighborhood search approaches to beam orientation optimization in intensity modulated radiation therapy treatment planning. J. Global Optim. 42, 587–607 (2008)
3. Bortfeld, T., Schlegel, W.: Optimization of beam orientations in radiation therapy: some theoretical considerations. Phys. Med. Biol. 38, 291–304 (1993)
4. Craft, D.: Local beam angle optimization with linear programming and gradient search. Phys. Med. Biol. 52, 127–135 (2007)
5. Das, S.K., Marks, L.B.: Selection of coplanar or non coplanar beams using three-dimensional optimization based on maximum beam separation and minimized non-target irradiation. Int. J. Radiat. Oncol. Biol. Phys. 38, 643–655 (1997)
6. Custódio, A.L., Rocha, H., Vicente, L.N.: Incorporating minimum Frobenius norm models in direct search. Comput. Optim. Appl. 46, 265–278 (2010)
7. Custódio, A.L., Vicente, L.N.: Using sampling and simplex derivatives in pattern search methods. SIAM J. Optim. 18, 537–555 (2007)
8. Davis, C.: Theory of positive linear dependence. Am. J. Math. 76, 733–746 (1954)
9. Deasy, J.O., Blanco, A.I., Clark, V.H.: CERR: A Computational Environment for Radiotherapy Research. Med. Phys. 30, 979–985 (2003)
10. Dias, J., Rocha, H., Ferreira, B.C., Lopes, M.C.: A genetic algorithm with neural network fitness function evaluation for IMRT beam angle optimization. Cent. Eur. J. Oper. Res. (in press) doi:10.1007/s10100-013-0289-4
11. Lee, E.K., Fox, T., Crocker, I.: Integer programming applied to intensity-modulated radiation therapy treatment planning. Ann. Oper. Res. 119, 165–181 (2003)
12. Li, Y., Yao, D., Yao, J., Chen, W.: A particle swarm optimization algorithm for beam angle selection in intensity modulated radiotherapy planning. Phys. Med. Biol. 50, 3491–3514 (2005)
13. Liu, H.H., Jauregui, M., Zhang, X., Wang, X., Dongand, L., Mohan, R.: Beam angle optimization and reduction for intensity-modulated radiation therapy of non-small-cell lung cancers. Int. J. Radiat. Oncol. Biol. Phys. 65, 561–572 (2006)
14. Pugachev, A., Xing, L.: Computer-assisted selection of coplanar beam orientations in intensity-modulated radiation therapy. Phys. Med. Biol. 46, 2467–2476 (2001)
15. Pugachev, A., Xing, L.: Incorporating prior knowledge into beam orientation optimization in IMRT. Int. J. Radiat. Oncol. Biol. Phys. 54, 1565–1574 (2002)
16. Rocha, H., Dias, J.M., Ferreira, B.C., Lopes, M.C.: Incorporating Radial Basis Functions in Pattern Search Methods: Application to Beam Angle Optimization in Radiotherapy Treatment Planning. In: Murgante, B., Gervasi, O., Misra, S., Nedjah, N., Rocha, A.M.A.C., Taniar, D., Apduhan, B.O. (eds.) ICCSA 2012, Part III. LNCS, vol. 7335, pp. 1–16. Springer, Heidelberg (2012)
17. Rocha, H., Dias, J.M., Ferreira, B.C., Lopes, M.C.: Beam angle optimization for intensity-modulated radiation therapy using a guided pattern search method. Phys. Med. Biol. 58, 2939–2953 (2013)
18. Rocha, H., Dias, J.M., Ferreira, B.C., Lopes, M.C.: Selection of intensity modulated radiation therapy treatment beam directions using radial basis functions within a pattern search methods framework. J. Glob. Optim. 57, 1065–1089 (2013)
19. Rocha, H., Dias, J.M., Ferreira, B.C., Lopes, M.C.: Pattern search methods framework for beam angle optimization in radiotherapy design. Appl. Math. Comput. 219, 10853–10865 (2013)

20. Romeijn, H.E., Ahuja, R.K., Dempsey, J.F., Kumar, A., Li, J.: A novel linear programming approach to fluence map optimization for intensity modulated radiation therapy treatment planing. Phys. Med. Biol. 48, 3521–3542 (2003)
21. Stein, J., Mohan, R., Wang, X.H., Bortfeld, T., Wu, Q., Preiser, K., Ling, C.C., Schlegel, W.: Number and orientation of beams in intensity-modulated radiation treatments. Med. Phys. 24, 149–160 (1997)

Maximizing Expectation on Vertex-Disjoint Cycle Packing

João Pedro Pedroso

INESC Porto and Faculdade de Ciências, Universidade do Porto, Portugal
jpp@fc.up.pt

Abstract. This paper proposes a method for computing the expectation for the length of a maximum set of vertex-disjoint cycles in a digraph where vertices and/or arcs are subject to failure with a known probability. This method has an immediate practical application: it can be used for the solution of a kidney exchange program in the common situation where the underlying graph is unreliable. Results for realistic benchmark instances are reported and analyzed.

Keywords: Kidney exchange programs, Cycle packing, Expectation optimization, Combinatorial optimization.

1 Introduction

In many countries, recent legislation allows patients needing a kidney transplant to receive it from a living donor ready to provide one of her kidneys. When the transplant from the donor is not possible, due to blood type or other incompatibilities, the patient-donor pair may enter a kidney exchange program (KEP). The KEP may allow two (or more) patients with incompatible pairs to exchange their donors, in such a way that each recipient receives a compatible kidney from the donor of another pair (see Fig. 1).

An instance of the KEP with five pairs is presented in Fig. 2. The set of patients, donors and their interconnections is often called a *market*, and is represented as a directed graph $G = (\mathcal{V}, \mathcal{A}, v)$. In this graph, $\mathcal{V}$ is the set of all incompatible patient-donor pairs; two vertices i and j are connected by arc (i, j)

Fig. 1. Left: incompatible pairs $P_1 - D_1$ and $P_2 - D_2$ exchange donors, allowing P_1 to receive a transplant from D_2 and vice versa. Right: graph representation of this situation, where vertices are patient-donor pairs, and arcs link a patient to compatible donors.

B. Murgante et al. (Eds.): ICCSA 2014, Part II, LNCS 8580, pp. 32–46, 2014.

if the donor of pair i is compatible with the patient of pair j; and weights v_i, v_{ij} may be used to rate vertices and/or arcs (for simplicity, we consider $v_i = 0$ for all vertices and $v_{ij} = 1$ for all arcs, in this initial graph). An *exchange* is defined as a cycle in the graph. In the situation of Fig. 2 the possible exchanges are $1-2-5-3-1$, $1-2-3-1$, $1-2-4-1$, $2-5-3-2$, $2-3-2$ and $2-5-2$. A feasible exchange can be defined as a set of vertex-disjoint cycles. The size of an exchange is the sum of the lengths of its cycles. The maximum exchange in this example has size four, corresponding to the cycle $1-2-5-3-1$. In many situations the length of each cycle in feasible exchanges is limited to a given value K, *e.g.*, due to limitations in the number of operation rooms that are simultaneously available. If cycles of length greater than $K = 3$ are not allowed, several solutions are possible, each with one cycle of length three; as we will see later, they may be non-equivalent in terms of their *expected* number of arcs.

Fig. 2. Instance with five pairs (left); the maximum number of transplants is four, corresponding to the cycle $1-2-5-3-1$ (right)

Standard approaches to this problem consist of finding a set of vertex-disjoint cycles that maximizes the weighted sum of chosen cycles, where the weight of a cycle is typically its length (*i.e.*, $v_{ij} = 1$); this corresponds to maximizing the number of transplants. In this work we will focus on maximizing the *expectation* of the size of an exchange, given that there may be withdrawal of vertices (patients and/or donors may back out) and arcs (*e.g.*, due to incompatibilities that are detected in the last minute). In the occurrence of a withdrawal, vertices in the cycle involved may be rearranged between them if another cycle is possible, but other vertices cannot be involved; this is called a *contingency plan* in [1].

A summary of models for solving a KEP has been presented in [2], where several formulations are presented, analyzed, and compared in terms of tightness and experimental performance. The topic to be developed in this paper — maximizing the expectation of the number of transplants in a given setting — has been initially raised in [1] and [3], which propose a model for maximizing the expected utility when arcs are subject to failure; a simulation system, where patient-donor pairs are generated and assigned to a cycle in a dynamic version of the KEP is also proposed there. The study of the dynamics in a KEP, concerning the question of how often and how exactly should the exchange be run so as to make the mechanism efficient, has also been addressed in [4] and in references therein; more recently, issues on market design for kidney exchange have been analyzed in [5]. An analysis of the efficacy of chains in kidney exchange models, indicating that their length should be small — best results were obtained for a maximum length of four, in a dynamic setting —, is provided in [6].

2 Background: Mathematical Programming Model

There are several possibilities for modeling the optimization problem of a KEP [2]. One of the most successful is the *cycle formulation* proposed in [7], where one needs to enumerate all cycles in the market graph $G = (\mathcal{V}, \mathcal{A})$ with length at most K. Notice that if $K = 2$ the problem can be formulated as a matching, and if K is unlimited as an assignment; in these cases, it is solvable in polynomial time (see [8] and [9], respectively); otherwise, it is proven to be NP-hard [7]. For each cycle c in the set $\mathcal{C}$ of cycles of the graph, let variable x_c be 1 if c is chosen for the exchange, 0 otherwise. The cycle model of the KEP is the following integer linear program:

$$\text{maximize} \quad \sum_{c} w_c x_c, \tag{1a}$$

$$\text{subject to} \quad \sum_{c:i \in c} x_c \leq 1, \qquad \forall i \in \mathcal{V}, \tag{1b}$$

$$x_c \in \{0, 1\}, \qquad \forall c \in \mathcal{C}. \tag{1c}$$

In the case of arcs with weight 1, the weight of a cycle is $w_c = |c|$, *i.e.*, its coefficient in the objective is its length; otherwise, it is typically the sum of the utilities assigned to arcs in the cycle. The objective function (1a) maximizes the weight of the exchange. Constraints (1b) ensure that every vertex is in at most one cycle (*i.e.*, each donor will donate, and the corresponding patient will receive, at most one kidney); a feasible solution is, thus, a vertex-disjoint cycle packing. The difficulty of this formulation is due to the number of variables; the number of cycles of length at most K may be very large, even for graphs with a modest size (see, *e.g.*, Table 1 or [2]). Very good results were mentioned in [7], obtained by implementing a column generation method within a branch-and-price scheme.

Enumerating cycles for this formulation (or for an associated column generation procedure) may be complemented with an appropriate assessment of the cycle's value, so as to encompass objectives other than the weighted sum of its arcs. Indeed, we may evaluate the expectation of the number of transplants (possibly weighted) in a given subgraph of G induced by the vertices in a cycle. In other words, we propose to assess the expectation of the weight of a maximum packing of vertex-disjoint cycles *in the subgraph induced by the vertex set of each cycle*, given the probability of failure of its components (vertices and arcs). The value of this expectation will be used as the cycle's coefficient in the objective of the model above.

The assignment problem that arises when K is unlimited corresponds to maximizing a packing of vertex-disjoint cycles with no restriction in cycle length. It will be used later; we call it the *optimization subproblem*, and for a given directed graph $G = (\mathcal{V}, \mathcal{A})$ it can be formulated as follows.

$$\begin{aligned}
\text{maximize} \quad & \sum_{(i,j)\in\mathcal{A}} x_{ij}, && && \text{(2a)}\\
\text{subject to} \quad & \sum_{(i,j)\in\mathcal{A}} x_{ij} = \sum_{(j,i)\in\mathcal{A}} x_{ji} && \forall i \in \mathcal{V}, && \text{(2b)}\\
& \sum_{(i,j)\in\mathcal{A}} x_{ij} \le 1 && \forall i \in \mathcal{V}, && \text{(2c)}\\
& x_{ij} \in \{0,1\}, && \forall (i,j) \in \mathcal{A}. && \text{(2d)}
\end{aligned}$$

3 Unreliable Graphs

3.1 Vertex Withdrawal

One of the problems in the implementation of the solution of a KEP instance is that patients or donors may become unavailable, *e.g.*, due to illness or to backing out. In such cases, the previously found solution cannot be implemented in its totality; the cycle where failure occurred cannot be implemented, though the vertices involved may be rearranged if another (shorter) cycle within them can be formed (it is usually accepted that other cycles cannot be changed; depending on the timing of the failure, vertices not assigned to any cycle could potentially be considered, but we are not tackling this case). In this situation, instead of using the cycle's length, the value of a cycle can be better assessed by the expectation of the number of transplants it will lead to, given the probability of failure of each of its vertices.

As an example, consider again the graph of Fig. 2. If $K = 3$ (i.e. the maximum cycle length is three), there are three different optimal solutions: $c_1 = 1-2-4-1$, $c_2 = 1-2-3-1$, and $c_3 = 2-5-3-2$, all of which are packings (*i.e.*, any other cycle leads to vertex overlap; see Fig. 3). Even though the length is the same for all these cycles, the expected length may be different: in c_1 withdrawal of any of the vertices leads to no transplants, whereas in c_2 there can be two transplants (cycle $2-3-2$) in case (only) vertex 1 fails. In c_3, either 3 or 5 may fail, and two

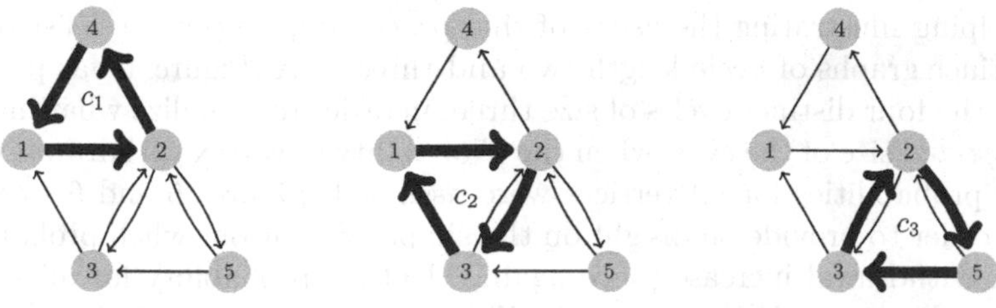

Fig. 3. Cycles c_1, c_2, c_3 of length 3 in the example graph

transplants are still possible. Given the probability of failure p_i for each vertex in the graph, the expected length of each of these cycle configurations can be computed as follows:

$$
\begin{aligned}
E[c_1] =& 3\,(1-p_1)\,(1-p_2)\,(1-p_4) \\
E[c_2] =& 3(1-p_1)(1-p_2)(1-p_3) + 2p_1(1-p_2)(1-p_3) \\
=& (3-p_1)(1-p_2)(1-p_3) \\
E[c_3] =& 3(1-p_2)(1-p_3)(1-p_5) + 2(1-p_2)p_3(1-p_5) + 2(1-p_2)(1-p_3)p_5 \\
=& (p_2-1)(p_3p_5+p_3+p_5-3)
\end{aligned}
$$

The expectation of the length of the maximum set of vertex-disjoint cycles of a (sub)graph can be computed for the general case with Algorithm 1.

Algorithm 1. Expected size for a cycle c (unreliable vertices)
Input: $G_c = (\mathcal{V}_c, \mathcal{A}_c), \quad p_i \; \forall i \in \mathcal{V}_c$
Output: Expectation for the input graph
(1) $E := 0$ ← *accumulator for expectation*
(2) $\mathcal{S} := 2^{\mathcal{V}_c}$ ← *set of all subsets of* $\mathcal{V}_c$
(3) **foreach** $\mathcal{R} \in \mathcal{S}$: ← *vertices remaining*
(4) $\quad \mathcal{Q} := \mathcal{V}_c \backslash \mathcal{R}$ ← *vertices quitting*
(5) $\quad \mathcal{A}' := \{(i,j) \in \mathcal{A}_c : i \in \mathcal{R}, j \in \mathcal{R}\}$ ← *arcs in subgraph induced by* $\mathcal{R}$
(6) $\quad z :=$ maximum of (2a)–(2d) for subgraph $G' = (\mathcal{R}, \mathcal{A}')$
(7) $\quad E := E + z \prod_{i \in \mathcal{R}} (1-p_i) \prod_{i \in \mathcal{Q}} p_i$
(8) **return** E

The purpose of this algorithm is to compute the expectation of a given cycle, to be used during the enumeration phase of the main optimization method when using the cycle formulation presented in Section 2. The evaluation of each cycle c (its coefficient w_c in the objective function) is, in this case, the *expected length* of the maximum set of vertex-disjoint cycle *in the subgraph induced by vertices in the current cycle*, $G_c = (\mathcal{V}_c, \mathcal{A}_c)$, which is given as a parameter to the algorithm. The key issue here is to check, for each subset of the set of vertices $\mathcal{V}_c$, if their withdrawal allows forming other cycles in G_c or not. This is done by solving the optimization subproblem on the subgraph $G' = (\mathcal{R}, \mathcal{A}')$, in line (6). Notice that we need not to be concerned with the size of the cycles here: the size of $\mathcal{V}_c$ has been limited already, as we are considering only vertices in a cycle found in the enumeration phase.

For helping illustrating the usage of this procedure, we present a list of possible distinct graphs of cycle length two and three in A. Figure 4 was prepared based on the four distinct cycles of size three, in order to visualize what happens to the *expected* size of the cycle when the probability of vertex withdrawal varies; identical probabilities for all vertices were assumed. Figures 5 and 6 were prepared in order to provide an insight on the shape of solutions when probabilities of vertex withdrawal increase (once again, identical probability for all vertices was assumed). As can be seen, very significant changes in the solution structure may occur.

3.2 Arc Withdrawal

Another source of unreliability comes from arc withdrawal: connections in the graph may become unavailable, mostly because of new sources of incompatibility between donors and patients discovered in a last, more thorough examination. As in the previous case, the cycle where failure occurred will have to be resolved. The expectation of the number of transplants corresponding to a given cycle

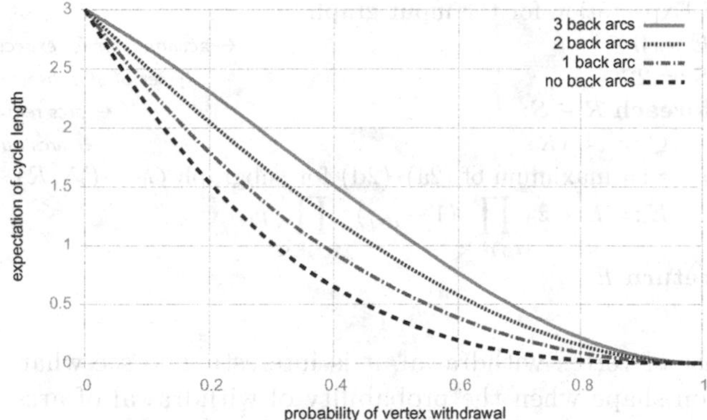

Fig. 4. Expectation of cycle length as a function of vertex failure probability (considered identical for all vertices) for the four distinct cycle configurations with 3 vertices

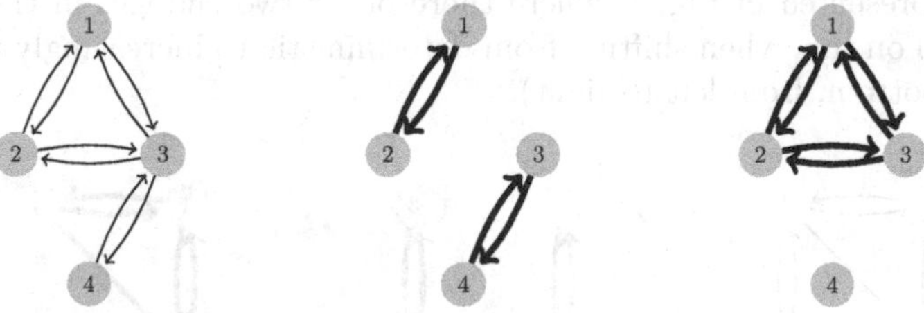

Fig. 5. Left: original graph. Center: solution for high vertex reliability ($p_i < 1/3$). Right: solution for low vertex reliability ($p_i > 1/3$).

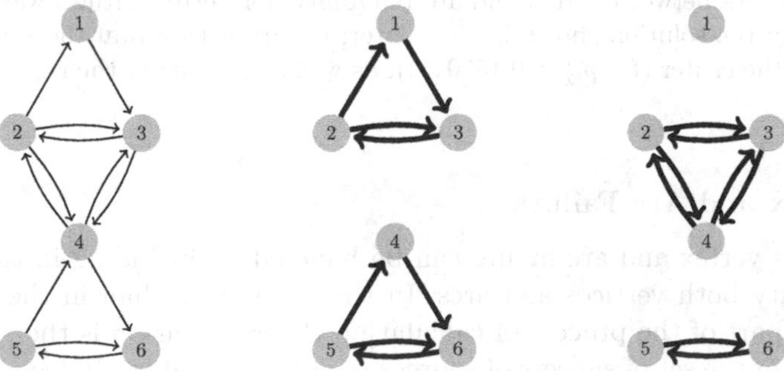

Fig. 6. Left: original graph. Center: solution for high vertex reliability ($p_i < 0.2$). Right: solution for low vertex reliability ($p_i > 0.2$).

configuration can be determined with an algorithm similar to Algorithm 1 but the sources of unreliability are arcs instead of vertices. For a given graph G, the data required for calculating expectation is, in this case, the probability of failure p_{ij} for each arc (i, j) in the graph; the expectation of the length of the maximum set of vertex-disjoint cycles of a (sub)graph can be computed with Algorithm 2.

Algorithm 2. Expected size for a cycle c (unreliable arcs).
Input: $G_c = (\mathcal{V}_c, \mathcal{A}_c), \quad p_{ij} \; \forall (i,j) \in \mathcal{A}_c$
Output: Expectation for the input graph

$E := 0$ *←accumulator for expectation*
$\mathcal{S} := 2^{\mathcal{A}_c}$ *←set of all subsets of $\mathcal{A}_c$*
foreach $\mathcal{R} \in \mathcal{S}$: *←arcs remaining*
  $\mathcal{Q} := \mathcal{A}_c \backslash \mathcal{R}$ *←arcs quitting*
  $z :=$ maximum of (2a)–(2d) for subgraph $G' = (\mathcal{V}_c, \mathcal{R})$
  $E := E + z \prod_{(i,j)\in\mathcal{R}} (1 - p_{ij}) \prod_{(i,j)\in\mathcal{Q}} p_{ij}$
return E

As in the case of vertex withdrawal, it is interesting to see what happens in terms of solution shape when the probability of withdrawal of arcs in a graph increase. Figures 7 and 8 show how solutions vary with this probability, here considered identical for all arcs. Figure 7 presents a situation where vertex withdrawal and arc withdrawal may lead to different solutions. A more complex situation is presented in Fig. 8, where there occur two changes in the solution for the graph on top, when shifting from deterministic to increasingly unreliable situations (bottom, from left to right).

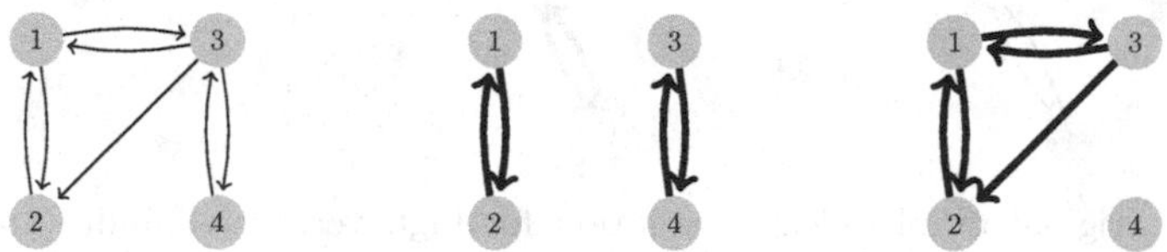

Fig. 7. Differences between vertex and arc reliability: for vertex withdrawal, left hand graph has only the solution shown in the center; for arc withdrawal the solution may be the one in the center (for $p_{ij} < 0.1670\ldots$), as well as the one in the right (for larger p_{ij})

3.3 Vertex and Arc Failure

Simultaneous vertex and arc failure can be handled by including in the sources of unreliability both vertices and arcs. In this case more than in the previous, the limiting part of the process of calculating the expectation is the generation and analysis of the set of subsets of sources of potential failure; indeed, the set of subsets of $\mathcal{S}$ in Algorithm 3 becomes extremely large, forbidding the direct use of this algorithm. An alternative to this will be presented in the next section.

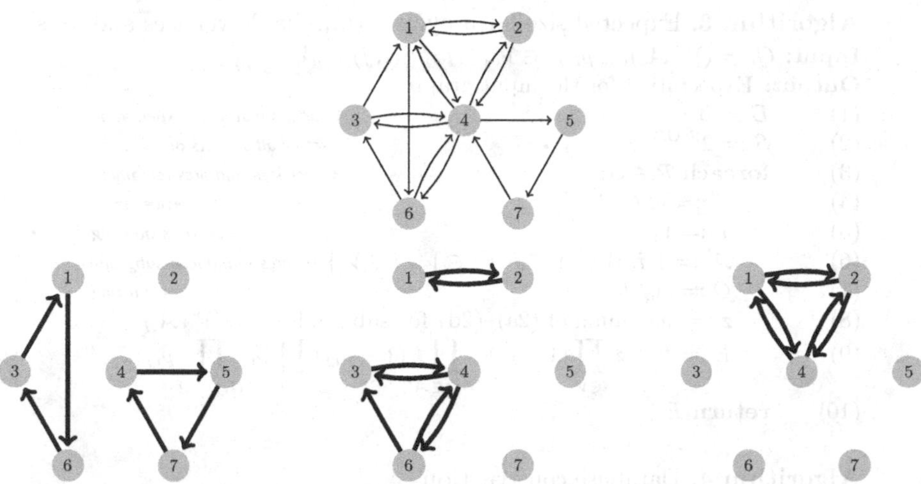

Fig. 8. Example of a more complex graph (top): solutions from high to low reliability (bottom, from left to right)

4 Solution Procedure

The limitations of algorithms 1 to 3 are due to the need of enumerating the set of all subsets of sources of unreliability (*i.e.*, $\mathcal{V}$, $\mathcal{A}$, of $\mathcal{V} \cup \mathcal{A}$). In other words, the limiting part is the very rapid growth of the set of subsets of $\mathcal{S}$ with the cardinality of $\mathcal{S}$, in line (2) of each algorithm. Indeed, this makes the method not viable except for very small graphs. However, here we are dealing with subgraphs induced by vertices in a cycle of length at most K, found when enumerating cycles in the original graph. As the length K of cycles of the original graph to consider is, in practice, limited to a small number (due to limitations on the number of simultaneous operations), the algorithms have practical relevance if properly used.

An alternative to directly using the previous algorithms, in settings where the output is required with minimum delay, is to have previously prepared a database of possible configurations of graphs up to a given size, and for each of them store the function calculating the expression given by the algorithms (possibly after simplifying this expression, since it may become very large). As the graphs that we are dealing with here are small (the maximum number of vertices is K) and their number is limited, with appropriate data structures it is computationally acceptable to search this database for a graph which is isomorphic to the one under consideration, *i.e.*, isomorphic to the subgraph induced by nodes in a cycle of the original graph, itself found during the enumeration phase when setting up the main optimization problem. A list of all possible graphs with cycle length up to three, and the expressions for computing the corresponding expectations, is presented in A. Notice that the number of different graphs grows very rapidly with the number of vertices; for graphs with four vertices there are 61 distinct,

Algorithm 3. Expected size for a cycle c (unreliable vertices and arcs)
Input: $G_c = (\mathcal{V}_c, \mathcal{A}_c), \quad p_i \, \forall i \in \mathcal{V}_c, \quad p_{ij} \, \forall (i,j) \in \mathcal{A}_c$
Output: Expectation for the input graph

$E := 0$ ← *accumulator for expectation*
$\mathcal{S} := 2^{\mathcal{V}_c \cup \mathcal{A}_c}$ ← *set of all subsets of $\mathcal{V}_c \cup \mathcal{A}_c$*
foreach $\mathcal{R} \in \mathcal{S}$: ← *vertices and arcs remaining*
  $\mathcal{V}' := \mathcal{V}_c \cap \mathcal{R}$ ← *vertices in $\mathcal{R}$*
  $\mathcal{T} := \mathcal{V}_c \backslash \mathcal{V}'$ ← *vertices quitting*
  $\mathcal{A}' := \{(i,j) \in \mathcal{A}_c \cap \mathcal{R} : i \in \mathcal{V}', j \in \mathcal{V}'\}$ ← *arcs in induced subgraph*
  $\mathcal{Q} := \mathcal{A}_c \backslash \mathcal{R}$ ← *arcs quitting*
  $z :=$ maximum of (2a)–(2d) for subgraph $G' = (\mathcal{V}', \mathcal{A}')$
  $E := E + z \prod_{i \in \mathcal{V}'} (1 - p_i) \prod_{(i,j) \in \mathcal{A}'} (1 - p_{ij}) \prod_{i \in \mathcal{T}} p_i \prod_{(i,j) \in \mathcal{Q}} p_{ij}$
return E

Algorithm 4. Database construction
(1) Modify algorithms 1 to 3 for returning the *expression* for computing expectation;
(2) Enumerate possible graph configurations up to the desired size;
(3) For all distinct graphs (*i.e.*, non-isomorphic to one already in the database):
(4) Apply the relevant algorithm (1, 2, or 3);
(5) Optionally, simplify the expression obtained with mathematical computation software;
(6) Store in a database the graph as a key and the expression as the data;
(7) *(Use the database for setting up instances of Problem (1a)–(1c))*

non-isomorphic graphs that contain (at least) a cycle of length 4; for graphs with five vertices, this number is 3725 (see [10]). Maximum cycle length of five may, indeed, be the practical limit for this method; interestingly, it is also a common limit to the number of simultaneous transplants in kidney exchange programs.

In order to use this approach, a preliminary step is the construction of a database of all possible graphs containing at least a cycle of relevant length (excluding isomorphic graphs) and their corresponding expressions, as detailed in Algorithm 4. It turns out that, even with a careful (though straightforward) implementation, the most time consuming operation in this algorithm is the precomputation of formulas for expectations for each of the graph configurations.

After the database is constructed, it can be used for setting up the optimization problem, using Algorithm 5. Here, the expectation formula determined with Algorithm 4 and stored in the database is used in line (7) for computing the objective coefficient w_c for each cycle c appearing in the instance. For this purpose, a cycle in the instance is matched with an isomorphic cycle previously enumerated and stored in the database; *i.e.*, a key that matches the current cycle is searched for, the corresponding expectation formula is extracted, and the probabilities from the instance's cycle are mapped to the cycle stored. Finally, a numeric value for the expectation is computed with that formula, and used as the cycle's coefficient in the objective.

Algorithm 5. Solution procedure
(1) Read instance: compatibility between pairs, withdrawal probabilities
(2) Prepare compatibility graph
(3) Enumerate cycles of relevant size
(4) Setup optimization model:
(5) Create one variable for each cycle
(6) Constraints: each vertex in at most one cycle
(7) Objective coefficients: for each cycle, the corresponding expectation
(8) Solve optimization model, Problem (1a)–(1c)

After the optimization instance is prepared, it is solved in line (8) of the algorithm; this is done by means of a general-purpose mixed-integer programming solver. Computationally, this turned out to be relatively inexpensive, compared to the effort required for setting up the instance. The solution of the optimization problem can afterwards be used for the actual implementation of the KEP: contacting the selected pairs, verifying if the solution is indeed possible (*i.e.*, checking if there are no back outs), making a last-minute compatibility check for each pair, possibly rearranging the solution, and finally executing the transplants.

4.1 Computational Experiment

Algorithm 5 has been used to optimize size expectation for vertex-disjoint cycle packing on benchmark instances used in kidney exchange programs, available in [2]. This set, named *blood-type test instances*, includes graphs constructed by means of a generator referred to in [11], which creates random graphs based on probabilities of blood type and donor–patient compatibility. Instances tested here vary on size, from 10 to 1000 vertices; hence, they correspond to realistic KEP situations. Sizes up to 100 vertices include 50 different instances each; for larger sizes there are 10 different instances per size. Each instance's data of [2] was complemented with probabilities for vertex and arc withdrawal, putting together the *non-deterministic* version of the instance[1]. The maximum cycle size considered was three; this allowed the complete enumeration of all relevant cycles for the instances considered, and thus the usage of the proposed algorithms in a straightforward setting (*i.e.*, there was no need of recourse to more sophisticated methods, such as decomposition and column generation, for solving the main optimization problem).

Table 1 presents average results for instances of each size. The average number of cycles of length three, for all instances of each size, is shown in column $\overline{|\mathcal{C}|}$. Values on column $\overline{z}^D$ are average objective values obtained when maximizing the number of transplants in a deterministic setting, *i.e.*, with $w_c = |c|$. Column $\overline{z}^N$ lists, for the same instances, the average objective values obtained for the non-deterministic case. This is the average, for each instance size, of the maximum

[1] Instances' data are available in `http://www.dcc.fc.up.pt/~jpp/code/KEP`

Table 1. Graph size of benchmark instances and average number of cycles with sizes 2 or 3 ($|\overline{\mathcal{C}}|$); average values for the objective value in deterministic and non-deterministic settings, and performance of the corresponding solutions in the reverse situation (mean and standard deviation of percentage of missed operations).

Size	$\lvert\overline{\mathcal{C}}\rvert$	$\overline{z}^D$	$\overline{y}^D$	$\overline{M}^D$	s^D	$\overline{z}^N$	$\overline{y}^N$	$\overline{M}^N$	s^M
10	5.62	2.720	0.170	18.6	28.3	0.243	2.420	5.2	11.2
20	46.02	8.200	0.396	52.2	30.3	1.013	7.000	13.1	13.8
30	109.1	12.26	0.712	47.7	25.1	1.495	10.58	12.3	11.4
40	221.8	17.28	1.046	50.2	20.6	2.213	14.76	13.9	9.1
50	442.2	23.48	1.312	59.4	17.2	3.352	20.04	14.0	6.9
60	722.6	27.64	1.658	59.9	19.4	4.256	24.46	11.0	7.4
70	1117.	34.04	1.886	61.6	10.9	4.932	30.02	11.4	5.4
80	1686.	39.54	2.279	63.5	11.5	6.260	34.00	13.6	6.3
90	2472.	46.18	2.626	63.4	9.1	7.467	40.36	12.4	5.2
100	3264.	51.28	2.864	65.1	9.6	8.314	44.96	12.3	4.8
200	24401.	107.1	5.431	74.7	3.3	21.40	97.20	9.3	2.6
300	87342.	163.6	8.570	75.7	2.9	35.21	150.1	8.4	3.5
400	224763.	221.3	13.06	75.3	3.3	52.93	206.8	6.6	1.7
500	419670.	280.6	16.00	77.4	2.7	71.00	264.0	6.0	1.3
600	791068.	341.8	20.00	77.5	1.6	88.73	322.4	5.7	1.5
700	1057876.	383.3	21.49	78.0	2.1	97.43	364.3	4.9	1.2
800	1586344.	441.8	23.84	79.5	2.1	116.22	425.1	3.8	1.1
900	2343290.	498.8	29.87	78.2	1.8	137.56	481.0	3.6	1.5
1000	2952990.	548.3	33.60	78.2	1.7	153.54	530.3	3.3	1.3

expected number of transplants with withdrawal probabilities on vertices and arcs, obtained with values for w_c as determined by algorithms 4 and 5.

Let us consider an optimal solution to the deterministic version of a given instance. Besides its objective value z^D, we can compute the expectation of the number of transplants that it would lead to in a corresponding non-deterministic setting; let us denote it by y^D. Similarly, let z^N be the objective value for the non-deterministic version of an instance, and y^N the number of transplants that the solution would lead to in the corresponding deterministic setting. The percentage of missed operations when the deterministic solution is used in a probabilistic setting can be calculated as $M^D = 100(z^N - y^D)/z^N$. Analogously, the percentage of missed operations if the non-deterministic solution is used in a completely reliable setting can be calculated as $M^N = 100(z^D - y^N)/z^D$. Averages of these values $\overline{M}^D$, $\overline{M}^N$, and the corresponding standard deviations s^D, s^M, are listed for each problem size in Table 1. These results show that solutions for maximum expectation are much more accurate and robust in a deterministic setting than the inverse. This tendency seems to be more pronounced on large instances (see also Fig. 9).

Concerning CPU times required for solving these problems, the limiting part was cycle enumeration and the evaluation of the corresponding expectation. Indeed, solving the integer optimization problems using Gurobi version 5.0.1 [12]

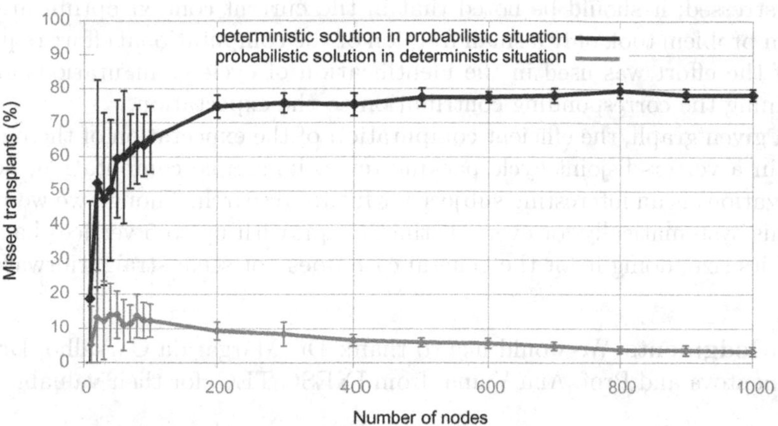

Fig. 9. Percentage of missed operations if deterministic and non-deterministic solutions are used in the reverse situation as a function of the instance size (mean values plus/minus standard deviation).

required less that 0.1 second for the 10-vertex instances, less than 1 second for 100 vertices, and about 400 seconds for the 1000-vertex instances. Cycle enumeration and evaluation, implemented in the Python language, also used less that one second for instances with 100 or fewer vertices, but the time required increased up to more that 5000 seconds for the largest, 1000-vertex instances. The system used was the following: a computer with an Intel Xeon processor at 3.0 GHz, running Linux version 2.6.32; only one thread was assigned to this experiment.

5 Conclusions

This paper proposes a method for computing the maximum expectation for the length of a set of vertex-disjoint cycles in a digraph where vertices and/or arcs are subject to failure with a known probability. This problem has a practical application on maximizing the expected number of transplants in a kidney exchange program. It makes use of a formulation where all cycles of feasible length are enumerated. An intermediate step in the solution procedure is the construction of a database associating cycle configurations to the corresponding expectations, for all small graphs of relevant size; this is employed for setting up the main optimization problem.

A set of well-known benchmark instances was used for testing the procedure, considering a maximum cycle-size of three; the experiment unveiled a superior robustness of models maximizing expectation, as opposed to models maximizing total cycle length in a deterministic setting. This is a clear indication that probabilistic information should be taken into account for the solution of this problem.

In usual situations the difficulty of the solution of NP-hard optimization problems is stressed; it should be noted that in the current context optimization of the main problem took only a small fraction of the computational effort required. Most of the effort was used in the identification of cycle configurations and in determining the corresponding contribution to the expectation.

For a given graph, the efficient computation of the expectation of the number of arcs in a vertex-disjoint cycle packing (even its simple computation, not its maximization) is an interesting subject for future research. Though we were able to do this systematically for every distinct graph with up to 5 vertices having a cycle of its size, doing it for the general case does not seem straightforward.

Acknowledgments We would like to thank Dr. Margarida Carvalho, Dr. Xenia Klimentova and Prof. Ana Viana, from INESC TEC, for their valuable comments.

This work is partially supported by the ERDF — European Regional Development Fund through the COMPETE Programme (operational programme for competitiveness) and by National Funds through the FCT — Fundação para a Ciência e a Tecnologia (Portuguese Foundation for Science and Technology) within project "KEP - New models for enhancing the kidney transplantation process" (FCT ref: PTDC/EGE-GES/110940/2009).

References

1. Li, Y., Kalbfleisch, J., Song, P.X., Zhou, Y., Leichtman, A., Rees, M.: Optimization and simulation of an evolving kidney paired donation (KPD) program. Working Paper Series 90, Department of Biostatistics, University of Michigan (May 2011), `http://www.bepress.com/umichbiostat/paper90`
2. Constantino, M., Klimentova, X., Viana, A., Rais, A.: New insights on integer-programming models for the kidney exchange problem. European Journal of Operational Research 231(1), 57–68 (2013)
3. Chen, Y., Li, Y., Kalbfleisch, J.D., Zhou, Y., Leichtman, A., Song, P.X.K.: Graph-based optimization algorithm and software on kidney exchanges. IEEE Trans. Biomed. Engineering 59(7), 1985–1991 (2012)
4. Ünver, M.U.: Dynamic kidney exchange. Review of Economic Studies 77(1), 372–414 (2010)
5. Sönmez, T., Ünver, M.U.: Market Design for Kidney Exchange. In: Oxford Handbook of Market Design. Oxford University Press (2013)
6. Dickerson, J.P., Procaccia, A.D., Sandholm, T.: Optimizing kidney exchange with transplant chains: theory and reality. In: van der Hoek, W., Padgham, L., Conitzer, V., Winikoff, M. (eds.) AAMAS, pp. 711–718. IFAAMAS (2012)
7. Abraham, D.J., Blum, A., Sandholm, T.: Clearing algorithms for barter exchange markets: enabling nationwide kidney exchanges. In: MacKie-Mason, J.K., Parkes, D.C., Resnick, P. (eds.) ACM Conference on Electronic Commerce, pp. 295–304. ACM (2007)
8. Edmonds, J.: Paths, trees, and flowers. Canadian Journal of Mathematics. Journal Canadien de Mathématiques 17, 449–467 (1965)

9. Kuhn, H.W.: The Hungarian Method for the assignment problem. Naval Research Logistics Quarterly 2, 83–97 (1955)
10. Harary, F., Palmer, E.M.: Graphical Enumeration. Academic Press, New York (1973)
11. Saidman, S.L., Roth, A.E., Sönmez, T., Ünver, M.U., Delmonico, F.L.: Increasing the opportunity of live kidney donation by matching for two- and three-way exchanges. Transplantation 81, 773–782 (2006)
12. Gurobi Optimization, Inc.: Gurobi Optimizer Reference Manual, Version 5.0 (2012), `http://www.gurobi.com`
13. Wolfram Research, Inc.: Mathematica. Version 8.0 edn. Wolfram Research, Inc., Champaign, Illinois (2010)

A Graphs for Vextex Failure

Given the probability p_i for failure of vertex i, and/or probability p_{ij} of failure of arc (i, j), expressions of the expectation for the size of the cycle is presented for all cycle configurations in two- and three-vertex graphs. The original, more lengthy expressions were simplified using the mathematical computation software *Mathematica*[13].

A.1 Graphs with Two Vertices (1 Configuration)

Graph	Expectation with failure on
	vertices: $2(p_1-1)(p_2-1)$ arcs: $2(p_{12}-1)(p_{21}-1)$ both: $2(p_1-1)(p_2-1)(p_{12}-1)(p_{21}-1)$

A.2 Graphs with Three Vertices (4 Configurations)

Graph	Expectation with failure on
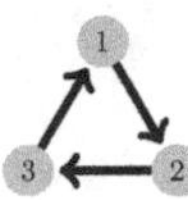	vertices: $-3(p_1-1)(p_2-1)(p_3-1)$ arcs: $-3(p_{12}-1)(p_{23}-1)(p_{31}-1)$ both: $3(p_1-1)(p_2-1)(p_3-1)(p_{12}-1)(p_{23}-1)(p_{31}-1)$
	vertices: $-(p_1-1)(p_2-3)(p_3-1)$ arcs: $-(p_{12}(2p_{13}+1)(p_{23}-1)-(2p_{13}+1)p_{23}+3)(p_{31}-1)$ both: $(p_1-1)(p_3-1)(p_2(p_{12}-1)(2p_{13}+1)(p_{23}-1)$ $-p_{12}(2p_{13}+1)(p_{23}-1)+2p_{13}p_{23}+p_{23}-3)(p_{31}-1)$
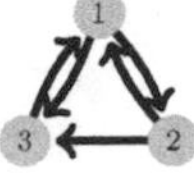	vertices: $(p_1-1)(p_3+p_2(p_3+1)-3)$ arcs: $3+(2p_{13}p_{21}+1)p_{23}(p_{31}-1)-2p_{21}p_{31}-p_{31}$ $-p_{12}(p_{23}(p_{31}-1)+2p_{13}(p_{21}p_{23}-1)(p_{31}-1)-2p_{21}p_{31}+p_{31}+1)$ both: (see A.3)
	vertices: $-p_3-p_2(p_3+1)+p_1(-p_3+p_2(3p_3-1)-1)+3$ arcs: (see A.3) both: (this expression would not fit in one A4 page)

A.3 Longer Expectation Expressions

First case: unreliable vertices and arcs are considered for graph

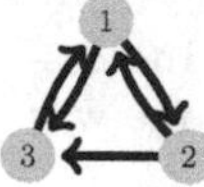

$$G = (\{1,2,3\}, \{(1,2),(1,3),(2,1),(2,3),(3,1)\}).$$

The expression for expectation when vertices and arcs are subject to failure is the following:

$$\begin{aligned}&(p_1-1)\left(2p_{13}p_{12}-2p_{13}p_{21}p_{23}p_{12}-p_{23}p_{12}-2p_{13}p_{31}p_{12}-2p_{21}p_{31}p_{12}+2p_{13}p_{21}p_{23}p_{31}p_{12}\right.\\&\quad+p_{23}p_{31}p_{12}+p_{31}p_{12}+p_{12}+2p_{13}p_{21}p_{23}+p_{23}-p_3\left(-p_{23}+p_{21}\left(2-2p_{13}p_{23}\right)\right.\\&\quad+p_{12}\left(-2p_{21}+p_{23}+2p_{13}\left(p_{21}p_{23}-1\right)+1\right)+1)\left(p_{31}-1\right)+2p_{21}p_{31}-2p_{13}p_{21}p_{23}p_{31}\\&\quad-p_{23}p_{31}+p_{31}+p_2\left(p_{12}-1\right)\left(-p_{31}p_{23}+p_{23}-2p_{13}\left(p_{21}p_{23}-1\right)\left(p_{31}-1\right)\right.\\&\qquad\left.\left.+p_3\left(-2p_{21}+p_{23}+2p_{13}\left(p_{21}p_{23}-1\right)+1\right)\left(p_{31}-1\right)+2p_{21}p_{31}-p_{31}-1\right)-3\right)\end{aligned}$$

Second case: reliable vertices and unreliable arcs are considered for graph

$$G = (\{1,2,3\}, \{(1,2),(1,3),(2,1),(2,3),(3,1),(3,2)\}).$$

The expression for expectation when only arcs are subject to failure is the following:

$$\begin{aligned}&-p_{21}p_{23}-p_{21}p_{31}p_{23}+p_{21}p_{32}p_{23}+p_{21}p_{31}p_{32}p_{23}+p_{31}p_{32}p_{23}-p_{32}p_{23}-p_{21}p_{31}\\&-p_{21}p_{31}p_{32}-p_{31}p_{32}+p_{13}\left(p_{23}\left(p_{31}-1\right)\left(-p_{32}+p_{21}\left(p_{32}+1\right)+1\right)-\left(p_{21}-1\right)p_{31}\left(p_{32}-1\right)\right)\\&+p_{12}\left(-\left(p_{23}\left(p_{31}-1\right)+p_{31}+1\right)p_{32}+p_{21}\left(p_{23}\left(p_{31}-1\right)\left(p_{32}-1\right)+p_{32}+p_{31}\left(p_{32}+1\right)-1\right)\right.\\&\quad+p_{13}\left(p_{23}\left(p_{31}+1\right)\left(p_{32}-1\right)+\left(p_{31}-1\right)\left(p_{32}+1\right)+p_{21}\left(\left(p_{31}-1\right)\left(p_{32}-1\right)\right.\right.\\&\qquad\left.\left.\left.+p_{23}\left(p_{31}\left(1-3p_{32}\right)+p_{32}+1\right)\right)\right)\right)+3\end{aligned}$$

On Modelling Approaches for Planning and Scheduling in Food Processing Industry

G.D.H. Claassen[1] and Eligius M.T. Hendrix[2]

[1] Operations Research and Logistics, Wageningen University
frits.claassen@wur.nl

[2] Computer Architecture, Universidad de Málaga
eligius@uma.es

Abstract. We consider developments in lot-sizing and scheduling, particularly relevant for problem settings arising in food processing industry. Food processing industry (FPI) reveals several specific characteristics which make integrated production planning and scheduling a challenge. First of all, setups are usually sequence-dependent and may include the so-called non-triangular setup conditions. Secondly, planning problems in FPI have to deal with product decay due to deterioration of inventory. We give an overview of lot-sizing and scheduling models, and assess their suitability for addressing sequence-dependent setups, non-triangular setups and product decay. We show that a trend exists towards so-called big bucket models. However, the advantage of these approaches may become a major obstacle in addressing the identified characteristics in FPI.

Keywords: Capacitated lot-sizing and scheduling, sequence-dependent setups, non-triangular setups, perishability, product decay, food processing industry.

1 Introduction

Adequate and efficient production planning and scheduling is one of the most challenging problems for present-days enterprises. Especially scheduling and sizing of production lots, is an area of increasing research attention within the wider field of production planning and scheduling, Clark et al. (2011). Although lot-sizing problems have been studied extensively, most of the literature is focused on discrete manufacturing. Moreover, there is an on-going research trend directed towards incorporating real-world issues and specificities of simultaneous lot-sizing and scheduling, Jans and Degraeve (2008); Quadt and Kuhn (2008).

Lot-sizing and scheduling in Food Processing Industry (FPI) is usually more complex than in other continuous and discrete processing environments. This is primarily due to issues like inevitable decline in quality of products, related quality requirements and safety regulations of products, market driven standards regarding shelf life and variability of demand and prices. Secondly, the diversity of products in FPI increased considerably in the past decades and global competition on the food market has forced manufacturers to participate in an on-going

B. Murgante et al. (Eds.): ICCSA 2014, Part II, LNCS 8580, pp. 47–59, 2014.

trend towards increased variety (i.e. ingredients and flavors, customized packaging, prints and/or labels) of (new) products. Soman et al. (2004b) state that the majority of research contributions do not address specific characteristics of food processing, e.g. high capacity utilization, sequence-dependent setups and limited shelf life due to product decay.

Production lines in FPI usually operate under tight capacity constraints. As products take the same route, a production line may be planned as a single resource. Changeovers between products that share the same line in a food processing environment often imply that both changeover costs and times depend on the production sequence of individual items. In order to avoid unnecessary changeovers and improve efficient use of available production capacity, customer demand has to be pooled in production orders (lots). When sequence-dependent setup times are predominant, available capacity for production depends on both the sequence and the size of the lots. In such a situation, lot-sizing and scheduling should be applied simultaneously, Meyr (2000).

In general practice, lot-sizing and scheduling problems are solved separately in successive hierarchical phases, (Soman et al., 2004a; Claassen and van Beek, 1993; Drexl and Kimms, 1997; Kreipl and Pinedo, 2004; Soman et al., 2007). First optimal lot-sizes for given product families are determined and afterwards production schedules of customer orders are generated. The generated schedules on the shop floor often fail to realize production targets because changeover losses are not correctly accounted for on a higher planning level. As a consequence, the planning process has to be redone (with or without over-time) and/or frequent rescheduling takes place in daily practice, Kreipl and Pinedo (2004). Currently, there exists a general consensus regarding a closer integration of lot-sizing and scheduling decisions, Clark et al. (2011); Jans and Degraeve (2008); Meyr (2000); Gupta and Magnusson (2005); Almada-Lobo et al. (2008); Menezes et al. (2011).

Although the survey of Drexl and Kimms (1997) already focused on the integration of lot-sizing and scheduling, Jans and Degraeve (2008) conclude after another decade in their review that the boundaries between lot-sizing and scheduling are fading, but further integration still constitutes a challenging research track. The latter may explain why even in present-days Advanced Planning and Scheduling (APS) systems, the planning and scheduling modules are seen as unusable, or unable to handle the complexity of the underlying capacitated planning problems, Pochet and Wolsey (2006).

Planning (i.e. lot-sizing) models differ from scheduling models in a number of ways. Kreipl and Pinedo (2004) give an extensive overview of practical issues for planning and scheduling processes. In a recent special issue on lot-sizing and scheduling, Clark et al. (2011) confirm the need for more realistic and practical variants of models for simultaneous lot-sizing and scheduling. Features such as (i) non-triangular setups, (ii) perishability, and (iii) delivery time windows are labelled by the authors as hot topics and open research opportunities.

In this paper we focus on two interrelated problem characteristics that argue the need for simultaneous planning and scheduling, particularly in FPI:

(i) Sequence-dependent setups and non-triangular setups. With respect to sequence-dependent setup costs and times under tight capacity constraints there is a complicating issue, referred to as the triangular setup conditions Clark et al. (2011); Gupta and Magnusson (2005); Almada-Lobo et al. (2008), that holds for FPI too, Menezes et al. (2011). Due to processing conditions of different product variants (e.g. several heating and/or cooling levels) and other product specific requirements (e.g. flavors, addition of specific additives and/or the danger of contamination between subsequent production runs), the common assumption regarding the triangular setup conditions often does not hold in FPI. If these conditions do not hold, it implies that changeover costs and times between two subsequent products i and j may become substantially less by processing another product k between i and j. As a consequence, applying models that assume triangular setup conditions may generate non-consistent solutions from a scheduling point of view.

(ii) Product decay. The quality or value of perishable food products usually deteriorates rapidly after production. Product decay may be delayed by conditioned storage, but quality depends on product age, and restricted shelf-lives are inevitable. Considering product decay in lot-sizing enforces smaller production quantities for perishable products. Consequently, individual products are produced i.e. scheduled at higher frequency. This increases the difficulty of sequencing.

This paper intends to contribute to the recognized need for more realistic variants of models for simultaneous lot-sizing and scheduling under tight capacity constraints Clark et al. (2011); Jans and Degraeve (2008); Almada-Lobo et al. (2008); Menezes et al. (2011); Almada-Lobo et al. (2007). We give an overview of model developments for simultaneous lot-sizing and scheduling directive for a problem formulation with the following characteristics: a multi-item, single machine lot-sizing and scheduling problem for FPI with sequence-dependent setup costs and times and product decay. The setup state of the machine should be preserved over period boundaries including idle time (i.e. setup carry-overs) and any additional assumption with respect to the changeover matrices should be relaxed (e.g. the triangular setup conditions).

We assess the proposed models for addressing sequence-dependent setups (including non-triangular setups) and product decay. The objective here is to focus on modelling developments in time that are directive for the identified problem characteristics, and to expose their shortcomings and disadvantages. For a general overview of lot-sizing problems we refer to several reviews of the past, Kuik et al. (1994); Drexl and Kimms (1997); Karimi et al. (2003) and two more recent overviews: Jans and Degraeve (2008); Quadt and Kuhn (2008).

The overview in this paper shows that a trend exists of preferred modelling approaches. However, these approaches may i) disrupt a crucial balance between total setup costs and inventory-holdings costs and ii) hamper a further integration between production and distribution planning. We state that crucial aspects for integrated planning and scheduling may unfoundedly disappear from sight.

One of the most important features of models for lot-sizing and scheduling is the segmentation of the planning horizon. From a modelling point of view it is convenient to distinguish two general classes of models (Eppen and Martin, 1987), i.e. small bucket (SB) and big (or large) bucket (BB) modelling approaches. In SB models, the planning horizon is divided into a finite number of small time periods such that in each period either at most two products can be produced, or there will be no production at all. Conversely, in BB approaches the planning horizon is divided into longer periods, usually of the same length. In each period, multiple products may be produced. As a consequence, SB models have been applied mostly over short time planning horizons and BB models are usually associated with medium term planning horizons.

Sections 2 and 3 provide an overview of model developments for SB and BB approaches respectively. Section 4 describes the state of affairs regarding issues of product decay for lot-sizing and scheduling. A summary can be found in Section 5 to analyze the literature overview. Section 6 concludes.

2 Small Bucket Approaches

Crucial for small bucket modelling approaches is that at most one set-up may occur in a period. In this class of models, the so-called all-or-nothing assumption usually holds. In most models only one item may be produced in a time interval and, if so, production uses (in most cases) full capacity. In SB models, lot-sizes include the production of the same product for one or several consecutive periods. Alternatively, if a setup is performed and when it comes to non-zero setup times, both setups and production runs comprise a number of time intervals. A lot includes the production of a single product for one or several consecutive periods. Next, we discuss development in time of SB-approaches.

2.1 DLSP: Discrete Lot-sizing and Scheduling Problem

The Discrete Lot-sizing and Scheduling Problem (DLSP) is a typical example within the class of small bucket approaches. The basic DLSP includes (sequence-independent) setup costs and setup carry-over but at zero setup times, Fleischmann (1990). Inclusion of setup carry-over implies that setup states of a machine are carried over between period boundaries. Porkka et al. (2003) compare models with and without setup carry-overs. The authors show that substantial savings (regarding costs and production time) can be derived from fundamentally different production plans enforced by carry-overs. Comparable results are found by Sox and Gao (1999). However, in the basic DLSP, setup states are not preserved over idle time. Sequence-dependent setup costs and times are neither considered in the DLSP. Many extensions of the (basic) DLSP have been described in literature. We refer to Drexl and Kimms (1997) and Salomon et al. (1991) for a broader view on variants of the DLSP.

2.2 Extensions of the DLSP

Fleischmann (1994) analyses the multi-item single machine DLSP with sequence-dependent setup costs. An artificial product (i=0) is introduced to represent idleness of the machine. Salomon et al. (1997) continue this work and reformulate the DLSP to capture the characteristic of sequence-dependent times (DLSPSD). However, the triangular setup conditions are assumed to hold. Machine idleness is represented by an artificial product. Jordan and Drexl (1998) present a comparable model in which idleness is indicated by an artificial product too. It should be mentioned that for models in which idleness is represented by an artificial product (i=0), the changeover matrix must comply with strict conditions to cope with sequence-dependent setup times. In all other cases the setup state of the machine is not correctly carried over across the boundaries of idleness.

Wolsey (1997) extended the work of Constantino (1996) for problems with sequence-independent setups to formulations with sequence-dependent setup times and costs. In this paper, the presented model will be referred to as (GSB), i.e. the general small bucket model. In the (GSB), idleness is not represented by an artificial product (i=0). However, the triangular setup conditions should hold.

2.3 CSLP: Continuous Setup Lot-sizing Problem

An early paper in which sequence-dependent costs are modelled is due to Karmarkar and Schrage (1985). Their model is called the Continuous Setup Lot-sizing Problem (CSLP). The CSLP is closely related to the DLSP. Main difference is that the CSLP allows production of quantities less than the available production capacity in a time period. Still, at most one product can be produced in each time interval.

2.4 PLSP: Proportional Lot-sizing and Scheduling Problem

The fundamental assumptions of the DLSP and the CSLP stimulated Drexl and Haase (1995) to study a new type of model, the Proportional Lot-sizing and Scheduling Problem (PLSP). The PLSP is based on a widening of the common all-or-nothing production principle in SB models. The PLSP assumes that at most one setup may occur within a period. Hence, at most two products can be produced in a period. Main difference between the PLSP and the DLSP is the possibility to compute continuous lot-sizes and to preserve the setup state over idle time. However, setup costs and times of (extended) PLSP formulations are considered to be sequence-independent, Suerie (2006).

3 Big Bucket Approaches

In contrast to small bucket models, the planning horizon of a big bucket (BB) model is usually divided into longer periods, mostly of equal length. Time intervals in a BB model may represent a time slot of one week (or more) in the real world, Drexl and Kimms (1997). In each period, multiple products can be manufactured. Relaxing the all-or-nothing production principle of (most) SB models implies that a BB model includes the possibility to determine continuous lot-sizes.

3.1 CLSP: Capacitated Lot-sizing Problem

The Capacitated Lot-Sizing Problem (CLSP) is a typical example of a big bucket model. It is closely related to the (small bucket) DLSP; decision variables, parameters and objective function are the same for both problems,(Drexl and Kimms, 1997). However, sequence-dependent setup costs and times, or more in general scheduling decisions, are not integrated into the CLSP. As a consequence, setup carry-overs between period boundaries are not included either. Suerie and Stadtler (2003) use the simple plant location problem to obtain a tight and new model formulation for setup carry-overs in the CLSP with sequence-independent setup costs and times.

3.2 GCLP: Generalized Capacitated Lot-sizing Problem

Sox and Gao (1999) introduce the Generalized Capacitated Lot-sizing Problem (GCLP). The GCLP uses less binary variables for including setup carry-overs in the CLSP with sequence-independent setup costs and no setup times. Sequence-independent setup times can be included; probably at the expense of additional computational effort. The authors also apply the network reformulation approach as proposed by Eppen and Martin (1987) and compare the behavior of a set of models. The results demonstrate that incorporating setup carry-over has a significant effect on both costs and lot-sizes.

In all aforementioned BB approaches, the emphasis is directed towards combining characteristics of a big bucket model like the CLSP (i.e. allow production of more products per period without setup carry-overs) with a small bucket model like the DLSP (production of only one product per period with setup carry-overs) in a single framework. Still, sequence-dependent setup costs and times are not considered in the BB models above.

3.3 GLSP: General Lot-sizing and Scheduling Problem

Fleischmann and Meyr (1997) proposed a combination of CLSP and DLSP, i.e. the General Lot-sizing and Scheduling Problem (GLSP). The GLSP is a big bucket model in which the planning horizon is divided into T macro-periods. To obtain the production sequence of the items, each macro-period is subdivided into a subset of micro-periods of variable length. The GSLP assumes all-or-nothing production for micro-periods. The number of micro-periods within each macro-period must be fixed in advance in the MIP model. As a consequence, a lot (i.e. a sequence of micro-periods assigned to the same item) may contain idle micro-periods. Sequence-dependent costs are considered, but setup times are disregarded in the (basic) GLSP. In order to cope with cases in which the triangular setup conditions of the cost matrix do not hold, the authors introduce minimum lot-sizes. Meyr (2000) extended the GLSP with sequence-dependent setup times. Again, minimum lot-sizes are used to avoid a wrong evaluation of setup costs (and setup time, respectively) if the setup matrices do not satisfy the triangular setup conditions. It should be mentioned that the introduction of minimum lot-sizes may have an impact on economical lot-sizes. Transchel et al. (2011) present a tailored hybrid mixed-binary model based on the GLSP for a practical problem from process industry and show that minimum production quantities affect the MIP performance for real world test instances. Ferreira et al. (2009) present a GLSP-based model too that integrates production lot-sizing and scheduling decisions for a Brazilian soft drink plant.

Block planning approaches can be regarded as a practical variant of the GLSP in which macro- (i.e. blocks) and micro periods are distinguished. An important assumption in block planning approaches is a predefined production sequence of (variable) batch-sizes, Lütke entrup et al. (2005); Bilgen and Günther (2010); Baumann and Trautmann (2012). In other words, there is a unique period-block assignment and each product occurs at the same given position (micro-period) in each block. As a consequence, within the planning horizon of T periods, each product i=1..N is scheduled T times. The number of production lots in the schedule equals N*T. We refer to Günther et al. (2006) for a complete description of block planning.

3.4 Extensions of the CLSP

A study to extend the CLSP was initiated by Gopalakrishnan et al. (1995). The authors developed a modelling framework for the (single machine) CLSP with setup carry-overs. Setup times and costs were assumed to be constant across all products and time periods. This assumption was relaxed in a modified framework that included product-dependent and sequence-independent setup costs and times by Gopalakrishnan (2000). Haase (1996) takes the CLSP as a starting point but extends the model with sequence-dependent setup costs. Moreover, the setup state of the machine can be preserved over idle times. The model formulation does not consider (sequence-dependent) setup times and it is assumed that the triangular setup conditions for setup costs hold. Haase and Kimms (2000) consider both sequence-dependent setup costs and times. It is assumed that setup times satisfy the triangular setup conditions. The authors formulate the problem by considering only efficient (predefined) production sequences. Efficient sequences are found by solving a travelling salesman problem.

Gupta and Magnusson (2005) extend the framework of Gopalakrishnan (2000) by including sequence-dependent setup times and setup costs. From a scheduling point of view, the CLSP with sequence-dependent setup times is closely related to the travelling salesman problem (TSP). In every period a connected tour (or sequence) between multiple products has to be determined. The distance matrix in the TSP corresponds to the matrix of setup costs in the (extended) CLSP. Almada-Lobo et al. (2008) show that the model formulation as proposed by Gupta and Magnusson (2005) does not eliminate disconnected sub tours. As a consequence, it may generate infeasible solutions. Almada-Lobo et al. (2007) present two correct model formulations for the identified problem characteristics, provided that the triangular setup conditions with respect to the setup matrices (costs and times), hold. In order to avoid disconnected sub tours, the authors add a polynomial set of sub tour elimination constraints. Menezes et al. (2011) present an extension of the CLSP which handles non-triangular setup costs and times while enforcing minimum lot-sizes.

Next, we focus on papers that discuss lot-sizing and scheduling of perishable products.

4 Product Decay

Although a vast body of literature exists on inventory management for perishable products, surprisingly little has been done to include product decay in traditional lot-sizing and scheduling models. One of the first contributions in this area is provided by Soman et al. (2004a). The authors focus on shelf life considerations in the economic lot scheduling problem (ELSP). Models of this class usually assume constant demand, do not account for sequence-dependent setup times and aim to generate production cycles for several products on a single resource. Lütke entrup et al. (2005) propose three MILP models that integrate shelf-life issues into production planning and scheduling for an industrial case study of yoghurt production. The authors apply a block planning approach (see Section 3.3) in which a block covers all products based on the same recipe. Shelf-life aspects are taken into account by considering a shelf-life-dependent pricing component that may also include inventory-holding costs. Chen et al. (2009) and Kopanos et al. (2012) argue the need to develop models for better coordination between production scheduling and vehicle routing for perishable food products. Lee and Yoon (2010) consider a coordinated production-and-delivery scheduling problem that incorporates different inventory-holding costs between production and delivery stages. The results may only apply to specific situations but the study can be regarded as a first attempt to allow different (stage-dependent) inventory-holding costs. Chen et al. (2009) conclude that papers discussing production scheduling and/or distribution of perishable goods are relatively rare. Amorim et al. (2011) state that papers discussing simultaneous lot-sizing and scheduling for perishable goods are even rarer. These authors deal with simultaneous lot-sizing and scheduling of perishable products using a multi-objective framework. The main idea is to separate economic production tangible costs from intangible value of having fresher products in two conflicting objectives.

5 Summary

Table 1 gives an overview of key publications that are directive for a problem formulation which is characterized as a multi-item, single machine lot-sizing and scheduling problem with sequence-dependent setup costs and times. The setup

Table 1. Overview of key publications. Seq-Dep: Sequence dependent, n-trian: Non-triangular, c-over: carry over.

					Setup	Seq - Dep		n-trian
Year	Author	model	SB	BB	c-over	cost	time	setup
DLSP			x		-	-	-	-
1985	Karmarkar & Schrage	CLSP	x		-	x	-	-
1994	Fleischmann	DLSP	x		-	x	-	-
1995	Drexl & Haase	PLSP	x		x	-	-	-
1997	Salomon et al.	DLSP	x		-	x	-	-
1997	Wolsey	GSB	x		x	x	x	-
CLSP				x	-	-	-	-
1996	Haase	CLSP		x	x	x	-	-
1997	Fleischmann & Meyr	GLSP	x	x	x	x	-	-
1999	Sox & Goa	GCLP		x	x	-	-	-
2000	Gopalakrishnan et al.	CLSP		x	x	-	-	-
2000	Meyr	GLSP	x	x	x	x	x	-
2003	Surie & Stadtler	CLSPL		x	x	-	-	-
2005	Gupta & Magnussen	CLSP		x	x	x	-	-
2007	Almada-Lobo et al.	CLSP		x	x	x	x	-
2011	Menezes et al.	CLSP		x	x	x	x	x

state of the machine should be preserved over period boundaries (including idle time) and any additional assumption with respect to the changeover matrix (e.g. the triangular setup conditions) is excluded.

Note that Table 1 only refers to the presented model formulations and not to the proposed solution approaches. The (GSB) is a SB formulation for the specified problem without product decay, provided that the triangular setup conditions hold. If these conditions do not hold, setup state changes will occur without production changes. Literature shows that there is a clear tendency to propose BB models for short time horizons too. Moreover, both the survey of Quadt and Kuhn (2008) and the results in Table 1 reveal an interesting trend in which BB approaches are preferred to SB models.

6 Conclusions

Purpose of this paper is to study how literature deals with realistic variants of models for simultaneous lot-sizing and scheduling. We consider developments in lot-sizing and scheduling particularly relevant for problem settings arising in food processing industry (FPI) and focus on i) sequence-dependent setups (including non-triangular setups), ii) product decay of inventory due to perishability, and (iii) a desired tuning of modules for production planning with physical distribution planning (i.e. delivery time windows). Although Big Bucket (BB) models are usually associated with medium term planning horizons, literature reveals an interesting trend in which these models are proposed for short term planning horizons too. From a computational point of view, models with large time intervals (i.e. a week) are preferred over Small Bucket (SB) approaches. However, we argue that segmentation of the planning horizon is a key issue for simultaneous lot-sizing and scheduling. The observed preference for segmentation in BB approaches implies that the following crucial aspects may disappear from sight:

- Main principle of optimality for lot-sizing models
  The general accepted principle of optimality for lot-sizing models is based on the best compromise between total production costs on the one hand and total inventory-holding costs on the other hand. Inventory costs in lot-sizing models are calculated from inventory levels at the end of each period. As time intervals in BB models represent long periods (e.g. a week) multiple batches can be produced in a single period. As a consequence, inventory costs for batches manufactured at the start of periods are assumed to be equal to inventory costs of batches produced at the end of the same period. As a result, total inventory-holding costs are underestimated and the crucial principle of optimality for lot-sizing problems may be disrupted. Segmentation of the planning horizon is the key in modelling this balance correctly. SB models offer the framework to calculate these costs more adequately.
- Decline of product quality and limited shelf life
  In FPI, product decay is primarily associated with the age of products. Incorporating perishability issues like product decay requires defining the moments of production for manufactured products, unambiguously. Segmentation of the planning horizon is the key to capture the age of manufactured products.
- Delivery time windows for physical distribution
  Obviously, a close coordination of production scheduling and delivery planning will become an important issue, Chen et al. (2009); Clark et al. (2011). Products in FPI usually include highly perishable items that must be delivered within allowable time frames. In order to contribute to improved logistical performance, production planning and scheduling modules for FPI should have, at least to a certain extent, the flexibility to take issues for physical distribution into consideration. In contrast to BB models, SB approaches offer the time frame to attune short term physical distribution planning to production planning and scheduling, e.g. by assigning demand to specific time slots in a 24-hours production environment.
- Scheduling lots
  A major advantage of small time intervals may be a better control of the sequence of lots. Using large time intervals implies that sequencing the lots within each period may become complex. Moreover, planned maintenance for production facilities can be scheduled much easier and accurately by applying SB models. In each period of a BB model, a sequencing problem (i.e. travelling salesman problem) has to be solved. Incorporating this feature may become complex, especially in case the triangular setup conditions do not hold.

Despite of a larger number of time periods in the planning horizon, the strengths of SB approaches can be used to develop a SB model that (i) can handle sequence-dependent setups (including non-triangular setups), (ii) addresses product decay by using age-dependent holding costs. Such a model offers (iii) a starting point to integrate solutions of a production planning module with delivery time windows for physical distribution of products to customers.

Acknowledgments. This work has been funded by grants from the Spanish Ministry (TIN2008-01117, TIN2012-37483-C03-01) and Junta de Andalucía (P11-TIC-7176), in part financed by the European Regional Development Fund (ERDF).

References

Almada-Lobo, B., Klabjan, D., Carravilla, M.A., Oliveira, J.F.: Single machine multi-product capacitated lot sizing with sequence-dependent setups. International Journal of Production Research 45(20), 4873–4894 (2007)

Almada-Lobo, B., Oliveira, J.F., Carravilla, M.A.: A note on "the capacitated lot-sizing and scheduling problem with sequence-dependent setup costs and setup times. Computers & Operations Research 35(4), 1374–1376 (2008)

Amorim, P., Antunes, C.H., Almada-Lobo, B.: Multi-objective lot-sizing and scheduling dealing with perishability issues. Industrial & Engineering Chemistry Research 50(6), 3371–3381 (2011)

Baumann, P., Trautmann, N.: A continuous-time milp model for short-term scheduling of make-and-pack production processes. International Journal of Production Research 51(6), 1707–1727 (2012)

Bilgen, B., Günther, H.O.: Integrated production and distribution planning in the fast moving consumer goods industry: a block planning application. OR Spectrum 32(4), 927–955 (2010)

Chen, H.-K., Hsueh, C.-F., Chang, M.-S.: Production scheduling and vehicle routing with time windows for perishable food products. Computers & Operations Research 36(7), 2311–2319 (2009)

Claassen, G.D.H., van Beek, P.: Planning and scheduling packaging lines in food-industry. European Journal of Operational Research 70(2), 150–158 (1993)

Clark, A., Almada-Lobo, B., Almeder, C.: Lot sizing and scheduling: industrial extensions and research opportunities. International Journal of Production Research 49(9), 2457–2461 (2011)

Constantino, M.: A cutting plane approach to capacitated lot-sizing with start-up costs. Mathematical Programming 75(3), 353–376 (1996)

Drexl, A., Haase, K.: Proportional lotsizing and scheduling. International Journal of Production Economics 40(1), 73–87 (1995)

Drexl, A., Kimms, A.: Lot sizing and scheduling - survey and extensions. European Journal of Operational Research 99(2), 221–235 (1997)

Eppen, G.D., Martin, R.K.: Solving multi-item capacitated lot-sizing problems using variable redefinition. Operations Research 35(6), 832–848 (1987)

Ferreira, D., Morabito, R., Rangel, S.: Solution approaches for the soft drink integrated production lot sizing and scheduling problem. European Journal of Operational Research 196(2), 697–706 (2009)

Fleischmann, B.: The discrete lot-sizing and scheduling problem. European Journal of Operational Research 44(3), 337–348 (1990)

Fleischmann, B.: The discrete lot-sizing and scheduling problem with sequence-dependent setup costs. European Journal of Operational Research 75(2), 395–404 (1994)

Fleischmann, B., Meyr, H.: The general lotsizing and scheduling problem. Or Spektrum 19(1), 11–21 (1997)

Gopalakrishnan, M.: A modified framework for modelling set-up carryover in the capacitated lotsizing problem. International Journal of Production Research 38(14), 3421–3424 (2000)

Gopalakrishnan, M., Miller, D.M., Schmidt, C.P.: A framework for modeling setup carryover in the capacitated lot-sizing problem. International Journal of Production Research 33(7), 1973–1988 (1995)

Günther, H.O., Grunow, M., Neuhaus, U.: Realizing block planning concepts in make-and-pack production using milp modelling and sap apo. International Journal of Production Research 44(18-19), 3711–3726 (2006)

Gupta, D., Magnusson, T.: The capacitated lot-sizing and scheduling problem with sequence-dependent setup costs and setup times. Computers & Operations Research 32(4), 727–747 (2005)

Haase, K.: Capacitated lot-sizing with sequence dependent setup costs. Or Spektrum 18(1), 51–59 (1996)

Haase, K., Kimms, A.: Lot sizing and scheduling with sequence-dependent setup costs and times and efficient rescheduling opportunities. International Journal of Production Economics 66(2), 159–169 (2000)

Jans, R., Degraeve, Z.: Modeling industrial lot sizing problems: a review. International Journal of Production Research 46(6), 1619–1643 (2008)

Jordan, C., Drexl, A.: Discrete lotsizing and scheduling by batch sequencing. Management Science 44(5), 698–713 (1998)

Karimi, B., Ghomi, S., Wilson, J.M.: The capacitated lot sizing problem: a review of models and algorithms. Omega-International Journal of Management Science 31(5), 365–378 (2003)

Karmarkar, U.S., Schrage, L.: The deterministic dynamic product cycling problem. Operations Research 33(2), 326–345 (1985)

Kopanos, G.M., Puigjaner, L., Georgiadis, M.C.: Simultaneous production and logistics operations planning in semicontinuous food industries. Omega 40(5), 634–650 (2012)

Kreipl, S., Pinedo, M.: Planning and scheduling in supply chains: an overview of issues in practice. Production and Operations Management 13(1), 77–92 (2004)

Kuik, R., Salomon, M., Vanwassenhove, L.N.: Batching decisions - structure and models. European Journal of Operational Research 75(2), 243–263 (1994)

Lee, I.S., Yoon, S.H.: Coordinated scheduling of production and delivery stages with stage-dependent inventory holding costs. Omega-International Journal of Management Science 38(6), 509–521 (2010)

Lütke Entrup, M., Günther, H.O., Van Beek, P., Grunow, M., Seiler, T.: Mixed-integer linear programming approaches to shelf-life-integrated planning and scheduling in yoghurt production. International Journal of Production Research 43(23), 5071–5100 (2005)

Menezes, A.A., Clark, A., Almada-Lobo, B.: Capacitated lot-sizing and scheduling with sequence-dependent, period-overlapping and non-triangular setups. Journal of Scheduling 14(2), 209–219 (2011)

Meyr, H.: Simultaneous lotsizing and scheduling by combining local search with dual reoptimization. European Journal of Operational Research 120(2), 311–326 (2000)

Pochet, Y., Wolsey, L.A.: Production Planning by Mixed Integer Programming. Series in Operations Research and Financial Engineering. Springer Science+Business Media, Inc., New York (2006)

Porkka, P.P., Vepsalainen, A.P.J., Kuula, M.: Multiperiod production planning carrying over set-up time. International Journal of Production Research 41(6), 1133–1148 (2003)

Quadt, D., Kuhn, H.: Capacitated lot-sizing with extensions: A review. 4or-a Quarterly Journal of Operations Research 6(1), 61–83 (2008)

Salomon, M., Kroon, L.G., Kuik, R., Vanwassenhove, L.N.: Some extensions of the discrete lotsizing and scheduling problem. Management Science 37(7), 801–812 (1991)

Salomon, M., Solomon, M.M., VanWassenhove, L.N., Dumas, Y., DauzerePeres, S.: Solving the discrete lotsizing and scheduling problem with sequence dependent setup costs and set-up times using the travelling salesman problem with time windows. European Journal of Operational Research 100(3), 494–513 (1997)

Soman, C.A., van Donk, D.P., Gaalman, G.J.C.: Combined make-to-order and make-to-stock in a food production system. International Journal of Production Economics 90(2), 223–235 (2004a)

Soman, C.A., van Donk, D.P., Gaalman, G.J.C.: A basic period approach to the economic lot scheduling problem with shelf life considerations. International Journal of Production Research 42(8), 1677–1689 (2004b)

Soman, C.A., van Donk, D.P., Gaalman, G.J.C.: Capacitated planning and scheduling for combined make-to-order and make-to-stock production in the food industry: An illustrative case study. International Journal of Production Economics 108(1-2), 191–199 (2007)

Sox, C.R., Gao, Y.B.: The capacitated lot sizing problem with setup carry-over. Iie Transactions 31(2), 173–181 (1999)

Suerie, C.: Modeling of period overlapping setup times. European Journal of Operational Research 174(2), 874–886 (2006)

Suerie, C., Stadtler, H.: The capacitated lot-sizing problem with linked lot sizes. Management Science 49(8), 1039–1054 (2003)

Transchel, S., Minner, S., Kallrath, J., Loehndorf, N., Eberhard, U.: A hybrid general lot-sizing and scheduling formulation for a production process with a two-stage product structure. International Journal of Production Research 49(9), 2463–2480 (2011)

Wolsey, L.A.: MIP modelling of changeovers in production planning and scheduling problems. European Journal of Operational Research 99(1), 154–165 (1997)

A Multiobjective Approach for a Dynamic Simple Plant Location Problem under Uncertainty

Joana M. Dias[1,2] and Maria do Céu Marques[2,3]

[1] Faculdade de Economia, Universidade de Coimbra, 3004-512 Coimbra, Portugal
[2] Inesc-Coimbra, Rua Antero de Quental, 199, 3000-033 Coimbra, Portugal
[3] Instituto Politécnico de Coimbra, ISEC, DFM, Rua Pedro Nunes, 3030-199 Coimbra
joana@fe.uc.pt, cmarques@isec.pt

Abstract. Location problems are, by nature, strategic decisions since facilities will usually be in operation in the medium and long terms. It is necessary to decide today, given the available information, knowing that the consequences of today's decisions will remain in time. Having to include in the decision making process data that will only be known with certainty in the future does not advise the use of deterministic models: the existing uncertainty should be explicitly included in the models. The notion of *optimal solution* becomes fragile: it will be difficult to find a single solution that is the best in all possible future realizations of uncertainty. In this paper we consider a dynamic simple plant location problem, where uncertainty is explicitly considered through the use of scenarios. We advocate the use of a multiobjective approach as a valuable tool in guiding the decision-making process, iteratively or as an *off-line* generation procedure.

Keywords: Location Problems, Uncertainty, Scenarios, Multiobjective, Pareto-efficient.

1 Introduction

Plant location problems are strategic problems by nature, usually associated with significant investments, and with medium to long term consequences. Whenever a decision-maker (DM) has to decide where to locate a warehouse, a plant, a store, the amount of data that should be considered in the decision making process is not only huge but often not known with certainty at the time the decision has to be made. Furthermore, location decisions are often associated with other problems like inventory management, transportation or assignment problems. As a simple example, let us consider the problem of locating a warehouse that will serve as supplier for a set of customers. There are only two potential locations for the warehouse: one location has very low construction and other fixed costs, but it is very far away from any of the customers (implying important transportation costs); the other location is near most customers, but the fixed costs incurred are very high. When deciding, we cannot split the location and transportation/assignment problems in two distinct problems, or we would end up with a suboptimal solution. They have to be considered simultaneously

B. Murgante et al. (Eds.): ICCSA 2014, Part II, LNCS 8580, pp. 60–75, 2014.

in the decision making process. However, dealing with those two problems at the same time is not as straightforward as it might seem, especially when we are considering that the facility to be located will be in operation during a given time horizon. Although it is not easy to change the location of warehouses or other facilities from one time period to the next, because of the direct and indirect costs incurred, it is not that difficult to change the way the demands of customers are being satisfied and the goods transported. If there is a change in the road infrastructure, or an important increase or decrease in fuels' prices, or there is a road that became tolled, the decision maker will be able to adjust the assignment/transportation decisions to the new reality. A similar reasoning could be made if we were thinking of other problems like inventory management or production scheduling.

In plant location problems we are thus faced with having to simultaneously make decisions that are deeply interconnected but that have very different natures: on one hand, location decisions that are strategic by nature and difficult to reverse, on the other hand, other related decisions that can be easily reverted whenever new information is available. The strategic nature of location problems is clearly described and emphasized in [1], where not only the classical static and deterministic models are defined, but also dynamic and stochastic location problems are discussed.

In this paper a dynamic simple plant location problem under uncertainty is considered. There is a set of potential locations for opening facilities, and there is a set of clients with a given demand that has to be satisfied by opened facilities. There is a planning horizon, and it will be necessary to decide which facilities to open and when to open them. The only data that is known for sure is the one related with the present moment: if the facilities are opened today, and the assignment of clients to facilities is made today, then all fixed location costs and assignment costs for the present time period are known with certainty. However, for all other future time periods, fixed opening costs are not known for sure, as well as assignment costs. In the future, even the set of clients or the potential locations for new facilities can be uncertain. In the present work, uncertainty is represented through a set of possible future scenarios. Two different stages for decision-making are considered: on one hand, location decisions determining when and where facilities should be opened during the planning horizon are set at the present moment and cannot be changed throughout the planning horizon (first stage); on the other hand, the assignment of clients to facilities that is made in each time period considering the existing opened facilities and the assignment costs for that particular period and for the scenario that came to occur (second stage).

Whenever uncertainty is explicitly considered, the notion of both feasible and optimal solution becomes very fragile (see, for instance, [2]). Regarding feasibility, it can be the case that it is not possible to guarantee feasibility under all possible realizations of the uncertain parameters. In the present work, as we are dealing with an uncapacitated problem, a solution that is feasible for one scenario will be feasible for all others. Regarding optimality, it is often the case that no single solution will be optimal under all possible future scenarios. A possible approach would be to consider a single objective function that would represent the expected objective function value over all scenarios. Optimizing the expected value of a given objective function can be the best thing to do if we are facing a decision making process that has to be repeated over and over again, always under uncertainty. Only in this case does the expected value represent the DM's payoff in the long run. Location problems do not fit this profile. Most of the times they are "once in a lifetime" decisions. Furthermore, we cannot talk about uncertainty without thinking about risk, and different DMs have different attitudes towards risk [3]. If a DM is neutral towards risk, then an expected value approach could be defendable. But if the DM is averse to risk, then he will probably be concerned with the worst of all scenarios only: he would be interested in making the worst possible outcome the best possible. This approach is known as the min-max approach. Many other possible ways of dealing with uncertainty can be disguised, some under an umbrella usually known as robust optimization. A review of facility location under uncertainty can be found in [4].

There is a whole set of publications dedicated to multiobjective stochastic programming, usually tackling the problem by reducing it to a single objective stochastic program or transforming it to a deterministic multiobjective program (see, for instance, [5, 6, 7, 8, 9, 10, 11]). The approach described in this paper is different: we tackle a single objective location problem under uncertainty by resorting to a multiobjective approach. The concept of Pareto-efficiency is thus applied in the context of a single objective problem under uncertainty. Without making any kind of assumptions regarding the attitude of the decision maker towards risk, it is possible to consider the dynamic plant location problem under uncertainty as a multiobjective problem, where each scenario will give rise to one objective. The DM will only be interested in Pareto-efficient solutions. A Pareto-efficient solution can be understood as a solution where it is not possible to improve one objective without deteriorating at least one other. In this context, a solution will be Pareto-efficient if it is not possible to improve the objective function value under one scenario without deteriorating its value under at least one other scenario. The multiobjective approach can be used embedded in an interactive decision-making process or as an *off-line* generation procedure, where the whole set of efficient solutions is calculated.

Actually, it is very difficult to find the concept of Pareto efficiency being applied in this context. Iancu and Trichakis [12] have, in a recent paper, introduced the Pareto efficiency concept in robust optimization, showing that some robust solutions are not efficient, and apply their methodology to portfolio optimization, inventory management and project management. Nevertheless, their focus is solely on Pareto efficient worst-case solutions. Kaläi et al [13] introduces the concept of lexicographic α-robustness, that can be seen as an extension of lexicographic programming used to tackle multiobjective problems.

This paper is organized as follows: the next section presents the mathematical model, and introduces the main concepts used throughout the text. Section 3 describes the multiobjective approach, and shows how some "robust solutions" can be, in fact, Pareto-efficient solutions. Section 4 illustrates the procedure through the use of some illustrative examples. In Section 5 the main conclusions are stated, and paths for future research are delineated.

2 Mathematical Modelling

Consider the set of potential locations for facilities denoted by $J = \{1,...,j,...,M\}$, and the set of possible customers' locations denoted by $I = \{1,...,i,...,N\}$. The planning horizon is defined as a set of time periods $\Gamma = \{1,...,t,...,T\}$. There are fixed costs associated with opening a facility at a given location and keeping this facility open until the end of the planning horizon. There are also assignment costs related to the assignment of customers to opened facilities during the planning horizon. The uncertainty considered in this problem is present not only in the fixed and assignment costs, but also in the existence of each possible client in each time period, and in the possibility of opening a facility in each potential location in the beginning of each time period. The uncertainty is represented by a set of possible future scenarios $\Phi = \{1,...,s,...S\}$. Suppose that each scenario s will occur with probability p_s such that $\sum_{s\in\Phi} p_s = 1$. The decision maker can be aware of these probabilities or not. Consider the following notation:

$$\delta_{its} = \begin{cases} 1, \text{ if customer } i \text{ has a demand to be fulfilled during} \\ \quad \text{period } t \text{ under scenario } s \\ 0, \text{ otherwise} \end{cases} \quad ,i \in I, t \in \Gamma, s \in \Phi;$$

$$\beta_{jts} = \begin{cases} 1, \text{ if it is possible to open facility } j \text{ at the beginning} \\ \quad \text{of period } t \text{ under scenario } s \\ 0, \text{ otherwise} \end{cases} \quad , j \in J, t \in \Gamma, s \in \Phi;$$

f_{jts} = fixed cost of opening facility *j* in the beginning of period *t* plus operation and maintenance costs until the end of the planning horizon (and possibly closing costs if the facility is to be closed after *T*) under scenario *s*, $j \in J, t \in \Gamma, s \in \Phi$;

c_{ijts} = cost of assigning customer *i* to facility *j* during period *t* under scenario *s*, $i \in I, j \in J, t \in \Gamma, s \in \Phi$.

The decision variables to be considered are defined as follows:

$$x_{ji} = \begin{cases} 1, \text{ if facility } j \text{ is opened at the beginning of period } t \\ 0, \text{ otherwise} \end{cases} \quad , j \in J, t \in \Gamma;$$

$$y_{ijts} = \begin{cases} 1, \text{ if customer } i \text{ is assigned to facility } j \text{ in period } t \\ \quad \text{under scenario } s \\ 0, \text{ otherwise} \end{cases} \quad ,i \in I, t \in \Gamma, s \in \Phi.$$

Notice that location decisions represented by x_{ji} do not depend on the realizations of the uncertain parameters, but the assignment decisions do: their values are only determined when the uncertainty is resolved for period *t*.

Representing location variables x_{ji} by a vector $\boldsymbol{x}$ and assignment variables by a vector $\boldsymbol{y}$, the simple dynamic location problem under uncertainty can be formulated as follows.

SDLPU:

$$\text{"min"}G(\boldsymbol{x},\boldsymbol{y}) \tag{1}$$

Subject to:

$$\sum_{j\in J} y_{ijts} = \delta_{its}, i\in I, t\in\Gamma, s\in\Phi \tag{2}$$

$$\sum_{\tau=1}^{t} x_{j\tau} \ge y_{ijts}, i\in I, j\in J, t\in\Gamma, s\in\Phi \tag{3}$$

$$\sum_{t\in\Gamma} x_{jt} \le 1, j\in J \tag{4}$$

$$x_{jt} \le \beta_{jts}, j\in J, t\in\Gamma, s\in\Phi \tag{5}$$

$$x_{jt} \in \{0,1\}, j\in J, t\in\Gamma \tag{6}$$

$$y_{ijts} \in \{0,1\}, i\in I, j\in J, t\in\Gamma, s\in\Phi \tag{7}$$

Constraints (2) guarantee that each client with demand will be assigned to a facility in each period t and scenario s. Constraints (3) guarantee that clients can only be assigned to opened facilities. Constraints (4) state that a facility can only be opened once during the whole planning horizon. Constraints (5) guarantee that it is only possible to open facilities in a given time period if that is allowed by all scenarios. If we consider that if β_{jts} is equal to 0 then f_{jts} is equal to $+\infty$, then constraints (5) can be dropped.

After fixing variables x_{jt}, a simple assignment problem has to be solved for each time period t and scenario s. We assume that the assignment problem is solved only when we know what was the scenario that came to occur. It is worth noticing that these assignment problems will always have a feasible solution as long as at least one facility is open at each time period, since the model does not consider capacity constraints.

Although the meaning of constraints is clear, this problem is still not well-defined. Since we are in the presence of possible future scenarios, how should we interpret "min"? Let us consider that we are interested in minimizing total costs. Even so, different decision makers will have different interpretations of (1).

When a decision has to be taken under uncertainty, and the consequences of that decision depend on the realization of uncertainty that comes to occur, the objective function to consider is not independent of the risk profile of the DM. DMs that are risk averse will tend to look for min-max solutions: solutions that minimize the worst possible outcome. DMs that like to take risks could prefer making a decision that could lead to the most advantageous payoff, even if this "best of all" payoff only has a tiny chance of occurring. One way of interpreting (1) could be to consider the expected value of a given objective function over all scenarios. If this is the case, and we are interested in the minimization of the expected total cost, then (1) could be interpreted as (8).

$$\min \sum_{s\in\Phi}\sum_{t\in\Gamma}\sum_{j\in J} p_s f_{jts} x_{jt} + \sum_{s\in\Phi}\sum_{t\in\Gamma}\sum_{j\in J}\sum_{i\in I} p_s c_{ijts} y_{ijts} \tag{8}$$

It is also possible to imagine that we are interested in guaranteeing that the worst payoff is the best possible (min-max approach):

$$\min_s \left\{ \max \sum_{t\in\Gamma}\sum_{j\in J} f_{jts} x_{jt} + \sum_{t\in\Gamma}\sum_{j\in J}\sum_{i\in I} c_{ijts} y_{ijts} \right\} \tag{9}$$

We can also think that what really matters is to consider the minimization of the total cost for the most probable scenario s^*:

$$\sum_{t\in\Gamma}\sum_{j\in J} f_{jts^*} x_{jt} + \sum_{t\in\Gamma}\sum_{j\in J}\sum_{i\in I} c_{ijts^*} y_{ijts^*} \tag{10}$$

Or minimize the maximum regret, where F_s^* represents the optimal objective function value when considering only scenario *s*:

$$\min_s \left\{ \max \left(\sum_{t\in\Gamma}\sum_{j\in J} f_{jts} x_{jt} + \sum_{t\in\Gamma}\sum_{j\in J}\sum_{i\in I} c_{ijts} y_{ijts} - F_s^* \right) \right\} \tag{11}$$

In reality, the DM will be interested in compromise solutions: it is seldom the case where a single solution will be the best under all possible realizations of uncertainty. Making a decision will always give a better result under some circumstances than under others, and the DM is not capable of controlling what these circumstances will be. If it is not possible to do any kind of assumptions regarding the risk profile of the decision maker, or about his preferences, then one possible approach could be to interpret (1) not as one single objective function but as a set of objective functions instead. Let $F_s(\boldsymbol{x},\boldsymbol{y})$ represent the total cost incurred under scenario *s* when solution $(\boldsymbol{x},\boldsymbol{y})$ is considered (12).

$$F_s(\boldsymbol{x},\boldsymbol{y}) = \sum_{t\in\Gamma}\sum_{j\in J} f_{jts} x_{jt} + \sum_{t\in\Gamma}\sum_{j\in J}\sum_{i\in I} c_{ijts} y_{ijts},\ s\in\Phi \tag{12}$$

SDLPU would then become a multiobjective problem (SDLPU_MO), where (1) is interpreted as (13).

SDLPU_MO:

$$\min\{F_1(\boldsymbol{x},\boldsymbol{y}),\cdots,F_s(\boldsymbol{x},\boldsymbol{y}),\cdots,F_S(\boldsymbol{x},\boldsymbol{y})\} \quad (13)$$

Subject to: (2)-(7)

Independently of the preferences or profile of the DM and assuming only that the DM is rational, he will only be interested in solutions such that it is not possible to improve the objective function of one given scenario without deteriorating the objective function of at least one other scenario. This means that the DM will only be interested in Pareto-efficient solutions (known as non-dominated solutions if we consider the objective space).

Definition 1: Consider $(\boldsymbol{x},\boldsymbol{y})$ an admissible solution for SDLPU_MO. $(\boldsymbol{x},\boldsymbol{y})$ is a Pareto efficient solution if and only if there is no other solution $(\boldsymbol{x}',\boldsymbol{y}')$ such that $F_s(\boldsymbol{x}',\boldsymbol{y}') \le F_s(\boldsymbol{x},\boldsymbol{y})$ and $F_s(\boldsymbol{x}',\boldsymbol{y}') < F_s(\boldsymbol{x},\boldsymbol{y})$ for at least one scenario s. The image of an efficient solution in the objective space is known as a *non-dominated solution.*

3 Multiobjective Approach

There are several different ways of dealing with a multiobjective programming problem. One such way is the so-called interactive approach. The interactive approach considers interchanging calculation and dialogue phases. In the calculation phase a non-dominated solution is calculated and showed to the DM. The DM will then react by giving some new information that will guide the calculation of the new non-dominated solution to be calculated in the next iteration. The process continues until the DM is satisfied with a given solution or the total set of non-dominated solutions is found (see, for instance, [14]). The major drawback of this approach has to do with the possibility of having calculation phases taking too much computational time, not promoting a real-time interaction and making the process not attractive to the DM. The main advantage has to do with the ability of searching areas of the solutions' surface that are interesting to the DM, not wasting time or resources calculating solutions that the DM will simply discard. Moreover, whenever a non-dominated solution is encountered, there is a region in the objective space that is no longer interesting (the one that is dominated by this solution), and another region where there cannot be any admissible solutions (or else this solution would not be non-dominated). So, it is possible, in each iteration, to eliminate regions from further searches.

Another way of dealing with multiobjective problems considers the *a priori* and *off-line* calculation of the whole set (or a significant number) of non-dominated solutions. The solutions can then be presented to the DM, all at the same time, or using an interactive approach similar to the one previously described. The set of non-dominated solutions can even be analyzed by using multicriteria decision-making techniques ([15, 16]). One of the advantages of this approach is that the computational burden of calculating the solutions is made *a priori*, promoting a faster action-reaction interaction with the decision maker since no optimizations will be done. Furthermore, it will be possible to show more information to the DM, like statistics or the fulfillment of other criteria not explicitly considered in the model.

The choice between an interactive or a generation approach should be done considering several aspects of the problem: What is the dimension of the problem? How long does it take to calculate a solution? Whatever is the choice in a particular situation, there has to be some kind of procedure to calculate non-dominated solutions. There are several auxiliary programming problems that can be used to calculate non-dominated solutions, mostly relying on the optimization of some mono-objective programming problem, being the most well known the consideration of a weighted sum of the objective functions. When dealing with integer or mixed-integer problems, care has to be taken to guarantee that the chosen procedure is capable of calculating non-supported non-dominated solutions (lying inside duality gaps).

In this paper we resort to a result due to Ross and Soland [17], considering an auxiliary mono-objective programming problem AUX, where $\lambda_s > 0$ is a weight associated with scenario s such that $\sum_{s\in\Phi}\lambda_s = 1$, and M_s is an upper bound to the objective function related to scenario s.

AUX:

$$\min \sum_{s\in\Phi}\sum_{t\in\Gamma}\sum_{j\in J}\lambda_s f_{jts} x_{jt} + \sum_{s\in\Phi}\sum_{t\in\Gamma}\sum_{j\in J}\sum_{i\in I}\lambda_s c_{ijts} y_{ijts} \tag{14}$$

Subject to: (2)-(7)

$$\sum_{t\in\Gamma}\sum_{j\in J} f_{jts} x_{jt} + \sum_{t\in\Gamma}\sum_{j\in J}\sum_{i\in I} c_{ijts} y_{ijts} \le M_s, s\in\Phi \tag{15}$$

Let $\boldsymbol{M} \in \Re^S$ represent the vector of M_s values.

Proposition 1. (based on [17]): For any vector $\boldsymbol{\lambda}$ such that $\sum_{s\in\Phi}\lambda_s = 1$ and $\lambda_s > 0, s\in\Phi$, $(\boldsymbol{x},\boldsymbol{y})$ is an efficient solution of SDLPU_MO if and only if it is the optimal solution of AUX for some vector $\boldsymbol{M} \in \Re^S$.

Based on Proposition 1, it is now possible to verify whether some of the most common interpretations of (1) will lead to efficient solutions or not. Considering objective function (8), if constraints (15) are added to SDLPU, we can guarantee that the corresponding solution will be an efficient solution.

Proposition 2: If $\lambda_s = p_s, \forall s$, then AUX will calculate a solution considering the minimization of the expected total cost over all scenarios. This solution is a non-dominated solution to SDLPU_MO.

Proof: The proof follows directly from Proposition 1. If $\lambda_s = p_s, \forall s$ and if M_s is made large enough, then we are actually minimizing the expected total cost. We can thus conclude that this solution is a non-dominated solution of SDLPU_MO.

Let us define a solution as α-robust , $0<\alpha<1$, if and only if the objective function value under any of the scenarios is not more than $\alpha\%$ worst than the best possible solution for that particular scenario [18]. If M_s are properly chosen, then AUX can be used to calculate an α-robust solution, and it is possible to guarantee that the α-robust solution is indeed efficient. Let F_s^* be the optimal objective function value if only scenario *s* is considered.

Proposition 3: The solution obtained by solving AUX such that $M_s=(1+\alpha)F_s^*, s\in\Phi$ and $0<\alpha<1$, is a α-robust non-dominated solution of SDLPU_MO.

Proof: The result follows directly from Proposition 1.

It has been proven that the solution obtained by considering the minimization of the worst outcome can be dominated [12]. Nevertheless, AUX could also be used to calculate an efficient min-max solution. In a first stage, it would be necessary to solve the problem of minimizing the maximum cost under all scenarios. This could be done by solving the following programming problem:

MIN-MAX:

$$\min\delta \tag{16}$$

Subject to:

(2)-(7)

$$\sum_{t\in\Gamma}\sum_{j\in J}f_{jts}x_{jt}+\sum_{t\in\Gamma}\sum_{j\in J}\sum_{i\in I}c_{ijts}y_{ijts}\leq\delta, s\in\Phi \tag{17}$$

After solving this problem, it will be possible to guarantee the calculation of an efficient solution that is also min-max optimal by considering appropriate values for M_s.

Proposition 4: Let δ represent the worst case objective function, calculated by solving MIN-MAX. If M_s is defined such that $M_s=\delta, s\in\Phi$, then AUX will generate an efficient min-max solution.

Proof: After solving MIN_MAX, δ will represent the worst case objective that we are willing to accept. This means that we will only be interested in solutions such that the objective function value for any scenario *s* will be less than or equal to δ. Comparing (17) with (15), it is easy to see that if M_s is defined such that $M_s=\delta, s\in\Phi$, then any efficient solution calculated will also be a min-max solution.

A similar reasoning could be applied if we are interested in minimizing the maximum regret. Consider the following restrictions:

$$\sum_{t\in\Gamma}\sum_{j\in J} f_{jts}x_{jt} + \sum_{t\in\Gamma}\sum_{j\in J}\sum_{i\in I} c_{ijts}y_{ijts} - F_s^* \leq \delta, s\in\Phi \tag{18}$$

Proposition 5: Let δ represent the maximum allowed regret calculated by solving MIN-MAX, with restriction (17) replaced by (18). If M_s is defined such that $M_s = F_s^* + \delta, s\in\Phi$, then AUX will generate an efficient solution that minimizes maximum regret.

Proof: It follows the previous proof.

It is quite easy to embed the use of AUX in both an interactive and an *off-line* generation procedure. In an interactive approach, the dialogue phase with the DM consists in defining new M_s values. Notice that, in reality, these values do no more no less than defining a region of search. In a generating approach, M_s values can be automatically generated in a way that guarantees that the whole objective space is explored. The automatic generation of vector $\boldsymbol{M}$ can be done resorting to two simple data structures: a binary tree, with as much levels as the number of scenarios, and a matrix. Each time a new solution is calculated based on a given vector $\boldsymbol{M}$, a binary tree is generated such that it will define all possible future vectors $\boldsymbol{M}$. These vectors are then saved in a matrix so that they can be retrieved in future iterations. To give a simple example of this procedure, imagine a situation with three scenarios. The initial vector $\boldsymbol{M}$ is equal to $\left[M_1^1, M_2^1, M_3^1\right]$. By using this vector, we obtain a non-dominated solution with the following values for each of the 3 objective function values: $\left[F_1^1, F_2^1, F_3^1\right]$, with $F_1^1 \leq M_1^1, F_2^1 \leq M_2^1, F_3^1 \leq M_3^1$. Based on these two vectors, a binary tree can be built (Fig. 1), where 8 possible search regions are defined.

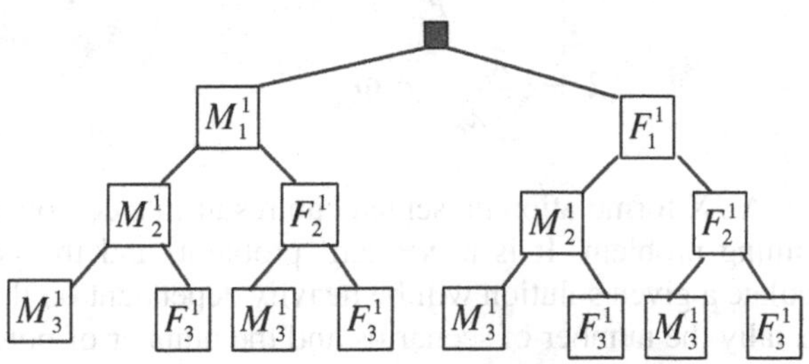

Fig. 1. Binary tree for automatic generation of vector $\boldsymbol{M}$

The path from the root to each node of the tree will define a new future vector $\boldsymbol{M}$. These values can be stored in a matrix, so that they can be retrieved in a future iteration of the algorithm (Table 1). Whenever a new solution is calculated, a new binary tree is built and the corresponding vectors added to the matrix. In each iteration, vector $\boldsymbol{M}$ will be determined by the next column of this matrix. Some of the problems will be unfeasible and should not be considered in the matrix. Using $\left[F_1^1, F_2^1, F_3^1\right]$, for instance, will not be interesting because it corresponds to an unfeasible problem (if that was not so, $\left[F_1^1, F_2^1, F_3^1\right]$ would not be a non-dominated solution). Other vectors will end up with optimal solutions that are already known ($\left[M_1^1, M_2^1, M_3^1\right]$, for instance, is not an interesting vector). Furthermore, knowing that one given problem is impossible will allow us to conclude that other $\boldsymbol{M}$ vectors will also lead to impossible problems (if $[M_1^1, F_2^1, M_3^1]$ leads to a problem that is impossible, then it is not worth to explore region $[M_1^1, F_2^1, F_3^1]$, for instance). This search method is easily implementable and will guarantee that the whole objective space is explored.

Table 1. Example of a table for automatic definition of vector $\boldsymbol{M}$

M_1^1	M_1^1	M_1^1	F_1^1	F_1^1	F_1^1
M_2^1	F_2^1	F_2^1	M_2^1	M_2^1	F_3^1
F_3^1	M_3^1	F_3^1	M_3^1	F_3^1	M_3^1

It should be stressed that the weights vector $\boldsymbol{\lambda}$ that is used in (14) does not represent any kind of DM's preferences. These weights can and should be changed in accordance with vector $\boldsymbol{M}$ in order to help decreasing the computational time needed to calculate a solution [14]. If $\boldsymbol{M}$ is more demanding for a given scenario, meaning that M_s is close to the best objective function value F_s^*, then the respective objective function weight should be increased. One simple way of doing this is setting $\boldsymbol{\lambda}$ as follows:

$$\begin{aligned} \lambda_s &= 1 - \frac{M_s - F_s^*}{F_s^*}, s \in \Phi \\ \lambda_s &= \frac{\lambda_s}{\sum_{s \in \Phi} \lambda_s}, s \in \Phi \end{aligned} \tag{19}$$

Actually, the AUX formulation presented can result in a computationally heavy integer programming problem. It is a NP-hard problem, and the computational time needed to calculate a given solution will be heavily dependent on the dimension of the problem, especially the number of scenarios and the number of potential locations for facilities.To solve AUX we can resort to general solvers or use dedicated procedures, both exact and heuristic procedures. Although the latter will not be able to guarantee the optimality of the calculated solution, they can be a very good choice especially in the presence of an interactive procedure, where the most important thing will be to define a region of interest for the DM. It is even possible to think of using a heuristic procedure in a first stage, and then an exact procedure to actually guarantee the optimality of the solution of interest. Although any number of scenarios can be considered, in many real life situations DMs look at small number of possible future realizations of uncertainty (usually the worst, best and most probable scenarios).

4 Illustrative Examples

In this section we will illustrate the multiobjective approach to the simple plant location problem under uncertainty by resorting to two small examples of randomly generated problems. The problems were generated using the algorithm described in [19]. All AUX problem instances were solved by a general solver (Cplex V12.6). In the first example, we consider a problem with 25 potential locations for facilities, 100 potential clients, 10 time periods and 2 scenarios. Let us consider the use of an interactive procedure based on [14]. The first thing to do is to calculate the optimal solution for both scenarios. The best possible value of the objective function for each scenario will delineate the region of interest. These solutions are shown in Fig. 2, in the objective space. The DM is free to set the vector $\boldsymbol{M}$ as he wishes. Let us assume that he does not want to explore any particular region, so he decides to define $M_1 = 218195$ and $M_2 = 153313$. These limits are defined by the two non-dominated solutions already calculated (that correspond to the optimum solution of each scenario). The weights are considered to be $\lambda_1 = 0.8$ and $\lambda_1 = 0.2$ (according to (19)). The solution reached is shown in Fig. 3.

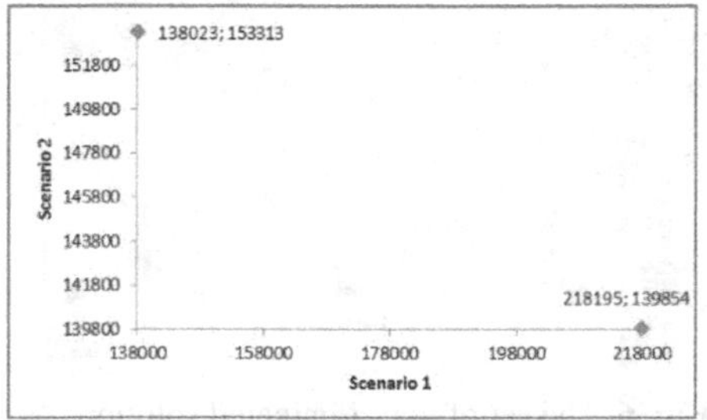

Fig. 2. Optimal solutions for each scenario

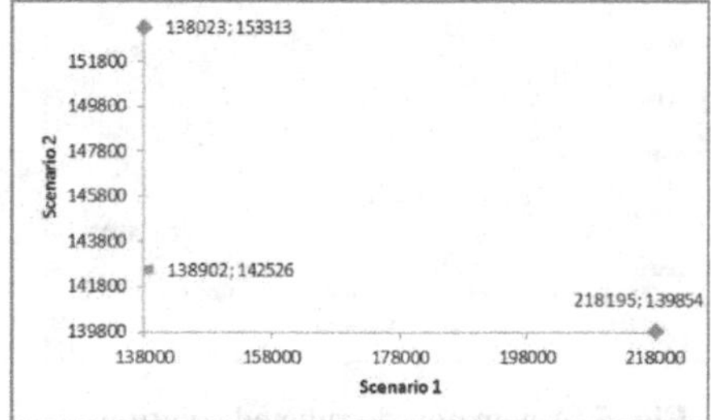

Fig. 3. The first non-dominated solution calculated

Let us look at more detail at Fig. 3. Considering the newly calculated non-dominated solution, it is easy to see that two regions of the objective space are no longer interesting. This is shown in Fig. 4. Region A will only have solutions that are dominated by the solution calculated. Region B has only non-admissible solutions. The decision-maker can now decide whether to explore region C or region D. Let us suppose that he would explore region D. Then M_1 will remain equal to 218195 and M_2 will be set to 142526 (given by the new non-dominated solution just calculated). Fig. 5 shows the new solution calculated. The procedure would be repeated until the DM is satisfied, or the whole objective space has been explored. The whole set of non-dominated solutions found is shown in Fig. 6. It is possible to observe the compromises that exist between the two scenarios. Moreover, it is also possible to calculate the non-dominated solution that minimizes the maximum regret or that minimizes the maximum cost under both scenarios. These solutions are highlighted in Fig. 7.

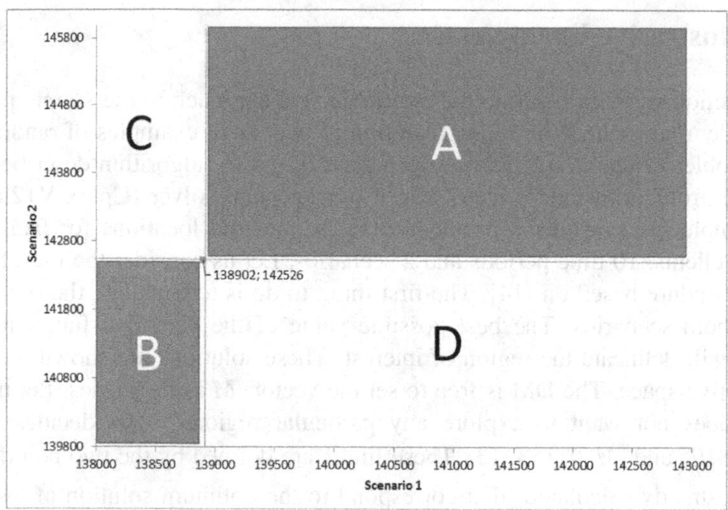

Fig. 4. The shaded areas A and B are no longer of interest

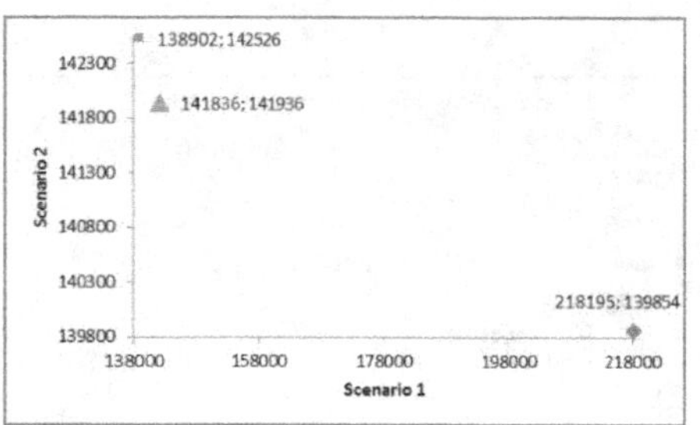

Fig. 5. A new non-dominated solution

Fig. 6. The set of non-dominated solutions

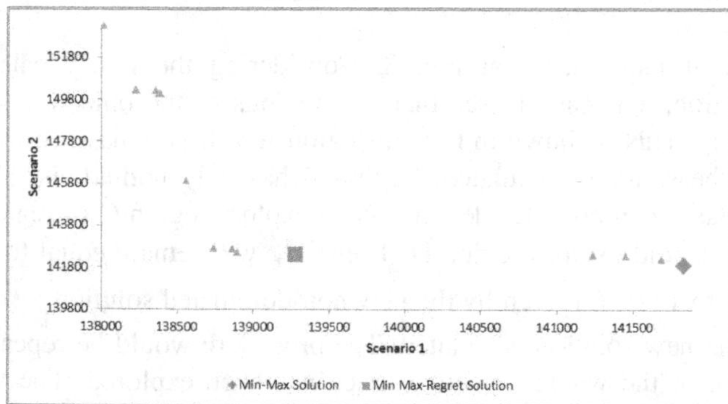

Fig. 7. The Min-Max and Min-Max Regret Solutions

It can also be interesting to analyze the solutions by looking at the opened facilities. In this problem, a set of 7 facilities is opened exactly in the same time period in all solutions calculated. Table 2 shows, for each non-dominated solution, which are the facilities to be opened and when should they be opened.

Table 2. Time period in which each facility is opened in each solution

Objective Function		Opened Facilities														
s=1	*s=2*	2	3	4	5	6	7	9	11	14	16	18	20	22	24	25
138023	153313	1	1	-	7	6	4	2	2	3	2	4	6	3	2	-
138228	150276	1	1	-	7	6	4	2	2	3	2	4	6	3	2	1
138237	150257	1	1	-	7	6	4	2	2	3	3	4	6	3	2	1
138360	150238	1	1	-	7	6	4	2	2	3	3	4	-	3	2	1
138384	150093	1	1	-	-	6	4	2	2	3	2	4	6	3	2	1
138393	150074	1	1	-	-	6	4	2	2	3	3	4	6	3	2	1
138564	145957	1	1	-	7	6	4	2	5	3	2	4	6	3	2	-
138720	145827	1	1	-	-	6	4	2	5	3	2	4	6	3	2	-
138746	142709	1	1	-	7	6	4	2	5	3	2	4	6	3	2	1
138869	142690	1	1	-	7	6	4	2	5	3	2	4	-	3	2	1
138902	142526	1	1	-	-	6	4	2	5	3	2	4	6	3	2	1
139281	142430	1	1	7	-	6	4	2	5	3	2	4	6	3	2	1
141238	142389	-	1	-	7	6	4	2	5	3	2	4	6	3	2	1
141457	142365	-	1	-	-	6	4	2	5	3	2	4	6	3	2	1
141695	142200	-	1	7	-	6	4	2	5	3	2	1	6	3	2	1
141836	141936	-	1	7	-	6	4	2	5	3	2	4	6	3	2	1
145742	140500	1	1	-	-	6	4	2	5	3	2	4	2	3	2	1
146121	140404	1	1	7	-	6	4	2	5	3	2	4	2	3	2	1
147507	140307	-	1	-	7	6	4	2	5	3	2	4	2	3	2	1
218195	139854	-	1	7	-	6	4	2	5	3	2	4	2	3	2	1

In Fig. 8, the non-dominated solutions calculated by using the *off-line* generation procedure are shown, considering a randomly generated problem with 10 potential locations for facilities, 50 potential clients, 5 time periods and 3 scenarios. With more than 3 scenarios, the visualization of non-dominated solutions in the objective space is no longer possible, and it could be better to resort to the use of multicriteria analysis tools.

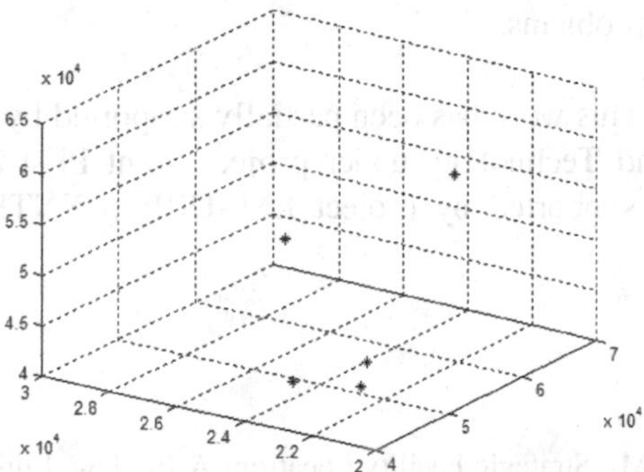

Fig. 8. Non-dominated solutions in 3D objective space

5 Conclusions

In this paper we describe a multiobjective approach for decision-making applied to a simple dynamic location problem under uncertainty. This approach has several advantages when compared with the more traditional robust approaches. Regarding the DM, the only assumption made is that the DM is rational, thus will only be interested in Pareto-efficient solutions. It is not necessary to estimate any kind of probabilities associated with the possible future scenarios. A set of solutions is calculated, instead of only one. The DM will thus have a much broader view of the compromises that exist among the possible scenarios. Many of the classic robust solutions, like minimizing the worst result or the maximum regret can be considered as special cases. Making available to the DM a set of solutions will make it possible to apply other robustness measures that are more difficult to incorporate directly in a mathematical programming problem (like, for instance, the *bw*-robustness measure [20]).

The major drawback of the described approach has to do with the fact that it implies the sequential optimization of NP-hard problems. Depending on the problem's dimension, this can be computationally expensive. One way of circumventing the problem is through the use of efficient heuristics in a first stage, and the use of exact procedures in a second phase where the preferred region is already delineated. Another possibility is the use of metaheuristics that work with populations of solutions (like genetic algorithms) and that are capable of generating sets of efficient solutions in each generation (see, for instance, [21]).

In the problem considered, any solution that is admissible for one given scenario will be admissible for all others. This situation does not always occur. Defining the concept of Pareto-efficiency for single objective dynamic location problems where feasibility can also be affected by uncertainty is an interesting path of research.

Although we have applied this approach to a simple dynamic location problem under uncertainty, it can easily be generalized to many other linear, integer or mixed-integer programming problems.

Acknowledgements. This work has been partially supported by the Portuguese Foundation for Science and Technology under project grant PEst-OE/ EEI/UI308/2014. This work has been supported by project EMSURE (CENTRO 07 0224 FEDER 002004).

References

1. Owen, S., Daskin, M.: Strategic Facility Location: A Review. European Journal of Operational Research 111, 423–447 (1998)
2. Mulvey, J.M., Vanderbei, R.J., Zenios, S.A.: Robust optimization of large-scale systems. Operations Research 43, 264–281 (1995)
3. Rieger, M.O., Wang, M., Hens, T.: Risk Preferences Around the World. Management Science. Published online in Articles in Advance (February 21, 2014)
4. Snyder, L.V.: Facility location under uncertainty: a review. IIE Transactions 38, 537–554 (2006)

5. Hulsurkar, S., Biswal, M.P., Sinha, S.B.: Fuzzy programming approach to multi-objective stochastic linear programming problems. Fuzzy Sets and Systems 88, 173–181 (1997)
6. Teghem Jr., J., Dufrane, D., Thauvoye, M., Kunsch, P.: STRANGE: an interactive method for multi-objective linear programming under uncertainty. European Journal of Operational Research 26, 65–82 (1986)
7. Urli, B., Nadeau, R.: PROMISE/scenarios: An interactive method for multiobjective stochastic linear programming under partial uncertainty. European Journal of Operational Research 155, 361–372 (2004)
8. Abdelaziz, F.B.: Solution approaches for the multiobjective stochastic programming. European Journal of Operational Research 216, 1–16 (2012)
9. Gutjahr, W.J.: Two metaheuristics for multiobjective stochastic combinatorial optimization. In: Lupanov, O.B., Kasim-Zade, O.M., Chaskin, A.V., Steinhöfel, K. (eds.) SAGA 2005. LNCS, vol. 3777, pp. 116–125. Springer, Heidelberg (2005)
10. Guillén, G., Mele, F.D., Bagajewicz, M.J., Espuna, A., Puigjaner, L.: Multiobjective supply chain design under uncertainty. Chemical Engineering Science 60, 1535–1553 (2005)
11. Cardona-Valdés, Y., Álvarez, A., Ozdemir, D.: A bi-objective supply chain design problem with uncertainty. Transportation Research Part C 19, 821–832 (2011)
12. Iancu, D.A., Trichakis, N.: Pareto Efficiency in Robust Optimization. Management Science 60, 130–147 (2014)
13. Lamboray, C., Vanderpooten, D.: Lexicographic alfa-robustness: An alternative to min-max criteria. European Journal of Operational Research 220, 722–728 (2012)
14. Dias, J., Captivo, M.E., Clímaco, J.: An interactive procedure dedicated to a bicriteria plant location model. Computers & Operations Research 30, 1977–2002 (2003)
15. Mladineo, N., Margeta, J., Brans, J.P., Mareschal, B.: Multicriteria Ranking of Alternative Locations for Small Scale Hydro Plants. European Journal of Operational Research 31, 215–222 (1987)
16. Barda, O.H., Dupuis, J., Lencioni, P.: Multicriteria Location of Thermal Power Plants. European Journal of Operational Research 45, 332–346 (1990)
17. Ross, T., Soland, R.: A Multicriteria Approach to the Location of Public Facilities. European Journal of Operational Research 4, 307–321 (1980)
18. Lim, G.J., Sonmez, A.D.: Y-Robust Facility Relocation Problem. European Journal of Operational Research 229, 67–74 (2013)
19. Marques, M.C., Dias, J.: Simple dynamic location problem with uncertainty: a primal-dual heuristic approach. Optimization 62, 1379–1397 (2013)
20. Roy, B.: Robustness in Operational Research and Decision Aiding: a multi-faceted issue. European Journal of Operational Research 200, 629–638 (2010)
21. Dias, J., Captivo, M.E., Clímaco, J.: A Memetic Algorithm for Multi-objective Dynamic Location Problems. Journal of Global Optimization 42, 221–253 (2008)

Joint Scheduling and Optimal Charging of Electric Vehicles Problem

Ons Sassi[1] and Ammar Oulamara[2]

[1] University of Lorraine - LORIA, Nancy, France
ons.sassi@loria.fr

[2] University of Lorraine, Ile de Saulcy, Metz, France
ammar.oulamara@loria.fr

Abstract. In this paper, we consider the joint scheduling and optimal charging of electric vehicles problem. This problem can be defined as follows: Given a fleet of Electric Vehicles - EV and Combustion Engine Vehicles - CV, a set of tours to be processed by vehicles and a charging infrastructure, the problem aims to optimize the assignment of vehicles to tours and to minimize the charging cost of EVs while considering several operational constraints mainly related to chargers, electricity grid, and EVs driving range. We prove the NP-hardness in the ordinary sense of this problem. We provide a mixed-integer linear programming formulation to model the joint Scheduling and Optimal Charging of EVs problem (EVSCP) and we use CPLEX to solve small and medium instances. A two-phase sequential heuristic, based on the Maximum Weight Clique Problem (MWCP) and the Minimum Cost Flow Problem (MCFP), is developed to solve large instances of the EVSCP. Computational results on a large set of real and randomly generated test instances show the efficiency and the effectiveness of the proposed approaches.

Keywords: Electric vehicle, Assignment problem, Charging problem, Mathematical programming, Heuristic, Greedy algorithm, Maximum weight clique, Minimum cost flow problem.

1 Introduction

The transport sector, heavily dependent on oil, is among the first-ranked sectors for energy consumption and greenhouse gas emissions. To combat environmental and energy challenges, Electric Vehicles (EVs) may provide a clean and safe alternative to the internal combustion engine vehicles (CV). However, electric car industry is still facing many weaknesses related to the battery management. The first one is the limited EV driving range. The second weakness is related to the long charging time of EVs (fully charging the battery pack can take around 8 hours with Level 1 chargers) and the availability of a charging infrastructure. The third weakness concerns the electricity grid on which EV charging may put a significant stress. A large scale adoption of EVs may then strongly affect the electricity grid which is already facing numerous challenges. Thus, the new

B. Murgante et al. (Eds.): ICCSA 2014, Part II, LNCS 8580, pp. 76–91, 2014.

ecosystem of EVs (vehicles - chargers - electricity grid) leads to new optimization challenges aiming to develop efficient models and decision tools to manage EV fleets and take into account all constraints related to the ecosystem.

This study is a part of a French industrial project led by a company that manages large captive fleets. This R&D project aims at designing, with a progressive approach, an intelligent system that manages the charging infrastructures and will allow an economical and ecological sustainable deployment of EVs fleet in the business context.

Optimization problems related to EVs can be classified into four classes: (i) EV used in the context of Smart Grid or Vehicle to Grid (V2G) problems, (ii) controlled EV charging problems, (iii) charging infrastructure network design problems and (iv) EV routing problems.

In the V2G context, EVs batteries are used to store energy. Those batteries can serve as an energy resource by sending electricity back into the grid in order to either prevent or postpone load shedding. In [1], the authors consider EV charging and V2G capabilities for congestion management. They propose a novel algorithm, based on power distribution factors, for optimal EV congestion management. Other studies related to the V2G can be found in [2, 3].
The controlled EV charging problems consist in a better management of the charging load in order to minimize the charging cost. A design of a simulation environment, which produces charging schedules using a multi-objective evolutionary optimization algorithm is presented in [4]. In [5], an energy consumption scheduler able to reduce peak power load in smart places based on genetic algorithms is exposed. A concept of real-time scheduling techniques for EV charging that minimizes the impact on the power grid and guarantees the satisfaction of consumers charging requirements is suggested in [6].
The charging infrastructure network design problem is considered in [7–10]. In [7], the design of charging infrastructures problem is addressed. The author develops a model using an integer program in order to determine the locations of charging stations. Furthermore, in [8], an integer programming model to optimize the locations and the number of battery exchange stations is presented. Recently, in [9], the authors use the concepts of set covering problems to model the allocation of multiple types of charging stations. In [10], a mathematical program with complementarity constraints is developed to determine an optimal allocation of a given number of public charging stations.
The EV routing problem is less considered in the literature. The problem of energy-optimal routing problem is addressed in [11]. In [12], the authors formulate the Green Vehicle Routing Problem (GVRP) as a Mixed Integer Linear Program (MIP). Two constructive heuristics are developed to solve this problem. An overview of the GVRP is given in [13].

In this paper, we address the Electric Vehicle Scheduling and Charging Problem (EVSCP) defined as follows: Given a set of tours, a fleet of electric and combustion engine vehicles and a charging infrastructure, the objective is to find the optimal way to assign tours to vehicles and to minimize the EVs charging cost while satisfying several constraints mainly related to the electricity grid, chargers and EVs batteries capacities. To the best of our knowledge, no previous study was devoted to tackle this problem in the literature.

The remainder of the paper is organized as follows. Section 2 presents the problem and describes the notation used in the remaining sections. In Section 3, we prove the NP-Hardness of the problem. Section 4 provides a detailed description of the mathematical formulation of the problem. In Section 5, exact and heuristic approaches are presented. Section 6 summarizes the computational results. Concluding remarks are given in Section 7.

2 Problem Description and Notation

We consider a set $M_1 = \{1, \ldots, m_1\}$ of EVs and a set $M_2 = \{1, \ldots, m_2\}$ of Combustion Engine Vehicles (CVs), needed to process a set $N = \{1, \ldots, n\}$ of tours. Each vehicle can process at most one tour at a time. Each EV j operates with a battery characterized by a nominal capacity of embedded energy CE_j(kWh) and a State of Charge (SoC) at time $t = 0$, denoted by SoC_j^0 and expressed as a percentage of CE_j (0% = empty; 100% = full). At low and high SoC's values, the battery tends to degrade faster [14]. In order to improve its lifetime after repeated use and to respect the security issues, at each time, SoC_j should be in the interval $[SoC_j^{\min}, SoC_j^{\mathrm{Max}}]$, where $SoC_j^{\min}$ and SoC_j^{Max} are the minimal and maximal allowable values of SoC, respectively. We assume that there are m chargers available to charge EVs during the time horizon $[0, T]$. This time horizon is divided into T equidistant time periods, $t = 1, \ldots, T$, each of length d, where t represents the time interval $[t-1, t]$ (in our case, $d = 15$ minutes). At each time period t, each charger can apply on EV j a charging power $p_{jt} \in [p^{\min}, p^{\mathrm{Max}}]$ where $p^{\min}$ and p^{Max} are the minimal and maximal powers that can be delivered by the charger, respectively. Thus an EV charged with a power p_{jt} during the time period t retrieves a total amount of energy equal to $d \times p_{jt}$(kWh). We denote by GP_t the electricity grid capacity available for EV charging at time t; i.e., at each time period t, the total grid power available to charge all EVs is limited to GP_t. Let C_t be the energy cost during the time interval t. There is a set of n tours to be processed by the vehicles. Each tour i is characterized by a start time st_i, a finish time ft_i, a weight w_i (length expressed in km) and the energy E_i(kWh) required to perform this tour using an EV. Two tours i and j overlap if their intersection is nonempty, i.e., $[st_i, ft_i] \cap [st_j, ft_j] \neq \emptyset$, otherwise they are disjoint. Given a tour i, let $V(i)$ be the set of tours that $\forall\ j \in V(i)$, i and j overlap. Note that all tours are already constructed. Thus the start and finish times of each tour are known in advance and, to satisfy operational constraints, preemption of tours is not allowed and the vehicles cannot charge while in tour. The objective consists in maximizing the use of EVs, i.e.; the total weight of tours processed with EVs, and minimizing the overall charging costs. This problem can be seen as a fixed interval scheduling problem [15, 16] with additional constraints of energy.

The fixed interval scheduling problem without additional constraints is characterized as the problem of scheduling a number of jobs, each with fixed

starting and finishing times, and it is stated as follows. Let n be the number of intervals of the form $[s_j, f_j)$ with $s_j < f_j$, for $j = 1, \ldots, n$ and m the number of machines. Each machine can process at most one job at a time and it is always available. The objective is to process a maximum number of jobs on the machines. The fixed interval scheduling problem has been largely considered in the literature, particularly the variant where the objective is to maximize the number of (weighted) jobs that can be feasibly scheduled [17–20]. This variant can be solved using a min-cost flow formulation (See, e.g., [17] and [18]). When jobs have unit weights, a greedy algorithm that finds the maximum number of jobs to process on machines is proposed in [19], and in [20]. If each job can only be carried out by an arbitrary given subset of machines, the problem becomes NP-hard [17]. However, the fixed interval scheduling problem with energy constraints has never been considered in the literature.
In the following we give an illustrative example of our problem.

Example. We consider a problem in which five tours have to be scheduled. The start times, the finish times, the weights and the energy consumptions of those tours are given in Table 1. These tours should be processed using two EVs each with a battery nominal capacity of 22 kWh. We consider that, initially, both batteries contain 12 kWh and after the completion of tours, each battery should contain 6 kWh. We assume that $p^{\min} = p^{\max} = 3$ kW and, at each time period t, $C_t = 1$ euro and $GP_t = 6$ kWh.

Table 1. Instance of EVSCP

Tour	Start time (h)	Finish time (h)	Weight (km)	Energy consumption (kWh)
T_1	1	3	20	5
T_2	2	4	20	5
T_3	5	7	20	5
T_4	5	8	30	10
T_5	8	10	20	5

A feasible solution to our problem is given in Figure 1.

3 NP-Hardness Result

In this section, we prove the weak NP-hardness of the EVSCP. We use the reduction to the Partition problem, which is known to be NP-complete in the ordinary sense [21]. This problem can be stated as follows:

Given a set $P = \{a_1, a_2 \ldots a_p\}$ of p integers such that $\sum_{i=1}^{p} a_i = 2B$, is there a partition of P into two subsets P_1 and P_2 such that $\sum_{i \in P_1} a_i = \sum_{i \in P_2} a_i = B$?

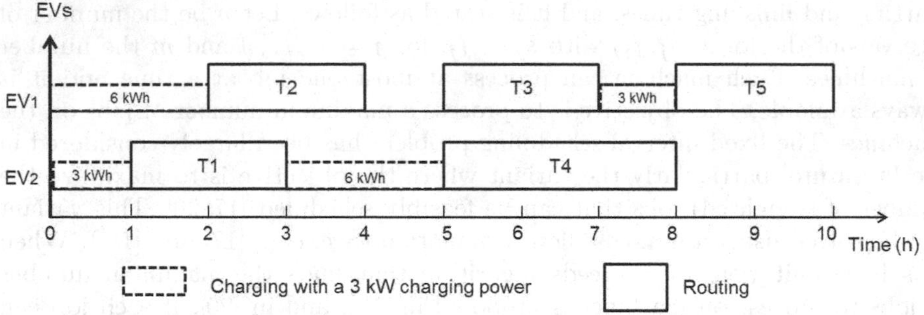

Fig. 1. Solution to EVSCP: EV_1 executes the set of tours $\{T_2, T_3, T_5\}$ and gets charged between $t = 0$ and $t = 2$ and between $t = 7$ and $t = 8$ with a charging power of 3 kW. EV_2 performs the tours T_1 and T_2 and is charged between $t = 0$ and $t = 1$ and between $t = 3$ and $t = 5$.

Theorem 1. *The problem EVRCP is NP-hard in the weak sense.*

Proof. Given an arbitrary instance of the Partition problem, we build an instance $(\mathcal{I})$ of the EVSCP with a set of $2p + 2$ tours. The start times, the finish times and the energy consumptions of those tours are given in Table 2. We assume that $a_0 = 0$. Without loss of generality, we can restrict our analysis to the case

Table 2. Characteristics of the instance $(\mathcal{I})$

Tour	Start time	Finish time	Energy consumption
$Ti; i = 1, \ldots, 2p$	$B + a_{i-1}$	$B + a_{i-1} + a_i$	a_i
T_{2p+1}	$3B$	$3B + 1$	B
T_{2p+2}	$3B$	$3B + 1$	B

when we have two available EVs with a battery capacity of $2B$kWh. We assume that there exist two chargers. Each charger can provide one discrete power of 1kW. The grid capacity available for EV charging is equal to 2kW at each time interval. At $t = 0$, we assume that the batteries of both EVs are empty. A feasible assignment of tours to EVs should verify the following constraints.

- The EVs should have a sufficient energy before processing tours.
- The maximum battery capacity of each EV should never be exceeded during the charging process.
- At each time, the total power used to charge EVs cannot exceed the grid's maximum capacity.

In what follows, we show that there exists a schedule of tours and a feasible charging planning of EVs if, and only if, the Partition problem admits a solution.

First, assume that the Partition problem has a solution and let P_1 and P_2 be the required subsets of P. The desired schedule of tours is constructed by building

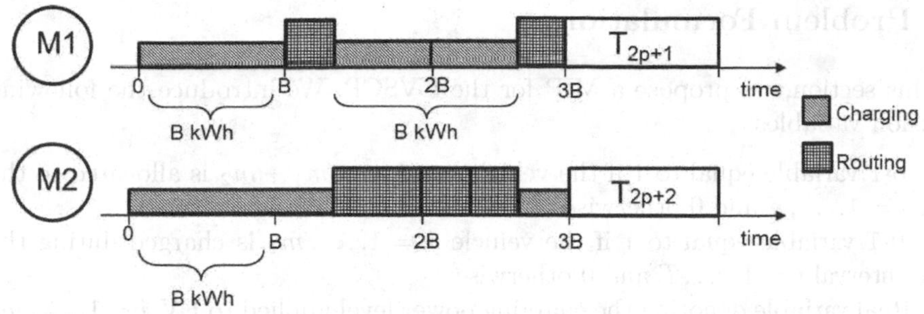

Fig. 2. Solution for the scheduling problem on two EVs M1 and M2

two sets L_1 and L_2 of tours $T_1, \ldots, T_{2p}$, where tours of L_1 and L_2 correspond to the sets P_1 and P_2, respectively; i.e., $T_i \in L_j$, if $a_i \in P_j$, $i = 1, \ldots, 2p$, $j = 1, 2$. Assign tours of L_1 to EV_1 and tours of L_2 to EV_2. The tours T_{2p+1} and T_{2p+2} are assigned to EV_1 and EV_2, respectively. In order to execute the assigned tours, the EVs are charged as follows. During the time interval $[0, B]$, each EV is charged with a power of 1kW, at $t = B$ each EV recovers then BkWh. During the time interval $[B, 3B + 1]$, EV_1 processes all tours of L_1, for B units of time, and it is charged when it is not processing any tour (i.e., EV_1 is charged during the intervals $[B, 3B] \bigcap_{a_i \in P_2} [B + a_{i-1}, B + a_{i-1} + a_i]$) and finally it processes the tour T_{2p+1}, whereas EV_2 processes all tours of L_2 and get charged during $[B, 3B] \bigcap_{a_i \in P_1} [B + a_{i-1}, B + a_{i-1} + a_i]$ before performing the tour T_{2p+2}. It is easy to see that, during the different charging phases, each EV recovers enough amount of energy to process all tours. Therefore, the obtained schedule is feasible.

Conversely, assume now that there exists a feasible schedule of tasks and a charging planning of EVs . Since T_{2p+1} and T_{2p+2} have the same start and finish times, we assume, without loss of generality, that T_{2p+1} and T_{2p+2} are executed by EV_1 and EV_2, respectively. Let R_1 and R_2 be the two sets of tours assigned to EV_1 and EV_2, respectively. Let $\sum_{a_i \in R_1} a_i = A_1$ and $\sum_{a_i \in R_2} a_i = A_2$ be the amounts of energy required by the EVs to execute the sets of tasks A_1 and A_2. Then, EV_1 and EV_2 need a total amount of energy equal to $B + A_1$ and $B + A_2$, respectively. Assume that $A_1 > B$. Then, EV_1 needs an amount of energy greater than $2B$. EV_1 can be charged only during $[0, 3B] \bigcap_{a_i \in R_2} [B + a_{i-1}, B + a_{i-1} + a_i]$ and it may recover at most a total amount of energy E equal to $3B - A_1 < 2B$kWh. Since $E < 2B$, there is not enough energy to process the set of tasks R_1 and T_{2p+1} . Then $A_1 \leq B$. Following the same reasoning, we have $A_2 \leq B$. Since $A_1 + A_2 = 2B$, we conclude that $A_1 = B$ and $A_2 = B$. Therefore, R_1 and R_2 represent a solution to the Partition problem.

4 Problem Formulation

In this section, we propose a MIP for the EVSCP. We introduce the following decision variables:

x_{ij} : 0-1 variable equal to 1 if the vehicle $j = 1, \ldots, m_1 + m_2$ is allocated to the tour $i = 1, \ldots, n$ and 0 otherwise.
y_{jt} : 0-1 variable equal to 1 if the vehicle $j = 1, \ldots, m_1$ is charged during the time interval $t = 1, \ldots, T$ and 0 otherwise.
p_{jt} : Real variable denoting the charging power level applied to EV $j = 1, \ldots, m_1$ at time interval $t = 1, \ldots, T$.
In order to reduce the complexity of the proposed MIP and without loss of generality, we assume that the energy E_i required to perform the tour i is consumed during the last period ft_i ($[ft_i - 1, ft_i]$) of the tour; i.e., $E_{i,ft_i} = E_i$ and $E_{i,l} = 0$, $l = st_i, st_i + 1, \ldots, ft_i - 1$. The EVSCP is formulated as a MIP. Its mathematical formulation $(\mathcal{P})$ is as follows:

$$\text{Lex} \quad (\text{Max} \sum_{i=1}^{n} \sum_{j=1}^{m_1} w_i \times x_{ij} \; ; \; \text{Min} \quad \sum_{j=1}^{m_1} \sum_{t=1}^{T} C_t \times p_{jt}) \tag{1}$$

$$\sum_{j=1}^{m_1+m_2} x_{ij} = 1, \quad \forall i \tag{2}$$

$$x_{ij} + \sum_{i' \in V(i)} x_{i'j} \leq 1, \quad \forall i, \forall j \tag{3}$$

$$\sum_{t=st_i}^{ft_i} y_{jt} + (ft_i - st_i + 1) \times x_{ij} \leq (ft_i - st_i + 1), \quad \forall i, \forall j \in M_1 \tag{4}$$

$$\sum_{j=1}^{m_1} p_{jt} \leq GP_t, \quad \forall t \tag{5}$$

$$p^{\min} \times y_{jt} \leq p_{jt}, \quad \forall t, \forall j \in M_1 \tag{6}$$

$$p_{jt} \leq p^{\text{Max}} \times y_{jt}, \quad \forall t, \forall j \in M_1 \tag{7}$$

$$SoC_j^0 + \frac{\sum_{t \leq st_i - 1} d \times p_{jt} - \sum_{l=1/ft_l \leq st_i - 1}^{n} E_{l,ft_l} \times x_{lj}}{CE_j} \leq SoC_j^{\text{Max}}, \quad \forall i, \forall j \in M_1 \tag{8}$$

$$SoC_j^0 + \frac{\sum_{t \leq ft_i} d \times p_{jt} - \sum_{l=1/ft_l \leq ft_i} E_{l,ft_l} \times x_{lj}}{CE_j} \geq SoC_j^{\min}, \quad \forall i, \forall j \in M_1 \tag{9}$$

Constraints (2) ensure that each tour is assigned to exactly one vehicle. Constraints (3) guarantee that no vehicle can be assigned to overlapping tours. Constraints (4) prohibit charging the EV when it is processing tours. Constraints (5) ensure that, at each time interval t, the total power used to charge the EVs does not exceed the grid's maximum capacity. Constraints (6) and (7) guarantee the respect of the minimum and the maximum powers of chargers when charging the EVs. Constraints (8) and (9) ensure that the SoC of each EV is in the interval $[SoC^{\min}, SoC^{\text{Max}}]$. Our goal is to optimize two lexicographical objective functions (Lex (Max $\sum_{i=1}^{n}\sum_{j=1}^{m_1} w_i \times x_{ij}$; Min $\sum_{j=1}^{m_1}\sum_{t=1}^{T} C_t \times p_{jt}$)), i.e., the objective functions are arranged in order of importance and the optimization problems are solved one at a time. The first objective consists in maximizing the EVs kilometers travelled ($\sum_{i=1}^{n}\sum_{j=1}^{m_1} w_i \times x_{ij}$). The second objective aims at minimizing the EVs charging costs ($\sum_{j=1}^{m_1}\sum_{t=1}^{T} C_t \times p_{jt}$).

5 Solving Approaches

5.1 Exact Approach

The proposed MIP is solved in two steps using CPLEX. In the first step, the objective is to maximize the EVs kilometers travelled ($\sum_{i=1}^{n}\sum_{j=1}^{m_1} w_i \times x_{ij}$). The optimal solution generated in the first step serves as the starting solution of the second step. A new constraint which ensures that the number of EVs kilometers travelled is greater than the objective function found in the first step, is added to the MIP in the second step. Then, the new MIP is solved with the second objective function ($\sum_{j=1}^{m_1}\sum_{t=1}^{T} C_t \times p_{jt}$). These two steps are described in Algorithm 1.

Algorithm 1. Exact method

1: **Input:** Data of EVSCP
2: **Output:** A solution to the EVSCP
3: **Step 1**: - Solve the MIP with the objective function $K = \sum_{i=1}^{n}\sum_{j=1}^{m_1} w_i \times x_{ij}$
4: - Let K^* be the value of the objective function of the optimal solution S^*
5: **Step 2**: - Add the constraint $\sum_{i=1}^{n}\sum_{j=1}^{m_1} w_i \times x_{ij} \geq K^*$ to the MIP formulation
6: - Solve the new MIP with the objective function $\sum_{j=1}^{m_1}\sum_{t=1}^{T} C_t \times p_{jt}$ starting from the solution S^*

5.2 Sequential Heuristic Approach

In this section, we describe a Sequential Heuristic - SH to solve the EVSCP. This heuristic is mainly based on the following idea. For each EV, a set of tours is selected and assigned to that EV. Then, a charging schedule for this EV is proposed while satisfying all constraints described in Section 4. This heuristic interleaves two sub-algorithms: tours selection algorithm and charging schedule algorithm.

Algorithm 2. Overall heuristic's algorithm

Input: n tours, m_1 EVs and m_2 CVs
Output:A solution to the EVSCP
for Each EV **do**
    Apply the tours selection algorithm to get a set of tours achievable by that EV
    Given the set of selected tours assigned to EV, apply the charging schedule algorithm to optimize the charging cost of that EV
end for
for Each CV **do**
    Apply the tours selection algorithm
end for

Tours Selection Algorithm. To assign a set of tours to an EV and maximize the kilometers travelled (i.e., maximize the weight of tours assigned to that EV), the Maximum Weight Clique Problem (MWCP) [22, 23] is exploited. Recall that a MWCP is defined as follows. Let $G = (V; E)$ be an arbitrary undirected and weighted graph, where V is the set of nodes and E is the set of edges. For each node i is associated a weight w_i. Two distinct nodes are said to be adjacent if they are connected by an edge. Given a subset S of nodes, the weight assigned to S will be denoted by $W(S) = \sum_{i \in S} w_i$. A clique of graph G is a subset of V in which all nodes are pairwise adjacent. A clique S is called maximal if no strict superset of S is a clique. A maximal weight clique S is a clique which is not contained in any other clique having weight larger than $W(S)$.

The tours selection problem is represented through an undirected weighted graph $G = (V, E)$, where V is the set of nodes and E is the set of edges. Each node $i \in V$ represents a tour which is characterized by a weight (number of kilometers). The nodes are indexed in a nondecreasing order of their starting times st_i. Two nodes i and j are connected by an edge if the tours i and j are disjoint. The tours selection problem is then equivalent to the MWCP in graph G with additional constraints. The MWCP is NP-hard problem [21]. Balas et al. [24] provide some polynomial solvable cases of MWCP in specific graphs. The interval graph is one of the specific graphs for which the the MWCP is a polynomial problem. However, even if our graph G is an interval graph, the MWCP in graph G should ensure that a feasible schedule for EVs charging exists. To solve the tours selection problem, we propose a heuristic that finds the maximum weight clique in graph G and makes sure that the charging schedule exists.

Overview of the Algorithm. Our algorithm consists of two steps. In the first step, a maximal weight clique is constructed using a greedy heuristic. In the second step, the weight of that clique is improved by removing and/or adding one or several nodes. The two steps are repeated till stop conditions are fulfilled. The best clique is updated at each iteration.

• **Initial Step.** Initially, the clique C is empty. The algorithm selects randomly a vertex i and adds it to C. To expand this clique, a list of possible nodes

$L = \{v \in V | v \text{ is connected to all nodes of } C \text{ and } v \notin C\}$ is created. Then, the algorithm picks up a node v^* from L and adds it to C. v^* is the node with the largest total sum of the weights of its adjacent vertices that respects the *admission condition* (described later) related to the feasibility of a charging schedule. This process is repeated until L is empty or all nodes of L do not satisfy the *admission condition.*

• **Improvement Step.** When there are no more nodes that could increase the clique weight, the algorithm computes, for each node $i \in C$, the list of nodes $N_i = \{v \notin C | v \text{ is connected to all nodes of } C \setminus \{i\}\}$, and removes from C the node i with $W(N_i) \geq W(N_j), \forall j \in C$, where $W(N_i) = \sum_{k \in N_i} w_k$. Then, a new list L is selected and the process of adding and/or removing nodes is repeated till a specific timeout is reached. When the improvement step is stopped, the current clique is cleared and the algorithm restarts with the initial step till the algorithm's stop condition is reached. Finally, the algorithm generates the best clique found among all constructed cliques and removes the best clique's nodes from the graph G.

• **Stop Condition.** The algorithm continues until either a fixed limited processing time is reached or the best clique weight reaches the target clique weight (i.e., if a clique is found that has the same weight as the total weight of the graph).

Admission Condition. The admission condition ensures that the tours corresponding to the nodes of the current clique are achievable by an EV (i.e., there is a feasible charging schedule to perform all tours of the current clique by the EV). The existence of a feasible schedule is guaranteed thanks to the metric *me*. *me* is the minimal amount of energy that the EV should recover before starting the first tour of the clique, i.e., $me = \sum_{k=1}^{|C|}(E_k - Y_k)$, where Y_k is the amount of energy that may be charged between successive tours $k-1$ and k; i.e., $Y_k = d \times \sum_{t=ft_{k-1}+1}^{st_k - 1} \min\{p^{\text{Max}}, p'_t\}$, where p'_t is the residual power grid at time interval t. We assume that $Y_1 = 0$. A node v is added to the current clique, if, following the chronological order, for each node j in the clique, $me_j = \sum_{k=1}^{j}(E_k - Y_k)$ is lower than the EV's battery capacity and $me_j \leq d \times \sum_{t=0}^{st_1 - 1} \min\{p^{\text{Max}}, p'_t\}$.

In the following we provide a small example to describe the *Tours Selection Algorithm.* Consider the EVRCP instance defined in Section 2. The different steps of the *Tours Selection Algorithm* are given in Figure 3.

Charging Schedule Optimization. Given a clique C^* of tours assigned to an EV j, the objective here is to provide a feasible charging schedule that minimizes the total charging cost and respects the constraints described in Section 4. Below a charging schedule algorithm is proposed. It is based on a Minimum Cost Flow Formulation and runs in $O(((T+|C|)log(T+|C|))^2)$ polynomial time.

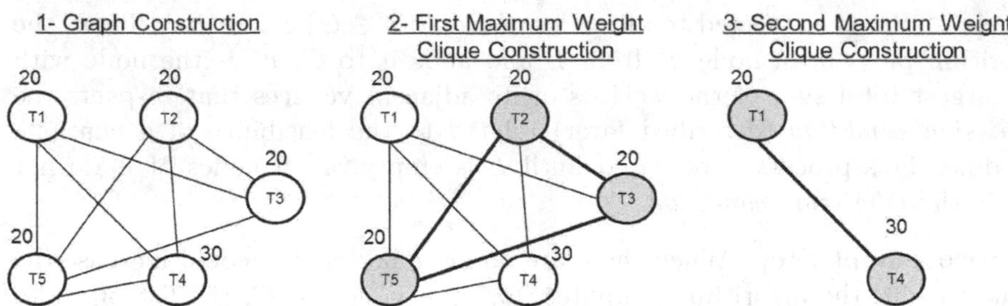

Fig. 3. Illustration of the Tours Selection Algorithm: The first maximum weight clique is composed of the nodes T_2, T_3 and T_5. Note that the clique $\{T_1, T_4, T_5\}$ does not satisfy the Admission Condition.

For a given set V of vertices of the clique C^*, a network $G = (W, A)$ is defined as follows:

- The set of nodes W consists of (i) a source s, (ii) the nodes t_i representing the time periods $[t_i, t_i+1]$, except the time periods when the EV is not available for charging, i.e., $i \in \{1, \ldots, T\} \setminus \cup_{v \in \{1,\ldots,|V|\}} \{st_v, st_v+1, \ldots, ft_v\}$, (iii) the nodes v_k, $k = 1, \ldots, |V|$ representing the tours indexed in the increasing order of their start times st_k, (iv) a sink p.
- The set of directed arcs with restricted capacities A consists of (i) the arcs (s, t_i) with bounded capacities $max_i = \min\{p^{Max}, p'_i\} \times d$ and a cost C_i corresponding to the energy cost during the time period t_i, (ii) the arcs (t_i, v_k) if $t_i < st_k$ and $t_i > st_{k-1}$ with a capacity ∞ and a cost equal to zero, (iii) the arcs (v_{k-1}, v_k), $k = 2, \ldots v_{|V|}$ with a maximum capacity equal to $CE_j - E_{k-1}$ and a cost equal to zero, (iv) the arcs (v_k, p), $k = 1, \ldots v_{|V|}$ with a lower bound and an upper bound equal to E_k and a cost equal to zero.

Let $f(i, j)$ be the flow on the arc $(i, j) \in A$ in a feasible solution to this MCFP. We define a feasible schedule to the corresponding charging problem as follows: at each t_i, where $i \in \{1, \ldots, T\} \setminus \cup_{v \in V} \{st_v, st_v+1, \ldots, ft_v\}$, apply on the EV j a charging power $p_{jt_i} = \frac{f(s,t_i)}{d}$. The MCFP can be solved in $O(((T+|V|)log(T+|V|))^2)$ time (See, e.g., [25]).

A feasible solution to the Charging Schedule Optimization problem is given in Figure 4.

6 Computational Experiments and Discussion

Extensive numerical experiments are conducted to evaluate the quality of solutions obtained through the exact method and the heuristic on real instances and randomly generated instances. To be able to asses the solution quality, we use real instances provided by the Group La Poste and other randomly generated instances. All the experiments are carried out on an Intel Xeon E5620 2.4GHz

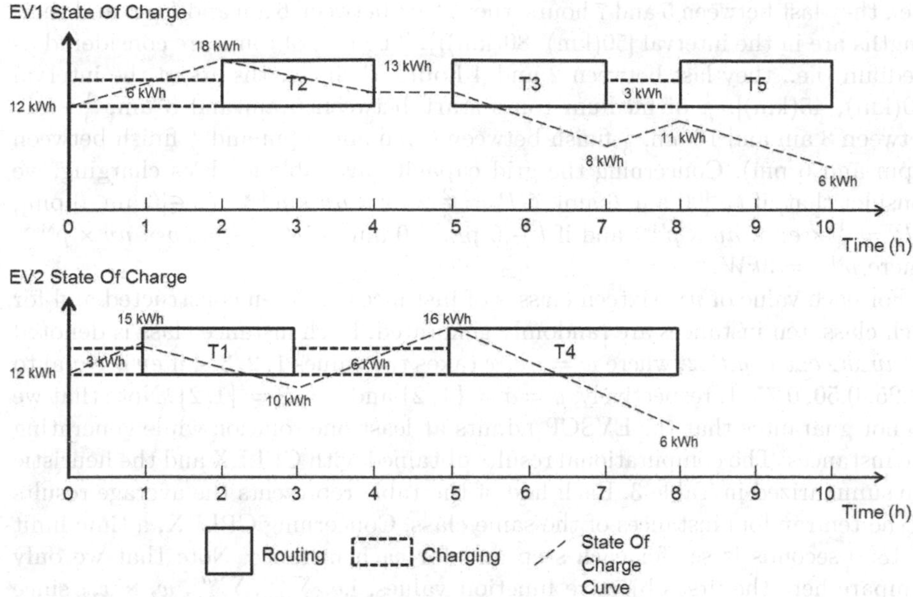

Fig. 4. Illustration of a solution to the Charging Schedule Optimization problem. The charging power is equal to 3 kW. The total charging cost of both EVs is equal to 18 euros.

processor, with 8GB RAM memory. The tested algorithms are coded in the C++ language. The commercial solver ILOG CPLEX 12.5 is used to solve the MIP. The results of the algorithms are compared in terms of quality of the generated solutions and run times.

The first real data instance is composed of 18 CVs and 8 EVs. All EVs considered have 22(kWh) battery packs. Forty-five tours have to be assigned to those vehicles. The second instance is composed of 14 CVs, 8 EVs with a battery capacity of 22(kWh) and 4 other EVs with a battery capacity of (23.5)kWh. Forty-six tours have to be assigned to those vehicles. For those two instances, both CPLEX and the heuristic succeeded to generate optimal solutions in a very short computational time. For reasons of confidentiality, we cannot provide the computational results related to those instances.

Random instances are generated as follows. The total number of vehicles $nv = m_1 + m_2$ takes its values from the set $\{40, 80, 120\}$. The number of EVs m_1 is a ratio of nv; i.e., $m_1 = ev \times nv$. ev takes its values from the set $\{0.25, 0.5, 0.75, 1.0\}$. A parameter a was used to distinguish the capacity of EVs batteries, namely, $a = 1$ means that all EVs have a capacity of 22(kWh) and $a = 2$ means that 50% of EVs have a capacity equal to 22(kWh) and the rest of EVs have a capacity of 16(kWh). The parameter tt is related to the number of tours. If $tt = 1$, then $n = nv \times c$, where $c \in [1.2, 1.3]$. If $tt = 2$, then $n = nv \times c$, where $c \in [1.5, 1.6]$. All tours start not earlier than 6 am and finish not later than 8 pm and they need between 0.15 and 0.35(kWh) per kilometer. 50% of those tours are long (i.e., they last between 5 and 7 hours, they start between 6 am and 1 pm and their lengths are in the interval [50(km), 80(km)]). The rest of tours are considered as medium (i.e., they last between 2 and 4 hours, their lengths are in the interval [20(km), 45(km)]. $\frac{1}{6}$ of medium tours start between 6 am and 8 am, $\frac{1}{3}$ start between 8 am and 10 am, $\frac{1}{6}$ finish between 6 pm and 8 pm and $\frac{1}{3}$ finish between 4 pm and 6 pm). Concerning the grid capacity available for EVs charging, we consider that, if $t \in$[00 am, 6 am], $GP_t = \frac{2}{3} \times ev \times nv \times p^{\text{Max}}$, if $t \in$[6 am, 6 pm], $GP_t = \frac{1}{5} \times ev \times nv \times p^{\text{Max}}$ and if $t \in$[6 pm, 00 am], $GP_t = \frac{1}{2} \times ev \times nv \times p^{\text{Max}}$, where $p^{\text{Max}} = 3kW$.

For each value of nv, sixteen classes of instances are then constructed and for each class, ten instances are randomly generated. Each instance class is denoted by $nv\_w\_ce\_x\_a\_y\_tt\_z$, where $w = nv$, x takes the values 1, 2, 3, 4 if ev is equal to : 0.25, 0.50, 0.75, 1, respectively, $y = a = \{1, 2\}$ and $z = tt = \{1, 2\}$. Note that we do not guarantee that the EVSCP admits at least one solution while generating the instances. The computational results obtained with CPLEX and the heuristic are summarized in Table 3. Each line of the table represents the average results of the ten random instances of the same class. Concerning CPLEX, a time limit of 1800 seconds is set for each step and for each instance. Note that we only compare here the first objective function values, i.e. $\sum_{i=1}^{n} \sum_{j=1}^{m_1} w_i \times x_{ij}$ since the charging costs depend on the EVs kilometers travelled.

The entries of Table 3 show, for each solving approach, (i) the success, i.e. the percentage of solved instances among the ten tested instances of the same class, by "solved instances", we mean the instances for which at least one solution was generated, within the time limit, even if it is not optimal, (ii) the Gap_{UB} of the best solution, denoted by s, (generated by either CPLEX or the heuristic) in relation to the Upper Bound (UB) (generated by CPLEX) computed as: $\text{Gap}_{\text{UB}}(s) = \frac{UB-s}{s}$, (iii) the Gap_{max} computed as $\text{Gap}_{\text{max}}(s) = \frac{\max(S_{\text{EM}}, S_{\text{H}})-s}{s}$, where S_{EM} and S_{H} are the values of the objective functions of the solutions generated by CPLEX and the heuristic, respectively, (iv) the average run time in seconds (s) needed to generate an assignment and charging solution to the EVSCP.

The computational results show that the heuristic exhibits an excellent performance both in terms of efficiency and effectiveness. Indeed, it consistently exhibits the best average time of resolution and it solves more instances. For only three classes of instances, CPLEX generated better solutions than the heuristic and the Gap_{max} was at most equal to 0.40%. For four classes of instances, both CPLEX and the heuristic succeeded to find the optimal solutions. In most cases, the heuristic succeeded in finding quickly enough approximate good solutions with an average Gap_{UB} of around 15% and of at most 25%, when CPLEX failed to find any solution. The percentage success of the heuristic was better than that of CPLEX for 28 classes of instances. Both exact and heuristic approaches failed to solve the instances composed of 80 and 100 EVs. This may be due to the fact that there is no feasible solutions for those instances.

Table 3. Performance results of CPLEX and the heuristic on generated random instances. Entries of '-' signify that the runs were terminated because of excessive computational time requirements. Entries of '*' signify that it was impossible to estimate the concerned values.

	Exact method				**Heuristic**			
Instance	Success	Gap_{UB}	Gap_{max}	Time(s)	Success	Gap_{UB}	Gap_{max}	Time(s)
nv_40_ce_1_a_1_tt_1	100%	8.06%	0%	1800	100%	8.36%	0.27%	0.12
nv_40_ce_1_a_1_tt_2	100%	13.66%	0.14%	1800	100%	13.5%	0%	0.14
nv_40_ce_1_a_2_tt_1	100%	14.72%	4.96%	1800	100%	9.30%	0%	0.09
nv_40_ce_1_a_2_tt_2	90%	20.89%	6.18%	1800	100%	13.85%	0%	0.12
nv_40_ce_2_a_1_tt_1	100%	12.73%	0.39%	1800	100%	12.29%	0%	0.14
nv_40_ce_2_a_1_tt_2	40%	21.53%	2.54%	1800	90%	18.52%	0%	0.20
nv_40_ce_2_a_2_tt_1	100%	20.67%	6.88%	1800	100%	12.90%	0%	0.14
nv_40_ce_2_a_2_tt_2	20%	24.18%	5.46%	1800	90%	17.75%	0%	0.20
nv_40_ce_3_a_1_tt_1	100%	5.88%	0%	362	100%	6.00%	0.12%	0.15
nv_40_ce_3_a_1_tt_2	40%	7.84%	0%	1800	80%	8.27%	0.40%	0.21
nv_40_ce_3_a_2_tt_1	100%	1.51%	0.30%	409	100%	1.20%	0%	0.174
nv_40_ce_3_a_2_tt_2	70%	14.61%	4.68%	1800	80%	9.48%	0%	0.24
nv_40_ce_4_a_1_tt_1	30%	0%	0%	1800	30%	0%	0%	0.21
nv_40_ce_4_a_1_tt_2	10%	0%	0%	254	10%	0%	0%	0.39
nv_40_ce_4_a_2_tt_1	30%	0%	0%	428	30%	0%	0%	0.19
nv_40_ce_4_a_2_tt_2	10%	0%	0%	223	10%	0%	0%	0.25
nv_80_ce_1_a_1_tt_1	100%	31.78%	10.96%	1800	100%	18.76%	0%	0.48
nv_80_ce_1_a_1_tt_2	80%	64.53%	32.65%	1800	100%	24.04%	0%	0.79
nv_80_ce_1_a_2_tt_1	100%	73.66%	15.30%	1800	100%	19.39%	0%	0.43
nv_80_ce_1_a_2_tt_2	80%	92.11%	54.66%	1800	100%	24.21%	0%	0.66
nv_80_ce_2_a_1_tt_1	100%	96.82%	70.66%	1800	100%	15.32%	0%	0.62
nv_80_ce_2_a_1_tt_2	50%	93.55%	65.93%	1800	100%	16.64%	0%	1.09
nv_80_ce_2_a_2_tt_1	90%	153.29%	119.47%	1800	100%	15.41%	0%	0.63
nv_80_ce_2_a_2_tt_2	20%	87.43%	63.67%	1800	100%	14.52%	0%	1.06
nv_80_ce_3_a_1_tt_1	70%	6.47%	0.50%	1800	100%	5.92%	0%	0.69
nv_80_ce_3_a_1_tt_2	0%	-	-	1800	100%	*	*	1.08
nv_80_ce_3_a_2_tt_1	30%	7.81%	2.63%	1800	100%	5.33%	0%	0.72
nv_80_ce_3_a_2_tt_2	0%	-	-	1800	100%	*	*	1.18
nv_80_ce_4_a_1_tt_1	10%	0%	0%	1800	0%	-	-	-
nv_80_ce_4_a_1_tt_2	0%	-	-	1800	0%	-	-	-
nv_80_ce_4_a_2_tt_1	0%	-	-	1800	0%	-	-	-
nv_80_ce_4_a_2_tt_2	0%	-	-	1800	0%	-	-	-
nv_120_ce_1_a_1_tt_1	70%	551.01%	435%	1800	100%	21.78%	0%	1.05
nv_120_ce_1_a_1_tt_2	0%	-	-	1800	100%	*	*	1.8
nv_120_ce_1_a_2_tt_1	60%	820.20%	667.94%	1800	100%	19.83%	0%	1.02
nv_120_ce_1_a_2_tt_2	0%	-	-	1800	100%	*	*	1.55
nv_120_ce_2_a_1_tt_1	50%	223.05%	178.95%	1800	100%	15.81%	0%	1.56
nv_120_ce_2_a_1_tt_2	0%	-	-	1800	100%	*	*	3.16
nv_120_ce_2_a_2_tt_1	80%	215.18%	169.88%	1800	100%	16.79%	0%	1.68
nv_120_ce_2_a_2_tt_2	0%	-	-	1800	100%	*	*	3.06
nv_120_ce_3_a_1_tt_1	20%	25.03%	19.88%	1800	100%	4.29%	0%	1.76
nv_120_ce_3_a_1_tt_2	0%	-	-	1800	100%	*	*	3.46
nv_120_ce_3_a_2_tt_1	0%	-	-	1800	100%	*	*	1.81
nv_120_ce_3_a_2_tt_2	0%	-	-	1800	100%	*	*	3.48
nv_120_ce_4_a_1_tt_1	0%	-	-	1800	0%	-	-	-
nv_120_ce_4_a_1_tt_2	0%	-	-	1800	0%	-	-	-
nv_120_ce_4_a_2_tt_1	0%	-	-	1800	0%	-	-	-
nv_120_ce_4_a_2_tt_2	0%	-	-	1800	0%	-	-	-

7 Conclusion

In this paper, we considered the problem of optimizing the allocation of tours to EVs and minimizing the charging costs simultaneously. This new problem was shown to be NP-hard in the ordinary sense using its reduction to the Partition problem. To solve the EVSCP to optimality, we proposed a mixed integer linear programming formulation to maximize the EVs use and to minimize the charging costs. An exact method based on two phases was developed to solve the MIP with CPLEX. We have also described a sequential heuristic to solve the EVSCP. It interleaves two sub-algorithms to assign a set of tours to EVs using the Maximum Weight Clique Problem and to plan its charging using a Minimum Cost Flow formulation. Experiments results showed the efficiency and effectiveness of the proposed heuristic.

References

1. Lopez, M.A., Martin, S., Aguado, J.A., De la Torre, S.: V2G strategies for congestion management in microgrids with high penetration of electric vehicles. Electr. Pow. Syst. Res. 104, 28–34 (2013)
2. Jian, J., Zhu, X., Shao, Z., Niu, S., Chan, C.C.: A scenario of vehicle-to-grid implementation and its double-layer optimal charging strategy for minimizing load variance within regional smart grids. Energ. Convers. Manage. 78, 508–517 (2014)
3. Soares, J., Sousa, T., Morais, H., Vale, Z., Canizes, B., Silva, A.: Application-Specific Modified Particle Swarm Optimization for energy resource scheduling considering vehicle-to-grid. Appl. Soft Comput. 13, 4264–4280 (2013)
4. Ramezani, M., Graf, M., Vogt, H.: A simulation environment for smart charging of electric vehicles using a multi-objective evolutionary algorithm. In: Kranzlmüller, D., Tjoa, A.M. (eds.) ICT-GLOW 2011. LNCS, vol. 6868, pp. 56–63. Springer, Heidelberg (2011)
5. Lee, J., Park, G.-L., Kwak, H.-Y., Jeon, H.: Design of an energy consumption scheduler based on genetic algorithms in the smart grid. In: Jędrzejowicz, P., Nguyen, N.T., Hoang, K. (eds.) ICCCI 2011, Part I. LNCS, vol. 6922, pp. 438–447. Springer, Heidelberg (2011)
6. Kang, J., Duncan, S.J., Mavris, D.N.: Real-time Scheduling Techniques for Electric Vehicle Charging in Support of Frequency Regulation. Procedia Comput. Sci. 16, 767–775 (2013)
7. Wang, Y.W.: An optimal location choice model for recreation-oriented scooter recharge stations. Transport. Res. Part D: Transport and Environment 12(3), 231–237 (2007)
8. Wang, Y.W.: Locating battery exchange stations to serve tourism transport. Transport. Res. Part D: Transport and Environment 13(3), 193–197 (2008)
9. Wang, Y.W., Lin, C.C.: Locating multiple types of recharging stations for battery-powered electric vehicle transport. Transport. Res. Part E 58, 76–87 (2013)
10. He, F., Wu, D., Yin, Y., Guan, Y.: Optimal deployment of public charging stations for plug-in hybrid electric vehicles. Transport. Res. Part B 47, 87–101 (2013)
11. Artmeier, A., Haselmayr, J., Leucher, M., Sachenbacher, M.: The optimal routing problem in the context of battery-powered electric vehicles. In: Workshop CROCS at CPAIOR 2010, 2nd International Workshop on Constraint Reasoning and Optimization for Computational Sustainability (2010)

12. Erdogan, S., Miller-Hooks, E.: A Green Vehicle Routing Problem. Transport. Res. Part E 48, 100–114 (2012)
13. Lin, C., Choy, K.L., Ho, G.T.S., Chung, S.H., Lam, H.Y.: Survey of Green Vehicle Routing Problem: Past and future trends. Expert. Syst. Appl. 41, 1118–1138 (2014)
14. Bashash, S., Moura, S.J., Forman, J.C., Fathy, H.K.: Plug-in hybrid electric vehicle charge pattern optimization for energy and battery longevity. J. Power Sources 196, 541–549 (2011)
15. Kolen, W.J.A., Lenstra, J.K., Papadimitriou, C.H., Spieksma, F.C.R.: Interval Scheduling: A Survey. Nav. Res. Logist. 54, 530–543 (2007)
16. Kovalyov, M.Y., Ng, C.T., Cheng, T.C.E.: Fixed interval scheduling: Models, applications, computational complexity and algorithms. Eur. J. Oper. Res. 178, 331–342 (2007)
17. Arkin, E.M., Silverberg, E.B.: Scheduling with fixed start and end times. Discrete Appl. Math. 18, 1–8 (1987)
18. Bouzina, K.I., Emmons, H.: Interval scheduling on identical machines. J. Global Optim. 9, 379–393 (1996)
19. Faigle, U., Nawijn, W.M.: Note on scheduling intervals on-line. Discrete Appl. Math. 58, 13–17 (1995)
20. Carlisle, M.C., Lloyd, E.L.: On the k-coloring of intervals. Discrete Appl. Math. 59, 225–235 (1995)
21. Garey, R., Johnson, S.: Computers and Intractability: A Guide to the Theory of NP-Completeness, A Series of Books in the Mathematical Sciences. William H. Freeman, United States (1979)
22. Brijnesh, J.J., Wysotski, F.: The Maximum Weighted Clique Problem and Hopfield Network. In: Proceedings - European Symposium on Artificial Neural Networks (ESANN), Bruges, pp. 331–336 (2004)
23. Yamaguchi, K., Masuda, S.: A New Exact Algorithm for the Maximum Weight Clique. In: The 23rd International Technical Conference on Circuits/Systems, Computers and Communications, Yamaguchi (2008)
24. Balas, E., Yu, C.S.: On graphs with polynomially solvable maximum-weight clique problem. Networks 19, 247–253 (1989)
25. Orlin, J.B.: A Faster Strongly Polynomial Minimum Cost Flow Algorithm. Oper. Res. 41(2), 338–350 (1993)

Automatic Clustering Using a Genetic Algorithm with New Solution Encoding and Operators

Carolina Raposo[1], Carlos Henggeler Antunes[2], and João Pedro Barreto[1]

[1] Institute of Systems and Robotics, Dept. of Electrical and Computer Engineering, University of Coimbra, Portugal
{carolinaraposo,jpbar}@isr.uc.pt

[2] INESC Coimbra, Dept. of Electrical and Computer Engineering, University of Coimbra, Portugal
ch@deec.uc.pt

Abstract. Genetic algorithms (GA) are randomized search and optimization techniques which have proven to be robust and effective in large scale problems. In this work, we propose a new GA approach for solving the automatic clustering problem, ACGA - Automatic Clustering Genetic Algorithm. It is capable of finding the optimal number of clusters in a dataset, and correctly assign each data point to a cluster without any prior knowledge about the data. An encoding scheme which had not yet been tested with GA is adopted and new genetic operators are developed. The algorithm can use any cluster validity function as fitness function. Experimental validation shows that this new approach outperforms the classical clustering methods K-means and FCM. The method provides good results, and requires a small number of iterations to converge.

Keywords: Genetic Algorithms, Clustering, Calinski-Harabasz index, K-means, Fuzzy C-Means.

1 Introduction

Clustering is the problem of classifying an unlabeled dataset into groups of similar objects. Each group, called a cluster, consists of objects that are similar between themselves, according to a certain measure, and dissimilar to objects of other groups, with respect to the same measure. Clustering has had an important role in areas such as computer vision [1], data mining and document categorization [2], and bioinformatics [3].

The major difficulty in the clustering problem is the fact that it is an unsupervised task, i.e., in most cases the structural characteristics of the problem are unknown. In particular, the spatial distribution of data in terms of number, volumes, and shapes of clusters is not known.

A classical clustering method is K-means [4], in which the data is divided into a set of K clusters. This division depends on the initial clustering, and is

B. Murgante et al. (Eds.): ICCSA 2014, Part II, LNCS 8580, pp. 92–103, 2014.

computed by minimizing the squared error between the centroid of a cluster and its elements. One variant of K-means is the Fuzzy C-Means (FCM), in which the elements of the dataset may belong to more than one cluster with different membership degrees [5]. FCM proceeds by updating the cluster centroids and the membership degrees until a convergence condition is attained. Both methods have proven to perform well in many situations. However, they can easily get stuck in local minima, depending on the initial clustering. Moreover, for problems where the number of clusters is unknown a priori, it is necessary to perform the clustering for different values of K and choose the best solution according to a certain evaluation function. This becomes impractical and tedious in the presence of a large number of clusters.

These issues (local minima and unknown number of clusters) can be tackled using GA [6], which have proven to be effective in search and optimization problems [7]. GA are high level search heuristics that mimic the process of natural evolution, based on the principle of survival of the fittest, by using selection, crossover and mutation mechanisms. Solutions (individuals) are encoded as strings, called *chromosomes*, and, at each iteration, the fitness of solutions in the population is computed for determining the evolutionary process.

In [8] a GA was proposed to find the optimal partition of a dataset, given the number of clusters. More recently, methods based on GA have been proposed to solve the automatic clustering problem, i.e., when no information about the dataset is known [9–13].

We propose a new GA approach for solving the automatic clustering problem, ACGA - Automatic Clustering Genetic Algorithm - which uses a solution encoding based on the one presented in [14]. Two new mutation and one new crossover operators have been developed in order to guarantee the existence of high diversity in the population. As fitness function any internal cluster validity measure can be used, for which the Calinski-Harabasz (CH) index [15] was chosen. An experimental validation using real [16] and synthetic [17] datasets was conducted, and the performance of ACGA was compared with the K-means and FCM algorithms. ACGA outperforms these two classical methods, converging to solutions with higher fitness function values.

The interest and motivation of this study have been provided in the introduction. An overview of the GA approach describing the encoding scheme, and genetic operators is presented in Section 2. Experiments and results are discussed in Section 3. Finally, some conclusions are drawn in Section 4.

2 Overview of the GA Approach

In this Section a new GA approach for solving the automatic clustering problem is presented. An encoding scheme based on [14] is used, and new genetic operators (mutation and crossover) are developed.

Solutions represent different points in the search space, to which quality values are assigned, according to a pre-defined fitness function. The best B solutions of the population are preserved, and pass to the next generation, thus endowing the evolutionary process with a certain degree of elitism. A fixed number of solutions is then selected to act as parents. Crossover is applied to the parent chromosomes, originating new solutions, which are then subject to mutation. The new population is evaluated, and this procedure is repeated until a stopping criterion is satisfied.

Algorithm 1 shows the main steps of the GA. In the next Subsections, the encoding scheme is described, as well as the procedures for generating the initial population, evaluating each solution, selecting the parent population, and performing the crossover and mutation operators.

Algorithm 1: Pseudo-Code of the GA

$t \leftarrow 0$;
Generate initial population $P(t)$;
Evaluate $P(t)$;
while *Stopping criterion not satisfied* **do**
  Select parent population $P'(t)$ from $P(t)$;
  Apply genetic operators to $P'(t) \rightarrow P(t+1)$;
  Replace random solutions in $P(t+1)$ with the best B solutions in $P(t)$;
  Evaluate $P(t+1)$;
  $t \leftarrow t+1$;
Result: Best solution (highest fitness value) of the population in the last generation.

2.1 Encoding Scheme

In this work, we adopted an encoding scheme based on the one used in [14]. Each solution i is a fixed-length string represented by activation values A_{ij} and cluster centroids $m_{ij}, j = 1, \ldots, K_{max}$, where K_{max} is the maximum number of clusters defined by the user. Then, solution i of the population is represented as:

A_{i1}	A_{i2}	...	$A_{iK_{max}}$	m_{i1}	m_{i2}	...	$m_{iK_{max}}$

where A_{ij} are activation values defined in the interval $[0, 1]$. An activation value A_{ij} larger than 0.5 means that the cluster with centroid m_{ij} is active, i.e., it is selected for partitioning the dataset. Otherwise, the respective cluster is inactive in chromosome i.

Since each centroid m_{ij} is a d-dimensional vector, where d is the number of features of the data, each solution is a string with total length equal to $K_{max} + dK_{max}$. As an example, for a maximum number of clusters $K_{max} = 3$ and a dataset with $d = 2$ features, the string

0.1	**0.8**	**0.6**	10	0.5	**11**	**0.9**	**14**	**0.3**

represents a solution i with two activated clusters with centroids $m_{i2} = [11, 0.9]$ and $m_{i3} = [14, 0.3]$.

This encoding scheme has advantages over the more popular label-encoding, where each position is an integer value corresponding to one cluster. These include the fact that it may lead to smaller solution vectors, in the presence of large datasets, and solutions with isolated one-element clusters are more difficultly generated, since the labeling of each data point only depends on the location of the centroids. However, this encoding has the disadvantage of requiring one label to be assigned to each data point, before computing the fitness of each solution. This is done by finding the active centroid closest to each point (smallest euclidean distance).

2.2 Initialization of the Population

Each candidate solution in the population is initialized independently, with the activation values and the cluster centroids initialized separately.

The K_{max} activation values A_{ij} are initialized as random values drawn from the standard uniform distribution in the interval $[0, 1]$. In order to guarantee that each initial solution has at least 2 active clusters, a minimum of 2 activation values are forced to be larger than 0.5.

For initializing the K_{max} centroids m_{ij}, the interval in which the data points are contained is determined. More specifically, for each dimension relative to each feature, the minimum and the maximum values in the dataset are determined. Then, for each dimension, K_{max} random values are sampled from the corresponding interval, originating K_{max} points that are the initial centroids.

The size of the population P (number of candidate solutions) is a user-defined value.

2.3 Solution Evaluation

A quality measure is assigned to each solution, which is computed using a fitness function. In this work we chose to define the fitness function as the Calinski-Harabasz (CH) index [15], which uses the quotient between the intra-cluster average squared distance and the inter-cluster average squared distance. For solution s, $\mathsf{CH}(s)$ is computed as in equation (1):

$$\mathsf{CH}(s) = \frac{\sum_{i=1}^{N} ||X_i - m_s||^2 - \sum_{j=1}^{K_s} \sum_{X \in C_{sj}} ||X - m_{sj}||^2}{\sum_{j=1}^{K_s} \sum_{X \in C_{sj}} ||X - m_{sj}||^2} \frac{N - K_s}{K_s - 1}, \quad (1)$$

where N is the number of objects X_i in the dataset, m_s is the centroid of the dataset, K_s is the number of clusters in solution s, and X are the objects in cluster C_{sj} with centroid m_{sj}. $|| \cdot ||$ denotes the L2 norm. Larger inter-cluster distances and smaller intra-cluster distances lead to larger CH values, i.e., higher quality solutions.

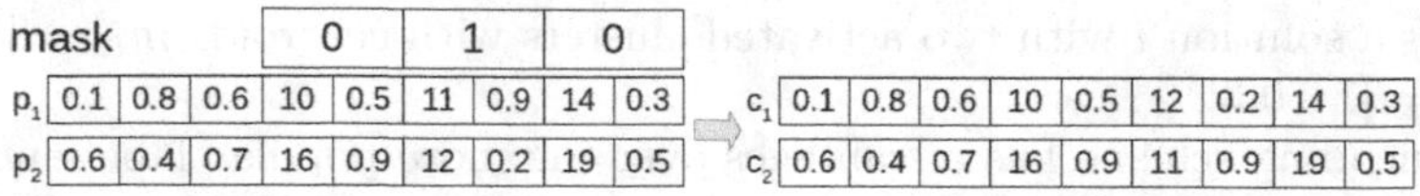

				mask	0		1		0	
p_1	0.1	0.8	0.6	10	0.5	11	0.9	14	0.3
p_2	0.6	0.4	0.7	16	0.9	12	0.2	19	0.5

c_1	0.1	0.8	0.6	10	0.5	12	0.2	14	0.3
c_2	0.6	0.4	0.7	16	0.9	11	0.9	19	0.5

Fig. 1. Example of the crossover between two parent solutions p_1 and p_2, originating two offspring c_1 and c_2. The solutions correspond to maximum number of clusters $K_{max} = 3$ and $d = 2$ features.

Note that for one-cluster solutions, the term $K_s - 1$ is substituted by 1, and we have the relation

$$\sum_{i=1}^{N} ||X_i - m_s||^2 = \sum_{j=1}^{K_s} \sum_{X \in C_{sj}} ||X - m_{sj}||^2, \tag{2}$$

leading to a CH value of 0. Thus, it is not necessary to explicitly deal with one-cluster solutions since they are automatically assigned a null quality value.

2.4 Selection of the Parent Solutions

Before applying the genetic operators, a set of parent solutions must be selected. This is done in a tournament scheme by selecting P_p random solutions from the population and finding the one with the highest fitness, which will act as a parent solution. This procedure is repeated P times, originating a parent population of the same size as the original population.

2.5 Genetic Operators

The genetic operators should be suited to the encoding scheme. Their objective is to ensure diversity in the population, such that better solutions can be produced through the evolutionary process. In this work, one new crossover scheme is presented, as well as two new mutation operators.

Crossover. For each pair of parent solutions in the population crossover is applied, originating two offspring. A binary mask of length K_{max} is created, where 1 occurs with a probability p_{cross}. The centroids in the parent solutions corresponding to the positions in the mask with a unitary value are exchanged, originating two children. Figure 1 shows an example of the crossover between parent solutions p_1 and p_2 yielding offspring c_1 and c_2.

Mutation. Each solution produced by the crossover operator goes through a mutation process. Two different types of mutation have been considered: mutation in the activation values, and mutation in the centroids.

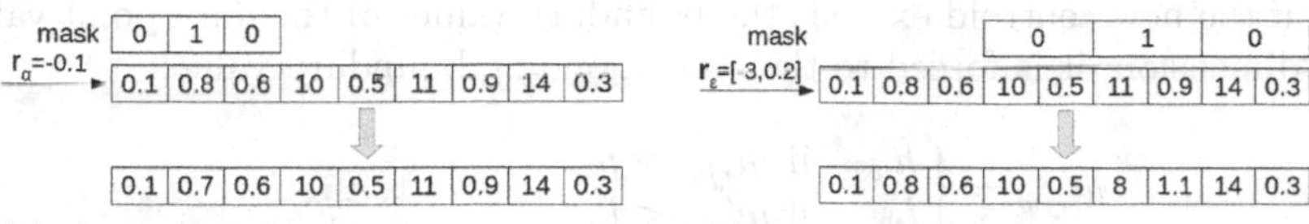

(a) An example of the activation value mutation.

(b) An example of the centroid mutation.

Fig. 2. Examples of the two different types of mutation proposed in this work: mutation in the activation values, and centroid mutation. The solution vectors correspond to maximum number of clusters $K_{max} = 3$, and $d = 2$ features.

Activation Value Mutation

Similarly to the crossover operator, a binary mask of length K_{max} is created, where 1 occurs with a probability p_{mut}. Let α be a user-defined parameter that controls the perturbation of the activation values. For each unitary value in the mask, a random value r_α is drawn in the interval $[-\alpha, \alpha]$. The activation value in the corresponding position is modified by adding r_α: $A'_{ij} = A_{ij} + r_\alpha$, where A'_{ij} is the modified activation value.

If the resulting activation value A'_{ij} does not belong to the interval $[0, 1]$, it is forced to 0 or 1:

$$A'_{ij} = \begin{cases} 1 & \text{if } A'_{ij} > 1 \\ 0 & \text{if } A'_{ij} < 0 \end{cases}. \quad (3)$$

Figure 2(a) shows an example of the application of mutation to the activation values. The centroid values remain unaltered, and only randomly chosen activation values, with probability p_{mut}, are altered.

Since the crossover operator does not influence the activation values, this type of mutation is the only possible way to achieve diversity in the number of clusters between solutions in consecutive iterations of the algorithm.

Centroid Mutation

The mutation operator applied to the centroids is similar to the activation value mutation, with the difference that the perturbation is a d-dimensional vector created according to the range of values of each feature of the data points.

Let ϵ be a user-defined parameter that controls the change in amplitude of the centroids. For each dimension w of the data points, $w = 1, \ldots, d$, let $g_w = h_w - l_w$ be the difference between the largest value relative to feature w found in the dataset, h_w, and the smallest value relative to the same feature, l_w. By defining $t_w = \epsilon g_w$ for each feature, a vector $\mathbf{r}_\epsilon = [r_{\epsilon 1}, \ldots, r_{\epsilon d}]$ can be created by drawing random values $r_{\epsilon w}$ from the intervals $[-t_w, t_w]$.

The centroids m_{ij} that correspond to unitary values in the binary mask (created as in the activation value mutation scheme) are modified by adding the perturbation vector $\mathbf{r}_\epsilon$: $m'_{ij} = m_{ij} + \mathbf{r}_\epsilon$, where m'_{ij} is the modified centroid.

Again, if the new centroid exceeds the boundary values of the data point values in any dimension, it is forced to the corresponding boundary value:

$$m'_{ijw} = \begin{cases} h_w & \text{if } m'_{ijw} > h_w \\ l_w & \text{if } m'_{ijw} < l_w \end{cases}, \forall w = 1, \ldots, d. \tag{4}$$

Figure 2(b) shows an example of the application of mutation to the centroids. In this case, only randomly chosen centroids are altered.

The activation value mutation may drastically change the number of clusters in a solution, originating child solutions considerably different from the parent solutions. This may be undesirable, since it easily leads to solutions with significantly lower fitness values. Thus, this type of mutation is applied less often than the centroid mutation. The algorithm proceeds until either a maximum number of iterations I_{max}, or a fixed number of iterations without improvement of the best solution in the population I_{ni} are reached.

3 Experiments and Results

In this Section, we compare the performance of ACGA with two classical clustering methods: K-means [4] and FCM [5]. Since these two methods require the user to input the number of clusters, we performed the clustering for a varying number of clusters, and chose the solution with the highest quality value, using the same fitness function used in the GA (CH index). We tested for $K = 2, \ldots, K_{max}$, with $K_{max} = 20$, clusters. Experiments were performed using the datasets in Table 1. The first three are synthetic datasets downloaded from [17], presenting different degrees of cluster overlapping, as can be seen in Figure 3. The remaining sets are real-life datasets, which were obtained from [16], and are commonly used for the validation of clustering methods. The parameters refer to the original labeling provided with the datasets. Table 1 shows the number of clusters K, the number of objects N, and the number of features d for each dataset.

Table 1. Description of the datasets used in the experiments. K is the number of clusters, N is the number of data points, and d is the number of features.

Dataset	K	N	d	Avg Intra Cluster Dist.	Avg Inter Cluster Dist.	CH value
S2	15	5000	2	42294.60	182459.12	12541
S3	10	3254	2	54507.97	174914.69	5002
S4	15	5000	2	55312.63	149595.93	5385
Iris	3	150	4	0.67	2.15	486
B. Cancer	2	638	9	2.66	4.65	303
Seeds	3	210	7	1.58	3.85	310

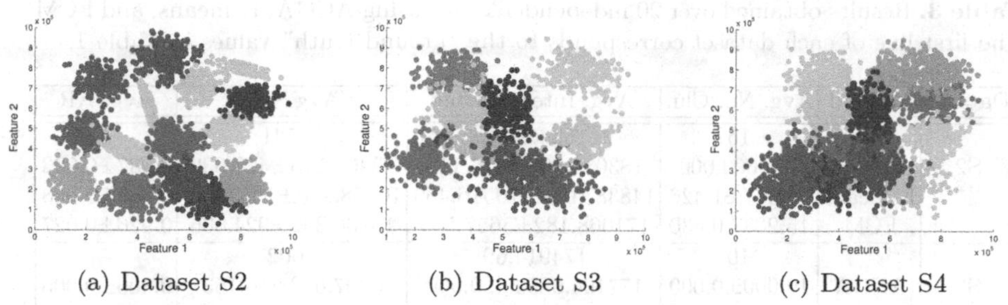

(a) Dataset S2 (b) Dataset S3 (c) Dataset S4

Fig. 3. The three synthetic datasets used in the experiments, presenting different degrees of cluster overlapping. Colors are used to identify the clusters.

Table 2. User-defined values used in the experiments

Maximum number of clusters, K_{max}	20
Population size, P	40
No. of solutions chosen for parent selection, P_p	3
Crossover probability, p_{cross}	0.3
Mutation probability, p_{mut}	0.1
Activation value mutation parameter, α	0.1
Centroid mutation parameter, ϵ	0.3
No. of elite solutions, B	4
Maximum no. of iterations, I_{max}	5000
Maximum no. of iterations without improvement, I_{ni}	500

Moreover, the average intra-cluster and inter-cluster distances were computed. For each cluster, the intra-cluster distance is the mean of the distances of all data points to the cluster centroid. The inter-cluster distance is computed as the minimal distance between cluster centroids, for each cluster. Table 1 also shows the CH value, computed using equation (1). For all the datasets, 20 independent runs of each method were performed. Since K-means and FCM depend on the initial clustering which is obtained by randomly sampling the dataset, and generating random membership values, respectively, slightly different results may be obtained in different runs. For ACGA, the values in Table 2 were assigned to the parameters described in Section 2. Table 3 shows the results obtained for all the datasets in Table 1, using K-means, FCM, and ACGA. The last column of Table 3 presents the values for the Adjusted Rand (AR) index [18], which is a measure of agreement between two partitions and is frequently used in cluster validation. We computed the AR index between the original labelings and the ones obtained with each method. This index indicates how similar the partition obtained is to the original one. For two random partitions, the expected value of the AR index is 0, and its maximum value is 1, obtained when comparing equal partitions.

Table 3. Results obtained over 20 independent runs using ACGA, K-means, and FCM. The first line of each dataset corresponds to the "ground truth" values in Table 1.

Dataset	Method	Avg. No. Clu.	Avg. Inter Cluster	Avg. CH	Avg. AR
S2	-	15	182459.12	12541	-
	ACGA	15.000±000	183081.207±316.112	13461.515±420.630	0.937±0.003
	K-means	16.967±1.426	148383.088±10516.459	10658.228±1697.369	0.859±0.048
	FCM	15.233±0.430	171068.182±5657.558	13106.362±924.506	0.926±0.027
S3	-	10	174914.69	5002	-
	ACGA	10.000±0.000	177104.685±420.960	6307.633±16.153	0.806±0.003
	K-means	10.267±1.112	173354.951±15347.364	5482.044±584.628	0.736±0.042
	FCM	10.133±0.434	170345.561±9489.542	6091.281±387.783	0.785±0.041
S4	-	15	149595.93	5385	-
	ACGA	15.000±0.000	152034.821±358.558	7865.943±12.925	0.724±0.003
	K-means	15.933±1.460	138291.741±7920.736	6788.132±605.469	0.664±0.035
	FCM	15.567±0.626	141600.223±4524.683	7570.782±339.299	0.713±0.019
Iris	-	3	2.15	486	-
	ACGA	3.00±0.000	2.185±0.000	506.297±0.000	0.886±0.000
	K-means	3.067±0.450	1.932±0.319	473.496±75.401	0.791±0.158
	FCM	3.00±0.000	1.940±0.000	506.297±0.000	0.886±0.000
Breast Cancer	-	2	4.65	303	-
	ACGA	2.000±0.000	11.793±1.117	1026.084±0.351	0.847±0.004
	K-means	2.000±0.000	13.816±0.000	1026.262±0.000	0.846±0.000
	FCM	2.000±0.000	13.886±0.000	1023.939±0.000	0.830±0.000
Seeds	-	3	3.85	310	-
	ACGA	3.000±0.000	3.107±0.153	329.191±3.108	0.704±0.012
	K-means	3.000±0.000	3.988±0.000	328.267±0.000	0.706±0.000
	FCM	3.000±0.000	3.970±0.000	329.918±0.000	0.694±0.000

By comparing the AR index in Table 3, it can be seen that ACGA generally outperforms both K-means and FCM. As an example, an AR value of 0.886 obtained for the iris dataset means that 6 data points were misclassified, corresponding to 4%. The superiority of ACGA is clear by comparing the results obtained for all synthetic datasets, where both K-means and FCM failed to partition the dataset in the correct number of clusters in every run. Figure 4 shows a partition of the S2 dataset obtained with the three methods, where each cluster is identified by a color. Comparing the results to the original clusters in Figure 3(a), it can be seen that our solution (Figure 4(a)) yielded the correct number of clusters, with only a few misclassified objects. K-means (Figure 4(b)) performed poorly since it was not capable of finding the optimal number of clusters. FCM (Figure 4(c)) produced a result considerably better than K-means. However, it still failed to correctly classify many data points. Moreover, ACGA leads to smaller intra-cluster distances and larger inter-cluster distances, and consequently larger CH values. Due to cluster overlapping, the CH values computed using the original partitions (Table 1) are, in some cases, lower than the ones obtained with ACGA. Also, it presents much smaller standard deviations than both K-means and FCM, especially in the synthetic datasets. This means that ACGA is capable of producing stable results, independently of the initialization, converging to very similar solutions in different runs.

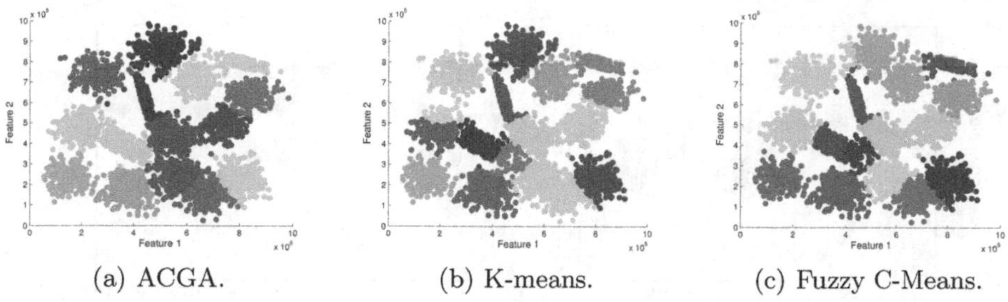

(a) ACGA. (b) K-means. (c) Fuzzy C-Means.

Fig. 4. Results obtained for dataset S2 (refer to Figure 4) using ACGA, K-means, and FCM. Clusters are identified by colors.

Table 4. Average number of iterations required to reach the results in Table 3

Dataset	S2	S3	S4	Iris	B. Cancer	Seeds
Iterations	2936.5	2199.5	3316.0	15.6	588.5	353.3

Table 4 shows the average number of iterations necessary to achieve the solutions in Table 3. The values were obtained by subtracting the number of iterations without improvement of the best solution in the population, I_{ni}, to the total number of iterations. It can be seen that for smaller and "well-behaved" datasets, ACGA is fast and converges after a small number of iterations. For larger datasets, such as S2, it requires approximately 3000 iterations.

Generally, using parameters such as the mutation probability that adapt to the evolution of the algorithm is a good technique since higher diversity in the population can be achieved, preventing the algorithm to get stuck in local minima. Thus, we tested the influence of an increase in p_{mut} triggered by reaching a certain number of iterations without improvement of the best solution in the population. Results obtained with dataset S2 are shown in Figure 5(a), which plots the average (blue) and maximum (red) quality of the population, as well as the number of clusters of the best solution (black), in each iteration. It can be seen that, despite increasing diversity in the population, applying mutation at higher rates leads to a decrease in the average quality of population, which is clear by observing the evolution in the last 500 iterations. Thus, we kept all parameters fixed after a tuning process, and the algorithm behaves as depicted in Figure 5(b), where the average quality curve presents always the same pattern. The maximum quality in the population does not decrease due to elitism: the best B solutions in each generation pass to the next generation unaltered. This moderate degree of selective pressure due to elitism contributed to improve the results without reducing exploration. Also, the high population diversity is evinced by the significant changes in the average fitness in consecutive iterations of the algorithm. The curves show that significant changes in the maximum quality are generally caused by changes in the number of clusters.

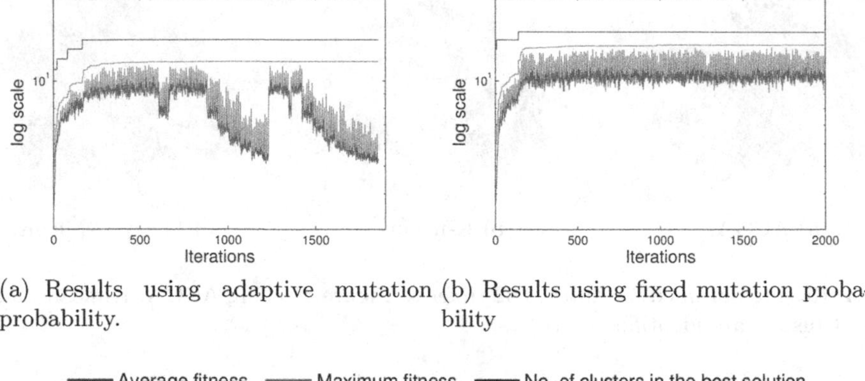

(a) Results using adaptive mutation probability.

(b) Results using fixed mutation probability

Fig. 5. Evolution of the fitness and number of clusters of the solutions in the population for dataset S2. The figures show the maximum (red) and average (blue) fitnesses of the population, as well as the number of clusters of the best solution in each iteration (black). Fitness values have been scaled by a factor of 10^{-3}.

4 Conclusion

We present a new GA approach, ACGA, for automatically finding the optimal number of clusters in real and synthetic datasets, with different degrees of overlapping, and correctly assigning each data point to a cluster. It uses an encoding scheme that, to the best of our knowledge, had never been incorporated into GA. This required the development of new genetic operators that ensure diversity in the population. ACGA does not require any information about the data, and is able to outperform two classical clustering methods: K-means and FCM. Experiments included three synthetic and three real datasets, and results show that ACGA leads to partitions very similar to the original ones, requiring a small number of iterations to converge.

Acknowledgments. This work was financially supported by a PhD grant (SFRH/BD/88446/2012) from the Portuguese Foundation for Science and Technology (FCT). The author Carlos Henggeler Antunes acknowledges the support of FCT project PEst-OE/EEI/UI0308/2014 and QREN Mais Centro Program iCIS project (CENTRO-07-ST24-FEDER-002003).

References

1. Belahbib, F., Souami, F.: Genetic algorithm clustering for color image quantization. In: 3rd European Workshop on Visual Information Processing (EUVIP), pp. 83–87 (2011)

2. Mecca, G., Raunich, S., Pappalardo, A.: A New Algorithm for Clustering Search Results. Data and Knowledge Engineering 62, 504–522 (2007)
3. Valafar, F.: Pattern Recognition Techniques in Microarray Data Analysis: A Survey. Annals of New York Academy of Sciences 980, 41–64 (2002)
4. Hartigan, J., Wong, M.: Algorithm AS 136: A K-Means Clustering Algorithm. Applied Statistics 28(1), 100–108 (1979)
5. Bezdek, J., Ehrlich, R., Full, W.: FCM: The fuzzy c-means clustering algorithm. Computers and Geosciences 10(2-3), 191–203 (1984)
6. Holland, J.: Genetic algorithms. Scientific American (1992)
7. Srinivas, M., Patnaik, M.: Genetic algorithm: A survey. IEEE Computer 27(6), 17–26 (1994)
8. Murthy, C., Chowdhury, N.: In search of optimal clusters using GA. Pattern Recognition Letters 17, 825–832 (1996)
9. Tseng, L., Yang, S.: A genetic approach to the automatic clustering problem. Pattern Recognition 34(2), 415–424 (2001)
10. Agustin-Blas, L., Salcedo-Sanz, S., Jimenez-Fernandez, S., Carro-Calvo, L., Del Ser, J., Portilla-Figueras, J.A.: A new grouping GA for clustering problems. Expert Systems with Applications 39(10) (2012)
11. Sheikh, R., Raghuwanshi, M., Jaiswal, A.: Genetic Algorithm Based Clustering: A Survey. In: First International Conference on Emerging Trends in Engineering and Technology, vol. 2(6), pp. 314–319 (2008)
12. Liu, Y., Wu, X., Shen, Y.: Automatic clustering using genetic algorithms. Applied Mathematics and Computation 218(4), 1267–1279 (2011)
13. He, H., Tan, Y.: A two-stage genetic algorithm for automatic clustering. Neurocomputing 81, 49–59 (2012)
14. Das, S., Abraham, A., Konar, A.: Automatic Clustering Using an Improved Differential Evolution Algorithm. IEEE Transactions on Systems, Man and Cybernetics, Part A: Systems and Humans 38(1), 218–237 (2008)
15. Calinski, R., Harabasz, J.: A dendrite method for cluster analysis. Communications in Statistics 3(1), 1–27 (1974)
16. Asuncion, A., Newman, J.: UCI Machine Learning Repository. University of California, Department of Information and Computer Science, Irvine, CA (2007), `http://www.ics.uci.edu/~mlearn/MLRepository.html`
17. Speech and Image Processing Unit. Clustering datasets, `http://www.cs.joensuu.fi/sipu/datasets/`
18. Hubert, L., Arabie, P.: Comparing Partitions. Journal of Classification (2), 193–218 (1985)

On Simplicial Longest Edge Bisection in Lipschitz Global Optimization

Juan F.R. Herrera[1], Leocadio G. Casado[1],
Eligius M.T. Hendrix[2], and Inmaculada García[2]

[1] Universidad de Almería (ceiA3), Almería, Spain
{juanfrh,leo}@ual.es
[2] Universidad de Málaga, Málaga, Spain
{eligius,igarciaf}@uma.es

Abstract. Simplicial subsets are popular in branch-and-bound methods for Global Optimization. Longest Edge Bisection is a convenient way to divide a simplex. When the number of dimensions is greater than two, irregular simplices (not all edges have the same length) may appear with more than one longest edge. In these cases, the first longest edge is usually selected. We study the impact of other selection rule of the longest edge to be bisected next on the development of a branch-and-bound algorithm to solve multidimensional Lipschitz Global Optimization instances. Experiments show a significant reduction in the number of evaluated simplices for most of the test problems.

Keywords: Longest Edge Bisection, Branching rule, Branch-and-bound, Lipschitz optimization.

1 Introduction

Branch-and-bound (BnB) methods are commonly used in Global Optimization (GO) problems when the solution requires a certain guaranteed accuracy. The method is based on the recursive division of the problem into sub-problems until a solution is reached. The method generates a search tree and avoids visiting regions where no feasible and non-optimal solution can be found. This is a deterministic method, it always returns the same solution for a given accuracy. GO problems belong to the complexity class of NP-hard problems [1]. BnB methods can be defined by four rules [2,3]:

Branching also known as division rule, determines the method used to divide a sub-problem.

Bounding determines an upper and/or lower bound of the solution contained in a sub-problem. These values are used in the rejection rule.

Selection of the sub-problem to be processed, and possibly divided, next. The selection rule determines how the search tree is built.

Rejection of a sub-problem that does not to contain the global optimal solution.

B. Murgante et al. (Eds.): ICCSA 2014, Part II, LNCS 8580, pp. 104–114, 2014.

Sometimes it is also necessary for several cases to define a fifth rule called the *Termination* rule. This rule determines if the BnB process continues on a non-rejected sub-problem or it belongs to the set of sub-problems that may contain the solution.

The set of rules determines the development of the search. The efficiency of the BnB algorithm can be increased by improving the rules, leading to a reduction of the execution time and facilitating the solution of more complex problems with the same computational resources.

This study focuses on Simplicial BnB algorithms, where the initial search space is an n-dimensional hyper-cube partitioned into a set of non-overlapping simplices. A n-simplex is a convex hull of $n+1$ affinely independent vertices. Recent studies show an interesting improvement in the number of generated sub-simplices when a different heuristic is applied in the iterative refinement of a regular n-simplex [4]. In that study, the complete binary tree is built by bisecting the longest edge of a sub-simplex until the width, determined by the length of the longest edge, is smaller or equal to a given accuracy ϵ. We study the effect of similar heuristics in a BnB algorithm applied to solve the Multidimensional Lipschitz Global Optimization problem (MLGO), where the shape of the binary tree is also determined by the rejection rule using several values of the accuracy.

Therefore, this study is focused on the branching rule. Section 2 briefly explains the MLGO algorithm. Section 3 describes the studied division heuristics. Section 4 shows the results and Section 5 concludes the work.

2 Multidimensional Lipschitz Global Optimization

We focus on the MLGO problems as described in [5]. The goal is to find at least one global minimum point x^* of

$$f^* = f(x^*) = \min_{x \in X} f(x),$$

where the feasible area $X \subset \mathbb{R}^n$ is a compact set. Function f is said to be Lipschitz-continuous on X if there exists a maximum slope L called the Lipschitz constant such that $|f(x) - f(y)| \leq L\|x-y\|$, $\forall x, y \in X$, see [6]. The norm $\|\cdot\|$ is usually taken as Euclidean. However, the generation of bounds also allows other norms.

Algorithm 1 introduces an overview of the BnB method to solve the MLGO problem. It uses explicitly simplicial partition sets that are bisected over the longest edge to generate new points to be evaluated and Lipschitzian lower bounds. It guarantees to find an ϵ-approximation x^U of the minimum point x^* such that $f(x^U) \leq f^* + \epsilon$.

In Global Optimization, a feasible region is usually box-constrained, i.e. the feasible region is a hyper-rectangle. Therefore, most BnB methods use hyper-rectangular partitions. However, other types of partitions may be more suitable for some optimization problems. For the use of simplicial partitions, the feasible region should be partitioned into simplices. The most preferable initial covering

Algorithm 1. Simplicial BnB algorithm, bisection

Require: X, f, L_2, L_1, L_∞, ϵ

Partition X into simplices C_k
Start the working list as $\Lambda := \{C_k\}$
The set of vertices $V := \{v_i \in C_k \in \Lambda\}$
Set $f^U := \min_{v \in V} f(v)$ and $x^U := \arg\min_{v \in V} f(v)$
Determine lower bounds $LB(C_k)$ based on L_2, L_1, L_∞ — *Bounding rule*
while $\Lambda \neq \emptyset$ **do**
  Extract a simplex $C = C_k$ from Λ — *Selection rule*
  Bisect C into $C1$ and $C2$ generating x — *Branching rule*
  if $x \notin V$ **then**
    Add x to V
    if $f(x) < f^U$ **then**
      Set $f^U := f(x)$ and $x^U := x$
      Remove all C_k from Λ with $LB_k > f^U - \epsilon$
    end if
  end if
  Determine lower bounds $LB(C1)$ and $LB(C2)$ — *Bounding rule*
  Store $C1$ in Λ if $LB(C1) \leq f^U - \epsilon$
  Store $C2$ in Λ if $LB(C2) \leq f^U - \epsilon$
end while
return x^U, f^U

is face-to-face vertex triangulation. It involves partitioning the feasible region into a finite number of n-dimensional simplices with vertices that are also the vertices of the feasible region. A standard method [7] is triangulation into $n!$ simplices. All simplices share the diagonal of the feasible region and have the same hyper-volume. Figure 1 depicts a hypercube of dimension three divided in six irregular simplices.

2.1 Selection Rule

For the selection rule in the algorithm, we focus on a depth-first search by selecting the best simplex among those generated in the last division, until no further division is possible. In general, depth-first search is used to reduce the memory requirement of the algorithm. In a previous work, a hybrid selection rule, combination of best-first and depth-first, was used [8]. Nevertheless, the memory requirement is high, leading to longer execution times [9]. A wide study on selection strategies for this type of algorithms can be found in [10].

2.2 Bounding Rule

The determination of a lower bound is an important computational step. Therefore, we study which calculations are involved. Consider simplex C with vertices $v_0, v_1, \ldots, v_n$. In our study, the Lipschitz constants are given *a priori*. Like in [5],

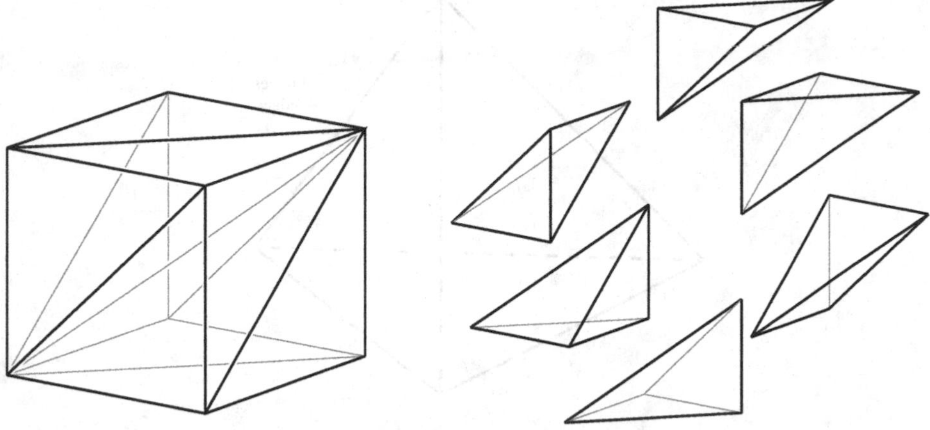

Fig. 1. Division of a hypercube into six irregular simplices

we may consider the maximum slope of a differentiable function based on several norms:

$$L_1 = \max_{x \in X} \|\nabla f(x)\|_1$$
$$L_2 = \max_{x \in X} \|\nabla f(x)\|_2$$
$$L_\infty = \max_{x \in X} \|\nabla f(x)\|_\infty$$

The basis of the lower bound is that for each vertex v_i we have a lower bounding function

$$\varphi_i(x) := f(v_i) - L\|x - v_i\| \leq f(x). \tag{1}$$

For the multi-dimensional case, researchers looked for the easy computational determination of

$$\min_{x \in C} \max_i \varphi_i(x) \tag{2}$$

where the objective function is in general neither convex nor concave. Summarizing, the lower bound is obtained using (2) and L_1, L_2 and L_∞ with the corresponding norms in (1). For more details see [5].

2.3 Rejection Rule

A simplex C is rejected if

$$LB(C) > f^U - \epsilon.$$

3 Longest Edge Bisection

In the literature we can find many ways of dividing a simplex [6]. One of them is the Longest Edge Bisection (LEB), which is a popular way of iterative refinement

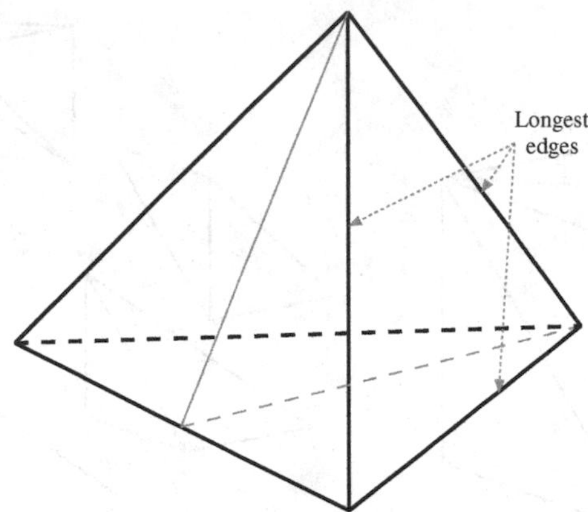

Fig. 2. Longest Edge Bisection, denoting a sub-simplex with three equal longest edges

in the finite element method, since it is very simple and can easily be applied in higher dimensions [11]. This method consists of splitting a simplex using the hyperplane that connects the middle point of the longest edge of a simplex with the opposite vertices as illustrated in Figure 2.

The LEB process is not uniquely determined for a 3-simplex [11]. For $n = 2$ the longest edge is either unique or the simplex is symmetric and the bisection of any longest edge produces similar sub-simplices. For $n \geq 3$ several longest edges may appear in an irregular simplex, see Figure 2. A heuristic for selecting the longest edge has been investigated in the refinement of a regular n-simplex [4]. The investigated longest edge selection rules are the following:

LEB$_1$ The natural way to select a longest edge is to take the first one found. The first one depends of course on the coding and storing of the vertices and edges, i.e. the index number assigned to each vertex of the simplex. When a simplex is split and two sub-simplices are created, the new vertex of each sub-simplex has the same index as the one it substitutes. This rule is used as a benchmark in measuring performance.

LEB$_\alpha$ For each vertex in a longest edge, the sum of the angles between edges ending at that vertex is determined and that longest edge that obtains the smallest sum is selected. This rule was elaborated and applied in [4].

Example 1. *For a tetrahedron with vertices v_i, $i = 1, 2, 3, 4$, the alpha value of the edge $v_2 - v_1$ is given by the sum of the angles with vertex v_1 ($\widehat{v_2 v_1 v_3}$, $\widehat{v_3 v_1 v_4}$, and $\widehat{v_2 v_1 v_4}$) and with vertex v_2 ($\widehat{v_1 v_2 v_3}$, $\widehat{v_1 v_2 v_4}$, and $\widehat{v_3 v_2 v_4}$) calculated using trigonometric identities.*

The research question is how a different selection among several longest edges will influence the development of the BnB algorithm for MLGO.

Table 1. Test functions

No.	Test problem	n	L_1	L_2	L_∞	Domain
1	Levy No. 15	4	1273.2	1196.4	1195.5	$[-10, 10]^4$
2	Shekel 5	4	204.1	102.4	56.1	$[0, 10]^4$
3	Shekel 7	4	300.1	151.5	86.1	$[0, 10]^4$
4	Shekel 10	4	408.2	204.5	110.8	$[0, 10]^4$
5	Schwefel 1.2	4	600	313.69	200	$[-5, 10]^4$
6	Powell	4	92216	48251.5	29270	$[-4, 5]^4$
7	Levy No. 9	4	26.1	14.4	8.3	$[-10, 10]^4$
8	Levy No. 16	5	422.93	370.66	369.68	$[-5, 5]^5$
9	Levy No. 10	5	34.375	16.56	8.25	$[-10, 10]^5$
10	Levy No. 17	6	421.7	358.6	357.4	$[-5, 5]^6$

4 Results

The set of problems used for the evaluation are taken from [12] (see Appendix). Table 1 summarizes the most important characteristics of each problem. Algorithm 1 has been executed for the set of test problems, using the strategies described in Section 3 (LEB_1 and LEB_α) for selecting the longest edge to be bisected. Experimental results have been obtained with an accuracy of $\epsilon = \frac{1}{8}L_2$ for four-dimensional instances, $\epsilon = \frac{3}{8}L_2$ for five-dimensional instances and $\epsilon = \frac{12}{8}L_2$ for the six-dimensional instance.

Numerical results obtained from these experiments are shown in Table 2. Columns NS_1 and NS_α provide the number of simplex evaluations performed by Algorithm 1 when the longest edge to be bisected is chosen by rules LEB_1 and LEB_α, respectively. Data in column NLE_1 represents the percentage of the divided sub-simplices which contains two or more longest edges when LEB_1 rule is used. The algorithm, coded in C, has been carried out on a node of Bullxual, consisting of two eight-core 2.00 GHz Intel® Xeon® E5-2650 processors and 64 GB of main memory.

Experimental results in column NLE_1 show that the number of simplices having more than one longest edge is relatively large, between 56% and 74%. Nevertheless, using LEB_α rule the number of divided simplices containing two or more longest edges with the minimum LEB_α value is lower: around 8% for four-dimensional problems and less than 3% for five and six-dimensional problems. Therefore, LEB_α rule is more effective than LEB_1.

Column *Diff* in Table 2 compares the number of evaluated simplices following the rule LEB_α with the number of evaluated simplices by LEB_1 as a reduction percentage measured as

$$\mathit{Diff} = \frac{\text{NS}_1 - \text{NS}_\alpha}{\text{NS}_1} \times 100.$$

Table 2. Comparison of LEB_α with LEB_1

No.	NS_1	$Time_1$	NLE_1	NS_α	$Time_\alpha$	Diff.
1	5,593,162,478	13,533	59%	1,731,072,258	6,434	69%
2	619,816,602	1,414	62%	409,265,106	1,459	34%
3	1,312,244,486	2,991	59%	480,214,824	1,723	63%
4	1,331,168,300	3,110	59%	507,048,676	1,804	62%
5	1,363,992,350	3,071	60%	731,110,834	2,526	46%
6	1,568,043,214	3,525	61%	848,498,976	2,777	46%
7	79,541,580	189	60%	40,431,804	148	49%
8	3,688,449,152	18,300	59%	3,353,894,904	22,805	9%
9	1,051,441,204	5,016	56%	975,336,498	6,447	7%
10	322,865,034	4,023	74%	255,783,394	4,193	21%

Experimental results in column *Diff* show that the number of evaluated simplices using LEB_α is considerably reduced with respect to the use of LEB_1, up to 69% in *Levy No. 15* test function.

LEB_α causes a computational overhead in the bounding rule. Nevertheless, this overhead is compensated due to the significant reduction of the number of evaluated simplices in most of the cases and thus the execution time is less. Test problems with a dimension $n > 4$ experience a slight reduction in the number of evaluated simplices, not enough to improve the execution time of the algorithm.

Experimental results for a set of three-dimensional problems taken from [13] are not shown in tables because the number of evaluated simplices are the same for LEB_1 and LEB_α.

5 Conclusion

Every rule in the BnB schema plays an important role in terms of efficiency. Selecting the correct edge leads to a reduction of the number of evaluated simplices as shown the experimental results of this work. The efficiency of two Longest Edge Bisection strategies, LEB_1 and LEB_α, in a real problem like Lipschitz Global Optimization is shown. For test problems with dimension greater than three, LEB_α provides a reduction in the number of evaluated simplices with respect to LEB_1 (the easiest and most frequently used).

As a future work, the application of new heuristics with less computational cost than LEB_α will be studied. Additionally, it will be interesting to develop a LEB heuristic which does not need to calculate the length of the edges. This would speed up considerably the execution time since a great percentage of this time is spent doing these calculations.

Acknowledgments. This work has been funded by grants from the Spanish Ministry (TIN2008-01117 and TIN2012-37483) and Junta de Andalucía (P11-TIC-7176), in part financed by the European Regional Development Fund (ERDF). J.F.R. Herrera is a fellow of the Spanish FPU program.

References

1. Christodoulos, A.F., Pardalos, P.M. (eds.): State of the Art in Global Optimization: Computational Methods and Applications. Nonconvex Optimization and Its Applications, vol. 7. Kluwer Academic Publishers (1996)
2. Mitten, L.G.: Branch and bound methods: general formulation and properties. Oper. Res. 18(1), 24–34 (1970)
3. Ibaraki, T.: Theoretical comparisons of search strategies in branch and bound algorithms. Int. J. Comput. Inf. Sci. 5(4), 315–344 (1976)
4. Aparicio, G., Casado, L.G., Hendrix, E.M.T., García, I., Toth, B.G.: On computational aspects of a regular n-simplex bisection. In: 2013 Eighth International Conference on P2P, Parallel, Grid, Cloud and Internet Computing (3PGCIC), pp. 513–518 (October 2013)
5. Paulavičius, R., Žilinskas, J.: Global optimization using the branch-and-bound algorithm with a combination of Lipschitz bounds over simplices. Technol. Econ. Dev. Econ. 15(2), 310–325 (2009)
6. Horst, R., Tuy, H.: Global Optimization (Deterministic Approaches). Springer, Berlin (1990)
7. Todd, M.J.: The computation of fixed points and applications. Lecture Notes in Economics and Mathematical Systems, vol. 24. Springer (1976)
8. Herrera, J.F.R., Casado, L.G., Paulavičius, R., Žilinskas, J., Hendrix, E.M.T.: On a hybrid MPI-Pthread approach for simplicial branch-and-bound. In: 2013 IEEE 27th International Parallel and Distributed Processing Symposium Workshops PhD Forum (IPDPSW), pp. 1764–1770 (May 2013)
9. Herrera, J.F.R., Casado, L.G., Hendrix, E.M.T., Paulavičius, R., Žilinskas, J.: Dynamic and hierarchical load-balancing techniques applied to parallel branch-and-bound methods. In: 2013 Eighth International Conference on P2P, Parallel, Grid, Cloud and Internet Computing (3PGCIC), pp. 497–502 (October 2013)
10. Paulavičius, R., Žilinskas, J., Grothey, A.: Investigation of selection strategies in branch and bound algorithm with simplicial partitions and combination of Lipschitz bounds. Optim. Lett. 4(2), 173–183 (2010)
11. Hannukainen, A., Korotov, S.: On numerical regularity of the face-to-face longest-edge bisection algorithm for tetrahedral partitions. Sci. Comput. Program. 90, Part A, 34–41 (2014)
12. Paulavičius, R., Žilinskas, J.: Simplicial global optimization. Springer Briefs in Optimization. Springer, New York (2014)
13. Hansen, P., Jaumard, B.: Lipschitz optimization. In: Horst, R., Pardalos, P. (eds.) Handbook of Global Optimization. Nonconvex Optimization and Its Applications, vol. 2, pp. 407–493. Springer US (1995)

Appendix: Function Definitions

Levy No. 15

$$f(x) = \sin^2 3\pi x_1 + \sum_{i=1}^{n-1} (x_i - 1)^2 (1 + \sin^2 3\pi x_{i+1}) + (x_n - 1)^2 \times (1 + \sin^2 2\pi x_n)$$

Shekel 5

$$f(x) = -\sum_{i=1}^{5} \frac{1}{(x - a_i)(x - a_i)^\mathsf{T} + c_i}$$

where

$$\begin{aligned} a_1 &= (4,4,4,4),\ c_1 = 0.1 \\ a_2 &= (1,1,1,1),\ c_2 = 0.2 \\ a_3 &= (8,8,8,8),\ c_3 = 0.2 \\ a_4 &= (6,6,6,6),\ c_4 = 0.4 \\ a_5 &= (3,7,3,7),\ c_5 = 0.4 \end{aligned}$$

Shekel 7

$$f(x) = -\sum_{i=1}^{7} \frac{1}{(x - a_i)(x - a_i)^\mathsf{T} + c_i}$$

where

$$\begin{aligned} a_1 &= (4,4,4,4),\ c_1 = 0.1 \\ a_2 &= (1,1,1,1),\ c_2 = 0.2 \\ a_3 &= (8,8,8,8),\ c_3 = 0.2 \\ a_4 &= (6,6,6,6),\ c_4 = 0.4 \\ a_5 &= (3,7,3,7),\ c_5 = 0.4 \\ a_6 &= (2,9,2,9),\ c_6 = 0.6 \\ a_7 &= (5,5,3,3),\ c_7 = 0.6 \end{aligned}$$

Shekel 10

$$f(x) = -\sum_{i=1}^{10} \frac{1}{(x-a_i)(x-a_i)^\top + c_i}$$

where

$$\begin{aligned}
a_1 &= (4.0, 4.0, 4.0, 4.0),\ c_1 = 0.1\\
a_2 &= (1.0, 1.0, 1.0, 1.0),\ c_2 = 0.2\\
a_3 &= (8.0, 8.0, 8.0, 8.0),\ c_3 = 0.2\\
a_4 &= (6.0, 6.0, 6.0, 6.0),\ c_4 = 0.4\\
a_5 &= (3.0, 7.0, 3.0, 7.0),\ c_5 = 0.4\\
a_6 &= (2.0, 9.0, 2.0, 9.0),\ c_6 = 0.6\\
a_7 &= (5.0, 5.0, 3.0, 3.0),\ c_7 = 0.6\\
a_8 &= (8.0, 1.0, 8.0, 1.0),\ c_8 = 0.7\\
a_9 &= (6.0, 2.0, 6.0, 2.0),\ c_9 = 0.5\\
a_{10} &= (7.0, 3.6, 7.0, 3.6),\ c_{10} = 0.5
\end{aligned}$$

Schwefel 1.2

$$f(x) = \sum_{i=1}^{4}\left(\sum_{j=1}^{i} x_j\right)^2$$

Powell

$$f(x) = (x_1 + 10x_2)^2 + 5(x_3 - x_4)^2 + (x_2 - 2x_3) + 10(x_1 - x_4)^4$$

Levy No. 9

$$f(x) = \sin^2 3\pi y_1 + \sum_{i=1}^{n-1}(y_i - 1)^2(1 + \sin^2 3\pi y_{i+1}) + (y_n - 1)^2$$

where

$$y_i = 1 + \frac{x_i - 1}{4}$$

Levy No. 16

$$f(x) = \sin^2 3\pi x_1 + \sum_{i=1}^{n-1} (x_i - 1)^2 (1 + \sin^2 3\pi x_{i+1}) + (x_n - 1)^2 (1 + \sin^2 2\pi x_n)$$

Levy No. 10

$$f(x) = \sin^2 3\pi y_1 + \sum_{i=1}^{n-1} (y_i - 1)^2 (1 + 10 \sin^2 \pi y_{i+1}) + (y_n - 1)^2$$

where

$$y_i = 1 + \frac{x_i - 1}{4}$$

Levy No. 17

$$f(x) = \sin^2 3\pi x_1 + \sum_{i=1}^{n-1} (x_i - 1)^2 (1 + \sin^2 3\pi x_{i+1}) + (x_n - 1)^2 (1 + \sin^2 2\pi x_n)$$

Heuristics to Reduce the Number of Simplices in Longest Edge Bisection Refinement of a Regular n-Simplex

Guillermo Aparicio[1], Leocadio G. Casado[2], Boglárka G-Tóth[3], Eligius M.T. Hendrix[4], and Inmaculada García[4]

[1] TIC 146: Supercomputing-Algorithms Research Group, Universidad de Almería, Agrifood Campus of International Excellence (ceiA3), 04120, Spain
guillermoaparicio@ual.es

[2] Department of Informatics, University of Almería, Agrifood Campus of International Excellence (ceiA3), 04120, Spain
leo@ual.es

[3] Department of Differential Equations, Budapest University of Technology and Economics, Egry J. u. 1., Budapest, 1111, Hungary
bog@math.bme.hu

[4] Department of Computer Architecture, Universidad de Málaga, Campus de Teatinos, 29017, Spain
{Eligius,igarciaf}@uma.es

Abstract. In several areas like Global Optimization using branch-and-bound methods, the unit n-simplex is refined by bisecting the longest edge such that a binary search tree appears. The refinement usually selects the first longest edge and ends when the size of the sub-simplices generated in the refinement is smaller than a given accuracy. Irregular sub-simplices may have more than one longest edge only for $n \geq 3$. The question is how to choose the longest edge to be bisected such that the number of sub-simplices in the generated binary tree is minimal. The difficulty of this Combinatorial Optimization problem increases with n. Therefore, heuristics are studied that aim to minimize the number of generated simplices.

Keywords: Regular Simplex, Longest Edge Bisection, Complete Binary Tree.

1 Introduction

Global Optimization deals with finding the minimum or maximum value of an objective function f on a closed set with a non-empty interior. We focus here on the so-called standard n-simplex defined in the $(n+1)$-dimensional space

$$S = \left\{ x \in \mathbb{R}^{n+1} \quad \sum_{i=1}^{n+1} x_i = 1;\ x_i \geq 0 \right\}, \tag{1}$$

B. Murgante et al. (Eds.): ICCSA 2014, Part II, LNCS 8580, pp. 115–125, 2014.

that is used, for instance, to describe blending problems. In blending problems a product is obtained as result of mixing of raw materials. The set of possible mixtures is the unit simplex, where x_i represent the fraction of raw material i in mixture x [4,5].

Branch-and-bound methods (B&B) are widely used to solve Global Optimization (GO) problems where the solution is required to have a guaranteed accuracy. A B&B performs an exhaustive search for optima based on successive decomposition of the search space into smaller sub-regions until a given precision is reached. Bounds on the objective function are calculated for each sub-region, allowing discarding sub-regions where the global optimum cannot be located. We focus on the binary tree implicitly generated by the branching (refinement) where the simplex division is defined by the Longest Edge Bisection rule (LEB) [1,7] without pruning taking place.

Aparicio at al. in [3] investigates the effect of the LEB rule on the number of generated simplices, their roundness and whether sub-simplices have similar shapes. A new selection method of the longest edge to be bisected, denoted by LEB_α, was presented. The use of LEB_α selection in the refinement of a 3-simplex produces eight shape classes. The goals of this paper are the following: i) to determine whether the LEB_α rule generates the minimum number of simplices for a 3-simplex at a precision high enough; ii) to develop new and easier to calculate longest edge selection heuristics and iii) to extend the studies to $n > 3$.

The paper is organized as follows. Section 2 describes the simplex refinement process based on the longest edge bisection. Section 3 describes several longest edge selection heuristics. Section 4 discusses the findings and Sect. 5 concludes.

2 Simplex Refinement Using Longest Edge Bisection

An n-simplex is defined by its set of vertices $V = \{v_1, \ldots, v_{n+1}\}$, $v_j \in \mathbb{R}^{n+1}$. $\omega(S)$ denotes the size (width, i.e. longest edge) of a simplex S, i.e. $\max_{i,j} \|v_i - v_j\|$. In this research, the set to be refined is a regular n-simplex, called S_1, scaled to an initial edge length of 1, $\omega(S_1) = 1$. Fig. 1 shows the 2-simplex instance.

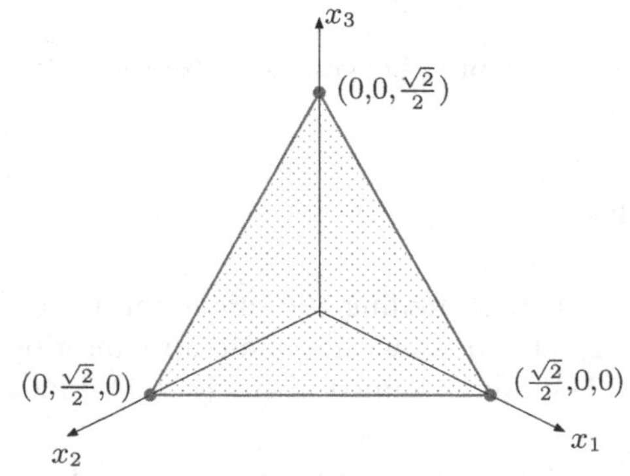

Fig. 1. A regular 2-simplex with edge length 1

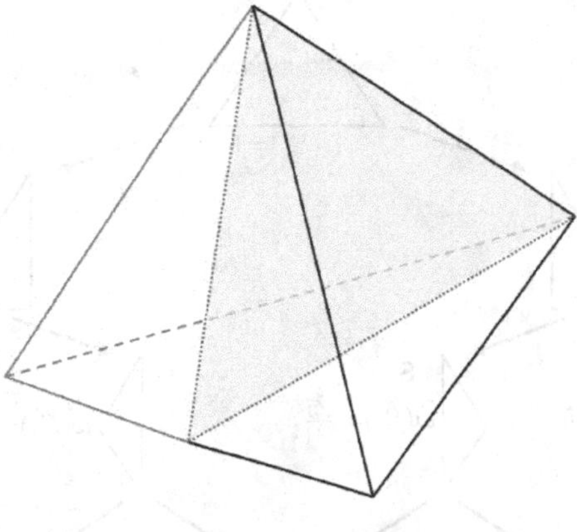

Fig. 2. Longest-edge bisection on a regular 3-simplex

Algorithm 1. Simplex Refinement algorithm

Procedure SR(S_1, ϵ)
$\Lambda := \{1\}$ {Set of leaf indices; simplices not yet split}
$ns := 1$ {Number of simplices}
while $\Lambda \neq \emptyset$ **do**
  Extract a simplex i from Λ
  if $\omega(S_i) > \epsilon$ **then** {Final accuracy not reached}
    $\{j,k\}$:=SelectLE(S_i)
    $\{S_{2i}, S_{2i+1}\}$:= Bisect(S_i, j, k)
    Store simplices $2i$ and $2i+1$ in Λ.
    $ns := ns + 2$.

The longest-edge bisection algorithm is a popular way of iterative refinement in the finite element method, since it is very simple and can easily be applied in higher dimensions [6]. It is based on splitting a simplex using the hyperplane that connects the mid point of the longest edge of a simplex with the opposite vertices, as illustrated in Fig.2.

Algorithm 1 provides the Simplex Refinement (SR) procedure which bisects the initial simplex iteratively. In principle, the refinement can continue indefinitely, but we describe the process here (like in B&B) such that there is a stopping criterion; the branching continues until the size of a node is less than or equal to the desired accuracy ϵ. To study the resulting binary tree, the index i is used to number the simplices. The set Λ provides the leaves of the tree corresponding to the simplices that have not been refined yet. Fig. 3 illustrates the result of SR on a 2-simplex with termination criterion $\omega(S) \leq 0.5$.

For a 2-simplex, selection among the longest edges in SelectLE is not required, as the longest edge is either unique, or the sub-simplex to be divided is regular and the choice does not alter the number and shapes of the generated

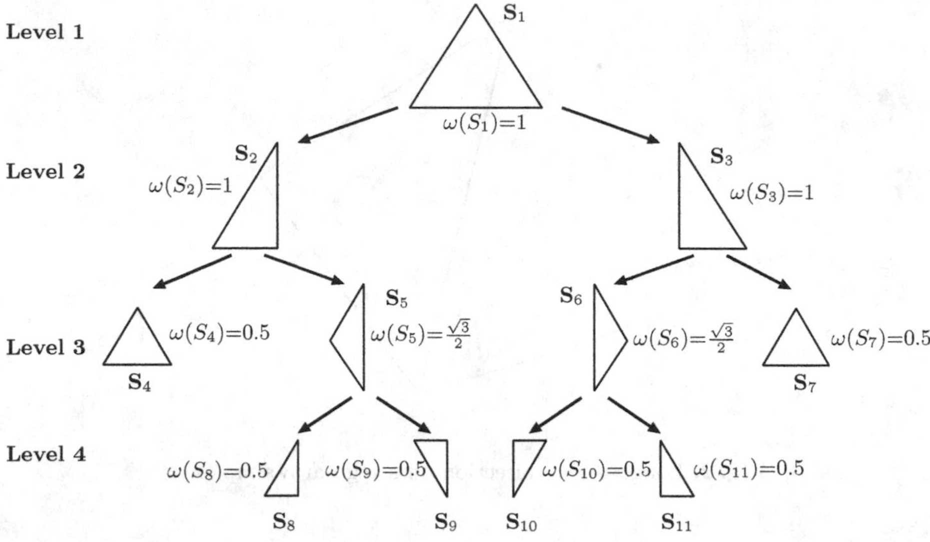

Fig. 3. Binary tree generated by the SR algorithm on a 2-simplex with $\epsilon = 0.5$

sub-simplices. Mitchell in [8] showed that one can avoid edge length calculations in a 2-simplex by always bisecting the edge of the two oldest vertices.

Fig. 2 shows the result of the first bisection for a regular 3-simplex. It does not matter which edge is selected first, because all generated sub-simplices are similar. Notice that there exist three longest edges for the sub-simplices generated from the first subdivision. So, one of the longest edges has to be selected for bisection, which in this case alter the shape of the new sub-simplices. How to choose among the longest edges is one of the main questions in this article.

The number of simplices in the finite binary tree generated by Algorithm 1 depends on how fast the size of the simplices decreases when we go deeper into the tree. [3] studies the number and classes of simplices for $n \leq 3$ when a specific longest edge selection method, denoted by LEB_α (see Sect. 3.2) is used as SelectLE() in Algorithm 1. The question is whether it can be done better, i.e. does a rule exist that generates less number of simplices in an easier way? We investigate that by studying new heuristic rules when there are several choices for SelectLE().

3 LE Selection Heuristics for Bisection

Longest Edge bisection is performed in simplex refinement in order to maintain the shape of the generated simplices as round as possible. This avoids the generation of needle shape simplices. In this way, the length of an edge in a simplex cannot be greater than two times the length of the other. Using LE bisection and $\omega(S) \leq \epsilon$ termination criterion, the convergence of the algorithm is assured.

As described before, during the refinement of a regular n-simplex ($n \geq 3$) sub-simplices may have several longest edges. The design of new heuristics for the LE selection should contribute to reduce the number of generated sub-simplices and to maintain or enhance the characteristics of the longest edge bisection. We have considered the following heuristics having different characteristics to be tested. Starting from a heuristic with low computational cost, more expensive heuristics are presented. The aim is to generate more round simplices, such that in total the tree contains less.

3.1 First Longest Edge, LEB$_1$

The natural way to select a longest edge is to take the first one found. Which is the first one? It depends of course on the coding and storing of the vertices and edges, i.e. the index number assigned to each vertex of the simplex. The new vertex usually has the same index as the one it substitutes. This rule, which we denote as LEB$_1$ is used as a benchmark in measuring performance.

3.2 Longest Edge with the Smallest Angles, LEB$_\alpha$

In an n-simplex, each vertex v is connected to n edges. In total there are $\binom{n}{2}$ associated angles between the edges at v. The idea is that one intends to obtain sub-simplices as round as possible. To do so, for each vertex in a longest edge, the sum of the angles is determined and the longest edge with smallest sum is selected. This rule has been elaborated and applied in [3]. Dividing the longest edge with the smallest sum of angles at its vertices causes splitting the biggest angles of the simplex, which are in fact at vertices not belonging to the selected longest edge and therefore resulting in rounder sub-simplices. For higher dimensions, i.e. $n > 3$, more than one longest edge with the smallest sum of vertex's angles may appear and one should add more criteria to decide on the edge to be selected. Here, we will simply evaluate what happens if one selects the first edge with the smallest sum.

3.3 Longest Edge, Midpoint Furthest from the Centroid, LEB$_C$

The distance of the midpoint of edge $\{v_i, v_j\}$ to the centroid C is given by

$$||\frac{v_i + v_j}{2} - C||. \tag{2}$$

The LEB$_C$ rule selects the longest edge $\{v_i, v_j\}$ that gives the optimum of

$$max_{i,j}||\frac{v_i + v_j}{2} - C||. \tag{3}$$

3.4 Longest Edge with Small Edge Lengths at the Vertices via Weights, LEB_W

New vertex v_n, generated by bisection has a small sum of edge lengths compared to the other vertices. Every vertex v_i is given a weight $w_i \in \{0,\ldots,n-1\}$ during the refinement which is increased if v_i is connected to a halved edge. Vertices of the initial simplex have weight 0 in this heuristic. A new vertex obtains a weight of 1 because it is connected to the smallest and just bisected edge. The weight of v_i is increased by one, if $\{v_i, v_j\}$ is split for any v_j. In that case, the weight of v_i is changed to $w_i := (w_i + 1) \mod n$. LEB_W selects the longest edge with minimal overall weights at its vertices.

We consider the computational effort for the determination of the heuristic rules. All presented methods need to determine the set of longest edges. In terms of computational complexity LEB_1 is relatively simple; just the first longest edge is bisected. Heuristic LEB_W requires updating and storing information for each edge and no complicated calculation are necessary for that. Heuristic LEB_C requires the calculation of the centroid and distances from it to mid-point of the longest edges. This means, no additional values have to be stored, but distances have to be calculated. The most complicated method is LEB_α, as it requires calculating angles after each bisection.

When a selection rule is applied to determine the longest edge to be bisected, one or more longest edges can satisfy the rule. The goal is to find an unambiguous selection rule producing the smallest binary tree.

4 Experimental Results

The heuristics described in Section 3 are evaluated on a regular n-simplex with edge length 1. Four different values of ϵ, ϵ_1=1/8, ϵ_2=1/16, ϵ_3=1/32 and ϵ_4=1/64 are used for n=3, 4 and 5 and only ϵ_1 for n=6 and 7. When there is more than one longest edge satisfying the criterion related to the heuristic, the first longest edge according to the indices of the vertices is bisected. To study those cases where several edges satisfy the selection criterion, we also evaluate for each heuristic a variant where the longest edge is chosen at random among them. For instance, variant $\text{LEB}_{\alpha,R}$ selects an edge at random among those longest edges that have the same minimum sum of angles value. $\text{LEB}_{1,R}$ denotes a random selection among all longest edges. For these randomized variants, the average of two executions is shown in tables. The aim of having randomized versions is just to check if they outperform the presented heuristics due to their ambiguity for different values of n.

4.1 3-Simplex

We first focus on the refinement of the 3-simplex (tetrahedron). Table 1 shows the number of simplices generated by each selection rule varying the value of ϵ. Rule LEB_1 generates the largest number of sub-simplices for all accuracy

Table 1. Number of simplices in a 3-simplex refinement

	ϵ_1	ϵ_2	ϵ_3	ϵ_4
LEB_1	4,027	32,167	259,701	2,114,687
$LEB_{1,R}$	3,607	31,763	237,709	1,975,207
LEB_α	2,751	21,887	174,847	1,398,271
$LEB_{\alpha,R}$	2,751	21,887	174,847	1,398,271
LEB_C	2,751	21,887	174,847	1,398,271
$LEB_{C,R}$	2,751	21,887	174,847	1,398,271
LEB_W	3,223	26,843	221,619	1,822,377
$LEB_{W,R}$	3,108	26,134	217,516	1,808,182

values ϵ. Even a random selection of the longest edge ($LEB_{1,R}$) performs better. The methods that generate the smallest number of sub-simplices are LEB_α and LEB_C followed by LEB_W. In terms of the number of generated simplices, LEB_α and LEB_C are 32%, 32%, 33% and 34% more efficient than LEB_1 for $\epsilon_1, \epsilon_2, \epsilon_3$ and ϵ_4, respectively.

A further investigation of the generated tree for $\epsilon = 0.25$ was made, considering all possible paths for the selection rules. It shows that the bisected longest edges leading to the minimum number of simplices are the same as those selected by LEB_α and LEB_C. The combinatorial explosion hinders obtaining such results for higher values of n. An earlier investigation [2] focused on so-called simplex shape classes; two simplices belong to the same class when they are equal after scale, translation and rotation/flip. For this specific case (3-simplex) has been shown that applying LEB_α leads to 8 shape classes for any termination precision. As the classes are repeated when going deeper into the tree (smaller value of ϵ) the rules LEB_α and LEB_C also generate the minimum number of simplices for smaller values of ϵ. Rules $LEB_{\alpha,R}$ and $LEB_{C,R}$ generate the same binary tree as using the rules LEB_α and LEB_C. Apparently, for a 3-simplex the generated tree is independent of the selected longest edge from the set with the same selection criterion value according to rules LEB_α and LEB_C, respectively.

We now consider the uniqueness of the selected edge by each rule, as several longest edges may have the same criterion value. Table 2 indicates the number of longest edges (NLE) with the same criterion value for a heuristic. For each rule the number of simplices having NLE$= 1, 2, 3, 6$ longest edges is given in the table. The values indicate the ambiguity degree of each heuristic. It is worth to highlight the absence of NLE $= 4$ and 5. Moreover, the table shows that the heuristics providing the minimum number of simplices do not generate regular simplices (NLE=6). Rules LEB_α and LEB_C are the least ambiguous heuristic.

Table 2 shows that LEB_α generates 3,040 simplices with three longest edges having the same LEB_α. Those simplices belong to a symmetric class where it does not matter which is the longest edge to be bisected.

Table 2. Number of divided simplices having NLE number of longest edges with the same minimum or maximum criterion value for a 3-simplex and $\epsilon = \epsilon_3$

NLE	LEB_1	LEB_α	LEB_C	LEB_W
1	115,093	84,382	84,382	102,059
2	10,458	0	0	7,113
3	4,002	3,040	3,040	1,634
6	297	1	1	3
Sum	129,850	87,423	87,423	110,809

4.2 4-Simplex

Table 3 shows the same information as Table 1 for a 4-simplex. The comparison of the heuristics follows the same pattern as in the 3-simplex case. It is interesting to see that now rule LEB_C leads to the smallest number of generated simplices. In terms of the number of generated simplices, LEB_C is 28%, 25%, 22% and 20% more efficient than LEB_1 for $\epsilon_1, \epsilon_2, \epsilon_3$ and ϵ_4, respectively. The earlier remarks about symmetric classes with respect to rule LEB_α seem to be valid; the random variant produces the same result as choosing the first edge with the same criterion value. However, this is not the case for $\text{LEB}_{C,R}$.

Table 3. Number of simplices in a 4-simplex refinement

	ϵ_1	ϵ_2	ϵ_3	ϵ_4
LEB_1	99,563	1,577,047	25,260,927	407,426,105
$\text{LEB}_{1,R}$	84,317	1,327,427	22,719,455	363,775,505
LEB_α	71,831	1,196,327	19,895,047	329,359,823
$\text{LEB}_{\alpha,R}$	71,831	1,196,327	19,895,047	329,359,823
LEB_C	71,547	1,189,213	19,722,933	325,901,453
$\text{LEB}_{C,R}$	71,795	1,193,753	19,780,355	327,013,131
LEB_W	75,049	1,291,883	21,997,793	369,323,649
$\text{LEB}_{W,R}$	74,283	1,278,983	21,903,465	370,036,154

Table 4 shows the same information as Table 2 for a 4-simplex and $\epsilon = \epsilon_2$. NLE $= 8$ and 9 do not occur. All heuristics that result in a reduction of the number of generated simplices over LEB_1 do not repeat the initial regular simplex shape. Now rule LEB_C produces the smallest number of ambiguity cases, followed closely by LEB_α.

Table 4. Number of divided simplices having NLE number of longest edges with the same minimum or maximum criterion value for a 4-simplex and $\epsilon = \epsilon_2$

NLE	LEB_1	LEB_α	LEB_C	LEB_W
1	631,632	562,008	554,465	567,051
2	126,962	34,828	38,602	70,761
3	23,700	922	1,146	7,443
4	4,945	404	388	632
5	379	0	4	19
6	660	0	0	84
7	172	0	0	0
10	73	1	1	1
Sum	765,060	598,163	594,606	645,991

4.3 5-Simplex

Table 5 shows the same information as Table 1 for a 5-simplex. Surprisingly, LEB_W provides the smallest binary tree for ϵ_1 and ϵ_2, while LEB_α is the best for ϵ_3 without taking into account random versions.

Table 5. Number of simplices in a 5-simplex refinement

	ϵ_1	ϵ_2	ϵ_3
LEB_1	2,510,297	79,614,191	2,536,097,441
$LEB_{1,R}$	2,052,713	67,795,829	2,144,283,073
LEB_α	1,988,255	64,783,583	2,120,863,607
$LEB_{\alpha,R}$	1,988,255	64,783,583	2,120,862,291
LEB_C	1,968,215	64,978,969	2,129,779,527
$LEB_{C,R}$	1,939,315	63,961,855	2,112,991,041
LEB_W	1,806,301	62,937,019	2,142,429,229
$LEB_{W,R}$	1,750,849	61,726,572	2,091,107,630

The best non-random heuristics generate 28%, 21% and 16% less simplices than LEB_1 for ϵ_1, ϵ_2 and ϵ_3, respectively. For 5-simplices we provide results only for $\epsilon = \epsilon_1$, ϵ_2 and ϵ_3. The argument on the symmetric classes for the LEB_α is only valid here for ϵ_1 and ϵ_2. For $LEB_{C,R}$ we always observe a small deviation with respect to LEB_C. $LEB_{W,R}$ always improves the results obtained by LEB_W. In general, random versions of the heuristics produce better results showing that the ambiguity degree of the studied heuristics increases with the dimension.

Table 6 shows the same information as Table 2 for a 5-simplex and $\epsilon = \epsilon_1$. One can observe that apparently the bisection process does not generate simplices with NLE=11, 13 and 14 edges with the same selection criterion value. All rules

Table 6. Number of non-leaf simplices having NLE options for longest edges selection heuristics, in a 5-simplex and $\epsilon = \epsilon_1$

NLE	LEB_1	LEB_α	LEB_C	LEB_W
1	936,148	967,940	940,178	770,768
2	240,444	25,392	41,760	111,600
3	54,213	488	1,854	17,508
4	16,722	288	272	2,770
5	5,174	0	0	429
6	1,859	18	42	74
7	253	0	0	0
8	209	0	0	0
9	24	0	0	0
10	76	0	0	0
12	11	0	0	0
15	15	1	1	1
Sum	1,255,148	994,127	946,523	903,150

apart from LEB_1 do not repeat the regular simplex. It is surprising to see that for LEB_α, LEB_C and LEB_W the smallest number of generated simplices comes with the largest ambiguity.

4.4 6-Simplex and 7-Simplex

Table 7 shows the number of simplices for a 6-simplex and a 7-simplex when $\epsilon = \frac{1}{8}$ is chosen.

Table 7. Number of simplices in a 6-simplex and 7-simplex refinement for $\epsilon = \epsilon_1$

n	LEB_1	LEB_α	LEB_C	LEB_W
6	71,915,489	54,297,713	55,275,973	47,362,449
7	2,306,293,327	1,429,247,747	1,554,393,609	1,326,290,317

LEB_W is the method providing the smallest binary tree for this precision but it is not known if the same occurs for a higher precision.

5 Conclusions

Selecting the first longest edge in the refinement of a regular n-simplex is an easy rule, but appears to generate a larger number of sub-simplices than other rules.

Even a random selection performs better. Several heuristic rules were studied. The tested rules give a fair reduction of the generated binary tree. The rule that gives the highest reduction depends on the dimension $n \in \{3, 4, 5, 6, 7\}$. It has been shown that the tested rules LEB_α and LEB_C lead to the smallest tree for n=3. Rule LEB_C gives the smallest tree for $n = 4$. Rule $\text{LEB}_{W,R}$ appears as the best for $n \geq 5$. For large dimensional simplices ($n \geq 5$), random version of the heuristics in general provides better results than the non-random versions. It means that the ambiguity degree of the studied heuristics increases with the dimension.

Acknowledgments. This work has been funded by grants from the Spanish Ministry of Science and Innovation (TIN2008-01117,TIN2012-37483), and Junta de Andalucía (P11-TIC-7176), in part financed by the European Regional Development Fund (ERDF).

References

1. Adler, A.: On the Bisection Method for Triangles. Mathematics of Computation 40(162), 571–574 (1983)
2. Aparicio, G., Casado, L.G., Hendrix, E.M.T., G.-Tóth, B., García, I.: On the minimum number of simplex classes in longest edge bisection refinement of a regular n-simplex. Informatica (submitted)
3. Aparicio, G., Casado, L.G., Hendrix, E.M.T., García, I., G.-Tóth, B.: On computational aspects of a regular n-simplex bisection. In: Proceedings of the 2013 Eighth International Conference on P2P, Parallel, Grid, Cloud and Internet Computing, Compiegne, France, pp. 513–518 (October 2013)
4. Ashayeri, J., van Eijs, A., Nederstigt, P.: Blending modelling in a process manufacturing: A case study. European Journal of Operational Research 72(3), 460–468 (1994)
5. Casado, L.G., García, I., G.-Tóth, B., Hendrix, E.M.T.: On determining the cover of a simplex by spheres centered at its vertices. Journal of Global Optimization 50(4), 645–655 (2011)
6. Hannukainen, A., Korotov, S., Křížek, M.: On numerical regularity of the face-to-face longest-edge bisection algorithm for tetrahedral partitions. Science of Computer Programming 90(A), 34–41 (2014)
7. Horst, R.: On generalized bisection of n-simplices. Mathematics of Computation 66(218), 691–698 (1997)
8. Mitchell, W.F.: A comparison of adaptive refinement techniques for elliptic problems. ACM Trans. Math. Softw. 15(4), 326–347 (1989)

Multiple Roots of Systems of Equations by Repulsion Merit Functions

Gisela C.V. Ramadas[1], Edite M.G.P. Fernandes[2], and Ana Maria A.C. Rocha[2]

[1] Department of Mathematics, School of Engineering, Polytechnic of Porto, 4200-072 Porto, Portugal
gcv@isep.ipp.pt

[2] Algoritmi Research Centre, University of Minho, 4710-057 Braga, Portugal
{emgpf,arocha}@dps.uminho.pt

Abstract. In this paper we address the problem of computing multiple roots of a system of nonlinear equations through the global optimization of an appropriate merit function. The search procedure for a global minimizer of the merit function is carried out by a metaheuristic, known as harmony search, which does not require any derivative information. The multiple roots of the system are sequentially determined along several iterations of a single run, where the merit function is accordingly modified by penalty terms that aim to create repulsion areas around previously computed minimizers. A repulsion algorithm based on a multiplicative kind penalty function is proposed. Preliminary numerical experiments with a benchmark set of problems show the effectiveness of the proposed method.

Keywords: System of equations, multiple roots, penalty function, repulsion, harmony search.

1 Introduction

In this paper, we aim to investigate the performance of a repulsion algorithm that is based on a multiplicative kind penalty merit function, combined with a metaheuristic optimization algorithm, the harmony search (HS), to compute multiple roots of a system of nonlinear equations of the form

$$f(x) = 0, \tag{1}$$

where $f(x) = (f_1(x), f_2(x), \ldots, f_n(x))^T$, each $f_i : \Omega \subset \mathbb{R}^n \to \mathbb{R}$, $i = 1, \ldots, n$ is a continuous possibly nonlinear function in the search space and Ω is a closed convex set, herein defined as $[l, u] = \{x : -\infty < l_i \le x_i \le u_i < \infty, i = 1, \ldots, n\}$. The functions $f_i(x)$, $i = 1, \ldots, n$ are not necessarily differentiable implying that analytical and numerical derivatives may not be used. The work herein presented comes in the sequence of the study published in [1,2]. To compute a solution of a nonlinear system of equations is equivalent to compute a global minimizer of the optimization problem

$$\min_{x \in \Omega \subset \mathbb{R}^n} \mathcal{M}(x) \equiv \sum_{i=1}^{n} f_i(x)^2, \tag{2}$$

B. Murgante et al. (Eds.): ICCSA 2014, Part II, LNCS 8580, pp. 126–139, 2014.

in the sense that they have the same solutions. Thus, a global minimizer and not just a local one, of the function $\mathcal{M}(x)$, known as merit function, in the set Ω, is required. Problem (2) is similar to the usual least squares problem for which many iterative methods have been proposed. They basically assume that the objective function is twice continuously differentiable. However, the objective $\mathcal{M}$ in (2) is only once differentiable if some, or just one, of the f_i, $(i = 1, \ldots, n)$ are not differentiable. Thus, the most popular Newton-type and Quasi-Newton methods should be avoided [3,4,5,6]. Furthermore, their convergence and practical performance are highly sensitive to the user provided initial approximation. Additionally, they are only capable of finding one root at each run of the algorithm. Since a global minimizer of problem (2) is required, classical optimization techniques with guaranteed convergence to local minimizers cannot be applied. When the optimization problem is nonlinear and non-convex, metaheuristics are able to avoid convergence to local minimizers and to generate good quality solutions in less time than most classical techniques. Metaheuristics are general heuristic methods which can be applied to a wide variety of optimization problems. In 2001 emerged the HS algorithm that relies on a set of points and is inspired by natural phenomena [7]. It draws its inspiration not from a biological or physical process like most metaheuristic optimization techniques, but from an artistic one – the improvisation process of musicians seeking a wonderful harmony. HS has efficient strategies for exploring the entire search space, as well as techniques to exploit locally a promising region to yield a high quality solution in a reasonable time. The dynamic updating of two important parameters in the HS algorithm has improved the efficiency and robustness of the metaheuristic [8]. Until today, the HS paradigm has been implemented in many areas, such as in engineering, robotics, telecommunications, health and energy [9,10], in scheduling problems [11], in transportation problems [12], and in seismic isolation systems [13].

Although finding a single root of a system of nonlinear equations is a trivial task, finding all roots is one of the most demanding problems. Multistart methods are stochastic techniques that have been used to compute multiple solutions to problems [14,15,16]. In a multistart strategy, a search procedure is applied to a set of randomly generated points of the search space to converge sequentially along the iterations to the multiple solutions of the problem, in a single run. However, the same solutions may be located over and over again along the iterations and the computational effort turns out to be quite heavy. Other approaches that combine metaheuristics with techniques that modify the objective function in problem (2) have been reported in the literature [17,18,19,20]. The technique in [20] relies on the assignment of a penalty term to each previously computed root so that a repulsion area around the root is created. In [19], an evolutionary optimization algorithm is used together with a type of polarization technique to create a repulsion area around each previously computed root. The repulsion areas force the algorithm to move to other areas of the search space and look for other roots thus avoiding repeated convergence to already located solutions.

In this study, we further explore this penalty-type approach to create repulsion areas around previously detected roots and propose a repulsion algorithm that is capable of computing multiple roots of a system of nonlinear equations through the invoking of the HS algorithm with modified merit functions. We propose a multiplicative kind penalty function based on the inverse of the 'erf' function, known as error function.

The proposed algorithm is tested on 13 benchmark systems of nonlinear equations and the obtained results are compared to the results produced by other penalty type functions that have been recently proposed in the literature. It is shown that the proposed 'erf' penalty function is competitive with other penalties in comparison.

The paper is organized as follows. Section 2 reports on penalty type functions and describes the proposed repulsion algorithm and Section 3 addresses the HS metaheuristic to compute global minimizers of merit functions with accuracy and efficiency. Then, some numerical experiments are shown in Section 4 and we conclude the paper in Section 5.

2 Repulsion Merit Functions

This section aims to discuss the implementation of penalty type functions to create repulsion areas around previously computed solutions of a system of nonlinear equations, thus avoiding the convergence to already located solutions. The proposed repulsion algorithm solves a sequence of global optimization problems by invoking a solver to locate a global minimizer of a sequentially modified merit function. The first call to the global solver considers the original merit function (2). Thereafter the merit function needs to be modified to avoid locating previously computed minimizers. Let the first located minimizer be ξ_1. The idea is to define a repulsion area around ξ_1 so that it will be no more a global minimizer of the modified merit function. The minimization problem is then based on the modified merit function $\bar{\mathcal{M}}$ with a repulsion area created around ξ_1 so that the solver will not find it again. We now show two modified objective functions available in the literature. The first one, presented in [20], is:

$$\min_{x \in \Omega \subset \mathbb{R}^n} \bar{\mathcal{M}}(x) \equiv \mathcal{M}(x) + \beta e^{-\|x-\xi_1\|} \tag{3}$$

where β is a large positive parameter and aims to scale the penalty for approaching the root ξ_1. Thus, after k roots of the problem (1) having been identified, herein denoted by $\xi_1, \xi_2, \ldots, \xi_k$, the repulsion modified merit function is

$$\bar{\mathcal{M}}(x) = \mathcal{M}(x) + \sum_{i=1}^{k} P^A(x; \xi_i, \beta, \rho), \tag{4}$$

where the superscript A stands for 'additive-type penalty' and each penalty term is given by

$$P^A(x; \xi_i, \beta, \rho) = \begin{cases} \beta e^{-\|x-\xi_i\|}, & \text{if } \|x - \xi_i\| \le \rho \\ 0, & \text{otherwise} \end{cases} \tag{5}$$

where $\rho = \min\{0.1, \|\xi_j - \xi_l\|/2 : j \neq l \text{ and } j, l = 1, \ldots, k\}$ is a small problem dependent parameter and defines the radius of the repulsion area, so that other solutions not yet identified, and outside the repulsion area, are not penalized in $\bar{\mathcal{M}}$ [20]. We note that an additive penalty term like (5) satisfies the following properties:

- $P^A(x; \xi_i, \beta, \rho) \to \beta$ when $x \to \xi_i$, and increases with parameter β, in a way that ξ_i is no longer a minimizer of $\bar{\mathcal{M}}(x)$;
- $P^A(x; \xi_i, \beta, \rho) \to 0$ when x moves away from ξ_i, such that the merit function is not affected outside the repulsion area.

Other type of penalty term aiming to create a repulsion area around a previously computed minimizer, say ξ_i, but with a distinctive behavior, is presented in [19]:

$$P^M(x; \xi_i, \alpha) = |\coth(\alpha\|x - \xi_i\|)|, \tag{6}$$

where coth is the hyperbolic cotangent function and α is a positive parameter greater or equal to one, called density factor and used to adjust the radius of the repulsion area. Here, the superscript M means that the penalty is of a 'multiplicative-type'. The main properties in this repulsion context are the following:

- $P^M(x; \xi_i, \alpha) \to \infty$ when $x \to \xi_i$, so that ξ_i is no longer a minimizer of $\bar{\mathcal{M}}(x)$;
- $P^M(x; \xi_i, \alpha) \to 1$ when x moves away from ξ_i, in a way that the merit function is not affected outside the repulsion area.

Similar arguments may be used to create a sequence of global minimization problems based on the modified merit function $\bar{\mathcal{M}}$ [19] which creates repulsion areas around the located global minimizers ξ_i, $i = 1, \ldots, k$ so that the solver will not converge again to the same solutions:

$$\bar{\mathcal{M}}(x) = \mathcal{M}(x) \prod_{i=1}^{k} P^M(x; \xi_i, \alpha). \tag{7}$$

We now illustrate the behavior of the above referred penalty terms (5) and (6), as the corresponding parameters β and α increase. See Figure 1. While the parameter α, in the penalty $|\coth(\alpha\|x - \xi_i\|)|$, aims to define the radius of the repulsion area (figure on the right), the parameter β in penalty $\beta e^{-\|x-\xi_i\|}$ aims to scale the penalty created by moving close to a previously located solution. The radius of the repulsion area is defined by the parameter ρ.

We now present the main ideas behind the new repulsion merit function. It uses the error function, denoted by 'erf', which is a mathematical function defined by the integral

$$\operatorname{erf}(x) = \frac{2}{\sqrt{\pi}} \int_0^x e^{-t^2}\, dt,$$

satisfies the following properties

$$\operatorname{erf}(0) = 0,\ \operatorname{erf}(-\infty) = -1,\ \operatorname{erf}(+\infty) = 1,\ \operatorname{erf}(-x) = -\operatorname{erf}(x), \tag{8}$$

and has a close relation with the normal distribution probabilities. When a series of measurements are described by a normal distribution with mean 0 and

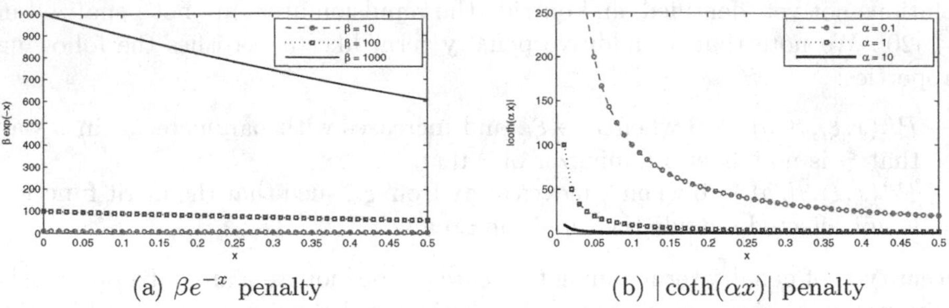

(a) βe^{-x} penalty (b) $|\coth(\alpha x)|$ penalty

Fig. 1. Penalty terms in $(0, 0.5]$, for different values of the parameters

standard deviation σ, the erf function evaluated at $\frac{x}{\sigma\sqrt{2}}$, for a positive x, gives the probability that the error of a single measurement lies in the interval $[-x, x]$. In a penalty function context aiming to prevent convergence to a located root ξ_i, thus defining a repulsion area around it, we propose the multiplicative inverse 'erf' penalty function:

$$P_e^M(x; \xi_i, \delta, \bar{\rho}) = \begin{cases} |\text{erf}(\delta \|x - \xi_i\|)|^{-1}, & \text{if } \|x - \xi_i\| \leq \bar{\rho} \\ 1, & \text{otherwise} \end{cases} \tag{9}$$

which depends on the parameter $\delta > 0$ to scale the penalty for approaching the already computed solution ξ_i, and on the parameter $\bar{\rho}$ to adjust the radius of the repulsion area, where $\bar{\rho} = 0.1 \min_{i=1,\ldots,n}(u_i - l_i)$. We note that the penalty term tends to $+\infty$ when x approaches the root ξ_i, meaning that ξ_i is no longer a minimizer of the modified penalty merit function. According to the properties in (8), as $x \to \infty$, the penalty $P_e^M(x; \xi_i, \delta, \bar{\rho}) \to 1$ meaning that the modified merit function is of 'multiplicative-type' and thus it is not affected when far from previous located roots:

$$\bar{\mathcal{M}}(x) = \mathcal{M}(x) P_e^M(x; \xi_i, \delta, \bar{\rho}). \tag{10}$$

We include Figure 2 to show how the penalty behaves with the parameter δ. Our proposal for the implementation of the penalty (9) is the following:

- the parameter δ is used to scale the penalty in the neighborhood of ξ_i, i.e., when $\|x - \xi_i\| \approx 0$, noting that the penalty increases as δ decreases;
- the parameter $\bar{\rho}$ is used to adjust the radius of the repulsion area, and this may depend closely on the problem at hand.

Algorithm 1 contains the main steps of the repulsion algorithm. The set Ξ, empty at the beginning of the iterative process, contains the roots that are computed and are different from the previous ones. To check if a computed root ξ has been previously located the following conditions

$$|\mathcal{M}(\xi) - \mathcal{M}(\xi_i)| \leq \epsilon \text{ and } \|\xi - \xi_i\| \leq \epsilon \tag{11}$$

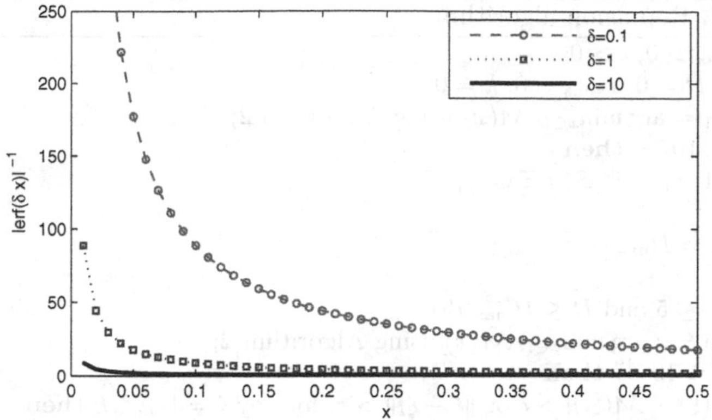

Fig. 2. $|\,\mathrm{erf}(\delta x)|^{-1}$ penalty in $(0, 0.5]$, for $\delta = 0.1$, $\delta = 1$ and $\delta = 10$

must hold, for all $\xi_i \in \Xi$ and a small positive ϵ. Although the repulsion strategy is rather successful in locating multiple roots and avoiding repeated convergence to previously located solutions, these may be occasionally recovered suggesting that the penalty could be increased. The algorithm stops when five unsuccessful iterations (counter It_{uns} in the algorithm) are encountered. An iteration is considered unsuccessful when the solution produced by Algorithm 2 is not a global minimizer of the merit function. In practice, and after an empirical study, we consider that a solution is not a global one if $\mathcal{M} > 10^{-10}$. Anyway, Algorithm 1 is allowed to run for $It_{\max}$ iterations, the user provided threshold.

We now consider an example to illustrate the behavior of the three above described penalty functions. This is a system with many roots in the considered search space Ω.

Example 1. Let the function $\mathcal{M}$ illustrated in Figure 3 be the merit function of the system

$$f(x) = \begin{cases} \cos(x_1) = 0 \\ \sin(x_2) = 0 \end{cases}.$$

In $[-5, 5]^2$, the merit function has 12 global minimizers [18]. Table 1 shows the average number of roots, $N.\text{roots}_{\text{avg}}$, the average number of function evaluations, NFE_{avg}, and time (in seconds), T_{avg}, found in five experimental runs produced by our algorithm, where we implemented:

- the 'exp' penalty term as described in (5) with $\beta = 1000$;
- the 'coth' penalty as shown in (6) with $\alpha = 10$;
- the 'erf' penalty with $\delta = 10$ but without the condition with the parameter $\bar{\rho}$ (aiming to work like the coth function) (erf_1);
- the 'erf' penalty using $\delta = 0.1$ and $\bar{\rho} = \rho$ as defined to be used in (5) (erf_2);
- and finally, the 'erf' penalty using $\delta = 0.1$ and $\bar{\rho}$ as proposed to be used in (9) (erf_3).

Algorithm 1. Repulsion algorithm

Require: $It_{\max} > 0$, $\epsilon > 0$;
Set $\Xi = \emptyset$, $It = 0$, $It_{\text{uns}} = 0$, $k = 0$;
Compute $\xi_1 = \arg\min_{x \in \Omega} \bar{\mathcal{M}}(x)$ using Algorithm 2;
if $\mathcal{M}(\xi_1) \leq 10^{-10}$ **then**
  Set $k = 1$, $r_1 = 1$, $\Xi = \Xi \cup \xi_1$;
else
  Set $It_{\text{uns}} = It_{\text{uns}} + 1$;
end if
while $It_{\text{uns}} \leq 5$ and $It \leq It_{\max}$ **do**
  Compute $\xi = \arg\min_{x \in \Omega} \bar{\mathcal{M}}(x)$ using Algorithm 2;
  if $\mathcal{M}(\xi) \leq 10^{-10}$ **then**
    if $|\mathcal{M}(\xi) - \mathcal{M}(\xi_i)| > \epsilon$ or $\|\xi - \xi_i\| > \epsilon$, for any $i = 1, \ldots, k$ **then**
      Set $k = k + 1$, $\xi_k = \xi$, $r_k = 1$, $\Xi = \Xi \cup \xi_k$;
    else
      Set $r_l = r_l + 1$ $(\xi_l \in \Xi)$;
    end if
  else
    Set $It_{\text{uns}} = It_{\text{uns}} + 1$;
  end if
  Set $It = It + 1$;
end while
return k (number of located roots), Ξ (located roots), r_i, $i = 1, \ldots, k$ (number of times each root was recovered)

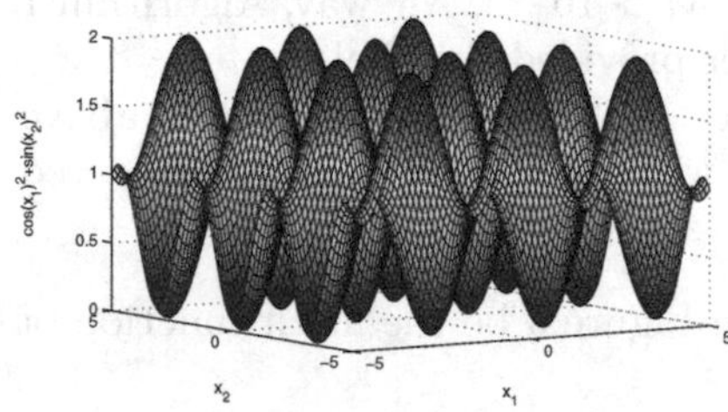

Fig. 3. $\mathcal{M}$ function of Example 1

Table 1. Comparison of penalty repulsion functions

penalty	$N.\text{roots}_{\text{avg}}$	NFE_{avg}	T_{avg}
exp	10.6	14204	1.562
coth	11.2	20695	1.860
erf_1	9.4	15426	2.344
erf_2	11.4	15334	1.894
erf_3	12.0	16728	1.313

Based on this example we conclude that the proposed methodology, summarized by variant erf_3, performed better for this experiment. For completeness, we report now the roots produced by one of the five experimental runs of erf_3:

(1.57080295e+00, -3.14159058e+00), (-4.71238170e+00, -3.14159195e+00), (-4.71239481e+00, 1.99367704e-06), (4.71238471e+00, 3.14158920e+00), (-1.57080249e+00, -3.14159349e+00), (1.57080522e+00, -3.68619544e-06), (-1.57079403e+00, 7.33762541e-06), (-1.57079621e+00, 3.14160097e+00), (-4.71239131e+00, 3.14159943e+00), (4.71239663e+00, -3.14159506e+00), (4.71237915e+00, 1.11265112e-06), (1.57079907e+00, 3.14158507e+00).

3 Improved Harmony Search

The HS algorithm was developed to solve global optimization problems in an analogy with the music improvisation process where music players improvise the pitches of their instruments to obtain better harmony [7,9]. An overview of the existing variants of the HS is presented by Alia and Mandava in [21]. Here, the improved harmony search (I-HS) variant [8] is used to compute a global solution of the problem (2).

At each iteration, the I-HS algorithm provides a set of solution vectors from which the best and the worst solutions, in terms of their fitness - objective function values - are selected. The candidate solutions are saved in the harmony memory (HM). Throughout the iterative process there are HMS (the size of the HM) solutions. After generating the HM randomly in the search space Ω, x^j, $j=1,\ldots,$HMS, the vectors are evaluated and the best harmony, x^{best}, and the worst, x^{worst}, in terms of objective/merit function value are selected. Thereafter, a new harmony is improvised meaning that a new vector y is generated using three improvisation operators:

- O_1: HM operator
- O_2: random selection operator
- O_3: pitch adjustment operator.

A harmony memory considering rate (HMCR) represents the probability of choosing the component of the new harmony/vector from the HM (operator O_1). Otherwise, the component is randomly generated in Ω (operator O_2):

$$y_i = \begin{cases} x_i^j, j \text{ random } \in \{1,\ldots,\text{HMS}\}, & \text{if } \tau_1 < \text{HMCR} \\ l_i + \tau_2(u_i - l_i), & \text{otherwise} \end{cases} \tag{12}$$

for $i=1,\ldots,n$, where τ_1, τ_2 are uniformly distributed random variables in $[0,1]$. Based on a pitch adjusting rate (PAR), the operator O_3 is subsequently applied to refine only the components i produced by O_1, as follows:

$$y_i = \begin{cases} y_i \pm \tau\text{BW}, & \text{if } \tau < \text{PAR} \\ y_i, & \text{otherwise} \end{cases} \tag{13}$$

where BW is an arbitrary distance bandwidth and τ is a random number in the range $[0,1]$. Finally, the HM is updated. The new harmony is compared with the worst harmony in the HM, in terms of $\mathcal{M}$ values. The new harmony is included in the HM, replacing the worst one if it is better than the worst harmony.

As shown in (13), the classical HS algorithm uses fixed value for both PAR and BW. However, small values of PAR with large values of BW can considerably increase the number of iterations required to converge to an optimal solution of (2). Experience has shown that BW must take large values at the beginning of the iterative process to enforce the algorithm to increase the diversity of solution vectors. However, small BW values in the final iterations increase the fine-tuning of solution vectors. Furthermore, large values of PAR combined with small values of BW usually cause the improvement of best solutions in the final stage of the process. To eliminate some of the drawbacks due to fixed values of PAR and BW, the I-HS variant [8] dynamically defines parameter values that depend on the iteration counter It_{HS} of the algorithm:

$$\text{PAR}(It_{\text{HS}}) = \text{PAR}_{\min} + It_{\text{HS}} \frac{(\text{PAR}_{\max} - \text{PAR}_{\min})}{It_{\text{HS}_{\max}}} \tag{14}$$

where $It_{\text{HS}_{\max}}$ represents the allowed maximum number of iterations, $\text{PAR}_{\min}$ and $\text{PAR}_{\max}$ are the minimum and maximum pitch adjusting rate respectively, and

$$\text{BW}(It_{\text{HS}}) = \text{BW}_{\max} e^{c\, It_{\text{HS}}}, \text{ for } c = \frac{\ln(\frac{\text{BW}_{\min}}{\text{BW}_{\max}})}{It_{\text{HS}_{\max}}} \tag{15}$$

where $\text{BW}_{\min}$ and $\text{BW}_{\max}$ are the minimum and maximum bandwidth respectively. The main steps of the I-HS algorithm are as represented in Algorithm 2 below:

Algorithm 2. I-HS algorithm

Require: k and ξ_i, $i = 1, \ldots, k$ (from Algorithm 1), HMS, Ω, $It_{\text{HS}_{\max}} > 0$;
Set $It_{\text{HS}} = 1$;
for $j = 1, \ldots,$HMS **do**
  Randomly generate $x^j \in \Omega$;
  Compute $\bar{\mathcal{M}}(x^j)$;
end for
Based on $\bar{\mathcal{M}}$, select x^{best} and x^{worst};
while $It_{\text{HS}} \leq It_{\text{HS}_{\max}}$ and $\bar{\mathcal{M}}(x^{\text{best}}) > 10^{-10}$ **do**
  Improvise a new harmony $y \in \Omega$;
  Compute $\bar{\mathcal{M}}(y)$;
  Based on $\bar{\mathcal{M}}$, update HM and select x^{best} and x^{worst};
  Set $It_{\text{HS}} = It_{\text{HS}} + 1$;
end while
return $\xi \leftarrow x^{\text{best}}$ (for Algorithm 1)

4 Computational Experiments

The experiments were carried out on a PC Intel Core 2 Duo Processor E7500 with 2.9GHz and 4Gb of memory. The algorithms were coded in Matlab Version 8.0.0.783 (R2012b). In this study, the thirteen problems used for benchmark [16,17,19,20,22,23] are listed in Table 2 that also contains the number of roots in the search space Ω. These are the values set to the parameters: $\epsilon = 0.005$, $It_{\max} = 30$, HMS=10 when $n = 1, 2$ and 12 for $n = 3, 4$, HMCR=0.95, $\text{PAR}_{\min}$=0.35, $\text{PAR}_{\max} = 0.99$, $\text{BW}_{\min} = 10^{-6}$ and $\text{BW}_{\max} = 5$ [8]. We remark that the maximum number of iterations allowed in Algorithm 2, $It_{\text{HS}_{\max}}$, varies

Table 2. Problems set

NonD2	$f_1 = x_1^2 - x_2^2$ $f_2 = 1 - \|x_1 - x_2\|$ 2 roots in $[-3, 3]^2$
Trans	$f_1 = x_1^2 - x_2 - 2$ $f_2 = x_1 + \sin(\pi x_2/2)$ 3 roots in $[-3, 3]^2$
Himmelblau	$f_1 = 4x_1^3 + 4x_1x_2 + 2x_2^2 - 42x_1 - 14$ $f_2 = 4x_2^3 + 4x_1x_2 + 2x_1^2 - 26x_2 - 22$ 9 roots in $[-5, 5]^2$
Geometry	$f_1 = x_1x_2 + (x_1 - 2x_3)(x_2 - 2x_3) - 165$ $f_2 = (x_1x_2^3)/12 - (x_1 - 2x_3)(x_2 - 2x_3)^3/12 - 9369$ $f_3 = (2(x_2 - x_3)^2(x_1 - x_3)^2x_3) \,/\, (x_1 + x_2 - 2x_3) - 6835$ 2 roots in $[0, 50]^3$
Floudas	$f_1 = (0.25/\pi)x_2 + 0.5x_1 - 0.5\sin(x_1x_2)$ $f_2 = (e/\pi)x_2 - 2ex_1 + (1 - 0.25/\pi)(e^{2x_1} - e)$ 2 roots in $[0.25, 1] \times [1.5, 2\pi]$
Merlet	$f_1 = -\sin(x_1)\cos(x_2) - 2\cos(x_1)\sin(x_2)$ $f_2 = -\cos(x_1)\sin(x_2) - 2\sin(x_1)\cos(x_2)$ 13 roots in $[0, 2\pi]^2$
Reactor	$f_1 = (1 - R)\,(D/(10(1 + B_1)) - x_1)\, e^{10x_1/(1+10x_1/\gamma)} - x_1$ $f_2 = (1 - R)\,(D/10 - B_1x_1 - (1 + B_2)x_2))e^{10x_2/(1+10x_2/\gamma)} + x_1$ $-(1 + B_2)x_2$ with $D = 22$, $B_1 = B_2 = 2$, $R = 0.960$ and $\gamma = 1000$ 7 roots in $[0, 1]^2$
P1syst	$f_1 = x_1 + x_2 - 3$ $f_2 = x_1^2 + x_2^2 - 9$ 2 roots in $[-3, 3]^2$
Papersys	$f_1 = x_1 - \sin(2x_1 + 3x_2) - \cos(3x_1 - 5x_2)$ $f_2 = x_2 - \sin(x_1 - 2x_2) + \cos(x_1 + 3x_2)$ 3 roots in $[-3, 3]^2$
Casestudy5	$f_1 = e^{x_1^2} - 8x_1\sin(x_2)$ $f_2 = x_1 + x_2 - 1$ $f_3 = (x_3 - 1)^3$ 2 roots in $[0, 1]^3$
Casestudy7	$f_1 = x_1^3 - 3x_1x_2^2 - 1$ $f_2 = 3x_1^2x_2 - x_2^3 + 1$ 3 roots in $[-1, 2]^2$
Manipulator	$f = 3.9852 - 10.039x^2 + 7.2338x^4 - 1.17775x^6$ $+(-8.8575x + 20.091x^3 - 11.177x^5)\sqrt{1 - x^2}$ 6 roots in $[-1, 1]$
Trigonometric	$f = \sin(0.2x)\cos(0.5x)$ 19 different roots in $[-50, 50]$

with the problem: 1000 in NonD2, 2000 in Merlet and P1syst, 5000 in Trans, Floudas, Casestudy5, Casestudy7, Manipulator and Trigonometric, and 10000 in Himmelblau, Geometry, Reactor and Papersys.

Tables 3 – 5 report the average results produced by the proposed Algorithm 1 using:

- the 'exp' penalty with $\beta = 1000$ and ρ, as defined in (5),
- the 'coth' penalty with $\alpha = 10$, as described in (6), and
- the 'erf' penalty with $\delta = 0.1$ and $\bar{\rho}$, as defined in (9),

respectively, where the columns show:

- the name of the problem, Prob.;

Table 3. Numerical results from 'exp' penalty with $\beta = 1000$, considering (4) and (5)

Prob.	SR	$N.\text{roots}_{\text{avg}}$	NFE_{avg}	T_{avg}	NFE_{root}	T_{root}
NonD2	100	2.0	1507	0.109	753	0.055
Trans	97	3.0	9688	0.684	3266	0.231
Himmelblau	0	5.9	37664	3.146	6420	0.536
Geometry	43	1.4	8865	0.633	6332	0.452
Floudas	100	2.0	6290	0.426	3145	0.213
Merlet	20	11.4	10311	1.570	902	0.137
Reactor	7	5.7	157388	12.857	27451	2.243
P1syst	97	2.0	2240	0.157	1139	0.080
Papersys	60	2.4	16479	1.135	6866	0.473
Casestudy5	100	2.0	6941	0.505	3471	0.252
Casestudy7	90	2.9	10207	0.713	3520	0.246
Manipulator	0	5.0	12505	0.889	2501	0.178
Trigonometric	0	8.7	12760	1.383	1467	0.159

Table 4. Numerical results from 'coth' penalty with $\alpha = 10$, considering (6) and (7)

Prob.	SR	$N.\text{roots}_{\text{avg}}$	NFE_{avg}	T_{avg}	NFE_{root}	T_{root}
NonD2	100	2.0	1734	0.141	867	0.070
Trans†	90	2.9	30817	2.416	10627	0.833
Himmelblau†	0	6.2	58940	4.909	9456	0.788
Geometry	63	1.6	16779	1.324	10273	0.811
Floudas†	100	2.0	16385	1.262	8193	0.631
Merlet†	10	10.1	11577	1.119	1146	0.111
Reactor†	3	6.0	467962	39.799	77563	6.597
P1syst†	100	2.0	12303	0.951	6151	0.476
Papersys	60	2.4	17523	1.309	7301	0.545
Casestudy5	23	3.1	13617	1.124	4393	0.363
Casestudy7	90	2.9	10509	0.816	3624	0.282
Manipulator†	100	6.0	105303	7.353	17551	1.225
Trigonometric†	0	13.2	33123	2.585	2509	0.196

† - problems where some or all roots were recovered more than once (not necessarily in all runs).

- percentage of runs (out of 30) where all the roots were located, SR (%);
- the average number of located roots per run, $N.\text{roots}_{\text{avg}}$;
- the average number of merit function evaluations per run, NFE_{avg};
- the average time in seconds per run, T_{avg};
- the average number of function evaluations required to locate a root, NFE_{root};
- the average time required to locate a root, T_{root}.

These preliminary results are very encouraging. Numerical results suggest that the new 'erf' penalty function clearly has some advantages over 'coth' penalty

Table 5. Numerical results from 'erf' function with $\delta = 0.1$, considering $\bar{\rho} = 0.1 \min_i(u_i - l_i)$ in (9)

Prob.	SR	N.roots$_{avg}$	NFE_{avg}	T_{avg}	NFE_{root}	T_{root}
NonD2	100	2.0	1495	0.114	748	0.057
Trans	90	2.9	10074	0.781	3474	0.269
Himmelblau	60	8.6	56018	4.319	6539	0.504
Geometry	53	1.4	8598	0.659	6291	0.482
Floudas	97	2.0	6306	0.457	3206	0.232
Merlet	100	13.0	12574	1.048	967	0.081
Reactor	10	5.9	165713	12.577	28087	2.132
P1syst†	100	2.0	4361	0.366	2181	0.183
Papersys	7	1.7	11955	0.863	6897	0.498
Casestudy5	83	2.1	7938	0.620	3780	0.295
Casestudy7	100	3.0	10523	0.768	3508	0.256
Manipulator	100	6.0	21563	1.689	3594	0.281
Trigonometric†	53	18.6	45499	4.781	2446	0.257

† - problems where some or all roots were recovered more than once.

function and the 'exp' penalty function. Overall, 'coth' and 'exp' penalties produce fairly similar results for most problems. We observed that the 'erf' penalty function demonstrated to be a promising and viable tool for computing multiple roots of systems of nonlinear equations.

5 Conclusions

A repulsion algorithm is presented for locating multiple roots of a system of nonlinear equations. The proposed algorithm relies on a multiplicative kind penalty merit function that depends on two parameters. One aims to scale the penalty and the other adjusts the radius of the repulsion area, so that convergence to previously located solutions is avoided. The algorithm has been successfully applied and tested with a benchmark set of problems. The numerical experiments also lead us to allege that the 'erf' penalty function is indeed more accurate, reliable, and efficient at locating multiple roots than the other alternatives in comparison.

Acknowledgments. The authors wish to thank two anonymous referees for their valuable comments and suggestions to improve the paper. This work has been supported by CIDEM (Centre for Research & Development in Mechanical Engineering, Portugal) and FCT - Fundação para a Ciência e Tecnologia within the Projects Scope: PEst-OE/EME/UI0615/2014 and PEst-OE/EEI/UI0319/2014.

References

1. Ramadas, G.C.V., Fernandes, E.M.G.P.: Solving systems of nonlinear equations by harmony search. In: Aguiar, J.V., et al. (eds.) 13th International Conference Computational and Mathematical Methods in Science and Engineering, vol. IV, pp. 1176–1186 (2013) ISBN: 978-84-616-2723-3
2. Ramadas, G.C.V., Fernandes, E.M.G.P.: Combined mutation differential evolution to solve systems of nonlinear equations. In: 11th International Conference of Numerical Analysis and Applied Mathematics 2013 AIP Conf. Proc., vol. 1558, pp. 582–585 (2013)
3. Dennis, J.E., Schnabel, R.B.: Numerical Methods for Unconstrained Optimization and Nonlinear Equations. Prentice-Hall Inc. (1983)
4. González-Lima, M.D., Oca, F.M.: A Newton-like method for nonlinear system of equations. Numer. Algorithms 52, 479–506 (2009)
5. Martínez, J.M.: Practical Quasi-Newton methods for solving nonlinear systems. J. Comput. Appl. Math. 124, 97–122 (2000)
6. Nowak, U., Weimann, L.: A family of Newton codes for systems of highly nonlinear equations. Technical Report Tr-91-10, K.-Z.-Z. Inf. Berlin (1991)
7. Geem, Z.W., Kim, J.H., Loganathan, G.: A new heuristic optimization algorithm: harmony search. Simulation 76, 60–68 (2001)
8. Mahdavi, M., Fesanghary, M., Damangir, E.: An improved harmony search algorithm for solving optimization problems. Appl. Math. Comput. 188, 1567–1579 (2007)
9. Lee, K.S., Geem, Z.W.: A new meta-heuristic algorithm for continuous engineering optimization: harmony search theory and practice. Comput. Method Appl. M. 194, 3902–3933 (2004)
10. Manjarres, D., Landa-Torres, I., Gil-Lopez, S., Del Ser, J., Bilbao, M.N., Salcedo-Sanz, S., Geem, Z.W.: A survey on applications of the harmony search algorithm. Eng. Appl. Artif. Intel. 26(8), 1818–1831 (2013)
11. Estahbanati, M.J.: Hybrid probabilistic-harmony search algorithm methodology in generation scheduling problem. J. Exp. Theor. Artif. Intell. (2014), doi:10.1080/0952813X.2013.861876
12. Hosseini, S.D., Shirazi, M.A., Ghomi, S.M.T.F.: Harmony search optimization algorithm for a novel transportation problem in a consolidation network. Eng. Optimiz. (2013), doi:10.1080/0305215X.2013.854350
13. Nigdeli, S.M., Bekdas, G., Alhan, C.: Optimization of seismic isolation systems via harmony search. Eng. Optimiz. (2013), doi:10.1080/0305215X.2013.854352
14. Tsoulos, I.G., Lagaris, I.E.: MinFinder: Locating all the local minima of a function. Comput. Phys. Commun. 174, 166–179 (2006)
15. Voglis, C., Lagaris, I.E.: Towards "Ideal Multistart". A stochastic approach for locating the minima of a continuous function inside a bounded domain. Appl. Math. Comput. 213, 1404–1415 (2009)
16. Tsoulos, I.G., Stavrakoudis, A.: On locating all roots of systems of nonlinear equations inside bounded domain using global optimization methods. Nonlinear Anal.-Real 11, 2465–2471 (2010)
17. Hirsch, M.L., Pardalos, P.M., Resende, M.: Solving systems of nonlinear equations with continuous GRASP. Nonlinear Anal.-Real 10, 2000–2006 (2009)
18. Parsopoulos, K.E., Vrahatis, M.N.: On the computation of all global minimizers through particle swarm optimization. IEEE Trans. Evol. Comput. 8(3), 211–224 (2004)

19. Pourjafari, E., Mojallali, H.: Solving nonlinear equations systems with a new approach based on invasive weed optimization algorithm and clustering. Swarm and Evolutionary Computation 4, 33–43 (2012)
20. Silva, R.M.A., Resende, M.G.C., Pardalos, P.M.: Finding multiple roots of a box-constrained system of nonlinear equations with a biased random-key genetic algorithm. J. Global Optim. (September 2013), doi:10.1007/s10898-013-0105-7
21. Alia, O.M., Mandava, R.: The variants of the harmony search algorithm: an overview. Artif. Intell. Rev. 36, 49–68 (2011)
22. Grosan, C., Abraham, A.: A new approach for solving nonlinear equations systems. IEEE T. Syst. Man. Cy. A 38, 698–714 (2008)
23. Jaberipour, M., Khorram, E., Karimi, B.: Particle swarm algorithm for solving systems of nonlinear equations. Comput. Math. Appl. 62, 566–576 (2011)

Branch and Bound Based Coordinate Search Filter Algorithm for Nonsmooth Nonconvex Mixed-Integer Nonlinear Programming Problems

Florbela P. Fernandes[1,3], M. Fernanda P. Costa[2,3], and Edite M.G.P. Fernandes[4]

[1] ESTiG, Polytechnic Institute of Bragança, 5301-857 Bragança, Portugal
fflor@ipb.pt

[2] Department of Mathematics and Applications, University of Minho, 4800-058 Guimarães, Portugal
mfc@math.uminho.pt

[3] Centre of Mathematics, University of Minho, 4710-057 Braga, Portugal

[4] Algoritmi Research Centre, University of Minho, 4710-057 Braga, Portugal
emgpf@dps.uminho.pt

Abstract. A mixed-integer nonlinear programming problem (MINLP) is a problem with continuous and integer variables and at least, one nonlinear function. This kind of problem appears in a wide range of real applications and is very difficult to solve. The difficulties are due to the nonlinearities of the functions in the problem and the integrality restrictions on some variables. When they are nonconvex then they are the most difficult to solve above all. We present a methodology to solve nonsmooth nonconvex MINLP problems based on a branch and bound paradigm and a stochastic strategy. To solve the relaxed subproblems at each node of the branch and bound tree search, an algorithm based on a multistart strategy with a coordinate search filter methodology is implemented. The produced numerical results show the robustness of the proposed methodology.

Keywords: nonconvex MINLP, branch and bound, multistart, coordinate search, filter method.

1 Introduction

A wide range of problems arising in practical applications, which involve both discrete decisions and nonlinear real-world phenomena, are modeled as mixed-integer nonlinear programming (MINLP) problems. Examples of practical applications modeled as MINLP appear in various areas, such as, process engineering [14], water, gas, energy and transportation networks [5]. For a review of a wide range of MINLP applications, see [3].

B. Murgante et al. (Eds.): ICCSA 2014, Part II, LNCS 8580, pp. 140–153, 2014.

MINLP problems combine the combinatorial difficulty of optimizing over discrete variable sets with the challenge of handling nonlinear functions. They are the most general optimization problems, containing as special cases the mixed-integer linear programming (MILP) problem and the nonlinear programming (NLP) problem. This generality allows the modeling of a wide range of practical applications. When all functions involved in the problem are convex, the problem is a convex MINLP problem; otherwise it is a nonconvex one. Although MINLP problems are in general NP-hard, convex MINLPs are much easier to solve than nonconvex ones. This is due to the fact that the continuous relaxation of a convex MINLP problem, which is obtained by considering all the variables continuous, is itself convex and the computed solution being a global one provides a lower bound for the optimal solution of the MINLP. A variety of effective exact solution methods for convex MINLPs have been devised based on this fact. Unfortunately, when the MINLP is nonconvex, the continuous relaxation of a nonconvex MINLP is itself a global optimization problem, and therefore likely to be NP-hard. There is no guarantee that the computed solution of the continuous relaxation is a global optimum. The situation gets even worse when the MINLP model involves nonsmooth functions. For example, when the objective and constraints are provided as black-boxes. In this case, the MINLP is a nonsmooth and nonconvex problem.

There are different strategies to solve MINLP problems and a great majority rely on a branch and bound paradigm. One of the most well-known solvers for MINLP problems is the BARON (Branch And Reduce Optimization Navigator) solver [23]. For a review of the available solvers, we refer the reader to [3]. Recently, the nonconvex MINLP research area became more interesting due to its range of applications and the new techniques for globally solving NLP problems [8]. In the last 10–15 years, innovative heuristic type algorithms have also appeared: genetic algorithm [15], ant colony [22], pattern search algorithms [1], multistart Hooke and Jeeves algorithm [12], and differential evolution [19].

In this work we develop a derivative-free methodology to solve nonsmooth nonconvex MINLP problems. The proposed method is based on a branch and bound (BB) scheme, and the NLP problems that appear in the BB tree search are solved to optimality by a derivative-free global method that is based on a multistart algorithm coupled with a coordinate search filter method, initially developed in [11]. The therein called MCSFilter method is able to find multiple minima of a nonconvex NLP problem, and consequently the global one. The MCSFilter method is appropriate to solve nonsmooth problems since neither analytical nor numerical derivatives are required.

Our BB paradigm for nonsmooth nonconvex MINLP problems, henceforth denoted by BBMCSFilter, has been implemented in MatLab and the numerical experiments with a benchmark set of problems show that the method is competitive with others of the same type.

The remaining part of the paper is organized as follows. In Section 2 the MINLP problem is described and in Section 3 the proposed BBMCSFilter algorithm is presented and discussed. Section 4 reports on the computational experiments carried out using a benchmark set of problems (from the engineering field) from the *MINLPLib* library available online [6] and we conclude the paper with Section 5.

2 The MINLP Problem

The problem to be solved is of the form

$$\begin{array}{c} \min f(x,y) \\ \text{subject to } g_j(x,y) \leq 0,\ j \in J \\ x \in X,\ y \in Y \end{array} \tag{1}$$

where f is the objective function $f : \mathbb{R}^n \longrightarrow \mathbb{R}$, g_j, $j \in J = \{1, \cdots, m\}$, are the constraint functions and J is the index set of the g functions. The continuous variables are represented by the vector x, with $X \subset \mathbb{R}^{n_c}$ being the set of simple bounds on x:

$$X = \{x \in \mathbb{R}^{n_c} : l_x \leq x \leq u_x\},$$

with $l_x, u_x \in \mathbb{R}^{n_c}$. The integer variables are represented by the vector y, where $Y \subset \mathbb{Z}^{n_i}$ is the set of simple bounds on y:

$$Y = \{y \in \mathbb{Z}^{n_i} : l_y \leq y \leq u_y\},$$

with $l_y, u_y \in \mathbb{Z}^{n_i}$. The parameter $n = n_c + n_i$ represents the total number of variables. We will assume that at least one of the functions is nonlinear since this study aims to focus on MINLP problems. In the present study we are particularly interested in nonsmooth and nonconvex MINLPs. With this kind of problems two major issues have to be addressed:

- one is related with the integrality of some variables;
- the other is concerned with the lack of smoothness and convexity of the functions.

To settle the first issue, a BB paradigm is used. A brief description of the main ideas behind this classical technique is presented in the next section. The second issue is crucial since the nonconvex NLP relaxed problems that arise in the nodes of the BB tree search have multiple minima, some global and others local. When a minimum is found it is not possible to know if it is a global or a local one, until all the minima are computed and compared. Unless, a global search exact method with guaranteed convergence to a global minimum is used when solving the nonconvex NLP problems. Some heuristic methods that do not use derivative information by methodically searching the space can also guarantee convergence to global solutions with probability one. Both real analysis pointwise convergence and stochastic convergence are appropriate when convergence to a global minimum is required.

3 The BBMCSFilter Method

As previously stated, the BB paradigm and the MCSFilter method constitute the two major parts of the method. First, a summary of the BB paradigm is presented.

3.1 BB Method

The BB method was initially devised for MILP problems but has been applied ever since to MINLPs too. The first reference to nonlinear MINLP problems appears in 1965 [7]. The extension to convex MINLP problems is an easy task since the minimum of the convex relaxed NLP problem is also a global one [18]. BB methodology can be explained accordingly with a tree search [3,4,20]. We first define a continuous relaxation.

Definition 1. *Consider the (convex or nonconvex) MINLP problem* (1). *A continuous relaxation of problem* (1) *is:*

$$\begin{array}{c} \min f(x,y) \\ \textit{subject to } g_j(x,y) \leq 0,\ j \in J \\ x \in X \subset \mathbb{R}^{n_c},\ y \in Y_{\mathbb{R}} \subset \mathbb{R}^{n_i} \end{array} \tag{2}$$

with $Y_{\mathbb{R}} = \{y \in \mathbb{R}^{n_i} : l_y \leq y \leq u_y\}$, *meaning that all variables are real numbers.*

At the beginning of the process all the integer variables are relaxed and the relaxed NLP problem (2) is solved. This is the first node (also known as root) of the tree search. After solving this problem, if all integer variables take integer values at the solution then this solution also is the solution of the MINLP problem. However, in general, at this stage, some integer variables take non-integer values. Then, a tree search is performed in the space of the integer variables. Branching generates new NLP subproblems by adding new simple bounds to the new relaxed NLP subproblems. Next, a new subproblem is selected and solved. The solution of each relaxed subproblem provides a lower bound, $\underline{f}$, for the descent nodes of the tree (or the child nodes). The integer solutions (at some nodes of the tree) provide upper bounds, $\overline{f}$, on the optimal integer solution. This process of branching at each node continues until:

- the lower bound exceeds the best known upper bound;
- the NLP subproblem is infeasible;
- the solution provides integer values for the integer variables, thus providing an upper bound.

The BB algorithm stops when there are no more nodes to explore. In a BB technique there are two crucial components. One is related with choosing a good branching variable. The main goal is to choose the branching variable that minimizes the size of the tree that needs to be searched. However, this is not practical and the selection of a branching variable that maximizes the increase in the lower bound at a node is a good alternative. The other component is related with the choice of which relaxed problem (node) should be solved next. The main goal here is to find a good feasible solution as quickly as possible in order to reduce the upper bound, and to prove optimality of the current best integer solution by increasing the lower bound as quickly as possible.

Our main contribution is related with solving the relaxed NLP subproblems that appear in the multiple nodes of the tree search by a derivative-free multistart method so that a global solution of the relaxed subproblem (2) is obtained. This is a challenging task that is accomplished by the MCSFilter method.

3.2 MCSFilter Method for the Relaxed NLP Subproblems

In each node of the tree search a nonsmooth nonconvex NLP subproblem is required to be globally solved. To obtain the global optimal solution of these relaxed problems is crucial since fathoming /eliminating nodes of the BB tree search can no longer be made if the solution is not a global one. The process of eliminating nodes is very important because if all subproblems need to be solved the BB scheme will be very time consuming. So, it is necessary to eliminate some nodes as the result of a bounding scheme with $\underline{f}$ and $\overline{f}$.

The MCSFilter method which is used to solve the relaxed nonsmooth nonconvex NLP subproblems at each node of the tree has nice features:

- does not make use of any derivative information;
- solves nonconvex NLP problems;
- finds multiple solutions, global as well as the local ones;
- is based on a multistart strategy with regions of attraction coupled with a coordinate search filter methodology;
- is, relatively, simple to implement.

Since the variables of the relaxed problem are all real, hereafter we use z to denote the vector of all the n variables, the vectors l and u of the set $\mathbb{R}^n$ to denote the lower and upper bounds respectively, of all the variables. Thus, problem (2) is now formulated as:

$$\begin{array}{c} \min f(z) \\ \text{subject to } g_j(z) \leq 0, \ j \in J\,. \\ z \in [l,u] \subset \mathbb{R}^n \end{array} \tag{3}$$

This method is based on a multistart strategy. To explore the search space more effectively, the multistart algorithm sequentially generates points

$$z_i = l_i + \lambda\left(u_i - l_i\right), \, i = 1, \ldots, n,$$

where λ is a random number uniformly distributed in $[0,1]$, and applies a local search aiming to converge to the optimal solutions of the problem (3). Since this simple strategy may converge to some or all the minimizers over and over again, the implemented multistart strategy incorporates the concept of regions of attraction of minimizers to avoid convergence to the already computed solutions [17,24]. The region of attraction of a minimizer, z_*^i, associated with a local search procedure $\mathbf{L}$, is defined as:

$$A_i \equiv \left\{z \in [l,u] : \mathbf{L}(z) = z_*^i\right\}, \tag{4}$$

where $\mathbf{L}(z)$ is the minimizer obtained when the local search procedure $\mathbf{L}$ starts at point z. Computing the region of attraction A_i of a minimizer z_*^i is not an easy task. Alternatively, a stochastic procedure may be used to estimate the probability, p, that a sampled point will not belong to the union of the regions of attraction of already computed minimizers. This probability is estimated similarly to [24] but using instead forward differences to estimate the gradient of f [10]. Different stopping rules have been tested in a multistart algorithm context, see [17]. The algorithm should stop when all minima have been identified with certainty, and it should not require a large number of calls to the local search procedure to decide that all minima have been found. The rule used in our implementation gives an estimate of the fraction of uncovered space [17]. A formal description of the multistart algorithm for solving the nonsmooth nonconvex NLP subproblem (2), based on the CSFilter method as the procedure for the local search (4), is presented in Algorithm 1.

Algorithm 1. Multistart algorithm

Set $Z^* = \emptyset$ (contains the computed minimizers), $k = 1$, $t = 1$;
Randomly generate $z \in [l, u]$;
Compute $z_*^1 = \mathbf{L}(z)$ using Algorithm 2, set $Z^* = Z^* \cup z_*^1$, define A_1;
repeat
    Randomly generate $z \in [l, u]$;
    if z has a high probability of being outside $\bigcup_{i=1}^{k} A_i$ **then**
        Compute $z_* = \mathbf{L}(z)$ using Algorithm 2, set $t = t + 1$;
        if $z_* \notin Z^*$ **then**
            Set $k = k + 1$, $z_*^k = z_*$, $Z^* = Z^* \cup z_*^k$, compute A_k;
        else
            Update A_l (region of attraction of the nearest to z_* minimizer)
        end if
    end if
until the stopping rule is satisfied

Local CSFilter Method. The CSFilter method that combines a derivative-free local search technique with a filter methodology [13] is used as the search procedure $\mathbf{L}$, to compute a minimizer z_* starting from a sampled point $z \in [l, u]$ [11]. The classical coordinate search, which is a direct search method [16], and the filter methodology are combined to construct a local search procedure that does not require any derivative information. The filter methodology is implemented to handle the constraints by forcing the local search towards the feasible region. The main idea behind the filter approach is to interpret problem (3) as a bi-objective optimization problem aiming to minimize both the objective function $f(z)$ and a nonnegative continuous aggregate constraint violation function $\theta(z)$ defined by

$$\theta(z) = \|g(z)_+\|^2 + \|(l - z)_+\|^2 + \|(z - u)_+\|^2 \tag{5}$$

where $v_+ = \max\{0, v\}$. Therefore, the problem is rewritten as a bi-objective optimization problem of the form

$$\min_{z} \left(\theta(z), f(z)\right). \tag{6}$$

The filter technique incorporates the concept of nondominance, present in the field of multi-objective optimization, to build a filter that is able to accept trial approximations if they improve the constraint violation or objective function value [1,2,13]. A filter $\mathcal{F}$ is a finite set of points z, corresponding to pairs $(\theta(z), f(z))$, none of which is dominated by any of the others. A point z is said to dominate a point z' if and only if $\theta(z) \leq \theta(z')$ and $f(z) \leq f(z')$.

We now discuss the implemented CSFilter algorithm for the local procedure [11]. The pseudo-code for the local filter-based coordinate search algorithm is presented below in Algorithm 2. At the beginning, and every time the procedure is invoked inside the multistart algorithm, the filter is initialized to $\mathcal{F} = \{(\theta, f) : \theta \geq \theta_{\max}\}$, where $\theta_{\max} > 0$ is an upper bound on the acceptable constraint violation.

Let $\mathcal{D}_\oplus$ denote the set of $2n$ coordinate directions, defined as the positive and negative unit coordinate vectors, $\mathcal{D}_\oplus = \{e_1, e_2, \ldots, e_n, -e_1, -e_2, \ldots, -e_n\}$. The search begins with a central point, the current approximation $\tilde{z}$, as well as $2n$

Algorithm 2. CSFilter algorithm

Require: z (sampled in the Multistart algorithm) and parameter values; set $\tilde{z} = z$, $z_{\mathcal{F}}^{inf} = z$, $t = \tilde{z}$;

Initialize the filter;
Set $\alpha = \min\{1, 0.05 \frac{\sum_{i=1}^{n} u_i - l_i}{n}\}$;
repeat
  Compute the trial approximations $t_c^i = \tilde{z} + \alpha d_i$, for all $d_i \in \mathcal{D}_\oplus$;
  repeat
    Check acceptability of trial points t_c^i using (7) and (8);
    if there are some t_c^i acceptable by the filter **then**
      Update the filter;
      Choose t_c^{best};
      Set $t = \tilde{z}$, $\tilde{z} = t_c^{best}$; update $z_{\mathcal{F}}^{inf}$ if appropriate;
    else
      Compute the trial approximations $t_c^i = z_{\mathcal{F}}^{inf} + \alpha d_i$, for all $d_i \in \mathcal{D}_\oplus$;
      Check acceptability of trial points t_c^i using (7) and (8);
      if there are some t_c^i acceptable by the filter **then**
        Update the filter;
        Choose t_c^{best};
        Set $t = \tilde{z}$, $\tilde{z} = t_c^{best}$; update $z_{\mathcal{F}}^{inf}$ if appropriate;
      else
        Set $\alpha = \alpha/2$;
      end if
    end if
  until new trial t_c^{best} is acceptable
until the stopping condition is satisfied

---

trial approximations $t_c^i = \tilde{z} + \alpha d_i$, for $d_i \in \mathcal{D}_\oplus$, where $\alpha > 0$ is a step size. The constraint violation value and the objective function value of all $2n+1$ points are computed. If some trial approximations improve over $\tilde{z}$, reducing θ or f by a certain amount (see equations (7) and (8)), and are acceptable by the filter, then the best of these non-dominated trial approximations, t_c^{best}, is selected and the filter is updated (adding the corresponding entries to the filter and removing any dominated entries). Then, this best approximation becomes the new central point in the next iteration, $\tilde{z} \leftarrow t_c^{best}$. If, on the other hand, all trial approximations t_c^i are dominated by the current filter, then all t_c^i are rejected, and a restoration phase is invoked.

To avoid the acceptance of a point t_c^i, or the corresponding pair $\left(\theta(t_c^i), f(t_c^i)\right)$, that is arbitrary close to the boundary of $\mathcal{F}$, the trial t_c^i is considered to improve over $\tilde{z}$ if one of the conditions

$$\theta(t_c^i) < (1-\gamma_\theta)\, \theta(\tilde{z}) \;\text{ or }\; f(t_c^i) \leq f(\tilde{z}) - \gamma_f\, \theta(\tilde{z}) \tag{7}$$

holds, for fixed constants $\gamma_\theta, \gamma_f \in (0,1)$. However, the filter alone cannot ensure convergence to optimal points. For example, if a sequence of trial points satisfies $\theta(t_c^i) < (1-\gamma_\theta)\, \theta(\tilde{z})$ then it could converge to an arbitrary feasible point. Therefore, when $\tilde{z}$ is nearly feasible, $\theta(\tilde{z}) \leq \theta_{\min}$, for a small positive $\theta_{\min}$, the trial approximation t_c^i has to satisfy only the condition

$$f(t_c^i) \leq f(\tilde{z}) - \gamma_f\, \theta(\tilde{z}) \tag{8}$$

instead of (7), in order to be acceptable.

The best non-dominated trial approximation is selected as follows. The best point t_c^{best} of a set $T = \{t_c^i : t_c^i = \tilde{z} + \alpha d_i, d_i \in \mathcal{D}_\oplus\}$ is the point that satisfies one of two following conditions:

i) if there are some feasible points in T, t_c^{best} is the point that has the smallest objective function value among the feasible points;
ii) if there are no feasible points in T, t_c^{best} is the point that has the smallest constraint violation among the non-dominated infeasible points.

We remark that the filter is updated whenever the trial approximations t_c^i verify conditions (7) or (8) and are non-dominated.

When it is not possible to find a non-dominated best trial approximation, and before declaring the iteration unsuccessful, a restoration phase is invoked. In this phase, the most nearly feasible point in the filter, $z_{\mathcal{F}}^{inf}$, is recuperated and the search along the $2n$ coordinate directions is carried out about it to find the set $T = \{t_c^i : t_c^i = z_{\mathcal{F}}^{inf} + \alpha d_i, d_i \in \mathcal{D}_\oplus\}$. If a non-dominated best trial approximation is found, this point becomes the central point of the next iteration and the iteration is successful. Otherwise, the iteration is unsuccessful, the search returns back to the current $\tilde{z}$, the step size α is halved, and new $2n$ trial approximations t_c^i are generated around $\tilde{z}$. If a best non-dominated trial approximation is still not found, the step size reduction is repeated since another unsuccessful iteration has occurred. When α falls below $\alpha_{\min}$ [16], the algorithm stops, where $\alpha_{\min}$ is a small positive tolerance.

4 Numerical Results

To analyze the performance of the BBMCSFilter algorithm, a set of 23 test problems is used (see Table 1). The set contains inequality and equality constrained problems. Almost all problems are selected from published literature in several different engineering fields [12,19,21]. The BBMCSFilter method was coded in MatLab. Crucial BB algorithm specifications are:

- the branching variable is chosen by a simple heuristic that picks up the variable which maximizes the increase in the lower bound at that node, i.e.,

$$\arg \max_{i=1,\dots,n_i} \{|\underline{f}_i - \underline{f}|\},$$

  where $\underline{f} = f(x', y')$ is the solution of the NLP subproblem ($x' \in X, y' \in Y_{\mathbb{R}}$) and $\underline{f}_i = f(x', y_i)$, for $i = 1, \dots, n_i$, being $y_i = [y'_i]_{\mathbb{R}}$ the scalar rounding of y'_i to the nearest integer;
- the next subproblem to be solved is selected by a depth-first strategy, also known as *last-in-first-out*, so that the upper bounds are found as early as possible.

The results were obtained in a PC with an Intel Core i7-2600 CPU (3.4GHz) and 8 GB of memory. In the CSFilter method, we set after an empirical study: $\gamma_\theta = \gamma_f = 10^{-5}$, $\alpha_{\min} = 10^{-3}$, $\theta_{\min} = 10^{-6}$ and $\theta_{\max} = 10^2 \max\{1, 1.25\theta(z)\}$, where z is the point on entry in the local search. A maximum of 10 points were generated in the multistart algorithm and each problem was solved 30 times.

Table 1 shows the numerical results produced by the proposed BBMCSFilter method. The columns in the table show:

- the known global optimum, f^*;
- the average value of the obtained objective function values (over the 30 runs), 'f_{avg}';
- the standard deviation of obtained function values, 'SD';
- the average number of function evaluations, 'nfe_{avg}';
- the average CPU time (in seconds), 'T_{avg}';
- the average number of nodes, '$Nodes_{avg}$';
- the successful rate (percentage of runs that found a solution within an error of 10^{-2} of the known global optimal solution), 'SR' (%).

We may conclude from Table 1 that our algorithm found the global optimal solution in all the 30 runs when solving 17 of the 23 problems. The computed solutions are of high quality and the f_{avg} for all problems are very close to the known minimum. We remark that our values of f_{avg} under f^* (on problems f4, f8 and f22) are due to the slight constraint violation allowed by the algorithm when it stops. The values of SD are equal to zero on problems f8, f10, f16 and f18, and are moderate, ranging from 2.6E-02 to 1.0E-07, on the remaining problems, showing the consistency of the algorithm. The number of function evaluations and the time required by the algorithm are higher than one could expect.

Table 1. Numerical results produced by BBMCSFilter

Problem	f^*	f_{avg}	SD	T_{avg}	nfe_{avg}	$Nodes_{avg}$	SR
f1	2	2.00082	3.6E-04	10.0	3530	1.1	100
f2	2.124	2.124590	1.4E-06	1.5	1259	1.0	100
f3	1.07654	1.081640	8.1E-03	3.8	5274	3.0	87
f4	99.245209	99.239635	1.0E-07	0.2	670	1.0	100
f5	3.557463	3.560848	2.0E-03	59.3	76775	11.0	97
f6	4.579582	4.582322	9.3E-04	54.9	75413	10.7	100
f7	-17	-16.998054	2.3E-03	9.5	4296	1.8	100
f8	-32217.4	-32217.42778	0.0E00	3.2	18051	5.7	0
f9	7.6671801	7.667583	9.5E-04	30.3	28090	3.9	100
f10	-2.4444	-2.444444	0.0E00	2.4	2736	5.0	100
f11	3.2361	3.236121	8.7E-05	18.8	41635	10.0	100
f12	1.125	1.125115	2.9E-04	42.3	7770	2.8	100
f13	87.5	87.507043	1.7E-02	252.5	41852	3.0	90
f14	-6.666667	-6.666131	1.8E-04	0.5	1122	1.0	100
f15	-5.6848	-5.651952	2.6E-02	3567.4	393345	59.8	30
f16	2.000	2.000000	0.0E00	52.7	29847	4.9	100
f17	3.4455	3.445808	2.1E-04	18.2	5469	6.0	100
f18	2.2000	2.200000	0.0E00	8.6	11182	3.7	100
f19	6.0098	6.010714	6.6E-04	21.1	37132	5.3	100
f20	-17.0000	-16.994605	5.5E-03	52.5	27149	1.1	80
f21	-4.514202	-4.513448	6.8E-04	84.4	50146	4.6	100
f22	-13.401904	-13.401930	3.6E-04	57.8	84790	14.0	100
f23	-1.08333	-1.083245	5.4E-05	2.6	2458	1.0	100

Table 2. Numerical results obtained with BBGA

Problem	f_{avg}	SD	T_{avg}	nfe_{avg}	SR
f1	2.00	6.8E-05	2.5	10481	100
f2	2.1246	2.1E-04	2.4	11527	100
f3	1.078992	2.4E-03	3.3	13635	20
f5	3.564265	7.6E-03	13.5	47282	23
f6	4.5987	2.9E-02	13.7	46678	53
f7	-17	1.8E-04	3.4	14292	100
f8	-32217.4	1.5E-11	1.3	5220	100
f15	-5.684	1.9E-03	65.0	247055	87
f16	2.000	8.9E-07	3.2	12808	67
f17	3.446	2.0E-05	7.8	23489	100
f18	2.200	5.8E-05	4.5	15290	87

To compare our results with those of a BB scheme that uses the genetic algorithm solver from MatLab^{TM} Optimization Toolbox, to solve the nonsmooth nonconvex NLP relaxed subproblems, we include Table 2 with the results available in [9]. The columns in the table list the values of f_{avg}, SD, T_{avg}, nfe_{avg} and SR. The comparison with the therein called BBGA method involves

Table 3. Numerical results obtained from [19]

Problem	MDE			MDE-LS			MDE-IHS		
	f_{avg}	SD	nfe_{avg}	f_{avg}	SD	nfe_{avg}	f_{avg}	SD	nfe_{avg}
f1	2.009348	4.4E-02	1075	2.00000	0.0E00	1430	2.000001	0.0E00	3297
f2	2.167894	1.3E-01	827	2.124574	7.1E-05	653	2.124604	7.6E-05	1409
f4	99.240933	1.4E-03	403	99.241271	1.8E-03	684	99.512250	1.5E00	449
f5	3.599903	5.9E-02	37739	3.564912	2.9E-02	20116	3.561157	8.4E-03	27116
f6	4.661414	3.1E-01	7688	4.579595	3.0E-06	13023	4.579599	5.0E-06	14518
f8	-32217.427262	2.8E-03	1240	-32217.427106	3.7E-03	1955	-32217.427780	0.0E00	493
f9	7.918619	4.8E-02	96718	7.883841	9.9E-02	93524	7.848896	1.2E-01	83442
f12	1.124453	7.6E-02	30986	1.099805	5.6E-02	25766	1.094994	5.3E-02	22146
f13	89.879034	2.8E00	7777	88.230145	1.9E00	4436	87.497550	2.1E-03	5359

11 problems from the previous set of 23 problems. We conclude from the tables that BBMCSFilter method outperforms BBGA in terms of criteria SD and SR, but is outweighted by BBGA in criterion T_{avg} on all tested problems but one. As far as nfe_{avg} is concerned, BBMCSFilter is better than BBGA in 6 of the tested problems.

Finally, we compare our results with those reported in [19]. This paper presents two hybrid differential evolution (DE) algorithms. One, enhances a modified DE (MDE) algorithm with a local search operator, therein denoted by `MDE-LS`; the other, adds a second metaheuristic – the harmony search algorithm – to cooperate with the MDE algorithm, and is denoted by `MDE-IHS`. The results obtained from the `MDE` algorithm from which the other two hybrids were created are also reported. These results are shown in Table 3. When a comparison is made between BBMCSFilter and `MDE` it is possible to state that our method performs better than `MDE`, relatively to f_{avg} and SD. The BBMCSFilter method requires in general more function evaluations than `MDE` but on the other hand the quality of the solutions is higher. When comparing with the results of `MDE-LS` algorithm, we observe that BBMCSFilter produces better values of f_{avg} and SD in 7 of the 9 common problems (see Table 3). However, the values of nfe_{avg} are larger with BBMCSFilter on 6 problems. Observing the results obtained by `MDE-IHS` algorithm in Table 3, we may conclude that this method wins over BBMCSFilter in number of function evaluations, although the quality of the solutions in terms of f_{avg} and SD are lower than that of BBMCSFilter in 5 of the 9 common problems.

5 Conclusions

We have presented a method to solve nonsmooth nonconvex MINLP problems, the BBMCSFilter method, and showed that the BB paradigm coupled with a stochastic multistart method based on the classical coordinate search for a local exploitation of a minimizer is effective and worthy of further research. The MCSFilter method is used to solve the nonsmooth nonconvex NLP relaxed subproblems that appear in the multiple nodes of the BB tree search. The filter methodology aims to promote convergence to feasible and optimal solutions.

A set of benchmark problems was used to test the algorithm performance and the results are very promising. We observed that the proposed method consistently located the global optimum of all problems. The quality of the solutions is good and one can state that the BBMCSFilter behaves generally better than the other methods in comparison.

One problematic issue of the proposed BBMCSFilter method is related to the required large number of function evaluations which is a consequence of the set of search directions $\mathcal{D}_{\oplus}$ inside the MCSFilter method. For large dimensional problems, the computational effort in terms of number of function evaluations and consequently CPU time greatly increases with n. This issue is to be deepen in the near future.

Acknowledgments. The authors wish to thank two anonymous referees for their valuable comments and suggestions to improve the paper.

This work has been supported by FCT (Fundação para a Ciência e Tecnologia, Portugal) in the scope of the projects: PEst-OE/MAT/UI0013/2014 and PEst-OE/EEI/UI0319/2014.

References

1. Abramson, M.A., Audet, C., Dennis Jr., J.E.: Filter pattern search algorithms for mixed variable constrained optimization problems. Pac. J. Optim. 3(3), 477–500 (2007)
2. Audet, C., Dennis Jr., J.E.: A pattern search filter method for nonlinear programming without derivatives. SIAM J. Optimiz. 14(4), 980–1010 (2004)
3. Belotti, P., Kirches, C., Leyffer, S., Linderoth, J., Mahajan, A.: Mixed-Integer Nonlinear Optimization. Acta Numer. 22, 1–131 (2013)
4. Belotti, P., Lee, J., Liberti, L., Margot, F., Wächter, A.: Branching and bound tightening techniques for non-convex MINLP. Opt. Methods Softw. 24, 597–634 (2009)
5. Burer, S., Letchford, A.: Non-convex mixed-integer nonlinear programming: a survey. Surveys in Operations Research and Management Science 17, 97–106 (2012)
6. Bussieck, M.R., Drud, A.S., Meeraus, A.: Minlplib – a collection of test models for mixed-integer nonlinear programming. INFORMS J. Comput. 15(1), 1–5 (2003), `http://www.gamsworld.org/minlp/minlplib/minlpstat.htm`
7. Dakin, R.J.: A tree search algorithm for mixed integer programming problems. Comput. J. 8, 250–255 (1965)
8. D'Ambrosio, C., Frangioni, A., Liberti, L., Lodi, A.: A storm of feasibility pumps for nonconvex MINLP. Math. Program., Ser. B 136(2), 375–402 (2012)
9. Fernandes, F.P., Costa, M.F.P., Fernandes, E.M.G.P.: Assessment of a hybrid approach for nonconvex constrained MINLP problems. In: Proceedings of the 2011 International Conference on Computational and Mathematical Methods in Science and Engineering, pp. 484–495 (2011)
10. Fernandes, F.P., Costa, M.F.P., Fernandes, E.M.G.P.: A derivative-free filter driven multistart technique for global optimization. In: Murgante, B., Gervasi, O., Misra, S., Nedjah, N., Rocha, A.M.A.C., Taniar, D., Apduhan, B.O. (eds.) ICCSA 2012, Part III. LNCS, vol. 7335, pp. 103–118. Springer, Heidelberg (2012)
11. Fernandes, F.P., Costa, M.F.P., Fernandes, E.M.G.P.: Multilocal programming: A derivative-free filter multistart algorithm. In: Murgante, B., Misra, S., Carlini, M., Torre, C.M., Nguyen, H.-Q., Taniar, D., Apduhan, B.O., Gervasi, O. (eds.) ICCSA 2013, Part I. LNCS, vol. 7971, pp. 333–346. Springer, Heidelberg (2013)
12. Fernandes, F.P., Costa, M.F.P., Fernandes, E.M.G.P., Rocha, A.M.A.C.: Multistart Hooke and Jeeves filter method for mixed variable optimization. In: Simos, T.E., Psihoyios, G. (eds.) International Conference of Numerical Analysis and Applied Mathematics. AIP Conf. Proc., vol. 1558, pp. 614–617 (2013)
13. Fletcher, R., Leyffer, S.: Nonlinear programming without a penalty function. Math. Program. 91, 239–269 (2002)
14. Gueddar, T., Dua, V.: Approximate multi-parametric programming based B&B algorithm for MINLPs. Comput. Chem. Eng. 42, 288–297 (2012)
15. Hedar, A., Fahim, A.: Filter-based genetic algorithm for mixed variable programming. Numerical Algebra, Control and Optimization 1(1), 99–116 (2011)
16. Kolda, T.G., Lewis, R.M., Torczon, V.: Optimization by Direct Search: New Perspectives on Some Classical and Modern Methods. SIAM Rev. 45(3), 385–482 (2003)

17. Lagaris, I.E., Tsoulos, I.G.: Stopping rules for box-constrained stochastic global optimization. Appl. Math. Comput. 197, 622–632 (2008)
18. Leyffer, S.: Integrating SQP and branch-and-bound for mixed integer nonlinear programming. Comput. Optim. Appl. 18, 295–309 (2001)
19. Liao, T.W.: Two hybrid differential evolution algorithms for engineering design optimization. Appl. Soft Comput. 10, 1188–1199 (2010)
20. Liberti, L., Mladenović, N., Nannicini, G.: A recipe for finding good solutions to MINLPs. Math. Prog. Comp. 3, 349–390 (2011)
21. Ryoo, H.S., Sahinidis, N.V.: Global optimization of nonconvex NLPs and MINLPs with applications in process design. Comput. Chem. Eng. 19(5), 551–566 (1995)
22. Schlüter, M., Egea, J.A., Banga, J.R.: Extended ant colony optimization for non-convex mixed integer nonlinear programming. Comput. Oper. Res. 36, 2217–2229 (2009)
23. Tawarmalani, M., Sahinidis, N.V.: A polyhedral branch-and-cut approach to global optimization. Math. Program. 103(2), 225–249 (2005)
24. Voglis, C., Lagaris, I.E.: Towards "Ideal Multistart". A stochastic approach for locating the minima of a continuous function inside a bounded domain. Appl. Math. Comput. 213, 1404–1415 (2009)

Solving Multilocal Optimization Problems with a Recursive Parallel Search of the Feasible Region

Ana I. Pereira[1,2] and José Rufino[1,3]

[1] Polytechnic Institute of Bragança,
5301-857 Bragança, Portugal
[2] Algoritmi R&D Centre, University of Minho,
Campus de Gualtar, 4710-057 Braga, Portugal
[3] Laboratório de Instrumentação e Física Experimental de Partículas,
Campus de Gualtar, 4710-057 Braga, Portugal
{apereira,rufino}@ipb.pt

Abstract. Stretched Simulated Annealing (SSA) combines simulated annealing with a stretching function technique, in order to solve multilocal programming problems. This work explores an approach to the parallelization of SSA, named PSSA-HeD, based on a recursive heterogeneous decomposition of the feasible region and the dynamic distribution of the resulting subdomains by the processors involved. Three PSSA-HeD variants were implemented and evaluated, with distinct limits on the recursive search depth, offering different levels of numerical and computational efficiency. Numerical results are presented and discussed.

Keywords: Multilocal Optimization, Global Optimization, Parallel Computing.

1 Introduction

A multilocal programming problem aims to find all the local solutions of the minimization problem defined as

$$\min_{x \in X} f(x) \tag{1}$$

where $f : \mathbb{R}^n \to \mathbb{R}$ is a given multimodal objective function and X is a compact set defined by $X = \{x \in \mathbb{R}^n : a_i \leq x_i \leq b_i, i = 1, ..., n\}$.

So, the purpose is to find all local solutions $x^* \in X$ such that

$$\forall x \in V_\epsilon(x^*),\ f(x^*) \leq f(x), \tag{2}$$

for a positive value ϵ.

These problems appear in practical situations like ride comfort optimization [2], Chemical Engineering (process synthesis, design and control [3]), and reduction methods for solving semi-infinite programming problems [11, 19].

The most common methods for solving multilocal optimization problems are based on evolutionary algorithms, such as genetic [1] and particle swarm [14] algorithms. Additional contributions may be found in [10, 21–23].

B. Murgante et al. (Eds.): ICCSA 2014, Part II, LNCS 8580, pp. 154–168, 2014.

Stretched Simulated Annealing (SSA) was also proposed [15–17] as a method to solve multilocal programming problems, combining simulated annealing with stretching function technique, to identify the local minimizers.

In previous work [20], a first approach to the parallelization of SSA was introduced (PSSA), based on a decomposition of the search domain (feasible region) in a fixed number of homogeneous subdomains (*homogeneous decomposition*), and a deterministic assignment of those subdomains among the processors involved (*static distribution*). This previous approach, hereafter named PSSA-HoS, proved to be an effective way to increase the number of optima found.

This paper explores a novel approach, PSSA-HeD, that generates a variable number of heterogeneous subdomains of the initial search domain (*heterogeneous decomposition*) which are then assigned, on-demand, to the working processors (*dynamic distribution*). The aim of this new approach is to further increase the numerical performance of the previously developed PSSA-HoS approach.

The paper is organized as follows. Section 2 introduces the basic ideas behind Stretched Simulated Annealing (SSA). Section 3 is devoted to the new Parallel Stretched Simulated Annealing approach, PSSA-HeD. Section 4 describes criteria to filter the optima candidate set found by PSSA-HeD. Section 5 presents some numerical results. Finally, Section 6 concludes and defines future work.

2 Stretched Simulated Annealing

The Stretched Simulated Annealing (SSA) method solves a sequence of global optimization problems in order to compute the local solutions of the minimization problem (1) that satisfy condition (2). The objective function of each global optimization problem comes by applying a stretching function technique [13].

Let x_j^* be a particular solution. The mathematical formulation of the global optimization problem is as follows:

$$\min_{a \le x \le b} \Phi_l(x) \equiv \begin{cases} \hat{\phi}(x) \text{ if } x \in V_{\varepsilon^j}(x_j^*),\, j \in \{1, \ldots, N\} \\ f(x) \text{ otherwise} \end{cases} \tag{3}$$

where $V_{\varepsilon^j}(x_j^*)$ represents the neighborhood of the solution x_j^* with a ray ε^j.

The $\hat{\phi}(x)$ function is defined as

$$\hat{\phi}(x) = \bar{\phi}(x) + \frac{\delta_2[\text{sign}(f(x) - f(x_j^*)) + 1]}{2\tanh(\kappa(\bar{\phi}(x) - \bar{\phi}(x_j^*))} \tag{4}$$

and

$$\bar{\phi}(x) = f(x) + \frac{\delta_1}{2}\|x - x_j^*\|[\text{sign}(f(x) - f(x_j^*)) + 1] \tag{5}$$

where δ_1, δ_2 and κ are positive constants and N is the number of minimizers already detected.

To solve the global optimization problems (3) the simulated annealing (SA) method is used [8]. The SSA algorithm stops when no new optimum is identified after l consecutive runs. For more details see [15, 18].

3 Parallel Stretched Simulated Annealing (PSSA)

3.1 General Parallel Approach

The search for optima of nonlinear optimization functions through the SSA method is easily parallelizable. SSA searches for solutions in a given feasible region (search domain) by following a stochastic algorithm. It is possible to improve the number of optima found using SSA by increasing its parameter l, but that comes at the cost of higher execution time. An alternative to ameliorate the hit rate of SSA is to keep l constant and split the initial search domain in several subdomains to which SSA will be applied independently, whether serially (one subdomain at a time) or in parallel (several subdomains at the same time).

With as much processors/CPU-cores available as subdomains, each core could run a single SSA instance, dedicated to a specific subdomain. Moreover, the time that would take to search all subdomains simultaneously (in parallel) would be approximately the same that would take to search the initial domain[1], once running SSA on one subdomain has no data dependencies on any other subdomain. On the other hand, if the decomposition of the initial domain is too fine with relation to the number of available CPU-cores, that would lead to the serial processing of several subdomains by each SSA instance, which would still offer better performance than a purely sequential search of all subdomains.

In short, the general approach followed for the parallelization of SSA (PSSA) is based on a Data Decomposition of the problem domain, coupled with a Single Program Multiple Data (SPMD) execution model (i.e., having several instances of the same SSA implementation, dealing with different subdomains).

3.2 Implementation Details

The base SSA code was originally developed in ANSI C [9] and so was the supplemental code necessary for the parallel SSA (PSSA) variants.

In order to allow transparent execution, both on multi-core shared memory systems and on distributed memory HPC clusters, PSSA was built on the *message passing* paradigm, in the framework of the Message Passing Interface (MPI) specification [12]. More specifically, PSSA was developed in a Linux environment, on top of MPICH2 [7], a high-performance portable MPI implementation.

In this context, all PSSA variants operate in a *master-slaves* configuration: *slave* MPI processes apply SSA to problem subdomains; a *master* process performs pre-processing, coordination and post-processing; if c CPU-cores are enrolled, one core is reserved for the *master* and the remaining $c-1$ cores are for the *slaves*, with one *slave* per core (this is the MPI process mapping that most effectively exploits the available parallelism of our experimental environment).

The overall number of *slaves* is definable independently of the overall number of subdomains. This is both necessary and convenient: if the number of *slaves*

[1] Ignoring the setup time needed to spawn all instances and to post-process results. This time, however, may be counterbalanced by having the SSA instances finishing sooner, once subdomains are smaller search regions than the initial search domain.

were to always match the number of subdomains then, with fine-grain decompositions, there would be too much slaves for the available CPU-cores, preventing an efficient execution of PSSA. Thus, by separating the definition of the number of *slaves* from the number of subdomains, each number may be tuned at will.

The way in which the initial problem domain is decomposed and *slaves* get subdomains assigned depends on the PSSA variant: the *master* may be the one that partitions the problem domain and assigns subdomains to *slaves*, like in the PSSA-HeD approach explored in this paper; or *slaves* may conduct themselves such tasks autonomously, like in the PSSA-HoS approach [20]; in all cases the *master* is responsible for a final post-processing phase in which all optima candidates found by *slaves* are filtered using the criteria described in Section 4.

The final optima filtering should be conducted efficiently: depending on the specific optimization problem, it may have to cope with a number of candidates in the order of thousands or even millions, that must be stored in efficient data containers. Because ANSI C has no built-in container data types (e.g., lists, sets, etc.), an external implementation is necessary. The choice was to use the GLIBC `tsearch` built-in function family [4], that provides a very efficient implementation of balanced binary trees (more precisely, of Red-Black-Trees [5]).

All PSSA variants save (if requested) the optima candidates in CSV raw files. These raw files may be later re-filtered, using the same criteria or newest/ refined ones, thus avoiding the need to repeat (possibly lengthy) PSSA executions.

3.3 Heterogeneous Decomposition, Dynamic Distribution (PSSA-HeD)

Initial Decomposition. The search domain (or feasible region) of problem (1) is an n-dimensional interval, I, defined by the cartesian product of n intervals, one per each problem dimension: $I = I_1 \times I_2 \times ... \times I_n$. The PSSA-HeD approach starts by performing an *homogeneous decomposition* of these initial intervals.

Each initial interval I_i ($i = 1, 2, ..., n$) is subdivided in 2^m subintervals, such that each subinterval has the same relative width or granularity g, as given by

$$g = \frac{1}{2^m}, \text{ with } m \in \mathbb{N}_0 \tag{6}$$

A subdomain is thus a particular combination of subintervals (with one subinterval per problem dimension). The overall number of initial subdomains, $s_{initial}$, with granularity g, that is generated for n dimension problems is given by

$$s_{initial} = \left(\frac{1}{g}\right)^n = 2^{m \times n} \tag{7}$$

For instance, if the search space of a two dimensional ($n = 2$) function $f(x, y)$ is $I = [-10.0, 10.0] \times [-5.0, 5.0]$, the *homogeneous decomposition* of this initial domain with $m = 2$ (or $g = 0.5$) originates $s_{initial} = 4$ subdomains: $[-10.0, 0.0] \times [-5.0, 0.0]$, $[-10.0, 0.0] \times [0.0, 5.0]$, $[0.0, 10.0] \times [-5.0, 0.0]$ and $[0.0, 10.0] \times [0.0, 5.0]$.

Table 1 shows the number of initial subdomains ($s_{initial}$) as a function of the number of dimensions (n) and the granularities (g) used during this study.

Table 1. Decomposition granularity and number of initial subdomains.

$n = 2$			$n = 3$		
m	g	$s_{initial}$	m	g	$s_{initial}$
0	1.0	1	0	1.0	1
1	0.5	4	1	0.5	8
2	0.25	16	2	0.25	64
3	0.125	64	3	0.125	512
4	0.0625	256	4	0.0625	4096
5	0.03125	1024	5	0.03125	32768

Recursive Decomposition. With PSSA-HeD, SSA is first applied to the initial (sub)domain(s), and then, if optima are eventually found, child subdomains will be generated around them. Because an optimum may be located anywhere in its hosting (sub)domain, the child subdomains will not only be smaller than their parent (sub)domain, but will also typically vary in width, thus leading to an *heterogeneous decomposition*. The new child subdomains will, in turn, be searched using SSA and, if optima are found, more subdomains will be generated, until a stop criteria is met. As such, the decomposition is both dynamic and recursive, and the generated subdomains may be seen as part of an expanding search tree where each node/leaf subdomain refines its ancestor.

The stop criteria for this recursive behavior is as follows:

1) if none real optimum is found in a subdomain, then no child subdomains will be generated;
2) otherwise, such generation will take place, but only if the current branch of the search tree has not yet achieved a maximum depth or height $h \in \mathbb{N}$;
3) all subintervals of a new subdomain must have a minimum distance of μ from their parent optimum, or the new subdomain will be ignored.

With regard to the height h, a generic value of $h \in N$, means that a search branch may progress as far as $h - 1$ levels bellow the root level. Thus, $h = 1$ means that the search will be confined to the root of the search tree (in which case PSSA-HeD would be no different than PSSA-HoS). When $h = \infty$ such means that only criteria 1) and 3) are applied.

In PSSA-HeD, the initial set of homogeneous subdomains is the root of a search tree. If the root is to be defined as the full original domain of the optimization function, such is simply achieved with $g = 1.0$ (or $m = 0$). The purpose of setting $g < 1.0$ (or $m > 0$), thus starting the search with a grid of homogeneous subdomains, is to increase the probability of finding already several optima in the 1st level of the search tree and thus trigger the generation of many additional new subdomains. Otherwise, with $g = 1.0$, the number of optima found will typically be very limited and their descendant subdomains will be too few and too large to trigger a sustained recursive search.

Subdomains are assigned to the MPI *slave* processes in PSSA-HeD through a *dynamic distribution* in which the *master* process pushes subdomains to the *slaves*. This is advantageous because it inhibits the premature termination of the *slaves*: there may be times when all available subdomains are being processed by *slaves*; in this scenario, if an idle *slave* asked the *master* for a subdomain, it would receive none; but that would not mean that the *slave* could terminate once, in the near future, more new unprocessed subdomains might become available, as a byproduct of the current working *slaves*; thus, it is better for the *master* to push subdomains to the *slaves* (when they become available), than having the *slaves* pulling them from the *master* (at the risk of none being available).

In order to achieve the above behavior, the *master* manages a work-queue with all subdomains yet to process, and a slave-status-array with the current status (idle/busy) of each *slave*. Initially, the work-queue is populated with the starting grid of homogeneous subdomains (or with the single full domain, if such is the case), and all *slaves* are marked as idle in the slave-status-array.

The distribution of subdomains by the *slaves* is then just a matter of iterating through the slave-status-array and, for each idle *slave*, dequeue a subdomain from the work-queue, send it to the *slave*, and mark the *slave* as busy. During this iteration, the *master* may find all *slaves* to be busy, in which case nothing is removed from the work-queue; it may also find the work-queue to be empty, in which case nothing is assignable to the possible idle *slaves*; if the work-queue is empty and if all *slaves* are idle, such means the overall recursive search ended.

After a subdomain distribution round, and assuming the overall search process hasn't yet finished, the *master* will block, waiting for a message from some *slave*; that message will be empty if the *slave* found no optima in its assigned subdomain; otherwise, it will carry a set of optima found by the *slave* (and already filtered by him); in the later case, the optima are added to a global set of solutions that is being assembled by the *master* (based on all the contributions of the *slaves*); the optima are also used to generate new-subdomains that will be added to the work-queue; in any case, the *slave* is marked as idle in the slave-status-array; the *master* then performs the next distribution round.

4 Filtering Criteria

All PSSA variants produce false minima - some examples are the points in the limits of the subdomains generated. Therefore, filtering criteria are needed to eliminate such false minima. This section presents three criteria, to be used in sequence. In PSSA-HeD they are applied in the *slaves*, right after running SSA in a subdomain; thus, the *master* only receives sets of validated optima.

Criterion 1

At a given moment, there are a total of s subdomains (with $s \geq s_{initial}$). Each subdomain v is defined by n intervals with left and right limits a_i^v and b_i^v, respectively, for $i = 1, ..., n$. Consider x^v (with coordinates x_i^v, for $i = 1, ..., n$) a minimum found by at subdomain v. Define the vector d with components d_i as

$$d_i = \min\left\{\left|a_i^v - x_i^v\right|, \left|x_i^v - b_i^v\right|\right\}, \text{ with } i = 1, ..., n$$

and define Δ_1 as

$$\Delta_1 = \sqrt{\sum_{i=1}^{n} d_i^2}, \text{ with } i = 1, ..., n.$$

Criterion 1 is then defined as follows:

1. Consider ϵ_1 a positive constant.
2. If $\Delta_1 < \epsilon_1$ then x^v is not a candidate to a minimum of problem (1).

The situation targeted by this criterion is the one in which a subdomain v doesn't have minimum values except in its interval limits.

Criterion 2

Consider the unit vector, 1_i, with all components null except the component i with unit value. Consider the vector e, with component e_i defined as

$$e_i = \frac{f(x^v + \delta 1_i) - f(x^v)}{\delta}, \text{ for } i = 1, ..., n$$

with δ a small positive value. Define also Δ_2 as

$$\Delta_2 = \sqrt{\sum_{i=1}^{n} e_i^2}.$$

Criterion 2 is thus defined as:

1. Consider x^v that satisfy the Criterion 1.
2. If $\Delta_2 > \epsilon_2$ then x^v is not a candidate to a minimum of problem (1).

Criterion 3

Consider the set $X^* = \{x^j, j = 1..., n^*\}$ of all solutions that satisfy the Criterion 2 and let n^* be the cardinality of the set X^*.

Criterion 3 is defined as follows:

1. Consider $x^i \in X^*$.
2. The point x^i is a possible minimum value of problem (1) if

$$\left\| x^i - x^j \right\| > \epsilon_3, \text{ for all } j = 1, ..., n^* \text{ and } j \neq i$$

After applying the three criteria it is obtained the optima set that will be presented in the next section.

5 Numerical Results

5.1 Experimental Setup

PSSA-HeD was evaluated in a small commodity cluster of 4 nodes (with one Intel Q9650 3.0GHz quad-core CPU per each node), running Linux ROCKS version 5.4, with the Gnu C Compiler (GCC) version 4.1.2 and MPICH2 version 1.4.

All PSSA executions spawned 16 MPI processes (1 *master* and 15 *slaves*, one MPI process per cluster core), even if there were a surplus of unused *slaves* in certain scenarios. The MPICH2 "machinefile" used was designed to place the first 4 MPI processes (the *master* and the first 3 *slaves*) in a single node and scatter (alternately) the remaining 12 *slaves* across the other 3 nodes. This particular configuration maximizes performance both for scenarios with very few subdomains (mostly handled by the *slaves* of the 1st node without network exchanges), and with lots of subsubdomains (requiring *slaves* from all the nodes, in which case network exchanges benefit from the dispersion of their endpoints).

Five problems were evaluated: Ackley, Branin, Griewank, Michalewicz and Shubert [6]. All have more than one local solution, thus suitable to a parallel search of the solutions. Important parameters used were $\delta = 5.0$ and $l = 5$ for SSA, and $\mu = 0.001$, $\epsilon_1 = 10^{-4}$, $\epsilon_2 = 10^{-3}$ and $\epsilon_3 = 10^{-2}$ for PSSA-HeD.

Moreover, in order to know the performance gains introduced by the PSSA-HeD parallel approach, it was also necessary to conduct the optima search by executing SSA in sequence (one subdomain at a time). The set of subdomains searched serially is not exactly the same as the one searched in parallel by PSSA-HeD, once SSA is a stochastic algorithm. However, the overall number of subdomains searched (s), and the overall number of optima found (n^*), are similar for the two approaches, thus making SSA a valid baseline to evaluate PSSA-HeD.

5.2 Experimental Results

The results of the evaluation are presented in Tables 2 to 4, for different values of the recursive search depth: $h = 1$, $h = 2$ and $h = \infty$. In the following tables,

- g is the granularity of the initial decomposition,
- s is the overall number of subdomains searched with PSSA-HeD,
- n^* is the overall number of optima found with PSSA-HeD,
- $T_{PSSA-HeD}$ is the parallel search time (in seconds) with PSSA-HeD,
- T_{SSA} is the sequential search time (in seconds) with SSA,
- $S = T_{SSA}/T_{PSSA-HeD}$ is the speedup of PSSA-HeD against SSA,
- $r = n^*/T_{PSSA-HeD}$ is the search rate (optima/second) of PSSA-HeD.

The tables show that decreasing the decomposition granularity (g) yields, in general, a higher number of optima found (n^*). The only exception is the Branin function, with only 3 optima, that are all found in the original feasible region (when $g = 1.0$), and so no additional optima will ever be found by searching with $g < 1.0$. Of course, the primary reason for having more optima being found with smaller granularities is that the number of search subdomains generated

Table 2. Experimental results with $h = 1$

Problem	g	s	n^*	$T_{PSSA-HeD}$	T_{SSA}	S	r
Ackley	1.0000	1	1	1.09	1.09	1.00	0.92
	0.5000	4	9	7.57	20.16	2.66	1.19
	0.2500	16	57	8.23	88.98	10.81	6.93
	0.1250	64	233	29.90	363.70	12.16	7.79
	0.0625	256	676	98.70	1372.94	13.91	6.85
	0.0313	1024	1425	333.57	4975.80	14.92	4.27
Branin	1.0000	1	3	3.66	3.67	1.00	0.82
	0.5000	4	3	4.15	11.87	2.86	0.72
	0.2500	16	3	5.54	37.75	6.81	0.54
	0.1250	64	3	10.30	119.73	11.62	0.29
	0.0625	256	3	27.45	347.57	12.66	0.11
	0.0313	1024	3	85.32	1256.01	14.72	0.04
Griewank	1.0000	1	18	5.74	5.72	1.00	3.14
	0.5000	4	23	5.82	17.15	2.95	3.95
	0.2500	16	34	8.93	79.96	8.95	3.81
	0.1250	64	69	21.46	251.65	11.73	3.22
	0.0625	256	187	58.77	836.70	14.24	3.18
	0.0313	1024	576	194.69	2819.53	14.48	2.96
Michalewicz	1.0000	1	1	4.68	4.68	1.00	0.21
	0.5000	4	13	6.46	11.58	1.79	2.01
	0.2500	16	92	5.81	33.89	5.83	15.83
	0.1250	64	123	8.87	79.04	8.91	13.87
	0.0625	256	163	22.07	291.38	13.20	7.39
	0.0313	1024	200	87.94	1207.94	13.74	2.27
Shubert	1.0000	1	18	14.03	14.02	1.00	1.28
	0.5000	4	18	10.53	38.00	3.61	1.71
	0.2500	16	22	10.12	121.82	12.04	2.17
	0.1250	64	84	38.49	489.16	12.70	2.18
	0.0625	256	379	157.71	2276.67	14.38	2.40
	0.0313	1024	707	572.49	8657.49	15.12	1.23

(s) also becomes larger with smaller granularities. Moreover, this growth on the number of optima found, and the number of subdomains, is amplified when the maximum search depth, h, increases. Figures 1 and 2 allow to compare, for each problem (except Branin), the values of n^* attained with different values h.

A conclusion inferred from Figures 1 and 2 is that, in general, increasing the search depth (h) finds more optima, although going from $h = 2$ to $h = \infty$ often leads to marginal gains: for the Ackley and Shubert functions, the gains are modest, and for the Michalewicz function there isn't, in fact, any significant benefit; the Griewank function, in turn, exhibits clear gains. These gains vary, depending on the granularity: for the Ackley function, $g = 0.25$ translates in deeper searches being more advantageous; for the Shubert function such happens when $g = 0.125$; for the Griewank function, there are gains with all granularities.

Table 3. Numerical results with with $h = 2$

Problem	g	s	n^*	$T_{PSSA-HeD}$	T_{SSA}	S	r
Ackley	1.0000	5	1	6.92	19.46	2.81	0.14
	0.5000	30	15	20.31	160.26	7.89	0.74
	0.2500	202	99	89.29	1133.92	12.70	1.11
	0.1250	714	294	282.64	4392.66	15.54	1.04
	0.0625	2151	739	811.51	12799.00	15.77	0.91
	0.0313	5725	1427	2011.74	30152.16	14.99	0.71
Branin	1.0000	17	3	6.53	23.90	3.66	0.46
	0.5000	16	3	5.93	22.71	3.83	0.51
	0.2500	28	3	5.54	47.06	8.49	0.54
	0.1250	76	3	10.05	128.13	12.75	0.30
	0.0625	268	3	27.49	365.16	13.28	0.11
	0.0313	1036	3	88.51	1269.57	14.34	0.03
Griewank	1.0000	78	24	37.91	385.29	10.16	0.63
	0.5000	76	35	32.72	375.71	11.48	1.07
	0.2500	116	50	41.95	526.15	12.54	1.19
	0.1250	288	106	88.03	1231.37	13.99	1.20
	0.0625	955	271	234.04	3484.63	14.89	1.16
	0.0313	3249	792	685.68	10012.68	14.60	1.16
Michalewicz	1.0000	5	1	12.55	19.90	1.59	0.08
	0.5000	17	13	10.19	26.42	2.59	1.28
	0.2500	164	91	12.64	185.29	14.66	7.20
	0.1250	260	123	18.76	233.52	12.45	6.56
	0.0625	545	163	38.92	488.48	12.55	4.19
	0.0313	1465	201	102.53	1442.62	14.07	1.96
Shubert	1.0000	50	18	43.1	396.75	9.21	0.42
	0.5000	68	18	47.97	559.17	11.66	0.38
	0.2500	110	31	64.98	806.74	12.42	0.48
	0.1250	434	108	241.97	3529.14	14.59	0.45
	0.0625	1803	467	1031.21	15063.96	14.61	0.45
	0.0313	3893	715	2183.50	32507.96	14.89	0.33

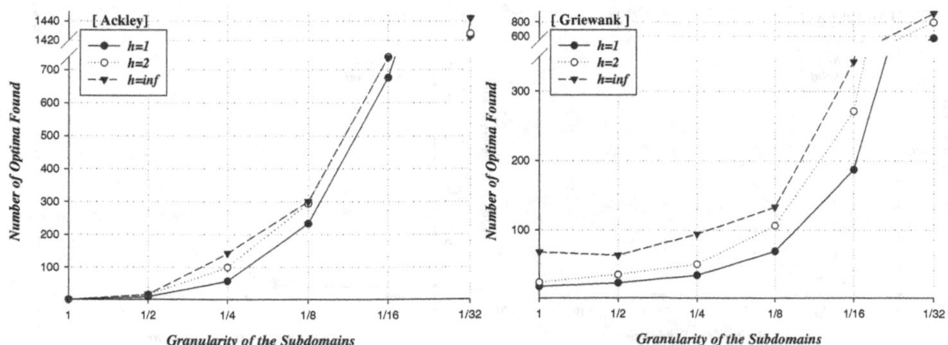

Fig. 1. Number of optima found for Ackley and Griewank

Table 4. Experimental results with with $h = \infty$

Problem	g	s	n^*	$T_{PSSA-HeD}$	T_{SSA}	S	r
Ackley	1.0000	5	1	6.95	19.02	2.74	0.14
	0.5000	36	16	18.15	160.06	8.82	0.88
	0.2500	390	142	165.18	2545.51	15.41	0.86
	0.1250	879	300	345.86	5449.28	15.76	0.87
	0.0625	2276	737	853.26	12721.95	14.91	0.86
	0.0313	5786	1443	2043.42	30329.25	14.84	0.71
Branin	1.0000	17	3	3.00	23.76	7.92	1.00
	0.5000	16	3	5.94	22.82	3.84	0.51
	0.2500	28	3	5.54	47.06	8.49	0.54
	0.1250	76	3	10.33	128.90	12.48	0.29
	0.0625	268	3	26.46	356.15	13.46	0.11
	0.0313	1036	3	88.24	1264.63	14.33	0.03
Griewank	1.0000	271	68	112.66	1025.34	9.10	0.60
	0.5000	220	63	108.45	1719.51	15.86	0.58
	0.2500	333	94	131.89	1866.75	14.15	0.71
	0.1250	526	133	175.29	2942.12	16.78	0.76
	0.0625	1533	342	425.85	6022.39	14.14	0.80
	0.0313	4529	919	1077.62	16120.09	14.96	0.85
Michalewicz	1.0000	5	1	9.6	20.07	2.09	0.10
	0.5000	17	13	10.01	26.37	2.63	1.30
	0.2500	164	91	12.12	186.93	15.42	7.51
	0.1250	260	123	22.14	232.62	10.51	5.56
	0.0625	550	166	36.57	488.42	13.36	4.54
	0.0313	1467	200	101.29	1444.65	14.26	1.97
Shubert	1.0000	50	18	47.32	425.21	8.99	0.38
	0.5000	72	19	49.44	591.67	11.97	0.38
	0.2500	150	32	89.68	1186.90	13.23	0.36
	0.1250	573	124	334.28	4614.55	13.80	0.37
	0.0625	2099	458	1215.03	18381.17	15.13	0.38
	0.0313	3915	716	2205.85	32726.42	14.84	0.32

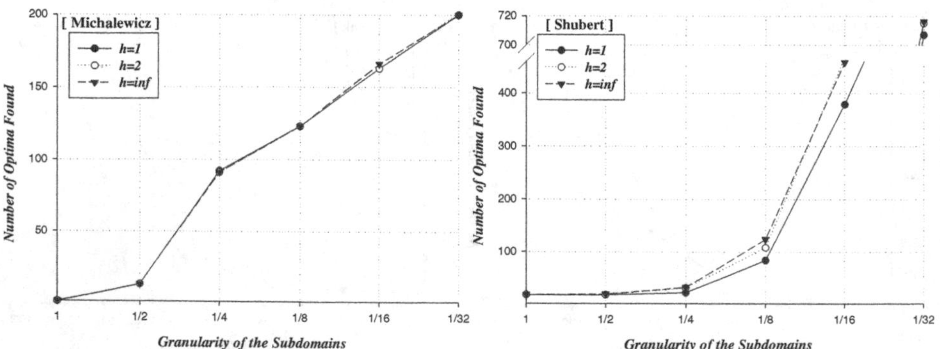

Fig. 2. Number of optima found for Michalewicz and Shubert

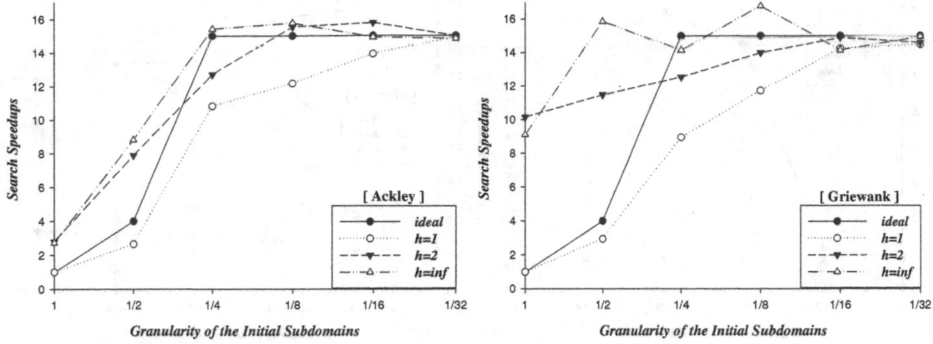

Fig. 3. Search speedups for Ackley and Griewank

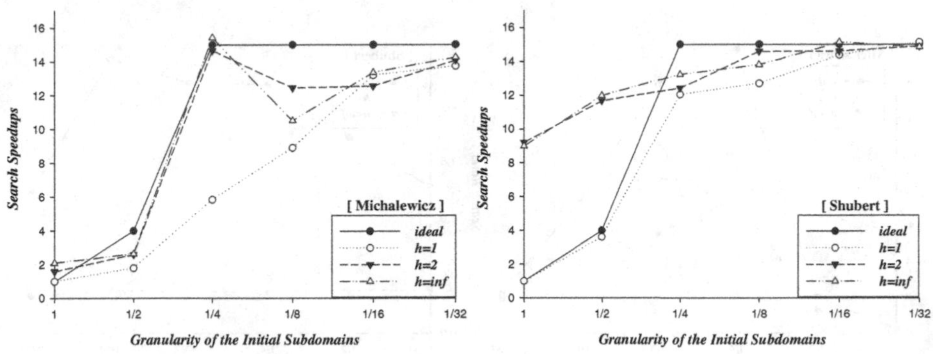

Fig. 4. Search speedups for Michalewicz and Shubert

With regard to the computational efficiency, the speedups (S) provided by PSSA-HeD against SSA executed serially over the same subdomain set, are not far from ideal values, denoted by S_{ideal}. This may be verified in the graphics from Figures 3 to 4 (again, the Branin function is omitted, for reasons already explained). The way in which the ideal speedup S_{ideal} is established is as follows:

- when $h = 1$, the number of subdomains is static, that is, $s = s_{initial}$ (as defined in Table 1); it becomes possible to define, *a priori*, the maximum expected speedup: with 15 MPI *slaves*, S_{ideal} will match the number of *slaves* actively engaged in optima search; if $s = 1$, then $S_{ideal} = 1$ once only 1 slave will be necessary (the other 14 will remain idle); if $s = 4$, then $S_{ideal} = 4$ once only 4 slaves will be needed[2]; when $s \geq 15$, all *slaves* will be necessary and the maximum theoretical speedup will be $S_{ideal} = 15$;

[2] The same rationale would be valid with $s = 8$, but the tested functions are all 2-dimensional and $s = 8$ emerges only with 3-dimensional problems – see Table 1.

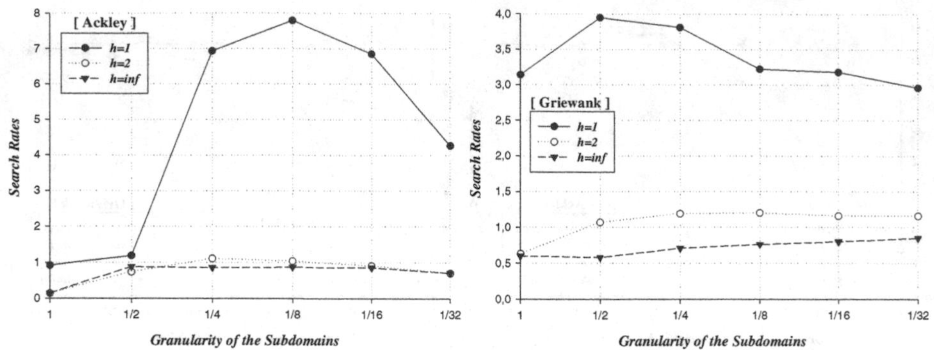

Fig. 5. Search rates for Ackley and Griewank (optima/second)

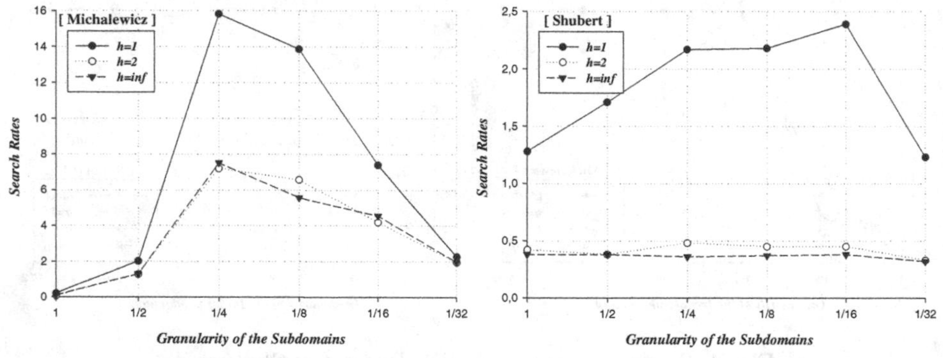

Fig. 6. Search rates for Michalewicz and Shubert (optima/second)

– if $h > 0$, then $s \geq s_{initial}$, and so the values 1, 4 and 15 of S_{ideal} with $h = 1$ are no longer upper bounds for the real speedup; instead, they become a lower bound for the ideal speedup (still, useful as reference for the real speedup).

The proximity between the measured (real) speedups and the theoretical ones (specially with small granularities or, conversely, with many subdomains) proves the merit, performance-wise, of the parallelization approach followed by PSSA-HeD. From a numerical point of view, the main advantage of PSSD-HeD was also already discussed: enabling the efficient finding of many more optima. However, the final decision on which granularity (g) and which search depth (h) to choose depends on the desired balance between i) number of optima found and ii) search time. In this regard, one way to combine both metrics into a single one is through the search rate r (optima/second), the last metric shown in the tables. Figures 5 to 6 present the graphics of r for all functions except Branin.

The search rate graphics support the following conclusions: a) $h = 1$ is the search depth limit that ensures the higher search rates, followed by $h = 2$ and then $h = \infty$; b) for each value of h, each function maximizes the search rate with a different granularity (*e.g.*, with $h = 1$, Ackley maximizes the rate with $g = 1/8$, Griewank maximizes with $g = 1/2$, Michalewicz maximizes with $g = 1/4$ and Shubert maximizes with $g = 1/16$); it is thus very difficult (if not impossible) to define a common granularity, that maximizes the search rate for all functions.

5.3 Comparison with PSSA-HoS

As initially stated, the new PSSA-HeD approach builds on a first attempt to parallelize SSA, then named PSSA [20]. This first approach, renamed as PSSA-HoS in the context of this paper, is based on a homogeneous decomposition of the search domain, a decomposition that is in fact identical to the one used in PSSA-HeD when $h = 1$; however, while PSSA-HoS performs a static distribution of the initial (and only) subdomain set by the MPI *slaves*, PSSA-HeD always performs a dynamic distribution irregardless of the parameter h. Although no detailed results are here supplied, PSSA-HoS was also executed under the same experimental conditions in which PSSA-HeD was evaluated. The conclusion was that the dynamic distribution performed by PSSA-HeD achieves better load balancing, providing to PSSA-HeD (with $h = 1$) marginally better search times ($\approx$10%, on average) than PSSA-HoS (the number of optima found is similar).

6 Conclusions and Future Work

This work expands previous investigation on the parallelization of the SSA stochastic algorithm, aimed at finding all local solutions of multimodal objective function problems. The computation experiments conducted showed that the new PSSE-HeD approach is capable of locating a large number of local optima, improving on the numerical efficiency of the previous PSSA-HoS approach. Moreover, PSSE-HeD may be tunned to achieve the desired compromise between search time and number of optima found. The speedups achieved by the new parallel code are also close to the experimental testbed ideal levels.

In the future, we intend to further refine PSSA and apply it to solve more complex constrained multilocal optimization problems.

Acknowledgments. This work was been supported by FCT (Fundação para a Ciência e Tecnologia) in the scope of the project PEst-OE/EEI/UI0319/2014.

References

1. Chelouah, R., Siarry, P.: A continuous genetic algorithm designed for the global optimization of multimodal functions. Journal of Heuristics 6, 191–213 (2000)
2. Eriksson, P., Arora, J.: A comparison of global optimization algorithms applied to a ride comfort optimization problem. Structural and Multidisciplinary Optimization 24, 157–167 (2002)

3. Floudas, C.: Recent advances in global optimization for process synthesis, design and control: enclosure of all solutions. Computers and Chemical Engineering, 963–973 (1999)
4. The GNU C Library, `http://www.gnu.org/software/libc/manual/`.
5. Guibas, L.J., Sedgewick, R.: A Dichromatic Framework for Balanced Trees. In: Proceedings of the 19th Annual Symposium on Foundations of Computer Science, pp. 8–21 (1978)
6. Hedar, A.R.: Global Optimization Test Problems, `http://www-optima.amp.i.kyoto-u.ac.jp/member/student/hedar/Hedar_files/TestGO.htm`
7. High-Performance Portable MPI, `http://www.mpich.org/`
8. Ingber, L.: Very fast simulated re-annealing. Mathematical and Computer Modelling 12, 967–973 (1989)
9. Kernighan, Ritchie, D.M.: The C Programming Language, 2nd edn. Prentice Hall, Englewood Cliffs (1988) ISBN 0-13-110362-8
10. Kiseleva, E., Stepanchuk, T.: On the efficiency of a global non-differentiable optimization algorithm based on the method of optimal set partitioning. Journal of Global Optimization 25, 209–235 (2003)
11. León, T., Sanmatías, S., Vercher, H.: A multi-local optimization algorithm. Top 6(1), 1–18 (1998)
12. Message Passing Interface Forum, `http://www.mpi-forum.org/`
13. Parsopoulos, K., Plagianakos, V., Magoulas, G., Vrahatis, M.: Objective function stretching to alleviate convergence to local minima. Nonlinear Analysis 47, 3419–3424 (2001)
14. Parsopoulos, K., Vrahatis, M.: Recent approaches to global optimization problems through particle swarm optimization. Natural Computing 1, 235–306 (2002)
15. Pereira, A.I., Fernandes, E.M.G.P.: A reduction method for semi-infinite programming by means of a global stochastic approach. Optimization 58, 713–726 (2009)
16. Pereira, A.I., Fernandes, E.M.G.P.: Constrained Multi-global Optimization using a Penalty Stretched Simulated Annealing Framework. In: Numerical Analysis and Applied Mathematics, AIP Conference Proceedings, vol. 1168, pp. 1354–1357 (2009)
17. Pereira, A.I., Fernandes, E.M.G.P.: Comparative Study of Penalty Simulated Annealing Methods for Multiglobal Programming. In: 2nd International Conference on Engineering Optimization (2010)
18. Pereira, A.I., Ferreira, O., Pinho, S.P., Fernandes, E.M.G.P.: Multilocal Programming and Applications. In: Zelinka, I., Snasel, V., Abraham, A. (eds.) Handbook of Optimization. Intelligent Systems Series, pp. 157–186. Springer (2013)
19. Price, C.J., Coope, I.D.: Numerical experiments in semi-infinite programming. Computational Optimization and Applications 6, 169–189 (1996)
20. Ribeiro, T., Rufino, J., Pereira, A.I.: PSSA: Parallel Stretched Simulated Annealing. In: Numerical Analysis and Applied Mathematics, AIP Conference Proceedings, vol. 1389, pp. 783–786 (2011)
21. Salhi, S., Queen, N.: A Hybrid Algorithm for Identifying Global and Local Minima When Optimizing Functions with Many Minima. European Journal of Operations Research 155, 51–67 (2004)
22. Tsoulos, I., Lagaris, I.: Gradient-controlled, typical-distance clustering for global optimization (2004), `http://www.optimization.org`
23. Tu, W., Mayne, R.: Studies of multi-start clustering for global optimization. International Journal Numerical Methods in Engineering 53, 2239–2252 (2002)

Stiction Detection and Quantification as an Application of Optimization

Ana S.R. Brásio[1,2], Andrey Romanenko[2], and Natércia C.P. Fernandes[1]

[1] CIEPQPF, Department of Chemical Engineering
University of Coimbra, Portugal
{anabrasio,natercia}@eq.uc.pt
[2] Ciengis, SA, Coimbra, Portugal
andrey.romanenko@ciengis.com

Abstract. Stiction is a major problematic phenomenon affecting industrial control valves. An approach for detection and quantification of valve stiction using an one-stage optimization technique is proposed. A Hammerstein Model that comprises a complete stiction model and a process model is identified from industrial process data. Some difficulties in the identification approach are pointed out and strategies to overcome them are suggested, namely the smoothing of discontinuity points. A simulation study demonstrates the application of the proposed technique.

Keywords: system identification, non-linear optimization, smoothing discontinuous functions, stiction detection and quantification.

1 Introduction

Stiction is one of the long-standing control valve problems in the process industry causing oscillations and, consequently, losses of productivity. Therefore, it is important to understand this phenomenon for its early detection and separation from other oscillation causes.

Because stiction is one of the major causes for oscillations in the controlled variable, numerous techniques for its detection and quantification in linear control loops have been suggested. Recently, stiction detection and quantification developments were proposed by means of system identification using the Hammerstein Model [1–11]. System identification deals with the building of mathematical models to describe dynamical systems based on observed data [12]. The use of optimization techniques for system identification has become more common, motivated by successful applications in different areas [13].

The Hammerstein Model is a model that consists of a static non-linear element in series with a linear dynamic part [14]. In the context of an industrial control loop, the non-linear element represents the sticky valve while the linear part models the process dynamics, as shown in Fig. 1. One of the identification strategies for such models is the decoupling of the non-linear and linear parts [2, 5, 11], with the process model parameters determined in the first place followed by the determination of the stiction model parameters.

B. Murgante et al. (Eds.): ICCSA 2014, Part II, LNCS 8580, pp. 169–179, 2014.

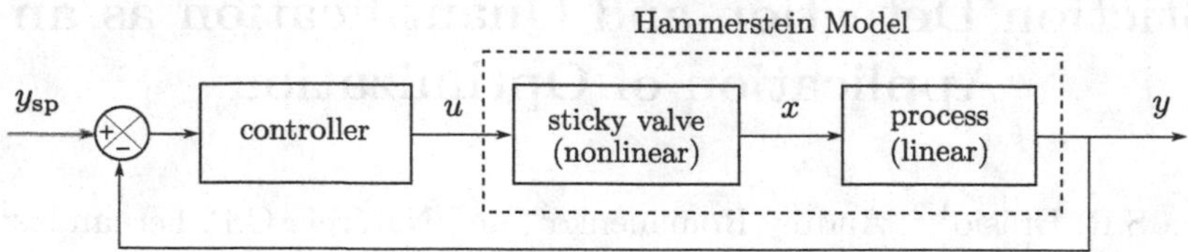

Fig. 1. Industrial control loop representing the Hammerstein Model, where y_{sp} is the variable setpoint, u is the controller output, x is the real valve position, and y is the controlled variable

The modeling of stiction may be carried out using first-principle or data-driven models. Data-driven models [1, 15–21] have been used more frequently due to their reduced number of parameters, simplicity and easy computational implementation. ARX [1, 2], ARMAX [6] models, or transfer functions [3] are examples of data-driven approaches. However, these models contain discontinuities in their formulation which is a significant disadvantage.

The present work explores the application of continuous optimization techniques to the system identification of a model with the stiction phenomenon. A strategy to deal with the discontinuities of the used Hammerstein Model in the context of the optimization procedure is proposed.

The paper is organized as follows: in Section 2, the proposed method is presented; Section 3 details a smoothing approach to avoid model discontinuities; in Section 4, the results are shown and discussed and, finally, in Section 5, main conclusions are drawn.

2 Method Formulation

A novel technique for detection and quantification of valve stiction in control loops based on one-stage identification is proposed in this section. The system to be identified is represented by the Hammerstein Model shown in Fig. 1.

After sufficiently rich process data is collected, the model parameters are determined such that the model response reproduces the observed response of the actual process. Mathematically, this is represented by the non-linear constrained optimization problem

$$\begin{aligned}
\underset{\mathbf{p}}{\text{minimize}} \quad & J(\mathbf{y}, \mathbf{u}, \mathbf{p}) && \text{(1a)}\\
\text{subject to} \quad & \dot{\mathbf{y}} = f(\mathbf{y}, \mathbf{x}, \mathbf{p}) && \text{(1b)}\\
& \mathbf{x} = g(\mathbf{u}, \mathbf{p}) && \text{(1c)}\\
& \mathbf{y}_{\mathrm{L}} \leq \mathbf{y} \leq \mathbf{y}_{\mathrm{U}} && \text{(1d)}\\
& \mathbf{u}_{\mathrm{L}} \leq \mathbf{u} \leq \mathbf{u}_{\mathrm{U}} && \text{(1e)}\\
& \mathbf{p}_{\mathrm{L}} \leq \mathbf{p} \leq \mathbf{p}_{\mathrm{U}} && \text{(1f)}\\
& h(\mathbf{p}) \leq 0, && \text{(1g)}
\end{aligned}$$

where J denotes the objective function, $\mathbf{p}$ is the parameters vector including both the stiction and the process models, $\mathbf{y}$ and $\mathbf{u}$ are the vectors of controlled variable and controller output (respectively), $\mathbf{x}$ is the vector of the real valve position, and the subscripts $_{\mathrm{L}}$ and $_{\mathrm{U}}$ stand for lower and upper bounds (respectively). The set of equations (1b) and (1c) defines a set of constraints arising from the Hammerstein Model dynamics. Inequalities (1g) enforce additional identification criteria.

As it may be seen in Fig. 1, the non-linear element scales the controller output and transforms it to the real valve position. The model expressed by the set of equations (1b) corresponds to this transformation and is represented by a stiction model existent in the literature. In contrast, the linear element whose output is the controlled variable is modeled by the linear model specified by (1c).

Given the Hammerstein Model and a set of n experimental data points $(t_i, y_{\mathrm{exp},i})$, the objective function J is written, according to the minimum least squares criterion, as

$$J = \left[\mathbf{y}_{\mathbf{exp}} - \mathbf{y}\right]^{\top} \mathbf{Q} \left[\mathbf{y}_{\mathbf{exp}} - \mathbf{y}\right], \tag{2}$$

where $\mathbf{Q}$ is a diagonal matrix containing the weights given to each observed variable. In this work, equal weight was given to all output variables.

From a practical point of view, the proposed technique only requires the controller output and the controlled variable data that may be accessed in the DCS (Distributed Control System) of industrial plants. Notice that the real valve position is an unmeasured intermediate variable.

3 Smoothing of Discontinuous Models

Stiction is essentially described by discontinuous non-linear models and that calls for mixed integer non-linear optimization problem formulation and a special class of optimizers. Alternatively, smoothing approaches for discontinuous models have been successfully applied. Some works introduced smoothing techniques in the context of exact penalty functions [22–24]. Others authors have suggested to express discontinuities by means of a step function and then to substitute this function by a continuous approximation [25–27]. This is the approach also adopted in the present work.

Consider the general discontinuous system

$$z(t) = \begin{cases} z_1(t), & \text{if } t \in T_1 \\ z_2(t), & \text{if } t \in T_2 \\ \vdots & \\ z_m(t), & \text{if } t \in T_m \end{cases}, \tag{3}$$

where $z_i(t)$, $i = 1, \cdots, m$, are continuously differentiable real functions over $\mathbb{R}^n$ subject to the conditions that define the subsets T_i. Assuming that the real expressions $e_k(t)$, $k = 1, \cdots, p$, are continuously differentiable over $\mathbb{R}^n$, the subsets T_i are defined as

$$T_i = \{t \in \mathbb{R}^n : e_k(t) < 0, \forall k \in L_i; e_k(t) \geq 0, \forall k \in G_i\} , \tag{4}$$

where L_i and G_i are, for branch i, the sets of indexes k for which $e_k(t) < 0$ and $e_k(t) \geq 0$, respectively.

The discontinuous function (3) may be expressed by means of the Heaviside function as

$$z(t) = \sum_{i=1}^{m} \prod_{k \in L_i} [1 - \mathcal{H}(e_k)] \prod_{k \in G_i} \mathcal{H}(e_k)\; z_i(t) , \tag{5}$$

with

$$\mathcal{H}(t) = \begin{cases} 1, & \text{if } t \geq 0 \\ 0, & \text{if } t < 0 \end{cases} , \tag{6}$$

that is,

$$\mathcal{H}(e_k) = \begin{cases} 1, & \text{if } e_k \geq 0 \\ 0, & \text{if } e_k < 0 \end{cases} . \tag{7}$$

It is possible to smooth the Heaviside function by approximating it by the hyperbolic function

$$\widetilde{\mathcal{H}}(t) = 0.5 + 0.5 \cdot \tanh{(r \cdot t)} , \tag{8}$$

where $\widetilde{\mathcal{H}}(t)$ is second-order continuously differentiable on $\mathbb{R}^n$ varying within the interval $[0, 1]$, and r is an accuracy parameter. Similarly to the approach of [27], the step function approximation here considered contains a single parameter. This parameter controls the accuracy of the approximation by adjusting the size of the neighborhoods around the discontinuity points over which the approximation has an effective effect.

Therefore, the continuous differentiable on $\mathbb{R}^n$ function that approximates function (3) may be written as

$$\widetilde{z}(t) = \sum_{i=1}^{m} \prod_{k \in L_i} \left[1 - \widetilde{\mathcal{H}}(e_k)\right] \prod_{k \in G_i} \widetilde{\mathcal{H}}(e_k)\; z_i(t) . \tag{9}$$

4 Application to a System Containing a Sticky Valve

As mentioned above, the Hammerstein Model comprises a non-linear model describing the sticky valve and a linear process model. The present paper uses the complete Chen Model, also called by the authors as two-layer binary tree model [18], to model the sticky valve.

In what concerns the process dynamics, it is modeled by the single-input single-output state-space model

$$\dot{\mathbf{y}}^* = a\, \mathbf{y}^* + b\, \mathbf{x}^* , \tag{10}$$

where a and b are state-space model constants, and the deviation variables vectors $\mathbf{y}^*$ and $\mathbf{x}^*$ are related to the original variables $\mathbf{y}$ and $\mathbf{x}$ through the simple translations $\mathbf{y}^* = \mathbf{y} - \bar{y}$ and $\mathbf{x}^* = \mathbf{x} - \bar{x}$, respectively.

In order to collect experimental data needed to perform a comparison between the smoothed and the original versions of the Chen Model and also needed for the identification process, a plant simulation was carried out using the Hammerstein Model containing the original discontinuous Chen Model. The parameters used in the simulation are: (i) for the stiction model: $f_S = 2.8\,\%$ and $f_D = 0.9\,\%$; (ii) for the process model: $a = 1$, $b = -1\,\%^{-1}$, $\bar{y} = 0$, and $\bar{x} = 0\,\%$. A sinusoidal excitation on the controller output with amplitude of 5 % and period of 40 min is applied to the system to generate a response, $\mathbf{y}_{\text{exp}}$, that includes sufficient dynamical information about the valve and the process. The obtained dataset contains $n = 101$ points covering an interval of 50 minutes with a sampling period of 0.5 minutes.

4.1 Smoothing of the Stiction Model

An enhanced flow diagram of the Chen Model is illustrated in Fig. 2. The diagram is complemented with some notes, relatively to the original model presented by its authors in [18], to better explain the model and the approach developed in the present paper.

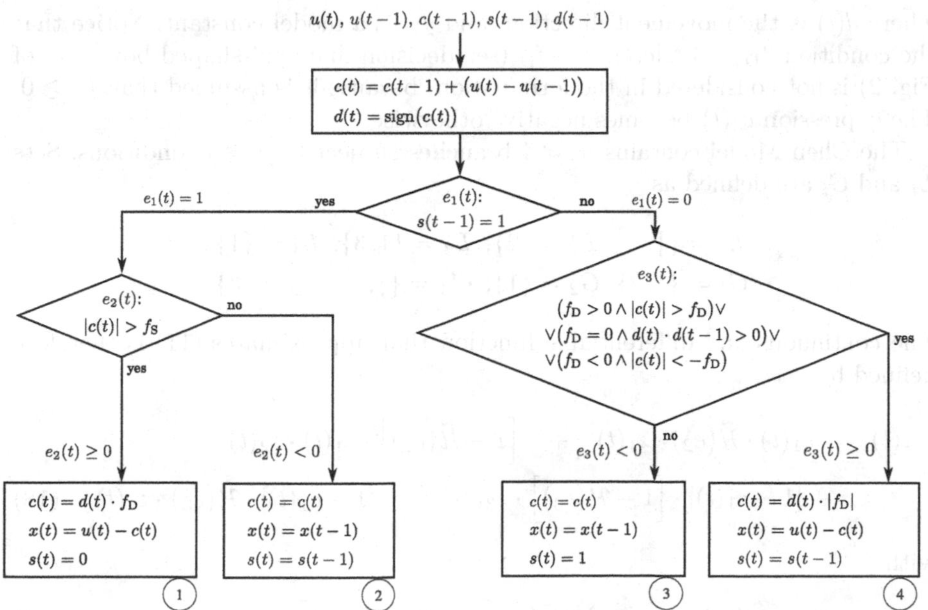

Fig. 2. Enhanced Chen Model flow diagram

The Chen Model may be rewritten as

$$z(t) = \begin{cases} z_1(t), & \text{if } e_1(t) = 1 \wedge e_2(t) \geq 0 \\ z_2(t), & \text{if } e_1(t) = 1 \wedge e_2(t) < 0 \\ z_3(t), & \text{if } e_1(t) = 0 \wedge e_3(t) < 0 \\ z_4(t), & \text{if } e_1(t) = 0 \wedge e_3(t) \geq 0 \end{cases}, \tag{11}$$

where $z(t)$ is a general variable used to represent the outputs $c(t)$, $x(t)$ and $s(t)$. Expressions $e_1(t)$ and $e_2(t)$ are given by

$$e_1(t) = s(t-1)\,, \tag{12}$$

$$e_2(t) = c(t) - f_{\mathrm{S}}\,, \tag{13}$$

where $s(t)$ is the valve status flag, $c(t)$ is the accumulated force compensated by friction, and f_{S} is a model constant. The expression $e_3(t)$ becomes positive or equal to zero when

$$\Big(e_{31}(t) \geq 0 \wedge e_{32}(t) \geq 0\Big) \vee \Big(e_{31}(t) < 0 \wedge e_{33}(t) \geq 0\Big)\,, \tag{14}$$

with

$$e_{31}(t) = f_{\mathrm{D}}\,, \tag{15}$$

$$e_{32}(t) = |c(t)| - f_{\mathrm{D}}\,, \tag{16}$$

$$e_{33}(t) = d(t) \cdot d(t-1)\,, \tag{17}$$

where $d(t)$ is the movement direction, and f_{D} is a model constant. Notice that the condition $f_{\mathrm{D}} < 0 \wedge |c(t)| < -f_{\mathrm{D}}$ (see decision diamond-shaped box $e_3(t)$ of Fig. 2) is not considered in the formulation, because it is assumed that $f_{\mathrm{D}} \geq 0$. The expression $e_3(t)$ becomes negative otherwise.

The Chen Model contains $m = 4$ branches subject to $p = 3$ conditions. Sets L_i and G_i are defined as

$$L_1 = \{\}, \quad L_2 = \{2\},\ L_3 = \{1,3\},\ L_4 = \{1\}\,,$$
$$G_1 = \{1,2\},\ G_2 = \{1\},\ G_3 = \{\}, \quad G_4 = \{3\}\,.$$

The continuous and differentiable function that approximates (11) is therefore defined by

$$\begin{aligned} \widetilde{z}(t) \; = \; & e_1(t) \cdot \widetilde{\mathcal{H}}(e_2) \cdot z_1(t) \; + \; \Big[1 - \widetilde{\mathcal{H}}(e_2)\Big] \cdot e_1(t) \cdot z_2(t) \\ & + [1 - e_1(t)] \cdot \Big[1 - \widetilde{\mathcal{H}}(e_3)\Big] \cdot z_3(t) \; + \; [1 - e_1(t)] \cdot \widetilde{\mathcal{H}}(e_3) \cdot z_4(t)\,, \end{aligned} \tag{18}$$

with

$$\widetilde{\mathcal{H}}(e_3) = \widetilde{\mathcal{H}}(e_{31}) \cdot \widetilde{\mathcal{H}}(e_{32}) + \Big[1 - \widetilde{\mathcal{H}}(e_{31})\Big] \cdot \widetilde{\mathcal{H}}(e_{33})\,. \tag{19}$$

It is noteworthy that $e_1(t)$ is used in (11) inside an equality condition which precludes the direct usage of the approximation approach described in Section 3. However, as shown in (12), this condition is given by the output $s(t-1)$ which is smoothed and valued between 0 and 1 similarly to $\widetilde{\mathcal{H}}(t)$. These facts enable the use of this variable directly in equation (18) in a similar way as the smoothed Heaviside functions, allowing to deal with the equality constraint.

Several simulations were performed in order to assess the performance of the developed approach. Fig. 3 depicts the simulation responses of the Hammerstein Model when the non-linear element is described by the original Chen Model (solid line) and also when the non-linear element is described by the smoothed Chen Model, proposed in the present paper (dashed lines and points). Being a measure of the quality of the approximation applied, the parameter r has a visible influence on the performance of the smoothed model. This influence is quantified in Table 1 by the mean squared error (MSE) associated with the simulations for different values of r. As it may be easily seen in Fig. 3, by using bigger values of r it is possible to reproduce better the data obtained by the original Chen Model. For bigger values of r, the approximation of the function occurs in smaller neighborhoods of the discontinuity points leading to a better

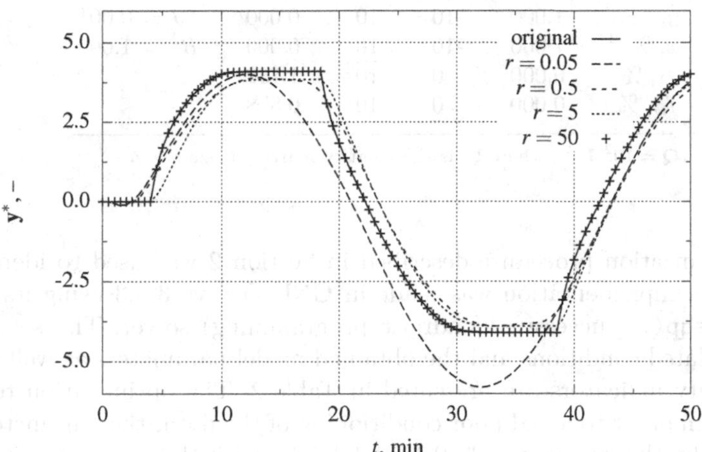

Fig. 3. Comparison between the Hammerstein Model using the original Chen Model and its smoothed version for different values of r

Table 1. MSE for different values of r

r	0.05	0.5	5	10	20	30	40	50
MSE	88.440	34.459	39.701	0.829	0.011	0.003	0.002	0.000

approximation. The value of $r = 50$ was selected based on a mean squared error tolerance of 10^{-3}.

4.2 Stiction Detection and Quantification

The Chen Model parameter f_S is linearly dependent on f_D through the mathematical relationship [18]

$$f_S = f_D + f_J\,. \tag{20}$$

Such dependence poses a difficulty in system identification, because it compromises the identifiability of the individual parameters. In order to overcome this problem, the model is reformulated to use parameters f_J instead of f_S for optimization purposes.

Table 2. System identification results

p	Initial	$\mathbf{p}_L$	$\mathbf{p}_U$	Fit	Indicators
a, –	-0.200	-10	10	-0.998	
b, %	0.020	-10	10	0.998	
$\bar{y}$, –	1.000	-10	10	0.000	$J = 0.001$
$\bar{x}$, $\%^{-1}$	1.000	-10	10	0.000	$R^2 = 1.000$
f_J, %	0.000	0	10	1.732	
f_D, %	0.000	0	10	0.898	

$\mathbf{Q} = 10^2\ \mathbf{I}_{101}$, where $\mathbf{I}_n$ is the identity matrix of size $n \times n$.

The optimization procedure described in Section 2 was used to identify the system. The implementation was made in GNU Octave 3.6.3 using its general non-linear `sqp()` (successive quadratic programming) solver. The set of optimization related conditions and the obtained model parameters as well as some fitting quality indicators are presented in Table 2. The optimization tolerance was 10^{-20}. In order to avoid poor conditioning of the data, the parameters were normalized by the vector $\boldsymbol{\alpha} = \left[-0.1\ 0.1\ 1\ 1\ 1\ 1\right]$, with the necessary changes in the model.

The profile predicted by the identified model, $\mathbf{y}^*_{\text{fit}}$, may be directly compared with the experimental profile, $\mathbf{y}^*_{\text{exp}}$, in Fig. 4. The initial profiles, $\mathbf{y}^*_{\text{init}}$ and $\mathbf{x}^*_{\text{init}}$, are also displayed revealing that the starting initial situation was significantly different from the experimental profiles. As it is possible to observe, the fitted Hammerstein Model is able to capture well the sticky valve and the process dynamics. The high correlation factor, R^2, and the lower objective function prove the effectiveness of the one-stage system identification technique.

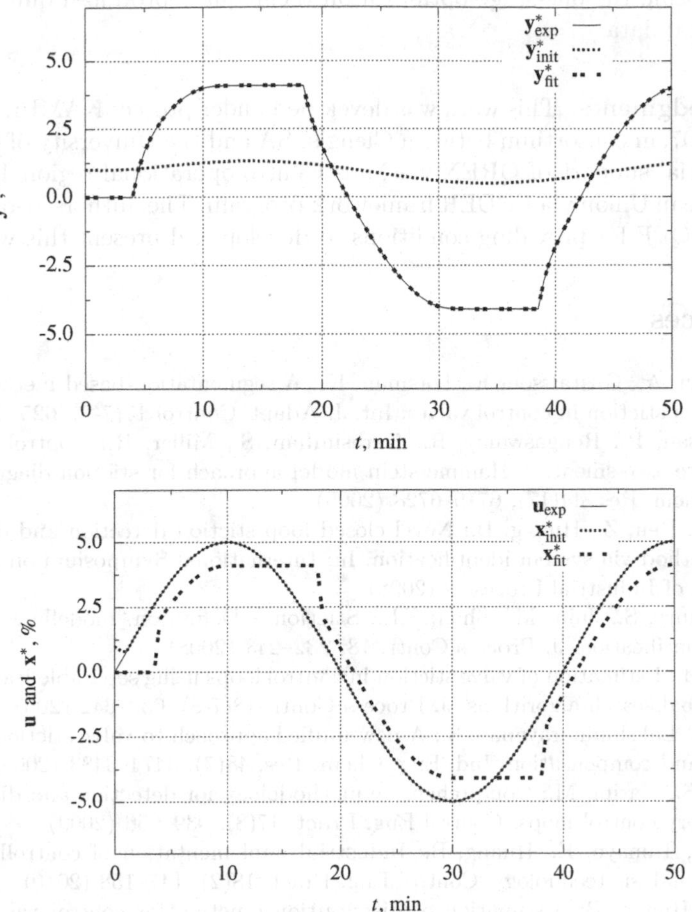

Fig. 4. System identification

5 Conclusions

A technique for detection and quantification of valve stiction in control loops based on one-stage optimization technique was developed using the Sequential Quadratic Programming algorithm.

The modeling approach adopted to describe a process containing a sticky valve is the so-called Hammerstein Model, which includes a non-linear element in series with a linear element. The former corresponds to the sticky valve and is modeled according to the Chen Model. The latter represents the process and is modeled by a single-input single-output state-space model.

Because the Chen Model discontinuities originate system identification difficulties, an approach based on the discontinuity points smoothing was suggested and successfully applied to the model. Finally, the smoothed Hammerstein Model identified using the one-stage optimization technique reproduced quite well the experimental data.

Acknowledgments. This work was developed under project NAMPI, reference 2012/023007, in consortium between Ciengis, SA and the University of Coimbra, with financial support of QREN via Mais Centro operational regional program and European Union via FEDER framework program. The authors also acknowledge CIEPQPF for providing conditions to develop and present this work.

References

1. Stenman, A., Gustafsson, F., Forsman, K.: A segmentation-based method for detection of stiction in control valves. Int. J. Adapt. Control 17(7-9), 625–634 (2003)
2. Srinivasan, R., Rengaswamy, R., Narasimhan, S., Miller, R.: Control loop performance assessment. 2. Hammerstein model approach for stiction diagnosis. Ind. Eng. Chem. Res. 44(17), 6719–6728 (2005)
3. Lee, K., Ren, Z., Huang, B.: Novel closed-loop stiction detection and quantification method via system identification. In: International Symposium on Advanced Control of Industrial Processes (2008)
4. Choudhury, S., Jain, M., Shah, S.L.: Stiction – Definition, modelling, detection and quantification. J. Process Contr. 18, 232–243 (2008)
5. Jelali, M.: Estimation of valve stiction in control loops using separable least-squares and global search algorithms. J. Process Contr. 18(7-8), 632–642 (2008)
6. Ivan, L., Lakshminarayanan, S.: A new unified approach to valve stiction quantification and compensation. Ind. Eng. Chem. Res. 48(7), 3474–3483 (2009)
7. Karra, S., Karim, M.: Comprehensive methodology for detection and diagnosis of oscillatory control loops. Control Eng. Pract. 17(8), 939–956 (2009)
8. Lee, K., Tamayo, E., Huang, B.: Industrial implementation of controller performance analysis technology. Control Eng. Pract. 18(2), 147–158 (2010)
9. Qi, F., Huang, B.: Estimation of distribution function for control valve stiction estimation. J. Process Contr. 21(8), 1208–1216 (2011)
10. Srinivasan, B., Spinner, T., Rengaswamy, R.: A reliability measure for model based stiction detection approaches. In: Symposium on Advanced Control of Chemical Processes, pp. 750–755 (2012)
11. Babji, S., Nallasivam, U., Rengaswamy, R.: Root cause analysis of linear closed-loop oscillatory chemical process systems. Ind. Eng. Chem. Res. 51(42), 13712–13731 (2012)
12. Ljung, L.: System identification: theory for the user. Prentice-Hall, New Jersey (1999)
13. Vandenberghe, L.: Convex optimization techniques in system identification. Technical report, University of California (2014)
14. Eskinat, E., Johnson, S., Luyben, W.: Use of Hammerstein models in identification of nonlinear systems. AIChE J. 37(2), 255–268 (1991)
15. Choudhury, A., Thornhill, N., Shah, S.: Modelling valve stiction. Control Eng. Pract. 13(5), 641–658 (2005)
16. Kano, M., Maruta, H., Kugemoto, H., Shimizu, K.: Practical model and detection algorithm for valve stiction. In: IFAC Symposium on Dynamics and Control of Process Systems, pp. 859–864. Elsevier, United Kingdom (2004)

17. He, Q., Wang, J., Pottmann, M., Qin, J.: A curve fitting method for detecting valve stiction in oscillating control loops. Ind. Eng. Chem. Res. 46(13), 4549–4560 (2007)
18. Chen, S., Tan, K., Huang, S.: Two-layer binary tree data-driven model for valve stiction. Ind. Eng. Chem. Res. 47(8), 2842–2848 (2008)
19. Zabiri, H., Mazuki, N.: A black-box approach in modeling valve stiction. J. Eng. Appl. Sci. 6(5), 277–284 (2010)
20. Wang, J., Sano, A., Chen, T., Huang, B.: A blind approach to identification of Hammerstein systems. In: IFAC World Congress on Block-oriented Nonlinear System Identification, pp. 293–312. Springer, London (2010)
21. Karthiga, D., Kalaivani, S.: A new stiction compensation method in pneumatic control valves. Int. J. Electron. Comput. Sci. Eng., 2604–2612 (2012)
22. Wu, Z., Bai, F., Yang, X., Zhang, L.: An exact lower order penalty function and its smoothing in nonlinear programming. Optimization 53(1), 51–68 (2004)
23. Wu, Z., Lee, H., Bai, F., Zhang, L.: Quadratic smoothing approximation to l_1 exact penalty function in global optimization. J. Ind. Manag. Optim. 1(4), 533–547 (2005)
24. Meng, K., Li, S., Yang, X.: A robust SQP method based on a smoothing lower order penalty function. Optimization 58(1), 23–38 (2009)
25. Goldfeld, S., Quandt, R.: Nonlinear methods in econometrics. North-Holland Publishing Company (1972)
26. Tishler, A., Zang, I.: A switching regression method using inequality conditions. J. of Econometrics 11(2-3), 259–274 (1979)
27. Zang, I.: Discontinuous optimization by smoothing. Math. Oper. Res. 6(1), 140–152 (1981)

On the Properties of General Dual-Feasible Functions

Jürgen Rietz, Cláudio Alves, José Manuel Valério de Carvalho, and François Clautiaux

Escola de Engenharia, Universidade do Minho, 4710-057 Braga, Portugal
Institut de Mathématiques de Bordeaux, Université de Bordeaux, 33405 Talence Cedex, France
juergen_rietz@gmx.de,
{claudio,vc}@dps.uminho.pt,
francois.clautiaux@math.u-bordeaux1.fr

Abstract. Dual-feasible functions have been used to compute fast lower bounds and valid inequalities for integer linear optimization problems. However, almost all the functions proposed in the literature are defined only for positive arguments, which restricts considerably their applicability. The characteristics and properties of dual-feasible functions with general domains remain mostly unknown. In this paper, we show that extending these functions to negative arguments raises many issues. We explore these functions in depth with a focus on maximal functions, *i.e.* the family of non-dominated functions. The knowledge of these properties is fundamental to derive good families of general maximal dual-feasible functions that might lead to strong cuts for integer linear optimization problems and strong lower bounds for combinatorial optimization problems with knapsack constraints.

Keywords: General dual-feasible functions, integer linear programming, valid inequalities, lower bounds.

1 Introduction

Dual-feasible functions were proposed first by Johnson in [4]. Although these functions have been used mainly to compute efficiently lower bounds for problems with knapsack constraints, Clautiaux *et al.* showed in [3] that they can be used also to compute valid inequalities for integer linear optimization problems. Almost all the functions proposed so far are restricted to very specific domains, containing only positive arguments. These functions have been studied extensively in the literature ([3, 6–8]). Properties such as *maximality* that characterizes the best (non-dominated) functions have been studied in [3, 6]. Their worst-case performance for the computation of lower bounds have been also explored in [7].

Restricting the domains of the dual-feasible functions to positive arguments limits clearly the applicability of these functions. The study of extensions of these functions to more general domains began very recently ([1, 9]). The first

B. Murgante et al. (Eds.): ICCSA 2014, Part II, LNCS 8580, pp. 180–194, 2014.

contribution in this sense focused on the extension to multidimensional domains [1]. The resulting functions were called *vector packing dual-feasible functions*, because they allow to compute lower bounds for the vector packing problem. In [9], the authors presented the first preliminary results concerning the extension of these functions to the domain of real numbers. In this paper, we explore in depth the properties of dual-feasible functions extended to the general domain of real numbers. As we will see throughout the paper, extending these functions to general domains is not trivial, and it clearly impacts on the properties observed for standard dual-feasible functions.

The paper is organized as follows. In Subsection 1.1, we introduce some principles related to superadditive functions and Integer Programming. In Subsection 1.2, we introduce the formal definitions underlying general dual-feasible functions, while in Subsection 1.3 we describe some general properties that will be used to simplify other proofs. In Section 2, we show some applications of dual-feasible functions, and we explore further the relation between these functions and valid inequalities (or cuts) for integer linear optimization problems. The properties of these general dual-feasible functions are analysed in Section 3, starting by the properties shared by all the general dual-feasible functions in Subsection 3.1, and developing in Subsection 3.2 the properties that hold specifically for the maximal functions.

1.1 Superadditive Functions in Integer Programming

When dealing with dual-feasible functions, references to notions as superadditivity and nondecreasing functions are frequent. For the sake of clarity, they are briefly recalled in the sequel for the general domains addressed in this paper. Let $Z \subseteq \mathbb{R}^m$. A function $F : Z \to \mathbb{R}$ is *superadditive* if the implication

$$\mathbf{x} + \mathbf{y} \in Z \Longrightarrow F(\mathbf{x}) + F(\mathbf{y}) \le F(\mathbf{x} + \mathbf{y}) \tag{1}$$

is true for all $\mathbf{x}, \mathbf{y} \in Z$. The function $F : Z \to \mathbb{R}$ is *nondecreasing* if for all $\mathbf{x}, \mathbf{y} \in Z$, it holds that

$$\mathbf{x} \le \mathbf{y} \Longrightarrow F(\mathbf{x}) \le F(\mathbf{y}).$$

The link between these superadditive and nondecreasing functions and integer linear optimization problems is set through the well-known *strong duality theorem*. Let $\mathbf{A} \in \mathbb{Q}^{m \times n}$, $\mathbf{b} \in \mathbb{Q}^m$ and $\mathbf{c} \in \mathbb{Q}^n$. Given an integer linear optimization problem

$$\mathbf{c}^\top \mathbf{x} \to \max! \text{ s.t. } \mathbf{A}\mathbf{x} \le \mathbf{b},\ \mathbf{x} \in \mathbb{Z}_+^n, \tag{2}$$

the dual consists in finding a nondecreasing and superadditive function $F : \mathbb{R}^m \to \mathbb{R}$ such that

$$F(\mathbf{b}) \to \min! \tag{3}$$

$$F(\mathbf{o}) = 0 \tag{4}$$

$$F(\mathbf{a}^j) \ge c_j \text{ for } j = 1, \ldots, n, \tag{5}$$

where $\mathbf{a}^j$ is the j-th column of the matrix $\mathbf{A}$, and $\mathbf{o}$ denotes the zero vector. As mentioned above, the relationship between the two problems relies on the *strong duality theorem*, *i.e.* if the problem (2) is solvable, then the optimal objective function values of the primal and the dual problem are the same. If (2) is infeasible, then $F(\mathbf{b})$ is unbounded from below $(-\infty)$, and if (2) yields an unbounded objective function value $(+\infty)$ then F does not exist. Moreover, if $\hat{\mathbf{x}}$ is an optimal solution of (2) then a necessary condition to the dual problem for F to be optimal is

$$F(\mathbf{Ax}) = \mathbf{c}^\top \mathbf{x} = F(\mathbf{b}) - F(\mathbf{b} - \mathbf{Ax})$$

for all $\mathbf{x} \in \mathbb{Z}_+^n$ with $\mathbf{x} \leq \hat{\mathbf{x}}$.

It is well-known that problem (2) is NP-hard, and hence, solving it exactly or finding an optimal superadditive and nondecreasing function $F : \mathbb{R}^m \to \mathbb{R}$ according to the above conditions (3)–(5) may be very difficult. However, as shown in [5], one may contribute to the resolution of (2) by deriving valid inequalities using a superadditive and nondecreasing function $F : \mathbb{R}^m \to \mathbb{R}$ fulfilling the demand (4). These functions yield the following valid inequalities:

$$\sum_{j=1}^{n} F(\mathbf{a}^j) * x_j \leq F(\mathbf{b}).$$

As we will show in Section 1.2, the best dual-feasible functions belong to the group of these superadditive and nondecreasing functions. The issue is that all the dual-feasible functions reported in the literature apply only to domains restricted to positive arguments. In this paper, we explore the properties of the general dual-feasible functions with no restrictions on their domain. These functions are formally defined next.

1.2 Dual-Feasible Functions

In the sequel, we introduce the definitions that support the notion of general dual-feasible functions, but first we recall the definition of the standard dual-feasible functions on which most of the literature on this topic is focused.

Definition 1. *A function $f : [0,1] \to [0,1]$ is a* dual-feasible function (DFF), *if for any finite set $\{x_i \in \mathbb{R}_+ : i \in I\}$ of nonnegative numbers, it holds that*

$$\sum_{i \in I} x_i \leq 1 \Longrightarrow \sum_{i \in I} f(x_i) \leq 1. \tag{6}$$

A general dual-feasible function is a generalization of the previous functions to any real arguments, *i.e.* arguments that are not restricted to positive values. This generalization extends the applicability of these functions, but it raises also many issues, leading to different properties as we will see in Section 3. Formally, the definition of a general dual-feasible function states as follows.

Definition 2. *A function $f : \mathbb{R} \to \mathbb{R}$ is a* general dual-feasible function, *if for any finite set $\{x_i \in \mathbb{R} : i \in I\}$ of real numbers, it holds that*

$$\sum_{i \in I} x_i \leq 1 \Longrightarrow \sum_{i \in I} f(x_i) \leq 1. \tag{7}$$

A general DFF f is a maximal general dual-feasible function (MDFF), *if there is no other general DFF g with $f(x) \leq g(x)$ for all $x \in \mathbb{R}$.*

In contrast to the situation of domain and range $[0, 1]$, the symmetry rule

$$f(x) + f(1 - x) = 1, \quad \text{for all} \quad x \leq 1/2, \tag{8}$$

is now not necessarily fulfilled by a MDFF, as happens for instance with the function $x \mapsto cx$ with any constant $c \in [0, 1]$ ([9]). More precisely, combining previous results introduced in [9], one may summarize as follows the conditions for a general DFF to be maximal.

Theorem 1. *Let $f : \mathbb{R} \to \mathbb{R}$ be a given function.*

(a) If f satisfies the following conditions, then f is a MDFF:
   1. $f(0) = 0$;
   2. f is superadditive, i.e. for all $x, y \in \mathbb{R}$, it holds that

$$f(x + y) \geq f(x) + f(y); \tag{9}$$

   3. there is an $\varepsilon > 0$, such that $f(x) \geq 0$ for all $x \in (0, \varepsilon)$;
   4. f obeys the symmetry rule (8).

(b) If f is a MDFF, then the above properties (1.)–(3.) hold for f, but not necessarily (4.).

(c) If f satisfies the above conditions (1.)–(3.), then f is monotonously increasing.

(d) If the symmetry rule (8) holds and f obeys the inequality (9) for all $x, y \in \mathbb{R}$ with $x \leq y \leq \frac{1-x}{2}$, then f is superadditive.

The monotonicity is now an important prerequisite for the assertion (a). Without that, the theorem would become false, as the following counter example of a function $f : \mathbb{R} \to \mathbb{R}$ shows (Figure 1).

$$f(x) := \begin{cases} 3x - 2, & \text{if } x < 0, \\ -x, & \text{if } 0 \leq x < 1/2, \\ 1/2, & \text{if } x = 1/2, \\ 2 - x, & \text{if } 1/2 < x \leq 1, \\ 3x, & \text{otherwise.} \end{cases} \tag{10}$$

The function f obeys only the 1st, 2nd and 4th condition of the theorem, and it is not a general DFF. $f(0) = 0$ is obviously fulfilled. The fourth condition is also checked easily. If $x < 0$, then $1 - x > 1$, and $f(x) + f(1 - x) = 3x - 2 + 3(1 - x) = 1$. If $0 \leq x < 1/2$, then $f(x) + f(1 - x) = -x + 2 - (1 - x) = 1$. To check the superadditivity, assume that $x \leq y \leq \frac{1-x}{2}$ according to point (d) of Theorem 1. Hence, we have $x + y \leq \frac{1+x}{2}$. It has to be verified that $d(x, y) := f(x + y) - f(x) - f(y)$ is not negative. The following cases arise:

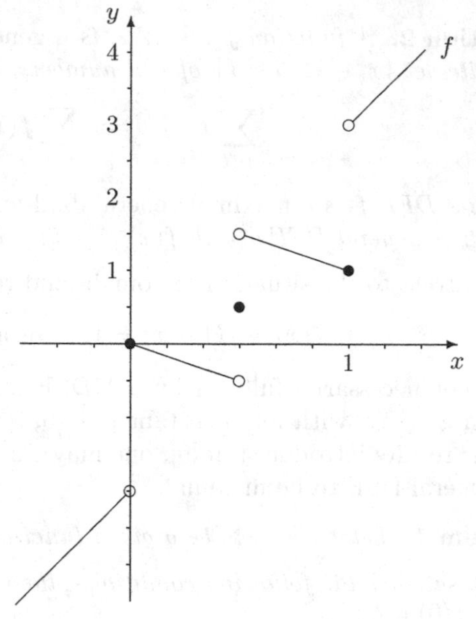

Fig. 1. Necessity of the monotonicity

1. $x + y < 0$: then $x < 0$, and hence, $d(x, y) = 3 * (x + y) - 2 - 3x + 2 - f(y) = 3y - f(y) \geq 0$.
2. $x < 0 \leq x + y$: then $x + y < 1/2$ and $y > 0$, and hence, $d(x, y) = -x - y - (3x - 2) - f(y) = 2 - 4x - y - f(y)$. If $y \leq 1$, then $d(x, y) \geq 0$ can easily be found. If $y > 1$, then $d(x, y) = 2 - 4(x + y) > 0$.
3. $x \geq 0$: if $y = 1/2$, then $y \leq \frac{1-x}{2}$ implies $x = 0$, and hence $d(x, y) = 0$. If $y < 1/2$, then $d(x, y) = f(x + y) + x + y \geq 0$.

The correctness of this example was only partially demonstrated in [9].

1.3 Introductory Properties

To simplify some of the proofs presented in this paper, we introduce next two lemmas that will be used respectively to prove superadditivity and to show that a given function F is appropriate for deriving valid inequalities for the problem (2). The former is inspired on Lemma 2 of [6], while the latter is based on the idea that showing nonnegativity for nonnegative arguments is usually simpler than proving monotonicity.

Lemma 1. *Let $f : \mathbb{R} \to \mathbb{R}$ and $b \in \mathbb{R}$ with $b > 0$ and $f(0) = 0$ be given. If f is convex on $[0, b]$, then the superadditivity (1) holds for all $x, y \in [0, b]$ with $x + y \leq b$. If f is convex on $[-b, 0]$, then the superadditivity (1) holds for all $x, y \in [-b, 0]$ with $x + y \geq -b$.*

Proof. Let $x_0, x_1, x_2, x_3 \in \mathbb{R}$ be given with $x_0 < x_1 < x_2 < x_3 \leq b$. The straight line through the points $(x_1; f(x_1))$ and $(x_2; f(x_2))$ can be described by $s(x) = f(x_1) + \frac{f(x_2)-f(x_1)}{x_2-x_1} * (x - x_1)$. We show $f(x_3) \geq s(x_3)$. The convexity of f in $[x_1, x_3]$ implies $f(x_2) \leq f(x_1) + \frac{f(x_3)-f(x_1)}{x_3-x_1} * (x_2 - x_1)$, and since $s(x_2) = f(x_2)$, we have $\frac{f(x_2)-f(x_1)}{x_2-x_1} \leq \frac{f(x_3)-f(x_1)}{x_3-x_1}$ and $s(x_3) = f(x_1) + \frac{f(x_2)-f(x_1)}{x_2-x_1} * (x_3 - x_1) \leq f(x_1) + \frac{f(x_3)-f(x_1)}{x_3-x_1} * (x_3 - x_1) = f(x_3)$. In the same way, $f(x_0) \geq s(x_0)$ is shown.

Let $s(x) = mx + n$ with certain $m, n \in \mathbb{R}$. If $x_1, x_2, x_1 + x_2 \in (0, b]$ or $x_1, x_2, x_1 + x_2 \in [-b, 0)$, then $f(x_1 + x_2) - f(x_1) - f(x_2) \geq s(x_1 + x_2) - s(x_1) - s(x_2) = -n = -s(0) \geq -f(0) = 0$. □

Lemma 2. *Let $F : \mathbb{R}^m \to \mathbb{R}$ be a superadditive function. Then $F(\mathbf{x}) \geq 0$ for all $\mathbf{x} \geq \mathbf{o}$, if and only if F is nondecreasing and fulfills the constraint (4).*

Proof. The superadditivity implies $F(2\mathbf{o}) \geq 2F(\mathbf{o})$, and hence, $F(\mathbf{o}) \leq 0$. The nonnegativity constraint yields $F(\mathbf{o}) \geq 0$. Therefore, we have $F(\mathbf{o}) = 0$. Additionally, if $\mathbf{x} \leq \mathbf{y}$, then $F(\mathbf{y} - \mathbf{x}) \geq 0$, such that the superadditivity yields $F(\mathbf{y}) = F(\mathbf{x} + \mathbf{y} - \mathbf{x}) \geq F(\mathbf{x}) + F(\mathbf{y} - \mathbf{x}) \geq F(\mathbf{x})$. On the other hand, if $\mathbf{x} \geq \mathbf{o}$, then the monotonicity and the condition (4) immediately imply $F(\mathbf{x}) \geq 0$. □

There are different functions as those of Lemma 2, as for example the following ones that might be seen as an extension to general domains of those presented in [1].

Proposition 1. *Let $\mathbf{u} \in \mathbb{R}^m_+$ and let $f : \mathbb{R} \to \mathbb{R}$ be a superadditive function with $f(x) \geq 0$ for all $x \geq 0$. Then $F : \mathbb{R}^m \to \mathbb{R}$, defined as $F(\mathbf{x}) := f(\mathbf{u}^\top \mathbf{x})$, fulfills the prerequisites of Lemma 2.*

Proof. If $\mathbf{x} \geq \mathbf{o}$, then $\mathbf{u}^\top \mathbf{x} \geq 0$, and hence, $F(\mathbf{x}) \geq 0$. To verify the superadditivity, choose any $\mathbf{x}, \mathbf{y} \in \mathbb{R}^m$. One obtains $F(\mathbf{x} + \mathbf{y}) = f(\mathbf{u}^\top(\mathbf{x} + \mathbf{y})) \geq f(\mathbf{u}^\top \mathbf{x}) + f(\mathbf{u}^\top \mathbf{y}) = F(\mathbf{x}) + F(\mathbf{y})$. □

2 Applications

2.1 Lower Bounds and Dual-Feasible Functions

A general dual-feasible function f yields feasible values to a certain dual linear optimization problem that corresponds to a 1-dimensional cutting stock problem (1D-CSP) defined as follows. Given a list of $m \in \mathbb{N}$ items of length $\ell_i \in (0, 1]$ $(i = 1, \ldots, m)$, each demanded b_i times, the objective is to put them into the minimal number of bins, such that in no used bin the total length of the packed items exceeds the bin size 1.

Comparing the situation for $f : \mathbb{R} \to \mathbb{R}$ with the case of domain and range $[0, 1]$ leads to the following similarity to the 1D-CSP. Let be given any fixed vectors $\mathbf{b}, \mathbf{l} \in \mathbb{R}^p$. All column vectors $\mathbf{a}^j \in \mathbb{Z}^p_+$ of the matrix $\mathbf{A}$ shall obey the inequality $\mathbf{1}^\top \mathbf{a}^j \leq 1$ (we assume the number n of the columns of $\mathbf{A}$ to be finite). Then, the dual of the continuous relaxation of the problem

$$\sum_{j=1}^{n} x_j \to \min! \text{ s. t. } \mathbf{Ax} = \mathbf{b},\ \mathbf{x} \in \mathbb{Z}_+^n \tag{11}$$

is

$$\mathbf{b}^\top \mathbf{u} \to \max! \text{ s. t. } \mathbf{u} \in \mathbb{R}^p, \mathbf{u}^\top \mathbf{a}^j \leq 1,$$

for $j = 1, \dots, n$. Given the general DFF $f : \mathbb{R} \to \mathbb{R}$, feasible values for this dual problem can be found by

$$u_i := f(\ell_i),\ i = 1, \dots, p, \tag{12}$$

because one gets for any $j \in \{1, \dots, n\}$ that $\mathbf{u}^\top \mathbf{a}^j = \sum_{i=1}^{p} a_{ij} * f(\ell_i) \leq 1$, since $\sum_{i=1}^{p} a_{ij} * \ell_i \leq 1$. If the problem (11) is feasible, a lower bound for the optimal objective function value of the problem (11) is $\mathbf{b}^\top \mathbf{u}$, with $\mathbf{u}$ obtained through (12). In other words, a lower bound for the 1D-CSP can be obtained simply by $\sum_{i=1}^{m} b_i \times f(\ell_i)$. The best possible bound by dual-feasible functions equals the optimal objective function value of the continuous relaxation of the corresponding column generation for the 1D-CSP, but it is obtained much faster.

2.2 Cuts and Dual-Feasible Functions

Apart from their use for computing lower bounds for problems with knapsack constraints and possibly negative coefficients, general dual-feasible functions can be used to derive valid inequalities (cuts) for integer linear optimization problems in a way that is similar to the one discussed in Section 1 for superadditive and nondecreasing functions. In this section, we show through an example that these functions can in fact provide cuts that are better than the well-known Chvátal-Gomory cuts.

Given the inequality system $\mathbf{Ax} \leq \mathbf{b}$, $\mathbf{x} \in \mathbb{N}^n$, one can use any nonnegative linear combination of the inequalities to derive cuts. Let $\mathbf{u} \geq \mathbf{o}$. Then $\mathbf{u}^\top \mathbf{Ax} \leq \mathbf{u}^\top \mathbf{b}$ yields the following scalar inequality

$$\mathbf{d}^\top \mathbf{x} \leq r, \tag{13}$$

and finally the Chvátal-Gomory cut

$$\sum_{j=1}^{n} \lfloor d_j \rfloor * x_j \leq \lfloor r \rfloor. \tag{14}$$

Suppose that $r > 0$. Dividing the inequality (13) by r and applying a maximal general dual-feasible function $f : \mathbb{R} \to \mathbb{R}$ with $f(1) = 1$ yields the valid inequality

$$\sum_{j=1}^{n} f\left(\frac{d_j}{r}\right) * x_j \leq 1. \tag{15}$$

Example 1. Consider the following maximal general dual-feasible function $g : \mathbb{R} \to \mathbb{R}$ introduced in [9], with the parameter $b \geq 1$:

$$g(x) := \begin{cases} b * \lfloor 2x \rfloor, & \text{if } x < 1/2, \\ 1/2, & \text{if } x = 1/2, \\ 1 - b * \lfloor 2 - 2x \rfloor, & \text{if } x > 1/2. \end{cases} \tag{16}$$

Additionally, consider the case where f is the function (16) with $b = 1$. One obtains the following simplified version of the function

$$f(d) = \begin{cases} \lfloor 2d \rfloor, & \text{if } d < 1/2, \\ 1/2, & \text{if } d = 1/2, \\ \lceil 2d \rceil - 1, & \text{if } d > 1/2. \end{cases} \tag{17}$$

If $r > 2$, then for any values of d_j, the possibility arises that the inequality (15) is stronger than (14), particularly if some d_j are a bit greater than $r/2$ and there is no $d_j < -r/2$. For example,

$$1.5x_1 - 1.4x_2 \leq 2.8$$

yields the Chvátal-Gomory cut

$$x_1 - 2x_2 \leq 2,$$

while the stronger inequality

$$x_1 - x_2 \leq 1$$

can be obtained by applying the maximal general dual-feasible function.

To compare the inequalities (14) and (15), the latter may be multiplied by $\lfloor r \rfloor$. In general, (15) can be stronger than (14), provided that $f(d_j/r) * \lfloor r \rfloor > \lfloor d_j \rfloor$ for some j. □

3 Properties of General Dual-Feasible Functions

In this section, we explore in depth the properties of general dual-feasible functions with a special focus on the maximal functions, *i.e.* those that are not dominated. First, general properties that apply to any general dual-feasible functions are discussed. In Subsection 3.2, we explore the properties of the maximal functions. The set of maximal general dual-feasible functions is naturally very large. The knowledge of these properties is fundamental to derive good families of general dual-feasible functions that might lead to strong cuts for integer linear optimization problems and strong lower bounds for combinatorial optimization problems with knapsack constraints. For the sake of clarity, the abbreviation $\mathsf{frac}(\cdot)$ will be used sometimes in the text below to denote the non-integer part of any real expression, *i.e.* $\mathsf{frac}(x) \equiv x - \lfloor x \rfloor$.

3.1 General Properties

The following proposition describes some necessary conditions for a function with possibly negative arguments to be a general dual-feasible function. The corresponding properties were first alluded partially in [9], but without proof. Here, we provide a formal proof of their validity. The estimations in this proposition will be used to derive other, non-obvious properties of maximal general dual-feasible functions in Proposition 4.

Proposition 2. *Let $f : \mathbb{R} \to \mathbb{R}$ be any function. If f is a general DFF, then it obeys the following two statements.*

1. *For any $x \in (0, 1]$, it holds that $f(x) \le 1/\lfloor 1/x \rfloor$.*
2. *For any finite set $\{x_i \in \mathbb{R} : i \in I\}$ of real numbers, the following holds:*

$$\sum_{i\in I} x_i \le 0 \Longrightarrow \sum_{i\in I} f(x_i) \le 0 \tag{18}$$

The converse is generally false.

Proof. If f is a general DFF and $x \in (0, 1]$, then let $n := \lfloor 1/x \rfloor \in \mathbb{N}\backslash\{0\}$. Setting $x_1 := x_2 := \ldots := x_n := x$ yields $\sum\limits_{i=1}^{n} x_i = nx \le 1$ and $\sum\limits_{i=1}^{n} f(x_i) = n * f(x) \le 1$ due to Definition 2, and hence we have $f(x) \le 1/n$.

Given $n \in \mathbb{N}$ and $x_i \in \mathbb{R}$ $(i = 1, \ldots, n)$ with $\sum\limits_{i=1}^{n} x_i \le 0$, it cannot happen that $\sum\limits_{i=1}^{n} f(x_i) = \varepsilon > 0$, because using each x_i in the quantity $p := \lfloor 1/\varepsilon \rfloor + 1$, *i.e.* setting $x_{n+1} := x_1, \ldots, x_{np} := x_n$, one gets $\sum\limits_{i=1}^{np} x_i = p * \sum\limits_{i=1}^{n} x_i \le 0$, but $\sum\limits_{i=1}^{np} f(x_i) = p * \varepsilon > 1$ in contradiction to Definition 2, and the fact that f is a DFF.

A counter-example to the assumption that the properties (1.) and (2.) imply that f would always be a dual-feasible function is this:

$$f(x) := \begin{cases} 0, & \text{if } 0 \le x \le 1, \\ 2x, & \text{otherwise.} \end{cases}$$

The property (1.) is obviously fulfilled, and (2.) too because of $f(x) \le 2x$ for all $x \in \mathbb{R}$, but $f(-1) + f(2) = 2 > 1$, in spite of $-1 + 2 \le 1$. □

Proposition 3 shows that if someone replaces a non-maximal general dual-feasible function by a dominating one and continues this process ad infinitum, then a general dual-feasible function is still obtained. Given a non-maximal general dual-feasible function $f_0 : \mathbb{R} \to \mathbb{R}$, one can first ensure the nonnegativity for nonnegative arguments by setting $f_1(x) := f_0(x)$ for $x < 0$ and $f_1(x) := \max\{0, f_0(x)\}$ for $x \ge 0$. After that, if the obtained function is not monotonous, then it can be replaced by a nondecreasing dominating dual-feasible function. If the resulting function is not superadditive, then it can be replaced by a superadditive dominating general dual-feasible function. If the corresponding function is still not maximal, then it can be improved further, and if always general dual-feasible functions were constructed, the final function will be also a general dual-feasible function.

Proposition 3. *The set of general dual-feasible functions $f : \mathbb{R} \to \mathbb{R}$ is closed, i.e. any converging sequence of general dual-feasible functions converges to a general dual-feasible function.*

Proof. Let (f_n) be a converging sequence of general dual-feasible functions, *i.e.* for each $x \in \mathbb{R}$, the limit $f(x) := \lim_{n\to\infty} f_n(x)$ exists. Let I be any finite index set. For any $x_i \in \mathbb{R}$ $(i \in I)$ with $\sum_{i\in I} x_i \leq 1$, one has $\sum_{i\in I} f(x_i) = \sum_{i\in I} \lim_{n\to\infty} f_n(x_i) = \lim_{n\to\infty} \sum_{i\in I} f_n(x_i) \leq 1$, because only a finite sum played a role in the limit. Therefore, f is a general dual-feasible function. □

Note that the limit $\lim_{n\to\infty} f_n(x)$ exists for any $x \in \mathbb{R}$, because each f_n was replaced by a dominating function. Hence, $f_n(x)$ grows monotonously with n, but it is bounded due to $f_n(x) \leq 1 - f_n(1-x)$.

Composing functions to obtain different functions is one of the ways of building dual-feasible functions as shown for example in [3]. In the following proposition, we show that composing general dual-feasible functions leads to functions that remain general dual-feasible functions.

Lemma 3. *The composition of general dual-feasible functions $f, g : \mathbb{R} \to \mathbb{R}$ is a general dual-feasible function.*

Proof. Let $\{x_i \in \mathbb{R} : i \in I\}$ be any finite set of real numbers with $\sum_{i\in I} x_i \leq 1$. One gets $\sum_{i\in I} g(x_i) \leq 1$, because g is a general dual-feasible function. Therefore, Definition 2 yields $\sum_{i\in I} f(g(x_i)) \leq 1$, because f is also a general dual-feasible function. □

3.2 Properties of Maximal General Dual-Feasible Functions

A first impression about the structure of maximal general dual-feasible functions is given in the following. One will see that these functions are affine-linearly bounded. The next proposition is motivated by the following functions $f_1, f_2, f_3 : \mathbb{R} \to \mathbb{R}$, which are not maximal general dual-feasible functions:

$$f_1(x) := \begin{cases} 1 - e^{-x}, & \text{if } x < 0, \\ x, & \text{if } 0 \leq x \leq 1, \\ e^{x-1}, & \text{if } x > 1. \end{cases} \qquad f_2(x) := \begin{cases} -\ln(1-x,) & \text{if } x < 0, \\ x, & \text{if } 0 \leq x \leq 1, \\ 1 + \ln x, & \text{if } x > 1. \end{cases}$$

$$f_3(x) := \begin{cases} \sqrt{2}x, & \text{if } x < 0, \\ \sqrt{2}x + 1 - \sqrt{x+1}, & \text{otherwise.} \end{cases}$$

The function f_1 violates the superadditivity condition, because

$$f_1(-2) = 1 - e^2 < 2 * (1 - e) = 2 * f_1(-1),$$

while f_2 is obviously not a general dual-feasible function because of $2*1+(-1) \leq 1$, but $2 * f_2(1) + f_2(-1) = 2 - \ln 2 > 1$. Regarding f_3, one has $f_3(0) = 0$ and $f_3(1) = 1$. Moreover, $f_3'(x) = \sqrt{2} - 1/(2\sqrt{x+1}) > 0$ for $x > 0$, such that f_3 is nondecreasing. The monotonicity of f_3' shows that f_3 is strictly convex for positive arguments, and therefore superadditive in $\mathbb{R}_+$. It is easy to see that the superadditivity of f_3 holds on entire $\mathbb{R}$. Therefore, f_3 is a general dual-feasible function fulfilling all the necessary conditions of Theorem 1, but it is not maximal as the next proposition states.

Proposition 4. *Let $f : \mathbb{R} \to \mathbb{R}$ be a maximal general dual-feasible function and $t := \sup\{f(x)/x : x > 0\}$. Then, $\lim\limits_{x\to\infty} \frac{f(x)}{x} = t \leq -f(-1)$, and for any $x \in \mathbb{R}$, it holds that $tx - \max\{0, t-1\} \leq f(x) \leq tx$, i.e. f is the sum of a linear and a bounded function.*

Proof. Since f is a maximal general dual-feasible function, it holds for every $x \in \mathbb{R}_+$ that

$$0 = f(0) \geq f(-x) \geq f(-\lceil x \rceil) \geq \lceil x \rceil * f(-1) \geq (x+1) * f(-1),$$

because $f(-1) \leq 0$. Proposition 2 (point 2.) implies $f(x) + f(-x) \leq 0$, and hence, we have $0 \leq f(x) \leq -(x+1) * f(-1)$. Therefore, $|f(x)| = \mathcal{O}(|x|)$ for $|x| \to \infty$.

Now, we show using Proposition 2 (point 1.) that $\lim\limits_{x\downarrow 0} \frac{f(x)}{x}$ is finite. For $x \in (0,1)$, one obtains

$$0 \leq f(x) \leq 1/\lfloor \frac{1}{x} \rfloor < 1/(\frac{1}{x} - 1),$$

and hence, $f(x)/x < \frac{1}{1-x}$ and $0 \leq \lim\limits_{x\downarrow 0} \frac{f(x)}{x} \leq 1$. Therefore, t remains finite.

The definition of t yields for every $c < t$ that there is a $y \in \mathbb{R}$, $y > 0$ with $f(y)/y \geq c$. Let $n := \lfloor x/y \rfloor$ with $x > 0$. Hence, we have $n \in \mathbb{N}$ and $ny \leq x < (n+1)y$. The monotonicity and superadditivity of f yield $f(x) \geq f(ny) \geq cny$ and $f(x)/x \geq c * \frac{n}{n+1}$. If $x \to \infty$, then $n \to \infty$, and hence, $\lim\limits_{x\to\infty} \frac{f(x)}{x} \geq c$. Since this inequality holds for all $c < t$, it follows that $\lim\limits_{x\to\infty} \frac{f(x)}{x} = t$, because this limit may not exceed t.

If $f \equiv 0$, then the proof is complete. Therefore assume $f \not\equiv 0$. Before the last assertion, first $f(x) < cx$ for all $c, x \in \mathbb{R}$ with $x < 0 < c < t$ is shown indirectly. The definition of t implies that for every $c \in (0,t)$, there is an $x_0 > 0$ with $f(x_0)/x_0 > c$. Suppose there is an $x_1 \in \mathbb{R}$ with $x_1 < 0$ and $f(x_1) \geq cx_1$. Since $c > 0 > x_1$ and $f(x_0)/c > x_0$, the open interval $\left(-\frac{x_0}{x_1}, -\frac{f(x_0)}{cx_1}\right)$ is not empty and contains a rational number p/q with $p, q \in \mathbb{N} \setminus \{0\}$. One gets

$$qx_0 < -px_1 < q * f(x_0)/c,$$

and hence $px_1 + qx_0 < 0$ and $q * f(x_0) > -cpx_1$. But $p * f(x_1) \geq cpx_1$, and therefore

$$p * f(x_1) + q * f(x_0) > cpx_1 - cpx_1 = 0,$$

contradicting Proposition 2 (point 2.). Therefore, no such x_1 exists, *i.e.* for any $c, x \in \mathbb{R}$ with $x < 0 < c < t$, it holds that $f(x) < cx$. For fixed $x < 0$, this inequality $f(x) < cx$ for all $c \in (0,t)$ implies finally $f(x) \leq tx$. For $x \geq 0$, the assertion $f(x) \leq tx$ is trivial. Setting $x := -1$ yields $f(-1) \leq -t$ or equivalently $t \leq -f(-1)$.

It remains to prove that $f(x) \geq tx - \max\{0, t-1\}$ for any $x \in \mathbb{R}$. If $t \leq 1$, then f is dominated by the function $x \mapsto tx$, which is a maximal general dual-feasible function as shown in [9]. Since f is a maximal general dual-feasible function, it must be identical to the dominating dual-feasible function, *i.e.* $f(x) = tx$ for all $x \in \mathbb{R}$.

If $t > 1$ then the dominating function is $g : \mathbb{R} \to \mathbb{R}$ with

$$g(x) := \max\{f(x), tx + 1 - t\}.$$

Obviously, $g(0) = 0$ and $g(x) \geq f(x)$ for all $x \in \mathbb{R}$. Moreover, g is monotonous, because both parts in the maximum expression are nondecreasing. The superadditivity of g has still to be checked. Choose for this purpose any $x, y \in \mathbb{R}$. Since

$$g(x + y) \geq f(x + y) \geq f(x) + f(y),$$

the desired result follows immediately, if $g(x) = f(x)$ and $g(y) = f(y)$. If $f(x) < g(x) = tx + 1 - t$, then $f(y) \leq ty$ yields $g(y) \leq ty$, and hence, we have

$$g(x) + g(y) \leq tx + ty + 1 - t \leq g(x + y).$$

The other case $f(y) < g(y)$ is similar, such that all cases are done. Since $g(1) = 1$ and g is nondecreasing and superadditive, it is a general dual-feasible function dominating f, but f is a maximal general dual-feasible function, and hence $f(x) = g(x) \geq tx + 1 - t$ for all $x \in \mathbb{R}$. □

These results are applied to prove partially symmetry in the following proposition. The idea to prove symmetry at the point $1/2$ under the given prerequisites is built on [2], but the approach by Carlier and Néron cannot be applied for the generalized dual-feasible functions.

Proposition 5. *If $f : \mathbb{R} \to \mathbb{R}$ is a maximal general dual-feasible function and not of the kind $x \mapsto tx$ with $0 \leq t < 1$, then $f(1) = 1$ and $f(1/2) = 1/2$.*

Proof. Let $t := \sup\{f(x)/x : x > 0\} \geq 0$. If $t < 1$, then $f(x) = tx$ for all $x \in \mathbb{R}$, such that the proof is complete. Otherwise $t \geq 1$ and $f(x) \geq tx + 1 - t$ for all $x \in \mathbb{R}$, and $f(1) \geq 1$, and therefore $f(1) = 1$. Since f is a maximal general dual-feasible function, it is nondecreasing and superadditive. Suppose $f(1/2) < 1/2$. Of course, $f(1/2) > 1/2$ is impossible. Define $g : \mathbb{R} \to \mathbb{R}$ by

$$g(x) := \begin{cases} f(x), \text{ if } x \neq 1/2, \\ 1/2, \text{ otherwise.} \end{cases}$$

Due to the assumption, g cannot be a general dual-feasible function, *i.e.* there are values $n \in \mathbb{N} \setminus \{0\}$ and $x_1, \ldots, x_n \in \mathbb{R}$ with $\sum_{i=1}^{n} x_i \leq 1$, but $\sum_{i=1}^{n} g(x_i) > 1$. Without loss of generality, assume $x_i = 1/2$ for $i \leq k$ and $x_i \neq 1/2$ for $i > k$ with a certain $k \in \mathbb{N}$, $k \leq n$. That yields $\sum_{i=k+1}^{n} x_i \leq 1 - k/2$ and

$$1 < \sum_{i=1}^{n} g(x_i) = k/2 + \sum_{i=k+1}^{n} f(x_i) \leq k/2 + f(\sum_{i=k+1}^{n} x_i) \leq k/2 + f(1 - k/2),$$

where empty sums equal zero. Since $f(1) = 1$, one gets further

$$\begin{aligned} 1 < \mathsf{frac}(k/2) + \lfloor k/2 \rfloor * 1 + f(1 - k/2) \leq \mathsf{frac}(k/2) + f(1 - k/2 + \lfloor k/2 \rfloor * 1) = \\ = \mathsf{frac}(k/2) + f(1 - \mathsf{frac}(k/2)). \end{aligned}$$

If k is even, then the contradiction $1 < 0 + f(1)$ arises, and if k is odd then $1 < 1/2 + f(1/2)$ yields also a contradiction. □

The following proposition provides further insight into the structure of maximal general dual-feasible functions. While Lemma 1 helps proving superadditivity, the following assertion demonstrates the limits of possible convexity for maximal functions.

Proposition 6. *If $f : \mathbb{R} \to \mathbb{R}$ is a maximal general dual-feasible function, then it cannot be strict convex in an environment of the point zero and also not concave (except linear) in an interval $[0, b]$ or $[-b, 0]$ with $b > 0$.*

Proof. We have that $f(0) = 0 \geq f(x) + f(-x)$ for any $x \in \mathbb{R}$, because f is a maximal general dual-feasible function. Therefore, f cannot be strict convex at 0. Suppose, f is concave in the interval $[0, b]$. Then $-f$ is convex in $[0, b]$, and hence, according to Lemma 1, $-f(x + y) \geq -f(x) - f(y)$ for all $x, y \in [0, b]$ with $x + y \leq b$. Therefore, we would have $f(x + y) = f(x) + f(y)$ for all these x, y, implying that f is linear in $[0, b]$ due to the monotonicity of f. The situation is similar for the interval $[-b, 0]$. □

Although the composition of general dual-feasible functions yield a general dual-feasible function, the same does not apply concerning maximality as shown in the following proposition.

Proposition 7. *The composition or convex combination of maximal general dual-feasible functions is not necessarily a maximal general dual-feasible function.*

Proof. As shown in [9], the following functions $f, g : \mathbb{R} \to \mathbb{R}$ are maximal general dual-feasible functions for any $b, c \in \mathbb{R}$ with $0 \leq c \leq 1 \leq b$:

$$f(x) := cx, \tag{19}$$

$$g(x) := \begin{cases} b * \lfloor 2x \rfloor, & \text{if } x < 1/2, \\ 1/2, & \text{if } x = 1/2, \\ 1 - b * \lfloor 2 - 2x \rfloor, & \text{if } x > 1/2. \end{cases} \tag{20}$$

Using them with parameters $c \in (0,1)$ and $b \geq 1/c$ yields the composition $f(g(\cdot)) = c * g(\cdot)$, *i.e.*

$$f(g(x)) = \begin{cases} bc * \lfloor 2x \rfloor, & \text{if } x < 1/2, \\ c/2, & \text{if } x = 1/2, \\ c - bc * \lfloor 2 - 2x \rfloor, & \text{if } x > 1/2, \end{cases}$$
$$\leq \begin{cases} bc * \lfloor 2x \rfloor, & \text{if } x < 1/2, \\ 1/2, & \text{if } x = 1/2, \\ 1 - bc * \lfloor 2 - 2x \rfloor, & \text{if } x > 1/2, \end{cases}$$

and for $x \geq 1/2$ the inequality is strict. The latter function is of the same type as the function (20) with parameter $bc \geq 1$, and hence it is a maximal general dual-feasible function. The function $f(g(\cdot))$ can also be seen as a convex combination of the constant zero-function $f \equiv 0$, which is also a maximal general dual-feasible function, with factor $1 - c$ and g with factor c. □

The following lemma is an extension of Lemma 3 of [6] to domain and range $\mathbb{R}$. This lemma gives a tool to prove that a symmetric function f, which is bounded by a maximal general dual-feasible function g for arguments less than $1/2$, is a maximal general dual-feasible function under some further conditions.

Lemma 4. *Let $f, g : \mathbb{R} \to \mathbb{R}$ be given with $f(1) = 1$, and fulfilling the conditions (3.) and (4.) of Theorem 1. If g is a maximal general dual-feasible function, and $f(x) \leq g(x)$ for all $x < 1/2$, and $f(x+y) \geq f(x) + f(y)$ for all $x, y \in \mathbb{R}$ with $x + y < 1/2$, then f is a maximal general dual-feasible function.*

Proof. The symmetry of f and g implies $f(0) = 0$ and $f(1/2) = 1/2 = g(1/2)$. Hence, $f(x) \leq g(x)$ for all $x \leq 1/2$. According to Theorem 1, it remains to be proved that f fulfills the superadditivity condition (9) for all $x, y \in \mathbb{R}$ with $x \leq y \leq \frac{1-x}{2}$ and not only if $x + y < 1/2$, as in the prerequisite demanded. Therefore, assume now $x + y \geq 1/2$. That yields $x \geq 0$ and $y \leq 1/2$, because $x + y \leq \frac{1+x}{2}$. Therefore, we have

$$\begin{aligned} f(x+y) &= 1 - f(1-x-y) \geq 1 - g(1-x-y) = \\ &= g(x+y) \geq g(x) + g(y) \geq f(x) + f(y). \end{aligned}$$

□

4 Conclusions

Building non-dominated general dual-feasible functions with no particular constraints on the values of their arguments is very relevant both for computing valid inequalities for integer linear optimization problems and strong lower bounds for combinatorial problems with knapsack constraints and general coefficients. For this purpose, it is fundamental to know the properties that characterize the set of maximal (non-dominated) general dual-feasible functions. In this paper, we showed that the properties of standard dual-feasible functions do not necessarily apply when one goes into the domain of real numbers. We identified and proved formally many relevant properties of maximal general dual-feasible functions. The knowledge of these properties will help in avoiding the use of weak general dual-feasible functions, and it will guide researchers into the development of useful families of strong functions.

Acknowledgements. This work was supported by FEDER funding through the Programa Operacional Factores de Competitividade - COMPETE and by national funding through the Portuguese Science and Technology Foundation (FCT) in the scope of the project PTDC/EGE-GES/116676/2010. Additionally, the work was supported by FCT through the postdoctoral grant SFRH/BPD/45157/2008 for Jürgen Rietz.

References

1. Alves, C., de Carvalho, J., Clautiaux, F., Rietz, J.: Multidimensional dual-feasible functions and fast lower bounds for the vector packing problem. Eur. J. Oper. Res. 233, 43–63 (2014)
2. Carlier, J., Néron, E.: Computing redundant resources for the resource constrained project scheduling problem. Eur. J. Oper. Res. 176, 1452–1463 (2007)
3. Clautiaux, F., Alves, C., de Carvalho, J.: A survey of dual-feasible and superadditive functions. An. Oper. Res. 179, 317–342 (2010)
4. Johnson, D.: Near optimal bin packing algorithms. Dissertation. Massachussetts Institute of Technology, Cambridge, Massachussetts (1973)
5. Nemhauser, G., Wolsey, L.: Integer and combinatorial optimization. Wiley-Interscience (1998)
6. Rietz, J., Alves, C., de Carvalho, J.: Theoretical investigations on maximal dual feasible functions. Oper. Res. Let. 38, 174–178 (2010)
7. Rietz, J., Alves, C., de Carvalho, J.: Worst-case analysis of maximal dual feasible functions. Opt. Let. 6, 1687–1705 (2012)
8. Rietz, J., Alves, C., de Carvalho, J.: On the extremality of maximal dual feasible functions. Oper. Res. Let. 40, 25–30 (2012)
9. Rietz, J., Alves, C., de Carvalho, J., Clautiaux, F.: Computing valid inequalities for general integer programs using an extension of maximal dual-feasible functions to negative arguments. In: 1st International Conference on Operations Research and Enterprise Systems, Vilamoura (2012)

A Global Optimization Approach Applied to Structural Dynamic Updating

Marco Dourado[1], José Meireles[1], and Ana Maria A.C. Rocha[2]

[1] Department of Mechanical Engineering,
Centre for Mechanical and Materials Technologies (CT2M), University of Minho, Portugal
{mdourado,meireles}@dem.uminho.pt
[2] Department of Production and Systems,
Algoritmi Research Centre, University of Minho, Portugal
arocha@dps.uminho.pt

Abstract. In this paper, the application of stochastic global optimization techniques, in particular the *GlobalSearch* and *MultiStart* solvers from MatLab®, to improve the updating of a structural dynamic model, are presented. For comparative purposes, the efficiency of these global methods relatively to the local search method previously used in a *Finite Element Model Updating* program is evaluated. The obtained solutions showed that the *GlobalSearch* and *MultiStart* solvers are able to achieve a better solution than the local solver previously used, in the updating of a structural dynamic model. The results show also that the *GlobalSearch* solver is more efficient than the *MultiStart*, since requires less computational effort to obtain the global solution.

Keywords: Finite Element Model Updating, Global Optimization, Structural Dynamic.

1 Introduction

Optimization problems can go from simple linear functions with few variables, until the most complex problems of non-linear functions, with many variables, with constraints on the variables and many optimal local solutions [1].

Depending on the problem under study, local or global optimization methods can be used to find the maximum or minimum of a function. The selection of a method for a particular application depends on the characteristics of the problem and what is desired, such as type of design variables, whether or not all local minima are desired, and availability of gradients of the functions. Many engineering optimization problems are multimodal and require the application of global search methodologies, in order to avoid the optimizer to be trapped in the first minimum or maximum local found. The global search methodologies allow the optimizer to evolve into other areas of the feasible region, being possible to obtain more and best solutions.

There are two major classes of methods depending on whether or not they incorporate any stochastic elements to solve the global optimization problem: deterministic and stochastic methods.

B. Murgante et al. (Eds.): ICCSA 2014, Part II, LNCS 8580, pp. 195–210, 2014.

Deterministic methods provide a theoretical guarantee of locating the global minimum. Stochastic methods only give guarantee in a probabilistic sense that the global minimum point will be found. On the other hand, stochastic methods are usually faster in locating a global optimum than deterministic ones. In most global optimization algorithms (both deterministic and stochastic) it is possible to identify two phases: a global phase and a local phase. The exhaustive exploration of the search space is delegated to the global phase, where the function is evaluated at a number of randomly sampled points. In the local phase, the sample points are manipulated, by means of local searches, to yield a candidate global minimum [2]. For an introduction to deterministic and stochastic methods in global optimization, see e.g. Horst and Tuy [3] and Törn and Zilinskas [4], respectively.

There are some examples of application of deterministic methods in structural engineering area, such as the work of Stolpe [5], that presents a *branch-and-bound* algorithm for global optimization of the minimum weight stress-constrained truss topology problem extended with displacement bounds and local buckling constraints. Later, using the same algorithm, Achtziger and Stolpe [6] developed a study to determine the optimal variables of a truss structure. However, Lin and Chen [7] emphasize that deterministic methods are based on assumptions of objective and constraint functions, and therefore the deterministic methods cannot be applied to general structural problems with satisfactory efficiency.

Thus, the stochastic methods became relevant to solve the most global optimization problems, since they adapt better to real problems or black-box formulations, and they prove to be very useful for applications in the field of structural engineering optimization problems, as for example, presented by Lucor [8]. They were inspired on natural environmental, biological, physical and chemical processes, composed by populations of individuals or elements that interact between them, and with their environment. The algorithms based on these natural phenomena are called *Swarm Intelligence* [9]. Those social behaviors have been crucial for the development of the random search methods and multi start methods. Some examples of application of these two subclasses of stochastic methods are the work of Lin and Chen [7] to study multistage optimization algorithms for simultaneously seeking multiple optimal solutions in a structural problem and Eriksson and Arora [10] to study the efficiency of three global stochastic optimization algorithms, with continuous variables, in the optimization of the ride comfort of a city bus. Within the stochastic methods are also the sub-class of evolutionary methods, such as *Genetic Algorithms* (GA), and *Simulated Annealing* (SA), used by Sonmez [11] to obtain multi optimal shapes for two-dimensional structures subject to quasi-static loads and restraints, and Venanzi e Materazzi [12] to optimize wind-excited structures. The hybridization of a genetic algorithm and a non-smooth proximal bundle method is used in Auvinen et al. [13] to minimize the weight of a forest machine, and Keller [14] applied evolutionary algorithms to a case study of an air-plane's side rudder.

This paper intends to show the application of global stochastic optimization methods, in the structural engineering field, namely in the optimization of structural dynamic models with resort to methods of improving finite element models, usually denoted by *Finite Element Model Updating*. These improvements can be conducted under two types of approach:

1. in the updating of simplified numerical models, representative of detailed physical models which present high computation time. The simplified model is submitted to updating by a *Finite Element Model Updating* methodology until obtain dynamic behavior similar to the physical model, also denominated as reference model [15]. Thus, it is possible to obtain a light computationally model and at the same time representative of the physical model. It is important to refer that, in these cases, the main interest is in the correlation of dynamic behavior, independently of the parameters values optimized;
2. in the structural modification to the optimization of the models. Detailed numerical models of physical models are built and submitted to optimization to: improve the dynamic behavior and/or achieve a model with similar behavior but with geometrical and/or physical parameters more advantageous from the design point of view [16].

The optimization methodologies help to fit on the control of updating process, nevertheless still constitute a developing task. Important works in the *Finite Element Model Updating* area, using global stochastic optimization methods, could be found in Levin and Lieven [17], that compares various implementations of two algorithms, the GA and SA, to find the global minimum, amongst many local minima, of an objective function that describes the finite element model updating of a flat plate wing structure in the frequency domain. Teughels et al. [18] use the *Coupled Local Minimizers* (CLM) method in the *Finite Element Model Updating* program for the damage identification of a reinforced concrete beam. The method combines the fast convergence of the local gradient-based algorithms with the global approach of GA, resulting in an efficient global optimization algorithm, able to find the global minimum of the objective function. The same method was used by Bakir et al. [19] to update the finite element model of a reinforced concrete frame, using 24 design variables. The authors compare the CLM method with different optimization local search methods, such as the *Gauss–Newton* method, *Levenberg–Marquardt* algorithm and *Sequential Quadratic Programming* (SQP) algorithm, and prove that the global method gave better results. Ameri et al. [20] used the *Globalized Bounded Nelder-Mead* method to find the optimal fiber orientation of laminated cylindrical panels based on natural frequencies by maximization of fundamental natural frequency. The obtained results show good accuracy and cost optimization when compared with results of GA.

Following the same principles of the cited authors, and in order to improve the efficiency of a *Finite Element Model Updating* program, two global stochastic optimization techniques, the *GlobalSearch* and *MultiStart* commands available in Matlab®, are used and compared with each other when applied to the updating process of a structural model. The aim is to compare the obtained solutions with the local solutions previously obtained in the *Finite Element Model Updating* program, developed by Meireles [21,22]. This *Finite Element Model Updating* program, has implemented in its optimizer a local search method that uses the SQP algorithm performed through the *fmincon* command from Matlab® to find the optimal global value. However, this implemented local search strategy, have difficulties to reach the global optimum, since it was developed to find local solutions.

The organization of this paper is as follows. Section 2 presents the mathematical formulation of the problem. Section 3 describes the optimization process and the models description used in the optimization process are presented in Section 4. Section 5 shows the computational experiments done with the local and the global solvers as well as a discussion of the obtained solutions. This paper is concluded in Section 6.

2 Problem Formulation

The optimization problem consists in the minimization of an objective function, related with the frequencies and respective mode shapes correlation between the reference model and the numerical model, defined as

$$\begin{array}{ll} \min & f(x) \\ \text{s.t.} & x_{LB} \leq x \leq x_{UB} \end{array} \tag{1}$$

where x is the vector with the updated parameters for the numerical model, and x_{LB}, and x_{UB} are the lower and upper bounds on the variables, respectively.

The objective function of the optimization problem is defined by the sum of three specific functions, as

$$f(x) = f_{\varphi C}(x) + f_{\varphi U}(x) + f_{\lambda}(x) \tag{2}$$

where $f_{\varphi C}(x)$ represents the quantification of the difference between numerical and reference correlated mode pairs, $f_{\varphi U}(x)$ represents the quantification of the difference between numerical and reference uncorrelated mode pairs and $f_{\lambda}(x)$ represents the quantification of the difference between numerical and reference frequencies.

Function $f_{\varphi C}(x)$ is given by

$$f_{\varphi C}(x) = -\frac{\sum_{i=1}^{N_C} MAC_{ii}(x)}{\sum_{i=1}^{N_C} MAC_{ii}(x^0)} \tag{3}$$

where N_C is the number of correlated mode pair values of the diagonal MAC matrix and the vector x^0 contains the initial updating parameters. The MAC matrix is defined by

$$MAC_{ij}(x) = \frac{\left(\left(\varphi_i^{Ref}\right)^T \varphi_j^{Num}\right)^2}{\left(\left(\varphi_i^{Ref}\right)^T \varphi_i^{Ref}\right)\left(\left(\varphi_j^{Num}\right)^T \varphi_j^{Num}\right)} \tag{4}$$

where, φ_i^{Ref} is the ith reference mode shape and φ_j^{Num} is the jth numerical mode shape [22].

Function $f_{\varphi U}(x)$ is given by

$$f_{\varphi U}(x) = \left(\frac{1}{N_U}\right) \frac{\sum_{j=1}^{N_U} \sum_{\substack{i=1 \\ j \neq i}}^{N_U} MAC_{ij}(x)}{\sum_{j=1}^{N_U} \sum_{\substack{i=1 \\ j \neq i}}^{N_U} MAC_{ij}(x^0)} \tag{5}$$

where N_U is the number of uncorrelated mode pairs values, outside of the diagonal MAC matrix.

Function $f_\lambda(x)$ represents the quantification of the difference between numerical and reference frequencies, given by

$$f_\lambda(x) = -\frac{\sum_{i=1}^{N_\lambda} \left(\omega_i^{Ref} - \omega_i^{Num}(x)\right)^2}{\sum_{i=1}^{N_\lambda} \left(\omega_i^{Ref} - \omega_i^{Num}(x^0)\right)^2} \tag{6}$$

where N_λ is the number of eigenvalues λ corresponding to the correlated mode pairs, ω^{Ref} is the reference frequency and ω^{Num} is the numerical frequency, respectively defined by $\omega^{Ref} = \sqrt{\lambda^{Ref}/2\pi}$ and $\omega^{Num} = \sqrt{\lambda^{Num}/2\pi}$. The quadratic term in (6) is used to accelerate the convergence and to obtain only positive differences between the frequencies of the two models. Numerical mode shapes φ^{Num} and numerical eigenvalues λ^{Num} are function of these updating parameters. The relationship between them can be written as

$$(\varphi^{Num}, \lambda^{Num}) = f(x_1, x_2, x_3, \dots, x_p) \tag{7}$$

where p is the number of updating parameters. The updated physical parameters x, that represent the best improvement of the numerical model, are obtained when the objective function (2) is minimized.

3 Optimization Process

The optimization process uses the interaction between Matlab® and Ansys® to improve the dynamic characteristics of the numerical model calculating the objective function value and finding the optimal value of the physical parameters.

The flowchart of the interaction algorithm between optimization method in Matlab® and Ansys® is represented in Fig.1.

The first step of the structural optimization process is to idealize the desired behavior of the dynamic model to develop, or collect experimental data of a physical model considered as the reference model. The next step is associated with the construction of a numerical model in the finite element program ANSYS®, that should describe the idealized dynamic model from which its dynamic characteristics are obtained. These dynamic characteristics are transferred to the optimizer of the *Finite Element Model Updating* program developed in MatLab®, in order to optimize the dynamic behavior of the numerical model when compared to the reference model. It is considered that the type of structure, in that this methodology is applicable, is sufficiently rigid that the damping can be neglected.

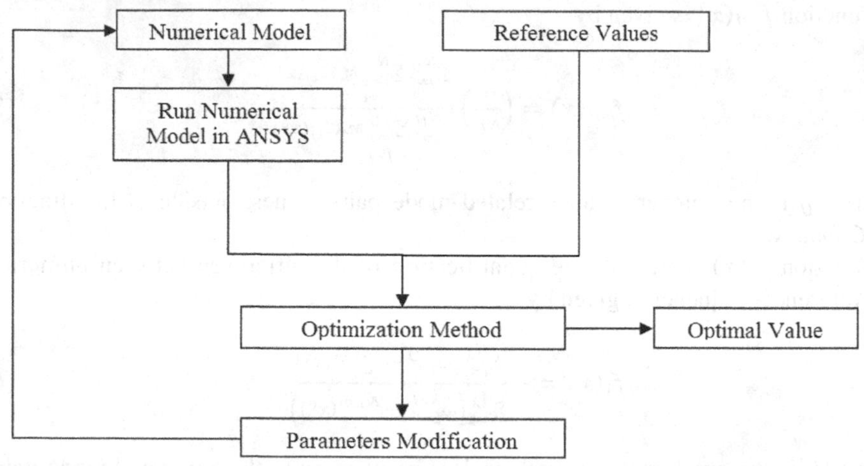

Fig. 1. Interaction flowchart between Matlab® and Ansys®

In this study, the Finite Element Model Updating program, implemented in MatLab®, uses a global solver, provided by the *Global Optimization Toolbox* [23] that searches for the optimal global value of the objective function (1). Two global solvers are used in the optimization process, performed by *GlobalSearch* and *MultiStart* commands, in order to test its efficiency and effectiveness in the updating process of a structural model. A prior version of *the Finite Element Model Updating* program uses the local solver, provided by the command *fmincon* from Matlab [21,22]. Following, the referred commands are briefly introduced.

The *fmincon* command aims to find a minimum of a constrained problem of multiple variables. Given an initial starting point, this solver can work with four algorithms type: *active-set*, *interior-point*, SQP and *trust-region-reflective*. As described in [23], the *active-set* and SQP algorithms work of similar way. In these algorithms, a *Quadratic Programming* subproblem is solved, where, at each iteration, the BFGS (*Broyden–Fletcher–Goldfarb–Shanno*) formulae is used to estimate the *Hessian* of the *Lagrangian* function. The *interior-point* algorithm is an approach to solve a sequence of approximate minimization problems. The *trust-region-reflective* algorithm is a subspace *trust-region* method and is based on the *interior-reflective Newton* method. Here, each iteration involves the approximate solution of a large linear system using the method of *Preconditioned Conjugate Gradients*. A work with application of this method can be found in Voormeren and Rixen [24].

GlobalSearch and *MultiStart* implement stochastic search methods and are similar when finding global or multiple solutions. Both algorithms use multiple start points to sample multiple basins of attraction and start a local solver, such as *fmincon*, from a variety of starting points and store local and global solutions found during the search process. Generally the starting points are random.

The *GlobalSearch* solver performs in two phases: a local phase and a global phase. In the local phase, the sampled points, randomly obtained, are manipulated by a local search to find candidates for local minimum. In the global phase the local minimum with best objective function value is used as an approximation to the global optimum. The solver uses a scatter search strategy in order to generate the trial points. Then, it analyzes the start points and rejects all of those that are unlikely to improve the best local minimum found so far.

The *MultiStart* solver uses uniformly distributed start points within bounds, or user-supplied start points. Then, it runs the local solver at all start points, or, optionally, all start points that are feasible with respect to bounds or inequality constraints.

4 Models Description

In this section the models description is presented, where a numerical model will be optimized taking into account a reference model from which are extracted the reference values of mode shapes and respective natural frequencies.

4.1 Reference Model

The reference model is a steel sheet with dimensions 200x300x10 mm^3, represented by width (w), height (h) and thickness (t), as shown in Fig.2.

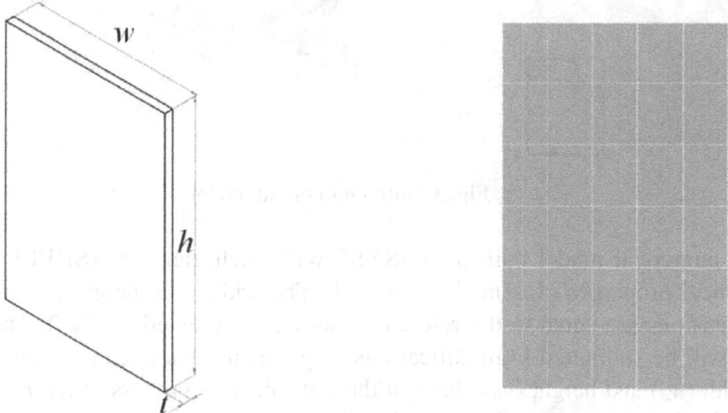

Fig. 2. Reference model

This model is built in ANSYS® with shell elements (SHELL63), and is submitted to modal analysis for extraction of mode shapes measured in 24 points and respective natural frequencies. The mechanical properties of the steel sheet are presented in Table 1.

Table 1. Mechanical properties of the reference model

Property	Symbol	Units	Value
Young's Module	E_x	Pa	$2.1\text{x}10^{11}$
Young's Module	$E_y = E_z$	Pa	$2.2\text{x}10^{11}$
Poisson's Ratio	$\upsilon_{xy} = \upsilon_{yz} = \upsilon_{zx}$	-	0.27
Density	ρ	kg/m^3	7847

4.2 Numerical Model

The numerical model to be optimized has a set of 240 areas of variable geometry, as shown in Fig.3. The areas are created from points, some with variable coordinates, enabling the change of all areas of the model. The points with variable coordinates are function of the geometrical parameters: width (w_a) and height (h_b). The coordinates of the points chosen for reading the mode shapes are kept constant in order to coincide with the readings of reference.

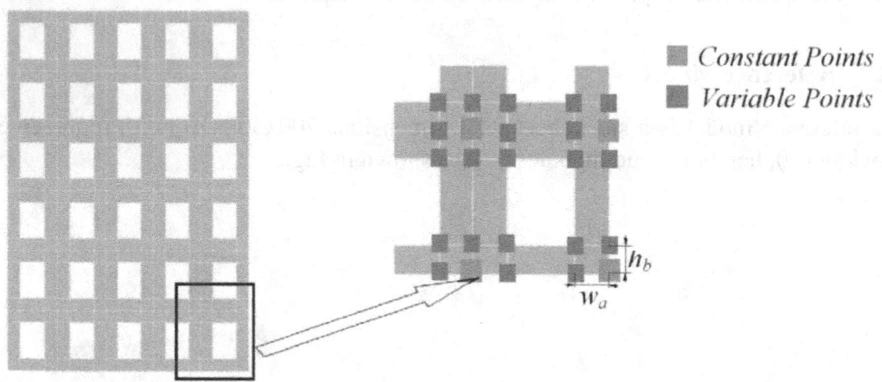

Fig. 3. Initial numerical model

The numerical model built in ANSYS® with shell elements (SHELL63), has the mechanical properties presented in Table 1. The width (w), height (h) and thickness (t) dimensions are equal to the reference model, represented in Fig.2. The numerical model will be submitted to modifications of geometric parameters, such as thickness (t), width (w_a) and height (h_b), through the optimization process in the *Finite Element Model Updating* program. The initial values of the parameters and their lower and upper bounds are indicated in Table 2.

Table 2. Parameters vector of the numerical model

Property	Variable	Units	Initial Value	Lower bound	Upper bound
Thickness	t	mm	10	1	20
Width a	w_a	mm	10	10	19
Height b	h_b	mm	10	10	24

It is expected that the optimal value of width (w_a) and height (h_b) variables, have a clear tendency to converge to the upper bound, in order to fill the empty spaces of the steel sheet.

5 Computational Experiments

In this section the numerical model is optimized and the computational results are presented. First, the solutions obtained with the local solver *fmincon* are showed, and then the ones with each of the global solver, *GlobalSearch* and *MultiStart*.

The local solver *fmincon* is performed on the supplied initial point x^0 with the *active-set, interior-point*, SQP and *trust-region-reflective* algorithms. The *GlobalSearch* solver is performed with 100 and 400 trial points, where the number of points analyzed in stage one is 100 and 400, respectively. So, the *GlobalSearch* applies the *fmincon* solver, first, in the supplied initial point x^0 and, second, in the starting points defined in the option *NumStageOnePoints*, making only an initial assessment of the score function of each one. Finally the *GlobalSearch* applies the *fmincon* solver in the point with best score. Therefore, the *GlobalSearch* solver makes complete evaluation in only two points, in the supplied initial point x^0 and in the best starting point among the trial points of stage one. The *MultiStart* solver is performed with 10 and 20 trial points, and *fmincon* solver is executed in all of them, and the evaluation is complete for all of them. All the others optional parameter values of the optimization solvers have the default values.

5.1 Local Solver Results

For the local solver *fmincon* analysis, the search is only performed on the starting point x^0 and theoptimization results are presented in Table 3.

Table 3. Optimization results for *fmincon* solver

Output		active-set	interior-point	SQP	trust-region
Nr. function evaluations		162	138	239	162
Optimization time [h]		~1.700	~1.500	~2.500	~1.700
	t	10.160	10.158	10.142	10.160
x local [mm]	w_a	14.582	14.677	15.318	14.582
	h_b	14.881	14.843	14.528	14.881
Optimal local value *f(x)*		4.402	4.401	4.393	4.402

The best optimal value found, 4.393, is achieved by SQP algorithm. With the other algorithms the optimal value of $f(x)$ is very similar between them. The solver requires 239 function evaluations and 2.5 hours to achieve the best optimal local value of the objective function.

5.2 *GlobalSearch* Solver Results

In the *GlobalSearch* solver analysis, there are, in general, improvements relatively to the local solver solution. In the first experiment, with 100 trial points, where optimization results are presented in Table 4, the SQP algorithm obtains the optimal value of 4.163, which means an improvement of 5.236% when compared with the optimal solution obtained with the local search method (4.393). With the other algorithms, only the interior-point achieves a slight improvement of 0.068% when compared with the value obtained with the same algorithm (4.401). With the active-set and trust-region-reflective algorithms do not found any improvement there. The *Globalsearch* solver with the SQP algorithm requires 532 function evaluations and 5.5 hours to achieve the best optimal global value, requiring approximately twice more (120%) of optimization time and function evaluations than with the local solver.

Table 4. Optimization results for *GlobalSearch* with 100 trial points

Output		active-set	interior-point	SQP	trust-region
Nr. function evaluations		526	505	532	526
Optimization time [h]		~5.500	~5.200	~5.500	~5.500
	t	10.160	10.152	9.990	10.160
x global [mm]	w_a	14.582	14.878	19.000	14.582
	h_b	14.881	14.762	10.000	14.881
Optimal global value $f(x)$		4.402	4.398	4.163	4.402

When using 400 trial points (see Table 5), the results are about the same although a large computational time (more than double).

Table 5. Optimization results for *GlobalSearch* with 400 trial points

Output		active-set	interior-point	SQP	trust-region-
Nr. function evaluations		1149	1224	1168	1215
Optimization time [h]		~12.500	~13.300	~12.800	~13.200
	t	10.145	10.135	9.990	10.138
x global [mm]	w_a	15.137	15.440	19.000	15.389
	h_b	14.645	14.517	10.000	14.511
Optimal global value $f(x)$		4.394	4.391	4.163	4.392

The *active-set* algorithm achieves a little improvement of 0.182% compared with the value obtained with the local solver (4.402). The *interior-point* and *trust-region-reflective* algorithms achieve a slight improvement of 0.227% compared with the value obtained with the local solver (4.401 and 4.402, respectively). The optimal global value obtained with the SQP algorithm remains in 4.163. The solver requires 1168 function evaluations and 12.8 hours to achieve the best optimal global value of the objective function, needing approximately five times more (412%) of optimization time and function evaluations than with the local solver.

5.3 *MultiStart* Solver Results

In the *MultiStart* solver analysis, when the search is performed with 10 trial points (see results in Table 6), the four algorithms have a favorable evolution, when compared to local solver results. The *active-set* and *trust-region-reflective* algorithms achieve a slight improvement of 0.250% compared with the value obtained with the local solver (4.402). The *interior-point* algorithm achieves a little improvement of 0.182% compared with the value obtained with the local solver (4.401). The SQP algorithm achieves an improvement of 5.236% compared with the value obtained with the local solver (4.393) and achieve the best optimal global value of function *f(x)* (4.163). The solver requires 1735 function evaluations and 19 hours to achieve the best optimal global value of the objective function, needing approximately 7.6 times more (660%) of optimization time and function evaluations than with the local solver.

Table 6. Optimization results with *MultiStart* for 10 trial points

Output		active-set	interior-point	SQP	trust-region-
Nr. function evaluations		1553	1628	1735	1665
Optimization time [h]		~17.000	~17.800	~19.000	~18.200
	t	10.135	10.140	9.990	10.137
x global [mm]	w_a	15.433	15.279	19.000	15.400
	h_b	14.519	14.596	10.000	14.522
Optimal global value *f(x)*		4.391	4.393	4.163	4.391

When executing *MultiStart* with 20 trial points, where results are presented in Table 7, the improvements are more evident relatively to the ones obtained with the local solver.

Table 7. Optimization results with *MultiStart* for 20 trial points

Output		active-set	interior-point	SQP	trust-region-
Nr. function evaluations		3588	2887	3804	3200
Optimization time [h]		~39.200	~31.500	~41.600	~35.000
	t	9.990	10.135	9.990	9.990
x global [mm]	w_a	19.000	15.444	19.000	19.000
	h_b	10.000	14.513	10.000	10.000
Optimal global value *f(x)*		4.163	4.391	4.163	4.163

The optimal value of function $f(x)$ improves 5.429% with *active-set* and *trust-region-reflective* algorithms compared with the value obtained with the local solver (4.402). The *interior-point* algorithm improves slightly 0.227% compared with the value obtained with the local solver (4.401). The optimal global value obtained with SQP algorithm remains in 4.163. The solver requires 3200 function evaluations and 35 hours to achieve the best optimal global value of the objective function with the *trust-region-reflective* algorithm, needing approximately 14 times more (1300%) of optimization time and function evaluations than with the local solver.

5.4 Discussion of Results

Following the discussion of results is presented for local and global solutions.

Local Solution Discussion

In general, the local solver *fmincon* converges to a good solution for the four algorithms, achieving a correlation of mode shapes and natural frequencies, between the reference and numerical model, with good quality.

The color graphs of Fig.4 represent the MAC matrix and frequencies matrix, and quantifies the correlation among the reference and numerical model. In MAC matrix the diagonal should be as dark as possible and bright outside of the diagonal, to represent a good correlation among mode shapes, and the frequencies matrix should be as bright as possible to represent a good correlation among the frequencies.

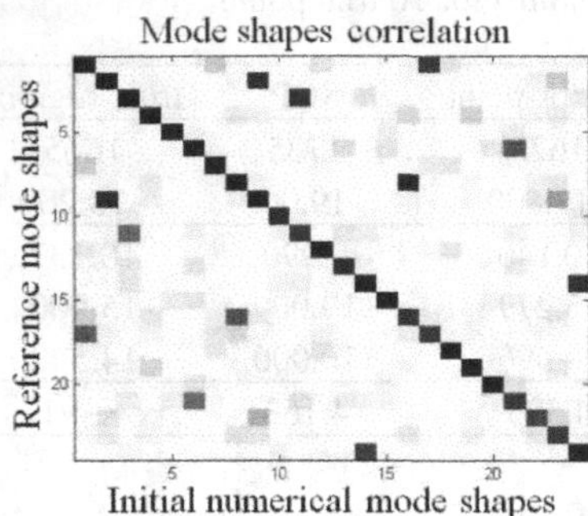

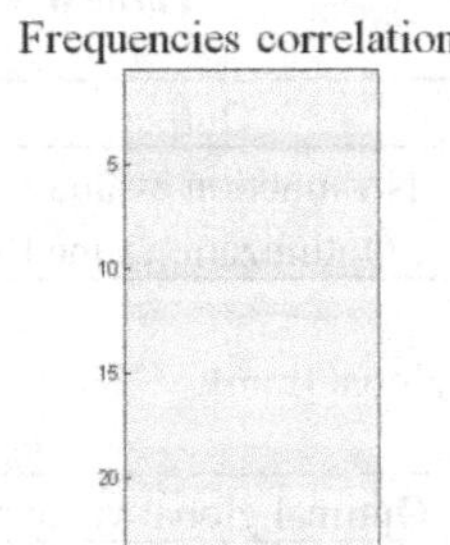

Fig. 4. Initial correlation

The value of first function evaluation, for any used algorithms, is 23.042, because the initial point x^0 is the same for all cases. This value has the meaning of the geometric distance between the reference model and initial numerical model, imposed by initial variables of point x^0.This originates a weak correlation between, mainly, the natural frequency values of the two models, since the correlation between all mode shapes in diagonal MAC matrix is quite close to the unit, as shown in Fig.4.

After the optimization is complete, the quality of the natural frequencies correlation improves considerably, and reveals a slight improvement in MAC matrix, as shown in Fig.5a. The SQP algorithm is the one that achieves the best optimal value of objective function, and consequently, the best correlation among the two models.

The final numerical model, presented in Fig.5b, suffers significant changes due to the convergence of width (w_a) and height (h_b) parameters to the upper bounds. As the thickness (t) parameter suffers a small change in relation to the initial value, the numerical model is now closer to the reference model, both geometrically and in terms of its dynamic behavior.

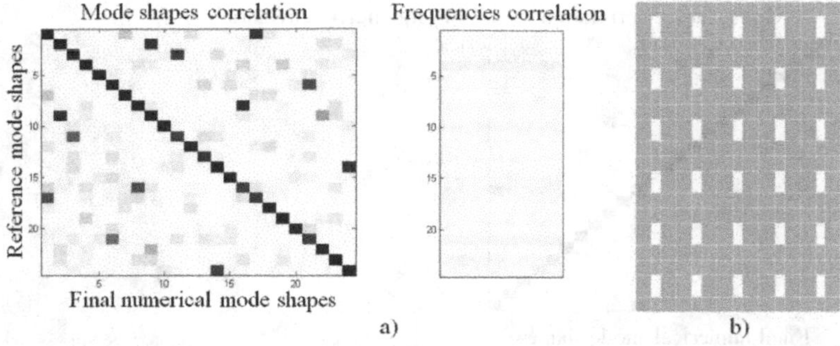

Fig. 5. a) Best correlation for local optimization; b) Best final numerical model for local optimization

Global Solution Discussion

The global solvers, *GlobalSearch* and *MultiStart* are able to converge to a better solution than with the local solver *fmincon*, since we are facing a multimodal problem and they are prepared to find global solutions.

The *MultiStart* solver is the one that reveals more robustness in the set of the four algorithms. With 10 trial points, just the SQP algorithm is able to achieve the best optimal value of objective function (4.163), but with 20 trial points just *interior-point* algorithm does not achieves this value. The *GlobalSearch* solver does not reveal as robust as the *MultiStart* solver, because just the SQP algorithm is able to obtain the best optimal value of objective function $f(x)$. Despite the higher number of trial points, the *GlobalSearch* solver has the advantage of being able to select the best trail points among the starting points defined in the option *NumStageOnePoints* and reject the others. The solution obtained with the *GlobalSearch* solver using 100 trial points saves optimization time and function evaluations in approximately 3.5 times (245.5%) face to *MultiStart* solver with 10 trial points and approximately 6.4 times (536.4%) face to *MultiStart* solver with 20 trial points. Hence, the *GlobalSearch* solver is more efficient than *Multistart* since requires less computational effort to obtain the global solution.

The quality of the correlation between mode shapes and natural frequencies of the two models is presented in Fig.6a and illustrates the improvement relatively to the local solver when using a global solver.

The final numerical model, presented in Fig.6b, is closer to the reference model because the width (w_a) parameter converges for the upper bound value. The height (h_b) parameter keeps the initial value, and the thickness (t) parameter suffers a small change, with regard to initial value, and converges to the lower bound. This parameter together with the other two, originates a final numerical model with very similar geometry and dynamic behavior in relation to the reference model.

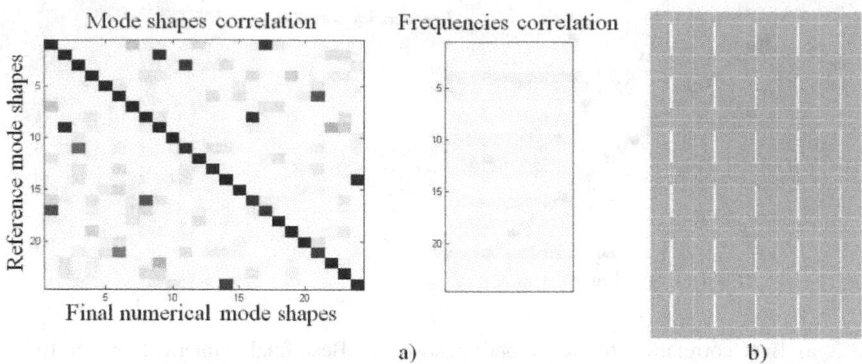

Fig. 6. a) Best correlation for global optimization; b) Best final numerical model for global optimization

6 Conclusions

The aim of this paper was to apply two stochastic global optimization techniques for the optimization of a dynamic structural finite element model, and to establish a comparison with the previously local search method used in the *Finite Element Model Updating* program. The global solvers have the advantage of being able to work with a higher number of trial points, and therefore, are more efficient than the local solver. The two global solvers tested work in a different way, and therefore the results may also be different. Both global solvers achieve the same optimal global value of the objective function, requiring, however, different optimization times and function evaluations. In this case, the *GlobalSearch* is the fastest solver to achieve the best optimal global value when working with the SQP algorithm. The *MultiStart* solver achieved the same best optimal global value with the active-set, SQP and *trust-region-reflective* algorithms however needed six times more computational effort in terms of execution time and number of function evaluations.

The example used can be considered too oriented, which may increase the possibility of convergence of the local method, and somehow reduce the ability of perception of higher capacity of global methods. However, it was evident the evolution of the final numerical model to get closer to the geometry of the reference model when applied global optimization techniques.

In the future, more complex models will be studied and the use of stochastic global optimization methods based on Swarm Intelligence will be investigated.

Acknowledgments. The authors gratefully acknowledge the Centre for Mechanical and Materials Technologies (CT2M) and the Portuguese Funds through FCT - "Fundação para a Ciência e a Tecnologia" under Project PEst-OE/EEI/UI0319/2014.

References

1. Nocedal, J., Wright, S.J.: Numerical Optimization. Series in Operations Research. Springer, Heidelberg (1999)
2. Horst, R., Pardalos, P.: Handbook of Global Optimization. Kluwer (1995)
3. Horst, R., Tuy, H.: Global Optimization Deterministic Approaches. Springer (1996)
4. Törn, A., Zilinskas, A.: Global Optimization. Springer (1989)
5. Stolpe, M.: Global optimization of minimum weight truss topology problems with stress, displacement, and local buckling constraints using branch-and-bound. Int. J. Numer. Meth. Eng. 61, 1270–1309 (2004)
6. Achtziger, W., Stolpe, M.: Truss topology optimization with discrete design variables - Guaranteed global optimality and benchmark examples. Struct. Multidisc. Optim. 34, 1–20 (2007)
7. Lin, C.-Y., Chen, W.-T.: Stochastic multistage algorithms for multimodal structural optimization. Comp. Struct. 24, 233–241 (2000)
8. Lucor, D., Enaux, C., Jourdren, H., Sagaut, P.: Stochastic design optimization: Application to reacting flows. Comput. Method Appl. M. 196, 5047–5062 (2007)
9. Birbil, S.I.: Stochastic Global Optimization Techniques. Faculty of North Carolina State University of Raleigh (2002)
10. Eriksson, P., Arora, J.S.: A comparison of global optimization algorithms applied to a ride comfort optimization problem. Struct. Multidisc. Optim. 24, 157–167 (2002)
11. Sonmez, F.O.: Shape optimization of 2D structures using simulated annealing. Comput. Method Appl. M. 196, 3279–3299 (2007)
12. Venanzi, I., Materazzi, A.L.: Multi-objective optimization of wind-excited structures. Eng. Struct. 23, 983–990 (2007)
13. Auvinen, P., Makela, M.M., Makinen, J.: Structural optimization of forest machines with hybridized nonsmooth and global methods. Struct. Multidisc. Optim. 23, 382–389 (2002)
14. Keller, D.: Global laminate optimization on geometrically partitioned shell structures. Struct. Multidisc. Optim. 43, 353–368 (2011)
15. Mottershead, J.E., Friswell, M.I.: Model updating in structural dynamics: A survey. J. Sound Vib. 167, 347–375 (1993)
16. Maia, N., Montalvão e Silva, J.: Theoretical and Experimental Modal Analysis. Research Studies Press Ltd., Hertfordshire (1997)
17. Levin, R., Lieven, N.: Dynamic Finite Element Model Updating Using Simulated Annealing and Genetic Algorithms. Mech. Syst. Signal Pr. 12, 91–120 (1998)
18. Teughels, A., De Roeck, G., Suykens, J.A.K.: Global Optimization by coupled local minimizers and its application to FE model updating. Comp. Struct. 81, 2337–2351 (2003)
19. Bakir, P.G., Reynders, E., De Roeck, G.: An improved finite element model updating method by the global optimization technique 'Coupled Local Minimizers'. Comp. Struct. 86, 1339–1352 (2008)
20. Ameri, E., Aghdam, M.M., Shakeri, M.: Global optimization of laminated cylindrical panels based on fundamental natural frequency. Compos. Struct. 94, 2697–2705 (2012)
21. Meireles, J.: Análise Dinâmica de Estruturaspor Modelos de Elementos Finitos Identificados Experimentalmente. PhD Thesis, University of Minho (2007) (in Portuguese)
22. Allemang, R.J., Brown, D.L.: A Correlation Coefficient for Modal Vector Analysis. In: Proceedings of the 1st International Modal Analysis Conference, Florida, Holiday Inn (1982)

23. MathWorks, Global Optimization Toolbox: User's Guide R2011b. The MathWorks Inc., Massachusetts (2011)
24. Voormeren, S., Rixen, R.: Updating component reduction bases of static and vibration modes using preconditioned iterative techniques. Comput. Method Appl. M. 253, 39–59 (2013)

A Hybrid Heuristic Based on Column Generation for Two- and Three- Stage Bin Packing Problems

Filipe Alvelos[1], Elsa Silva[2], and José Manuel Valério de Carvalho[1]

[1] Centro Algoritmi / Departamento de Produção e Sistemas, Universidade do Minho, 4710-057 Braga, Portugal

[2] INESC TEC (formerly INESC Porto), Campus da FEUP, Rua Dr. Roberto Frias, 378, 4200-465 Porto, Portugal

{falvelos,vc}@dps.uminho.pt, emsilva@inescporto.pt

Abstract. We address two two-dimensional bin packing problems where the bins are rectangular and have the same size. The items are also rectangular and all of them must be packed with the objective of minimizing the number of bins. In the first problem, the two-stage problem, the items must be packed in levels. In the second problem, the restricted 3-stage problem, items can be grouped in stacks which are packed in levels.

We propose a new decomposition model where subproblems are associated with the item that initializes a bin. The decomposition is solved by a heuristic which combines (perturbed) column generation, local search, beam branch-and-price, and the use of a general purpose mixed integer programming solver. This approach is closely related with SearchCol, a framework for solving integer programming / combinatorial optimization decomposition models.

Computational results with 500 instances from the literature show that the proposed hybrid heuristic is efficient in obtaining high quality solutions. It uses more 8 and 17 bins than the 7364 and 7340 bins of a compact model from the literature for the 2 and 3-stage problems, respectively, while the sum of the time spent for all instances is 35% and 58% of the time spent by the compact model.

Keywords: Integer programming / combinatorial optimization, Column generation, bin packing.

1 Introduction

Cutting and packing problems have been studied for more than half a century. In this type of problems, it is intended to choose an assignment of small items to large objects which corresponds to a minimum (or maximum) value of some objective measure. In a feasible assignment, the small items can not overlap and each one must lie entirely within one large object.

Cutting and packing problems are particularly relevant in industries where those operations are required in the production process (e.g., furniture) and have

B. Murgante et al. (Eds.): ICCSA 2014, Part II, LNCS 8580, pp. 211–226, 2014.

also played an important role in the development of optimization models and methods, judging from the number and diversity of approaches and publications devoted to this type of combinatorial optimization problems [9,7,22].

We address two 2-dimensional bin packing problems. In these problems, a virtually infinite number of rectangular identical bins is available for packing a set of rectangular items with the aim of minimizing the number of used bins. According to the packing and cutting typology of Wäscher et al. [22], we address a two-dimensional rectangular Single Bin Size Bin Packing Problem (SBSBPP).

We consider two variants of the above mentioned problems. The first is the two-stage bin packing problem (2BPP), where the items must be packed in levels. A level is a set of items placed horizontally without overlap and not exceeding the width of the bin. The height of a level is the height of the tallest item of all the items in the level. Levels can not overlap and therefore, in a feasible packing, the sum of the heights of the levels in a bin must be less than or equal to the height of the bin. An example of a solution to the 2BPP in which three levels are used is given in the left side of Figure 1.

The second variant is the restricted three-stage bin packing problem (r3BPP). In the r3BPP, items with the same width can be grouped vertically to form stacks which are packed in levels. Being the restricted version, in each level, at least one item is packed alone (its stack only has one item - itself) and the height of a level is given by the tallest item packed alone in the level. The right side of Figure 1 illustrates a restricted 3-stage packing.

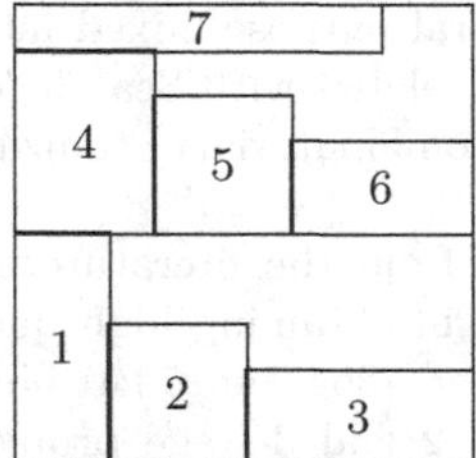

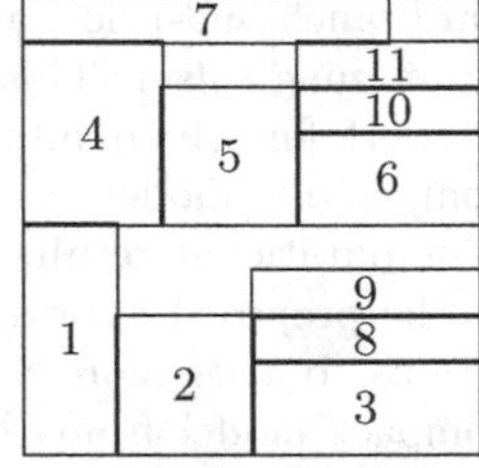

Fig. 1. The left side bin consists in a 2-stage packing of 7 items in one bin. Items 1,2,3 are packed in level 1; items 4,5,6 are packed in level 2; item 7 is packed in level 3. The right side bin shows a restricted 3-stage packing of 11 items. Items 3,8,9 are packed in one stack which is packed in level 1; items 6, 10, and 11 are packed in stack which is packed in level 2. The remaining items can be seen as stacks of only one.

Despite the difference of the actual operations in a given application, the 2BPP and r3BPP are equivalent to the two-stage and restricted three-stage guillotine cutting stock problems (CSP). However, it is common in the cutting and packing literature, to give a different meaning for each designation. In BP instances almost all items have different sizes, while in CS instances many items have the same sizes and therefore can be grouped in item types with a demand [22]. This difference has significant implications when solving BPPs/CSPs.

Most solution methods for CSPs take advantage of the concept of pattern (a set of items and, possibly, their layout in a stock sheet) which allows representing a solution as a set of patterns with multiplicities. Examples are the seminal approach of Gilmore and Gomory [9,10] and the approaches in [6,8]. In [1] a comparison of different approaches of this type is made. Exceptions are the pseudo-polynomial models of [21] and [15].

In BPPs items are treated individually. Due to their more combinatorial structure (when compared with cutting stock), most approaches for BPP are heuristic. Although linear programming based approaches (including column generation) exist, they are not so dominant. A survey on 2-dimensional bin packing is presented in [13]. One of the first heuristic approaches for the 2BPP is due to Berkey and Wang [5] which developed fast algorithms based on packing the items by a sequence, deciding their location according to a pre-established criterion (e.g. first location where the current item fits). In [12] these ideas are extended to other criteria and other fast heuristics are introduced and incorporated in a tabu search approach.

Among the fast heuristics of [12] is the knapsack heuristic which is used in the proposed hybrid heuristic for the 2BPP as the first incumbent solution (and also for initializing column generation). In the knapsack heuristic, binary knapsack problems are solved sequentially to form levels. In each of those problems, the tallest unpacked item is included in a level, and each remaining unpacked item is considered in the binary knapsack problem with a weight equal to its width and a profit equal to its area. The capacity of the knapsack is the width of the bin minus the width of the tallest item. A algorithm for the one dimensional bin packing problem packs the levels into bins.

Approaches based on branch-and-price were developed in [18] and [19]. In the former, the free (with no stages) BPP is addressed by solving the column generation (CG) subproblems by constraint programming. In the latter, restricted and unrestricted 3-stage variants are addressed, being the CG subproblems solved by a hierarchy of methods (including a genetic algorithm and a MIP solver to assure optimality). Stabilization procedures are also used.

A heuristic based on CG but not using the usual linear programming driven algorithm is described in [16]. The subproblem solutions are generated by problem specific fast algorithms and the resulting set covering model is tackled by a Lagrangean heuristic.

In this paper, an approach also based on CG is presented. When all the bins have the same size, in CG algorithms for cutting and packing (in any dimension) the common approach is to have a master variable associated with a feasible packing. In each CG iteration, a single subproblem is solved identifying the most attractive packing (in a single bin).

We propose a new decomposition model where each master variable is associated with a feasible packing initialized with a given item. A packing is initialized by an item if the item is placed in the first position of the first level of the bin. There are n subproblems where n is the number of items. Assuming the items are sorted by non-increasing heights, it is easy to show that in subproblem k, only items $j = k, ..., n$ must be considered. A bin is of type k if it is initialized by item k.

The potential advantages of using the new decomposition model include: (i) the CG subproblems are simpler as they already include one item (the one that initializes the bin) and have less variables, (ii) in each iteration of CG, many columns (up to n) can be included in the restricted master problem without any additional effort than solving the subproblem.

The decomposition model just introduced is solved by a hybrid heuristic approach which has two sequential phases: (i) beam branch-and-price with local search and (ii) SearchCol with MIP solver.

Beam [17] branch-and-price [4] combines beam search with branch-and-price by selecting only the most promising subset of nodes of each level. In each node, a local search is conducted.

SearchCol, "Metaheuristic search by column generation", is a general framework for decomposable integer programming / combinatorial optimization problem [2,3]. There are three main components in SearchCol. The first is CG which is responsible for generating subproblem solutions and information about them (e.g. their value in the linear relaxation). The second is a search algorithm which uses the subproblem solutions provided by CG as the search space and is responsible to improve the incumbent while updating information on subproblem solutions (e.g. the number of times each subproblem solution appeared in a "good" global solution). The third is responsible for, based on the available information provided by CG and by the searcher, generating perturbations, i.e. additional constraints to be incorporated in CG. In the proposed hybrid heuristic the searcher is a MIP solver.

One more potential advantage of the proposed decomposition model is related with the solution method just introduced: given that more subproblems are solved per iteration, it is expected that more columns will be generated by CG and the search space provided to the MIP will be larger.

We compare the computational results of the hybrid heuristic with the compact integer programming model from [14] (2BPP) and [19] (r3BPP) solved by the commercial MIP solver Cplex 12.4 [11].

In the next section, we describe the usual decomposition for BPPs and introduce the new decomposition (the item decomposition) and detail some of the issues involved in its solution by CG. In section 4, the hybrid heuristic is detailed. In section 5, computational results are reported and discussed. The last section refers to the main conclusions.

2 Decomposition Models

2.1 Bin Decomposition Model

In the two-dimensional BPPs addressed in this paper, there is an unlimited number of rectangular bins with width W and height H in which a set of n rectangular items must be packed. Item i, $i = 1, ..., n$, has width w_i and height h_i. The objective is to minimize the number of bins used to pack all the items.

A straightforward decomposition model, suitable for column generation based approaches, is based on associating decision variables with packings in single bins. Formally, and representing the set of all feasible packings in one bin by S, we define the decision variables $y_s = 1$, if the $s-th$ feasible packing is used, $y_s = 0$ otherwise, for all $s \in S$.

We also define parameters $a_{is}, i = 1, ..., n; s \in S$, corresponding to the number of times item i is included in the $s-th$ feasible packing,

These parameters could be defined as binary, but as discussed at the end of this subsection, that would imply a less efficient solution approach.

The bin decomposition model is

$$\min \sum_{s \in S} y_s \tag{1}$$

$$\text{subject to} \sum_{s \in S} a_{is} y_s \geq 1 \qquad i = 1, ..., n \tag{2}$$

$$y_s \in \{0, 1\} \qquad s \in S \tag{3}$$

The objective function (1) minimizes the number of used bins. Constraints (2) assure all items are packed. The linear relaxation of the decomposition model (obtained by replacing constraints (3) by $y_s \geq 0$) can be solved by CG.

CG iterates between a restricted master problem (RMP) and a subproblem. The RMP is a smaller version (not all decision variables are considered) of the decomposition model and, when optimized, provides a feasible solution to the restricted problem and a dual solution. The subproblem uses the dual solution to identify decision variables that may improve the current solution (the ones with negative reduced cost). When the subproblem does not identify any attractive decision variable, the optimal solution of the RMP is also an optimal solution to the linear relaxation of the decomposition model.

Representing the dual variables of the constraints (2) by π_i, $i = 1, ..., n$, the reduced cost of variable y_s is given by $\bar{c}_s = 1 - \sum_{i=1}^{n} a_{is}\pi_i$

The subproblem of CG consists in selecting which items to include in a bin given that each item i has a profit of π_i and taking into account the geometric constraints of the problem being solved.

For the 2BPP the CG subproblem can be solved in two steps, as first proposed in [10]. In the first step, the maximum value of the levels initialized by each item are obtained. For each level i, $i = 1, ..., n$, the problem is to decide which items should be packed in order to maximize the profit of the level. We define the decision variables $x_j = 1$ if item j is included in the level and $x_j = 0$ otherwise.

We represent the value of an optimal solution for level i by Z_i, $i = 1, ..., n$. This value is obtained by solving the following binary knapsack for level i:

$$\begin{aligned} Z_i = \pi_i + \max & \sum_{j=i+1}^{n} \pi_j x_j \\ \text{subject to} & \sum_{j=i+1}^{n} w_j x_j \leq W - w_i \\ & x_j \in \{0, 1\} \qquad j = i+1, ..., n \end{aligned} \tag{4}$$

Constraints (4) state that the width of the level i (already with item i) cannot exceed the width of the bin.

In the second step, it is decided which levels obtained in the first step are packed in the bin in order to maximize the total profit (equivalently, minimize the reduced cost). We define the decision variables $z_i = 1$, if level i is included in the bin, $z_i = 0$, otherwise. The binary knapsack problem with the objective function of maximizing $\sum_{z=1}^{n} Z_i z_i$ subject to the constraint $\sum_{i=1}^{n} h_i z_i \leq H$ (the height of the packing cannot exceed the height of the bin) is solved to obtain the single bin packing corresponding to the solution of the subproblem.

In solving the subproblem for the r3BPP problem, we first obtain the most attractive stack for each item. The candidates for being included with item i are the items $j, j > i$ and $w_j = w_i$. Note that given that the widths of the items in a stack are the same, the order of the items within the stack is irrelevant and items with $j < i$ can be excluded from any stack with item i.

The advantage of solving the subproblem by successive knapsacks is the availability of very efficient algorithms for knapsack problems. However, there is also a disadvantage. For example, in the 2BPP, one item appears at most once in a level but it may appear more than once in a packing in a single bin if it is included in two different selected levels. This justifies the use of (general) integers coefficients a_{is} (i.e., the decomposition model is a generalised set covering model). The repetition of items in a single bin is not desirable because the lower bound provided by the linear relaxation of the decomposition model is worst than the one obtained by forcing parameters a_{is} to be binary. However, avoiding this repetition would imply additional constraints in the subproblem, which would destroy the (efficient) method based on solving binary knapsack problems.

Note that replacing each $a_{is} > 1$ by $a_{is} = 1$ does not solve the problem. If that replacement is made before the reduced cost calculation, the optimality of the solution obtained by column generation is no longer assured. If that replacement is made after the reduced cost calculation, the same subproblem solution may appear in a subsequent iteration and the convergence of column generation is not assured.

2.2 Item Decomposition Model

In this section, we propose a different decomposition model. The main difference is that one subproblem is defined for each item. Although the number of subproblems is much larger than in the decomposition of the previous subsection, in each column generation iteration, several subproblem solutions can be introduced in the RMP. We define the subproblem of item k as the subproblem responsible for generating solutions in which the corresponding bins are initialized (by the level initialized) by item k.

More formally, and representing the set of all feasible packings in a single bin initialized by item k or with no items (a null packing) by S^k, we define the decision variables $y_s^k = 1$, if the $s - th$ feasible packing initialized by item k is used, $y_s^k = 0$, otherwise, for $s \in S^k; k = 1, ..., n$.

We also define the parameters $a_{is}^k, k = 1, ..., n; i = k, ..., n; s \in S^k$, corresponding to the number of times item i is included in the $s-th$ feasible packing initialized by item k.

The item decomposition model is

$$\min \sum_{k=1}^{n} \sum_{s \in S^k} y_s^k \tag{5}$$

$$\text{subject to} \sum_{k=1}^{i} \sum_{s \in S^k} a_{is}^k y_s^k \geq 1 \qquad i = 1, ..., n \tag{6}$$

$$y_s^k \in \{0, 1\} \qquad k = 1, ..., n; s \in S^k$$

The objective function (5) minimizes the number of used bins. Constraints (6) assure all items are packed.

The subproblem of item k is to decide which items, from the ones with height smaller than k, are packed in a bin initialized by k or if the item k should not initialize a bin. The subproblem of each item is the subproblem described in the previous subsection but with fewer items. Accordingly, the two steps (for 2BPP, three for r3BPP) algorithm based on binary knapsacks described before can be applied for solving every subproblem of the item decomposition.

The linear relaxations of the bin and item decompositions provide the same lower bound to the value of an integer optimal solution. Each decomposition can be seen as a Dantzig-Wolfe decomposition of [14] (2BPP) and [19] (r3BPP) where the subproblem does not have the integrality property. Therefore the decompositions provide equal or better lower bounds than the linear relaxations of the compact integer programming models of [14] (2BPP) and [19] (r3BPP).

3 The Hybrid Heuristic

We solve the item decomposition by a hybrid heuristic with two sequential phases, both based on column generation. We introduce the hybrid heuristic based on the item decomposition, although it can also rely on the bin decomposition.

3.1 Beam Branch-and-Price with Local Search

In the first phase of the hybrid heuristic, beam branch-and-price with local search is applied. At each level of the branch-and-price tree, only the *bw* nodes (not fathomed) with lower values of the relaxation, where *bw* is a parameter known as the beam width, are explored.

In each node of the branch-and-price tree, after the linear relaxation of the node is solved by CG, if the solution is fractional and the node is not fathomed, a local search heuristic is applied. The local search relies on the search space, representation and evaluation of solutions, and neighbourhood structure, defined in the SearchCol framework.

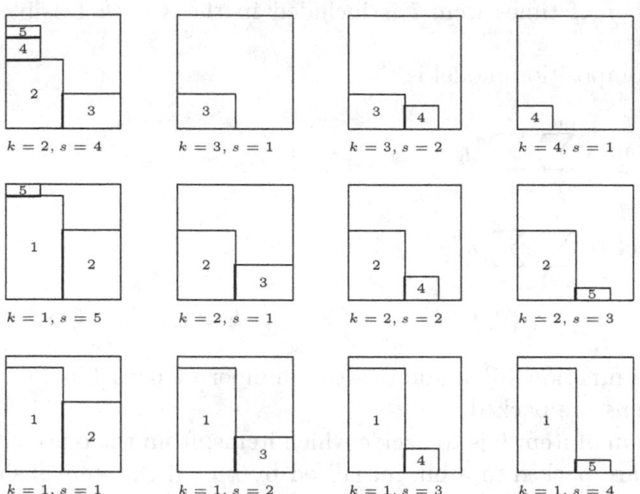

Fig. 2. Example of a set of subproblem solutions

In Figure 2, an example of subproblem solutions generated by CG is given (null solutions are not shown but it is assumed they were generated and have index $s = 0$). An initial solution is obtained by rounding the solution provided by CG. The search space is made of all global solutions that can be obtained by the combination of one solution from each subproblem (including the null). A global solution is represented as a solution from each subproblem k, i.e. a feasible packing for each bin initialized by k (or the empty bin). For example, $u = (5, 0, 1, 1, 0)$, where the k-th position contains the index of the subproblem solution (s) of subproblem k. Solution u is feasible because all items are packed. A neighbour solution is obtained by modifying the solution of one and only one subproblem. For example, $v = (2, 0, 1, 1, 0)$ is a neighbour of u. v is infeasible given that item 2 is not packed.

3.2 SearchCol with MIP

In the second phase of the hybrid heuristic, a SearchCol algorithm using a MIP solver for its search phase is applied. As shown in Figure 3, each iteration of a SearchCol algorithm has three main steps: (i) apply CG to the linear relaxation of a perturbed (restricted) master problem, (ii) conduct a search in the space provided by subproblem solutions obtained so far, and (iii) define perturbations for the next iteration. A perturbation is a constraint that forces a subproblem variable to take value 1 or 0, and a perturbed (restricted) master problem is the original master problem with additional constraints fixing subproblem variables.

In the proposed hybrid heuristic, CG of step 1 is initialized with all the subproblem solutions generated by the beam-branch-and-price. The search of steps 2 and 6 is conducted by a (general purpose) MIP solver in all the subproblem

```
1: Column generation
2: Search
3: repeat
4:     Define column generation perturbation
5:     Optimize perturbed column generation
6:     Search
7: until Stopping criterion fulfilled
```

Fig. 3. SearchCol algorithm

solutions generated by CG. A time limit of $0.2n$ seconds, where n is the number of items, is used as a stopping criterion for each search. In step 4, the set of perturbations used is updated. A perturbation (k, i) forces an item i to be packed in a bin initialized by a given item k. Perturbations are translated to the RMP variables. For example, using the instance of Figure 2, a perturbation $(1, 5)$ would correspond to the additional constraint $y_4^1 + y_5^1 \geq 1$ in the RMP. The dual variables of the perturbation constraints are taken into account in the subproblems.

Perturbations can be grouped to fix the solution of a subproblem. For example, using again the instance of Figure 2, the set of perturbations $\{(2, 2), (2, 3), (2, 4), (2, 5)\}$ forces subproblem solution 4 of subproblem 2 to be selected. We name this type of perturbations as subproblem perturbations. Each subproblem has at most one subproblem perturbation. A subproblem for which a subproblem perturbation is defined is named a perturbed subproblem.

In each iteration 10% of the subproblems are perturbed. Once a subproblem is perturbed, it remains perturbed until the end of the execution of the hybrid heuristic. The selection of the subproblems to be perturbed is made as follows.

For each subproblem not yet perturbed, the score of all solutions are calculated and the solution with higher score is selected. The score of a subproblem solution is obtained through

$$score = PresenceInc + CurWeight + 0.1 SearchRecency$$

where

- $PresenceInc$ is 1 if the subproblem solution is in the incumbent; and 0 otherwise;
- $CurWeight$ is the value of the variable of the decomposition model associated with the subproblem solution the last time CG was applied;
- $SearchRecency$ is the number of times the subproblem solution belonged to global solutions obtained by the MIP solver (normalized to a value between 0 and 1).

The unperturbed subproblems are sorted in descending order of the score of their selected solutions. The 10% subproblems with higher score are perturbed.

The algorithm terminates when optimality is proven (based on the lower bound provided by CG the first time it is applied), CG becomes infeasible, or all subproblems are fixed.

4 Computational Results

4.1 Implementation and Experimental Setting

The described algorithms were implemented in SearchCol++ [20], a general framework coded in C++ for the implementation of column generation based algorithms. When using SearchCol++, only information on the decomposition and problem specific components must be coded. More precisely, the user must provide the number of subproblems, the sense and right hand side of the master constraints, a subproblem solver, the way the duals of the global constraints modify the coefficients in the objective function of the subproblem variables, and the way a solution of a subproblem maps to a column. Solely based on that information, branch-and-price, beam branch-and-price, metaheuristics based on CG and SearchCol algorithms are applied with all their other components (as the search algorithms, search trees, or perturbations) hidden from the user and controlled through input parameters.

SearchCol++ uses Cplex 12.4 [11] as the linear programming solver for the restricted master problems and also as the MIP solver in the search step. The hybrid heuristic based on the bin decomposition was implemented by, in each iteration of CG, adding to the RMP only the most attractive column from all the subproblems, while when based on the item decomposition, the most attractive column from each subproblem is added to the RMP.

Computations were performed on a desktop computer with an Intel Core i7 - 3.3 GHz processor and with 32 GB RAM. The compact integer programming models from [14] (2BPP) and [19] (r3BPP) were solved by the commercial MIP solver Cplex 12.4 [11].

We used the 500 instances described in [18,19]. These instances are divided in 10 classes. Each class has subsets of 10 instances each one with 20, 40, 60, 80, and 100 items. A time limit of 3600 seconds was set to all tested methods.

4.2 2 Stage Problem

Tables 1 and 2 present global computational results for the 2BPP. In appendix A more detailed results are shown. The quality of the lower bound is significantly better for the decomposition models (13.6% in average, in some instances more than 50%) when compared with the compact model. In Table 1 average results for 100 instances with the same number of items are shown. The relative variation in the number of used bins in percentage between the two solutions of consecutive components of the proposed hybrid heuristic applied to the item decomposition are shown in columns two to five. Columns six to eight correspond to comparison of the hybrid heuristic based on the item decomposition with the knapsack heuristic, with the hybrid heuristic based on the bin decomposition and with the compact model.

The following notation is used.

- K - solution obtained by the Knapsack heuristic;
- R - Root solution when item decomposition is used (obtained after column generation and local search are applied for the first time, i.e. in the root of the beam branch-and-price);
- T - Tree solution when item decomposition is used (obtained at the end of the beam branch-and-price with local search);
- G - (first) Global solution when item decomposition is used (obtained by the MIP solver the first time it is used - first SearchCol iteration);
- I - final solution when Item decomposition is used (obtained after SearchCol with MIP is applied);
- B - final solution when the *Bin* decomposition is used;
- C - solution of the Compact model.

Table 1. 2-stage problem: Relative variation (in percentage) between components of the hybrid heuristic and other methods. Δ A to B corresponds to $100(ValueB - ValueA)/ValueA$.

n	Δ K to R	Δ R to T	Δ T to G	Δ G to I	Δ K to I	Δ B to I	Δ C to I
20	-0.93	0.00	0.00	-0.19	-1.12	0.00	0.19
40	-0.59	0.00	-0.30	-0.69	-1.57	0.00	0.20
60	-0.33	0.00	-0.33	-1.14	-1.80	-0.14	0.14
80	-0.25	0.00	-0.40	-0.90	-1.53	-0.15	0.15
100	-0.21	-0.12	-0.54	-0.71	-1.58	-0.17	0.00

The root solution (R) (column generation plus local search) improves the knapsack heuristic (K) solution in the five sets of instances. In the 4 sets of instances with less items, the nodes other than the root of the beam branch-and-price do not improve the root solution. However, columns generated by this tree may be used later by the MIP solver. The first time the MIP solver is used (providing the first global solution after local search) significant improvements occur. However, the most significant improvements are due to the SearchCol step of the hybrid heuristic (column Δ G to I).

The last three columns show the relative variation between the hybrid heuristic used in the item decomposition and the knapsack heuristic, the bin decomposition, and the compact model. A negative value means the hybrid heuristic produces a better result. The item decomposition approach is better than the bin decomposition approach for the three set of instances with more items (and equal for the other two). The compact model obtained the best solutions.

The absolute differences considering the 500 instances are given in Table 2 and confirm that the step where the quality of the solution of the hybrid heuristic most improves is the SearchCol step. The hybrid heuristic uses less 9 bins and spends less 30 minutes when based on the item decomposition than when based in the bin decomposition. The hybrid heuristic uses more 8 bins than the compact model but in less seven hours and a half.

Table 2. 2-stage problem: Global comparisons. Values correspond to the number of used bins in all 500 instances. Times (given in seconds) are for all the 500 instances. $\Delta Time$ and $\Delta Value$ are the absolute difference between the method in the column and the compact model. Times of the item decomposition are cumulative.

	R	T	G	I	K	B	C
Time	1854	7006	7849	14720	0	13036	41998
Value	7464	7461	7432	7372	7490	7381	7364
$\Delta Time$	-40145	-34992	-34150	-27279	-41998	-28962	0
$\Delta Value$	100	97	68	8	126	17	0

4.3 3 Stage Problem

Tables 3 and 4 present global computational results for the r3BPP. In appendix B more detailed results are shown. The notation used and the results presented are similar to the ones in the tables for the 2BPP, with the exception that for the r3BPP no knapsack heuristic was implemented. As in the 2BPP, the quality of the lower bound is significantly better for the decomposition models (13.7% in average, in some instances more than 50%). The comparative results for the r3BPP are similar to the ones obtained for the 2BPP. The hybrid approach based on the item decomposition achieves solutions with almost the same quality much faster.

Table 3. 3-stage problem: Relative variation (in percentage) between components of the hybrid heuristic and other methods

n	Δ R to T	Δ T to G	Δ G to I	Δ B to I	Δ C to I
20	0.00	-0.19	-0.38	-0.19	0.00
40	0.00	-0.30	-0.99	0.00	0.10
60	0.00	-0.40	-1.21	-0.07	0.62
80	-0.05	-0.40	-1.34	-0.10	0.25
100	-0.29	-0.71	-1.05	-0.21	0.08

The absolute differences considering the 500 instances are given in Table 4 and confirm that the step where the quality of the solution of the hybrid heuristic most improves is the SearchCol step. The hybrid heuristic based on the item decomposition uses less 9 bins and spends less than half of the time than when based on the bin decomposition. The hybrid heuristic uses more 17 bins than the compact model but in less four hours.

Table 4. 3-stage problem: Global comparisons

	R	T	G	I	B	C
Time	5014	12185	13369	21480	46795	37233
Value	7465	7457	7422	7340	7349	7323
$\Delta Time$	-32219	-25048	-23864	-15752	9562	0
$\Delta Value$	142	134	99	17	26	0

5 Conclusions

In this paper, a heuristic based on column generation was applied to two 2-dimensional bin packing problems: the two-stage and the restricted three-stage problems. In these problems, the packing of the items must be done in levels (two-stage problem) or in stacks and levels (three-stage problem).

We introduced a new decomposition model which gave promising results, since its use as the base of the proposed heuristic produced better results than the usual decomposition model. This new decomposition can be extended to other bin packing problems (as the geometric details are hidden in the subproblem).

The proposed hybrid heuristic is based on several components. Besides column generation, it includes concepts and/or components from beam search, branch-and-price, local search, SearchCol, perturbations, a problem specific solver for single bin subproblems, and a general purpose MIP solver.

The hybrid heuristic was compared against compact models for 500 instances from the literature with 20 to 100 items, with very good results. Considering the sum for all instances, the heuristic used 8 bins more (0.11% of the number of bins used by the compact model) in less 7.5 hours approximately for the 2BPP and more 17 bins (0.23%) in less 4 hours approximately for the r3BPP.

Acknowledgments. This work was financed by Fundação para a Ciência e a Tecnologia (Portuguese Foundation for Science and Technology) within projects "SearchCol: Metaheuristic search by column generation" (PTDC/EIAEIA/100645/2008) and PEst-OE/EEI/UI0319/2014.

References

1. Alvarez-Valdes, R., Parajon, A., Tamarit, J.M.: A computational study of lp-based heuristic algorithms for two-dimensional guillotine cutting stock problems. OR Spektrum 24, 179–192 (2002)
2. Alvelos, F., de Sousa, A., Santos, D.: SearchCol: Metaheuristic search by column generation. In: Blesa, M.J., Blum, C., Raidl, G., Roli, A., Sampels, M. (eds.) HM 2010. LNCS, vol. 6373, pp. 190–205. Springer, Heidelberg (2010)
3. Alvelos, F., de Sousa, A., Santos, D.: Combining column generation and metaheuristics. In: Talbi, E.-G. (ed.) Hybrid Metaheuristics. SCI, vol. 434, pp. 285–334. Springer, Heidelberg (2013)
4. Barnhart, C., Johnson, E.L., Nemhauser, G.L., Savelsbergh, M.W.P., Vance, P.H.: Branch-and-price: column generation for solving huge integer programs. Operations Research 46, 316–329 (1998)
5. Berkey, J., Wang, P.: Two-Dimensional Finite Bin-Packing Algorithms. The Journal of the Operational Research Society 38, 423–429 (1987)
6. Cintra, G.F., Miyazawa, F.K., Wakabayashi, Y., Xavier, E.C.: Algorithms for two-dimensional cutting stock and strip packing problems using dynamic programming and column generation. European Journal of Operational Research 191, 61–85 (2008)
7. Dyckhoff, H.: A typology of cutting and packing problems. European Journal of Operational Research 44, 145–159 (1990)

8. Furini, F., Malaguti, E., Durán, R.M., Persiani, A., Toth, P.: A column generation heuristic for the two-dimensional two-staged guillotine cutting stock problem with multiple stock size. European Journal of Operational Research 218, 251–260 (2012)
9. Gilmore, P.C., Gomory, R.E.: A Linear programming approach to the cutting-stock problem. Operations Research 9, 849–859 (1961)
10. Gilmore, P.C., Gomory, R.E.: Multistage cutting stock problems of two and more dimensions. Operations Research 13, 94–120 (1965)
11. ILOG, ILOG CPLEX 12.4 - User's Manual (2011)
12. Lodi, A., Martello, S., Vigo, D.: Heuristic and metaheuristic approaches for a class of two-dimensional bin packing problems. INFORMS Journal on Computing 11, 345–357 (1999)
13. Lodi, A., Martello, S., Monaci, M.: Two-dimensional packing problems: A survey. European Journal of Operational Research 41, 241–252 (2002)
14. Lodi, A., Martello, S., Vigo, D.: Models and bounds for two dimensional level packing problems. Journal of Combinatorial Optimization 8, 363–379 (2004)
15. Macedo, R., Alves, C., Valério de Carvalho, J.M.: Arc-flow model for the two-dimensional guillotine cutting stock problem. Computers and Operations Research 37, 991–1001 (2010)
16. Monaci, M., Toth, P.: A Set-covering based heuristic approach for bin-packing problems. INFORMS Journal on Computing 18, 71–85 (2006)
17. Ow, P.S., Morton, T.E.: Filtered beam search in scheduling. International Journal of Production Research 26, 35–62 (1988)
18. Pisinger, D., Sigurd, M.: Using decomposition techniques and constraint programming for solving the two-dimensional bin-packing problem. INFORMS Journal on Computing 19, 1007–1023 (2007)
19. Puchinger, J., Raidl, G.: Models and algorithms for three-stage two-dimensional bin packing. European Journal of Operational Research 183, 1304–1327 (2007)
20. SearchCol++, `http://searchcol.dps.uminho.pt/`
21. Silva, E., Alvelos, F., Valério de Carvalho, J.M.: An integer programming model for two- and three-stage two-dimensional cutting stock problems. European Journal of Operational Research 205, 699–708 (2010)
22. Wäscher, G., Haussner, H., Schumann, H.: An improved typology of cutting and packing problems. European Journal of Operational Research 183, 1109–1130 (2007)

A Detailed Results for the Two-Dimensional Two-Stage Bin Packing Problem

Averages for each 10 instances with the same number of items and of the same class. Values (columns V) corresponds to the number of used bins. Times (columns T) are expressed in seconds.

Columns "Compact LR" and "Decs LR" provide information about the linear relaxations (B - bound, T - time in seconds). The time to solve the linear relaxation of the compact model and the time spent by the knapsack heuristic were always almost 0 and therefore are not included in the table.

Remaining columns are for the solution obtained with each method.

Class	n	Compact LR	Decs LR		Compact Integer		Knps Heur	Bin Decomposition				Item Decomposition							
								Root		Final		Root		Tree		Global		Final	
		B	B	T	T	V	V	T	V	T	V	T	V	T	V	T	V	T	V
	20	6.4	7.1	0	0	7.3	7.5	0	7.4	0	7.3	0	7.3	0	7.3	0	7.3	0	7.3
	40	12.1	13.7	0	0	13.8	14	0	14	9	13.8	0	14	0	14	1	13.9	7	13.8
1	60	18.5	20.3	0	0	20.3	20.7	0	20.7	1	20.3	0	20.7	0	20.7	0	20.6	10	20.3
	80	25.3	27.7	0	0	27.7	28.5	0	28.5	2	27.7	0	28.3	0	28.3	2	28.1	15	27.7
	100	30.5	32.4	0	0	32.4	33.2	1	33.2	44	32.5	0	33.1	1	33.1	7	32.9	62	32.4
	20	1	1	0	0	1	1	0	1	0	1	0	1	0	1	0	1	0	1
	40	2	2	0	0	2	2	0	2	0	2	0	2	0	2	0	2	0	2
2	60	2.5	2.6	3	363	2.8	2.8	2	2.8	5	2.8	3	2.8	10	2.8	12	2.8	19	2.8
	80	3.1	3.1	8	8	3.3	3.3	6	3.3	13	3.3	8	3.3	26	3.3	29	3.3	51	3.3
	100	3.9	3.9	20	13	4	4	15	4	22	4	20	4	60	4	62	4	80	4
	20	4.4	5.4	0	0	5.4	5.4	0	5.4	0	5.4	0	5.4	0	5.4	0	5.4	0	5.4
	40	8.4	9.8	0	0	9.8	10.1	0	10	0	9.8	0	9.9	0	9.9	0	9.9	0	9.8
3	60	12.7	14	0	1	14	14.7	0	14.7	12	14.2	0	14.6	0	14.6	1	14.4	12	14.1
	80	17.3	19.7	0	1	19.7	20	0	20	1	19.7	0	20	0	20	0	19.9	0	19.7
	100	20.7	22.7	0	83	22.8	23.7	2	23.7	124	22.9	0	23.7	1	23.6	9	23.2	102	22.8
	20	1	1	0	0	1	1	0	1	0	1	0	1	0	1	0	1	0	1
	40	1.9	2	0	0	2	2	0	2	0	2	1	2	1	2	1	2	1	2
4	60	2.5	2.5	3	405	2.7	2.7	3	2.7	7	2.7	3	2.7	11	2.7	14	2.7	20	2.7
	80	3.1	3.3	8	246	3.3	3.4	10	3.4	24	3.4	8	3.4	14	3.4	16	3.4	23	3.4
	100	3.7	3.8	24	1001	4.1	4	31	4	71	4	24	4	70	4	75	4	106	4
	20	5.5	6.7	0	0	6.7	6.8	0	6.7	0	6.7	0	6.7	0	6.7	0	6.7	0	6.7
	40	10.4	12.3	0	0	12.3	12.5	0	12.5	0	12.3	0	12.4	0	12.4	0	12.4	0	12.3
5	60	15.8	18.3	0	0	18.3	18.9	0	18.8	10	18.4	0	18.7	0	18.7	0	18.7	0	18.3
	80	21.6	25	0	1	25	25.6	0	25.5	17	25.2	0	25.4	0	25.4	2	25.1	15	25
	100	26.1	28.8	0	9	28.8	29.8	1	29.7	26	28.9	0	29.6	0	29.4	1	29	16	28.8
	20	1	1	0	0	1	1	0	1	0	1	0	1	0	1	0	1	0	1
	40	1.8	1.9	1	0	1.9	2	1	2	1	2	1	2	1	2	1	2	1	2
6	60	2.1	2.2	4	66	2.2	2.3	4	2.3	6	2.3	4	2.3	7	2.3	8	2.3	11	2.3
	80	3	3	10	3	3	3	13	3	13	3	10	3	10	3	10	3	10	3
	100	3.2	3.4	33	438	3.5	3.5	43	3.5	70	3.5	33	3.5	53	3.5	56	3.5	71	3.5
	20	4.9	5.7	0	0	5.7	5.9	0	5.9	0	5.7	0	5.8	0	5.8	0	5.8	0	5.7
	40	9.9	11.5	0	0	11.5	11.8	0	11.8	0	11.5	0	11.7	0	11.7	0	11.7	0	11.5
7	60	14.1	16.2	0	1	16.2	16.4	0	16.4	1	16.2	0	16.4	0	16.4	0	16.3	0	16.2
	80	20	23.3	0	3	23.3	23.7	1	23.7	34	23.3	0	23.7	2	23.7	6	23.5	36	23.3
	100	23.8	27.5	20	7	27.5	27.9	21	27.9	160	27.5	20	27.9	169	27.9	173	27.7	212	27.5
	20	5	6	0	0	6.1	6.1	0	6.1	0	6.1	0	6.1	0	6.1	0	6.1	1	6.1
	40	9.8	11.4	0	0	11.5	11.9	0	11.8	7	11.5	0	11.9	0	11.9	1	11.7	12	11.5
8	60	14.3	16.4	1	1	16.4	16.9	1	16.9	16	16.4	1	16.9	2	16.9	4	16.8	12	16.4
	80	19.7	22.6	6	5	22.6	23.3	12	23.3	67	22.6	6	23.3	37	23.3	37	23.3	37	22.6
	100	24.1	28	41	369	28.2	28.5	45	28.5	271	28.1	41	28.5	219	28.5	227	28.4	274	28.1
	20	9.4	14.3	0	0	14.3	14.4	0	14.3	0	14.3	0	14.3	0	14.3	0	14.3	0	14.3
	40	18.2	27.8	0	0	27.8	28	0	27.8	0	27.8	0	27.8	0	27.8	0	27.8	0	27.8
9	60	27.9	43.7	0	0	43.7	43.9	0	43.7	0	43.7	0	43.7	0	43.7	0	43.7	0	43.7
	80	37.2	57.7	0	0	57.7	57.8	0	57.7	0	57.7	0	57.7	0	57.7	0	57.7	0	57.7
	100	45.1	69.4	0	0	69.5	69.7	0	69.5	0	69.5	0	69.5	0	69.5	0	69.5	0	69.5
	20	4.1	4.5	0	0	4.5	4.6	0	4.6	0	4.6	0	4.6	0	4.6	0	4.6	0	4.6
	40	7.1	7.6	0	0	7.7	7.8	0	7.8	5	7.8	0	7.8	0	7.8	2	7.8	14	7.8
10	60	9.9	10.4	0	362	10.5	10.7	1	10.7	25	10.5	0	10.7	0	10.7	3	10.7	25	10.5
	80	12.3	13.2	0	377	13.3	13.7	2	13.7	96	13.6	0	13.7	2	13.7	10	13.7	89	13.5
	100	15.4	16	1	439	16.3	16.6	7	16.6	142	16.6	1	16.6	4	16.6	16	16.6	127	16.5

B Detailed Results for the Two-Dimensional Restricted Three-Stage Bin Packing Problem

Averages for each 10 instances with the same number of items and of the same class. Values (columns V) corresponds to the number of used bins. Times (columns T) are expressed in seconds.

Columns "Compact LR" and "Decs LR" provide information about the linear relaxations (B - bound, T - time in seconds). The time to solve the linear relaxation of the compact model was always almost 0 and therefore the related column is not included in the table.

Remaining columns are for the solution obtained with each method.

Class	n	Compact LR	Decs LR		Compact Integer		Bin Decomposition				Item Decomposition							
							Root		Final		Root		Tree		First Global		Final	
		B	B	T	T	V	T	V	T	V	T	V	T	V	T	V	T	V
1	20	6.4	7.1	0	0	7.2	0	7.5	0	7.2	0	7.3	0	7.3	0	7.3	0	7.2
	40	12	13.6	0	0	13.7	1	14	1	13.7	0	13.9	0	13.9	0	13.9	8	13.7
	60	18.5	20	1	0	20.1	4	20.7	24	20.1	1	20.5	1	20.5	2	20.5	14	20.1
	80	25.3	27.5	2	0	27.5	12	28.5	42	27.5	2	28.3	2	28.3	6	28.1	38	27.5
	100	30.5	31.7	6	2	31.8	66	33.2	282	31.8	6	33.2	20	32.8	32	32.3	156	31.8
2	20	1	1	0	0	1	0	1	0	1	0	1	0	1	0	1	0	1
	40	1.9	2	2	0	2	4	2	4	2	2	2	2	2	2	2	2	2
	60	2.5	2.5	6	7	2.5	21	2.8	30	2.8	6	2.8	18	2.8	21	2.8	28	2.8
	80	3.1	3.1	15	337	3.1	62	3.3	83	3.3	15	3.3	36	3.3	39	3.3	55	3.3
	100	3.9	3.9	38	18	3.9	194	4	232	4	38	4	71	4	73	4	91	4
3	20	4.4	5.4	0	0	5.4	0	5.4	0	5.4	0	5.4	0	5.4	0	5.4	0	5.4
	40	8.2	9.8	0	0	9.8	1	10	1	9.8	0	10	0	10	0	9.9	1	9.8
	60	12.5	14	1	2	14	5	14.7	36	14.3	1	14.6	2	14.6	5	14.3	34	14.1
	80	17.3	19.3	2	1	19.4	20	20.0	63	19.4	2	20	5	19.9	7	19.7	24	19.4
	100	20.5	22.7	8	22	22.7	82	23.7	286	22.9	8	23.6	24	23.6	32	23.1	85	22.7
4	20	1	1	0	0	1	0	1	0	1	0	1	0	1	0	1	0	1
	40	1.9	1.9	2	1	2	4	2	4	2	2	2	2	2	2	2	2	2
	60	2.3	2.5	9	22	2.5	33	2.7	44	2.7	9	2.7	22	2.7	24	2.7	30	2.7
	80	3	3.1	29	744	3.3	141	3.4	221	3.4	29	3.4	88	3.4	93	3.4	122	3.4
	100	3.7	3.7	83	1148	3.9	548	4	684	4	83	4	253	4	260	4	308	4
5	20	5.4	6.7	0	0	6.7	0	6.7	0	6.7	0	6.7	0	6.7	0	6.7	0	6.7
	40	10.4	12.3	0	0	12.3	1	12.5	1	12.3	0	12.4	0	12.4	0	12.4	0	12.3
	60	15.7	18.3	1	0	18.3	4	18.9	6	18.3	1	18.8	1	18.8	1	18.7	7	18.3
	80	21.5	24.9	2	1	24.9	14	25.6	31	24.9	2	25.5	3	25.5	3	25.3	4	24.9
	100	26.0	28.7	7	6	28.7	75	29.8	173	28.7	7	29.8	15	29.7	16	29.5	32	28.7
6	20	1	1	0	0	1	0	1	0	1	0	1	0	1	0	1	0	1
	40	1.7	1.9	3	0	1.9	4	2	4	2	3	2	3	2	3	2	3	2
	60	2.1	2.2	13	4	2.2	37	2.3	42	2.3	13	2.3	17	2.3	19	2.3	21	2.3
	80	3	3	37	2	3	167	3	167	3	37	3	37	3	37	3	37	3
	100	3.2	3.4	102	728	3.6	636	3.5	703	3.5	102	3.5	143	3.5	145	3.5	161	3.5
7	20	4.7	5.7	0	0	5.7	0	5.9	0	5.7	0	5.8	0	5.8	0	5.8	0	5.7
	40	9.7	11.5	0	0	11.5	1	11.8	4	11.5	0	11.7	0	11.7	0	11.7	0	11.5
	60	14	16.1	1	0	16.1	6	16.4	22	16.1	1	16.4	1	16.4	1	16.4	1	16.1
	80	19.7	23.2	3	2	23.2	22	23.7	36	23.2	3	23.6	4	23.6	4	23.6	4	23.2
	100	23.8	27.1	24	44	27.2	124	27.9	275	27.2	24	27.9	61	27.7	86	27.4	152	27.2
8	20	5	6	0	0	6.1	0	6.1	0	6.1	0	6.1	0	6.1	0	6.1	1	6.1
	40	9.8	11.3	0	0	11.4	1	11.8	8	11.4	0	11.9	0	11.9	1	11.7	13	11.4
	60	14.3	16.4	1	1	16.4	6	16.9	27	16.4	1	16.9	4	16.9	5	16.8	14	16.4
	80	19.6	22.6	11	9	22.6	30	23.3	115	22.6	11	23.3	51	23.3	55	23.1	83	22.6
	100	24.1	28	59	3	28.1	107	28.5	378	28.1	59	28.5	237	28.5	244	28.4	288	28.1
9	20	9.4	14.3	0	0	14.3	0	14.3	0	14.3	0	14.3	0	14.3	0	14.3	0	14.3
	40	18.1	27.8	0	0	27.8	0	27.8	0	27.8	0	27.8	0	27.8	0	27.8	0	27.8
	60	27.9	43.7	1	0	43.7	1	43.7	1	43.7	1	43.7	1	43.7	1	43.7	1	43.7
	80	37.2	57.7	3	0	57.7	6	57.7	6	57.7	3	57.7	3	57.7	3	57.7	3	57.7
	100	45.1	69.4	8	0	69.5	21	69.5	22	69.5	8	69.5	8	69.5	8	69.5	9	69.5
10	20	4.1	4.5	0	0	4.5	0	4.6	0	4.6	0	4.6	0	4.6	0	4.5	0	4.5
	40	7.1	7.6	0	0	7.7	1	7.8	3	7.7	0	7.8	0	7.8	1	7.8	8	7.7
	60	9.7	10.3	2	46	10.3	13	10.7	34	10.4	2	10.7	3	10.7	6	10.6	37	10.5
	80	12.2	13	5	32	13	52	13.7	156	13.4	5	13.7	19	13.7	24	13.7	74	13.2
	100	15.3	16	15	540	16.1	198	16.6	430	16.5	15	16.6	60	16.6	73	16.5	197	16.2

Experiments with Firefly Algorithm

Rogério B. Francisco[1,2], M. Fernanda P. Costa[2], and Ana Maria A.C. Rocha[3]

[1] Escola Superior de Tecnologia e Gestão de Felgueiras, 4610-156 Felgueiras, Portugal
rbf@estgf.ipp.pt

[2] Centre of Mathematics, University of Minho, 4710-057 Braga, Portugal
mfc@math.uminho.pt

[3] Algoritmi Research Centre, University of Minho, 4710-057 Braga, Portugal
arocha@dps.uminho.pt

Abstract. Firefly Algorithm (FA) is one of the recent swarm intelligence methods developed by Xin-She Yang in 2008 [12]. FA is a stochastic, nature-inspired, meta-heuristic algorithm that can be applied for solving the hardest optimization problems. The main goal of this paper is to analyze the influence of changing some parameters of the FA when solving bound constrained optimization problems. One of the most important aspects of this algorithm is how far is the distance between the points and the way they are drawn to the optimal solution. In this work, we aim to analyze other ways of calculating the distance between the points and also other functions to compute the attractiveness of fireflies.

To show the performance of the proposed modified FAs a set of 30 benchmark global optimization test problems are used. Preliminary experiments reveal that the obtained results are competitive when comparing with the original FA version.

Keywords: Firefly algorithm, Unconstrained Optimization.

1 Introduction

The main objective of the optimization methods is to determine the maximum or the minimum of mathematical functions, called objective functions, which may be, or may not be, subject to constraints on its variables. Due to the wide variety of practical applications, optimization algorithms have been increasingly studied in the area of engineering and applied mathematics. The optimization methods can be divided into two major groups: deterministic and stochastic methods, which may use or not the derivatives of the objective and constraint functions.

The most of deterministic methods are local search methods. These methods are characterized for producing always the same set of solutions (optimal points) if the algorithm start under the same initial conditions. In turn, the stochastic methods are characterized by having one or more components of randomness, called stochastic components. This implies that for the same problem, and subject to the same initial conditions, these algorithms may not generate the same optimal solutions. There is a

B. Murgante et al. (Eds.): ICCSA 2014, Part II, LNCS 8580, pp. 227–236, 2014.

range of possibilities of how to form this stochastic component. For example, one way is to make a simple randomization by randomly sampling the search space or by making random walks. The majority of this type of methods is considered as metaheuristic.

Firefly Algorithm (FA) is an algorithm that belongs to the second group, that is, a stochastic and metaheuristic algorithm, and it was developed by Yang [13, 14]. It is a recent nature inspired optimization algorithm, inspired by the social behavior of fireflies, and is based on their flashing and attraction characteristics.

The firefly algorithm is one example among many, of the so called bio-inspired algorithms. Among such methods, the Genetic Algorithms [7], the Particle Swarm Optimization [5], the Ant Colony Optimization [3], the Artificial Fish Swarm Algorithm [11], are the best known algorithms that are inspired by phenomena and/or behaviors of nature and the animal world.

The first studies on bio-inspired algorithms are due to Reynolds [10] and Heppner and Grenander [8], which are based on the behavior and movement of flocks of birds. The Ant Colony Optimization algorithm, developed by Dorigo et al. [4], is based on the social behavior of insects and their form of communication to find the optimal path between their colony and its power supply. The Particle Swarm Optimization based on the movement of flocks of birds is due to Eberhart and Kennedy [6].

Following the trend of the study of natural collective behavior, Yang [13] introduced a new algorithm, known as Firefly Algorithm, inspired by the collective behavior of fireflies, specifically in how they attract each other. Previous studies have demonstrated that the FA obtained good results, indicating its superiority over some bio-inspired methods [9, 12, 14].

The paper is organized as follows. Section 2 briefly presents the key ideas of the original FA and describes the proposals to change attractiveness function in the FA. The numerical experiments are reported in Section 3 and the paper concludes in Section 4.

2 Firefly Algorithm for Bound Constrained Problems

In this study we are interested in solving the bound constrained global optimization problem by FA. The mathematical formulation of the optimization problem to be addressed in this paper is stated as follows:

$$\begin{array}{ll} \min\limits_{x \in \mathbb{R}^n} & f(x) \\ \text{subject to} & l \leq x \leq u \end{array} \tag{1}$$

where $f(x)$ is a continuous nonlinear objective function, l and u are the lower and upper bounds of the variables.

Following the firefly algorithm and the proposal changes to the attractiveness function are presented.

2.1 Original FA

The firefly algorithm is based on three main principles:

1. All fireflies are unisex, implying that all the elements of a population can attract each other.
2. The attractiveness between fireflies is proportional to their brightness. The firefly with less bright will move towards the brighter one. If no one is brighter than a particular firefly, it moves randomly. Attractiveness is proportional to the brightness which decreases with increasing distance between fireflies.
3. The brightness or light intensity of a firefly is related with the type of function to be optimized. In practice, the brightness of each firefly can be directly proportional to the value of the objective function.

This algorithm is based on two key ideas: the light intensity emitted and the degree of attractiveness that is generated between two fireflies.

The light intensity of firefly i, I_i, depends on the intensity I_0 of light emitted by firefly i and the distance r between firefly i and j. In [13], the light intensity I_i varies with the distance r_{ij} monotonically and exponentially. That is

$$I_i = I_0 e^{-\gamma r_{ij}}$$

where γ is the light absorption coefficient. In theory, $\gamma \in [0; +\infty[$, but in practice γ can be taken as 1.

Since the attractiveness β_{ij} of the firefly i depends on the light intensity seen by an adjacent firefly j and its distance r_{ij}, then the attractiveness β_{ij} is given by:

$$\beta_{ij} = \beta_0 e^{-\gamma r_{ij}{}^2} \tag{2}$$

where β_0 is the attractiveness at $r_{ij} = 0$.

In the original method, the distance r_{ij} between any two fireflies i e j, at x_i and x_j, could be given by the Cartesian distance (or 2-norm):

$$r_{ij} = \|x_i - x_j\|_2. \tag{3}$$

The movement of a firefly i towards another brightest firefly j is given by:

$$x_i = x_i + \beta_{ij}(x_j - x_i) + \alpha\epsilon_{ij} \tag{4}$$

where ϵ_{ij} is a random parameter generated by a uniform distribution and α is a parameter of scale.

In this paper, the light intensity of a firefly i, I_i, is determined by its objective function value.

The pseudo code of the Firefly Algorithm for bound constrained optimization problems can be summarized as follows.

Algorithm 1: Firefly Algorithm

Initialize population of m fireflies, $x_i,\ i = 1,2,\dots,m$.
Compute Light intensity $f(x_i)$, for all $i = 1,2,\dots,m$.

While (stopping criteria is not met) **do**
  for $i = 1$ to m
    for $j = 1$ to m
      if $f(x_i) > f(x_j)$ **then**
        Move firefly i towards j using (4)
      end if
    end for
  end for
  Update Light intensity $f(x_i)$ for all $i = 1,2,\dots m$.
  Rank the fireflies and find the current best
end while

2.2 Modified Attractiveness Approach

In this paper we want to study the effect of changing the calculation of the attractiveness function β (see (2)), either by changing the way of calculating the distance r between two points, either by changing its algebraic expression.
Let p-norm of $x \in \mathbb{R}^n$, for $p \in \mathbb{N}$, be the norm defined by:

$$\|x\|_p = \left(\sum_{i=1}^{n} |x_i|^p\right)^{\frac{1}{p}}$$

Let the infinity-norm, $\|x\|_\infty$, be defined by:

$$\|x\|_\infty = \max\{|x_i|\}_{i=1,\dots,n}.$$

In a first approach, a numerical experiment based on the change of the use of 2-norm, to compute the distance between two fireflies, is analyzed (see (3)). In this experiment, the 1-norm, 4-norm, 10-norm and infinity-norm are considered. The goal is to investigate these modifications in order to improve the efficiency and robustness of the FA (see (2)).

In a second approach, two new attractiveness functions, β^1 and β^2, are considered. Thus, a new attractiveness function, β^1 , that is monotonically decreasing as (2) is proposed:

$$\beta_{ij}^1 = \begin{cases} \beta_0 e^{-\gamma r_{ij}^2} & \text{if } r_{ij} \le d \\ 0.5 & \text{if } r_{ij} > d \end{cases} \tag{5}$$

Here, for a given value of d, if the distance between two fireflies is lesser or equal than d, then β^1 exponentially decreases with the distance. Whenever the distance between two fireflies is greater than d, the attractiveness takes a constant value. The motivation for this approach is due to the fact that in (2) the value of attractiveness can be considered negligible for certain values of distance.

Finally, another attractiveness function that takes into account the average rate of change of the firefly brightness is considered. This function, denoted by β^2, is defined by:

$$\beta_{ij}^2 = \begin{cases} \beta_0 e^{-\frac{f(x_i)-f(x_j)}{r_{ij}}} & \text{if } f(x_i) > f(x_j) \\ 0 & \text{otherwise.} \end{cases} \tag{6}$$

3 Numerical Experiments

The FA algorithm was coded in Matlab and the computational tests were performed on a PC with a 2.53 GHz Intel(R) Core (Tm) i5 Processor M460 and 4 Gb of RAM. We used a collection of 30 benchmark bound constrained global optimization test problems of dimension 2-30, described in full detail in the Appendix B of [1]. Each problem is solved 30 times.

The population of $m = 20$ fireflies is used and the algorithm stops when a maximum of 20000 function evaluations is reached. The parameters used are: $\beta_0 = 0.8$, $\gamma = 1$, $\alpha = 0.5$ and $d = 1$.

In the first set of experiments, different norms in (2) are used for the calculation of the distance r. In the second set of experiments, a comparison with three attractiveness functions is made. For comparison purposes, performance profiles are used. Finally, a table with the results obtained with the best strategy is presented.

To compare the results obtained by different algorithms (or solvers), we used a statistical tool called performance profiles, created by Dolan and Moré [2]. Performance profiles are depicted by a plot of a cumulative distribution $\rho(\tau)$ representing a performance ratio for the different solvers, based on a chosen metric. Generally, performance profiles can be defined representing a statistic computed for a given quality indicator value obtained in several runs. Performance profiles provide a good visualization and easiness of making inferences about the performance of the algorithms. The concept of performance profiles requires minimization of a performance metric and can be described as follows.

Let $\wp$ and $\mathcal{S}$ be the set of problems and the set of solvers in comparison, respectively. Let $m_{p,s}$ be the performance metric found by solver $s \in \mathcal{S}$ on problem $p \in \wp$ after a fixed number of function evaluations. Here, a metric that measures the relative improvement of the function values, a scaled distance to the optimal function value, is defined by:

$$m_{p,s} = \frac{favg_{p,s} - f_{opt}}{f_{worst} - f_{opt}} \tag{7}$$

where f_{opt} denotes the known optimal function value for a problem p, f_{worst} denotes the worst function value found among all solvers on the problem p and $f_{avg_{p,s}}$ denotes the average of the best function values found by a solver s on problem p, after a certain number of runs.

As $\min\{m_{p,s}: s \in \mathcal{S}\}$ can be zero for a particular problem, the comparison used in our study is based on the performance ratios defined by:

$$r_{p,s} = \begin{cases} 1 + m_{p,s} - \min\{m_{p,s}: s \in \mathcal{S}\}, & \text{if } \min\{m_{p,s} : s \in \mathcal{S}\} < \varepsilon \\ \dfrac{m_{p,s}}{\min\{m_{p,s}: s \in \mathcal{S}\}}, & \text{otherwise} \end{cases}$$

for $p \in \wp$, $s \in \mathcal{S}$ and $\varepsilon = 0.00001$.

The overall assessment of the performance of a particular solver s is given by

$$\rho_s(\tau) = \frac{\#\{p \in \wp : r_{p,s} \le \tau\}}{\#\mathcal{S}}, \tau \in \mathbb{R}.$$

The value of $\rho_s(1)$ gives the probability that the solver s will win over the others in comparison. Thus, to see which solver has the least value of the performance metric mostly, then $\rho_s(1)$ should be compared for all the solvers. The higher the ρ_s the better the solver is. On the other hand, for large values of τ, the solver robustness is measured by $\rho_s(\tau)$.

Figure 1 plots the performance profile for the average value of obtained best function values concerning FA with different norms in the attractiveness function (2). From Fig. 1 we may conclude that the average best solutions found for computing attractiveness based on 1-norm, over the 30 runs, outperforms the other versions in comparison. In particular, this one gives the best average solution in about 50% of the tested problems and dominates the other four solvers for all τ values.

Figure 2 shows the plots of the solvers in comparison that are related with the implementation of the two attractiveness functions ((5) and (6)) and the original attractiveness function (2). In all attractiveness functions the 1-norm is considered to compute the distance between fireflies. The attractiveness function defined by β^1 reveals as efficient as the original beta function. Their efficiency is shown in Fig. 2 at $\tau = 1$. However, for τ greater than approximately 2, the solver with β^1 as the attractiveness function wins against the others and dominates with respect to robustness.

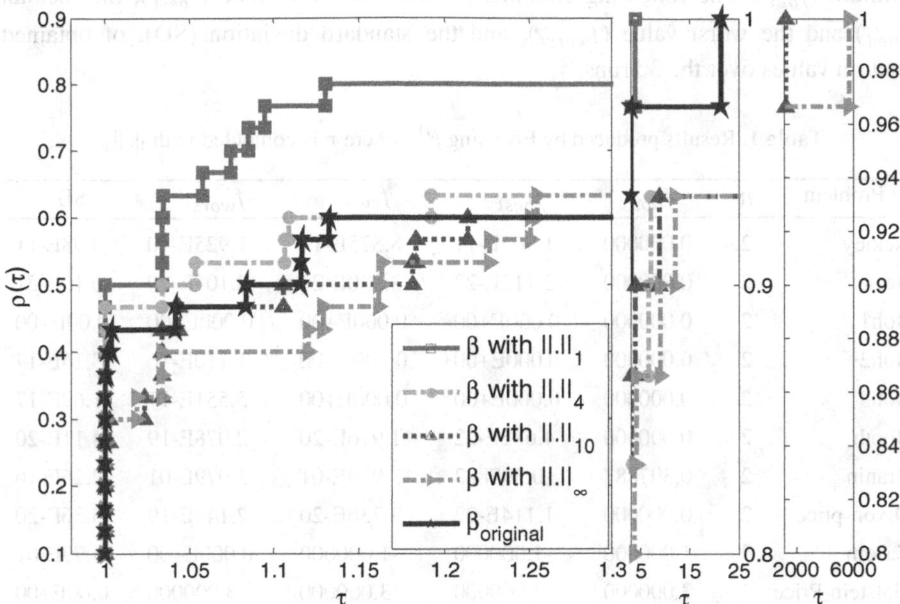

Fig. 1. Performance profiles on f_{avg} with different norms for computing distance

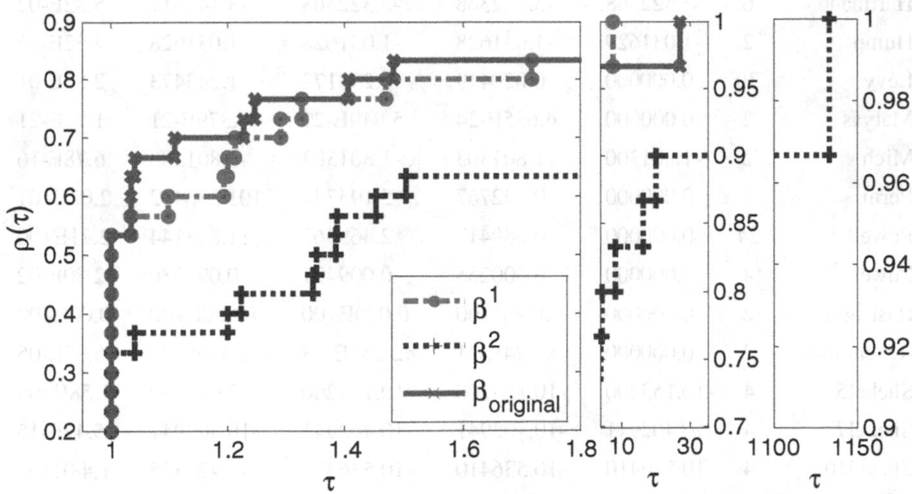

Fig. 2. Performance profiles on f_{avg} with different attractiveness functions

Table 1 summarizes some of the numerical results produced by version of FA with the attractiveness function β^1 and the distance r computed by 1-norm, within 20000 function evaluations among 30 runs. The first column shows the problems names, followed by the dimension of the problem and in the third column the known global optimum (f_{opt}). The following columns present: the best value (f_{best}), the median (f_{med}) and the worst value (f_{worst}), and the standard deviation (SD), of obtained function values over the 30 runs.

Table 1. Results produced by FA using β^1, where r is computed with $\|.\|_1$

Problem	n	f_{opt}	f_{best}	f_{med}	f_{worst}	SD
Ackley	2	0.000000	1.965E-11	8.375E-11	1.925E-10	3.98E-11
Beale	2	0.000000	2.112E-22	3.330E-21	2.105E-20	6.10E-21
Boh1	2	0.000000	0.000E+00	0.000E+00	0.000E+00	0.00E+00
Boh2	2	0.000000	0.000E+00	0.000E+00	1.110E-16	3.15E-17
Boh3	2	0.000000	0.000E+00	0.000E+00	5.551E-17	1.69E-17
Booth	2	0.000000	4.682E-22	1.916E-20	2.078E-19	4.11E-20
Branin	2	0.397887	0.397887	3.979E-01	3.979E-01	2.26E-16
Dixon-price	2	0.000000	1.114E-22	2.736E-20	2.149E-19	5.36E-20
Easom	2	-1.000000	-1.000000	-1.000000	0.000E+00	3.79E-01
Golstein-Price	2	3.000000	3.000000	3.000000	3.000000	0.00E+00
Griewank	2	0.000000	0.000E+00	0.007396	0.027125	5.84E-03
Hartman3	3	-3.862782	-3.862782	-3.862782	-3.861768	2.39E-04
Hartman6	6	-3.322368	-3.322368	-3.322368	-3.141611	5.82E-02
Hump	2	-1.031629	-1.031628	-1.031628	-1.031628	4.52E-16
Levy	30	0.000000	0.020419	0.233172	1.243473	2.80E-01
Matyas	2	0.000000	6.655E-24	5.949E-22	6.379E-21	1.21E-21
Mich	2	-1.801300	-1.801303	-1.801303	-1.801303	6.78E-16
Perm	4	0.000000	0.022767	5.195744	101.536442	2.02E+01
Powell	24	0.000000	0.689417	2.469367	11.843144	2.71E+00
Power	4	0.000000	0.000255	0.009421	0.092230	2.80E-02
Rastrigin	2	0.000000	0.000E+00	0.000E+00	0.000E+00	0.00E+00
Rosenbrock	2	0.000000	8.124E-21	2.284E-18	2.469E-04	5.87E-05
Shekel5	4	-10.153200	-10.153200	-10.153200	-2.682860	2.58E+00
Shekel7	4	-10.402941	-10.402941	-10.402941	-10.402941	5.42E-15
Shekel10	4	-10.536410	-10.536410	-10.536410	-2.427335	1.48E+00
Shubert	2	-186.730909	-186.730909	-186.730909	-180.251858	1.19E+00
Sphere_30	30	0.000000	0.000964	0.002590	0.003891	8.07E-04
Sum-squares	20	0.000000	0.013425	0.360771	3.942155	1.06E+00
Trid	10	-200.000000	-208.708837	-174.090109	-95.873874	2.91E+01
Zakharov	2	0.000000	2.773E-23	3.904E-21	1.818E-20	5.35E-21

The computed solutions are of high quality and the obtained f_{best} solutions obtained for the problems are very close to the known minimum, except for Perm, Powell, Power, Sphere_30, Sum_squares and Trid problems. The values of SD are in general, for all problems, reasonably small showing the consistency of this variant of FA algorithm.

4 Conclusions

In this paper the Firefly Algorithm, a stochastic global optimization algorithm, inspired by the social behavior of fireflies and based on their flashing and attraction, is presented for solving the bound constrained optimization problems.

In order to improve the efficiency of FA, other ways of calculating the distance between the points and other functions to compute the attractiveness of fireflies were tested and analyzed.

A set of benchmark global optimization test problems were used to show the performance of the proposed modified FAs and preliminary results revealed competitive when comparing with the original FA version.

In the future, an extension of FA algorithm, to solve constrained problems, based on penalty techniques will be addressed.

Acknowledgements. This work has been supported by FCT (Fundação para a Ciência e Tecnologia, Portugal) in the scope of the projects: PEst-OE/MAT/UI0013/2014 and PEst-OE/EEI/UI0319/2014.

References

1. Ali, M.M., Khompatraporn, C., Zabinsky, Z.B.: A numerical evaluation of several stochastic algorithms on selected continuous global optimization test problems. J. Global Optim. 31, 635–672 (2005)
2. Dolan, E.D., Doré, J.J.: Benchmarking Optimization Software with Performance Profiles. Preprint ANL/MCS-P861-1200 (2001)
3. Dorigo, M., Stützle, T.: Ant Colony Optimization. MIT Press (2004)
4. Dorigo, M., Caro, G.D., Gambardella, L.M.: Ant algorithms for discrete optimization. Université Libre de Bruxelles, Belgium (1999)
5. Eberhart, R.C., Kennedy, J., Shi, Y.: Swarm optimization. Academic Press (2001)
6. Eberhart, R.C., Kennedy, J.: Particle Swarm optimization. In: Proc. of IEEE International Conference on Neural Networks, Piscataway, NJ, pp. 1942–1948 (1995)
7. Goldber, D.E.: Genetic Algorithms in Search, Optimization and Machine Learning. Addison Wesley, Reading (1989)
8. Heppner, F., Grenander, U.: A stochastic nonlinear model for coordinated bird flocks. The Ubiquity of Chaos. AAAS Publications, Washington DC (1990)

9. Łukasik, S., Żak, S.: Firefly algorithm for continuous constrained optimization tasks. In: Nguyen, N.T., Kowalczyk, R., Chen, S.-M. (eds.) ICCCI 2009. LNCS (LNAI), vol. 5796, pp. 97–106. Springer, Heidelberg (2009)
10. Reynolds, C.W.: Flocks, herds and schools: a distributed behavioral model. Comp. Graph., 25–34 (1987)
11. Rocha, A.M.C., Fernandes, E.M.G.P., Martins, T.F.M.C.: Novel Fish swarm heuristics for bound constrained global optimization problems. J. Comput. Appl. Math. 235(16), 4611–4620 (2011)
12. Yang, X.-S.: Firefly Algorithm, Stochastic Test Functions and Design Optimization. Int. J. Bio-Inspired Computation 2(2), 78–84 (2010)
13. Yang, X.-S.: Nature-Inspired Metaheuristic Algorithms, 2nd edn. Luniver Press, Beckington (2010)
14. Yang, X.-S.: Firefly algorithms for multimodal optimization. In: Watanabe, O., Zeugmann, T. (eds.) SAGA 2009. LNCS, vol. 5792, pp. 169–178. Springer, Heidelberg (2009)

A New Branch-and-Price Approach for the Kidney Exchange Problem

Xenia Klimentova[1], Filipe Alvelos[2], and Ana Viana[1,3]

[1] INESC TEC,
Campus da FEUP, Rua Dr. Roberto Frias, 378, 4200-465 Porto, Portugal
[2] Centro Algoritmi / Departamento de Produção e Sistemas,
Universidade do Minho,
Campus de Gualtar, 4710-057 Braga, Portugal
[3] ISEP - School of Engineering, Polytechnic of Porto,
Rua Dr. António Bernardino de Almeida, 431, 4200-072 Porto, Portugal
xenia.klimentova@icc.ru, falvelos@dps.uminho.pt, aviana@inescporto.pt

Abstract. The kidney exchange problem (KEP) is an optimization problem arising in the framework of transplant programs that allow exchange of kidneys between two or more incompatible patient-donor pairs. In this paper an approach based on a new decomposition model and branch-and-price is proposed to solve large KEP instances. The optimization problem considers, hierarchically, the maximization of the number of transplants and the minimization of the size of exchange cycles. Computational comparison of different variants of branch-and-price for the standard and the proposed objective functions are presented. The results show the efficiency of the proposed approach for solving large instances.

Keywords: Kidney Exchange Problem, Integer Programming, Column Generation, Branch-and-price.

1 Introduction

The Kidney Exchange Problem arises in the framework of kidney exchange programs which were organized in different countries in recent years [18,6,26,27,1]. Patients with kidney failure can participate in these programs if they have a donor willing to donate him/her a kidney, but the pair is not physiologically compatible and because of that the transplantation can not be performed. In the most basic structure the idea is to organize exchanges between a number of such pairs so that a patient in one pair receives a kidney from a donor in the other pair and vice versa (see Fig. 1). The objective for optimization in a kidney exchange program is generally to maximize the collective benefit for a given pool of incompatible pairs, measured by the number of possible kidney exchanges [18,27], but other objectives may also be considered [19,10,22,15]. We will refer to this optimization problem as the *Kidney Exchange Problem* (KEP). Basically an exchange can be considered as a cycle in which the donor from one

B. Murgante et al. (Eds.): ICCSA 2014, Part II, LNCS 8580, pp. 237–252, 2014.

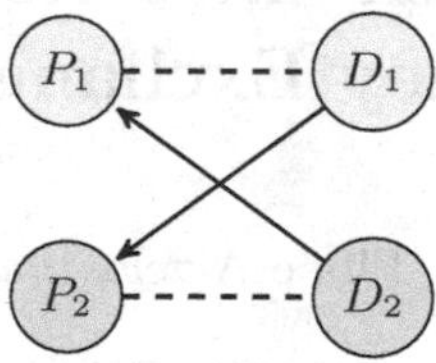

Fig. 1. The donor of the first pair D_1 gives a kidney to the patient P_2 of the second pair, and the patient P_1 gets a kidney from donor D_2

incompatible pair gives a kidney to a compatible recipient in another pair and so on, finally forming an alternating directed cycle (Fig. 1 illustrates an exchange cycle involving 2 pairs).

An important question for the KEP is the definition of an upper bound on the number of pairs that can be involved in a cycle. There are two main reasons that make this upper bound setting inevitable. First, because all operations in an exchange have to be performed simultaneously, the number of personnel and facilities needed for simultaneous operations bring several logistics problems. Second, because last-minute testing of donor and patient can elicit new incompatibilities that were not detected before, causing the donation and all possible exchanges in that cycle to be cancelled, it is preferred that cycles are of limited size. If only 2 pairs can be involved in an exchange cycle the KEP is a maximum cardinality matching problem which can be solved in polynomial time [27,21]. However when the maximum number of pairs is bounded by an integer number k, $3 \leq k < n$, where n is number of pairs in pool, the problem is NP-hard [6,1,17].

Integer Programming (IP) is a natural framework for modeling the KEP and develop methods for seeking an optimal solution. There are two known non-compact formulations for the KEP [25], so called "edge formulation" and "cycle formulation". The former has exponential number of constraints and the later exponential number of variables. As an alternative in [7] compact formulations with the number of variables and constraints bounded by polynomials were proposed. The most effective from a computational point of view is the so called "extended edge formulation". In [1] for the cycle formulation a branch-and-price scheme (combination of column generation with branch-and-bound) is implemented. The authors carried out computational experiments on data generated by current state-of-the-art generator [26] with $k = 3$. A branch-and-price approach for the cycle formulation was also studied in [15].

In this paper we propose a new decomposition model entitled disaggregated cycle decomposition (DCD) which is related to the compact IP formulation proposed in [7] and to the cycle formulation. The proposed model is solved by a branch-and-price algorithm. We also introduce a new objective function with two components, hierarchically related: the first one is the commonly used maximization of the number of transplants, and the second is related with the minimization of the number of pairs involved in exchange cycles. Through use of weights we can represent this hierarchy in a single objective function where the first component is made the most relevant one. Computational experience was carried out to test the proposed DCD and compare the standard and the newly proposed objective function on test instances of different size, for different values on the maximum length of the cycles (k). This new objective function is of practical interest as it increases the probability of the actual number of transplants i.e. the transplants that can still be performed after the last compatibility tests, that are only performed after pairs are matched (since the size of the cycles will tend to be smaller). It also makes branch-and-price much faster at reaching an optimal solution.

2 Problem Statement and Formulations

Graph theory can provide a natural framework for representing the KEP models. Given a directed graph $G(V, A)$, the set of vertices V is the set of incompatible donor-patient pairs. Two vertices i and j are connected by arc $(i, j) \in A$ if a donor from pair i is compatible with the patient of pair j. To each arc (i, j) is associated a weight w_{ij}. If the objective is maximizing total number of transplants $w_{ij} = 1$, $\forall (i, j) \in A$.

The Kidney Exchange Problem can be defined as follows:

Find a maximum weight packing of vertex-disjoint cycles with length at most k.

In the remaining of this document, without loss of generality, we will consider $w_{ij} = 1$, $\forall ij \in A$.

The work presented in this paper is related with two formulations previously presented in the literature, the cycle formulation and the reduced extended edge formulation, that can be described as follows.

2.1 Cycle Formulation

Let $\mathcal{C}$ be the set of all cycles in G with length at most k. We assume that a cycle is an ordered set of arcs. Define a variable $z_c = 1$ if cycle c is selected for the exchange $c \in \mathcal{C}$. Denote by $V(c) \subseteq V$ the set of vertices which belong to cycle c and the number vertices in the cycle is $|c| = |V(c)|$. The cycle formulation is written as follows:

$$\text{maximize} \quad \sum_{c \in \mathcal{C}} |c| z_c \tag{1}$$

$$\text{Subject to:} \quad \sum_{c: i \in V(c)} z_c \leq 1 \qquad \forall i \in V, \tag{2}$$

$$z_c \in \{0, 1\} \qquad \forall c \in \mathcal{C}. \tag{3}$$

The objective function (1) maximizes the number of transplants. Constraints (2) ensure that every vertex is in at most one of the selected cycles (so the donor will only donate one kidney and the patient will receive only one). The difficulty of this formulation is induced by exponential number of variables defined by exponential (in general case) number of cycles of length at most k.

2.2 Reduced Extended Edge Formulation

Let L be an upper bound on the number of cycles in any solution (e.g. the number of vertices). The cycles in the solution can be represented by an index l, with $1 \leq l \leq L$, the problem variables for the extended edge formulation being x_{ij}^l: $x_{ij}^l = 1$ if arc (i,j) is selected to be in a cycle with index l, $x_{ij}^l = 0$, otherwise.

To eliminate redundant variables and avoid symmetry of extended edge formulation some reduction procedures were implemented in [7]. It is imposed that a cycle l in the solution must have node l, and any other nodes in that cycle must have an index larger than l. Moreover, based on this restriction the variables can be eliminated when l cannot be in the same cycle with a node i or an arc $(i,j) \in A$. Denote by $\tilde{V}^l = \{i \in V \ : \ i \geq l\}$ and by d_{ij}^l the shortest path distance in terms of number of arcs in graph G from i to j for $i, j \in \tilde{V}^l$ such that the path passes only through vertices of set $\tilde{V}^l$. Let $d_{ij}^l = +\infty$ if there is no such path from i to j. For each vertex $l \in V$, let us build the set of vertices:

$$V^l = \{i \in V | i \geq l \text{ and } d_{li}^l + d_{il}^l \leq k\}.$$

It can happen that for some $l \in \{1, \ldots, L\}$, $V^l = \emptyset$. Denote by $\mathcal{L} \subseteq \{1, \ldots, L\}$ the set of indices l such that $V^l \neq \emptyset$. Define the set of arcs as:

$$A^l = \{(i,j) \in A \mid i, j \in V^l \text{ and } d_{li}^l + 1 + d_{jl}^l \leq k\}$$

Considering subgraphs $G^l(V^l, A^l)$ for all $l \in \mathcal{L}$, the reduced extended edge formulation is given as:

$$\text{maximize} \quad \sum_{l \in \mathcal{L}} \sum_{(i,j) \in A^l} x_{ij}^l \tag{4}$$

$$\text{subject to} \quad \sum_{j:(j,i) \in A^l} x_{ji}^l = \sum_{j:(i,j) \in A^l} x_{ij}^l \qquad \forall i \in V^l, \forall l \in \mathcal{L}, \tag{5}$$

$$\sum_{l \in \mathcal{L}} \sum_{i:(i,j) \in A^l} x_{ij}^l \leqslant 1 \qquad \forall j \in \bigcup_{l \in \mathcal{L}} V^l, \tag{6}$$

$$\sum_{(i,j) \in A^l} x_{ij}^l \leqslant k \qquad \forall l \in \mathcal{L} \tag{7}$$

$$\sum_{j:(i,j) \in A^l} x_{ij}^l \leqslant \sum_{j:(l,j) \in A^l} x_{lj}^l \qquad \forall i \in V^l, \forall l \in \mathcal{L} \tag{8}$$

$$x_{ij}^l \in \{0,1\}. \qquad \forall (i,j) \in A, \forall l \in \mathcal{L} \tag{9}$$

The objective (4) is to maximize the total number of arcs in the set of all subgraphs of the graph. Constraints (5) guarantee that in each cycle l the number of kidneys received by patient i is equal to number of kidneys given by donor i. Constraints (6) make sure a donor donates only once. Constraints (7) state that in each cycle a maximum number of k arcs is allowed. Finally constraints (8) assure that whenever an arc (i, j) is in cycle l a node with index l is also included in this cycle.

3 Disaggregated Cycle Decomposition

Exact methods based on the decomposition of the cycle formulation were implemented in [26,1,15]. Here we propose a different decomposition based on disaggregation of the cycles according to the subgraphs $G^l(V^l, A^l)$ defined above. For each $l \in \mathcal{L}$ let $\mathcal{C}^l$ be the set of cycles in subgraph G^l with length at most k arcs and including vertex l. For each cycle c in set $\mathcal{C}^l$ define a decision variable $y^{lc} = 1$ if cycle c of set $\mathcal{C}^l$ is chosen and $y^{lc} = 0$, otherwise. Similarly to the cycle formulation denote by $V^l(c) \subseteq V^l$ the set of vertices which belong to cycle $c \in$, $\mathcal{C}^l$ and the number of vertices in the cycle is $|c|^l = |V^l(c)|$.

The disaggregated cycle decomposition (DCD) is given as follows:

$$\text{maximize} \quad \sum_{l \in \mathcal{L}} \sum_{c \in \mathcal{C}^l} |c|^l y^{lc} \tag{10}$$

$$\text{subject to} \quad \sum_{l \in \mathcal{L}} \sum_{c \in \mathcal{C}^l : i \in V^l(c)} y^{lc} \leq 1 \quad \forall i \in V \tag{11}$$

$$y^{lc} \in \{0, 1\} \qquad \forall l \in \mathcal{L}, \forall c \in \mathcal{C}^l. \tag{12}$$

As in the cycle formulation, the number of variables of the DCD grows exponentially with the size of the underlying graph. Therefore approaches based on column generation (e.g. branch-and-price) are the most suitable for obtaining optimal solutions.

Column generation (CG) method allows obtaining an optimal solution of the linear relaxation of a decomposition without all the decision variables. CG alternates between solving a problem with a restricted set of variables (the so called restricted master problem - RMP), and solving subproblems where variables that may potentially improve the current solution of the RMP are identified by using their linear programming reduced costs. In the DCD a subproblem SP(l) is associated with each vertex $l \in \mathcal{L}$ and a solution of SP(l) is a cycle in graph G^l of length at most k and including vertex l. Using the variables x^l_{ij}, $(i, j) \in A^l$, $l \in \mathcal{L}$ defined above for the extended edge formulation, subproblem SP(l), $l \in \mathcal{L}$ is written as follows:

$$\text{maximize} \quad \sum_{(i,j)\in A^l} x^l_{ij} - \sum_{j\in V^l} \lambda_j \sum_{(i,j)\in A^l} x^l_{ij}, \tag{13}$$

$$\text{subject to} \quad \sum_{j:(j,i)\in A^l} x^l_{ji} = \sum_{j:(i,j)\in A^l} x^l_{ij} \qquad \forall i \in V^l, \tag{14}$$

$$\sum_{i:(i,j)\in A^l} x^l_{ij} \leq 1 \qquad \forall j \in V^l,\ j \neq l, \tag{15}$$

$$\sum_{i:(i,j)\in A^l} x^l_{il} = 1, \tag{16}$$

$$\sum_{(i,j)\in A^l} x^l_{ij} \leq k \tag{17}$$

$$x^l_{ij} \in \{0,1\} \qquad \forall (i,j) \in A^l,$$

Here λ_j, $j \in V$ are the dual variables for the packing constraints (11) of the *DCD*.

By construction of subgraphs G^l there always exists a feasible cycle including node l for each $l \in \mathcal{L}$ (otherwise $l \notin \mathcal{L}$) and the SP(l) always has an optimal solution. Hence the constraints (6) and (8) can be modified into (15) and (16). The other constraints are similar to constraints of the formulation (4)–(9).

A major issue when using relaxations in integer programming models is the quality of the bound they provide. It can be derived that the optimal value of the linear relaxation of the *DCD* is the same as the optimal value of the linear relaxation of the cycle formulation. Moreover the *DCD* is stronger (the optimal value of its linear relaxation is equal or better) than the extended edge formulation. This can be proved by noting that the *DCD* can be seen as a Dantzig-Wolfe decomposition [8] of the extended edge formulation represented by (4)–(9) where constraints (6) are the only constraints kept in the master problem (10)-(12), and convexity constraints are redundant. Since the subproblems do not have the integrality property (i.e. the polyhedrons of their linear relaxations have fractional extreme points), the bound provided by the linear relaxation of the decomposed model dominates the bound provided by the linear relaxation of the original model (as proved in general in [14], and by the equivalence between Lagrangean Relaxation and Dantzig-Wolfe decomposition).

Remark 1. State-of-the-art kidney exchange programs include altruistic donors, i.e., donors that are not associated to any patient, but willing to donate a kidney to someone in need. Included into a kidney exchange program an altruistic donor gives a kidney to a patient and the recipient's donor is "dominoed" to add another incompatible pair to the chain and so on. The last donor in the chain normally gives a kidney to the next compatible patient on the deceased donors waiting list [20,13,24]. European programs consider bounded chains with length at most k' (k' can be different from k) [19,15]. In programs running in the USA so called Never-Ending-Altruistic-Donor chains are considered, where a kidney from the last donor in a chain is not assigned to a patient in the deceased donor list [9,23,3,12]. Instead, the last donor acts as an altruistic donor in future matches. The cascading donor chain may continue indefinitely and the length of the chain is unbounded, unless a donor whose related recipient has already been transplanted drops out of the program. To take into account the European variant of the KEP with altruistic donors some constraints similar to (7) are to be added into the extended edge formulation. For the cycle formulation an extension on the set of cycles C with respect to inclusion of chains needs to be performed (see [7] for more details). As for DCD the sets C^l where l are altruistic donors have to be constructed with respect to constant k' (similarly to the cycle formulation), and k' substitutes k in the constraints (17) of the corresponding subproblems SP(l). These modifications do not change significantly the structure of the problem. Therefore the inclusion of altruistic donor chains is not considered in this paper. However all the implemented techniques can be adapted to the problem with bounded length chains as well.

4 A New Objective Function

As mentioned above the bound k on the number of incompatible pairs involved in exchange cycles is important because last-minute tests may find out incompatibilities that were not detected before, causing all possible exchanges in that cycle to be cancelled. Hence among all possible optimal solutions of the problem the ones having smaller cycles are preferable. This goal can be achieved by introducing an additional objective function related with the number of cycles (which is desired to be high) and considering the objectives hierarchically. Given that the number of cycles is $\sum_{l\in\mathcal{L}}\sum_{c\in C^l} y^{lc}$, the problem with the second objective can be written as:

$$\text{maximize} \quad \sum_{l\in\mathcal{L}}\sum_{c\in C^l} y^{lc} \tag{18}$$

$$\text{subject to} \quad \sum_{l\in\mathcal{L}}\sum_{c\in C^l: i\in V^l(c)} y^{lc} \le 1 \quad \forall i \in V, \tag{19}$$

$$\sum_{l\in\mathcal{L}}\sum_{c\in C^l} |c|^l y^{lc} \ge v^*, \tag{20}$$

$$y^{lc} \in \{0,1\} \qquad \forall l \in \mathcal{L}, \forall c \in C^l. \tag{21}$$

where v^* is an optimal value of the problem DCD. Denote the hierarchical problem defined by DCD and (18)–(21) as H. It can be derived that the objective functions (10) and (18) may be combined into a single one, keeping their hierarchy, with help of weights as follows:

$$\sum_{l\in\mathcal{L}}\sum_{c\in C^l}(|V||c|^l + 1)y^{lc}. \tag{22}$$

The weight $|V|$ is given to maximize the number of transplants, implicitly stating that this objective is more important than the maximization of the number of cycles. An optimal value for hierarchical problem H may be obtained by solving the DCD with objective function (22) (we will denote this problem (22), (11)–(12) as DCD_{nc}). Indeed, as the number of cycles in a solution cannot exceed $|V|$, an optimal solution of the DCD_{nc} will lead to the same number of transplants obtained when solving DCD. However, thanks to the second component of (22), the solution for the DCD_{nc} will have equal or bigger number of cycles, implying that cycles are of equal or smaller sizes. Therefore, it can be stated that by solving the problem DCD_{nc}, we also solve the problem DCD.

5 Solution Methods

As the DCD has exponential number of variables branch-and-price based on CG is an adequate method to obtain optimal solutions for the problem. Implementation details are presented in this section. They are preceded by a preprocessing phase, that generates subgraphs $G^l(V^l, A^l)$. The shortest paths d^l_{ij} used for construction of the subgraphs are calculated using Floyd-Warshall algorithm [11].

5.1 Column Generation

The general scheme of the implemented column generation is given by Fig 2.

```
Step 0. Initialize the RMP.
Step 1. If a heuristic to solve subproblems is used
        {
           Do {
                Solve the RMP with a linear programming solver.
                Solve the subproblems SP(l) heuristically.
              }
           while(at least one subproblem provides one attractive column)
        }
        Else
           Go to Step 2.
Step 2. Do {
           Solve the RMP with a linear programming solver.
           Solve the subproblems SP(l) exactly.
           }
        while(at least one subproblem provides one attractive column)
STOP
```

Fig. 2. Column generation general scheme

Initialization of RMP

Although it is reported in [1] that it is effective to initialize RMP with some set of initial columns, our preliminary experiments showed that for the implementation proposed in this work the most effective methods are the ones which do not use any initial columns. So in the computational experiments presented in section 6 the RMP was initialized with solutions of sub-problems SP(l) obtained with 0 values of dual variables λ_j, $j \in V^l$, $l \in \mathcal{L}$.

Solving Sub-problems

In this paper two special procedures are implemented for finding "good" feasible and optimal solutions for each subproblem. The objective of each SP(l) is to find a cycle including vertex l with maximum weight. This problem can be solved by finding maximum weight paths from vertex l to all other vertices and adding the weight of the arc from each vertex to l (if it exists). By following this procedure the cycle with maximum weight can be discovered. It is known that Belman-Ford's algorithm is effective to find a shortest path in a graph with weighted arcs. A general scheme of the algorithm was adopted for finding a cycle of length up to k.

A heuristic to find a feasible cycle including node l for subproblem SP(l) is presented in Figure 3. Let h^l_{ij}, $(i,j) \in A^l$ be the current coefficients of the objective function of SP(l): $h^l_{ij} = 1 - \lambda_j$. The heuristic performs a random walk on subgraph G^l, starting from vertex l (Step 0). The trace of the performed walk is kept in δ which is an ordered set of nodes (Step 5). The method chooses a new current vertex u in each iteration from the set of active nodes U such that the corresponding arc has the maximum weight (Step 2). Afterwards it checks if there is a back arc to vertex l (Step 4). The method stops whenever the weighted cycle is found. The performed walk δ will define a solution of SP(l) with weight equal to the sum of coefficients h_{ij} for corresponding arcs.

Step 0. Initialize $U = V^l \backslash \{l\}$, $u = l$, $\delta = (l)$.
Step 1. If $U = \emptyset$ STOP.
Step 2. If $\{i \in U : (u,i) \in A^l\} = \emptyset$, then $U = U \backslash \{u\}$,
u =vertex in walk δ previous to u, GoTo Step 1.
Step 3. Choose node $i \in U$, such that $h_{ui} = \max_{k:(u,k) \in A^l} h^l_{uk}$,
$U = U \backslash \{i\}$
Step 4. If $(i,l) \in A^l$ then δ is a solution of SP(l), STOP
Step 5. If (length of δ equal to $k-1$), then
GoTo Step 2,
Else add i to walk: $\delta = (\delta, i)$, $u = i$, GoTo Step 1

Fig. 3. Heuristic for solving SP(l)

5.2 Branch and Price Based Approaches

Branch-and-price (B&P) is the combination of column generation and branch-and-bound methods used to obtain optimal solutions to decomposition models [4]. In this work we explored two ways of potentially increasing the efficiency of a pure branch-and-price scheme: (i) to solve the integer RMP of the root node by a general purpose solver with different time limits, and (ii) to apply local search based heuristics in the root or in all nodes of the search tree.

The second approach is related with the framework *SearchCol: Metaheuristic search by column generation* [2]. Within that framework global solutions to a problem are considered to be formed by selecting one solution from each subproblem (in some cases the null solution). SearchCol defines many problem independent metaheurisitic components such as constructive procedures, neighbourhoods, and evaluation functions. When combined with B&P, (meta)heuristics may be applied in chosen nodes of the tree after column generation solves the relaxation of the node. In the current work we used Local Search (LS) and Variable Neighbourhood Search (VNS) [16] based on the definition of n-neighbour — a solution which differs from the current one by n cycles. For LS $n = 1$ was used. The shaking phase of VNS was accomplished by using higher values of n. The LS and VNS were attempted to be used in the root node and in all nodes of the search tree. Initial solutions were obtained by selecting, for each subproblem, the cycle with higher value in the optimal RMP of the CG of the node.

6 Computational Tests

Computational experiments were carried out to evaluate the proposed decomposition for the KEP and to investigate the difference from a computational point of view of two objective functions of the problem. CPU times were obtained on a computer with Intel Xeon processor at 3.00 GHz, 8 Gb of RAM and running Linux OS. MIP solver CPLEX 12.4 was used.

We used SearchCol++ (http://searchcol.dps.uminho.pt/) to implement all the approaches developed. SearchCol++ is a C++ set of classes which implements column generation based algorithms (B&P and SearchCol) in a general way. Only problem specific components such as the exchange of information between the RMP and subproblems solvers must be implemented by the user. Thus branching, column management, the application of metaheuristics, and perturbations are hidden and controlled through parameters. For the sake of generality branching is performed in the original subproblem variables.

To generate the instances for the computational study current state-of-the-art generator [26] was used. It creates random graphs based on probability of blood type and of donor–patient compatibility. Instances with different number of nodes (starting from 100) were generated, 10 instances of each size. The problem was studied for different values of parameter k ranging from 3 to 6[1].

[1] Currently implemented kidney exchange programs work in general with a maximum value of $k = 3$. However this value has already been exceeded by large: the maximum number of simultaneous transplantations performed up to now being 6 [5].

Average times to find an optimal solution (excluding instances where a given CPU time limit was reached) with different methods for the problem with the standard and the proposed objective function (see (22) and section 4) are shown in tables 1 to 4; maximum CPU time was set to 1800 seconds for all methods. The following notation is used in the tables:

- $|V|$ is the number of nodes in the graph;
- t_{prep} is the average CPU time (in seconds) for preprocessing the instances;
- $B\&P$ is the average CPU time to find an optimal solution of the KEP with standard objective function (the DCD) by the pure Branch-and-Price method;
- $rMIP$ is the average CPU time for solving DCD by CG combined with MIP solver in root node;
- $rVNS$ is the average CPU time for the CG combined with VNS in root node;
- $rMIP_H$ is the average CPU time for the CG combined with MIP solver in root node, using an heuristic for solving Sub-Problems;
- $rMIP_{LS}$ is the average CPU time for the CG combined with MIP solver, with LS running in all nodes;
- $B\&P_{nc}$ is the average CPU time to find an optimal solution for the problem with objective function (22) (the DCD_{nc}) by the pure Branch-and-price method;
- $rMIP_{nc}$ is the average CPU time to find an optimal solution of the DCD_{nc} by the $rMIP$ method.

In some cases there were instances that could not be solved within the CPU time limit. For such cases we show in parenthesis for each method the number of instances out of 10 that were solved.

Note that, as mentioned in section 4, the optimal value of the problem with standard objective function may be derived from the optimal value of the problem with the new function (22). Thus methods $B\&P_{nc}$ and $rMIP_{nc}$ for the latter may be considered as solution methods for the problem with standard objective and compared with other approaches for this problem.

Computational results for $k = 3$ to $k = 6$ are presented in Tables 1 to 4.

Results show that, for $k = 3$, preprocessing times are almost equal or even exceed the time to solve a problem with e.g. $rMIP_{nc}$. Still, they do not exceed 2 minutes and for the harder instances, with bigger values of k, they are smaller than the time required to solve the problem. It can also be seen that, for $k = 3$, $rMIP$ outperforms $B\&P$ and $rVNS$. Moreover, no improvements are found when heuristics are used for solving SPs and local search is run in each node of the tree (see $rMIP$ versus $rMIP_H$ and $rMIP_{LS}$). In fact CPU times increase. Therefore these variants of $rMIP$ as well as $rVNS$ are not used in experiments with bigger values of k.

When the new objective is considered, $rMIP_{nc}$ cannot solve some instances within available CPU time. This method will also be discarded for computational tests with $k > 3$.

Table 1. CPU time (in seconds) for $k = 3$

$\|V\|$	t_{prep}	$B\&P$	$rMIP$	$rVNS$	$rMIP_H$	$rMIP_{LS}$	$B\&P_{nc}$	$rMIP_{nc}$
100	0.0	0.0	0.0	0.0	0.0	0.0	0.0	0.0
200	0.1	0.0	0.1	0.1	0.1	0.1	0.0	0.0
300	0.3	0.3	0.2	0.4	0.3	0.2	0.2	0.2
400	0.7	0.6	0.6	0.9	0.5	0.4	0.5	1.1 (9)
500	1.4	1.4	1.6	2.0	1.9	1.0	1.0	0.8
600	2.5	3.7	1.5	4.9	3.3	2.5	2.3	1.2 (7)
700	3.6	5.5	4.0	7.5	3.9	2.5	3.9	1.8 (7)
800	5.5	10.6	4.0	13.4	6.6	5.2	4.5	2.6 (8)
900	8.0	10.3	5.8	14.8	10.3	6.3	8.3	4.1 (8)
1000	10.4	18.5	9.3	24.3	21.1	13.1	11.7	5.2
1500	35.6	105.4	20.9	127.4	71.1	47.9	57.1	16.3 (7)
2000	87.6	513.2	89.8	574.4	299.8	91.6	206.0	51.4 (7)

Results for $k = 4$ are presented in Table 2. With this value of k the problem becomes harder to solve. Therefore the time spent by MIP solver in the $rMIP$ method was limited to a number of seconds obtained by multiplying a parameter by the number of vertices of the network. Extra computational tests were carried out for values of 0.1 and 0.01 of that parameter. They are presented in columns $rMIP_{0.1}$ and $rMIP_{0.01}$, respectively.

Table 2. CPU time (in seconds) for $k = 4$

$\|V\|$	$B\&P$	$rMIP$	$rMIP_{0.1}$	$rMIP_{0.01}$	$B\&P_{nc}$
100	0.0	0.1	0.1	0.1	0.0
200	0.7	2.5	2.5	1.5	0.2
300	5.0	34.8	19.0	7.8	1.1
400	15.7	214.4	79.7	20.7	3.8
500	46.3	369.3 (8)	131.5	47.9	6.8
600	118.0	296.4 (3)	201.3	109.1	15.5
700	176.6	-	333.6	198.2	30.1
800	375.5	-	729.7	392.4	50.5
900	746.4	-	903.6 (9)	748.3	144.4
1000	1315.5 (8)	-	1388.7 (6)	1375.2 (7)	82.4

It is obvious that $rMIP$ is not efficient. $rMIP_{0.01}$ with the tightest bound on the time spent by the MIP solver performs almost as $B\&P$ and these two are the most efficient among the methods for the problem with the standard objective function. However it is interesting that all those methods lose to the Branch-and-Price run for the problem with the newly proposed objective function (column $B\&P_{nc}$). It handles all instances with up to 1000 nodes in significantly shorter time.

Table 3. CPU time (in seconds) for $k = 5$

$\|V\|$	$B\&P$	$rMIP_{0.1}$	$rMIP_{0.01}$	$B\&P_{nc}$
100	0.1	0.3	0.3	0.0
200	2.0	23.0	5.8	0.4
300	17.9	71.0	27.3	3.3
400	66.8	146.5	80.1	10.1
500	255.0	355.8	211.7	14.6
600	618.1	741.5	579.8	36.8
700	1040.0	1107.5 (9)	906.3	87.5
800	1694.6 (2)	-	1453.1 (5)	117.5
900	-	-	-	125.4 (6)

Table 4. CPU time (in seconds) for $k = 6$

$\|V\|$	$B\&P$	$rMIP_{0.01}$	$B\&P_{nc}$
100	0.3	0.6	0.1
200	9.4	9.8	0.6
300	96.6	63.0	3.9
400	187.3	252.6	12.5
500	994.1	619.7	21.5
600	1085.5	1276.4	63.8
700	1688.6 (2)	1706.8 (1)	114.1
800	-	-	155.6
900	-	-	214.6 (4)

Similar conclusions can be made when $k = 5$ and $k = 6$ (see Tables 3 and 4). $rMIP_{0.01}$ with stronger limit on time spent by MIP solver and $B\&P$ work similar, with slight dominance of $rMIP_{0.01}$ for bigger graphs and $k = 5$ (5 instances with 800 nodes solved by $rMIP_{0.01}$ versus 2 instances solved by $B\&P$). Furthermore, the $B\&P_{nc}$ is significantly more efficient than the others, solving 6 instances with 900 nodes, none of which having been solved by any other method.

For $k = 6$ the Branch-and-Price method applied to the problem with the proposed objective function is again the most efficient method. It was able to solve 4 instances with 900 nodes, while the other methods could only handle at most 2 problems with 700 nodes.

It is worth reminding that although problems DCD and DCD_{nc} are different – they differ in the objective function – the latter also solves the former and therefore it is reasonable to compare CPU times between the formulations. This allows us to conclude that using the proposed decomposition with the new objective function DCD_{nc}, problems of significantly bigger size can be solved especially for larger k. When compared to computational experiments carried out in [7] on a computer with a Quad-Core Intel Xeon processor at 2.66 GHz, 16 Gb of RAM and running Mac OS X 10.6.6 with the same version of MIP solver the maximum size of solved instances was 1000 nodes for $k = 3$, and only 200 nodes for $k = 4$ and 100 nodes for $k = 5, 6$.

7 Conclusions

This paper proposes a new efficient approach for seeking an optimal solution for large-scale instances of the kidney exchange problem. The approach is based on a new decomposition model entitled disaggregated cycle decomposition, two component objective function and branch-and-price. An optimal solution for the proposed objective function has maximum number of transplants, trying also to reduce the size of the cycles involved in the exchanges.

Computational results show that the addition of the second objective component related with minimizing the sizes of cycles to the standard objective function of maximizing the number of transplants allows solving smaller instances faster and also handling larger instances. Within the proposed approach instances with 2000 pairs and a limit of 3 pairs involved in an exchange cycle were solved in 200 seconds in average. Instances with 1000 pairs and a limit of 4 pairs in a cycle were solved in 80 seconds, and instances with 800 pairs and a limit of 5 and 6 pairs in 120 and 155 seconds respectively. These results clearly outperform those obtained when using the standard objective function of maximizing the number of transplants, as well as the ones obtained by using a general purpose solver for the compact model in [7]. Thus, by solving the problem with the new objective function we can indirectly find a solution of the original problem, with standard objective function, with much less computational effort.

Acknowledgements. We would like to thank Dr. Nicolau Santos from the INESC TEC, Porto, Portugal for useful comments. This work is financed by the ERDF — European Regional Development Fund through the COMPETE Programme (operational programm for competitiveness), by National Funds through the FCT — Fundação para a Ciência e a Tecnologia (Portuguese Foundation for Science and Technology) within project "KEP - New models for enhancing the kidney transplantation process. /FCT ref: PTDC/EGE-GES/110940/2009", by the North Portugal Regional Operational Programme (ON.2 O Novo Norte), under the National Strategic Reference Framework (NSRF), through the European Regional Development Fund (ERDF), and by national funds through FCT within project "NORTE-07-0124-FEDER-000057", and has also been supported by FCT through projects "SearchCol: Metaheuristic search by column generation" (PTDC/EIAEIA/100645/2008) and PEst-OE/EEI/UI0319/2014.

References

1. Abraham, D., Blum, A., Sandholm, T.: Clearing algorithms for Barter exchange markets: Enabling nationwide kidney exchanges. In: Proceedings of the 8th ACM Conference on Electronic Commerce, June 13-16, pp. 295–304 (2007)

2. Alvelos, F., de Sousa, A., Santos, D.: Combining Column Generation and Metaheuristics. In: Talbi, E.-G. (ed.) Hybrid Metaheuristics. SCI, vol. 434, pp. 285–334. Springer, Heidelberg (2013)
3. Ashlagi, I., Gilchrist, D., Roth, A., Rees, M.: Nonsimultaneous chains and dominos in kidney paired donation - revisited. American Journal of Transplantation 11(5), 984–994 (2011)
4. Barnhart, C., Johnson, E., Nemhauser, G., Savelsbergh, M., Vance, P.: Branch-and-price: column generation for solving huge integer programs. Operations Research 46, 316–329 (1998)
5. BBC: BBC news website. six-way kidney transplant first (9/04/2008) (2008), `http://news.bbc.co.uk/1/health/7338437.stm` (last accessed in December 2012)
6. Biro, P., Manlove, D., Rizzi, R.: Maximum weight cycle packing in directed graphs, wiht application to kidney exchange programs. Discrete Mathematics, Algorithms and Applications 1(4), 499–517 (2009)
7. Constantino, M., Klimentova, X., Viana, A., Rais, A.: New insights on integer-programming models for the kidney exchange problem. European Journal of Operational Research 231(1), 57–68 (2013)
8. Dantzig, G., Wolfe, P.: Decomposition principle for linear programs. Operations Research 8, 101–111 (1960)
9. Dickerson, J., Procaccia, A., Sandholm, T.: Optimizing kidney exchange with transplant chains: Theory and reality. In: AAMAS 2012: Proc. 11th Intl. Joint Conference on Autonomous Agents and Multiagent Systems (June 2011)
10. Dickerson, J., Procaccia, A., Sandholm, T.: Failure-aware kidney exchange. In: EC 2013: Proc. 14th ACM Conference on Electronic Commerce (June 2013)
11. Floyd, R.: Algorithm 97: Shortest path. Communications of the ACM 5(6), 345 (1962)
12. Gentry, S., Montgomery, R., Segev, D.: Kidney paired donation: Fundamentals, limitations, and expansions. American Journal of Kidney Disease 57(1), 144–151 (2010)
13. Gentry, S., Montgomery, R., Swihart, B., Segev, D.: The roles of dominos and nonsimultaneous chains in kidney paired donation. American Journal of Transplantation 9, 1330–1336 (2009)
14. Geoffrion, A.: Lagrangean relaxation for integer programming. Mathematical Programming Study 2, 82–114 (1974)
15. Glorie, K., Wagelmans, A., van de Klundert, J.: Iterative branch-and-price for large multi-criteria kidney exchange. Econometric Institute report (2012-11) (2012)
16. Hansen, P., Mladenovic, N., Perez, J.: Variable neighbourhood search: methods and applications. Annals of Operations Research 175, 367–407 (2010)
17. Huang, C.: Circular stable matching and 3-way kidney transplant. Algorithmica 58, 137–150 (2010)
18. de Klerk, M., Keizer, K., Claas, F., Haase-Kromwijk, B., Weimar, W.: The Dutch national living donor kidney exchange program. American Journal of Transplantation 5, 2302–2305 (2005)
19. Manlove, D.F., O'Malley, G.: Paired and altruistic kidney donation in the UK: Algorithms and experimentation. In: Klasing, R. (ed.) SEA 2012. LNCS, vol. 7276, pp. 271–282. Springer, Heidelberg (2012)
20. Montgomery, R., Gentry, S., Marks, W., Warren, D., Hiller, J., Houp, J., Zachary, A., Melancon, J., Maley, W., Simpkins, H.R.C., Segev, D.: Domino paired kidney donation: a strategy to make best use of live non-directed donation. The Lancet 368(9533), 419–421 (2006)

21. Nemhauser, G., Wolsey, L.: Integer and Combinatiorial Optimization. A Wiley-Interscience Publication (1999)
22. Pedroso, J.: Maximizing expectation on vertex-disjoint cycle packing. Technical Report DCC-2013-5 (2013)
23. Rees, M., Kopke, J., Pelletier, R., Segev, D., Rutter, M., Fabrega, A., Rogers, J., Pankewycz, O., Hiller, J., Roth, A., Sandholm, T., Ünver, M., Montgomery, R.: A nonsimultaneous, extended, altruistic-donor chain. The New England Journal of Medicine 360, 1096–1101 (2009)
24. Roth, A., Sönmez, T., Ünver, M.: Kidney exchange. Quarterly Journal of Economics 119(2), 457–488 (2004)
25. Roth, A., Sönmez, T., Ünver, M.: Efficient kidney exchange: Coincidence of wants in markets with compatibility-based preferences. The American Economic Review 97(3), 828–851 (2007)
26. Saidman, S., Roth, A., Sönmez, T., Ünver, M., Delmonico, F.: Increasing the opportunity of live kidney donation by matching for two- and three-way exchanges. Transplantation 81, 773–782 (2006)
27. Segev, D., Gentry, S., Warren, D., Reeb, B., Montgomery, R.: Kidney paired donation and optimizing the use of live donor organs. The Journal of the American Medical Association 293(15), 1883–1890 (2005)

A Study of the Complexity of an Infeasible Predictor-Corrector Variant of Mehrotra Algorithm

Ana Paula Teixeira[1] and Regina Almeida[2]

[1] University of Trás-os-Montes e Alto Douro, UTAD,
Science and Technology School
Department of Mathematics
Quinta de Prados, 5000-801, Vila Real, Portugal
CIO. Faculty of Sciences, University of Lisbon, Portugal
ateixeir@utad.pt

[2] University of Trás-os-Montes e Alto Douro, UTAD,
Science and Technology School
Department of Mathematics
Quinta de Prados, 5000-801, Vila Real, Portugal
CM-UTAD. University of Trás-os-Montes e Alto Douro, Portugal
ralmeida@utad.pt

Abstract. An infeasible version of the algorithm developed by Bastos and Paixão is analyzed and its complexity is discussed. We prove the efficiency of this infeasible algorithm by showing its complexity and Q-linear convergence. We start by demonstrating that, at each iteration, the step size computed by this infeasible predictor-corrector variant algorithm is bounded below by $\frac{1}{250\,n^4}$ and has $O(n^4|\log(\epsilon)|)$ iteration complexity; thus, proving that, similarly to what happens with the feasible version, this infeasible version of Bastos and Paixão algorithm has polynomial iteration complexity and is Q-linearly convergent.

Keywords: Infeasible predictor-corrector variant, linear programming, Mehrotra-type algorithm, polynomial complexity.

1 Introduction

The fundamentals of the Interior Point Methods have been established in the last two decades of the 20th century, in particular by Karmarkar [5]. These methods have been widely used and analyzed, for example, in the works [3,4,6–14,17–24], and more recently in [1,15,16], being Mehrotra's predictor-corrector algorithm [12] one of the most implemented.

The great success of Mehrotra's approach in practice justified not only the development of new variants of this method, but also additional theoretical studies on the properties of these new algorithms, including their computational complexity; as can be seen, for example, in [1,7,13–16,19,21–23].

B. Murgante et al. (Eds.): ICCSA 2014, Part II, LNCS 8580, pp. 253–266, 2014.

Bastos and Paixão [2] presented specialized versions of a feasible predictor-corrector variant of Mehrotra's algorithm [12] for transportation and assignment problems. The major differences between this new variant and the classical predictor-corrector for Linear Programming are: the predictor direction is computed as in the primal-dual methods, it uses the same duality measure both for the predictor and the corrector directions and no line search is needed to obtain the duality measure, thus having the advantage of decreasing the running time per iteration. Extensive computational experiments (described in [2]), have indicated that this new version was computationally more efficient than the original one for both class of problems studied in that work. Based on the good computational results of this new feasible algorithm, Teixeira and Almeida [16] proved that the algorithm has $O(n^4|\log(\epsilon)|)$ iteration complexity.

Feasible interior point methods start and remain in the interior of the feasible set during all the process. However, many times, finding such a starting point is complicated. Theoretically, this difficulty can be overcome by reformulating the problem and embedding the original feasibility set into a new set in a higher dimensional space. The disadvantage of this approach is not only the fact that large unknown parameters are required, which, depending on their size, may introduce instability in the resolution process, but also the possibility of a change on the structure of the original problem that can make the problem harder to solve. Because of this, embedding techniques are not very used. A practical alternative to the embedding is to use an infeasible interior point method, which requires no more than the positivity of the variables and try to achieve feasibility and optimality simultaneously.

The complexity of infeasible interior point methods was studied, for example, by Al-Jeiroudi and Gondzio [1], Kojima, Meggiddo and Mizuno [6], Korzak [7], Mizuno [13], Potra [14], Wright [17], Wright and Y. Zhang [19], Zhang [21] and Zhang and Zhang [23].

Taking into account the results obtained in [2] and [16], we considered relevant to examine the theoretical efficiency of an infeasible version of the algorithm developed by Bastos and Paixão [2], in order to understand if it would be worthwhile its future implementation and application to transportation and assignment class of problems.

In this work we discuss the theoretical efficiency of an infeasible version of the algorithm developed by Bastos and Paixão [2], establishing the complexity bound of $O(n^4|\log(\epsilon)|)$ and, thus, proving that both feasible and infeasible versions of Bastos and Paixão algorithm are Q-linearly convergent.

Let us now present some concepts and notation that will be used. Consider $M_{m\times n}(\mathbb{R})$ as the set of the real $m \times n$ matrices. The standard primal-dual pair of Linear Programming problems is

$$\begin{array}{llll} \text{(P)} \min\limits_{x} c^T x & & \text{(D)} \max\limits_{u,v} b^T u & \\ \text{s.t.}\ Ax = b & & \text{s.t.}\ A^T u + v = c & \\ \qquad x \geq 0 & & \qquad v \geq 0 & \end{array}$$

with $A \in M_{m\times n}(\mathbb{R})$, $c, x, v \in \mathbb{R}^n$ and $b, u \in \mathbb{R}^m$.

The predictor direction of the algorithm in [2], $(\triangle x^a, \triangle u^a, \triangle v^a)$, is obtained by solving the system

$$\begin{cases} A\triangle X^a e & = -r_b \\ A^T \triangle U^a e + \triangle V^a e & = -r_c \\ V^k \triangle X^a e + X^k \triangle V^a e = \mu_k e - X^k V^k e \end{cases}. \quad (1)$$

The corrector direction, $(\triangle x, \triangle u, \triangle v)$, is obtained by solving the system

$$\begin{cases} A\triangle X e & = 0 \\ A^T \triangle U e + \triangle V e & = 0 \\ V^k \triangle X e + X^k \triangle V e = \mu_k e - \triangle X^a \triangle V^a e - X^k V^k e \end{cases}, \quad (2)$$

where $r_b = Ax - b$ and $r_c = A^T u + v - c$ are the residuals for the two linear equations, μ_k is the duality measure define by

$$\mu_k = \frac{c^T x^k - b^T u^k}{\theta(n)}, \quad \text{with} \quad \theta(n) = \begin{cases} n^2, & n \leq 5000 \\ n\sqrt{n} & n > 5000 \end{cases} \quad (3)$$

and e denotes the vector with all components equal to one and $X^k = diag\left(x^k\right)$ is a diagonal matrix with the elements of the vector x^k in the diagonal. Analogously, the matrices $V^k, \triangle X, \triangle U, \triangle V, \triangle X^a, \triangle U^a$ and $\triangle V^a$ are obtained by using the elements of the correspondent vectors in the diagonal.

This predictor-corrector algorithm also uses the infeasible infinity norm neighborhood defined by

$$\mathcal{N}_\infty^-(\gamma,\beta) = \Big\{(x^k,u^k,v^k) \in \mathcal{F}^\circ : \ \|(r_b^k, r_c^k)\| \leq \frac{\|(r_b^0, r_c^0)\|}{\mu_0}\,\beta\mu_k, \quad (4)$$
$$x_i^k v_i^k \geq \gamma\mu_k, \ i \in \mathcal{I}\Big\},$$

where $\gamma \in\,]0,1[$, $\beta \geq 1$ are given constants, (x^0, u^0, v^0) with $(x^0, v^0) > (0,0)$ is the initial point, (r_b^0, r_c^0) is the pair of residuals evaluated at the starting point, $\mathcal{I} = \{1, 2, \cdots, n\}$ and

$$\mathcal{F}^\circ = \{(x,u,v) \in \mathbb{R}^n \times \mathbb{R}^m \times \mathbb{R}^n : (x,v) > (0,0)\}.$$

This algorithm can be formalized in *Algorithm 1*.

Throughout this paper, not only are valid the definitions and notation previously mentioned but, we also consider the index sets

$$\mathcal{I}_+ = \{i \in \mathcal{I} : \triangle x_i^a \triangle v_i^a > 0\} \qquad \text{and} \qquad \mathcal{I}_- = \{i \in \mathcal{I} : \triangle x_i^a \triangle v_i^a < 0\}.$$

For simplicity of notation, we omit the iteration index of the triples that represent the predictor and the corrector directions.

In this paper, we use $\gamma = \frac{1}{2}$ in the infeasible infinity norm neighborhood, i.e., $\mathcal{N}_\infty^-\left(\frac{1}{2},\beta\right)$ and we consider $\lambda_k = \min\{\alpha_p^k, \alpha_d^k\}$.

Algorithm 1.

Require: $\gamma \in]0,1[$, $\beta \geq 1$, $(x^0, u^0, v^0) \in \mathcal{F}^\circ$ and $\mathcal{N}_\infty^-(\gamma,\beta)$ defined in (4);

Set $k = 0$;

while the termination criteria is not satisfied **do**

(a) Compute μ_k using (3);

(b) Obtain the predictor direction $(\triangle x^a, \triangle u^a, \triangle v^a)$ by solving (1);

(c) Obtain the corrector direction $(\triangle x, \triangle u, \triangle v)$ by solving (2);

(d) Compute primal and dual step sizes, α_p^k and α_d^k respectively,

$$(\alpha_p^k, \alpha_d^k) = \max_{(\alpha_1,\alpha_2)>(0,0)} \left\{ (\alpha_1, \alpha_2) : \ (x^k + \alpha_1 \triangle x, u^k + \alpha_2 \triangle u, v^k + \alpha_2 \triangle v) \in \mathcal{N}_\infty^-(\gamma,\beta) \right\};$$

(e) Compute

$$x^{k+1} = x^k + 0.9995\alpha_p^k \triangle x,$$
$$(u^{k+1}, v^{k+1}) = (u^k, v^k) + 0.9995\alpha_d^k (\triangle u, \triangle v);$$

(f) Set $k = k+1$;

end while

---

2 Technical Results

In order to establish the main results of this paper, we develop in this section some technical results. Lemmas 1, 2 and 3 give upper and lower bounds for the predictor direction of *Algorithm 1* .

Lemma 1. *Let $(\triangle x^a, \triangle u^a, \triangle v^a)$ be the predictor direction of Algorithm 1. Then, for all $i \in \mathcal{I}_+$,*

$$\triangle x_i^a \triangle v_i^a \leq \frac{5}{4} x_i^k v_i^k.$$

Proof. Using the third condition of (1), we have

$$\frac{\triangle x_i^a}{x_i^k} + \frac{\triangle v_i^a}{v_i^k} = \frac{\mu_k}{x_i^k v_i^k} - 1. \tag{5}$$

Since

$$0 \leq \left(\frac{\triangle x_i^a}{x_i^k} - \frac{\triangle v_i^a}{v_i^k} \right)^2 = \left(\frac{\triangle x_i^a}{x_i^k} \right)^2 + \left(\frac{\triangle v_i^a}{v_i^k} \right)^2 - 2\frac{\triangle x_i^a \triangle v_i^a}{x_i^k v_i^k},$$

then

$$\left(\frac{\triangle x_i^a}{x_i^k} \right)^2 + \left(\frac{\triangle v_i^a}{v_i^k} \right)^2 \geq 2\frac{\triangle x_i^a \triangle v_i^a}{x_i^k v_i^k}. \tag{6}$$

Therefore, using

$$\left(\frac{\triangle x_i^a}{x_i^k}+\frac{\triangle v_i^a}{v_i^k}\right)^2=\left(\frac{\triangle x_i^a}{x_i^k}\right)^2+\left(\frac{\triangle v_i^a}{v_i^k}\right)^2+2\frac{\triangle x_i^a\triangle v_i^a}{x_i^k v_i^k},$$

equality (5) and inequality (6), we have

$$\begin{aligned}\frac{\triangle x_i^a\triangle v_i^a}{x_i^k v_i^k}&=\frac{1}{4}\left(\frac{\mu_k}{x_i^k v_i^k}-1\right)^2\\&\leq\frac{1}{4}\left(\left(\frac{\mu_k}{x_i^k v_i^k}\right)^2+1\right).\end{aligned}$$

Furthermore, hence $x_i^k v_i^k \geq \frac{1}{2}\mu_k$ and $n\sqrt{n} \leq \theta(n) \leq n^2$, we have

$$\frac{\triangle x_i^a\triangle v_i^a}{x_i^k v_i^k}\leq\frac{5}{4}.$$

Lemma 2. *Let* $(\triangle x^a, \triangle u^a, \triangle v^a)$ *be the predictor direction of Algorithm 1. Then,*

$$\sum_{i\in\mathcal{I}_+}\triangle x_i^a\triangle v_i^a\leq\frac{5}{4}\theta(n)\mu_k.$$

Proof. Using Lemma 1, we have

$$\sum_{i\in\mathcal{I}_+}\triangle x_i^a\triangle v_i^a\leq\frac{5}{4}\sum_{i\in\mathcal{I}_+}x_i^k v_i^k\leq\frac{5}{4}\,x^{k^T}v^k\leq\frac{5}{4}\theta(n)\mu_k.$$

From this point on, we consider the predictor direction of Algorithm 1 satisfying $\triangle x^{a^T}\triangle v^a \geq 0$.

Lemma 3. *Let* $(\triangle x^a, \triangle u^a, \triangle v^a)$ *be the predictor direction of Algorithm 1. Then, for* $\triangle x^{a^T}\triangle v^a \geq 0$,

$$\sum_{i\in\mathcal{I}_-}|\triangle x_i^a\triangle v_i^a|\leq\sum_{i\in\mathcal{I}_+}\triangle x_i^a\triangle v_i^a\leq\frac{5}{4}\theta(n)\mu_k.$$

Proof. Using the definition of inner product, $\mathcal{I}_-$ and $\mathcal{I}_+$, we have

$$\begin{aligned}\triangle x^{a^T}\triangle v^a&=\sum_{i\in\mathcal{I}_+}\triangle x_i^a\triangle v_i^a+\sum_{i\in\mathcal{I}_-}\triangle x_i^a\triangle v_i^a\\&=\sum_{i\in\mathcal{I}_+}\triangle x_i^a\triangle v_i^a-\sum_{i\in\mathcal{I}_-}|\triangle x_i^a\triangle v_i^a|\end{aligned}$$

Using the fact that $\triangle x^{aT} \triangle v^a \geq 0$ and Lemma 2, we obtain

$$\sum_{i \in \mathcal{I}_-} |\triangle x_i^a \triangle v_i^a| \leq \sum_{i \in \mathcal{I}_+} \triangle x_i^a \triangle v_i^a$$
$$\leq \frac{5}{4}\, \theta(n)\, \mu_k.$$

Using the above technical lemmas, we obtain the following proposition which gives an upper bound estimate of the 2-norm of the inner product of the vectors $\triangle v$ and $\triangle x$.

Proposition 1. *Let $(x^k, u^k, v^k) \in \mathcal{N}_\infty^- \left(\frac{1}{2}, \beta\right)$ be the current iterate and $(\triangle x, \triangle u, \triangle v)$ be the solution of (2) . Then, for $\triangle x^{aT} \triangle v^a \geq 0$,*

$$\|\triangle V \triangle X e\| \leq 5n^4 \mu_k.$$

Proof. Multiplying the third equation of (2) by $\left(X^k V^k\right)^{-\frac{1}{2}}$, we obtain

$$\begin{aligned} &\left(X^k\right)^{-\frac{1}{2}} \left(V^k\right)^{\frac{1}{2}} \triangle x + \left(V^k\right)^{-\frac{1}{2}} \left(X^k\right)^{\frac{1}{2}} \triangle v = \\ &= \left(X^k V^k\right)^{-\frac{1}{2}} \left(\mu_k e - \triangle X^a \triangle V^a e - X^k V^k e\right). \end{aligned} \tag{7}$$

Taking the first term of (7) and

$$D = \left(V^k\right)^{-\frac{1}{2}} \left(X^k\right)^{\frac{1}{2}},$$

we have

$$\left(X^k\right)^{-\frac{1}{2}} \left(V^k\right)^{\frac{1}{2}} \triangle x + \left(V^k\right)^{-\frac{1}{2}} \left(X^k\right)^{\frac{1}{2}} \triangle v = D^{-1} \triangle x + D \triangle v.$$

Let us denote

$$w_1 = D^{-1}\triangle x, \quad w_2 = D\triangle v, \quad W_1 = \mathrm{diag}(w_1), \quad W_2 = \mathrm{diag}(w_2).$$

Using Lemma 5.3 of [18, p.88] (see Appendix) follows

$$\begin{aligned} \|W_1 W_2 e\| &= \|\triangle V \triangle X e\| \\ &\leq 2^{-\frac{3}{2}} \|w_1 + w_2\|^2 \\ &= 2^{-\frac{3}{2}} \left\| \left(X^k V^k\right)^{-\frac{1}{2}} \left(\mu_k e - \triangle X^a \triangle V^a e - X^k V^k e\right) \right\|^2 \\ &= 2^{-\frac{3}{2}} \left(\mu_k^2 \sum_{i \in \mathcal{I}} \frac{1}{x_i^k v_i^k} + \theta(n)\mu_k + \sum_{i \in \mathcal{I}} \frac{(\triangle x_i^a \triangle v_i^a)^2}{x_i^k v_i^k} - 2n\mu_k \right. \end{aligned}$$

$$- 2\mu_k \sum_{i\in\mathcal{I}} \frac{\triangle x_i^a \triangle v_i^a}{x_i^k v_i^k} + 2\sum_{i\in\mathcal{I}} \triangle x_i^a \triangle v_i^a \Big).$$

For $x_i^k v_i^k \geq \frac{1}{2}\mu_k$, we have

$$\mu_k^2 \sum_{i\in\mathcal{I}} \frac{1}{x_i^k v_i^k} \leq 2n\mu_k.$$

Now, let us analyze the term

$$\sum_{i\in\mathcal{I}} \frac{(\triangle x_i^a \triangle v_i^a)^2}{x_i^k v_i^k} = \sum_{i\in\mathcal{I}_+} \frac{(\triangle x_i^a \triangle v_i^a)^2}{x_i^k v_i^k} + \sum_{i\in\mathcal{I}_-} \frac{(\triangle x_i^a \triangle v_i^a)^2}{x_i^k v_i^k}. \tag{8}$$

Using Lemma 1 and the fact that (x^k, u^k, v^k) belongs to $\mathcal{N}_\infty^-\left(\frac{1}{2}, \beta\right)$, we get, for the case $i \in \mathcal{I}_+$

$$\begin{aligned}
\sum_{i\in\mathcal{I}_+} \frac{(\triangle x_i^a \triangle v_i^a)^2}{x_i^k v_i^k} &\leq \sum_{i\in\mathcal{I}_+} \frac{25}{16} x_i^k v_i^k \\
&\leq \frac{1}{16} \sum_{i\in\mathcal{I}} x_i^k v_i^k \\
&= \frac{25}{16}\theta(n)\mu_k.
\end{aligned}$$

For $i \in \mathcal{I}_-$, using Lemma 3, we have

$$\begin{aligned}
\sum_{i\in\mathcal{I}_-} \frac{(\triangle x_i^a \triangle v_i^a)^2}{x_i^k v_i^k} &\leq \frac{2}{\mu_k} \sum_{i\in\mathcal{I}_-} (\triangle x_i^a \triangle v_i^a)^2 \\
&\leq \frac{2}{\mu_k} \left(\sum_{i\in\mathcal{I}_-} |\triangle x_i^a \triangle v_i^a| \right)^2 \\
&\leq \frac{2}{\mu_k} \left(\frac{5}{4}\theta(n)\mu_k \right)^2 \\
&\leq \frac{25}{8} (\theta(n))^2 \mu_k.
\end{aligned}$$

Therefore, for (8), we have

$$\sum_{i\in\mathcal{I}} \frac{(\triangle x_i^a \triangle v_i^a)^2}{x_i^k v_i^k} \leq \left(\frac{25}{16}\theta(n) + \frac{25}{8}(\theta(n))^2 \right) \mu_k.$$

Using Lemma 3, the fact that (x^k, u^k, v^k) belongs to $\mathcal{N}_\infty^-\left(\frac{1}{2}, \beta\right)$ and the definition of $\mathcal{I}_-$ and $\mathcal{I}_+$, we obtain

$$\begin{aligned}
-2\mu_k \sum_{i \in \mathcal{I}} \frac{\triangle x_i^a \triangle v_i^a}{x_i^k v_i^k} &= -2\mu_k \sum_{i \in \mathcal{I}_+} \frac{\triangle x_i^a \triangle v_i^a}{x_i^k v_i^k} - 2\mu_k \sum_{i \in \mathcal{I}_-} \frac{\triangle x_i^a \triangle v_i^a}{x_i^k v_i^k} \\
&\leq -2\mu_k \sum_{i \in \mathcal{I}_-} \frac{\triangle x_i^a \triangle v_i^a}{x_i^k v_i^k} \\
&= 2\mu_k \sum_{i \in \mathcal{I}_-} 2 \frac{|\triangle x_i^a \triangle v_i^a|}{\mu_k} \\
&= 4 \sum_{i \in \mathcal{I}_-} |\triangle x_i^a \triangle v_i^a| \\
&\leq 4\mu_k \frac{5}{4} \theta(n) \\
&= 5\theta(n)\mu_k.
\end{aligned}$$

Furthermore, using Lemma 2, we get

$$\sum_{i \in \mathcal{I}} \triangle x_i^a \triangle v_i^a \leq \sum_{i \in \mathcal{I}_+} \triangle x_i^a \triangle v_i^a \leq \frac{5}{4} \theta(n)\, \mu_k.$$

In conclusion, estimating $\theta(n)$ by n^2, we have

$$\|\triangle V \triangle X e\| \leq l(n)\mu_k,$$

where

$$l(n) = 2^{-\frac{3}{2}} \left(\frac{161}{16} \theta(n) + \frac{25}{8} (\theta(n))^2 \right) \leq 5n^4.$$

By a straightforward calculation, we obtain the following recursive relation.

Lemma 4. *Let μ_k satisfy (3). Then*

$$\mu_{k+1} = \Big(1 - \delta_k(\lambda_k)\Big)\mu_k,$$

where

$$\delta_k(\lambda_k) = \lambda_k \left(1 - \frac{n}{\theta(n)} + \frac{\triangle v^{aT} \triangle x^a}{v^{kT} x^k} \right).$$

Proof. Since

$$c^T x^{k+1} - b^T u^{k+1} = (v^{k+1})^T x^{k+1},$$

where $x^{k+1} = x^k + \lambda_k \triangle x$ and $v^{k+1} = v^k + \lambda_k \triangle v$, we get

$$\begin{aligned} c^T x^{k+1} - b^T u^{k+1} &= (v^{k+1})^T x^{k+1} \\ &= (v^k + \lambda_k \triangle v)^T (x^k + \lambda_k \triangle x) \\ &= {v^k}^T x^k + \lambda_k \left(\triangle v^T x^k + {v^k}^T \triangle x \right) + \lambda_k^2 \triangle v^T \triangle x \end{aligned}$$

which is equivalent to

$$c^T x^{k+1} - b^T u^{k+1} = {v^k}^T x^k + \lambda_k \left({x^k}^T \triangle v + {v^k}^T \triangle x \right) + \lambda_k^2 \triangle v^T \triangle x. \tag{9}$$

Using the second equation of (2), $\triangle u^T A = -(\triangle v)^T$, we obtain

$$\triangle u^T A \triangle x = -(\triangle v)^T \triangle x. \tag{10}$$

Therefore, from the first equation of (2) and (10), we get

$$\triangle v^T \triangle x = 0. \tag{11}$$

Introducing the third equation of (2), we have

$$\begin{aligned} {x^k}^T \triangle v + {v^k}^T \triangle x &= \left[V^k \triangle X e + X^k \triangle V e \right]^T e \\ &= \left[\mu_k e - \triangle X^a \triangle V^a e - X^k V^k e \right]^T e \\ &= n\mu_k - {\triangle v^a}^T \triangle x^a - {v^k}^T x^k. \end{aligned} \tag{12}$$

Applying (11) and (12) to (9), we obtain

$$\begin{aligned} {v^{k+1}}^T x^{k+1} &= {v^k}^T x^k + \lambda_k \left(n\mu_k - {\triangle v^a}^T \triangle x^a - {v^k}^T x^k \right) \\ &= {v^k}^T x^k \left(1 - \lambda_k \left(1 - \frac{n}{\theta(n)} + \frac{{\triangle v^a}^T \triangle x^a}{{v^k}^T x^k} \right) \right). \end{aligned} \tag{13}$$

Using (3) and considering

$$\delta_k(\lambda_k) = \lambda_k \left(1 - \frac{n}{\theta(n)} + \frac{{\triangle v^a}^T \triangle x^a}{{v^k}^T x^k} \right),$$

we obtain the desired relation

$$\mu_{k+1} = \left(1 - \delta_k(\lambda_k) \right) \mu_k.$$

3 Polynomial Complexity

In this section we prove that *Algorithm 1* is Q-linearly convergent. More precisely, we demonstrate that, at each iteration, the step size computed by this algorithm is bounded below by $\frac{1}{250n^4}$. Consequently, proving that *Algorithm 1* has $O(n^4|\log(\epsilon)|)$ iteration complexity.

In the next result we discuss the step size estimate of the algorithm to establish its worst case iteration complexity, in order to analyze the asymptotic behavior of *Algorithm 1*.

Theorem 1. *Suppose that the current iterate* (x^k, u^k, v^k) *belongs to* $\mathcal{N}_\infty^-\left(\frac{1}{2}, \beta\right)$. *Let* $(\triangle x^a, \triangle u^a, \triangle v^a)$ *be the solution of (1) satisfying* $\triangle x^{aT} \triangle v^a \geq 0$ *and* $(\triangle x, \triangle u, \triangle v)$ *be the solution of (2). Then the maximum step size* λ_k*, that keeps* $(x^{k+1}, u^{k+1}, v^{k+1})$ *in* $\mathcal{N}_\infty^-\left(\frac{1}{2}, \beta\right)$, *satisfies*

$$\lambda_k \geq \frac{1}{250n^4}, \qquad k \geq 0, \quad n \geq 2.$$

Proof. In order to find the maximum nonnegative λ_k for which

$$x_i^{k+1} v_i^{k+1} \geq \frac{1}{2}\mu_{k+1},$$

with $i \in \mathcal{I}$, we define

$$t = \max_{i \in \mathcal{I}_+} \left\{ \frac{\triangle x_i^a \triangle v_i^a}{x_i^k v_i^k} \right\}. \tag{14}$$

Since $\triangle x^{aT} \triangle v^a \geq 0$, we have $\mathcal{I}_+ \neq \emptyset$. Let $i \in \mathcal{I}_+$, then

$$\begin{aligned} x_i^{k+1} v_i^{k+1} &= x_i^k v_i^k + \lambda_k \left(\mu_k - x_i^k v_i^k - \triangle x_i^a \triangle v_i^a\right) + \lambda_k^2 \triangle x_i \triangle v_i \\ &= (1 - \lambda_k) x_i^k v_i^k + \lambda_k \mu_k - \lambda_k \triangle x_i^a \triangle v_i^a + \lambda_k^2 \triangle x_i \triangle v_i. \end{aligned}$$

Applying the above defined t, we obtain

$$\begin{aligned} x_i^{k+1} v_i^{k+1} &\geq (1 - \lambda_k) x_i^k v_i^k + \lambda_k \mu_k - \lambda_k t x_i^k v_i^k + \lambda_k^2 \triangle x_i \triangle v_i \\ &= \left(1 - \lambda_k (1 + t)\right) x_i^k v_i^k + \lambda_k \mu_k + \lambda_k^2 \triangle x_i \triangle v_i \\ &\geq \left(1 - \lambda_k (1 + t)\right) x_i^k v_i^k + \lambda_k \mu_k - \lambda_k^2 \left(5n^4\right) \mu_k. \end{aligned}$$

Since each iterate must belong to $\mathcal{N}_\infty^-\left(\frac{1}{2}, \beta\right)$, if we take $1 - \lambda_k(1 + t) > 0$, we obtain

$$x_i^{k+1} v_i^{k+1} \geq \left(1 - \lambda_k (1 + t)\right) \frac{1}{2}\mu_k + \lambda_k \mu_k - 5\lambda_k^2 n^4 \mu_k.$$

By Lemma 1, we get $t \leq \frac{5}{4}$ and, consequently, $\frac{1}{1+t} \geq \frac{4}{9}$. Therefore, we take $\lambda_k \in \left[0, \frac{4}{9}\right]$. In order to guaranty that the next iterate belongs to $\mathcal{N}_\infty^-\left(\frac{1}{2}, \beta\right)$, we consider

$$(1 - \lambda_k(1+t)) \frac{1}{2}\mu_k + \lambda_k \mu_k - 5\lambda_k^2 n^4 \mu_k \geq \frac{1}{2}\mu_{k+1}.$$

Using Lemma 4 and the fact that $\triangle x^{a^T} \triangle v^a \geq 0$, we have

$$(1 - \lambda_k(t-1)) \frac{1}{2}\mu_k - 5\lambda_k^2 n^4 \mu_k \geq \frac{1}{2}\left(1 - \lambda_k \left(1 - \frac{n}{\theta(n)}\right)\right)\mu_k \geq \frac{1}{2}\mu_{k+1},$$

which is equivalent to

$$\lambda_k \left(2 - t - \frac{n}{\theta(n)}\right)\frac{\mu_k}{2} \geq 5\lambda_k^2 n^4 \mu_k. \tag{15}$$

Using Lemma 1 and the definition of $\theta(n)$ presented in (3), we obtain for $n \geq 2$

$$\left(2 - t - \frac{n}{\theta(n)}\right) \geq 2 - \frac{5}{4} - \frac{1}{\sqrt{n}} \geq \frac{3\sqrt{n} - 4}{4\sqrt{n}} \geq \frac{1}{25}. \tag{16}$$

From (15) and (16), we conclude,

$$\lambda_k \geq \min\left\{\frac{4}{9}, \frac{1}{250n^4}\right\} = \frac{1}{250n^4}.$$

The following theorem gives an upper bound for the number of iterations in which *Algorithm 1* stops and allows us to conclude that this infeasible version of the algorithm developed by Bastos and Paixão has Q-linear convergence.

Theorem 2. *Let $\epsilon \in]0,1[$ and $\triangle x^{a^T} \triangle v^a \geq 0$. Algorithm 1 stops after at most $O(n^4 |\log(\epsilon)|)$ iterations with a solution for which $x^T v \leq \epsilon$.*

Proof. Considering the definition of μ_k presented in (3) and using Lemma 4, we have

$$\mu_{k+1} = \Big(1 - \delta_k(\lambda_k)\Big)\mu_k,$$

where

$$\delta_k(\lambda_k) = \lambda_k \left(1 - \frac{n}{\theta(n)} + \frac{\triangle v^{a^T} \triangle x^a}{v^{k^T} x^k}\right).$$

In order to obtain an upper bound for μ_{k+1} we need to determine a lower bound for $\delta_k(\lambda_k)$. Since $\triangle x^{a^T} \triangle v^a \geq 0$, then

$$\delta_k(\lambda_k) \geq \lambda_k \left(1 - \frac{n}{\theta(n)}\right).$$

Furthermore, since $n\sqrt{n} \leq \theta(n) \leq n^2$, we have

$$\delta_k(\lambda_k) \geq \lambda_k \left(1 - \frac{1}{\sqrt{n}}\right).$$

Using Theorem 1, we have for $n \geq 2$

$$\mu_{k+1} \leq \left(1 - \lambda_k \left(1 - \frac{1}{\sqrt{n}}\right)\right) \mu_k \leq \left(1 - \frac{1.1 \times 10^{-3}}{n^4}\right) \mu_k.$$

Applying Theorem 3.2 of [18, p. 61] (see Appendix) we complete the proof.

4 Concluding Remarks

In this section, we briefly present some comments, summarize the results of the current work and present some future trends.

Infeasible interior-point algorithms empirically have proved to be efficient, see for example [6], [17] and [19]. Although the study of worst-case complexity has been one of the main topics of research on interior-point methods, the best measure of efficiency is still numerical performance. The inadequacy of worst-case complexity as a measure of computational efficiency is due to the fact that the worst-case complexity bounds obtained both for feasible and infeasible interior-point algorithms are often too conservative or too far from average cases. As a result, it seems more appropriate to use worst-case complexity as a qualitative, rather than quantitative, measure of reliability. Nevertheless, we have to have in mind that, an algorithm with a polynomial complexity bound is likely to be more reliable than one without it.

In this paper, we established an $O(n^4|\log(\epsilon)|)$ iteration complexity bound for an infeasible version of Bastos and Paixão Mehrotra-type predictor-corrector variant algorithm [2], thus, demonstrating that both feasible and infeasible versions of Bastos and Paixão algorithm have polynomial iteration complexity and are Q-linearly convergent. It is worth pointing out that we also proved that, at each iteration, the step size computed by this infeasible algorithm is bounded below by $\frac{1}{250n^4}$.

Some further improvement in this regard is certainly possible, but these results allow us to conclude that it might be worth to implement an infeasible version of Bastos and Paixão algorithm and analyse the results obtained from its application to transportation and assignment class of problems.

The strengthening of the presented results and the study of the convergence results of the infeasible version of Bastos and Paixão algorithm are subjects of future research.

Acknowledgments. A. Teixeira was financially supported (under the project PEst-OE/MAT/UI0152/2013) by the Research Unit Centro de Investigação Operacional, based at the Faculty of Sciences, University of Lisbon, financed by the Portuguese Foundation for Science and Technology (FCT - Fundação para a Ciência e a Tecnologia).

References

1. Al-Jeiroudi, G., Gondzio, J.: Convergence Analysis of the Inexact Infeasible Interior-Point Method for Linear Optimization. J. Optim. Theory Appl. 141, 231–247 (2009)
2. Bastos, F., Paixão, J.: Interior-point approaches to the transportation and assignment problems on microcomputers. Investigação Operacional 13(1), 3–15 (1993)
3. Freund, R.W., Jarre, F.: A QMR-based interior-point algorithm for solving linear programs. Mathematical Programming 76(1), 183–210 (1997)
4. Güler, O., Ye, Y.: Convergence behavior of interior-point algorithms. Mathematical Programming 60, 215–228 (1993)
5. Karmarkar, N.K.: A new polinomial-time algorithm for linear programming. Combinatorica 4, 373–395 (1984)
6. Kojima, M., Megiddo, N., Mizuno, S.: A primal-dual infeasible-interior point algorithm for linear programming. Mathematical Programming, Series A 61, 261–280 (1993)
7. Korzak, J.: Convergence analysis of inexact infeasible-interior-point algorithms for solving linear programming problems. SIAM Journal on Optimization 11(1), 133–148 (2000)
8. Lustig, I.J., Marsten, R.E., Shanno, D.F.: Computational experience with a primal-dual interior point method for linear programming. Linear Algebra and Its Applications 152, 191–222 (1991)
9. Lustig, I.J., Marsten, R.E., Shanno, D.F.: On implementing Mehrotra's predictor-corrector interior-point method for linear programming. SIAM Journal on Optimization 2(3), 435–449 (1992)
10. Lustig, I.J., Marsten, R.E., Shanno, D.F.: Interior point method for linear programming: Computational state of the art. ORSA Journal on Computing 6, 1–14 (1994)
11. McShane, K.A., Monma, C.L., Shanno, D.F.: An implementation of a primal-dual interior point method for linear programming. ORSA Journal on Computing 1, 70–83 (1989)
12. Mehrotra, S.: On the implementation of a primal-dual interior point method. SIAM Journal on Optimization 2, 575–601 (1992)
13. Mizuno, S.: Polynomiality of infeasible interior-point algorithms for linear programming. Mathematical Programming 67, 109–119 (1994)
14. Potra, F.: A quadratically convergent predictor-corrector method for solving linear programs from infeasible starting points. Mathematical Programming 67, 383–406 (1994)
15. Salahi, M., Peng, J., Terlaky, T.: On Mehrotra-Type Predictor-Corrector Algorithms. SIAM Journal on Optimization 18(4), 1377–1397 (2007)
16. Teixeira, A.P., Almeida, R.: On the Complexity of a Mehrotra-Type Predictor-Corrector Algorithm. In: Murgante, B., Gervasi, O., Misra, S., Nedjah, N., Rocha, A.M.A.C., Taniar, D., Apduhan, B.O. (eds.) ICCSA 2012, Part III. LNCS, vol. 7335, pp. 17–29. Springer, Heidelberg (2012)
17. Wright, S.: An infeasible interior point algorithm for linear complementarity problems. Mathematical Programming 67, 29–52 (1994)
18. Wright, S.J.: Primal-Dual Interior-Point Methods. SIAM, Philadelphia (1997)
19. Wright, S., Zhang, Y.: A superquadratic infeasible interior point method for linear complementarity problems. Mathematical Programming 73, 269–289 (1996)

20. Ye, Y.: Interior Point Algorithms, Theory and Analysis. John Wiley and Sons, Chichester (1997)
21. Zhang, Y.: On the convergence of a class of infeasible interior-point methods for the horizontal linear complementarity problem. SIAM Journal on Optimization 4(1), 208–227 (1994)
22. Zhang, Y., Zhang, D.: On polynomiality of the Mehrotra-type predictor-corrector interior-point algorithms. Mathematical Programming 68, 303–318 (1995)
23. Zhang, Z., Zhang, Y.: A Mehrotra-type predictor-corrector algorithm with polynomiality and Q-subquadratic convergence. Annals of Operations Research 62, 131–150 (1996)
24. Zhang, Y., Tapia, R.A., Dennis, J.E.: On the superlinear and quadratic convergence of primal-dual interior point linear programming algorithms. SIAM Journal on Optimization 2, 304–324 (1992)

Appendix

In this appendix we transcribe Theorem 3.2 and Lemma 5.3 of [18].

Theorem 3.2 [18, p.61]. *Let* $\epsilon \in]0,1[$ *be given. Suppose that our algorithm for solving the Karush-Kuhn-Tucker conditions associated with the standard primal-dual pair of Linear Programming problems generates a sequence of iterates that satisfies*

$$\mu_{k+1} \leq \left(1 - \frac{\delta}{n^\omega}\right)\mu_k, \quad k = 0, 1, \ldots$$

for some positive constants δ *and* ω. *Suppose too that the starting point* (x^0, u^0, v^0) *satisfies*

$$\mu_0 \leq \frac{1}{\epsilon^\kappa}$$

for some positive constant κ. *Then there exists an index* K *with*

$$K = O\left(n^\omega \left|\log(\epsilon)\right|\right)$$

such that

$$\mu_k \leq \epsilon \quad \textit{for all } \kappa \geq K.$$

Lemma 5.3. [18, p.88] *Let* u *and* v *be any two vectors in* $\mathbb{R}^n$ *with* $u^T v \geq 0$. *Then*

$$\|UVe\| \leq 2^{-\frac{3}{2}} \|u + v\|^2,$$

where $U = diag\,(u_1, \ldots, u_n)$ *and* $V = diag\,(v_1, \ldots, v_n)$.

A Multi-start Tabu Search Approach for Solving the Information Routing Problem

Hela Masri[1], Saoussen Krichen[2], and Adel Guitouni[3]

[1] LARODEC Laboratory, Institut Supérieur de Gestion de Tunis,
University of Tunis,
2000 Bardo, Tunisia
Faculté des Sciences Economiques et de Gestion de Tunis, Université de Tunis El Manar, 1068 Rommana, Tunisia
masri_hela@yahoo.fr

[2] LARODEC Laboratory, Faculty of Law, Economics and Management,
University of Jendouba,
Avenue de l'UMA, 8189 Jendouba, Tunisia
saoussen.krichen@isg.rnu.tn

[3] Peter B. Gustavson School of Business, University of Victoria, Victoria, Canada
adelg@uvic.ca

Abstract. In this paper, we propose a new global routing algorithm supporting advance reservation. A set of flows are shared across a communication network. Each flow has a source node, a destination node and a predetermined traffic demand. The design goal of the routing is to minimize the overall network congestion under the constraint that each flow should be sent along a single path without being bifurcated. We model this optimization problem as a single path multicommodity flow problem (SPMFP). As the complexity of the SPMFP is NP-Hard, a Multi-start Tabu Search (MTS) is proposed as a solution approach. The empirical validation is done using a simulation environment called Inform Lab. A comparison to a state-of-the-art ant colony system (ACS) approach is performed based on a real case of maritime surveillance application. The same instances are optimally solved using CPLEX. The experimental results show that the MTS produces considerably better results than the ACS to the detriment of the CPU time.

1 Introduction

Communication networks are becoming of greater importance as they are playing a vital role in distributed systems. The evolution of the technology combined to new requirements of efficient communication infrastructures to support information exchange is the driver for an emergent novel routing algorithms in communication systems. Despite that modern communication networks support low latency and high bandwidth communication services, information routing is still an issue, in particular for large volumes of data. For Instance, large scale collaborative applications, such as surveillance systems and grid computing, produce generally very large data collections, which can be on the order of terabytes. It

B. Murgante et al. (Eds.): ICCSA 2014, Part II, LNCS 8580, pp. 267–277, 2014.

would be reasonable, with such applications, to assume that receivers will schedule their requests ahead of time to avoid the overload of the network. Assuming that an advance reservations are made, the router can efficiently manage the transmission by generating the optimal routing plan to be run in later stages.

In this paper, we study the routing optimization problem in networks supporting advance reservation. The problem consists of sending various messages from a set of sources to different destinations across a network. We assume there is a centralized routing coordinator managing all information exchange requests. Each flow has a source node, a destination node and a traffic demand known beforehand. The design goal of the routing is to minimize the overall network congestion under the constraint that each flow should be sent along a single path without being bifurcated. Although a multi-path design may optimize the utilization of network bandwidth, the requirement that the data flow is transmitted using a single path arises in variety of contexts especially for real-time applications [13]. For example, a video streaming application requires keeping the traffic intact, that is, without demultiplexing at the source, independent switching at intermediate nodes, and multiplexing at the destination. Also, some types of links such as optical and cellular networks require a single path routing.

The solution of the considered routing problem consists of defining a single path for each pair of source-destination nodes such that it minimizes the traffic in the most congested arc. Much of the motivation of this work comes from the possible application to the case of routing in the backbone of a maritime surveillance network. A surveillance mission is characterized by the collaboration of dispersed entities processing diverse information exchange in order to achieve a global goal. These interactions rely on a communication network that is generally composed by mobile and fixed surveillance assets. The stationary nodes represent the network backbone. Given that such network is private and configurable, a centralized global routing algorithm for the backbone can be designed prior to the mission execution. Furthermore, with such surveillance application the routing protocol should guarantee the end-to-end delay between the different nodes by choosing a single path and an advance reservation design.

This routing problem can be modeled as a single path multicommodity flow problem (SMCFP) [3]. SMCFP was first introduced by Kleinberg [12]. Different exact solution methods were proposed in the literature, but given the NP-hardness of the SMCFP a heuristic approach is a better choice specially to solve real-life large instances. Recently, Li et al. [13] proposed an ant colony system metaheuristic to solve the SMCFP for optical switching networks, by optimizing the most congested arc. Minimizing network congestion is a crucial concern in traffic grooming over wavelength division multiplexing (WDM). In this paper, we propose to adapt a new metaheuristic based on multi-start tabu search (MTS) to test the strength of a local search method on solving SMCFP. The routing objective is to guarantee having a load balance over the network resources, by minimizing the most congested arc traffic value. Series of experiments are conducted based on a real application of maritime surveillance. The experimental results show that MTS performs consistently better than the ACS for large sized problems in terms of solution quality.

This paper is organized as follows. Section 2 provides the literature overview of multicommodity flow problems. Section 3 presents the problem description and states its mathematical formulation. Section 4 details the proposed solution approach. Section 5 provides some experimental results using different testbeds.

2 Literature Review

Multicommodity network flow (MCNF) is a known optimization problem that arises in variaty of communication network contexts. MCNF model can be defined as follows: Given a set of commodities and an underlying network structure consisting of a number of nodes and arcs, the algorithm should find an optimal routing plan to transmit the commodities through the network without violating the capacity limits [1]. A commodity in a communication network represents a traffic demand between a source node and a destination node. For a solution to be feasible it must comply with the flow conservation constraint for each node, i.e. the flow going into a node must equal the flow out of the node. Several objectives were considered in the MCFPs such as the optimization of the transmission delay, bandwidth consumption or packet loss [3,13,6]. The reader can refer to the comprehensive survey of Ahuja et al. [1] for more details about MCNF problems and the different solution approaches.

If the flow of each source-destination pair is restricted to lie on a single path, an NP-hard integer programming problem results. This problem is denoted the SMCFP or the unsplittable MCFP, first introduced by Kleinberg [12]. This assumption of having a single path is necessary in real time applications. It may be also applicable to circuit-oriented technologies such as optical networks employing WDM. In such network, to be able to send data from one access node to another, one needs to establish a light path a (i.e. a single path), between the two nodes and to allocate a free wavelength on all of the arcs on the path. Bandyopadhyay [2] studied the non-bifurcated traffic grooming problem modeled as an the SMCFP. The objective is to minimize the cost of the overall network by minimizing the number of lightpaths and maximizing the network throughput. Decomposition method was used to solve this problem for large networks. Barnhart et al. [3] proposed an exact method using branch and price and cut algorithm. Such exact algorithms can only solve small instances. More recently, to tackle larger problems, a metaheuristic approach based on ant colony system method was developed [13]. Two versions were considered. The first problem minimizes the network congestion, in order to solve the problem of traffic grooming over WDM. In the second problem, the authors considered the general case of the minimum cost SMCFP.

3 Problem Formulation

In this paper, we study the problem of information routing in communication networks. This problem consists of exchanging a set of messages K

(i.e. commodities) between pairs of source-destination nodes. The network can be modeled by a directed graph (N, A) where $N = \{v_1, .., v_|N|\}$ is the set of nodes and $A = \{e_1, .., e_|A|\}$ is the set of arcs. Each node in the network can be an information sender or/and a receiver requiring an information or a neutral relay node. Each arc from a node i to node j is represented by (i, j). The traffic demand $size_k$ of each flow k is also known beforehand. The solution of this problem consists of defining how to exchange messages from pairs of source-destination nodes by generating a single path for each flow, so that it ensures having a load balance in all the network. Therefore the objective function is the minimization of the traffic in the most congested arc. Given these statements, the SMCFP can be described using a mathematical model [13] as follows:

Notation

N	the set of nodes $\{v_1, .., v_{\|N\|}\}$
A	the set of arcs $\{e_1, .., e_{\|A\|}\}$
K	the set of commodoties
s_k	the source node of commodity k
d_k	the destination node of commodity k
$size_k$	the size of the supply of commodity k
λ_{max}	network congestion that equals the maximum traffic load on network's arcs
x_k	the flow vector of commodity k
x_{ij}^k	a binary decision variable $\begin{cases} 1 \text{ if the entire quantity of commodity } k \text{ uses arc } (i,j) \\ 0 \text{ otherwise} \end{cases}$

$$min\ \lambda_{max} \tag{1}$$

$s.t$

$$\sum_{k \in K} size_k\ x_{ij}^k \leq \lambda_{max} \qquad (i,j) \in A \tag{2}$$

$$\sum_{j:(i,j) \in A} x_{ij}^k - \sum_{j:(j,i) \in A} x_{ji}^k = \begin{cases} 1 \text{ if } i = s_k \\ \text{-1 if } i = d_k \\ 0 \text{ otherwise} \end{cases} \quad i \in N \ \ k \in K \tag{3}$$

$$x_{ij}^k \in \{0, 1\} \qquad (i,j) \in A \ \ k \in K \tag{4}$$

The objective of SMCFP is to minimize network congestion value λ_{max}. Constraints (2) define the maximum congestion value in the network. Constraints (3) and (4) are the flow k conservation constraints, they ensure that each commodity is sent along a single path linking its source s_k to its destination d_k.

4 Solution Approach: A Multi-start Tabu Search Algorithm (MTS)

The NP-hardness of the SMCFP makes the use of exact methods impractical to tackle large instances. Therefore, we propose to solve it using a metaheuristic method, based on MTS. The successful application of tabu search approach for solving similar routing problems [10] motivated the choice of the metaheuristic. The algorithm starts from an initial solution, and then it explores iteratively its neighborhood. Tabu search approach, first designed by Glover et al. [8], is similar to greedy search method. Its key idea is to introduce a memory notion in order to keep track of information related to the exploration. It uses an iterative schema; it explores a single solution at a time from which it generates a number of moves leading to other neighbors solutions. To escape from a local minimum, tabu search method can accept a neighbor solution that doesn't improve the current objective function. As soon as non- improving moves are possible, the risk of revisiting a solution is present. This is the point where the use of memory is helpful to forbid moves which might lead to recently visited solutions. Therefore, the neighborhood of a solution s, denoted as $Ng(s)$ depend upon the contents of the *tabuList*. The definition of $Ng(s)$ implies that some recently visited solutions are disregarded as they are considered as tabu solutions. The tabu list is organized as a queue. Hence, when the list is full it will forget about the oldest solutions. As the search process depends inherently of the initial solution, we propose a multi-start tabu search in order to explore a wider zone of the feasible solution space F.

Two highly important components of tabu search are intensification and diversification strategies. Intensification strategies are based on modifying choice rules to encourage solution features historically found good. It aims to intensify the search in some region of F because it may contain some acceptable solutions. This step should be done during a few iterations, after this it may be fruitful to explore another region, diversification will then tend to spread the exploration effort over different regions of F [8].

Before detailing the proposed algorithm, the main components of MTS heuristic are defined:

- **A coded solution:** A solution s is coded as a vector of $|K|$ paths ($|K|$ is the number of commodities). Each path contains the indexes i of the used arcs e_i.
- **Tabu list:** The tabu list contains the indexes of the last modified commodities. The size of this list is a parameter to be tuned in the experiments.
- **Neighborhood:** exploring the neighborhood of a solution s consists of changing one of the paths, using a greedy procedure, in order to find a new solution s'. The path to be changed corresponds to one of flows that cross the most congested link. Given the traffic loads in different links, the main idea is to generate a new path that minimizes the network congestion. We propose to use a probabilistic path construction strategy that starts from

the source node and moves until reaching its corresponding destination. To move from node i to j. The choice of a adjacent node depends on a local information $\theta_{ij} = 1/l_{ij}$, where l_{ij} is the traffic along the link (i, j). The roulette wheel selection is then applied to choose the next adjacent node. The probability distribution is:

$$p_{ij} = \frac{\theta_{ij}}{\sum_{h \in G(i)} \theta_{ih}} \quad \forall j \in G(i) \tag{5}$$

Where $G(i)$ defines the set of adjacent nodes of i.

- **Diversification:** A multi-start TS is implemented in order to explore a wider space and to avoid being trapped in a local minimum. An initial solution is generated by setting randomly a path for each commodity. We assume that the algorithm first generate the k-shortest paths for every source-destination pair using Yen's algorithm [11]. Hence, an initial solution is build by randomly picking a path to each commodity from the list of k-shortest paths.
- **Intensification:** Starting from a solution s, an iterative exploration of its neighborhood is carried until no improvement is noticed after a given number of iterations.

In what follwos, he basic outline of the MTS algorithm:

Muti-start Tabu Search Algorithm

Initialization

$i = 0$

Set the parameters values: number of local iterations NB, Number of starts $NB - start$, number of iterations without improvement max, the size of the $TabuList$

Generate the k-shortest path for the commodities

Iterative process

for $z = 0$ to $NB - start$ **do**

Generate first a solution s

$no_i mprovement = 0$, $x = 0$

while $x \leq NB$ and $no_i mprovement \leq max$ **do**

$s' = s$

Choose a path P_k of a commodity k to be modified ($k \notin TabuList$)

Add k to the $TabuList$

Reconstract P_k according to (5)

if s' better than s **then**

$no_i mprovement = 0$

else

$no_i mprovement$ $++$

```
      end if
      s = s'
      if best better than s then
         best = s
      end if
   end while
end for
```

Output: the optimal solution *best*

---

5 Experimental Study

To test the efficiency of the MTS in solving the SMCFP, a set of 40 real instances of routing problem in a maritime surveillance problem is used. A simulation testbed called Inform-Lab [14,7] is used. This simulation environment enables to execute different algorithms for distributed information fusion and dynamic resource management. An empirical comparison with the Ant Colony System (ACS) method [13] is performed. CPLEX 12.2 is also implemented to solve the instances optimally using the (1)-(4) formulation. Furthermore, the analysis of the results is supported by statistical nonparametric Wilcoxon signed-rank tests, with a 95% confidence level. In the subsequent, we first explain we implemented an automatic procedure to tune the MTS parameters and get the best configuration. Then, we discuss how the simulation environment to generate the set of instances. Finally, we report and discuss the results of the different generated instances.

5.1 Parameter Tuning

As the parameter setting of the MTS may influence its performance, we propose to apply an automatic procedure called F-Race to determine the best configuration. The F-Race is a statistical automatic procedure proposed by Birattari [5] to enable parameter configuring. In Table 1, a set of candidate configurations are first defined, the optimal configuration are highlighted in boldface. Note that 3 or 5 values for each parameter were first suggested; this means that there are $5 \times 5 \times 5 \times 3 = 375$ possible configurations from which we need to choose the best one.

Table 1. Parameter setting

Parameter	Value
NB	100, 150, **200**, 250, 300
$NB - start$	2, **5**, 8, 11, 14
max	**30**, 50, 70, 100
Size of $TabuList$	**10**, 30, 50

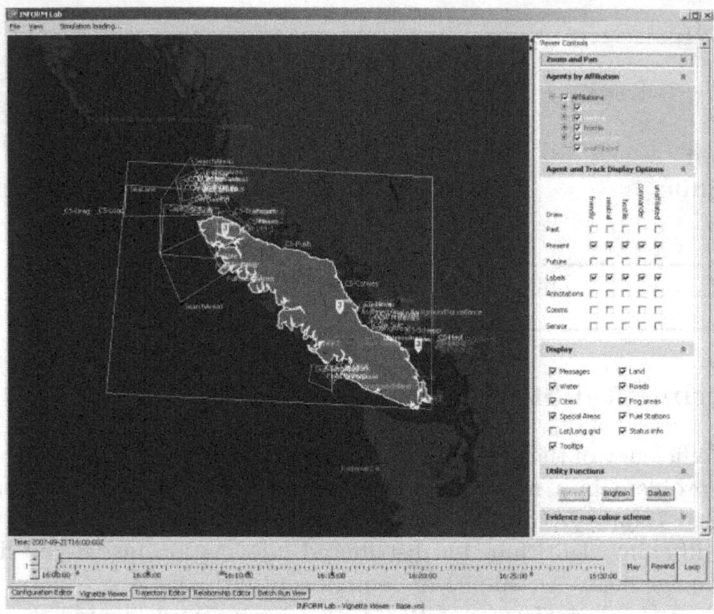

Fig. 1. Inform Lab testbed [14]

5.2 Experimental Results on Real Cases of Maritime Surveillance Problem

Inform lab is a testbed supporting the development of two fundamental groups of algorithms applied for wide-area surveillance applications: distributed dynamic information fusion and distributed dynamic resource management. It simulates different coastal surveillance scenarios (i.e. vignettes), such as the detection of sinking or smuggling entity. In a vignette several cooperative platforms and agents are cooperating in order to fulfill a mission. The interactions between the agents rely on a communication network that supports the surveillance mission by guaranteeing a good performance. The surveillance network includes mobile and fixed surveillance assets deployed in a large geographic area. The backbone of this network contains the stationary nodes.

In order to optimize the messages routing in the backbone of the surveillance network, The MTS is used. The considered real case study supports the mission "is smuggling" in the coastal area of Vancouver Island, contains a network of 50 nodes and 421 arcs. The nodes represent the potential sites in the pacific coast of Vancouver. This networks is heterogeneous composed by both optical and cellular mediums. Hence, a single path routing is necessary. Considering this network, different vignettes were generated to vary the number of commodities. Traffic estimation is performed prior to the execution of the algorithm to get the traffic demands. A screenshot of the testbed running is depicted in figure 1. The comparison was supported by the statistical nonparametric Wilcoxon.

Table 2. Experimental results

Instance	$\|k\|$	Cplex		MTS				ACS				Wilc
		OPT	CPU(s)	Best	Std	CPU(s)	*GAP*	Best	Std	CPU(s)	*GAP*	
1	100	**521**	342	**521**	0	5.2	0	**521**	0	6.4	0	≈
2	150	**572**	364	**572**	0	6.3	0	**572**	0	6.5	0	≈
3	200	**584**	379	**584**	0	6.5	0	**584**	0	7.1	0	≈
4	250	**593**	390	**593**	0	7.4	0	**593**	0	7.8	0	≈
5	300	**654**	532	**654**	0	8.7	0	**654**	0	8.9	0	≈
6	250	**681**	502	**681**	0	9.3	0	**681**	0	10.2	0	≈
7	400	**738**	487	**738**	0	13.7	0	**738**	5.7	11	0	≈
8	450	**784**	492	789	4.8	12.6	0.0063	**784**	3.2	11.7	0	≈
9	500	**806**	327	814	10.7	14.2	0.0098	**806**	4.9	13.5	0	−
10	550	**834**	402	**834**	1.2	17.2	0	**834**	5.7	13.3	0	≈
11	600	**876**	549	**876**	2.7	19.7	0	886	5.6	15.3	0.011	≈
12	650	**895**	698	**895**	4.1	22.4	0	**895**	4.8	14.3	0	≈
13	700	**1021**	1863	**1021**	5.4	20.2	0	1034	10.2	17.8	0.013	≈
14	750	**996**	423	**996**	3.7	24.5	0	1003	7.4	20.7	0.007	+
15	800	**1086**	742	**1086**	10.7	25.4	0	1099	12.7	18.9	0.012	+
16	850	**1116**	674	1119	8.5	23.7	0.0027	**1116**	2.6	19.5	0	≈
17	900	**1119**	802	1137	12.5	28.3	0.0158	1187	11.5	16.9	0.057	+
18	950	**1156**	845	1167	14.2	32.5	0.0094	1159	9.2	19.1	0.003	−
19	1000	**1235**	637	1246	8.6	35.6	0.0088	1256	12.8	22.3	0.017	+
20	1050	**1137**	1023	1139	11.2	41.3	0.0018	1142	10.5	23.4	0.004	≈
21	1100	**1298**	1568	**1298**	32.5	46.2	0	1346	14.9	28.6	0.036	+
22	1150	**1341**	718	1351	11.3	26.4	0.0074	1368	6.2	36.1	0.02	+
23	1200	**1421**	1084	1425	17.4	58.7	0.0028	1435	11.6	35.5	0.01	+
24	1250	**1479**	1148	1482	9.8	69.3	0.002	1502	12.7	31.6	0.015	+
25	1300	**1527**	893	1542	16.3	45.2	0.0097	1566	15.2	43.2	0.025	+
26	1350	**1486**	2435	1496	9.2	85.1	0.0067	1507	13.7	58.6	0.014	+
27	1400	**1647**	2155	**1647**	3.7	81.6	0	1702	20.9	50.4	0.032	+
28	1450	**1732**	1678	1741	15.4	102.4	0.0052	1759	17.6	56.1	0.015	+
29	1500	**1754**	1428	1762	12.6	92.3	0.0045	1798	17.5	59.6	0.024	+
30	1550	**1749**	1762	1768	18.4	113.8	0.0107	1775	16.3	57.3	0.015	≈
31	1600	**1884**	1964	1921	12.3	150.7	0.0193	1996	18.6	64.4	0.056	+
32	1650	-	-	**2087**	5.6	184.6	-	2155	15.9	79.2	-	+
33	1700	-	-	**2007**	23.1	140.2	-	2081	19.4	80.1	-	+
34	1750	-	-	**2112**	14.7	196.3	-	2123	22.5	109.4	-	≈
35	1800	**2178**	2866	2214	16.2	207.8	0.0163	2254	21.4	112.8	0.034	+
36	1850	**2238**	2314	2287	9.9	256.7	0.0214	2361	16.7	130.9	0.052	+
37	1900	-	-	**2273**	13.5	221.3	-	2311	25.9	129.8	-	+
38	1950	-	-	**2324**	18.1	214.8	-	2450	19.3	128.5	-	+
39	2000	-	-	**2465**	17.4	287.6	-	2502	24.3	134.8	-	+
40	2050	-	-	**2287**	19.2	313.2	-	2469	39.2	167.5	-	+
Average	-	1186.000	1045.030	1373.775	9.660	81.723	0.005	1400.100	11.915	46.975	0.014	

We report in table 2 the results of Cplex , MTS and ACS (the best generated solution value *best*, the standard deviation value *Std*, the CPU time in seconds and the *GAP* value) over 30 independent runs of each problem instance. The results of the Wilcoxon test are given in the last column *wilc*, where +, −, and ≈ denote respectively that the MTS performed significantly better than, significantly worse than, or statistically equivalent to the ACS.

Based on the experimental results one can notice that:

- The MTS outperformed the ACS in 28 instances in terms of the solution quality. The average *GAP* values are respectively 0.5% and 1.4% for MTS and ACS. This fact proves that MTS gives a better approximation of the optimal solution.
- Based on the Wilcoxon signed-rank test, MTS surpasses significantly the ACS in 22 instances. While, the ACS outperforms the MTS in only two instances. In the remaining 16 instances, the results of the statistical test are inconclusive.
- For instances 32, 33, 34, 37, 38, 39 and 40 CPLEX terminates and no optimal solution is returned. While, both metaheuristics continue to generate solutions in a reasonable CPU time.
- The ACS was able to converge to the best solution more rapidly than the MTS approach. The average required CPU time is 46.97 s for ACS and 81.72 for the MTS. These differences are more noticeable in the large instances, which questions the scalability of the MTS.
- Given the *Std* values, the MTS gives a slightly more accurate approximation of the global optimum.

6 Conclusion

In this paper, we studied the global routing problem in surveillance networks. As the backbone of such network is stationary and configurable, the routing can be modeled as a static optimization problem. Given the traffic demand between pairs of nodes, the routing plan defines for each flow a single path such that it minimizes the overall congestion. We proposed to model this problem as a SMCFP. A MTS is designed in order to solve such an $\mathcal{NP}$-hard problem. An experimental study is conducted on a large testbed of instances. A comparison with an existing ACS approach is performed. The numerical results, supported by a statistical test, prove the efficiency of MTS on generating better approximation of the global optimum. However, MTS requires is computionaly less efficient than the ACS, this may be explained by its constructive nature. As a future work, we are considering the dynamic routing problem as it models a more realistic situations.

References

1. Ahuja, R., Magnanti, T., Orlin, J.: Network Flows: Theory, Algorithms, and Applications. Prentice Hall Inc., Upper Saddle (1993)
2. Bandyopadhyay, S.: Dissemination of Information in Optical Networks: From Technology to Algorithms. Springer, Berlin (2007)
3. Barnhart, C., Hane, C.A., Vance, P.H.: Using branch-and-price-and-cut to solve origin-destination integer multicommodity flow problems. Operations Research 48(2), 318–326 (2000)

4. Baier, G., Köhler, E., Skutella, M.: On the k-splittable flow problem. In: Möhring, R.H., Raman, R. (eds.) ESA 2002. LNCS, vol. 2461, pp. 101–113. Springer, Heidelberg (2002)
5. Birattari, M.: The problem of tuning metaheuristics as seen from a machine learning perspective. PhD thesis, Universite Libre De Bruxelles (2005)
6. Bley, A.: Approximability of unsplittable shortest path routing problems. Networks 54(1), 23–46 (2009)
7. Dridi, O., Krichen, S., Guitouni, A.: A multi-objective optimization approach for resource assignment and task scheduling problem: Application to maritime domain awareness. In: IEEE Congress on Evolutionary Computation (CEC), pp. 1–8 (2012)
8. Glover, F., Taillard, E., Laguna, M., de Werra, D.: A user's guide to tabu search. Annals of Operations Research 41(1), 1–28 (1993)
9. Holmberg, K., Yuan, D.: A Multicommodity Network-Flow Problem with Side Constraints on Paths Solved by Column Generation. INFORMS Journal on Computing 15(1), 42–57 (2003)
10. Hyppolite, J.M., Galinier, P., Pierre, S.: A tabu search heuristic for the routing and wavelength assignment problem in multigranular optical networks. Photonic Network Communication 15(2), 123–130 (2008)
11. Jiménez, V.M., Marzal, A.: Computing the k Shortest Paths: A new algorithm and experimental Comparison. In: Vitter, J.S., Zaroliagis, C.D. (eds.) WAE 1999. LNCS, vol. 1668, pp. 15–29. Springer, Heidelberg (1999)
12. Kleinberg, J.M.: Approximation algorithms for disjoint paths problems, Ph.D. Thesis, MIT, Cambridge, MA (1996)
13. Li, X.Y., Aneja, Y.P., Baki, F.: An ant colony optimization metaheuristic for single-path multicommodity network flow problems. Journal of the Operational Research Society 61(9), 1340–1355 (2010)
14. MacDonald, Dettwiler and Associates Ltd., Inform Lab Wiki
15. Truffot, J., Duhamel, C., Mahey, P.: Using branch-and-price to solve multicommodity k-splittable flow problems. In: Proceedings of the International Network Optimization Conference (INOC) (2005)

IMRT Beam Angle Optimization Using Electromagnetism-Like Algorithm

Humberto Rocha[1], Ana Maria A.C. Rocha[2], Joana M. Dias[1,3], Brigida Ferreira[4,5], and Maria do Carmo Lopes[4,5]

[1] INESC-Coimbra, Rua Antero de Quental, 199
3000-033 Coimbra, Portugal
[2] Departmento de Produção e Sistemas,
Algoritmi Research Centre,
Universidade do Minho,
4710-057 Braga, Portugal
[3] Faculdade de Economia, Universidade de Coimbra,
3004-512 Coimbra, Portugal
[4] I3N, Departamento de Física,
Universidade de Aveiro,
3810-193 Aveiro, Portugal
[5] Serviço de Física Médica, IPOC-FG, EPE,
3000-075 Coimbra, Portugal
hrocha@mat.uc.pt, arocha@dps.uminho.pt,
joana@fe.uc.pt, brigida@ua.pt,
mclopes@ipocoimbra.min-saude.pt

Abstract. The selection of appropriate beam irradiation directions in radiotherapy – beam angle optimization (BAO) problem – is very important for the quality of the treatment, both for improving tumor irradiation and for better organs sparing. However, the BAO problem is still not solved satisfactorily and, most of the time, beam directions continue to be manually selected in clinical practice which requires many trial and error iterations between selecting beam angles and computing fluence patterns until a suitable treatment is achieved. The objective of this paper is to introduce a new approach for the resolution of the BAO problem, using an hybrid electromagnetism-like algorithm with descent search to tackle this highly non-convex optimization problem. Electromagnetism-like algorithms are derivative-free optimization methods with the ability to avoid local entrapment. Moreover, the hybrid electromagnetism-like algorithm with descent search has a high ability of producing descent directions. A set of retrospective treated cases of head-and-neck tumors at the Portuguese Institute of Oncology of Coimbra is used to discuss the benefits of the proposed algorithm for the optimization of the BAO problem.

Keywords: Electromagnetism-like mechanism, Descent Search, IMRT, Beam Angle Optimization.

B. Murgante et al. (Eds.): ICCSA 2014, Part II, LNCS 8580, pp. 278–289, 2014.

1 Introduction

Cancer is one of the most significant health problems worldwide with respect to its incidence and mortality alike. Radiation therapy is, with surgery and chemotherapy, one of the three main treatment approaches for cancer. More than 50% of all patients diagnosed with cancer, which corresponds to more than 7.6 million patients worldwide, benefit from radiation therapy, either to cure the disease or to palliate symptoms. With this therapy, several beams of ionizing radiation pass through the patient, sent at different incidence directions and centered at the tumor, attempting to sterilize all cancer cells while minimizing the collateral effects on the surrounding healthy organs and tissues. A modern type of radiation therapy is intensity modulated radiation therapy (IMRT), where the radiation beam is modulated by a multileaf collimator. Multileaf collimators enable the transformation of the beam into a grid of smaller beamlets of independent intensities allowing the irradiation of the patient using non-uniform radiation fields from selected angles. A common way to solve the IMRT optimization problems is to use a beamlet-based approach leading to a large-scale mathematical programming problem. Due to the complexity of the whole optimization problem, computation of mathematical algorithms is required to obtain improved solutions.

Typically, radiation is generated by a linear accelerator mounted on a gantry that can rotate along a central axis and is delivered with the patient immobilized on a movable couch. Irradiation from almost any angle, around the tumor, is assured by the combination of the movement of the couch with the rotation of the gantry. However, despite the fact that almost every angle is possible for radiation delivery, the use of coplanar angles is predominant. This is a way to simplify an already complex problem, and the angles considered lay in the plane of the rotation of the gantry around the patient. Furthermore, regardless of the evidence presented in the literature that selecting appropriate radiation beam incidence directions – beam angle optimization (BAO) problem – can lead to a plan's quality improvement [7,14], in clinical practice, most of the time, beam directions continue to be either manually selected by the treatment planner in a time-consuming trial and error iterative process or patients are irradiated using evenly spaced coplanar beams. The difficulty of solving the BAO problem, a highly non-convex problem with many local minima on a large search space [6], is one possible justification for the current clinical practice.

In this paper we present the benefits of using a hybrid electromagnetism-like algorithm with descent search for the optimization of the highly non-convex BAO problem. A set of clinical cases of head-and-neck tumors treated at the Portuguese Institute of Oncology of Coimbra is used to discuss the potential of this approach in the optimization of the BAO problem. The paper is organized as follows. In the next section we describe the BAO problem formulation. Section 3 briefly presents the hybrid electromagnetism-like algorithm with descent search. Computational tests using clinical examples of head-and-neck cases are presented in Section 4. In the last section we have the conclusions.

2 Beam Angle Optimization in IMRT Treatment Planning

The BAO problem is a quite difficult problem to solve since it is a highly non-convex optimization problem with many local minima – see Fig. 1. Except for rare exceptions, where the BAO problem is addressed as a non-convex nonlinear problem, for the vast majority of previous work on beam angle optimization, the continuous $[0°, 360°]$ gantry angles are discretized into equally spaced beam directions with a given angle increment, such as 5 or 10 degrees, where exhaustive searches are performed directly or guided by a variety of different heuristics including simulated annealing [5], genetic algorithms [9], particle swarm optimization [13] or other heuristics incorporating a priori knowledge of the problem [16]. Although those global heuristics can theoretically avoid local optima, globally optimal or even clinically better solutions can not be obtained without a large number of objective function evaluations. On the other hand, the use of single-beam metrics has been a popular approach to address the BAO problem as well, e.g., the concept of beam's-eye-view [15]. The concept is similar to a bird's eye view, where the object being viewed is the tumor as seen from a beam. The bigger the area of the tumor and the smaller the area of the surrounding organs is seen by the beam, the better candidate the beam is to be used in the treatment plan. Despite the computational time efficiency of these approaches, the quality of the solutions proposed cannot be guaranteed since the interplay between the selected beam directions is ignored. Many other attempts to address the BAO problem can be found in literature. Ehrgott et al. [10] discuss a mathematical framework that unifies the approaches found in literature. Aleman et al. [2] propose a response surface approach and include non-coplanar angles in beam orientation optimization. Lee et al. [12] suggest a mixed integer programming (MIP) approach for simultaneously determining an optimal intensity map and optimal beam angles for IMRT delivery. Schreibmann et al. [23] propose a hybrid multiobjective evolutionary optimization algorithm for IMRT inverse planning and apply it to the optimization of the number of incident beams, their orientations and intensity profiles. Other approaches include maximal geometric separation of treatment beams [7] or gradient searches [6].

The BAO problem is the first problem that arises in treatment planning, but its optimal solution is highly dependent on the optimal solution of the fluence map optimization (FMO) problem – the problem of deciding what are the optimal radiation intensities associated with each set of beam angles. Many of the previous BAO studies are based on a variety of scoring methods or approximations of the FMO to gauge the quality of the beam angle set leading to beam angle sets with no guarantee of optimality and questionable reliability since it has been extensively reported that optimal beam angles for IMRT are often non-intuitive [24]. The optimal solution of the FMO problem has been used to drive the BAO problem [1,6,23] including in our works [9,19,20,21]. Our approach for modeling the BAO problem uses the optimal solution value of the FMO problem as the measure of the quality for a given beam angle set. Thus, we will present the formulation of the BAO problem followed by the formulation of the FMO

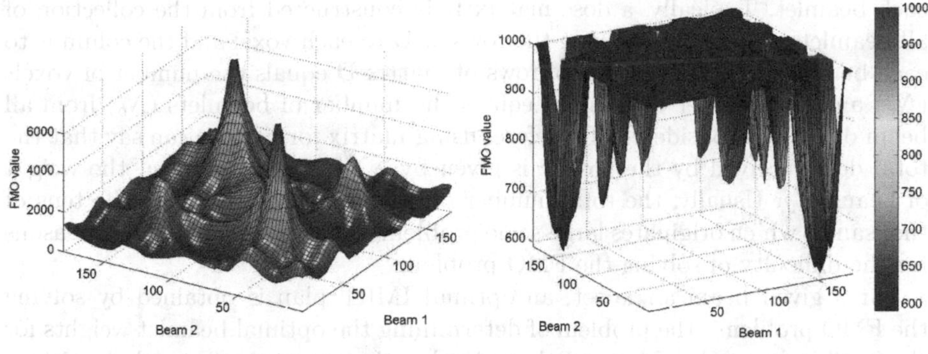

Fig. 1. 2-beam BAO surface (left) and truncated surface (right) to highlight the many local minima

problem we used. Here, we will assume that the number of beam angles is defined a priori by the treatment planner and that all the radiation directions lie on the same plane.

2.1 BAO Model

Let us consider n to be the fixed number of (coplanar) beam directions, i.e., n beam angles are chosen on a circle around the CT-slice of the body that contains the isocenter (usually the center of mass of the tumor). In our formulation, instead of a discretized sample, all continuous $[0°, 360°]$ gantry angles will be considered. Since the angle $-5°$ is equivalent to the angle $355°$ and the angle $365°$ is the same as the angle $5°$, we can avoid a bounded formulation. A simple formulation for the BAO problem is obtained by selecting an objective function such that the best set of beam angles is obtained for the function's minimum:

$$\begin{aligned} &\min\ f(\theta_1, \ldots, \theta_n) \\ &\text{s.t.}\ \ (\theta_1, \ldots, \theta_n) \in \mathbb{R}^n. \end{aligned} \tag{1}$$

Here, for the reasons stated before, the objective $f(\theta_1, \ldots, \theta_n)$ that measures the quality of the set of beam directions $\theta_1, \ldots, \theta_n$ is the optimal value of the FMO problem for each fixed set of beam directions. The FMO model used is presented next.

2.2 FMO Model

In order to solve the FMO problem, i.e., to determine optimal fluence maps, the radiation dose distribution deposited in the patient needs to be assessed accurately. Each structure's volume is discretized into small volume elements (voxels) and the dose is computed for each voxel considering the contribution of each beamlet. Typically, a dose matrix D is constructed from the collection of all beamlet weights, by indexing the rows of D to each voxel and the columns to each beamlet, i.e., the number of rows of matrix D equals the number of voxels (N_v) and the number of columns equals the number of beamlets (N_b) from all beam directions considered. Therefore, using matrix format, we can say that the total dose received by the voxel i is given by $\sum_{j=1}^{N_b} D_{ij}w_j$, with w_j the weight of beamlet j. Usually, the total number of voxels is large, reaching the tens of thousands, which originates large-scale problems. This is one of the main reasons for the difficulty of solving the FMO problem.

For a given beam angle set, an optimal IMRT plan is obtained by solving the FMO problem - the problem of determining the optimal beamlet weights for the fixed beam angles. Many mathematical optimization models and algorithms have been proposed for the FMO problem, including linear models [22], mixed integer linear models [11] and nonlinear models [1]. Here, we will use this latter approach that penalizes each voxel according to the square difference of the amount of dose received by the voxel and the amount of dose desired/allowed for the voxel. This formulation yields a quadratic programming problem with only linear non-negativity constraints on the fluence values [22]:

$$\min_w \sum_{i=1}^{N_v} \frac{1}{v_S} \left[\underline{\lambda}_i \left(T_i - \sum_{j=1}^{N_b} D_{ij}w_j \right)_+^2 + \overline{\lambda}_i \left(\sum_{j=1}^{N_b} D_{ij}w_j - T_i \right)_+^2 \right]$$

$$\text{s.t.} \quad w_j \geq 0, \ j = 1, \ldots, N_b,$$

where T_i is the desired dose for voxel i of the structure v_S, $\underline{\lambda}_i$ and $\overline{\lambda}_i$ are the penalty weights of underdose and overdose of voxel i, and $(\cdot)_+ = \max\{0, \cdot\}$. This nonlinear formulation implies that a very small amount of underdose or overdose may be accepted in clinical decision making, but larger deviations from the desired/allowed doses are decreasingly tolerated [1].

The FMO model is used as a black-box function and the conclusions drawn regarding BAO coupled with this nonlinear model are valid also if different FMO formulations are considered.

3 Electromagnetism-Like Algorithm

The Electromagnetism-like (EM) algorithm, developed by Birbil and Fang [3], is a population-based stochastic search method for global optimization that mimics the behavior of electrically charged particles. The method uses an attraction-repulsion mechanism to move a population of points towards optimality. The EM algorithm is designed for solving optimization problems in the following form:

$$\begin{array}{ll} \min & f(x) \\ \text{s.t.} & x \in \Omega, \end{array} \tag{2}$$

where $f : \mathbb{R}^n \to \mathbb{R}$ is a nonlinear continuous function and $\Omega = \{x \in \mathbb{R}^n : lb \leq x \leq ub\}$ is a bounded feasible region. We do not assume the objective function is convex and we consider it may have many local minima in the feasible region.

The EM algorithm simulates the electromagnetism theory of physics by considering each point in the population as an electrical charge that is released to the space. The charge of each point is related to the objective function value and determines the magnitude of attraction of the point over the others in the population. The better the objective function value, the higher the magnitude of attraction. The charges are used to find a direction for the movement of each points. The regions that have higher attraction will signal other points to move towards them. In addition, a repulsion mechanism is also introduced to explore new regions for even better solutions [3,4].

The EM algorithm is described in Algorithm 1 and comprises four main procedures: "Initialization", "Compute Force", "Move Points" and "Local Search".

Algorithm 1. EM algorithm

```
Initialization;
while stopping criterion in not met do
    Compute Force
    Move Points
    Local Search
end while
```

The "Initialization" procedure starts by randomly generating a sample of m points. Each point is uniformly distributed between the lower and upper bounds. Then, the objective function value for each point is calculated and the best point of the population, x^{best}, is identified as well as its corresponding objective function value f^{best}. After the initialization of the population, and identification of the best point of the population, the other procedures are executed until the stopping criterion is met.

In the "Compute Force" procedure, each particle's charge is calculated by the following formula:

$$q^i = \exp\left(-n\frac{f(x^i)-f(x^{best})}{\sum_{k=1}^{m}(f(x^k)-f(x^{best}))}\right), \quad i=1,\ldots,m. \tag{3}$$

that determines the power of attraction or repulsion for the point x^i. In this way the points that have better objective function values possess higher charges. After the charge calculation, the total force vector F^i on each point x^i is then calculated by adding the individual component forces, F^i_j, between any pair of points x^i and x^j,

$$F^i = \sum_{j=1, j\neq i}^{m} F^i_j = \begin{cases} (x^j - x^i)\frac{q^i q^j}{\|x^j-x^i\|^2} \text{ if } f(x^j) < f(x^i) \text{ (attraction)} \\ (x^i - x^j)\frac{q^i q^j}{\|x^j-x^i\|^2} \text{ if } f(x^j) \geq f(x^i) \text{ (repulsion)} \end{cases}, \tag{4}$$

for $i = 1, 2, \ldots, m$.

The "Move Points" procedure uses the total force vector, F^i, to move the point x^i in the direction of the force by a random step length λ. The best point, x^{best}, is not moved. To maintain feasibility, the force exerted on each point is normalized and scaled by the allowed range of movement towards the lower bound or the upper bound of the set Ω, for each coordinate. Thus, for $i = 1, 2, \ldots, m$ and $i \neq best$

$$x_k^i = \begin{cases} x^i + \lambda \frac{F^i}{\|F^i\|}(u - x^i) \text{ if } F^i > 0 \\ x^i + \lambda \frac{F^i}{\|F^i\|}(x^i - l) \text{ otherwise} \end{cases}. \tag{5}$$

The step length λ is assumed to be an uniformly distributed random variable in $[0, 1]$.

Finally, a local refinement around the best point of the population is done by the "Local search" procedure in order to improve the accuracy of EM [3,4,17,18]. This procedure implements a simple random line search algorithm, using the maximum feasible step length $s_{max} = \delta_{local}(\max[u - l])$, with $\delta_{local} > 0$, to guarantee that the local search always generates feasible points. A trial point y, componentwise defined by $y = x^{best} + \lambda s_{max}$, with $\lambda \sim U[0, 1]$, is computed. If y improves over x^{best} within a maximum number of local iterations, x^{best} is updated. See [3,17,18] for details.

4 Computational Results for Head-and-Neck Clinical Examples

The hybrid EM algorithm was tested using two clinical examples of retrospective treated cases of head-and-neck tumors at the Portuguese Institute of Oncology of Coimbra (IPOC). In general, the head-and-neck region is a complex area to treat with radiotherapy due to the large number of sensitive organs in this region (e.g., eyes, mandible, larynx, oral cavity, etc.). For simplicity, in this study, the organs at risk (OARs) used for treatment optimization were limited to the spinal cord, the brainstem and the parotid glands. The spinal cord and the brainstem are some of the most critical OARs in the head-and-neck tumor cases. These are serial organs, i.e., organs such that if only one subunit is damaged, the whole organ functionality is compromised. Therefore, if the tolerance dose is exceeded, it may result in functional damage to the whole organ. Thus, it is extremely important not to exceed the tolerance dose prescribed for these type of organs. Other than the spinal cord and the brainstem, the parotid glands are also important OARs. The parotid gland is the largest of the three salivary glands. A common complication due to parotid glands irradiation is xerostomia (the medical term for dry mouth due to lack of saliva). This decreases the quality of life of patients undergoing radiation therapy of head-and-neck, causing difficulties to swallow. The parotids are parallel organs, i.e., if a small volume of the organ is damaged, the rest of the organ functionality may not be affected. Their tolerance dose depends strongly on the fraction of the volume irradiated. Hence, if only a small fraction of the organ is irradiated the tolerance dose is much higher than if a larger fraction is irradiated. Thus, for these parallel structures, the organ

Table 1. Prescribed doses for all the structures considered for IMRT optimization

Structure	**Mean dose**	**Max dose**	**Prescribed dose**
Spinal cord	–	45 Gy	–
Brainstem	–	54 Gy	–
Left parotid	26 Gy	–	–
Right parotid	26 Gy	–	–
PTV1	–	–	70.0 Gy
PTV2	–	–	59.4 Gy
Body	–	80 Gy	–

mean dose is generally used instead of the maximum dose as an objective for inverse planning optimization. The tumor to be treated plus some safety margins is called planning target volume (PTV). For the head-and-neck cases in study it was separated in two parts with different prescribed doses: PTV1 and PTV2. The prescription dose for the target volumes and tolerance doses for the OARs considered in the optimization are presented in Table 1.

Our tests were performed on a 2.66Ghz Intel Core Duo PC with 3 GB RAM. In order to facilitate convenient access, visualization and analysis of patient treatment planning data, as well as dosimetric data input for treatment plan optimization research, the computational tools developed within MATLAB and CERR – a freeware computational environment for radiotherapy research [8] – are used widely for IMRT treatment planning research. The patients' CT sets and delineated structures are exported via Dicom RT to CERR. We used CERR 3.2.2 version and MATLAB 7.4.0 (R2007a). An automatized procedure for dose computation for each given beam angle set was developed, instead of the traditional dose computation available from IMRTP module accessible from CERR's menubar. This automatization of the dose computation was essential for integration in our BAO algorithm. To address the convex nonlinear formulation of the FMO problem we used a trust-region-reflective algorithm (*fmincon*) of MATLAB 7.4.0 (R2007a) Optimization Toolbox. In order to obtain a solution in a clinically acceptable computational time (one night), the population size of the EM algorithm was limited to 10 and the maximum number of function evaluations was set to 200. The equispaced solution was always included in the initial population.

Treatment plans with five to nine equispaced coplanar beams are used at IPOC and are commonly used in practice to treat head-and-neck cases [1]. Therefore, treatment plans of seven coplanar orientations were obtained using the hybrid EM algorithm. For each patient, ten runs of the EM algorithm were performed. Each run took 8 to 10 hours of computational time. Table 2 presents the computational results, considering the improvement of the objective function value of the final solution (f*EM*) when compared with the equidistant solution (f*equi*), the beam angle configuration typically used in clinical practice.

Table 2. FMO value (mean) improvement obtained by *EM* compared with the typical equispaced coplanar treatment plans *equi*

Case	f*equi*	Mean f*EM*	% decrease	Std
1	73.75	68.67	6.88%	0.95
2	161.22	151.94	5.76%	0.49

Despite the improvement in FMO value, the quality of the results can be perceived considering a variety of metrics. A metric usually used for plan evaluation is the volume of PTV that receives 95% of the prescribed dose. Typically, 95% of the PTV volume is required. The occurrence of coldspots, less than 93% of PTV volume receives the prescribed dose, and the existence of hotspots, the percentage of the PTV volume that receives more than 110% of the prescribed dose, are other measures usually used to evaluate target coverage. Mean and/or maximum doses of OARs are usually displayed to verify organ sparing.

Table 3. Target coverage obtained by treatment plans

Case	Target coverage	*EMbest*	*EMworst*	*equi*
1	PTV1 at 95 % volume	67.28 Gy	67.28 Gy	67.43 Gy
	PTV1 % > 93% of Rx (%)	99.33	99.59	99.53
	PTV1 % > 110% of Rx (%)	0.00	0.00	0.00
	PTV2 at 95 % volume	56.98 Gy	56.73 Gy	56.42 Gy
	PTV2 % > 93% of Rx (%)	96.75	96.49	96.11
	PTV2 % > 110% of Rx (%)	5.66	5.80	5.93
2	PTV1 at 95 % volume	65.08 Gy	65.03 Gy	65.08 Gy
	PTV1 % > 93% of Rx (%)	94.68	94.73	94.59
	PTV1 % > 110% of Rx (%)	0.00	0.00	0.00
	PTV2 at 95 % volume	56.28 Gy	56.38 Gy	56.08 Gy
	PTV2 % > 93% of Rx (%)	95.97	96.08	95.86
	PTV2 % > 110% of Rx (%)	19.14	18.97	19.04

The results regarding targets coverage are presented in Table 3. In clinical practice, in order to obtain a solution in a clinical acceptable time frame, we can only run the EM algorithm once. Although the small standard deviation obtained, results are presented for the best solution achieved and also for the worst, denoted *EMbest* and *EMworst*. We can verify that optimized treatment plans consistently obtained slightly better target coverage numbers compared to equidistant treatment plans, denoted *equi*. Organ sparing results are shown in Table 4. All the treatment plans fulfill the maximum dose requirements for the spinal cord and the brainstem. However, as expected, the main differences reside in parotid sparing. The optimized treatment plans clearly improve the usually

Table 4. OARs sparing obtained by treatment plans

Case	OAR	Mean Dose (Gy)			Max Dose (Gy)		
		EMbest	*EMworst*	*equi*	*EMbest*	*EMworst*	*equi*
1	Spinal cord	–	–	–	42.53	44.42	44.27
	Brainstem	–	–	–	52.98	53.93	53.18
	Left parotid	26.64	26.74	27.21	–	–	–
	Right parotid	25.41	25.59	27.03	–	–	–
2	Spinal cord	–	–	–	41.43	41.03	42.63
	Brainstem	–	–	–	47.53	48.78	48.08
	Left parotid	25.89	26.99	27.38	–	–	–
	Right parotid	28.34	28.91	29.14	–	–	–

clinically used equispaced treatment plans. Most important is to remark that, despite the best run achieved better results, the worst run managed to obtain, as well, a better parotid sparing than the typical equispaced configuration.

5 Conclusions

The BAO problem is a continuous global highly non-convex optimization problem known to be extremely challenging and yet to be solved satisfactorily. This paper proposes an alternative approach to the BAO problem using a hybrid electromagnetism-like algorithm with descent search, which is yet another step on the quest that may take us closer to find a better solution in a clinical acceptable time. The hybrid electromagnetism-like algorithm with descent search has already proved to be a suitable approach for the resolution of non-convex problems due to its faster progress towards optimality together with a higher consistency solution. For the clinical cases retrospectively tested, the use of this approach lead to solutions with better objective function value than the typical equispaced solution using a limited number of function evaluations. The improvement of the solutions in terms of objective function value corresponded, for the head-and-neck cases tested, to high quality treatment plans with better target coverage and with improved organ sparing.

Acknowledgments. This work was supported by QREN under Mais Centro (CENTRO-07-0224-FEDER-002003) and FEDER funds through the COMPETE program and Portuguese funds through FCT under project grant PTDC/EIA-CCO/121450/2010. This work has also been partially supported by FCT under project grant PEst-OE/EEI/UI308/2014 and in the scope of the project PEst-OE/EEI/UI0319/2014. The work of H. Rocha was supported by the European social fund and Portuguese funds from MCTES.

References

1. Aleman, D.M., Kumar, A., Ahuja, R.K., Romeijn, H.E., Dempsey, J.F.: Neighborhood search approaches to beam orientation optimization in intensity modulated radiation therapy treatment planning. J. Global Optim. 42, 587–607 (2008)
2. Aleman, D.M., Romeijn, H.E., Dempsey, J.F.: A response surface approach to beam orientation optimization in intensity modulated radiation therapy treatment planning. INFORMS J. Comput.: Computat. Biol. Med. Appl. 21, 62–76 (2009)
3. Birbil, S.I., Fang, S.-C.: An electromagnetism-like mechanism for global optimization. J. Global Optim. 25, 263–282 (2003)
4. Birbil, S.I., Fang, S.C., Sheu, R.L.: On the convergence of a population-based global optimization algorithm. J. Global Optim. 30, 301–318 (2004)
5. Bortfeld, T., Schlegel, W.: Optimization of beam orientations in radiation therapy: some theoretical considerations. Phys. Med. Biol. 38, 291–304 (1993)
6. Craft, D.: Local beam angle optimization with linear programming and gradient search. Phys. Med. Biol. 52, 127–135 (2007)
7. Das, S.K., Marks, L.B.: Selection of coplanar or non coplanar beams using three-dimensional optimization based on maximum beam separation and minimized non-target irradiation. Int. J. Radiat. Oncol. Biol. Phys. 38, 643–655 (1997)
8. Deasy, J.O., Blanco, A.I., Clark, V.H.: CERR: A Computational Environment for Radiotherapy Research. Med. Phys. 30, 979–985 (2003)
9. Dias, J., Rocha, H., Ferreira, B.C., Lopes, M.C.: A genetic algorithm with neural network fitness function evaluation for IMRT beam angle optimization. Cent. Eur. J. Oper. Res., doi:10.1007/s10100-013-0289-4 (in press)
10. Ehrgott, M., Holder, A., Reese, J.: Beam selection in radiotherapy design. Linear Algebra Appl. 428, 1272–1312 (2008)
11. Lee, E.K., Fox, T., Crocker, I.: Integer programming applied to intensity-modulated radiation therapy treatment planning. Ann. Oper. Res. 119, 165–181 (2003)
12. Lee, E.K., Fox, T., Crocker, I.: Simultaneous beam geometry and intensity map optimization in intensity-modulated radiation therapy. Int. J. Radiat. Oncol. Biol. Phys. 64, 301–320 (2006)
13. Li, Y., Yao, D., Yao, J., Chen, W.: A particle swarm optimization algorithm for beam angle selection in intensity modulated radiotherapy planning. Phys. Med. Biol. 50, 3491–3514 (2005)
14. Liu, H.H., Jauregui, M., Zhang, X., Wang, X., Dongand, L., Mohan, R.: Beam angle optimization and reduction for intensity-modulated radiation therapy of non-small-cell lung cancers. Int. J. Radiat. Oncol. Biol. Phys. 65, 561–572 (2006)
15. Pugachev, A., Xing, L.: Computer-assisted selection of coplanar beam orientations in intensity-modulated radiation therapy. Phys. Med. Biol. 46, 2467–2476 (2001)
16. Pugachev, A., Xing, L.: Incorporating prior knowledge into beam orientation optimization in IMRT. Int. J. Radiat. Oncol. Biol. Phys. 54, 1565–1574 (2002)
17. Rocha, A.M.A.C., Fernandes, E.M.G.P.: Modified movement force vector in an electromagnetism-like mechanism for global optimization. Opt. Methods Softw. 24, 253–270 (2009)
18. Rocha, A.M.A.C., Fernandes, E.M.G.P.: Numerical study of augmented Lagrangian algorithms for constrained global optimization. Optimization 60, 10–11 (2011)
19. Rocha, H., Dias, J.M., Ferreira, B.C., Lopes, M.C.: Beam angle optimization for intensity-modulated radiation therapy using a guided pattern search method. Phys. Med. Biol. 58, 2939–2953 (2013)

20. Rocha, H., Dias, J.M., Ferreira, B.C., Lopes, M.C.: Selection of intensity modulated radiation therapy treatment beam directions using radial basis functions within a pattern search methods framework. J. Global Optim. 57, 1065–1089 (2013)
21. Rocha, H., Dias, J.M., Ferreira, B.C., Lopes, M.C.: Pattern search methods framework for beam angle optimization in radiotherapy design. Appl. Math. Comput. 219, 10853–10865 (2013)
22. Romeijn, H.E., Ahuja, R.K., Dempsey, J.F., Kumar, A., Li, J.: A novel linear programming approach to fluence map optimization for intensity modulated radiation therapy treatment planing. Phys. Med. Biol. 48, 3521–3542 (2003)
23. Schreibmann, E., Lahanas, M., Xing, L., Baltas, D.: Multiobjective evolutionary optimization of the number of beams, their orientations and weights for intensity-modulated radiation therapy. Phys. Med. Biol. 49, 747–770 (2004)
24. Stein, J., Mohan, R., Wang, X.H., Bortfeld, T., Wu, Q., Preiser, K., Ling, C.C., Schlegel, W.: Number and orientation of beams in intensity-modulated radiation treatments. Med. Phys. 24, 149–160 (1997)

Improving Branch-and-Price for Parallel Machine Scheduling

Manuel Lopes[1], Filipe Alvelos[2,3], and Henrique Lopes[2]

[1] CIDEM-ISEP, 4200-072 Porto, Portugal
[2] Centro Algoritmi, Universidade do Minho, Campus de Gualtar, 4710-057 Braga, Portugal
[3] Departamento de Produção e Sistemas, Universidade do Minho, 4710-057 Braga, Portugal
falvelos@dps.uminho.pt, mpl@isep.ipp.pt

Abstract. In this paper we present a hybrid exact-heuristic method to improve a branch-and-price algorithm to solve the unrelated parallel machines with sequence-dependent setup times scheduling problem. As most of the computational time in the column generation (CG) process is spent in subproblems, two new heuristics to solve the subproblems are embedded in the branch-and-price (BP) framework with the aim to improve the efficiency of the process in obtaining optimal solutions. Computational results show that the proposed method improves a state-of-the-art BP algorithm from the literature, providing optimal solutions for large instances (e.g. 50 machines and 180 jobs) of the parallel machine scheduling problem with sequence dependent setup times, in significantly less time. One of the proposed approaches reduces, in average, to a half the time spent in the root of the branch-and-price tree and to a quarter the time spent in the full branch-and-price algorithm.

Keywords: parallel machine scheduling, sequence dependent setup times, column generation, branch-and-price.

1 Introduction

Scheduling is of crucial importance in today's business environment, as the efficient use of production resources and the effectiveness of deliveries has significant impacts on companies financial results and market share. Recent theoretical and computational developments have enabled the possibility to deal with more complex systems which can better represent the reality and lead to better decision making. We can find many examples of the use of parallel machines in the real world: manufacturing lines, multiprocessor computers, hospital assistance, public services, crane scheduling in port container terminals, and many others. The parallel machines scheduling problem also appears as a relaxation of more complex problems like the hybrid flow shop scheduling problem or the resource-constrained project scheduling problem. This is the main reason why

B. Murgante et al. (Eds.): ICCSA 2014, Part II, LNCS 8580, pp. 290–300, 2014.

parallel-machine scheduling has received much attention from researchers in recent years (for a literature review on parallel machines scheduling problems see, for example, [1]).

The majority of the intense research work on different scheduling problems that have appeared in the literature, since the first systematic approach to scheduling problems in the mid-1950s, assumed that the setup time is negligible or is part of the job processing time. The significant savings obtained when setup times/costs are explicitly considered for scheduling decisions in real world problems [3] induced an increasing research trend on this subject in the more recent decades. For a literature review on scheduling problems with setup times or costs see, for example [2].

We consider the problem of scheduling a set of independent jobs, with release dates and due dates, on several unrelated parallel machines with availability dates and sequence-dependent setup times, to minimize the total weighted tardiness. Using the $\alpha|\beta|\gamma$ Graham classification [4], this problem is classified as $R|a_k, r_j, s_{ij}|\sum w_j T_j$. This problem is strongly NP-hard, because it is a special case of $P||\sum w_j T_j$ which is strongly NP-hard, even when the number of machines is one [5].

The branch-and-price metod (BP) method is based on column generation (for a survey on the use of column generation to solve integer programming problems, see [6] and has been also successfully applied on solving parallel machines scheduling problems.

The first application of the BP method to this problem was presented by Lopes and Valério de Carvalho [7]. The BP algorithm applied to machine scheduling exploits the combinatorial structure of large problems by associating a subproblem to the decision of sequencing jobs on each machines. For an overview on decomposition methods, see [8,9].

This paper presents two new heuristics to solve the subproblems of the column generation process, with the aim to improve the efficiency of the BP algorithm, for the unrelated parallel machines with sequence-dependent setup times scheduling problem. The heuristics are embedded in the BP framework, within a hybrid procedure which iteratively uses the heuristic and the exact subproblem algorithms to obtain the optimal solution. In the next section we formulate the problem and present the decomposition model. The two heuristics developed to solve the subproblems are presented in Section 3. In Section 4 computational experiments are presented, and in Section 5 we conclude the paper.

2 Problem Definition and a Decomposition Model

The application of branch-and-price (BP) method to the unrelated parallel machines problem with sequence-dependent setup times, to minimize the total weighted tardiness of the jobs, was first presented by Lopes and Valério de Carvalho [7]. They considered a class of scheduling n independent jobs $N = 1, 2, , n$ on m unrelated parallel machines $M = 1, 2, , m$ to minimize the total weighted tardiness of the jobs. Each machine $k \in M$ has an availability date a_k and a configuration status l_k which represents the last job processed by machine k. Each job $j \in N$ must be processed by exactly one of the machines and its processing time depends on the machine, being represented by p_{jk}, $k \in M$. Each job j also has a weight w_j, a release date r_j, and a due date d_j. A sequence dependent setup time s_{ij} is incurred whenever job j is processed immediately after job i. There is an initial setup time, $s_{l_k}j$, if job j is the first one processed by machine k and $l_k \neq j$. If a job is completed after date d_j it will incur a tardiness penalty w_j per unit of tardiness time. Preemption of jobs is not allowed and a machine can only process a job at a time.

A decomposition model (DM) can be obtained by enumerating the set of all feasible schedules for all machines [7]. A machine schedule can be defined as a path p in a network represented by the sequence of nodes $(j_0, j_1, ..., j_H, j_{n+1})$, such that $H \leq n$, and where j_0 and j_{n+1} represent the source node (machine start) and the sink node (machine end), respectively, and the other nodes represent the jobs to be processed. Jobs $j_1, j_2, ..., j_H$ are processed in this order and v_{jp}^k represents the number of times job j is processed in path p, for machine k. The set of all feasible schedules for machine k is denoted by P^k. Representing the completion time of job j_h in the machine schedule p by $C_{j_h,p}^k$, the completion time of the next job in the sequence, $C_{j_{h+1},p}^k$, can be computed by the following relations:

$$C_{j_{h+1},p}^k = max(r_{j,h+1}, C_{j_h,p}^k + s_{j_h,j_{h+1}}) + p_{j_{h+1},k}$$

with

$$C_{j_0,p}^k = a_k \text{ and } C_{j_{H+1},p}^k = C_{j_H,p}^k, \forall k \in M, 1 \leq h \leq H$$

We define the cost of an arc (j_h, j_{h+1}) of machine schedule p, as the cost of completing job j_{h+1} immediately after job j_h at time $C_{j_h,p}^k$, on machine k,

$$c_{(j_h,j_{h+1})p}^k = w_j(max(max(r_{jh+1}, C_{j_h,p}^k + s_{j_h,j_{h+1}}) + p_{j_{h+1},k} - d_{j_{h+1}}, 0))$$

The cost of a machine schedule p in a machine k, c_p^k, is equal to the sum of the cost of the arcs belonging to p, and can be obtained by the expression,

$$c_p^k = \sum_{h=0}^{H} c_{j_h,j_{h+1},p}^k$$

We define decision variables as

$$y_p^k = \begin{cases} 1, \text{ if schedule } p \text{ of machine } k \text{ is selected for that machine} \\ 0 \text{ otherwise.} \end{cases}$$

The decomposition model is:

$$Min \sum_{k=1}^{m} \sum_{p \in P^k} c_p^k y_p^k \tag{1}$$

$$subject to:$$

$$\sum_{k=1}^{m} \sum_{p \in P^k} v_{jp}^k y_p^k = 1; j = 1, ..., n \tag{2}$$

$$\sum_{p \in P^k} y_p^k \leq 1; k = 1, ..., m \tag{3}$$

$$y_p^k \in \{0, 1\}; k = 1, ..., m; p \in P^k$$

The objective function (1) minimizes the total weighted tardiness. The first set of constraints (2) ensures that each job is processed exactly once; the second set of constraints (3) ensures that each machine is used at most once.

Because the number variables, y_p^k, corresponding to the machine schedules, grows exponentially with the number of jobs and the number of machines, in practice, the decomposition model cannot be solved directly, i.e., by approaches involving exhaustive enumeration. Rather, this problem is solved by column generation based approaches. Column generation allows obtaining an optimal solution to the linear relaxation of the decomposition model.

Column generation solves the linear relaxation of the decomposition model by considering a reduced set of decision variables (columns), called the restricted master problem (RMP). In each iteration of the CG algorithm, new columns are generated in the subproblems, each one associated with one machine, and added to the RMP. The simplex method supplies the set of prices (dual variables values) $\pi_j, j \in N$ and $\omega_k, k \in M$, associated with constraints 2 and 3, respectively, needed to solve the subproblems. The subproblem of machine k consists in determining the minimum reduced cost of a machine schedule p, on machine k, $\bar{c}_p^k$, which can be obtained by the following expression:

$$\bar{c}_p^k = \sum_{h=0}^{H} c_{j_h, j_{h+1}, p}^k - \omega_k - \pi_{j_{h+1}}$$

In the dynamic programming algorithms used in the BP algorithm of Lopes and Valério de Carvalho [7] to solve the subproblems, as the variables that define the state space ((j,t) - partial machine schedules going from machine node 0 to job j, having completed job j at time t) do not contain information about the jobs already processed, it may happen that, for a given machine schedule, a job is processed more than once, generating cyclic solutions (columns).

The branching strategy followed by [7] modifies the underlying problem network by the imposition of restrictions on possible paths. Each iteration of the branching process creates two new nodes in the branching tree which will be solved by the column generation process using the best-lower-bound selection rule. The branching is implemented on the flow value of arcs, x^k_{ij}, of the networks that represent the respective machine schedules:

1. For each non zero basic variable ($y^k_p > 0$), the total value of the flow on the respective arcs, x^k_{ij}, is calculated. Then, a score, which is higher for the most fractional flow values ($min|x^k_{ij} - 0.5|$) and for the arcs closer to the machine node 0, is given to each x^k_{ij}. The fractional flow value variable with the higher score is selected for branching.
2. For the variable selected two branches are created: one imposing that the total flow value on the arc (i, j), at machine k, is equal to one - $x^k_{ij} = 1$, i.e., the, arc (i, j) can only be used on machine k subproblem network, and the other imposing that the total flow value on the arc (i, j), at machine k, is equal to zero - $x^k_{ij} = 0$, i.e., the arc (i, j) cannot be used on machine k subproblem network.

The respective problem network must be modified to reflect the restrictions imposed by the branching decisions:

- If x^k_{ij} is fixed at zero, the arc (i, j) is simply removed from the subproblem network of machine k, eliminating the generation of schedules for machine k which include arc (i, j), and the cost of all the columns in the RMP that use arc (i, j) and machine k are penalized, i.e, take a very high cost compared to best integer solution found so far.
- If x^k_{ij} is fixed at one, the arcs $(i, v) \in A|v \neq j$ and $(l, j) \in A|l \neq i$ are removed from the subproblem network of machine k and the cost of all the columns in the RMP that use machine k and arcs $(i, v) \in A|v \neq j$ and $(l, j) \in A, l \neq i$ are penalized.

After optimization over the current columns, new columns are generated by the column generation process to obtain a new solution.

3 Improving Column Generation Efficiency

In the column generation algorithm, a very substantial part (in average, more than 80%) of the computational time is spent on solving the subproblems [7]. Furthermore, the solutions (columns) generated by the subproblems can be cyclic, i.e., jobs with $v^k_{jp} > 1$. We now propose two heuristics developed to try to overcome this issue.

These heuristics are based on the original dynamic programming model of [7] and both generate only acyclic (feasible) solutions. This results in constrained dynamic programming state spaces for which the optimal solution is not guaranteed.

In the first heuristic, Set of Jobs Heuristic (SJH), the main objective is to eliminate the cyclic solutions. They are eliminated by recording the information of the jobs already processed in the partial schedules generated by the dynamic programming algorithm. The dynamic programming state space is still defined by the variables last job processed (j) and time (t), but the cyclic paths are eliminated by recording, for each state (j,t), the set of jobs already processed in the partial schedule going from machine node 0 to the current state node (j,t) - $S_{j,t}$. The original dynamic programming [7] state space is constrained by considering only the acyclic schedules.

This process to eliminate the cyclic solutions, is different from the original optimal acyclic dynamic programming algorithm (Model 1) referred in [7] because in the latter the set of jobs already processed, S, is a variable of the dynamic programming state $((S,j,t))$, while in the former the same information is recorded in an auxiliary variable just to assure that a job is not processed more than once. For example, consider the partial schedules $(1,2,3,4)$ and $(3,5,6,4)$ with the same total processing time t. In the heuristic, these two partial schedules correspond to the same state $(j=4,t)$, while in the original optimal acyclic dynamic programming algorithm they would correspond to two different states, $(S=(1,2,3,4), j=4,t)$ and $(S=(3,4,5,6), j=4,t)$.

In the second heuristic, Number of Jobs Heuristic (NJH), the cyclic solutions are eliminated by the same process of SJH heuristic and the states of the original dynamic programming model of [7] are redefined to be the variables job (j) and number of jobs ($njob$) already processed in the partial machine schedule going from machine node 0 to job j, independently of the specific jobs included in the partial schedule. This results in a more constrained state space than heuristic SJH as, on top of the elimination of the cyclic schedules, only the most attractive of the different paths to the same job (j) and equal number of jobs visited ($njob$) is considered. For example, machine schedules $(2,3,1)$, $(4,5,1)$ and $(3,2,1)$, with different t values, represent three different states for the original dynamic programming model [7], while for NJH heuristic represent the same state $(j=1, njob=3)$ and, therefore, the constrained state space will only have one of them, the most attractive of the three (with the lowest reduced cost value).

3.1 SJH Heuristic

Let $F^k(j,t)$ be the minimum reduced cost of the partial machine schedule going from node 0 to job j processing only once all jobs in set $S_{j,t}$ and having completed the processing of job j at time t, at machine k. $F^k(j,t)$ can be computed by solving the following recurrence relations:

$$F^k(0,a_k) = -\omega_k$$

$$S_{0,a_k} = \{\}$$

$$F^k(j,t) = min_{i \in N} \begin{cases} F^k(j,t') + max(t-d_j,0)w_j - \pi_j| \\ t = max(t'+s_{ij}+p_{jk}, r_j+p_{jk}) \wedge j \notin S_{i,t'} \end{cases}$$

$$S_{j,t} = S_{i,t'} \cup j$$

for all j,k such that $j \in N, k \in M$.

3.2 NJH Heuristic

Define $S_{j,njob}$ as the set of jobs already processed in the partial schedule going from machine node 0 to the current state node $(j, njob)$. Let $G^k(j, njob)$ be the minimum reduced cost of the partial machine schedule of $njob$ jobs, going from node 0 to job j processing only once all jobs in set $S_{j,njob}$, at machine k. $G^k(j, njob)$ can be computed by solving the following recurrence relations:

$$G^k(0,0) = -\omega_k;$$

$$S_{0,0} = \{\}; t_{0,0} = a_k$$

$$G^k(n, njob) = min_{i \in N} \begin{cases} G^k(i, njob-1) + max(t_{j,njob} - d_j, 0)w_j - \pi_j| \\ t_{j,njob} = max(t_{i,njob-1} + s_{ij} + p_{jk}, r_j + p_j k) \wedge j \notin S_{i,njob_1} \end{cases}$$

$$S_{j,njob} = S_{i,njob-1} \cup j$$

for all j, k such that $j \in N, k \in M$.

The following example show how these two heuristics work.

Example. To demonstrate how the two heuristics perform in constraining the state space of the dynamic programming algorithm, we are going to use an example of 4 jobs and one machine (the behavior of the heuristics in one machine can be extrapolated to m machines), randomly generated accordingly to scheme proposed in [7]. Tables 1 and 2 show the instance.

Table 1. Setup times

s_{ij}	1	2	3	4
1		22	32	37
2	33		32	29
3	21	28		20
4	21	37	24	

Table 2. Jobs processing times, release times, due dates and penalties

j	p_j	r_j	d_j	w_j
1	65	42	189	7
2	23	0	88	20
3	57	6	83	5
4	73	11	150	74

The machine has an availability date $a_k = 0$ and an initial configuration status $l_k = 1$.

Tables 3, 4 and 5, show the state spaces for the original dynamic programming algorithm [7], the SJH heuristic, and the NJH heuristic, respectively.

Table 3. State space of original dynamic programming algorithm

Stage	0	1	2	3		4			
States (j,t)	(0,0)	(1,107)	(2,152)	(3,241)	(2,220)	(1,327)	(2,283)	(3,309)	(3,330)
		(2,45)	(3,196)	(4,254)	(4,285)	(1,340)	(4,325)	(4,322)	(4,343)
		(3,89)	(4,217)	(2,247)	(1,238)	(3,335)	(2,313)	(2,345)	(2,336)
		(4,110)	(1,143)	(4,289)	(4,242)	(1,345)	(3,334)	(3,327)	(4,378)
			(3,134)	(2,277)	(1,268)	(4,349)	(2,265)	(4,348)	(4,352)
			(4,147)	(3,298)	(2,242)	(1,375)	(4,330)	(1,328)	(2,322)
			(1,175)	(3,232)	(2,241)	(2,349)	(1,313)	(3,323)	(4,387)
			(2,140)	(4,253)	(3,285)	(3,366)	(2,287)	(3,357)	(4,344)
			(4,182)	(1,220)	(3,259)	(1,384)	(2,278)		
			(1,196)	(4,227)	(1,277)		(3,322)		
			(2,170)	(1,233)			(1,314)		
			(3,191)	(3,228)			(2,279)		

Table 4. State space of SJH heuristic

Stage	0	1	2	3		4			
States (j,t)	(0,0)	(1,107)	(2,152)	(3,241)	(2,220)				(3,330)
		(2,45)	(3,196)	(4,254)	(4,285)		(4,325)	(4,322)	
		(3,89)	(4,217)	(2,247)	(1,238)			(2,345)	(2,336)
		(4,110)	(1,143)	(4,289)	(4,242)	(3,335)	(3,334)		(4,378)
			(3,134)	(2,277)	(1,268)				
			(4,147)	(3,298)	(2,242)	(4,349)	(4,330)	(4,348)	
			(1,175)	(3,232)	(2,241)		(1,313)	(1,328)	
			(2,140)	(4,253)	(3,285)	(2,349)			
			(4,182)	(1,220)	(3,259)			(2,313)	(2,322)
			(1,196)	(4,227)	(1,277)	(3,366)	(3,322)		
			(2,170)	(1,233)			(1,314)	(1,340)	
			(3,191)	(3,228)					

Table 5. State space of NJH heuristic

Stage	0	1	2	3	4
States (j,ntar)	(0,0)	(1,1)	(1,2)	(1,3)	(1,4)
		(2,1)	(2,2)	(2,3)	(2,4)
		(3,1)	(3,2)	(3,3)	(3,4)
		(4,1)	(4,2)	(4,3)	(4,4)

From tables 3, 4, and 5, we can see that the original dynamic programming algorithm creates 76 states, and that the state space reduction performed by SJH and NJH heuristics is 22% (59 states) and 78% (17 states), respectively. The optimal solution is presented in Gantt chart of Figure 1.

Despite the dynamic programming constrained space of heuristics SJH and NJH, both returned the optimal solution.

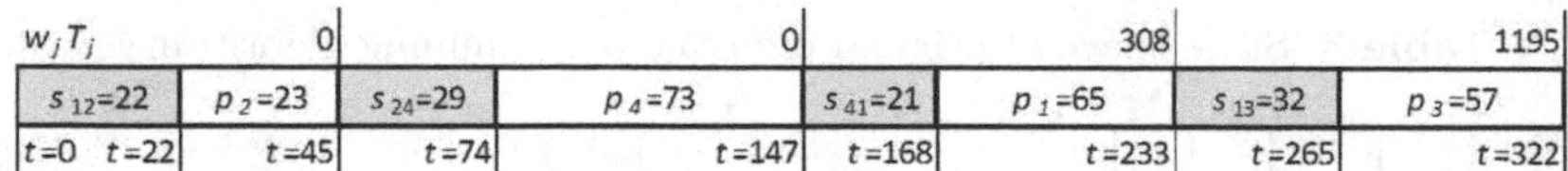

Fig. 1. Optimal solution of the example

The two heuristics were used in the BP algorithm of [7] to solve the subproblems in the column generation of all nodes of the BP tree according to the following scheme:

1. First phase, the heuristic (SJH or NJH) is used to solve the subproblems (one subproblem per each machine) until no attractive solutions (with negative reduced cost) are found;
2. Second phase, the exact dynamic programming algorithm is used until no attractive solutions (with negative reduced cost) are found.

Phase two is used to assure the optimality of the solution found. The performance of the two heuristics was tested by evaluating the time spent in the linear relaxation and in the BP algorithm. The results are shown in the next section.

4 Computational Results

The performance of the two heuristics was tested by evaluating the time and the number of columns as performance indicators. For the computational tests, we used the test instances $m - n - q$ referred in [10], which are classified according to the number of machines (m), number of jobs (n) and the congestion level ($q = 1, 2, 3, 4, 5$; the larger the value of q, the more congested the system will be, and the more tardy jobs will result). The computational tests were performed on a notebook pc with a 2.4 GHz i3-3110M Intel processor and 4 Gb of RAM. To evaluate and compare the performance of the two heuristics, the following set of test problems ($m - n$) was used: $10 - 50, 10 - 70, 10 - 90, 10 - 100, 20 - 100, 20 - 120, 30 - 150, 50 - 150$ and $50 - 180$. For each of the test problems, 10 random instances of congestion levels 3 and 4 (the most difficult [7]) were tested, summing up to 180 tested instances. The results of the computational tests are presented in Table 6 and Table 7.

The results show that both heuristics outperform the original exact algorithm for solving the subproblems. The average reduction of the SJH (the average of the $(t_{original} - t_{SJH})/t_{original}$) is 2% and 11% for the linear relaxation and BP, respectively. The average reduction of the NJH is really impressive since it reduces almost to a half (reduction of 48%) the time of lienar relaxation nd almost to a quarter the BP time (reduction of 24From the results of the comparison of the two heuristics performance, we conclude that:

- Heuristic NJH is faster than heuristic SJH in the LR phase (45%) and in total time (7%);
- Heuristic SJH generates less columns in the linear relaxation phase (24%) and in total (full BP) (10%) than heuristic NJH.

Table 6. Linear relaxation and branch-and-price computation times (in seconds) for the three alternative approaches

	Linear relaxation			Branch-and-price		
Instance	Exact	SJH	NJH	Exact	SJH	NJH
10-50-3	14	14	7	27	24	17
10-50-4	18	17	10	31	27	23
10-70-3	115	102	54	796	636	579
10-70-4	166	160	93	561	495	431
10-90-3	342	305	157	1359	1169	880
10-90-4	643	672	446	4502	3893	4160
10-100-3	627	588	268	4659	4199	3622
10-100-4	1439	1545	935	17871	13114	16109
20-100-3	68	67	32	634	544	425
20-100-4	94	99	48	635	597	474
20-120-3	216	179	67	1979	1657	1235
20-120-4	367	355	170	5159	4599	3970
30-150-3	173	163	69	2204	2101	1594
30-150-4	317	325	147	15948	15315	13274
50-150-3	35	39	24	721	746	610
50-150-4	35	39	28	552	557	449
50-180-3	118	103	53	4131	3809	2708
50-180-4	126	126	70	10894	10886	9104
Total	4913	4898	2678	72663	64368	59664

Table 7. Number of columns generated in the linear relaxation and branch-and-price for the three alternative approaches

	Linear relaxation			Branch-and-price		
Instance	Exact	SJH	NJH	Exact	SJH	NJH
10-50-3	3779	3697	4817	1438	1304	1570
10-50-4	4409	4161	5194	2862	2699	3379
10-70-3	6528	6293	8935	12156	11238	12586
10-70-4	7165	7041	10186	6701	5980	7118
10-90-3	8721	8570	13465	11616	12486	14645
10-90-4	9942	10132	16517	18834	16663	19463
10-100-3	10352	10166	17047	30281	28659	37079
10-100-4	12184	12887	21635	39415	31405	40130
20-100-3	7590	7685	9120	10390	9917	9943
20-100-4	8114	8217	9993	8362	7697	8111
20-120-3	9758	9689	12173	15067	15159	15618
20-120-4	11218	10976	14829	19505	18617	19201
30-150-3	11317	11203	13613	17063	16684	16659
30-150-4	12570	12534	16008	28393	27676	28255
50-150-3	9669	9782	10339	7491	7608	7522
50-150-4	9943	10090	10830	7045	7007	6926
50-180-3	12918	12851	13980	15569	15271	15407
50-180-4	12913	12986	14498	22265	22511	22477
Total	169090	168960	223179	274453	258581	286089

5 Conclusions

In this paper, a hybrid exact-heuristic method to improve a branch-and-price algorithm was proposed to solve the unrelated parallel machines with sequence-dependent setup times scheduling problem. Two new heuristics to solve the subproblems associated to the CG process are embedded in the branch-and-price framework, within a hybrid procedure which iteratively uses the heuristic and the exact subproblem algorithms to obtain the optimal solution. The computational results show that this new method outperforms a state-of-the-art branch-and-price algorithm from the literature, providing optimal solutions for large instances (e.g. 50 machines and 180 jobs) of the parallel machine scheduling problem with sequence dependent setup times, in significantly less time.

Acknowledgments. This work was financed by Fundação para a Ciência e a Tecnologia (Portuguese Foundation for Science and Technology) within projects "SearchCol: Metaheuristic search by column generation" (PTDC/EIAEIA/100645/2008) and PEst-OE/EEI/UI0319/2014.

References

1. Mokotoff, E.: Parallel Machine Scheduling Problems: a survey. Asia-Pacific Journal of Operational Research 18, 193–243 (2001)
2. Allahverdi, A., Ng, C.T., Cheng, T.C.E., Kovalyov, M.Y.: A survey of scheduling problems with setup times or costs. European Journal of Operational Research 187, 985–1032 (2008)
3. Allahverdi, A., Soroush, H.M.: The significance of reducing setup times/setup costs. European Journal of Operational Research 187, 978–984 (2008)
4. Graham, R.L., Lawler, E.L., Lenstra, J.K., Rinnooy Kan, A.H.G.: Optimization and approximation in deterministic sequencing and scheduling: a survey. Annals of Discrete Mathematics 5, 287–326 (1979)
5. Lenstra, J.K., Rinnooy Kan, A.H.G., Brucker, P.: Complexity of Machine Scheduling Problems. Annals of Discrete Mathematics 1, 343–362 (1977)
6. Barnhart, C., Johnson, E.L., Nemhauser, G.L., Savelsbergh, M.W.P., Vance, P.H.: Branch-and-price: column generation for solving huge integer programs. Operations Research 46, 316–329 (1998)
7. Lopes, M.J.P., de Carvalho, J.M.V.: A branch-and-price algorithm for scheduling parallel machines with sequence dependent setup times. European Journal of Operational Research 176, 1508–1527 (2007)
8. Desaulniers, G., Desrosiers, J., Solomon, M.M.: Column generation. Springer (2005)
9. Wilhelm, W.E.: A technical review of column generation in integer programming. Optimization and Engineering 2, 159–200 (2001)
10. Lopes, M.: Resolução de Problemas de Programação de Máquinas Paralelas pelo Método de Partição e Geração de Colunas. PhD thesis, Universidade do Minho (2004) (in Portuguese)

Fast Parallel Triangulation Algorithm of Large Data Sets in E^2 and E^3 for In-Core and Out-Core Memory Processing

Michal Smolik and Vaclav Skala

Faculty of Applied Sciences,
University of West Bohemia,
Univerzitni 22, CZ 30614 Plzen,
Czech Republic

Abstract. A triangulation of points in E^2, or a tetrahedronization of points in E^3, is used in many applications. It is not necessary to fulfill the Delaunay criteria in all cases. For large data (more then $5 \cdot 10^7$ points), parallel methods are used for the purpose of decreasing time complexity. A new approach for fast and effective parallel CPU and GPU triangulation, or tetrahedronization, of large data sets in E^2 or E^3, is proposed in this paper. Experimental results show that the triangulation/tetrahedralization, is close to the Delaunay triangulation/tetrahedralization. It also demonstrates the applicability of the method presented in applications.

1 Introduction

Today's applications need to process large data sets using several processors with shared memory in parallel processing, or/and on systems using distributed processing. In this paper we describe an approach applicable for effective triangulation in E^2 and E^3 (tetrahedralization) using CPU and/or GPU parallel or distributed systems, e.g. on computational clusters, for large data sets.

Many algorithms for triangulation in E^2 and E^3 have been developed and described with different criteria [1], [2], [5], [8]; mostly Delaunay triangulation in E^2 is used due to the duality with the Voronoi diagrams. The Delaunay triangulation maximizes the minimum angle; on the other hand, it does not minimize the maximum angle, which is required in some fields, like CAD systems etc. Moreover, if the points form a squared mesh, algorithms are sensitive to the numerical precision of computation.

It is well known that the Delaunay triangulation (DT) contains $O(N^{\lceil d/2 \rceil})$ simplicities where d is dimensionality. The computational complexity of the DT is $(N^{\lceil d/2 \rceil+1})$, i.e. for $d = 2$ is $O(N^2)$ and for $d = 3$ is $O(N^3)$.

B. Murgante et al. (Eds.): ICCSA 2014, Part II, LNCS 8580, pp. 301–314, 2014.

1.1 Motivation

However, in many cases we do not need exact Delaunay triangulation nor another specific triangulation, as triangulation "close enough" to the required type is acceptable. Weakening this strict requirement enables us to formulate a simple algorithm based on "divide and conquer (D&C)" strategy and the approach is independent from the triangulation property requirements.

There are the following critical issues to be solved if triangulation is to be applicable for large data sets:

- how to store data especially to have fast access to data on parallel/distributed system,
- how the triangulation is made on a data subset – we expect that each processor will process the given data subset resulting in a triangulated subset,
- how to join triangulated subsets in order to get the final large triangulation in E^2 or E^3.

Of course, implementation on CPU should be simple and implementation on GPU should be simple as well.

2 Proposed Algorithm

In this section, we will introduce a new fast parallel triangulation algorithm in E^2 and E^3. The main idea of this algorithm is to divide all input points into several subsets, perform a triangulation on each subset of them and then join them together.

First, in sections 2.1-6, we will introduce the proposed algorithm for parallel triangulation. In section 2.7, we will show how to divide data between multiple GPUs and/or cluster PCs. Finally, in section 2.8, we will propose an approach for large data processing.

2.1 Points Division

The approach proposed is based on D&C strategy and therefore input data set has to be split to several subsets. In our case, we will use rectangular grid of size $n \times m$ domains in E^2 (see Fig. 1), resp. $n \times m \times p$ domains in E^3. The grid *does not have to be necessarily regular* and we can adjust it according to the properties of the input data set. However, we will use orthogonal grid in our approach: it is not necessary because domains can be triangular or tetrahedral, etc.

In the case when a domain does not contain any point, we have to generate a random one and place it into this domain. This restriction is necessary because of the joining procedure which will be introduced later.

The virtual corner points of the grid are included in the domains. It means that now each data subset contains the original points plus the virtual corner points of the appropriate domain.

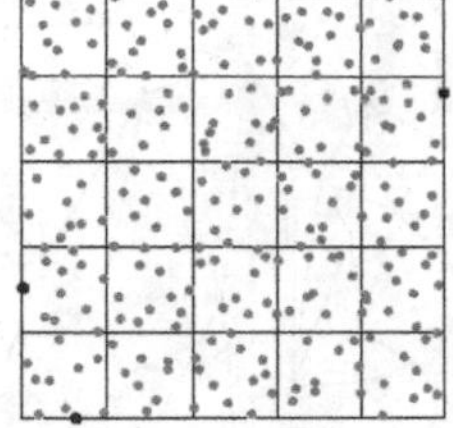

Fig. 1. Division of points into a rectangular grid

2.2 Domains Triangulation

Now, each domain can be triangulated using any triangulation library. Properties of the final triangulation will depend on which triangulation will be used. It should be noted that in some applications, it is inappropriate to use DT, as some other triangulations are more appropriate.

Each domain contains added virtual "corner" points. This is a great advantage because the convex hull of domain triangulation will only contain these virtual points (see Fig. 2).

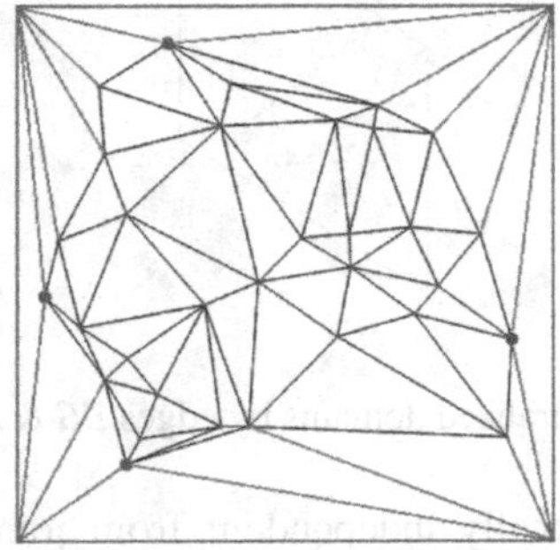

Fig. 2. Domain triangulation (in E^2)

Each domain is triangulated by a selected triangulation method, not necessarily DT. It should be noted that domain triangulations are totally independent and thus can be done in parallel. We have $n \times m$ independent processes in E^2, resp. $n \times m \times p$ in E^3.

2.3 Domains Joining

After domains triangulation, we have $n \times m$ triangulated domains in E^2, or $n \times m \times p$ in E^3, and we have to join them to only one triangulation. The process of joining two domains triangulations is in our case very simple. We only have to swap common edge EF to edge AB (see Fig. 3). Situation in E^3 is similar to situation in E^2.

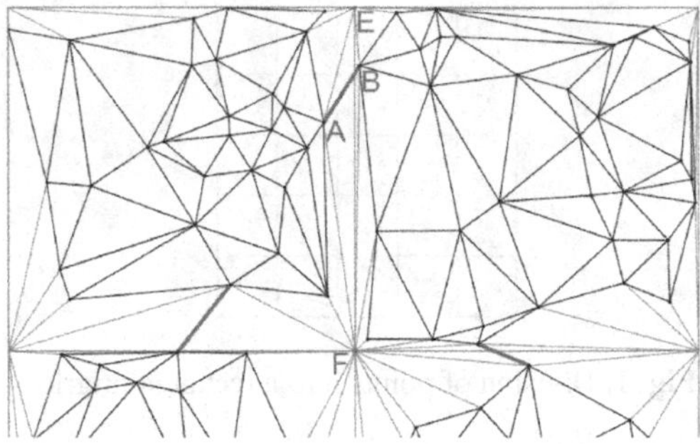

Fig. 3. Joining triangulated domains by edges $EF \rightarrow AB$ swapping

Two domains share one common side with vertices E, F, G and H, and thus we only have to swap edges EG and FH to edge AB (see Fig. 4). It can be seen that the connection of triangulated subsets is extremely simple in the E^2 case. In the E^3 case the situation is straightforward and not complicated as well.

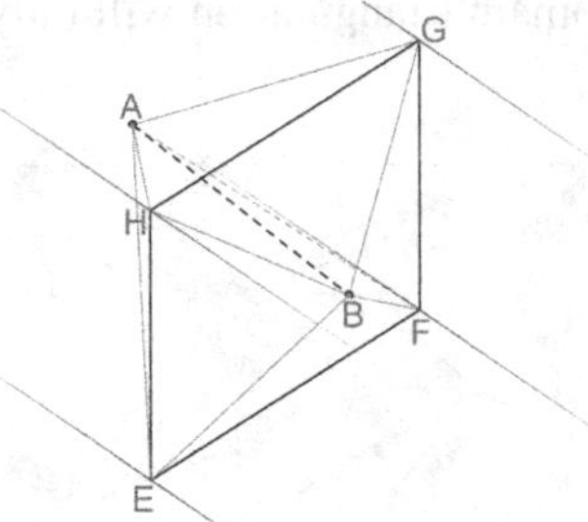

Fig. 4. Joining tetrahedralized domains by edges EG & $FH \rightarrow AB$ swapping

Joining two domains is totally independent from joining another two domains. Therefore, joining of all triangulations to one triangulation can be done in parallel without any conflicts.

2.4 Removing or Retaining of Virtual Corner Points

If the triangulation is used for scalar potential field in E^2 or E^3, or $2\&1/2D$ applications in GIS systems, the value in the virtual corner points can be approximated from the neighbors using Radial Basis Function Interpolation (RBF) [7]. Virtual corner points can be retained in the triangulation and thereby the triangulation is done. Otherwise the corner points have to be removed.

If the corner points have to be removed, there are several algorithms to manage deletion of vertices from triangulation/tetrahedralization [3], [4], [6]. Simply removing a vertex together with its incident simplices leaves a star-shaped hole in the triangulation, which is not necessarily convex. This approach will be described in the next subsection. Another approach is to move the vertex towards its nearest neighbor in several steps; each followed by a sequence of flips restoring the triangulation until the simplices between the two vertices are very flat and can be clipped out of the triangulation [9]. This approach will be described in the second subsection.

The process of removing one virtual corner point from triangulation is totally independent from removing any other virtual corner point. Thus removing of virtual corner points in the middle part of triangulation can be done totally in parallel.

Star-Shape Polygon Re-Triangulation. This algorithm removes a vertex from the triangulation and thus creates a star-shape hole (polygon/polyhedron) which has to be re-triangulated. The polygon can be divided into several parts (see Fig. 5). We have one center part and four "arms" in E^2, resp. six "arms" in E^3.

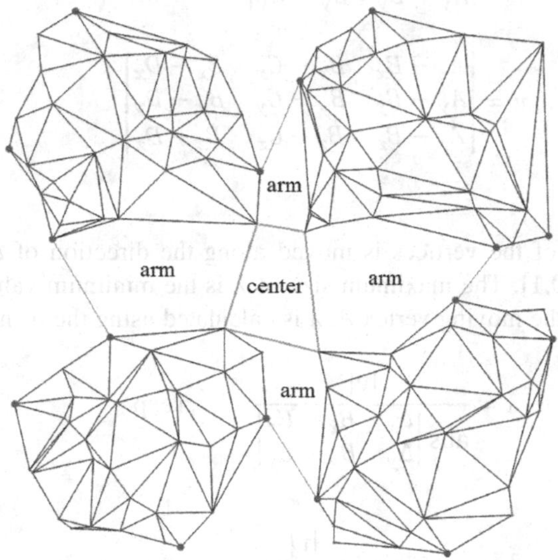

Fig. 5. Star-shape polygon (hole in triangulation)

The center part of the star-shape polygon contains the closest vertex from each surrounding domains. However, the number of vertices is usually four, or eight in E^3; more vertices can be included, e.g. the situation in Fig. 6. The center polygon can be triangulated using ear clipping algorithm, which is of computational complexity $O(N^2)$, but the number of vertices N is very small.

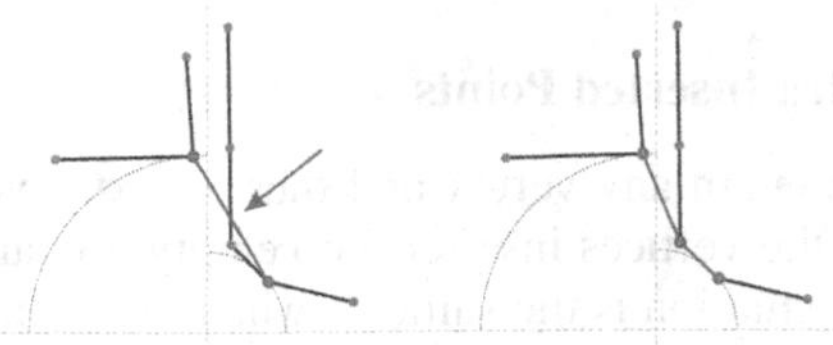

Fig. 6. Intersection of two edges (left) and the solution (right) in E^2

The arms of the star-shape polygon are monotone polygons in respect to axis x or y, resp. x, y or z. Monotone polygon can be triangulated in $O(N)$ time and thus triangulation of the star-shape hole is a really fast process.

Moving and Deleting Vertex. This algorithm moves the virtual corner point towards its nearest neighbor vertex in triangulation ($A \to A'$) [9]. The main question is how far a vertex v can be moved into a certain direction without invalidating the triangulation, i.e. without creating overlapping simplices. We can define the pseudo-orientation of a simplex $S = (A, B, C)$, resp. $S = (A, B, C, D)$, as follows:

$$v = \begin{vmatrix} A_x - B_x & B_x - C_x \\ A_y - B_y & B_y - C_y \end{vmatrix} \quad \text{, resp.}$$

$$v = \begin{vmatrix} A_x - B_x & B_x - C_x & B_x - D_x \\ A_y - B_y & B_y - C_y & B_y - D_y \\ A_z - B_z & B_z - C_z & B_z - D_z \end{vmatrix}. \tag{1}$$

Now suppose one of the vertices is moved along the direction of Δ, i.e. $A \to A' = A + \lambda\Delta$ with $\lambda \in [0,1]$. The maximum size of λ is the minimum value of all λ for all simplices incident the moving vertex A. λ is calculated using the formula:

$$\lambda = \frac{|v|}{\text{abs}\begin{vmatrix} \Delta_x & B_x - C_x \\ \Delta_y & B_y - C_y \end{vmatrix}} \quad \text{, resp.}$$

$$\lambda = \frac{|v|}{\text{abs}\begin{vmatrix} \Delta_x & B_x - C_x & B_x - D_x \\ \Delta_y & B_y - C_y & B_y - D_y \\ \Delta_z & B_z - C_z & B_z - D_z \end{vmatrix}}. \tag{2}$$

If $\lambda \geq 1$, then the vertex can simply be moved along the complete path Δ, whereas if $\lambda < 1$, the vertex A can only be moved by a fraction $\lambda\Delta$ and the triangulation has to be validated using a sequence of flips. After triangulation validation, we have to recalculate parameters λ and repeat the algorithm until vertex A is equal to A'.

2.5 Removing of Extra Inserted Points

Some domains did not contain any vertex and thus an extra vertex was inserted into such each domain. Now the vertices inserted into empty domains have to be removed from triangulation. This situation is the same as when removing virtual corner points. We can use an algorithm for "star shape polygon re-triangulation" or "moving and deleting vertex" algorithm. These algorithms for removing extra inserted "virtual" vertices were presented in the previous section.

2.6 Convex Hull Creation

The union of all simplices forms a convex hull. To create a convex hull of triangulation, we have to remove all virtual corner points at the border of the created grid. The vertices have to be removed and preserve the convex hull. There are many algorithms how to do it. One of them is the ear clipping algorithm. We remove all simplices containing one virtual corner point and then re-triangulate the border. Another way how to do it is to use the approach presented in section "Moving and Deleting a Vertex".

The process of triangulation from input vertices is done after removing all remaining virtual grid points.

2.7 Multiple GPUs or PCs

Today's applications need to process data sets in a short time. Therefore we may use several processors with shared memory, i.e. in parallel processing, or/and on systems using distributed processing, or/and systems using multiple GPUs.

When using several PCs, or/and GPUs, we have to find out how to divide the work and how to join results into one triangulation. Triangulations of domains are totally independent so there is no problem with work distribution. Joining of domains triangulations is, again, totally independent. In the case of retaining virtual corner points in final triangulation, there is no challenge in work distribution between PCs, or/and GPUs. Otherwise in the case of removing virtual corner points, we have to distribute work between PCs, or/and GPUs, according to Fig. 7. Both GPUs need triangulations of yellow domains for removing virtual corner points.

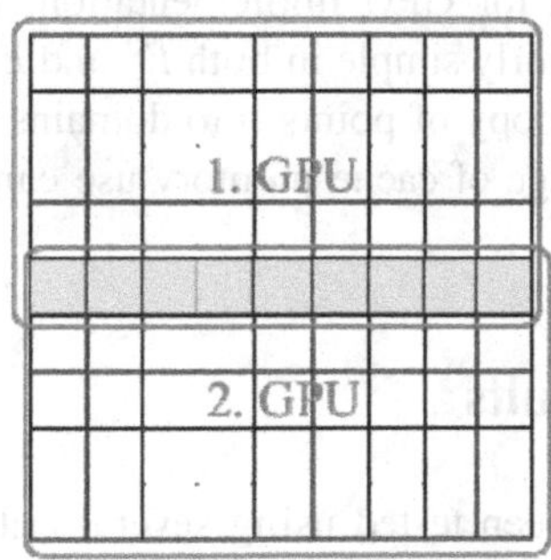

Fig. 7. Distribution of domains per GPU

Division of work between more PCs, or/and GPUs is no problem and can be easily implemented. Triangulation time can be easily reduced while using more PCs, or/and GPUs.

2.8 Large Data Processing

There are many algorithms for triangulation/tetrahedralization and only a few of them can be used for large data processing. The main problem is size of available memory, which is usually no more than tens of gigabytes. Therefore the number of points which can be triangulated/tetrahedralized, is limited by the available memory.

The approach proposed does not have this restriction on the maximal number of points. We can triangulate large data sets which cannot fit at once into the available memory. For one domain triangulation, we do not need any information about other domains. The situation in joining is almost the same. We only need information about two domains which will be joined. And finally, when removing one virtual corner point, we only need information about domains which contain this virtual corner point, i.e. four domains in E^2, or eight domains in E^3.

The input data set can be processed by parts. We can load input data only for some domains, perform parallel triangulation according the approach proposed, and save resulting triangulation/tetrahedralization, in a file. Then we can load data for following domains and perform the same operations. This is a very small change in the approach proposed and is easy to implement. Using this approach, we are able to perform a triangulation/tetrahedralization, on large input data sets with more than 10^7 vertices. The most important feature is that we are not restricted by the limited size of the maximal available memory.

3 Implementation

We implemented the approach proposed in C++ with using OpenMP for parallelization and in CUDA for GPU implementation. The implementation of the approach proposed has been fairly simple in both E^2 and E^3.

It is appropriate to save a copy of points into domains rather than only references to points. Then a full advantage of cache memory use can be taken, and speedup the implementation.

4 Experimental Results

The approach proposed has been tested using several criteria. First of all, we tested the optimal number of points per domain for the purpose of low time requirements. In the second part, we tested time performance of triangulation/tetrahedralization, for a different number of input points. After that, we tested the quality of triangulation/tetrahedralization. Finally, we tested our approach on both synthetic and real data sets.

The approach proposed has been tested on data sets using PC with the following configuration:

- CPU: Intel(R) Core(TM) i7 920 (4 × 2,67GHz) with 8 HyperThreads,
- GPU: 2 × GeForce GTX 295
  - 30 multiprocessors × 8 CUDA Cores per multiprocessor 1,38GHz
  - memory 896MB 1,05GHz
- memory: 12 GB RAM,
- operating system Microsoft Windows 7 64bits

4.1 Number of Points per Domain

The first part in the approach proposed is the division of all vertices into a grid. We need to know what the average number of points per domain is. According to that, we can compute parameters n and m, or n, m and p, to split input vertices into $n \times m$ domains in E^2, or $n \times m \times p$ domains in E^3.

We measured time complexity of triangulation/tetrahedralization, for different numbers of input vertices with uniform distribution and different numbers of points per domain. One example of the time measured for 10^7 points and a different number of points per domain can be seen in Graph 1. It can be seen that with an increasing number of points per domain time complexity decreases. This happens up to an optimal number of points per domain where the time complexity is minimal. From this number of points, the time complexity increases with an increasing number of points per domain.

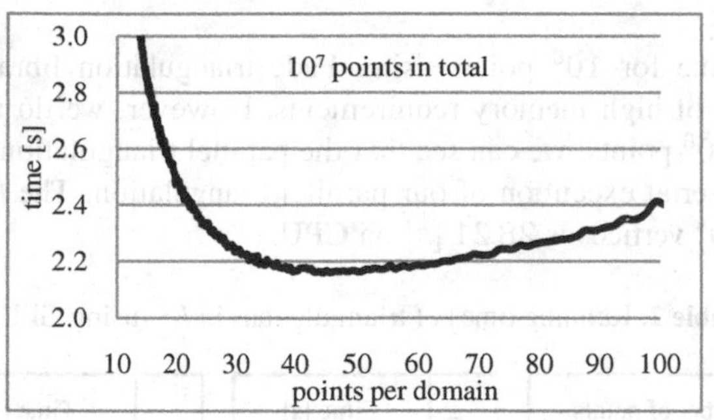

Graph 1. Number of points per domain for 10^7 points (in E^2) with uniform distribution

An optimal number of points per grid depends on the exact implementation of triangulation/tetrahedralization, which is used for domains triangulation. The next factor is the number of threads used during parallel triangulation.

In our case, we used eight hyper-threads and two different implementation of triangulation/tetrahedralization. In the case of using a brutal-force implementation, the optimal number of points per domain is 45 in E^2, or 171 in E^3. In the case of using an optimized implementation, the optimal number of points per domain is 2 000 in E^2, resp. 400 in E^3.

4.2 Time Performance

In some applications, time performance is one of an important criterion. We measured running times for triangulation/tetrahedralization, for different numbers of points with uniform distribution. Running times were measured for 8 threads running and for only 1 thread. The times of 1 thread running, have been compared with running times of publicly available serial library for triangulation called Fade[1], or serial library for tetrahedralization called TetGen[2].

Triangulation. Tab. 1 presents running times of triangulationon on CPU. Running times of triangulation on GPU in comparison with running times of publicly available GPU library GPU DT[3] can be seen in Tab. 2.

Table 1. Running times of triangulations in E^2 (using CPU)

	8 threads (4 cores)		1 thread			
Number of points		Time [s]		Time [s]		Time [s]
316 227	Parallel triangulation	0.06	Parallel triangulation	0.20	Fade library	0.27
1 000 000		0.18		0.67		0.88
3 162 277		0.65		2.23		2.96
10 000 000		2.16		7.33		9.58
31 622 776		7.99		24.88		35.66
100 000 000		28.21		81.94		

The running time for 10^8 points using Fade triangulation library could not be measured because of high memory requirements. However, we do not have time of triangulation for 10^8 points: we can see that the parallel triangulation is always faster, even when using serial execution of our parallel triangulation. The time required for triangulation of 10^8 vertices is $28.21\ [s]$ on CPU.

Table 2. Running times of triangulations in E^2 (using GPU)

Number of points		Time [s]		Time [s]
1 000	GPU parallel triangulation	0.010	GPU DT library	0.149
3 162		0.012		0.173
10 000		0.015		0.186
31 622		0.019		0.260
100 000		0.034		0.317
316 227		0.093		0.620
1 000 000		0.253		1.625

[1] Kornberger, B., Fade2D & Fade2.5D, Geom e.U. Software Development.

[2] Si, H., TetGen: A Quality Tetrahedral Mesh Generator and a 3D Delaunay Triangulator, Weierstrass Institute for Applied Analysis and Stochastics.

[3] GPU-DT: A 2D Delaunay Triangulator using Graphics Hardware, National University of Singapore.

According to the results from Tab. 2, it can be seen that our GPU triangulation is much faster than publicly available library for GPU triangulation called GPU DT. The time required for triangulation of 10^6 vertices is $0.253\,[s]$ on GPU.

Tetrahedralization. Running times of tetrahedralization in comparison with publicly available serial library TetGen can be seen in Tab. 3.

Table 3. Running times of tetrahedralizations in E^3

	8 threads (4 cores)	1 thread	
	Parallel tetrahedraliza.	Parallel tetrahedraliza.	TetGen library
Number of points	Time [s]	Time [s]	Time [s]
100 000	0.29	0.97	1.78
316 227	0.85	2.92	5.75
1 000 000	2.52	8.77	18.97
3 162 277	8.11	28.35	60.60
10 000 000	25.72	88.69	196.00
31 622 776	81.70	278.45	

The running time for $\sqrt{10} \cdot 10^7$ points using TetGen tetrahedralization library could not be measured because of high memory requirements. Although we do not have time of tetrahedralization for $\sqrt{10} \cdot 10^7$ points, we can see that the parallel tetrahedralization is always faster, even when using serial execution of our parallel tetrahedralization, and the speed-up is increasing with the increasing number of input vertices. The time required for tetrahedralization of $\sqrt{10} \cdot 10^7$ vertices is $81.7\,[s]$.

Speed-Up. Using Tab. 1 and Tab. 3, we can calculate the speed-up of parallel triangulation/tetrahedralization when using only one thread, in respect to the publicly available serial library for triangulation called Fade, or for tetrahedralization called TetGen, see Graph 2.

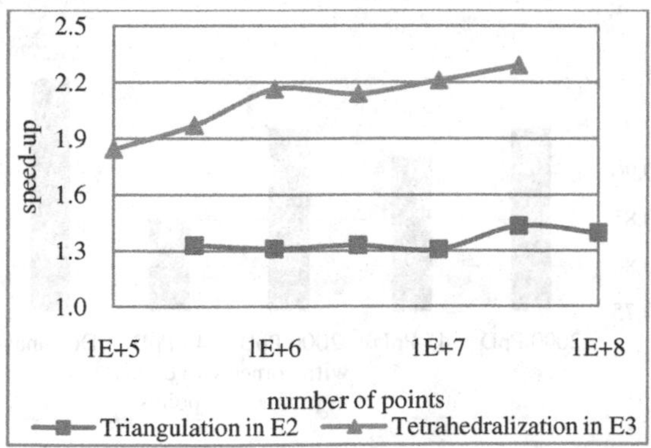

Graph 2. Speed-up of parallel triangulation (using 1 thread) to Fade library and speed-up of parallel tetrahedralization (using 1 thread) to TetGen library

4.3 Triangulation Quality

Delaunay triangulation maximizes the minimal internal angle in triangulation. Therefore, one test of triangulation quality is the distribution of minimal internal angles in triangulation. We measured the internal angle in degrees in E^2 and the internal solid angle in steradians in E^3 (see Graph 3). The data set contained 10^7 uniformly distributed points in E^2/E^3.

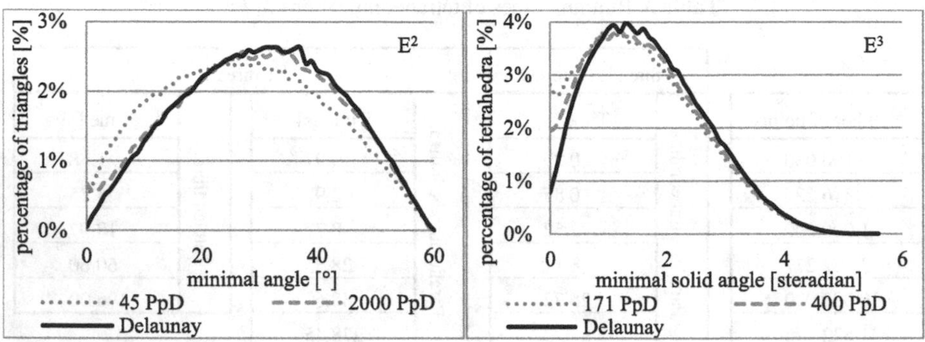

Graph 3. Distribution of minimal internal angles (PpD = Points per Domain) for triangulation/tetrahedralization of 10^7 uniformly distributed points

According to the results, a triangulation/tetrahedralization created with the algorithm proposed is very close to Delaunay triangulation, see Graph 2. Moreover, the inner parts of the domains are Delaunay's. The more points per domain are used, the closer to the Delaunay triangulation the triangulation proposed is.

The Delaunay triangulation maximizes the mean incircle radii. Due to this criterion, we calculated Graph 4. We can see a similar behavior like in Graph 3. The more points per domain are used, the closer to the Delaunay triangulation our triangulation is. If we retain corner points in triangulation, then the quality of triangulation is a bit worse. However, for 2 000 vertices there is almost no difference in the mean inradius for triangulation with and without virtual corner points.

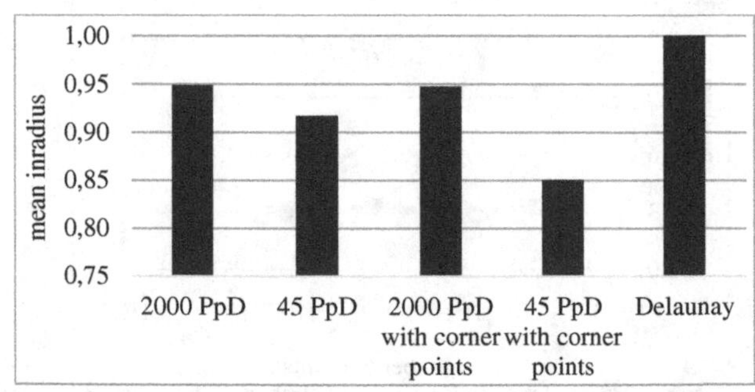

Graph 4. Mean inradius of triangles for different triangulations (mean inradius of Delaunay triangulation was normalized to size 1.0)

4.4 Synthetic and Real Data Sets

In many applications, we do not need to triangulate only uniformly distributed data sets, but real data sets. An example of real data sets may be sets for geographic information system applications. Triangulation of one such set is shown in Fig. 8. It is a GIS data set of South America. The set contains $1.1 \cdot 10^6$ points and the triangulation time of our parallel triangulation proposed was $0.42\ [s]$.

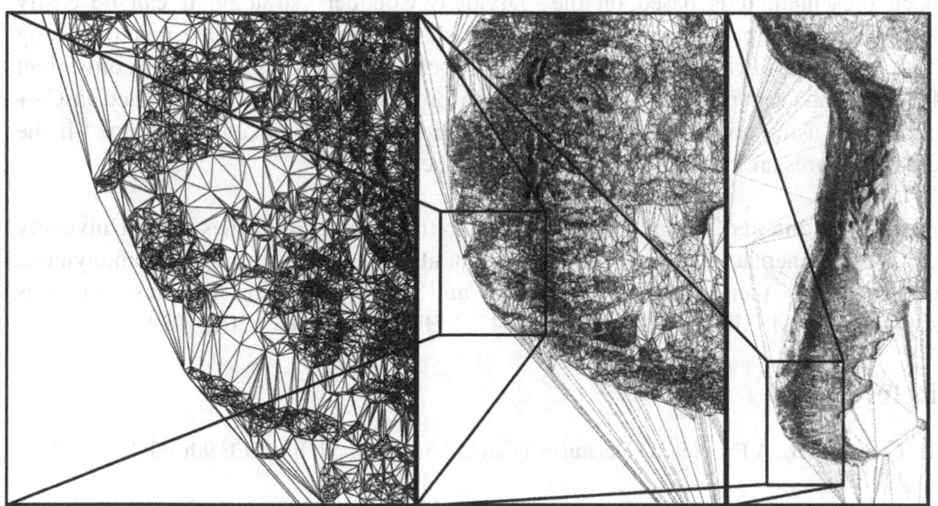

Fig. 8. Triangulation of South Americas GIS data set

We compared the obtained triangulation from Fig. 8 with the Delaunay triangulation of the same data set. We used the distribution of minimal internal angles in triangulation for comparison (see Graph 5).

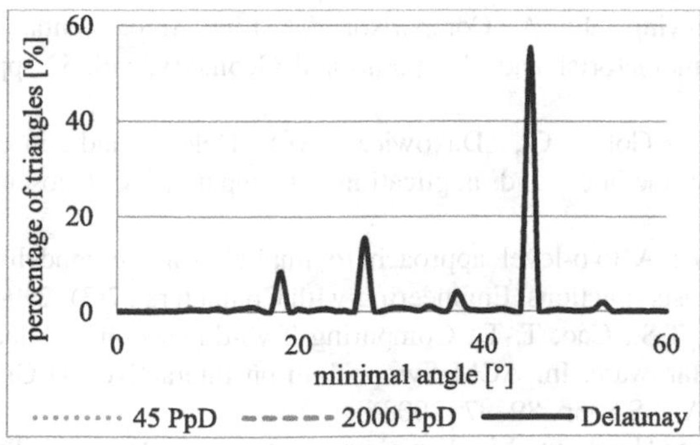

Graph 5. Distribution of minimal internal angle (PpD = Points per Domain)

It can be seen that there is almost no difference and all graphs are overlapped over each other. The highest maxima are for angles $45°$ (edges of the triangle in the proportion of $1:1:\sqrt{2}$), $26.6°$ (edges of the triangle in the proportion of $1:2:\sqrt{5}$) and $18.4°$ (edges of the triangle in the proportion of $1:3:\sqrt{10}$).

5 Conclusion

A new fast and simple to implement parallel triangulation algorithm in E^2 and E^3 has been presented. It is based on the "Divide & Conquer" strategy. It can be easily implemented on parallel environments with shared and/or distributed memory using both CPU and GPU. As it is scalable; the proposed algorithm is especially convenient for large data sets processing. The approach proposed has been implemented in C++ and tested using both CPU and GPU. An additional speed-up can be expected if the data structures are carefully implemented for the given HW.

Acknowledgments. The authors would like to thank their colleagues at the University of West Bohemia, Plzen, for their comments and suggestions, and anonymous reviewers for their valuable comments and hints provided. The research was supported by MSMT CR projects LG13047, LH12181 and SGS 2013-029.

References

1. Chen, M.-B.: A Parallel 3D Delaunay Triangulation Method. In: IEEE 9th ISPA, pp. 52–56 (2011)
2. Cignoni, P., Montani, C., Scopigno, R.: DeWall: A Fast Divide & Conquer Delaunay Triangulation Algorithm in Ed. Computer Aided Design 30(5), 333–341 (1998)
3. Guibas, L., Russel, D.: An Empirical Comparison of Techniques for Updating Delaunay Triangulations. In: ACM Symposium on Computer Geometry, pp. 170–179 (2004)
4. Ledoux, H., Gold, C.M., Baciu, G.: Flipping to Robustly Delete a Vertex in a Delaunay Tetrahedralization. In: Gervasi, O., Gavrilova, M.L., Kumar, V., Laganá, A., Lee, H.P., Mun, Y., Taniar, D., Tan, C.J.K. (eds.) ICCSA 2005. LNCS, vol. 3480, pp. 737–747. Springer, Heidelberg (2005)
5. Liu, Y.-X., Snoeying, J.: A Comparison of Five Implementations 3D Delaunay Tessellations. Combinatorial and Computational Geometry, vol. 52, pp. 439–458. MSRI Publ. (2005)
6. Mostafavi, M.A., Gold, C., Dakowicz, M.: Delete and insert operations in Voronoi/Delaunay methods and applications. Computers & Geosciences 29, 523–530 (2003)
7. Pan, R., Skala, V.: A two-level approach to implicit surface modeling with compactly supported radial basis functions. Engineering with Computers 27(3), 299–307 (2011)
8. Rong, G.D., Tan, T.S., Cao, T.-T.: Computing Two-dimensional Delaunay Triangulation Using Graphics Hardware. In: ACM Symposium on Interactive 3D Graphics and Games, Redwood City, CA, USA, pp. 89–97 (2008)
9. Schaller, G., Meyer-Hermann, M.: Kinetic and Dynamic Delaunay Tetrahedralization in Three Dimensions. Computer Physics Communications 162(1), 9–23 (2004)

A Robust Key Management Scheme Based on Node Hierarchy for Wireless Sensor Networks

A.S.M. Sanwar Hosen, Gideon, and Gi-hwan Cho*

Div. of Computer Science and Engineering,
Chonbuk University, Jeonju, S. Korea
{sanwar,ghcho}@jbnu.ac.kr,
pepsigidy@gmail.com

Abstract. Quality-of-Service (QoS) requirements of a wireless sensor network mainly rely on the reliable and secure communication. To meet these, we have designed a robust key management scheme to ensure security for wireless sensor networks. This scheme presents a hierarchical network topology in terms of key generation and distribution to incorporate secure data routing in-between the source and the destination. Thus, it emphasizes on key management tasks distributed at different levels of node, minimized key storage at each node, even minimized re-key messages for data encryption and decryption in preparing against any node compromised in the network. In the evaluation, our scheme significantly reduces the cost of key generation and distribution, even the computational overhead.

Keywords: Wireless Sensor Network, Hierarchy, Key Management, Security.

1 Introduction

A wireless sensor network (WSN) consists of a large number of micro electro-mechanical systems based tiny sensor devices. A sensor node has several resource constraints such as; limited computational capabilities, lack of battery supply, low memory space, and transmission limitation. A sensor node can be deployed anywhere and worked without human assistance [1]. These features introduce a reliable network design including critical issues like security, network access control, authentication, confidentiality, and compromising node. Here, it is well-known that key management is crucial to the secure operation of WSNs.

The network topology in WSN is the fundamental issue in order to increase the secured data received ratio with minimum network cost. It is generally agreed that the hierarchical data forwarding strategies encourage the different levels of group of sensor nodes, that is, a kind of group communication [2]. When the number of security keys may become quite large in the network, an efficient key management is a non-trivial problem. So it is important to generate the security keys and distribute them

* Corresponding author.

B. Murgante et al. (Eds.): ICCSA 2014, Part II, LNCS 8580, pp. 315–329, 2014.

among the constituted nodes with minimum computational overhead as well as the number of keys stored at each node should be minimized.

Meanwhile, security is the main aspect in the key distribution management. Attacker's main purpose is to either capture the key or the data that is transmitted from the source to the destination. In generally, an attacker can use active, passive and even physical attacks to compromise any node or capture a key in the network [3]. In active attacks, attackers have the ability to change the communication course happens between a sender and the receiver. Attacker can create, forge, alter, replace, block or re-route the messages. Attackers can trick both the sender and the receiver in communicating with them (Man-in-the-middle attacks), both the sender and the receiver think they were communicating between each other. In other case, attacker can modify and replace any message with anything he wants (Rewrite attacks). In passive attacks, the attacker only can eavesdrop the messages sent from a sender to the receiver, but he does not alter any messages. In physical attacks, attacker gain access over the sensor devices to capture the cryptographic keys of neighbor sensors.

In the aspect of the above issues, this work motivated to split the network into clusters. The elected cluster headers (CHs) are the data gathering points among their constituent cluster member (CM) nodes. All CHs are built a group. A group could be a set of sub-groups and each sub-group contains elected CH(s) nodes in a designated forwarding path from a source to a destination. To mitigate the security issues in the network, we are proposing a new key management technique to reduce the number of key stores and re-key messages in the situation of a node compromised in the network. This scheme presents a hierarchical network topology in terms of key generation and distribution to incorporate secure data routing in-between the source and the destination.

The key generation and distribution process are initiated by the base station (BS) and cluster header (CH) node in the network. In this, BS sends the keys to CHs along with individual seed encrypted with the personal keys of CHs in order to communicate with BS. Seeds play a vital role in generating keys for the CM nodes and neighboring CHs for sending data in-between them without interfering with the BS. There is a possibility of key theft during the data communication among the nodes, in this case, we use the re-keying scheme and thus the keys will be periodically changed by the BS as well as CHs to avoid vulnerability of some attacks in the network.

2 Related Works

In recent years, several key management techniques have been proposed. [4] proposed a q-composite pre-distribution, in which permits two sensors to setup a pairwise key only when they share at least q common keys for improving resiliency against node compromised. [5] developed a method name PIKE for key formation by using peer sensor nodes as trusted intermediates. The location-aware schemes which advance the security of the key pre-distribution scheme are proposed in [6][7]. [8] proposed a key management scheme based on matrix clustering, in which each cluster pointed in a matrix and each CH node generated and distributed group keys shared among their member nodes.

SEKM [9] proposed an efficient key management scheme, in which a key distribution server generates a very large size of key pool according to the number of sensor nodes. Before deployment phase, each sensor node has been equipped with a pseudo random function f and a special key K_S and each sensor calculated original key by using its own unique ID and a special key in order to establish pairwise key among nodes. [10] developed a method named NSKM for key management by using three categories of key corresponding to different types of node, namely, super cluster head (SCH), cluster head (CH), and cluster member node (CM). After established the keys, the BS assigned the K_{SC} as a shared key to the CHs and SCH node in order to communicate with BS in the network.

Here, our scheme differs from the previous works as we are proposing a simple and robust key management scheme by introducing a new grouping mechanism. That is, our method is interested to split a large group into some small sub-groups in order to maintain the data routing and the security keys. As a result, the proposed method offers a simple and robust network with minimizing key storage at the constituent node, by simplifying data encryption and decryption and new key generation in terms of node compromised in the network.

3 Network Model

The network model comes mainly from the traditional WSN models. BS is assumed as the destination for receiving sensed data from a source sensor node. Additionally, we can assume that, it could be needed to communicate in-between sensor nodes in the network. Generally, BS is secure and has no resource constraints compared to sensor nodes and it can communicate directly to any deployed sensor node in the network. In order to distribute the security keys to encrypt and decrypt data between sensor nodes and BS or in-between sensor nodes, the network model is based on the following assumptions;

- The entire network partitioned into clusters among the randomly deployed sensor nodes with unique IDs and personal keys K_P.
- Initially, a CH node is elected by the group of CM nodes in each cluster.
- The CH nodes are data gathering point for the constituted nodes in a cluster. They are responsible in the data forwarding process from a source nodes (a group of sensor nodes) to the BS or to communicate with other cluster through the intermediate CH(s) node or directly.
- We assume that an attacker will not be able to compromise any node for a small interval initially after the node is deployed. After this initial interval an attacker might be able to compromise any node.
- The different subset of CHs node is called as a sub-group from a group of CHs node toward the BS or any cluster. The first element (CH node) of a sub-group is the ingress node in the data routing path from a specific cluster to BS or another cluster of internal communication in the network.

- The sub-groups contain a set of CH(s) node could be based on the minimum data routing cost from a source to destination. So, multiple sub-groups could be built towards the BS or different clusters. For instance, let's assume that the data forwarding path contains the CHs; CH_1, CH_2, and CH_3. Then a sub-group will be formed with CH_1, CH_2 and CH_3, named as sub-group_1 that can communicate with each other even direct to BS through gateway CH_1 node, as shown an example of a network in Fig. 1. We assume that a CH node only knows the other CH(s) in a sub-group, whereas, the CMs node aware of their corresponding CH node only.

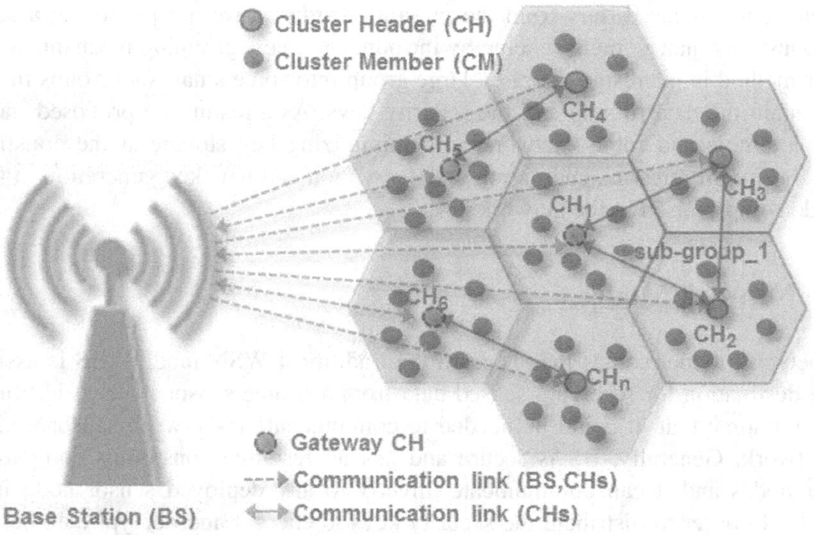

Fig. 1. An example of the proposed network model

- After gathering data at ingress node CH_i of a particular sub-group in order to forward the data to a destination BS or cluster, the ingress CH_i node will push the information of intermediate node(s) of the particular sub-group inside the forwarding packet. This path is called as a designated path. If the link setup between two CHs node in a sub-group is failed, then the CH_i node will select another sub-group (as bypassed path).
- On receiving the information of intermediate node(s) of the routing path/sub-group, the received CH node will select the next minimum routing cost path obtained intermediate CH node for further data forwarding towards any destination either BS or any cluster.

We make use of the following notations to describe our key management scheme and involved cryptographic operation in this paper.

$\{seed_1, seed_2,\ldots,seed_n\} \in S$: number of seeds used by the entire network
$K_{Bs\text{-}m} \in seed_i$: session key used by the BS and all gateway CH(s) node

$K_{B\text{-}Ad\text{-}m} \in seed_i$: administrative key used by the BS and all CHs node
$K_{Cs\text{-}i} \in seed_j$: session key used in sub-group_i
$K_{C\text{-}Ad\text{-}i} \in seed_j$: administration key in sub-group_i
$K_{Cms\text{-}i} \in seed_j$: session key used by all in a cluster
$K_{Cm\text{-}Ad\text{-}i} \in seed_j$: administration key used by CM node and CH_i node
K_P: personal key
$E_K(M)$: encrypt message M by key K
$D_K(M)$: decrypt message M by key K

4 The Proposed Robust Key Management Scheme

To enable a secure transmission, the most common way is through the use of data encryption and decryption. This requires that each authorized member node of a secure cluster or a sub-group has knowledge of a session key for sending data to the BS, shared by the entire cluster or the sub-group. The source node uses this session key for encrypting data packets before sending to the corresponding CH node.

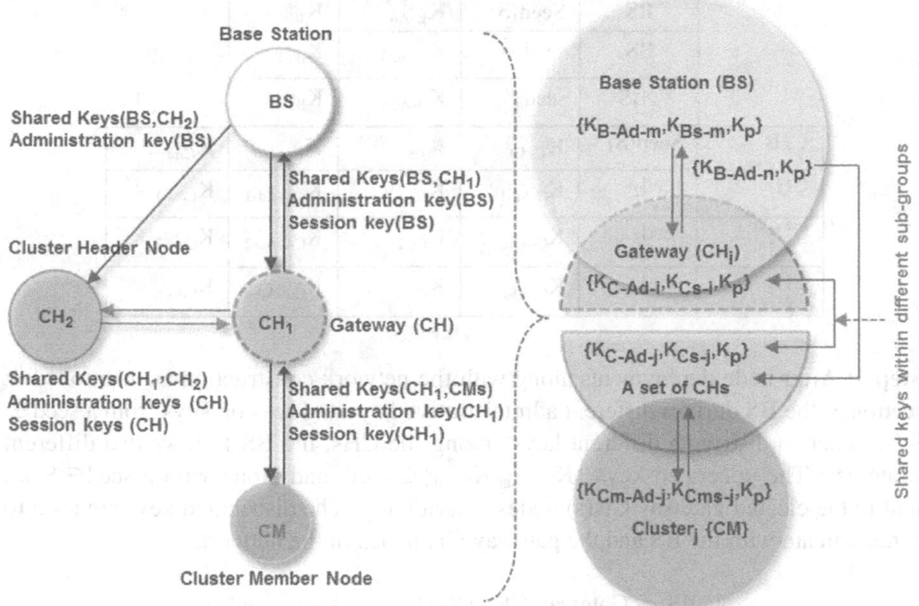

Fig. 2. The different types of key sharing in different node levels

We assume the existing of a key server functioned BS, whose aims to manage session and administration keys used by the gateway CH(s) node and BS. All CH nodes are partially involved in managing the session and administration keys to use within a sub-group and with their corresponding cluster. The administrative keys are used only for re-keying operation that takes place when the membership changes. A personal key is priorly stored at each node, known to their corresponding CH node and BS, respectively.

Fig. 2 shows the required different types of keys sharing in different levels of sub-group in our proposed scheme. The entire network is covered hierarchically in the data routing as well as in the key sharing. In the case, the BS is interlinked with the gateway CHs node of the sub-groups in the secure data routing using secret shared keys (e.g., administration and session key). The CH nodes are interlinked between themselves in a particular sub-group and share the secret keys between them. Consequentially, the CM nodes are connected with their corresponding CH node and used the shared keys in secure data routing between them.

4.1 Key Generation and Distribution

The proposed key generation and the distribution process in different stages in the following Table 1.

Table 1. Key pool of the proposed scheme

	BS	**Seed(S)**	$\mathbf{K_{B\text{-}Ad}}$	$\mathbf{K_{Bs}}$	
	BS	$seed_1$	$K_{B\text{-}Ad\text{-}1}$	$K_{Bs\text{-}1}$	
	BS	$seed_n$	$K_{B\text{-}Ad\text{-}n}$	$K_{Bs\text{-}n}$	
CHs	**Seed(S)**	$\mathbf{K_{C\text{-}Ad}}$	$\mathbf{K_{Cs}}$	$\mathbf{K_{Cm\text{-}Ad}}$	$\mathbf{K_{Cms}}$
CH_1	$seed_2$	$K_{C\text{-}Ad\text{-}1}$	$K_{Cs\text{-}1}$	$K_{Cm\text{-}Ad\text{-}1}$	$K_{Cms\text{-}1}$
CH_2	$seed_3$	$K_{C\text{-}Ad\text{-}2}$	$K_{Cs\text{-}2}$	$K_{Cm\text{-}Ad\text{-}2}$	$K_{Cms\text{-}2}$
CH_n	$seed_n$	$K_{C\text{-}Ad\text{-}n}$	$K_{Cs\text{-}n}$	$K_{Cm\text{-}Ad\text{-}n}$	$K_{Cms\text{-}n}$

Step 1: After node deployments along with the network construction as mentioned in section 3, the BS derives different administrative keys and session keys from a seed$\in$ S. In order to distribute different keys among the CHs, the BS follows two different strategies: The generated keys $\{K_{B\text{-}Ad\text{-}m}, K_{Bs\text{-}m}\} \in seed_i$ and along with a seed$\in$ S are sent to the elected gateway CH(s) nodes individually. The distributed keys are used to communicate with the BS and the gateway CH nodes in the network.

$$BS \rightarrow Gateway\ CH_i: (K_P[K_{B\text{-}Ad\text{-}m}, K_{Bs\text{-}m}, seed_j])$$

The generated keys $\{K_{B\text{-}Ad\text{-}n}\} \in seed_i$ along with the different seeds$\in$ S are sent to the remaining CHs node which are not the gateway node in the network.

$$BS \rightarrow CH_j: (K_P[K_{B\text{-}Ad\text{-}n}, seed_k])$$

Step 2: Each CH node receive the encrypted message and decrypts it using their K_P and store the keys as well as the assigned seed. The received administration and session keys from the BS are used in the control and data message encryption and decryption process between BS and CH(s) node (which are the gateway node) in the network, respectively. The CHs node derive keys from their assigned seed: $\{K_{C\text{-}Ad\text{-}i}, K_{Cs\text{-}i}, K_{Cm\text{-}Ad\text{-}i}, K_{Cms\text{-}i}\} \in seed_j$, where these keys are used for control and the data message encryption and decryption with a particular sub-group and within its own cluster, respectively. Therefore, each CH node derives four different keys from their received seed. Among them, two keys are distributed to the CHs node within a sub-group. Initially, the keys $\{K_{C\text{-}Ad}, K_{Cs}\}$ are used in their priority basis within a sub-group which obtains the different CHs node towards the BS/different clusters. For instance, the generated keys $\{K_{C\text{-}Ad\text{-}i}, K_{Cs\text{-}i}\}$, by the ingress node of a sub-group could be used first based on the negotiation, and then the next CH node in that sub-group and so on.

$$CH_i \text{ -> } CHs \in \text{sub-group\_i}: (K_P[K_{C\text{-}Ad\text{-}i}, K_{Cs\text{-}i}])$$

Step 3: The other two keys $\{K_{Cm\text{-}Ad\text{-}i}, K_{Cms\text{-}i}\}$ generated by a CH_i node are distributed among the CMs node in a particular cluster in order to communicate with CH_i node in the network.

$$CH_i \text{ -> } CMs \in CH_i: (K_P[K_{Cm\text{-}Ad\text{-}i}, K_{Cms\text{-}i}])$$

4.2 Encryption and Decryption Stages

In the secure communication of control and data messages between BS and CMs node as well as in-between CHs node, the encryption and decryption procedures are initiated at different levels of node (e.g., BS, CHs, and CMs node) hierarchically. Let's assume, BS initiates the key generation from a large key pool (S) and distribute to the corresponding CHs node. In the case, the BS encrypts the keys $K_{B\text{-}Ad}$, K_{Bs} and a seed using the individual K_P of gateway CH nodes. On receiving encrypted messages from BS, the gateway CH nodes decrypt the messages using their individual K_P and store the keys $K_{B\text{-}Ad}$, K_{Bs}. The corresponding keys are used to communicate with the BS for sending control and data messages encryption and decryption respectively. On the other hand, the BS generates only the $K_{B\text{-}Ad}$ and sends it to the normal CH node(s) along with an individual seed encrypted by the K_P of the respective CH(s) node. In terms of decrypt the key and the assigned seed from the BS, the normal CH(s) (which are not the gateway nodes) use their individual K_P and store the decrypted keys.

The generated four different keys at CH node: $K_{C\text{-}Ad}$, K_{Cs}, $K_{Cm\text{-}Ad}$, K_{Cms} from the received seed is encrypted using the corresponding K_P of CH(s) node of a particular sub-group and CM nodes of individual cluster respectively. On receiving encrypted messages from the CH node, the corresponding received nodes decrypt it using their individual K_P and store the keys shown in Fig. 3.

In the consequence, after the key establishment, the CM nodes will start to transmit data messages encrypted by K_{Cms} to CH node as an intermediate node towards the BS or any cluster. The CH node decrypts the encrypted message using the shared session key of CMs node. After gathering the messages at a gateway CH node, the CH nodes have transmitted the message to BS encrypted by the shared K_{Bs}. In order to transmit the gathered data to BS/another cluster, the CH node (which is not a gateway node) will choose a designated path (e.g., a sub-group) and forward the message encrypted by the shared K_{Cs} between CH(s) node in that sub-group. This scheme proposes the encryption and decryption in the Fig. 3.

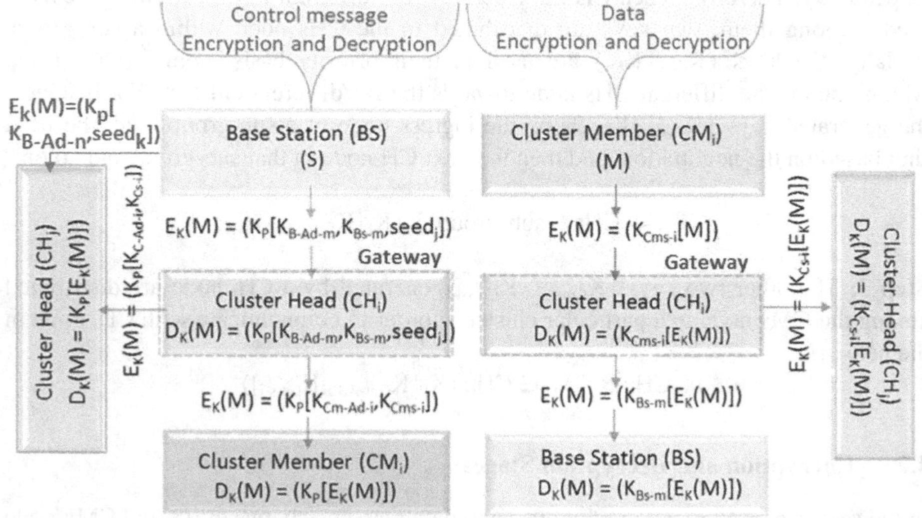

Fig. 3. Data encryption and decryption processes

4.3 Robustness in Terms of Nodes Compromised

In a group communication, if a node of any group is compromised by the adversary, it is required to change their communication keys (session key even though administration key) in order to secure the further data transmission among them. Therefore, it needs the re-keying message with new session/administrative key within that group. For a large group, the number of re-keying message could be high; consequently the network is required to dissipate more additional cost.

Our work makes use of a cluster based data routing as well as a key management scheme, where the entire network splits into clusters; the lower level of nodes form a cluster and the intermediate CHs node built different sub-groups. Different levels of sub-groups use different administration and session keys. Therefore, if a node is compromised, it is needed to change the administration and session key within that particular sub-group only. Because the constituted nodes are involved in a small number of nodes within a sub-group only, this approach is competent to reduce the number of re-keying message in order to change the compromised key(s) in the network.

Our proposed system provides a robust key management scheme that also will mitigate the attacks by the attacker. As soon as the initial key distribution is finished, the BS changes all the personal key K_P of all the nodes. Therefore, the new K_P of all the nodes are only known to the BS and the respective nodes in the network.

The changing of the K_P provides an additional security to the network. For instance, K_P could be used for the authentication in the network construction, reconstruction as well as in the new node joining process.

4.4 The Re-keying Process in Terms of CH Node Compromised

If an attacker compromises any CH node in a cluster, the keys in the CH node for communicating with the corresponding CM nodes and the other CH node(s) (within a sub-group) will be known to the attacker. The attacker then has the ability to listen or change the data. Therefore, when an attacker tries to compromise any CH nodes, the CH node will sends an alert message to the BS. If a CH node has already been compromised by an attacker, then the neighboring CH node(s) (sub-group members) will detect the attack and send an alert message to the BS denoting that the particular CH node is under an attack. If an attacker takes control of a CH node, then there is a possibility that the attacker may know the K_P of the entire CMs node in the cluster.

Once the BS is alerted with an attack of a particular CH node, the BS will checks with the respective CH node and then it will search for the next possible CH node

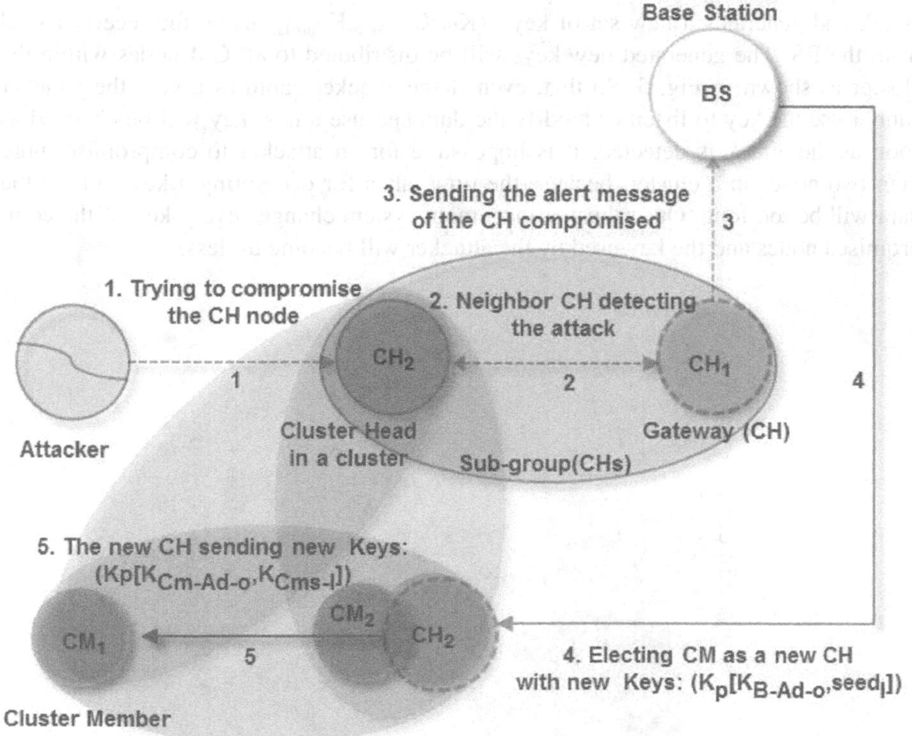

Fig. 4. The re-keying process in terms of CH node compromised

inside the same cluster. It selects a next possible node from a group of CM nodes, suitable to be as a CH node. After a node is selected, the BS will elect the particular CM node as the new CH node. Then the BS responds by sending the re-keys ($K_P[K_{B-Ad-o}$,$seed_l]$) to the newly elected CH node. The re-keys consists of a whole new set of administration, session keys along with a new seed.

From the new seed, the CH node will generate new administration and the session keys ($K_P[K_{Cm-Ad-o}$,$K_{Cms-l}]$) for the corresponding CM nodes. As soon as the new CH is elected, the compromised CH will get expired. The new administration and session keys (for gateway CH node) generated by the BS will be sent to the newly elected CH node. Since the attacker may possibly know the keys of the CMs, the newly elected CH will generate a group of administrative and session keys from the new seed and will distribute all the keys to the respective CH node(s) of the sub-group and corresponding CM nodes shown in Fig. 4. Therefore, the entire CMs node will use a new set of keys, that the attacker cannot compromise any node.

4.5 The Re-keying Process in Terms of CM Node Compromised

If the attacker tries to compromise any CM node in a cluster, the CM node detects the attack and sends an alert to the respective CH node. The CH node will also detect the attack and generates a new set of keys: ($K_P[K_{Cm-Ad-o}$,$K_{Cms-l}]$) using the received seed from the BS. The generated new keys will be distributed to all CM nodes within the cluster as shown in Fig. 5. So that, even if the attacker captures a key, the attacker cannot use the key to listen or modify the data because a new key will be changed as soon as the attack is detected. It is impossible for an attacker to compromise more than two nodes in a cluster, because the time taken for decrypting a key and get the data will be too long. Our robust management system changes every key of the compromised nodes and the key used by the attacker will become useless.

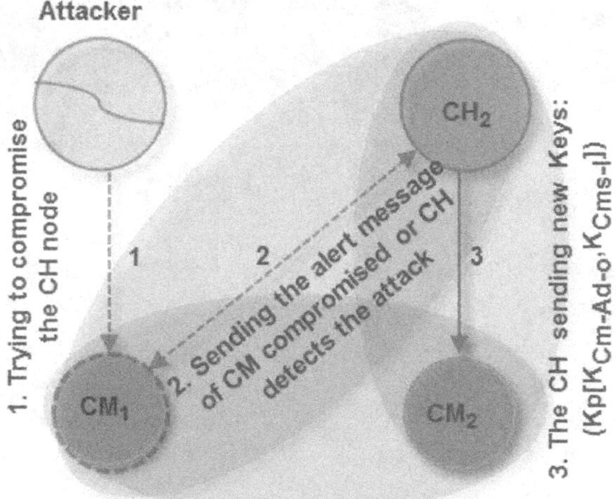

Fig. 5. The re-keying process in terms of CM node compromised

5 Performance Analysis and Discussion

It is well known that the practical aspects of secure WSNs implementation like secure key generation and distribution tasks at every sensor node is not reasonable due to the constraints of the sensor devices. This task should go under some higher capability nodes (i.e., BS, CH node) rather than every node in the networks. Priorly stored personal key at individual nodes are used in the initial key distribution process to decrypt the control message. Thus, the nodes could be feasible to protect against an unknown attacker at the very beginning of the network.

5.1 An Evaluation with Simulation

To evaluate the performance at different issues of this proposed key management scheme, the simulation were carried out in MATLAB 6.5 in a small scale with the following parameters: Processing cost 5mJ, Size of a seed 2000 bits, Size of key 128 bits. By taking the above parameters, the analysis were performed on a network of 180 sensor nodes deployed corresponding a BS on a network area of 800m × 800m. The entire network was split into a number of clusters based on the predefined value 9. Therefore, each individual cluster obtained 19 CM nodes and the sub-groups are formed with at most 9 CH nodes. In order to analysis the performances of this proposed scheme in terms of comparison with other key management schemes, we considered only the processing costs of key overhead as well as number of messages captured in terms of session key compromised at different sessions.

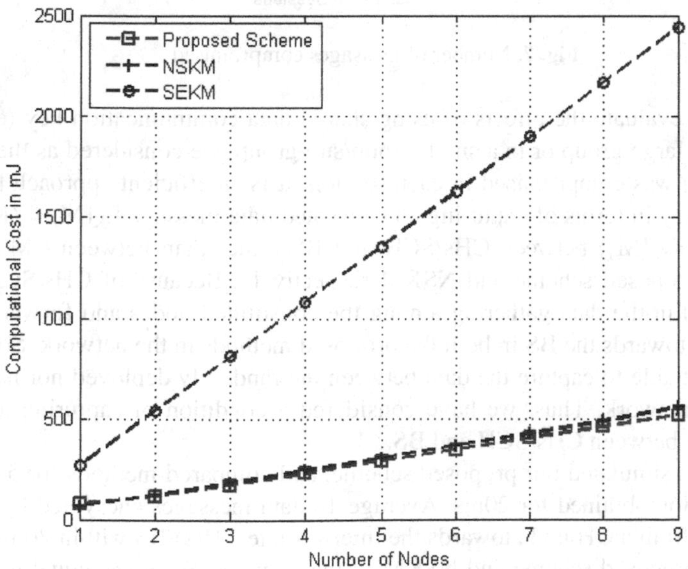

Fig. 6. Computational costs of key generation and distribution

The computational overhead in terms of key generation and distribution depends on the processing of the distributed key size at received node compared with the proposed schemes SEKM [9], NSKM [10]. So, minimizing different types of key and concatenating in the distribution messages could achieve minimized computational overhead at individual nodes shown in Fig. 6. To establish pairwise key in-between nodes could be the reason of the number of keys in the network, and consequentially required high computational cost.

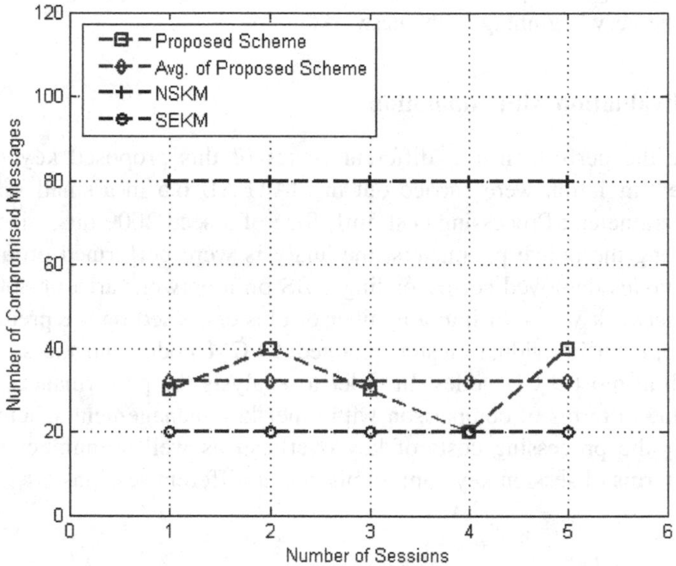

Fig. 7. Number of messages compromised

In order to evaluate the effects of using shared data communication key (e.g., session key) in a large group or in a small group/sub-group, we considered as the session key ($k_{Cs\text{-}i}/K_{SC}$) was compromised in each session. It is an efficient approach to attack on a session key in terms of capturing an encrypted data message $E_K(M) = (K_{Cs\text{-}i}[M])$ or $E_K(M) = (K_{SC}[M])$ between CHs/SCH and BS rather than between CM and CH node in our proposed scheme and NSKM respectively. Because of CHs/SCH nodes were involved in the data gathering among the constituted nodes and forwarding the gathered data towards the BS in both the proposed methods in the network. So that, an attacker might able to capture the data between the randomly deployed normal sensor nodes in the network. Thus, we have considered a condition of capturing a session key traversing between CHs/SCH and BS.

For this, we simulated our proposed scheme, and compared methods for 5 sessions and each session obtained for 20ms. Average 10 data messages encrypted by the session key sent from a CH node towards the intermediate CH(s)/BS within 20ms in both cases of our proposed scheme and NSKM in the network. Whereas, equal number of data messages were sent from a normal sensor node towards the neighboring node in the case of SEKM. We considered the session key $k_{Cs\text{-}i}/K_{SC}$ encryption and decryption messages between CH/SCH and BS in each session was known by the attacker in our proposed scheme and NSKM respectively. In case of our proposed scheme, we considered the session key of different sub-groups at different sessions was known by the attacker where the sub-groups were formed with member CHs node {3,4,3,2,4} respectively. Meanwhile, in SEKM, we assumed the session key (pairwise key) in-between two neighboring nodes were known by an attacker in each session.

In Fig. 7, it is shown that, our proposed scheme outperforms by NSKM, where our scheme significantly reduced the number of captured messages due to the session key captured by the attacker than that of NSKM, because of the session key is shared in a small number of CHs node in a sub-group only. Therefore, the attacker captured all of the sent data messages from the constituted CH nodes in that particular sub-group of that particular session. Whereas, in NSKM, the attacker captured all the messages from the entire constituted CHs node encrypted by the shared session key. But, SEKM performed better than that of NSKM and our proposed scheme because of the pairwise key are established at the most of two neighboring node only, even though the computational and storage over head is high at each node in the network.

5.2 Discussion

As mentioned above, attackers use active attacks to alter or replace the messages (session key $k_{Cs\text{-}i}/K_{SC}$) sent between CHs/SCHs node and the CMs node. The possibility of gaining access to the CHs node in NSKM is very accessible. If a CHs node in NSKM is compromised using the shared key, the whole network will become vulnerable and the attacker can compromise any node using the shared key in the network. The passive attack in NSKM is performed by the attacker by eavesdropping the messages (session key $k_{Cs\text{-}i}/K_{SC}$) exchanged between the CHs node and CMs node to determine the routing protocols and the structure of the network. This information can be used by the attacker in the future to perform any active attacks.

Our proposed scheme outperforms NSKM in both active and passive attacks. During active attacks, our proposed system will detects the attack happened between the CHs/SCHs and the CMs node by either the CHs/SCHs or BS and take the action by changing the new keys as soon as possible to prevent the attacker from compromising further nodes. In the case of passive attacks, the attacker will listen to the messages of a particular node or transaction, in our proposed system BS has the ability to change the role of CHs node from any CMs node in the case of CHs node compromised. Even if the attacker uses passive attacks and listen to messages, our proposed system will elect new CH node and distribute new keys to the new node and thus preventing passive attacks. Our proposed system plays a vital role in the robustness of the WSN network and proves to be high other than the NSKM method preventing from active and passive attacks.

From the point of key management schemes, WSN still have more research opportunities in the future. An effective key management scheme should provide a secure communication, emphasizing on fundamental issues of a well-consolidate WSNs design. Hence, our scheme can be used for solving the above issues.

Table 2. A comparison of the proposed scheme

Security Features	The Proposed Scheme	SEKM [9]	NSKM [10]
Resilience against sensor compromised before key establishment	Yes	Yes	Yes
Key generation and distribution task distributed at	BS, CH	BS, SN	BS, SCH, CH, CM
Encrypted message size of key distribution (avg.)	Medium	Medium	Medium
Key path establishment	Sub-group	Pairwise	Sub-group
Computational overhead	Negligible	High	Negligible
Storage overhead at sensor node	Minimized	High	Minimized
Re-key messages	Minimized	High	High
Robustness	High	Medium	Medium

As a summary, Table 2 shows an informal comparison among our proposed scheme, SEKM, and NSKM in terms of the security features and robustness. The key generation and distribution tasks are disposed to the BS and CH in our scheme, whereas in SEKM it is from BS and sensor nodes, and in NSKM is from BS, super cluster header, CH and CM.

6 Conclusion

In this paper, we provide a robust key management scheme based on the hierarchically defined network topology. With forming different groups based on the node involvement in the routing domain, it is well suited in terms of implementing data security issues to keep the network secure from the adversary. Splitting a large group into sub-groups could be the most efficient approach in order to reduce the number of rekeying messages in terms of a node compromised within a sub-group in the network. We considered the key generation and distribution tasks are disposed to BS and CH nodes. A CH node is selected from the constituted normal sensor nodes, so that the constituted CHs node's energy could be exhausted frequently in terms of key generation and distribution tasks. Therefore, the network needs to recall the reconstruction processes to select other normal sensor nodes as CHs node and it dissipates more additional cost. It is better to use different type of nodes as CH nodes which have no energy constraints along with this proposed scheme for various WSN applications rather than select normal sensor nodes as CH nodes. Meanwhile, an efficient method is required to select a session key among the session keys of CHs in a sub-group in order to communicate within that sub-group. We are going to demonstrate this key management scheme along with the proposed network model with considering other different issues in practice.

Acknowledgement. This research was supported by Basic Science Research Program through the National Research Foundation of Korea (KRF) funded by the Ministry of Education, Science and Technology (2012R1A1A2042035).

References

1. Sakthidharan, G., Chitra, S.: A Survey on Wireless Sensor Network: An Application Perspective. In: International Conference on Computer Communication and Information (ICCCI), pp. 1–5 (2012)
2. Eltoweissy, M., Heydari, M., Morales, L., Sudborough, I.: Combinatorial Optimization of Group Key Management. Journal of Network and Systems Management 12, 33–50 (2004)
3. Kadri, B., Moussaoui, D., Feham, M., Mhammed, A.: An Efficient Key Management Scheme for Hierarchical Wireless Sensor Networks. Wireless Sensor Network 4(6), 155–161 (2012)
4. Chan, H., Perrig, A., Song, D.: Random Key Predistribution Schemes for Sensor networks. In: IEEE Symposium Security and Privacy, pp. 197–213 (2003)
5. Chan, H., Perrig, A.: PIKE: Peer Intermediates for Key Establishment in Sensor Networks. In: IEEE INFOCOM, vol. 1, pp. 524–535 (2005)
6. Liu, D., Ning, P.: Location-based Pairwise Key Establishment for Static Sensor Networks. In: The 1st ACM Workshop on Security of Ad Hoc and Sensor Networks, pp. 72–82 (2003)
7. Zhang, Y., Liu, W., Lou, W., Fang, Y.: Securing Sensor Networks with Location-based Keys. In: IEEE Conferenceon Wireless Communications and Networking, vol. 4, pp. 1909–1914 (2005)
8. Cui, X., Zhang, Y.: A Key Management Scheme Based on Cluster Radiation Matrix in WSN. In: International Conference on Computer Science and Electrical Engineering (ICCSEE), vol. 3, pp. 719–722 (2012)
9. Nanda, R., Tiwari, S., Krishna, V.: Secure and Efficient Key Management Scheme for Wireless Sensor Networks. In: International Conference on Electronics Computer Technology (ICECT), vol. 5, pp. 58–61 (2011)
10. Gawdan, I., Chow, C., Zia, T., Sarhan, Q.: A Novel Secure Key Management Module for Hierarchical Clustering Wireless Sensor Networks. In: The 3rd International Conference on Computational Intelligence, Modelling and Simulation (CIMSiM), pp. 312–316 (2011)

A Method to Triangulate a Set of Points in the Plane*

Taras Agryzkov[1], José L. Oliver[2], Leandro Tortosa[1], and José F. Vicent[1]

[1] Departamento de Ciencia de la Computación e Inteligencia Artificial
Universidad de Alicante
Ap. Correos 99, E–03080, Alicante, Spain
[2] Departamento de Expresión Grafica y Cartografia
Universidad de Alicante
Ap. Correos 99, E–03080, Alicante, Spain

Abstract. Given a set of points S in the plane, we propose a triangulation process to construct a triangulation for the set S. Triangulating scattered point-sets is a a very important problem in computational geometry; hence the importance to develop new efficient algorithms and models to triangulate point-sets. This process presents two well distinguished phases; Phase 1 begins with the construction of an auxiliar triangular grid containing all the points. This auxiliar mesh will help us in the triangulation process, since we take this mesh as a reference to determine the points that may be joined by edges to construct triangles that will constitute the triangulation. Phase 2 takes as a starting point the auxiliar triangular mesh obtained in Phase 1. The objective now is to determine which of the vertices of the initial set S must be joined to form the triangles that will constitute the triangulation of the points. Some examples are shown in detail to understand the behaviour of the triangulation process.

Keywords: Triangulation, grid, lattice, Delaunay triangulation, mesh.

1 Introduction

We can say, in a simple way, that a triangulation is a decomposition into triangles (see [7,13,16]). We are interested in triangulations of point sets in the Euclidean plane, where the input is a set of points in the plane, denoted by S, and a triangulation is defined as a subdivision of the convex hull of S into triangles whose vertices are the points in S.

It is a well-known fact that n points in the plane can have many different triangulations. Three important fields in which triangulations are frequently used are finite-element methods, terrain modeling and social science research. In the first case, triangulations are used to subdivide a complex domain by creating a mesh of simple elements (triangles), over which a system of differential equations

* Partially supported by University of Alicante grant GRE10-34 and Generalitat Valenciana grant GV/2012/111.

B. Murgante et al. (Eds.): ICCSA 2014, Part II, LNCS 8580, pp. 330–341, 2014.

can be solved more easily (see [1,9,10]). In the second case, the points represent sampled from a terrain, and the triangulation provides a bivariate interpolating surface, providing an elevation model of the terrain (see [5,17]). In the third case the concept of triangulation refers to a process by which a researcher wants to verify a finding by showing that independent measures of it agree with or, at least, do not contradict it ([14,15]).

In all these cases, the shapes of the triangles can have serious consequences on the result. For example, in finite-element methods, the aspect ratio of the triangles is particularly important, since elements of large aspect ratio can lead to poorly conditioned systems (see [2]).

In most applications, the need for well-shaped triangulations is usually addressed by using the Delaunay triangulation (see [3,4,6,8,11,12]). The Delaunay triangulation of a point set S is defined as a triangulation in which the vertices are the points in S and the circumcircle of each triangle (that is, the circle defined by the three vertices of each triangle) contains no other point from S. The Delaunay triangulation has many known geometric properties and there exists an extense bibliography with several efficient and relatively simple algorithms to compute it. Intuitively, we can say that its triangles are considered well shaped, that is, they maximizes the minimum angle among all triangle angles, which implies that its angles are, in a sense, as large as possible.

2 Generation of a Triangular Mesh from a Set of Points

The primary objective that we set is the generation of a flat triangular mesh from a set of points S in the plane. That mesh, consisting of vertices and edges forming triangles, must have two main characteristics: may contain a number of points (vertices) greater than the initial set and its convex hull must contain all the points of the initial set.

In this process, we begin constructing a lattice graph; more exactly, we will draw a square grid graph and, afterwards, we will create from this grid a triangular grid graph that will be used as a starting point in the triangulation process. For the construction of this mesh we start from the cloud points. Let S be a set of n points in the plane $S = \{P_1, P_2, \ldots, P_n\}$, whose coordinates are given by $P_i = (x_i, y_i)$, for $i = 1, 2, \ldots, n$.

2.1 Construction of a Square Grid from the Set of Points

The first step is to construct a square grid graph containing l vertices or nodes, such that $l \approx 2n$; that is, the number of vertices of the new mesh should be, approximately, twice the number of points of S. To carry out this task we are going to construct a lattice grid in a process that can be summarized as follows:

- We start computing the maximum and minimum values of the coordinates (x_i, y_i), for the points P_i, for $i = 1, 2, \ldots, n$. Using these parameters a rectangle must be created with all the points in its interior.

- To be sure that the rectangle does not have points on the border we can expand the width and height.
- For the creation of the lattice, it is necessary to divide the rectangle into parts equal in height and in width, considering that the number of divisions must be a natural number. We will carry out appropriate rounding.
- After obtaining the divisions in length and width of the rectangle, the lattice is set by assigning one point to the different intersections.

The process described above can be implemented by the following algorithm.

ALGORITHM 1: *An algorithm to create a mesh from a set of points in the plane.*

INIT: We start with an initial set S of n points in the plane $P_i = (x_i, y_i)$, for $i = 1, 2, \ldots, n$. Let us assume that the coordinates x_i and y_i are positive (otherwise a translation in the plane can be applied).

OUTPUT: We obtain a lattice with l points, where $l \approx 2n$.

We perform the following steps:

Step 1. We compute the parameters $maxx = \max x_i$, $maxy = \max y_i$, $minx = \min x_i$, $miny = \min y_i$, for $i = 1, 2, \ldots, n$.

Step 2. We determine the rectangle that contains the points of the new mesh.

1. Take $nmaxx = maxx + 0.1 \cdot maxx$ and $nmaxy = maxy + 0.1 \cdot maxy$.
2. If $minx - 0.5 \cdot minx < 0$, then $nminx = 0$.
   Otherwise, $nminx = minx - 0.5 \cdot minx$.
3. If $miny - 0.5 \cdot miny < 0$, then $nminy = 0$.
   Otherwise, $nminy = miny - 0.5 \cdot miny$.
4. Width of the rectangle: $distx = nmaxx - nminx$.
5. Height of the rectangle: $disty = nmaxy - nminy$.

Therefore, the dimensions of the rectangle containing the n points P_i are $distx \times disty$.

Step 3. We create the grid in the rectangle.

We determine the distance d as

$$d = \sqrt{\frac{distx \cdot disty}{2n}}. \tag{1}$$

Then, the number of divisions of the grid is:

$$d_x = \lceil \frac{distx}{d} \rceil, \quad d_y = \lceil \frac{disty}{d} \rceil, \tag{2}$$

where d_x and d_y are the number of divisions of $distx$ and $disty$, respectively.

Step 4. We create a lattice $L_{d_x \times d_y}$, according to the values d_x and d_y obtained in the step before. Each of the intersections of the lattice represent a point in the new mesh. So, the new mesh has $l = d_x \times d_y$ points. The points of the lattice are K_{ij}, for $i = 1, 2, \ldots, d_x$ and $j = 1, 2, \ldots, d_y$.

Algorithm 1 allows us to construct an auxiliar square grid consisting of l points that we are going to denote by K_{ij}, for $i = 1, 2, \ldots, d_x$ and $j = 1, 2, \ldots, d_y$.

Remark that we take the origin of this grid at point $K_{11} = (nminx, nminy)$ located at the lower left corner. Using this notation, we can set the coordinates of each of the l points of the grid, according to the parameter d. In general,

$$K_{ij} = (nminx + d(j-1), nminy + d(i-1)).$$

In Figure 1 we have represented a square grid with the coordinates of the vertices in the corners of the rectangle, as well as the parameters *distx*, *disty* and d involved in its construction. We have also represented a generic point K_{ij}.

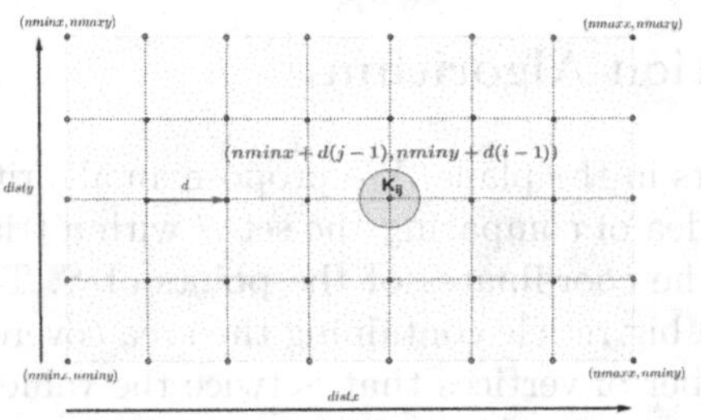

Fig. 1. The rectangular grid and the parameters involved in its construction

Note that the square grid is contained in the rectangle of dimensions $distx \times disty$ obtained in Step 2 of Algorithm 1.

When running Algorithm 1, we obtain a square grid (see Figure 1). However, our objective was to create a triangular mesh from a set of points; consequently, it is necessary to find a method of decomposition of a quadrilateral in triangular elements. This method is simple; it is enough to create an edge along a diagonal of the square and divide it into two triangles. That is, we create $(d_x - 1)(d_y - 1)$ edges joining K_{ij} with K_{i-1j-1}, for $i = 2, 3, \ldots, d_x$ and $j = 2, 3, \ldots, d_y$, (see Figure 2).

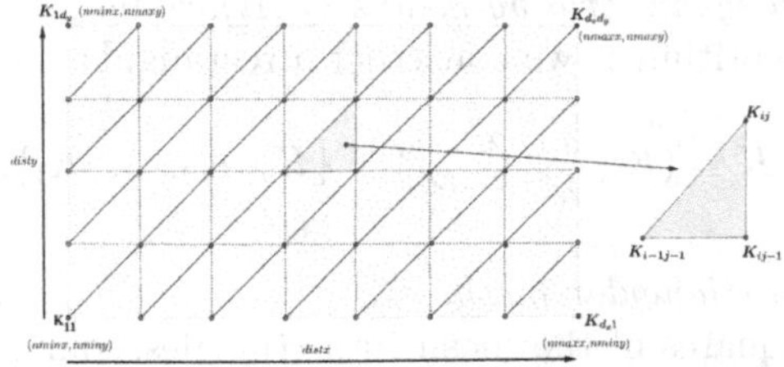

Fig. 2. The triangular lattice created from the square grid

As a consequence of this, if the dimensions of the square grid are such that the number of divisions is $d_x \times d_y$, then the triangle mesh constructed this way has the following characteristics: the number of nodes is $d_x \times d_y$, the number of horizontal edges is $(d_x - 1)d_y$, the number of vertical edges is $d_x(d_y - 1)$, and the number of diagonal edges is $(d_x - 1)(d_y - 1)$. Finally, the number of triangles is $2(d_x - 1)(d_y - 1)$.

Computing the parameters d, d_x and d_y by expressions (1) and (2), we obtain a number of points in the triangular mesh that is twice the number of points in S. We can perform a numerical simulation if we consider a rectangle with fixed dimensions and check the construction of the lattice varying the number of points n. We fix the parameters $distx$ and $disty$, and compute the parameters d, d_x, d_y and l for different values of n, using expressions (1) and (2).

3 The Triangulation Algorithm

Let S be a set of n points in the plane. We propose an algorithm to triangulate the set S based on the idea of comparing the set S with a triangular mesh that can be generated from the coordinates of the points of S. The new mesh may be seen (intuitively) as a big reticle containing the area covered by the points of S and containing a number of vertices that is twice the value of n.

The algorithm consists of two phases.

Phase 1. The first phase of the algorithm consists on the construction of an auxiliar triangular mesh by running Algorithm 1, which can be stated as follow: subdivide a bounding box of the points into squares, then divide each square into two triangles.

Phase 2. The second phase consists on a triangulation process based on the implementation of an algorithm that compares the mesh obtained in Phase 1 with the original set of points S, trying to determine the points that must be joined to construct the different triangles in the triangulation.

Phase 1. *Construction of an auxiliar triangular mesh.*

Let S be a set of n points in the plane $S = \{P_1, P_2, \ldots, P_n\}$, whose coordinates are given by $P_i = (x_i, y_i)$, for $i = 1, 2, \ldots, n$.

1. *Construction of a square grid by means of Algorithm 1.*
   After running Algorithm 1, we construct a rectangular mesh whose vertices are
   $$K = (K_{ij})_{i=1,2,\ldots,d_x}^{j=1,2,\ldots,d_y} = \{K_1, K_2, \ldots, K_l\},$$
   where $l \approx 2n$.
2. *Construction of a triangular mesh.*
   Decompose the squares of the mesh into triangles, joining the vertices K_{ij} and K_{i-1j-1} by edges. Therefore, a triangular mesh of $r = 2(d_x - 1)(d_y - 1)$ triangles is created. We denote this set of triangles $T = \{T_1, T_2, \ldots, T_r\}$. With the aim to simplify the notation, we say that the triangle T_i is given by its three vertices, that is, we represent a triangle as $T_i = \{K_{i1}, K_{i2}, K_{i3}\}$. It is noteworthy that this is only a criterion of notation to facilitate comprehension of the algorithm. When we write the subscripts $\{i1, i2, i3\}$ we do not mean or indicate that that the three vertices are in the same row. Just

what we must interpret is that $\{K_{i1}, K_{i2}, K_{i3}\} \in K$ are the vertices of the triangle T_i.

Phase 2. *Triangulation process*

To determine the triangles of the triangulation, we run the following algorithm:

ALGORITHM 2: *An algorithm to triangulate a set of points in the plane.*

INIT: Let S be the set of points; consider the set K of the l nodes or points of the triangle mesh generated by Algorithm 1, and T the triangles of the triangle mesh.

1. Associate each node of the triangular mesh with a point of the set S. For every node K_i, for $i = 1, 2, \ldots, l$, find $j \in \{1, 2, \ldots, n\}$ such that
$$d(K_i, P_j) \leq d(K_i, P_s), \text{ with } P_s \in S,$$
where $d(K_i, P_j)$ represents the euclidean distance from K_i to P_j. Save (K_i, P_j). We say that P_j is the node associated to K_i.
2. Change the nodes of the original triangles by their associated nodes. For every $T_i = \{K_{i_1}, K_{i_2}, K_{i_3}\} \in T$, substitute
$$\{K_{i_1}, K_{i_2}, K_{i_3}\} \longrightarrow \{P_{j_1}, P_{j_2}, P_{j_3}\},$$
where $P_{j_1}, P_{j_2}, P_{j_3}$ are the associated nodes of $K_{i_1}, K_{i_2}, K_{i_3}$, respectively.
   - If $P_{j_1} \neq P_{j_2} \neq P_{j_3}$, then save T_i.
   - If $P_{j_1} = P_{j_2}$, or $P_{j_1} = P_{j_3}$, or $P_{j_2} = P_{j_3}$, then continue.
3. Graph the set
$$\mathcal{C} = \{T_i = \{P_{j_1}, P_{j_2}, P_{j_3}\}, \text{ with } P_{j_1} \neq P_{j_2} \neq P_{j_3}\}.$$
4. After graphing the set $\mathcal{C}$ check the following conditions:
   - If some node is isolated we add new triangles, linking this node with their adjacent nodes.
   - Let T_i be a triangle from $\mathcal{C}$. If $T_i \bigcap T_j \neq \Phi$, for any $T_j \in \mathcal{C}$, then we proceed to eliminate T_i from the triangulation.

Broadly speaking, the reconstruction algorithm consists of 3 steps. First, to identify each of the vertices of the auxiliar triangular mesh with a point of S according to its proximity. We will refer to it as its representative. After that, we create a list of triangles whose vertices have different representatives; finally, we draw the triangles of the list (we draw edges between representatives), removing the triangles intersecting any other.

The idea underlying our approach is to establish a special equivalence relation between the original set of points and the new triangular mesh constructed from Algorithm 1. The equivalence relation consists on assigning to every node of the triangular mesh K_i a point of the set S which is nearest to it. After this association, we will have some equivalence sets because some nodes of the auxiliar triangular mesh will have as a representative the same node of the set of points S (remember that the cardinal of S is, approximately, half the number of points in the auxiliar mesh). After this concordance process, we only have to determine the triangles with different representatives to determine the nodes of the triangles that must be linked to graph the final triangulation.

The last item of the algorithm has a clear objective: to avoid the possibility to create some holes or unconnected regions in the final triangulation due to some special or rare configurations of the n points of S. When we say that a vertex is isolated we mean that only two edges leave from it. This is avoided connecting this vertex with the adjacent vertices to it.

Obviously there are many efficient algorithms to triangulate point-sets. One of the most interesting aspects of this model is the low computational cost required for its implementation. Let us note that the only step of the whole model in which arithmetic operations are required is in the triangulation algorithm. More exactly, at the beginning of the triangulation process, when the association between the nodes of the triangular mesh and the points of the original set of points is performed.

4 Some Examples of Triangulations

To assess the performance of the triangulation process described by means of Algorithm 2, we are going to see an example in detail. For this example we take $S = \{P_1, P_2, \dots, P_{30}\}$, where

$P_1 = (1,6)$, $P_2 = (6,3)$, $P_3 = (12,7)$, $P_4 = (18,12)$, $P_5 = (8,5)$, $P_6 = (3,11)$, $P_7 = (12,4)$, $P_8 = (15,3)$, $P_9 = (19,8)$, $P_{10} = (8,13)$, $P_{11} = (4,9)$, $P_{12} = (7,9)$, $P_{13} = (16,12)$, $P_{14} = (11,6)$, $P_{15} = (13,10)$, $P_{16} = (18,4)$, $P_{17} = (4,4)$, $P_{18} = (14,14)$, $P_{19} = (10,10)$, $P_{20} = (17,7)$, $P_{21} = (20,5)$, $P_{22} = (2,5)$, $P_{23} = (12,3)$, $P_{24} = (8,8)$, $P_{25} = (18,10)$, $P_{26} = (6,7)$, $P_{27} = (12,14)$, $P_{28} = (10,2)$, $P_{29} = (6,12)$, $P_{30} = (16,5)$.

According to the process described for the triangulation, we perform the two phases.

Phase 1. *Construction of an auxiliar triangular mesh.*

We construct an auxiliar triangular mesh from the set of points S. The first part of Algorithm 1 consists on creating a rectangular grid that contains the points of S. Then, $\max x_i = 20$, $\min x_i = 1$, $\max y_i = 14$, $\min x_i = 2$, therefore, $nmaxx = 22$, $nmaxy = 15.4$, $nminx = 0.9$, $nminy = 1.8$. Then,

$$distx = 21.1, \quad disty = 13.6.$$

In Figure 3(a) we have represented the rectangle containing the 30 points of S. Now, according to the step 3 in the algorithm, we create the grid from the dimensions of the rectangle. The distance d given by (1) is

$$d = \lceil \sqrt{\frac{distx \cdot disty}{2n}} \rceil = 2.19.$$

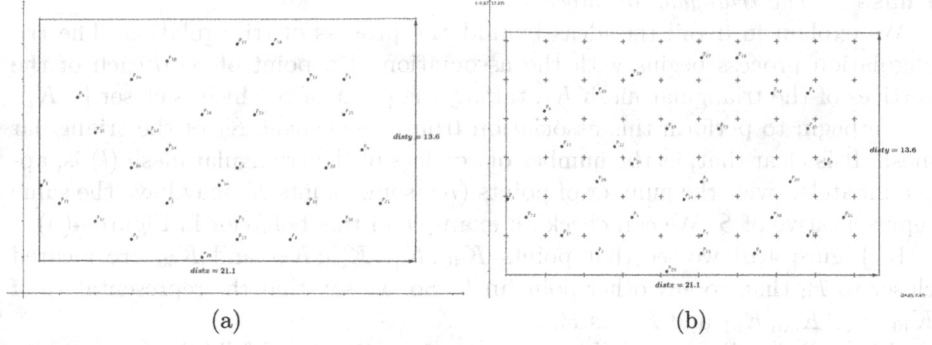

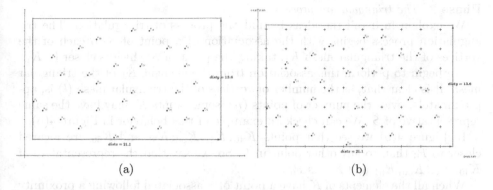

(a) (b)

Fig. 3. The image (a) represents the points of S in 2D with the rectangle containing them. In the right side (b) we superpose the grid constructed from the values of d, d_x and d_y.

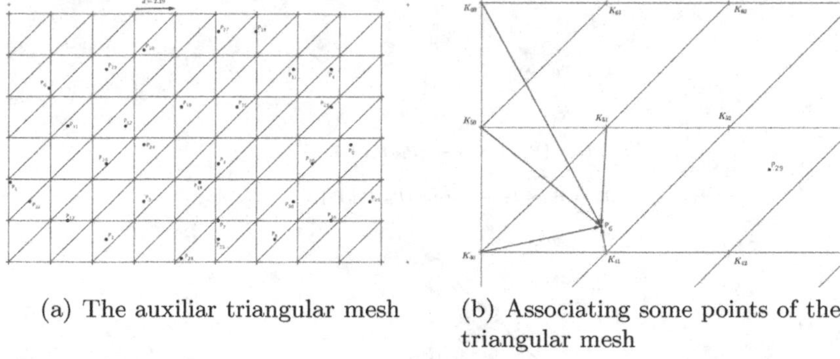

(a) The auxiliar triangular mesh

(b) Associating some points of the triangular mesh

Fig. 4. Some images of the example we are running

Then, from (2),

$$d_x = \lceil \frac{distx}{d} \rceil = 10, \quad d_y = \lceil \frac{disty}{d} \rceil = 7.$$

As a consequence of this, we obtain a lattice $L_{d_x \times d_y}$ with $d_x . d_y = 70$ points located in the intersections of the grid, as we can see in Figure 3(b).

The next step consists on generating a triangular mesh from the lattice. Joining the vertices of the lattice in an appropriate manner, a triangular mesh is obtained as we see in Figure 4(a).

As a result of the application of Phase 1 of the triangulation process, we have a triangular mesh with 70 points

$$K = (K_{ij})_{i=1,2,\ldots,10}^{j=1,2,\ldots,7} = K = \{K_1, K_2, \ldots K_{70}\}.$$

The number of triangles r is $r = 2(d_x - 1)(d_y - 1) = 108.$

Phase 2. *The triangulation process*

We explain in detail the ideas behind the process of triangulation. The triangulation process begins with the association of a point of S to each of the vertices of the triangular mesh K_i, taking the point of S which is closer to K_i.

We begin to perform this association from the element K_1 of the triangular mesh. It is clear that, as the number of vertices of the triangular mesh (l) is, approximately, twice the number of points (n), some points K_i may have the same representative of S. We can check an example of this behavior in Figure 4(b).

In Figure 4(b) we see that points $K_{40}, K_{41}, K_{50}, K_{51}$ and K_{60} are located closer to P_6 than to any other point in S. So, we say that the representative of $K_{40}, K_{41}, K_{50}, K_{51}$ and K_{60} is P_6.

When all the elements of K have a point of S associated following a proximity criterion, we proceed to check the vertices of the set of triangles T. When we find a triangle whose vertices have different representatives, we store it in order to be drawn afterwards. Once we have a list with all the triangles verifying this condition, we just only have to draw the triangles joining the representatives.

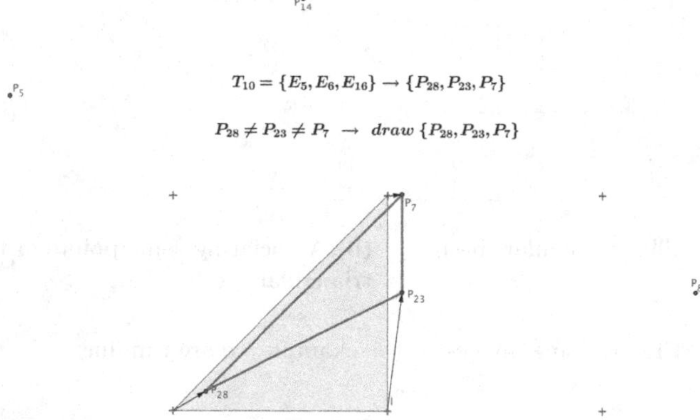

Fig. 5. A triangle whose vertices have different representatives

In Figure 5 we show a triangle in the auxiliar mesh with the property that each of the vertices has a different representative, so we have to draw a triangle with vertices P_7, P_{23}, P_{28}. This triangle will be part of the triangulation. All the triangles with this characteristic must be stored and represented in the last part of the algorithm.

We have completed the triangulation process as it can be seen in Figure 6. On the left (image (a)) we see a graphical representation of all the triangles that must be drawn, that is, the triangles of $\mathcal{C}$. In the image on the right (b), we have represented the triangles to complete the triangulation.

In this example, we only have one isolated vertex (P_2; therefore we create an edge from this vertex to the adjacent ones, creating the triangle P_2, P_5, P_{28}.

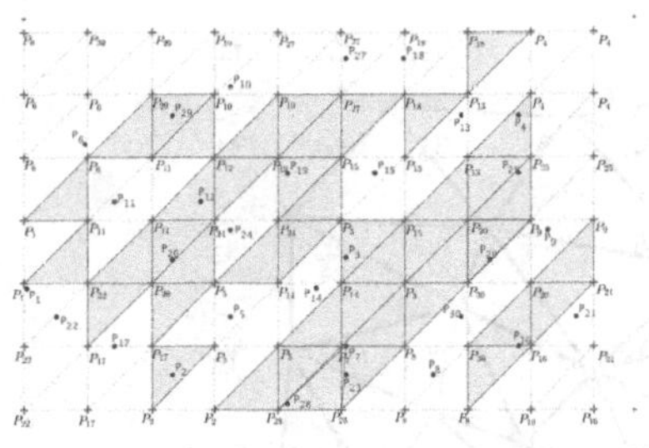

(a) The set of triangles $\mathcal{C}$

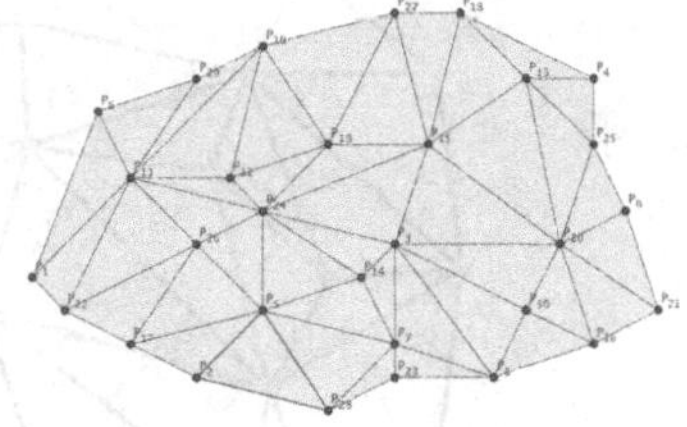

(b) The triangulation of the points

Fig. 6. Some images of the example we are running

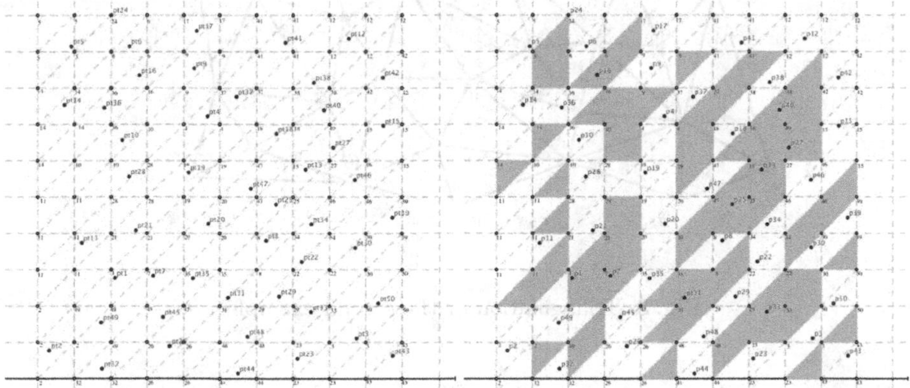

(a) The auxiliar triangular mesh and the set of points S

(b) The set $\mathcal{C}$ of the triangles that must be drawn

Fig. 7. An example with 50 points in the plane

Moreover, it is necessary to remove two triangles that intersect with other ones in the triangulation. More exactly, the triangles with vertices (P_3, P_{14}, P_{15}) and (P_9, P_{20}, P_{30}) must be removed since they intersect with some others triangles. This is a consequence of a fact that sometimes occurs as it is that for some configurations it is difficult to determine the representative of a node in the triangular mesh because the distances are very similar.

We visualize in Figure 7 another example of a triangulation for a set S of 50 points in the plane. Following a similar reasoning as in the previous example studied, we create an auxiliar triangular mesh from the coordinates of the points in S. Now, we have that $d_x = 10$ and $d_y = 10$. We create a triangular lattice $L_{d_x \times d_y} = L_{10\times 10}$ with 100 vertices and containing all the points in S. Joining the vertices diagonally, an auxiliar triangular mesh is created with vertices $K = \{K_1, K_2, \ldots, K_{100}\}$ and triangles $T = \{T_1, T_2, \ldots, T_{162}\}$ (see Figure 7(a)).

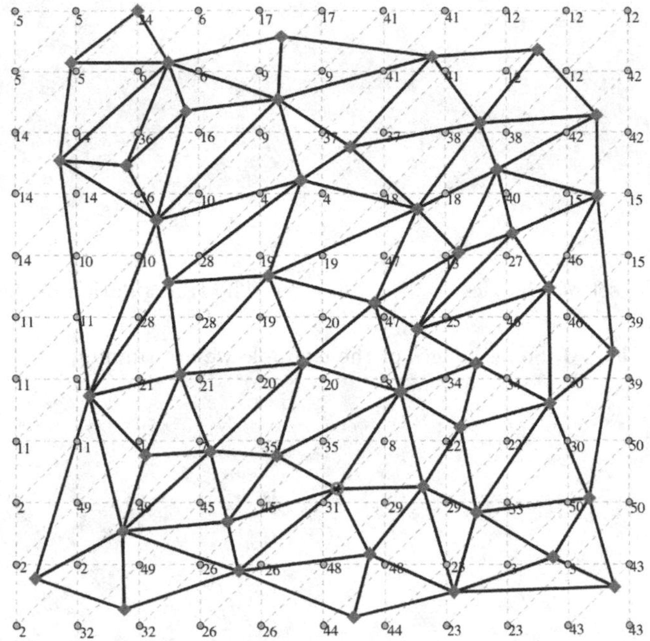

Fig. 8. Triangulation of the second example

Now, we only have to complete the Phase 2 of the triangulation process to obtain the triangulation. The set of triangles $\mathcal{C}$ for this example is represented in Figure 7(b).

After determining the set $\mathcal{C}$ of the triangles that must be represented to complete the triangulation, we proceed to draw all the triangles with the result that can be seen in Figure 8. In this example no vertex is isolated, so we do not need to complete the triangulation creating new edges between adjacent vertices. In this case, any triangle in $\mathcal{C}$ intersects with any other in this set, so it is not necessary to remove any triangle from the graphical representation.

5 Conclusions

We have introduced a triangulation process to perform triangulations of a set of points in the plane. This algorithm presents two well distinguished phases; Phase 1 begins with the construction of an auxiliar triangular grid containing all the points. This auxiliar mesh will help us in the triangulation process, since we take this mesh as a reference to determine the points that may be joined by edges to construct triangles that will constitute the triangulation.

Phase 2 determines the vertices of the initial set S that must be joined to form the triangles that will constitute the triangulation of the points. The main advantages of the triangulation process described in this paper are twofold; firstly, it constitutes a simple and quick process based on simple mathematical concepts as it is the euclidean distance. Secondly, this process may be easily adapted to three-dimensional space, since we only have to construct a lattice in 3D. The Phase 2 of the triangulation process is clearly independent of the number of coordinates of the points.

The numerical experiments we have performed, for different values of n show us the characteristics of the model in terms of adaptability and simplicity.

References

1. Briechle, K., Hanebeck, U.: Localization of a mobile robot using relative bearing measurements. IEEE Transactions on Robotics and Automation 20(1), 36–44 (2004)
2. Byröd, M., Josephson, K., Åström, K.: Fast optimal three view triangulation. In: Yagi, Y., Kang, S.B., Kweon, I.S., Zha, H. (eds.) ACCV 2007, Part II. LNCS, vol. 4844, pp. 549–559. Springer, Heidelberg (2007)
3. Carmichael, G.: A survey of Delaunay Triangulations: Algorithms and Applications (2008)
4. Chazelle, B., Devillers, O., Hurtado, F., Mora, M., Sacristán, V., Teillaud, M.: Splitting a Delaunay triangulation in linear time. Algorithmica 34, 39–46 (2002)
5. Cohen, C., Koss, F.: A comprehensive study of three object triangulation. Mobile Robots VII 1831, 95–106 (1993)
6. Delaunay, B.: Sur la sphere vide. Bull. Acad. Sci. USSR VII: Class. Scil., Mat. Nat., 793–800 (1934)
7. Devillers, O., Pion, S., Teillaud, M.: Walking in a Triangulation. International J. Found. Comput. Science 13, 181–199 (2002)
8. Ely, J.L., Leclerc, A.P.: Correct Delaunay triangulation in the presence of inexact inputs and arithmetic. Reliable Computing 6, 23–38 (2000)
9. Esteves, J., Carvalho, A., Couto, C.: Position and orientation errors in mobile robot absolute self-localization using an improved version of the generalized geometric triangulation algorithm. In: IEEE International Conference on Industrial Technology (ICIT), pp. 830–835 (2006)
10. Font-Llagunes, J., Batlle, J.: Consistent triangulation for mobile robot localization using discontinuous angular measurements. Robotics and Autonomous Systems 57(9), 931–942 (2009)
11. Khanban, A.A., Edalat, A.: Computing Delaunay triangulation with imprecise input data. In: Proc. 15th Canadian Conference Computer Geom., pp. 94–97 (2003)
12. Lischinski, D.: Incremental Delaunay Triangulation. Graphic Gems IV (1994)
13. Loera, J.A.: Triangulations: Structures and Algorithms. In: Mark de Berg, O.C. (ed.) Computational Geometry: Algorithms and Applications. Springer (2008)
14. Miles, M., Huberman, A.: Qualitative Data Analysis: An Expanded Sourcebook, Thousand Oaks. Sage Publications, California (1994)
15. Nordberg, K.: The Triangulation Tensor. Computer Vision and Image Understanding 113, 935–945 (2009)
16. Sharir, M., Welzl, E.: Random triangulations of planar point sets. In: 22nd Annual Symposium on Computational Geometry, pp. 273–281. ACM (2006)
17. Silveira, R.I.: Optimization of polyhedral terrains. Ph.D. Thesis, Utrecht University (2009)

3D Network Traffic Monitoring Based on an Automatic Attack Classifier

Diego Roberto Colombo Dias[1], José Remo Ferreira Brega[2], Luis Carlos Trevelin[1], Bruno Barberi Gnecco[3], João Paulo Papa[2], and Marcelo de Paiva Guimarães[4]

[1] Computer Science Department - Federal University of São Carlos, São Carlos, SP, Brazil
[2] Computer Science Department - UNESP, Bauru, SP, Brazil
[3] Corollarium Technologies, São Paulo, SP, Brazil
[4]Open University of Brazil – Federal University of São Paulo/Faccamp's Master Program, São Paulo, SP, Brazil

{diegocolombo.dias,trevelin}@dc.ufscar.br,
{remo,papa}@fc.unesp.br, brunobg@corollarium.com,
marcelodepaiva@gmail.com

Abstract. In the last years, the exponential growth of computer networks has created an incredibly increase of network data traffic. The management becomes a challenging task, requesting a continuous monitoring of the network to detect and diagnose problems, and to fix problems and to optimize performance. Tools, such as Tcpdump and Snort are commonly used as network sniffer, logging and analysis applied on a dedicated host or network segment. They capture the traffic and analyze it for suspicious usage patterns, such as those that occur normally with port scans or Denial-of-service attacks. These tools are very important for the network management, but they do not take advantage of human cognitive capacity of the learning and pattern recognition. To overcome this limitation, this paper aims to present a visual interactive and multiprojection 3D tool with automatic data classification for attack detection.

1 Introduction

Network data traffic in computer networks is originated by devices such as TVs, tablets, mobile phones, computers, sensors, personal digital assistants and camera throw applications as webmail, video-conference and internet banking. In last decade, with the growing of users and applications, the amount of data traffic has increased dramatically. Consequently, it is necessary a network management, in order to protect the data against improper access and usage. It is essential an ongoing monitoring to detect and diagnose vulnerabilities and threats.

In the most of networking protocols, the data are split into small segment, or packets, to be transmitted. These packets contain information - such as user identification, passwords and confidential data, and some time isn´t encrypted in some way - that let interested the attackers. These data can be also useful for other purposes - network administrator uses it to diagnose network faults. To accomplish this task is necessary to take care of heterogeneous networks, which involve sharing and

B. Murgante et al. (Eds.): ICCSA 2014, Part II, LNCS 8580, pp. 342–351, 2014.

transmitting data but also voice and video. It is complex because lacks tools to provide a sense of network situational awareness. Traditionally, this task accomplished using sniffers tools, such as TCPdump [1], Ngrep [2] and Snort [3], which work in "promiscuous mode" grabbing a copy of every packet that goes past the segment. Sniffers analyze the packets looking for suspicious usage patterns, such as those that occur normally with port scans or Denial-of-service (DoS) attacks. These tools are very important for the network management, but they do not take advantage of human cognitive capacity of the learning and pattern recognition. The network administrator has to analyze almost manually a huge data set generated by these sniffers.

This paper aims to present a solution to facilitate the detection of new network attacks automatically throw a multi-view 3D and interactive tool, amplifying the cognition, which means facilitate the perception of the events concerned (attacks). Two modules compose it: the automatic pattern classification based on optimum-path forest [4]; and a 3D users interface. This visual tool becomes necessary because the amount of data generated by the sniffers is enormous. Further, this data is normally saved in text log file. There are visual tools that have the objective of the tool present, but they don´t offer an automatic attack classifier in real time and are not interactive. Figure 1 depicts the architecture of Sniffer 3D applied to an academic environment, where many attacks occur.

Considering this, the main contributions of this work is to present an automatic attack classifier with an interactive interface with multiprojection. It is a powerful real time tool for the network administrator to detect the attacks pleasantly; with 3D, interaction and navigability. Furthermore, this tool can provide information, which might otherwise be missed during textual analysis. For example, it shows alerts when an attack is detected. The tool developed is an IDS (Intrusion Detection Systems).

This paper is organized as follows: section 2 is an overview of related work. Section 3 describes how the tool was developed, explaining the automatic attack classifier, the interactive interface and the tests. Section 4 presents the conclusions and future researches.

2 Related Work

Many network management tools uses information visualization techniques to present the data in a visual representation to get an easy way of perception, access and manipulation of the data. The main issue is how represent, visualize and select important information in these massive data to detect attacks.

EtherApe is tool that displays network activity graphically in 2D [5]. It uses lines to represent the connection between hosts. Hosts and links change in size with traffic. Each protocol is associated with a color. It supports Ethernet, FDDI, Token Ring, ISDN, PPP, SLIP and WLAN devices, and some encapsulation formats. The visual representation is not good when there is a huge number of hosts.

VISUAL (Visual Information Security Utility for Administration Live) is another tool which uses 2D representation [6]. Its mains purpose is to show the traffic

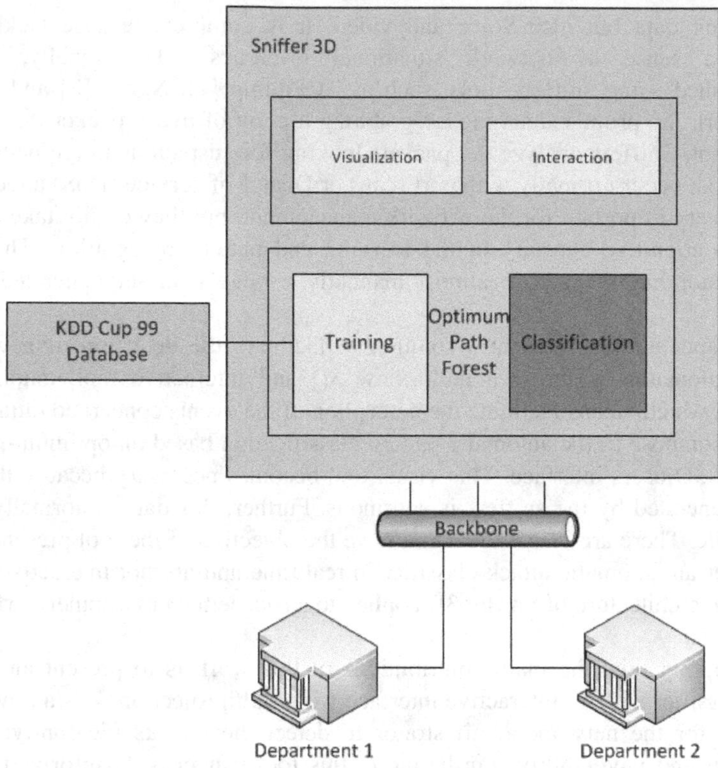

Fig. 1. The Architecture of Sniffer 3D

between local and external network. VISUAL does not provide analyze in real time, it relies on a preprocessor to treat network packet trace. It suffers from similar limitation of EtherApe, it is not easy to detect a attack when there is a large amount of data.

Spinning Cube of Potencial Doom is a 3D network visualizer that listen a network interface or stdin, extracts new connections, and maps them onto a cube [7]. The coordinates of each point are determined by mapping the source, destination, and port to the cube axis. The cube emphasizes the connection attempts and port scans using colored lines. Nevertheless, it is very hard to identify the origins of attempts. Additionally, it offers just a simple interaction, allowing only cube rotation.

NVisionIP is another 2D tool that focuses on the representation of the network traffic on an entire class-B IP network [8]. It has a single user interface that allows the network administrator to have an overview of the current state of network. It allows filtering the traffic based on upon a number of attributes useful in categorized security incidents.

Tcpdump prints out a description of the contents of packets on a network interface that match in a boolean expression [1]. The network administrator uses command line to interact with him. Snort is another tool that has the ability to perform real-time traffic analysis and packet logging on Internet Protocol (IP) networks [9]. It performs protocol analysis, content searching, and content matching. This tool can also be used to detect probes or attacks, including, operating system fingerprinting attempts, common gateway interface, buffer overflows, server message block probes, and stealth port scans. Tranfshow is another tool that continuously shows the information regarding packet traffic on the configured network interface [10]. It periodically sorts and updates this information. It may be useful for locating suspicious network traffic on the net. Tcpdump, Snort and Trafshow offer a poor visual representation. They are not appropriate to visualize huge amount of data.

Table 1 compares several tools. Each one has the own features, differing visualization form (1D, 2D or 3D), number of available traffic filters and the results that can present. In the last line is the tool developed in this work that differs from the others in the interactive interface and in the automatic attack classifier.

Table 1. Comparison tools related

Tool	Interface	Filter	Results presented
EtherApe	3D	Protocol	Traffic volume
Visual	2D	Port, Protocol	Traffic volume
Spinning Cube of Pontential Doom	3D	IP range	Attack quantity
NvisionIP	2D	Protocol	Traffic volume
Tcpdump	1D	Header parameters	Packet header
Snort	1D	Rules	Packet header
Trafshow	1D	IP range	Traffic volume
Tool developed	3D	Header parameters; network interface; quantity	Traffic volume; relationship; details on- demand

3 Tool Developed

Nowadays, all organization rely on computer network to run their business, generating a critical dependency of this infrastructure. Meanwhile Symantec report shows that [11]:

- They blocked a total of over 5.5 billion malware attacks in 2011, an 81% increase over 2010;
- Web based attacks increased by 36% with over 4,500 new attacks each day;

- 403 million new variants of malware were created in 2011, a 41% increase of 2010;
- SPAM volumes dropped by 34% in 2011 over rates in 2010.39% of malware attacks via email used a link to a web page;
- Mobile vulnerabilities continued to rise, with 315 discovered in 2011;
- In 2011, 232 million identities were exposed. An average of 82 targeted attacks take place each day;
- Mobile threats are collecting data, tracking users and sending premium text messages.

Facts like these boost investment in research in IDS development. These tools should find the threats and show to the network administrator, which will take the appropriate action. The ideal is that the administrator take care of the attacks while it is happing, requiring an automatic and real-time detection.

The tool developed aims facilitate the network administrator work, alerting him in real-time of attempted attacks. Two modules compose it: the automatic attack classifier; and the 3D visual interface. It was implemented using: Linux (operational system), C++ (programming language), Python (programming language), QT interface (design), Ogre3D (graphical engine) and LibpCap (sniffer).

3.1 Automatic Attack Classifier

The automatic pattern classification based on optimum-path forest implemented uses samples mode as nodes of a complete graph [4]. The most representative elements of each class (attacks) in the training set (prototypes) were selected as being the ones belonging to the border regions between classes. Each prototype competes with another looking for samples, offering least-cost paths and their respective labels. This process results in a training set partitioned into optimum-path trees. The union of these leads us to an optimum-path forest. This approach has several benefits over other methods of supervised pattern classification: is parameters free; treats natively multiclass problems; and it does not use form and/or class separation.

The classifier-training phase has to find a set of prototypes. Several heuristics can be applied, such as randomized choice of prototypes. However, a heuristic based randomized choice can affect the classifier performance, becoming it unstable and with a high sensitivity degree during the process of selecting the prototype. We aim to provide prototypes that overlap samples regions and class boundaries, since they are regions more susceptible to misclassification. We used the Minimum Spanning tree algorithm.

The classifier requests an adaptation to be implemented. We use the database KDD (Knowledge Discovery and Data Mining) CUP 99 to train the classifier. This database contains information about traffic data and their classification. It has 41 features with 22 types of classification of attacks. Initially, we don´t implemented all types of attacks. We applied a filter, generating a new database. It was added a new tables to connect source to the destine. It was composed by the following fields:

- duration: this field specifies duration of a connection;
- protocol_type: this field specifies the network protocol type (tcp, udp,...);
- service: this field specifies the network service used in the connection (http, telnet,..);
- src_bytes: this field specifies length of bytes sent from the source node;
- dst_bytes: this field specifies his field specifies length of bytes sent from the destine node;
- flag: link status (1 - normal ; 0- error);
- land : 1 - if source and destine are the same;
- wrong_fragment: quantity of wrong fragments;
- urgent: quantity of urgent packets.

The connection status (flag) is not recovered straight from the packets. This process is done using an automata with final states representing the connection status; the transaction states shows the packet type that can reach the state.

The Figure 2 depicts, in order, the steps present in the training/test phase of attacks in Sniffer 3D.

During the test phase the Sniffer 3D captures a packet (sample) and compares it with the optimum paths tree. The tree that matches with the sample is rotuled, otherwise, it is not classified as an attack.

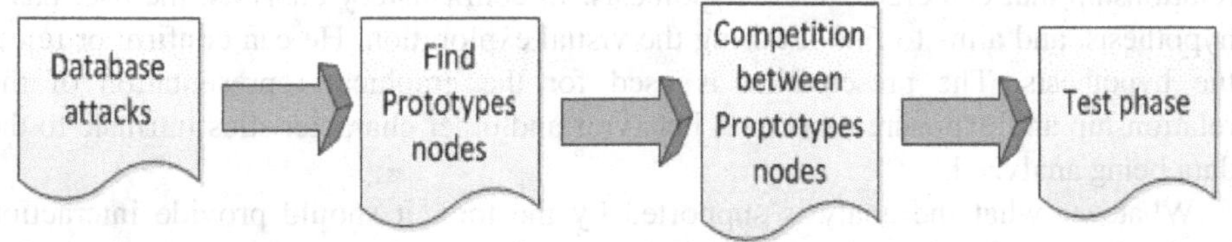

Fig. 2. Classification and Test Phase of Sniffer 3D

The Table 2 present five tests applied to the classifier. Test I and II classified two classes of attack: normal and smurf. Test III classified three classes: normal, smurf and back. Test IV classified five classes: normal, smurf, back, spy and nmap. Test V involved different classification types: not attack and attack, with the following classes: smurf, back, spy and nmap. For each test, we used a specific percentage of databases KDD to train the optimum-path forest algorithm. For example, Test I used 60%, then requesting more time for training (3863s). With 30%, it was possible to achieve a close hit rate that utilized 60%. The number of classes of attacks classified also affects the hit rate.

The traffic data used during the test was captured from a backbone network of a university.

Table 2. Evaluating the classifier

	Test I	Test II	Test III	Test IV	Test V
Classes	2	2	4	5	2
Training (% of database used)	60%	30%	30%	30%	30%
Test (% of database used)	40%	70%	70%	70%	70%
Training duration	3863s	942s	942s	980s	983s
Test duration	1138s	865s	976s	1017s	1015s
Hit rate	100%	100%	99,99%	89,81%	99,99%

3.2 3D Visual Interface

One of the main ideas of Information Visualization is help the user during the large amount of data interpretation. It can basically support tree activities: exploratory analysis, confirmatory analysis and presentation.

During exploratory analysis, the user don´t have idea of what knowledge is part of the data, and using an analytic process, he explores the visual presentation looking for relationship that can create some hypothesis. In confirmatory analysis, the user has a hypothesis, and aims to find it during the visual exploration. He can confirm or reject the hypothesis. The presentation is used for the graphical representation of the relationship and exposure, structure, behavior and other characteristics intrinsic to the data being analyzed.

Whatever what the analysis supported by the tool, it should provide interaction mechanisms to facilitate the data visualization from various perspectives, including [12]:

- Mapping: it is the process which determines how to visualize information or how to encode information into visual form;
- Selection: it means to select data among those data which is available according to the given task;
- Presentation: it treats of how to manage, organize information in the available space on the screen effectively;
- Scale and dimensionality: it treats of how to manage huge visualization in available space; and
- Rearrangement: it treats of how to organize, explore, and rearrange the visualization.

The main goal of this work is that the administrator can easily detect attacks over the computer network. We maps in real time the network traffic into a cube. It shows information such as source IP, destine IP, source port and destine port. This interface is similar to the Spinning Cube of Potencial Doom, but it is interactive, the user can rotate, translate the cube, zoom-in and zoon-out. Each point in the cube represents a connection. While the network traffic is captured, it is added colored points inside of the cube. The automatic classifier determinates the colors, such as red color is associated with a possible attack or virus.

The cube is configured using a special interface. The main settings are:

- Network interface: the administrator can visualize the traffic of a specific interface; or the traffic can be read from a log file;
- Axes: it allows to configure which information will be associated with each edge of the cube;
- Filters: the administrator can customize filters following LibpCap functionalities; and
- Indicators: it allows defining classification for each connection found and colors (packets types, source port, destine port, alerts...).

Furthermore, there is a graphical interface with details about the connections, that is showed when the administrator clicks over an object (point) in the 3D cube. Figure 3 depicts the cube and detail about a packet (one point inside the cube). Packets that are identified as attack are displayed in different colors according to the level of attack.

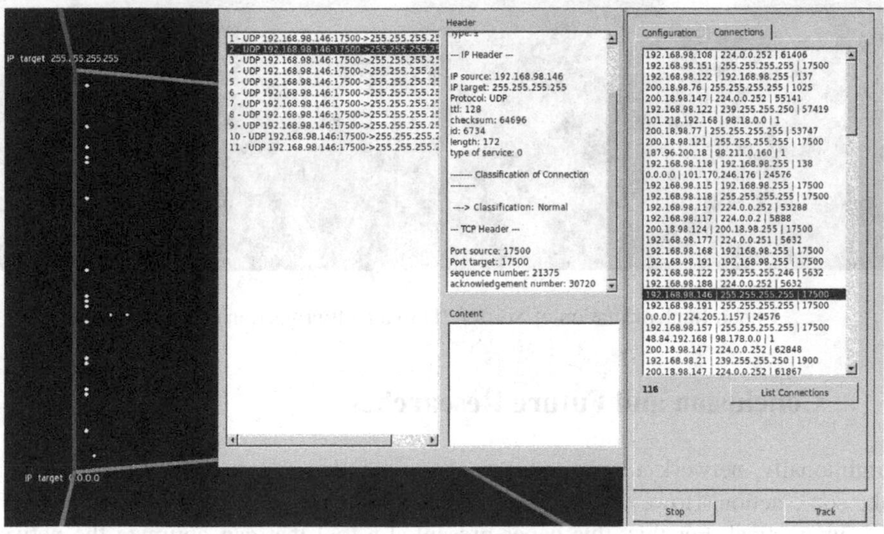

Fig. 3. Details about a connection

Interaction is critical to the success of our tool, since it affects directly the network administrator satisfaction and their efficiency in performing tasks. It can be performed through conventional devices, such as mouse and keyboard, and also by non-conventional devices, such as data gloves and motion trackers. In this project we used two devices originally developed for game consoles, Wii Remote, through the WiiUseJ library [13]; and Microsoft Kinect, throught the OpenNI library [14]. The network administrator can execute operations as zoom in, zoom out, rotation, pan and detail view.

It is possible use the Sniffer 3D in multiprojection environments. Figure 4 depicts an example where three Sniffer 3D instances are used. The purpose is to divide display in several cubes by ip ranges, allowing that large networks, which has many connections, can be displayed in a separate manner, eg, three university department, which are represented by three different cube.

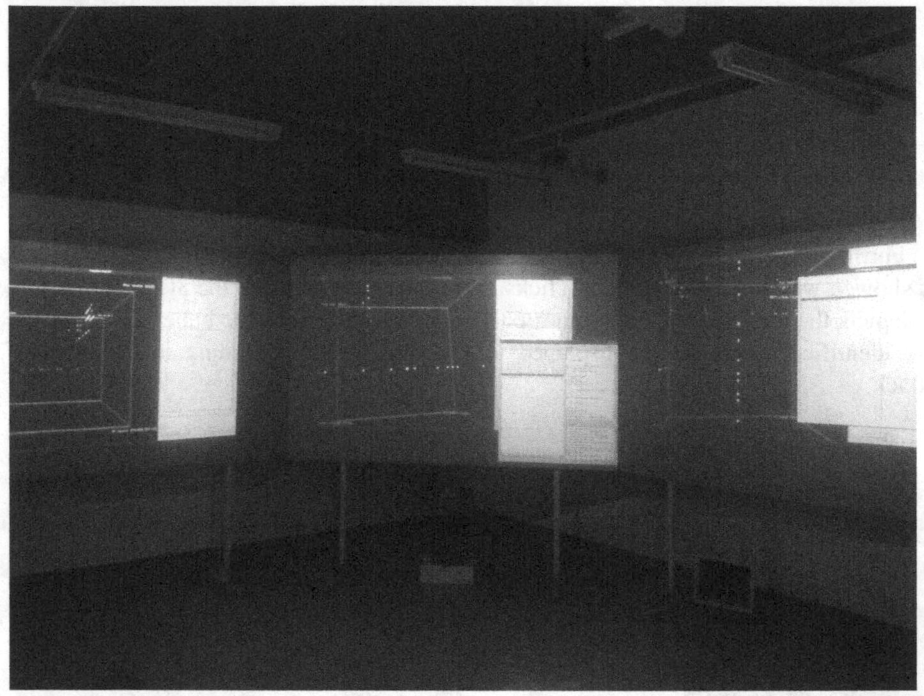

Fig. 4. Implementation of Sniffer 3D in a multiprojection environment

4 Conclusion and Future Researches

Traditionally, network administrator analyses text log files to detect attacks and to take some action. However, the growing constant of network computer becomes this task impractical. For this, this paper presented a tool that can optimize the network administrator job showing the analyze of the traffic data in an effective way.

The use of log files is not excluded in this work, because it can be utilized to later analysis, allowing comparison and improvement of tool developed. The tests was done with real data and showed efficiency of the tool developed.

The automatic classifier is to support the user, giving him the idea of what might be an attack. However, we believe only in user perception. How future research, it is expected that the Sniffer 3D can automatically take some decisions, such as closing a connection that may be trying to attack the network. Confirmations of activities to be taken by Sniffer 3D can also be generated, not excluding the user cognition.

We are aiming to compare our classifier with others based on neural networks. We believe that we can achieve a lower training time, and an improvement of the hit rate. Usability tests are being conducted as a next step of the research.

References

[1] TCPDUMP. TCPDUMM & LIBPCAP, http://www.tcpdump.org/ (accessed September 2012)

[2] NGREP, Ngrep – networl grep, http://ngrep.sourceforge.net/ (accessed September 2012)

[3] SNORT. Snort:Home Page, http://www.snort.org/ (accessed September 2012)

[4] Papa, J.P., Falcão, A.X., Suzuki, C.T.N.: Supervised Pattern Classification based on Optimum-Path Forest. Journal of Imaging Systems and Technology 19(2), 120–131 (2009) ISSN: 0899-9457

[5] ETHERAPE. EtherApe, a graphical network monitor, http://etherape.sourceforge.net/ (accessed September 2012)

[6] Ball, R., Fink, G.A., North, C.: Home-Centric Visualization of Network Traffic for Security Administration. In: VizSEC/DMSEC 2004: Proceedings of the 2004 ACM Workshop on Visualization and, pp. 55–64. ACM Press (2004)

[7] Lau, S.: The Spinning Cube of Potential Doom. Communications of the ACM 47(6) (June 2004)

[8] Lakkaraju, K., Yurcik, W., Lee, A.J.: NVisionIP: netflow visualizations of system state for security situational awareness. In: Proceedings of the 2004 ACM Workshop on Visualization and Data Mining For Computer Security, VizSEC/DMSEC 2004, Washington DC, USA, pp. 65–72. ACM, New York (2004), doi:http://doi.acm.org/10.1145/1029208.1029219

[9] SNORT. Snort network intrusion prevention and detection system, http://www.snort.org (accessed September 2012)

[10] TRAFSHOW. Network traffic monitoring utility, http://linux.maruhn.com/sec/trafshow.html (accessed September 2012)

[11] SYMANTEC. Symantec – Confidence in a connected world, http://www.symantec.com/threatreport/topic.jsp?id=highlights (accessed September 2012)

[12] Khan, M., Khan, S.S.: Data and Information Visualization Methods, and Interactive Mechanisms: A Survey. International Journal of Computer Applications 34(1), 0975–8887 (2011)

[13] WIIUSEJ. Java Api for Wiimotes: WiiUseJ, http://code.google.com/p/wiiusej/ (accessed September 2012)

[14] OPENNI. OpenNI – Introducing OpenNI, http://openni.org/ (accessed September 2012)

Topology Preserving Algorithms for Implicit Surfaces Simplifying and Sewing

Aruquia Peixoto[1] and Carlos A. de Moura[2]

[1] CEFET/RJ, Rio de Janeiro RJ 08544, Brazil
aruquia@gmail.com,
[2] Rio de Janeiro State University, Rio de Janeiro RJ 22550-900, Brazil
demoura@ime.uerj.br

Abstract. Two discretization methods for implicit surfaces are presented: the Non-Compact Dual Simplification and the Sewing Octree. They work with surfaces polygonalized with Dual Contouring, an adaptive method which uses an octree. The Non-Compact Dual Simplification (NDS) preserves the topology of simplified non-compact surfaces. This method can be used for non-compact as well as for compact surfaces in the case the polygonalization region does not contain the latter ones. The Sewing Octree is a method to glue two or more octrees that share faces or edges and contain portions of the surface polygonalized with Dual Contouring. These methods can be employed either independently or coupled, by dividing the original cube in two or more cubes, making the polygonalization, simplifying these regions with NDS, if necessary, and glueing the resulting surfaces with the Sewing Octree. We assure that, with this procedure, the resulting surface and the original one share the same topology.

Keywords: Computational Geometry, Computational Topology, Implicit Surface.

1 Introduction

Implicit surfaces appear in many areas of computer graphics, from geometric modeling to medical images. Their data can be obtained from an algebraic function or from sampling of a function in a spatial grid. Among many books on implicit surfaces we quote [1997], [1998b] and [2013].

A compact surface is defined as a closed and bounded surface, cf. [2009] and [2010]. Being bounded, a sphere is a compact surface, as it has no boundary. A bounded cube can completely enclose a given sphere, while any plane is a non-compact surface, since no cube can contain it. In this work we deal with compact surfaces as well as with non-compact ones.

In an implicit description, a surface is defined as the collection of points that satisfy $f(x, y, z) = c$, were c is the isolevel. The position of a point from the whole space with respect to an implicit, compact surface is characterized by one of the following three options:

B. Murgante et al. (Eds.): ICCSA 2014, Part II, LNCS 8580, pp. 352–367, 2014.

- If $f(x, y, z) < c$, the point lies in the region inside the surface;
- If $f(x, y, z) > c$, then the point lies outside it;
- And if $f(x, y, z) = c$, the point belongs to the surface.

An implicit compact surface thus defines a partition for the space: its two distinct sides (interior and exterior) and the surface itself.

Polygonalization of implicit surfaces can generate meshes with too many points, which can be hard to deal with. To avoid this we propose a method, the Non-Compact Dual Simplification (NDS), to simplify the mesh without loosing information on the original surface. We show that its results are never like the ones shown in Fig. 1, where the simplification of a plane leads to a single point.

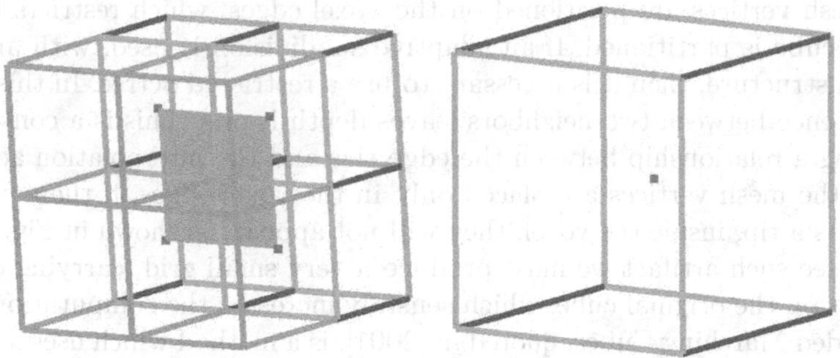

Fig. 1. Original portion of a plane and its simplification, which leads to a single point, therefore changing the original surface topology

Another way to deal with large amount of data is to divide the original region in smaller regions, make the polygonalization for each sub-region, then glue the resulting meshes. To have these regions glued, we introduce the Sewing Octree algorithm. Observe that even when dealing with a compact surface, it can overpass the border of the polygonalization region, just like when dealing with non-compact surfaces. A simplification may be performed by employing the Sewing Octree with the Non-Compact Dual Simplification. This can avoid changes in the original surface topology.

2 Related Work

There exist some well known methods to polygonize implicit surfaces, like Marching Cubes [1987], Extended Marching Cubes [2001], SurfaceNets [1998a] and Dual Contouring [2002]. Many applications are based on these methods, from Games to Medical Images, and some of them are implemented in GPU [2007b].

All these methods take off on a cube on which a partitioning is performed, so as to generate smaller cubes, called voxels. To know each voxel position with respect to the surface, its eight vertices must be checked out. Two cases clearly hold: either the surface intersects the voxel or not.

If all function values at the eight vertices are smaller (or bigger) than the isolevel, then the cube is considered to be entirely on one side of the surface, and no intersection holds. On the other hand, when the (so-called implicit) function yields some values at the vertices that are not smaller (or greater) than the isovalue, while they are strictly smaller (greater) at the remaining ones, we consider that the surface intersects the cube, and we must thus generate polygons to approximate the surface portion inside the cube.

Due to its simplicity to be understood and implemented, Marching Cubes algoritm [1987] is very frequently used. This method generates a partition for the cube, usually a uniform one, and makes the polygonalization to cross the voxels. For every voxel that intersects the surface, a table is analyzed and triangles are generated in order to approximate the portion of the surface inside that voxel.

All mesh vertices are positioned on the voxel edges, which restricts the way that the cube is partitioned. If an adaptive subdivision is used, with an octree as a data structure, then it is necessary to use a restricted octree. In this octree, the difference between two neighbors leaves depth is one. This is a consequence of existing a relationship between the edge size and the interpolation accuracy.

Since the mesh vertices are placed only in the voxel edges, if there exists an artifact, as a tip, inside the voxel, they will not appear, as shown in Fig. 2 a). In order to see such artifact we must produce a very small grid, carrying on more partitions on the original cube, which sensibly increases the computational cost.

Extended Marching Cubes, quoted in [2001], is a method which uses a uniform grid, and the way it works is similar to that of Marching Cubes. It is carried in two steps, introducing a change that allows to polygonize surfaces with more details with no need for too many cube partitions.

For every voxel that intersects the surface, its vertices are analyzed and the polygonalization edges are placed on its edges, as does the Marching Cubes algorithm, but it also analyzes the normal vectors at the voxel vertices. These normal vectors are then used to test the need to place a new mesh vertex inside the voxel, due to the existence of some artifact. This extra vertex is marked, and at the second step all voxels are crossed, testing if another extra vertex, that belongs to neighbor voxels, must be connected to this vertex, thus changing the mesh defined at the first step. With this method we can polygonize surfaces with artifacts that lie inside the voxels, as shown in Fig. 2 b).

The Surface Nets, presented in [1998a], still uses a uniform grid, but it changes the way the polygonalization points are positioned and connected. It is carried out through three steps. At the first one all voxels are crossed, and for each voxel that intersects the surface, one polygonalization vertex is placed at its center. At the second step those vertices are connected.

The final step consists in using the implicit function, in an iterative way, to place the vertices in their correct position, respecting the boundaries of the voxels to which it belongs, and preserving the surface topology. This step can be done many times for a better vertices placement. In Fig. 2 c), we can see that this method can generate better surface details inside the voxel.

With all mesh vertices positioned inside the voxel, and no more at its edges, it is possible to have an adaptive polygonalization method. This leads to the Dual Contouring Algorithm, presented in [2002]. To deal with these voxels created in an adaptive way, it is necessary to use a data structure, in this case an octree.

An octree is a tree data structure whose root corresponds to the original cube, each node corresponding to one intermediate cube on the original cube subdivision, while the leaf nodes correspond to the final and smallest cubes generated by this subdivision. In Fig. 2 d) there are ten leaves that correspond to the cubes which failed to be subdivided.

The vertices of every cube that corresponds to a leaf node are tested. If those vertices indicate that the surface intersects this cube, an interpolation is made placing a point inside such cube and close to the surface.

The connection between the mesh vertices is made using the Minimal Edge, which is defined as the smallest edge on the intersection of three or four cubes that meet at an edge. These edges vertices are tested, so as to determine whether the surface crosses this edge, in which case these mesh vertices are connected, generating a triangle or a quadrangle. In Fig. 2 d) the horizontal line, in the center of the cube, has four Minimal Edges.

With this adaptive method, we can generate more details, getting a more refined subdivision only in regions that have more details which are worth to be considered, like regions with a stronger curvature change. And since these mesh

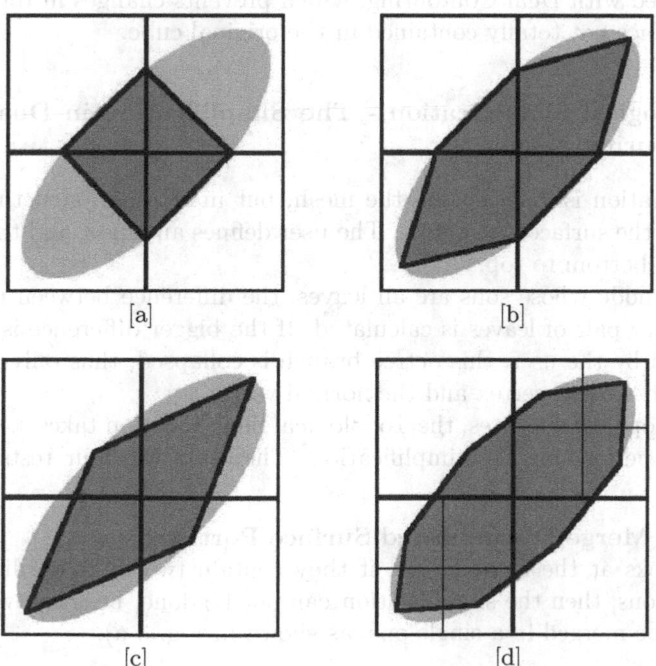

Fig. 2. Implicit surfaces polygonization methods. a) Marching Cubes. b) Extended Marching Cubes. c) Surface Nets. d) Dual Contouring.

vertices are placed inside the voxel, artifacts as peaks can be generated with more accuracy.

Even an adaptive polygonalization method, as Dual Contouring, can generate a large number of triangles, which leads to a simplification on the mesh. This simplification must be carried out rather carefully in order to avoid unwanted changes on the mesh.

The paper [2002], which introduces the Dual Contouring algorithm, also introduces the Topological Simplification, which will be explained with more details in section 3. It prevents a topological change in the mesh, if the surface is compact and totally contained on the original cube.

Other papers introduce methods that prevent topological changes when we simplify the mesh. The work presented in [2004] makes the simplification for two or more distinct surfaces, maintaining them separated, while the work presented in [2007a] prevents that a tubular region gets flattened.

3 Non-Compact Dual Simplification

The Non-Compact Dual Simplification (NDS), herein presented, is a method that prevents topology changes when we simplify a mesh that either fails to be compact or that is not totally contained in the original cube.

In order to understand NDS, we must understand the Topological Simplification, presented with Dual Contouring, which prevents changes in topology of a compact surface not totally contained in the original cube.

3.1 Topological Simplification - The Simplification in Dual Contouring

The simplification is done not on the mesh, but in the data structure used to polygonalize the surface, the octree. The user defines an angle, and the octree is crossed from bottom to top.

For every node whose sons are all leaves, the difference between the normal angles of every pair of leaves is calculated. If the bigger difference is above the angle defined by the user, this octree branch is collapsed, thus only the parent remains, with a mesh vertex and the normal vector.

To avoid topology changes, the Topological Simplification takes some precautions before performing the simplification. These are the four tests described below.

Prevent to Merge Disconnected Surface Portions
This test looks at the eight leaves; if they contain two or more disconnected surface portions, then the simplification can not be done, because two different regions will be merged in a single one, as shown in Fig. 3 a).

Test the Cube Center
If the cube center is in a position different from the eight vertices of the parent cube, then the simplification can not be done, because in this case there is

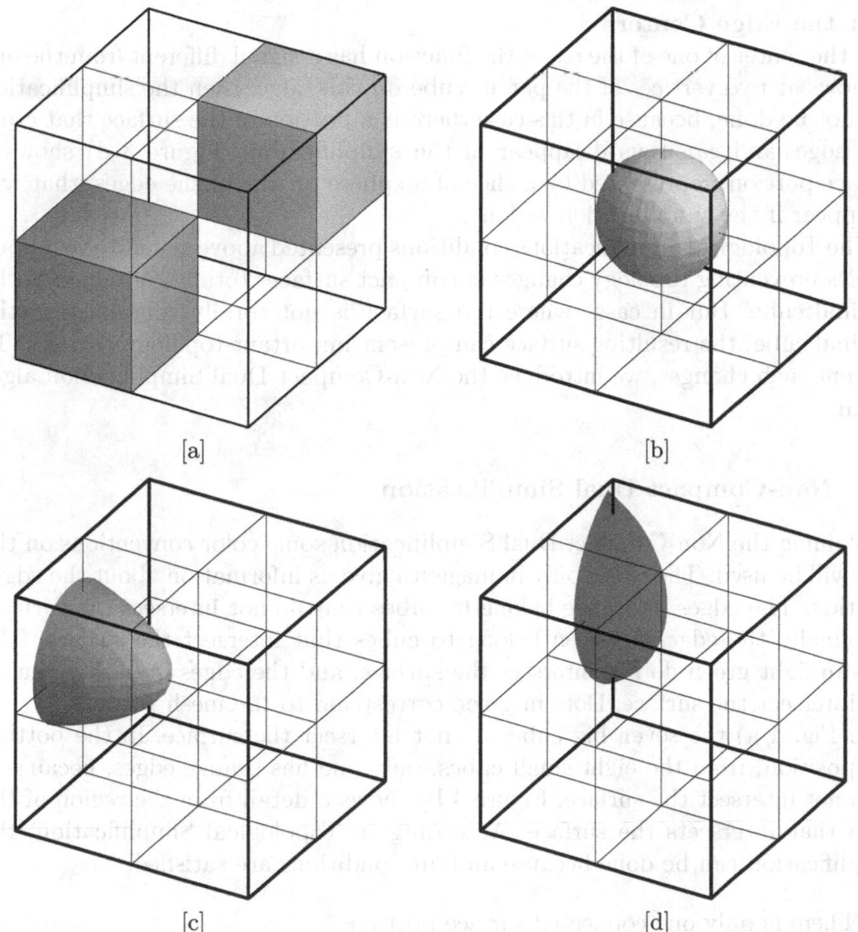

Fig. 3. Topological Simplification and the cases where the simplification can not be done. a) Two different surfaces. b) A surface in the cube center. b) A surface in one cube face. c) A surface on one of the cube edges.

a surface portion inside that cube which can disappear in the simplification. Figure 3 b) shows a portion of the surface, represented by the sphere, that will disappear if the simplification is carried on.

Test the Face Center
If at the center of a face the function has a signal different from its signal at the four vertices of the parent cube in this face, then the simplification can not be done, because in this case there is a portion of the surface that intersects this face and can thus disappear in the simplification. Figure 3 c) shows a portion of the surface, represented by the half sphere on one face, that will disappear if the simplification is carried on.

Test the Edge Center
If at the center of one of the edges the function has a signal different from the one it shows at two vertices of the parent cube on this edge, then the simplification can not be done, because in this case there is a portion of the surface that cross this edge, and can thus disappear in the simplification. Figure 3 d) shows a surface portion, represented by a slice of a sphere on one of the edges, that will disappear if the simplification is done.

The Topological Simplifications conditions presented above generate very good results preventing topology changes in compact surfaces totally contained in the original cube. But in cases where the surface is not totally contained on the original cube, the resulting surface can present important topology changes. To prevent such changes, we introduce the Non-Compact Dual Simplification algorithm.

3.2 Non-Compact Dual Simplification

Explaining the Non-Compact Dual Simplification some color conventions on the grid will be used. The edges only in magenta give us information about the edges position. The edges in orange belong to cubes that do not intersect the surface, and finally the edges in green belong to cubes that intersect the surface. The ones in light green do not intersect the surface, and the edges in dark green do not intersect the surface. Dots in green correspond to the mesh vertices.

In Fig. 4 a) the seven big cubes do not intersect the surface. In the bottom left position, from the eight small cubes, only one has orange edges, because it does not intersect the surface. Figure 4 b) shows a detail from the region of the cube that intersects the surface. According to Topological Simplification, the simplification can be done because all four conditions are satisfied:

- There is only one connected surface portion.
- At the cube center the function has the same signal as at the four parent vertices on the back face.
- At all six faces center the function has the same signal as at least at one of their faces vertex, in the parent cube.
- At all twelve edges centers the function has the same signal as at least at one of their edges vertex, in the parent cube.

If the simplification is done, the result will be a single point, as shown in Fig. 4 c), which is topologically different from the portion of the surface in Fig. 4 a). To prevent this, another condition is needed in order to ensure that the resulting surface will not degenerate in a point or disconnected surface portions.

The Non-Compact Dual Simplification works after the Topological Simplification. It tests if the connections inside the cube, from the edges that will disappear or be merged in another edge, are maintained in a remaining edge away from the border of the original cube. Notice that if an edge is in the border of the original cube, it connects only two cubes, with a maximum of two mesh vertices, and it can not generate a polygon from the mesh.

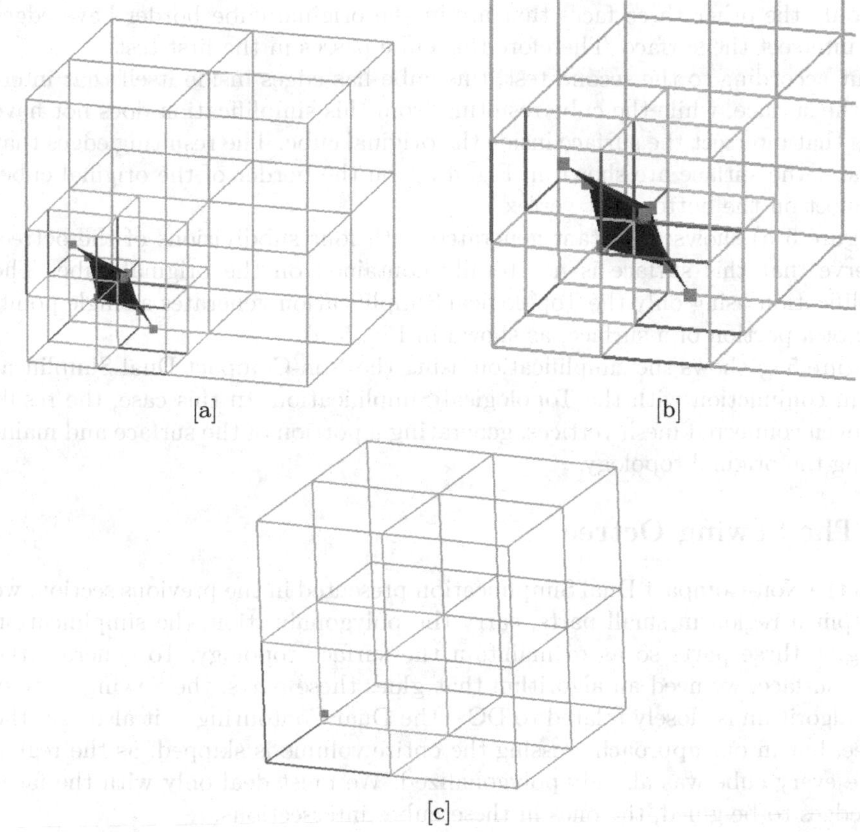

Fig. 4. The grid. a) Grid resulting from the polygonalization of a surface. b) Detail of this grid. c) The only point resulting from the simplification with the Topological Simplification.

The method makes then two tests:

Test Edges on Faces
For every cube face that intersects the surface, it tests if one of the edges that intersects the surface is internal, or if they are all on the original cube border. If there is an internal edge that intersects the surface, it tests if the corresponding face, resulting from the simplification, has an internal edge that intersects the surface, so as to maintain the connection.

Test Edges Inside the Cube
Test if the cube has edges inside itself that intersect the surface, and if so, test if the cube resulting from this simplification has internal edges that intersect the surface.

In Fig. 4 b) we have a cube divided in eight smaller cubes. We can see that all three faces that are internal to the original cube do not intersect the surface, and only the other three faces that are in the original cube border have edges that intersect the surface. Therefore this cube passes in the first test.

But according to the second test, this cube has edges inside itself that intersect the surface, while the cube resulting from this simplification does not have edges that intersect the surface inside the original cube. The resulting edges that intersect the surface are shown in Fig. 4 c), on the border of the original cube, and meet on the bottom left vertex.

Figure 5 a) shows an octant generated with four subdivisions of the octree. Observe that this surface is not totally contained on the original cube. The simplification using only the Topological Simplification generates a single point, and not a portion of a surface, as shown in Fig. 5 b).

Figure 5 c) shows the simplification using the Non-Compact Dual Simplification in conjunction with the Topological Simplification. In this case, the result are seven connected mesh vertices, generating a portion of the surface and maintaining the original topology.

4 The Sewing Octree

With the Non-Compact Dual Simplification presented in the previous section, we can split a region in small parts, carry the polygonalization, the simplification and glue these parts so as to maintain the surface topology. To generate the entire surface, we need an algorithm that glues these parts, the Sewing Octree. This algorithm is closely related to DC – the Dual Contouring –, it also uses the octree, but in our approach crossing the entire volume is skipped, as the region inside every cube was already polygonalized. We must deal only with the faces and edges to be glued, the ones in these cubes intersections.

A face can be defined as the encounter of two cubes. In the left of Fig. 6 we have four faces that result from the meeting of four cubes. In the right side of the same figure, we have three faces that originated from the encounter of three cubes, two faces come from the encounter of the bigger cube with the two small ones, while the remaining one comes from the encounter of the two small cubes.

An edge is the encounter of three or four cubes. We can see in Fig. 6 that the edge in blue results from the meeting of the cubes. The figure on the left shows this edge as the result of four cubes, while at right there is another edge which is generated from the encounter of three cubes. Observe that the real edges do not exceed the boundaries of the smallest cube, but in this case this edge was drawn bigger just to make it easier to locate them.

Since all portions of the surface inside the cubes were polygonalized, only the regions on the cube border must then be polygonized. That is the reason why the Sewing Octree works only for faces and edges. It is a recursive algorithm with only two calls, one for the edges and another for the faces.

Faces

When two cubes meet on a face, if one of them has a subdivision, they generate:

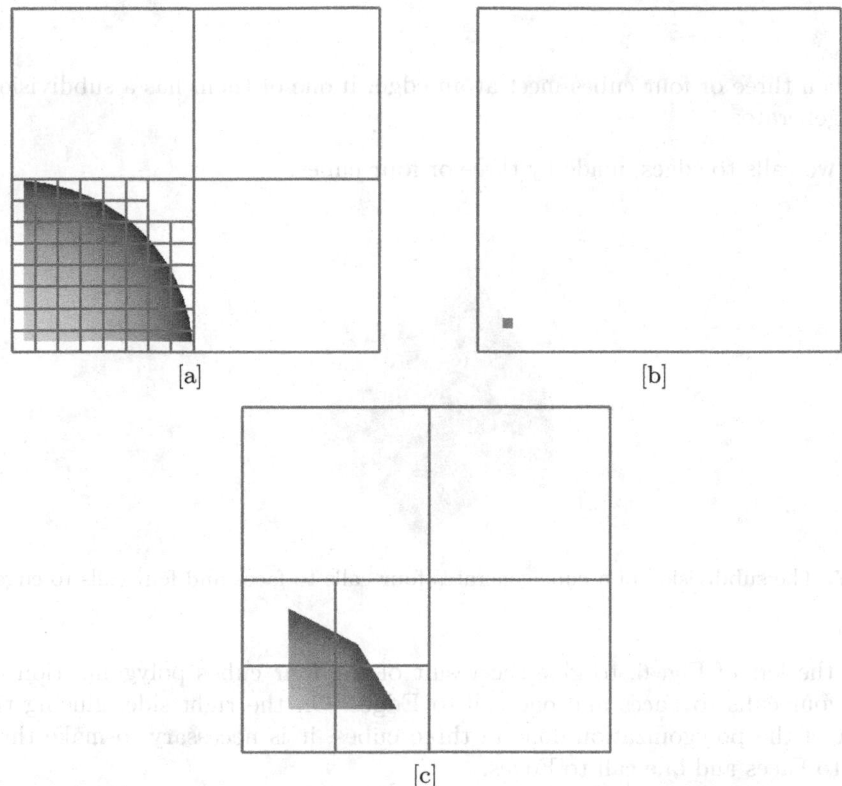

Fig. 5. Simplification. a) Octant of a sphere polygonalized with an octree with four subdivisions. b) The only point resulting of the simplification using only the Topological Simplification. c) The surface resulting of the simplification with the Non-Compact Dual Simplification.

Fig. 6. Edges defined by four and by three cubes

- Four calls to faces, made by two cubes.
- Four calls to edges, made by three or four cubes.

Edges

When three or four cubes meet at an edge, if one of them has a subdivision, they generate:

- Two calls to edges, made by three or four cubes.

Fig. 7. The subdivision of a cube generates four calls to faces and four calls to edges

In the left of Fig. 6, to glue the result of the four cubes polygonization we need four calls to Faces and one call to Edges. On the right side, glueing the result of the polygonization done in three cubes, it is necessary to make three calls to Faces and one call to Edges.

Figure 8 shows a sphere as an example. The original cube is divided in two cubes, on the left and right side, and the polygonization is performed in each cube separately. The result of the polygonalization of each separated cube is shown in Fig. 8 a), while b) shows the surface and the octree. ????

The Sewing Octree is used to glue these two disconnected portions of the surface, maintaining the polygonalization done inside the cubes. Figure 8 c) shows the final result.

5 Examples

In this section we show some results generated with the Non-Compact Dual Simplification, with the Sewing Octree, as well as from coupling both algorithms.

5.1 Example from Non-Compact Dual Simplification

We start with the simplification of the surface presented in Fig. 9 a). In Fig. 9 b) to d) we have the simplification only with the Topological Simplification, the only difference between these cases being the angle constraint required. Notice that in all three cases we have an undesirable change on the surface topology.

In Fig. 9 b) we have an open region degenerated in a single vertex, namely in the figure center. Figure 9 c) shows a worse case for the same region, it

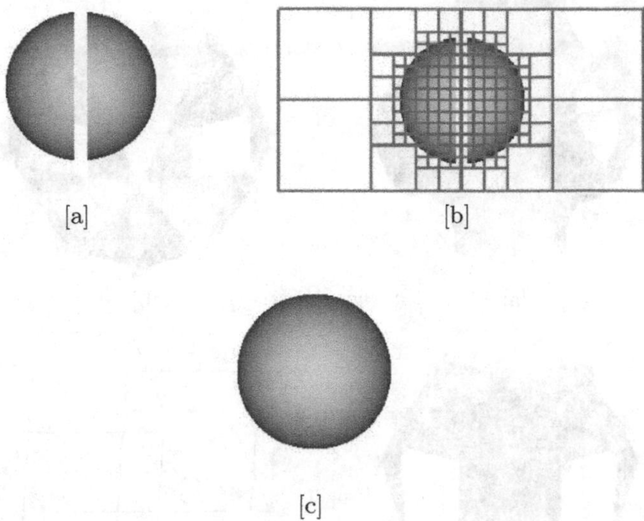

Fig. 8. A cube divided in two cubes. a) The two octrees resulting from the polygonalization. b) The surface resulting from the polygonalization. c) The surface resulting from the Sewing Octree.

degenerates in a disconnected single vertex; and Fig. 9 d) shows a simplification which leads to eight disconnected mesh vertices. Notice that in Fig. 9 d) the only edges in dark green, indicating that they connect vertex meshes around them, lie in the cube border.

Finally in Fig. 9 e) we have the simplification made with the same angle of Fig. 9 d), but in this case we use the Non-Compact Dual Simplification. The resulting surface has the same topological characteristics as the original surface.

5.2 Examples from Sewing Octree

In this section we show results from the Sewing Octree algorithm. Figure 10 shows a surface polygonalized with two octrees, in the right and left side. Figure a) shows at its center a vertical band where the surface failed to be polygonalized. This region corresponds to the face where these two octrees meet.

In Fig. 10 b) we used the Sewing Octree to connect these two surface portions and the result is the hyperboloid, which is the original surface.

The second example shows a torus polygonalized with two octrees, which is shown in Fig. 11. In this case we can see at the left side of Fig. a) that this portion of the torus has two holes and a connected region at its center.

When we use the Sewing Octree to connect these two surface portions these two holes are correctly connected to the corresponding hole, as shown in Fig. 11 b). In this figure we can see the entire torus polygonized.

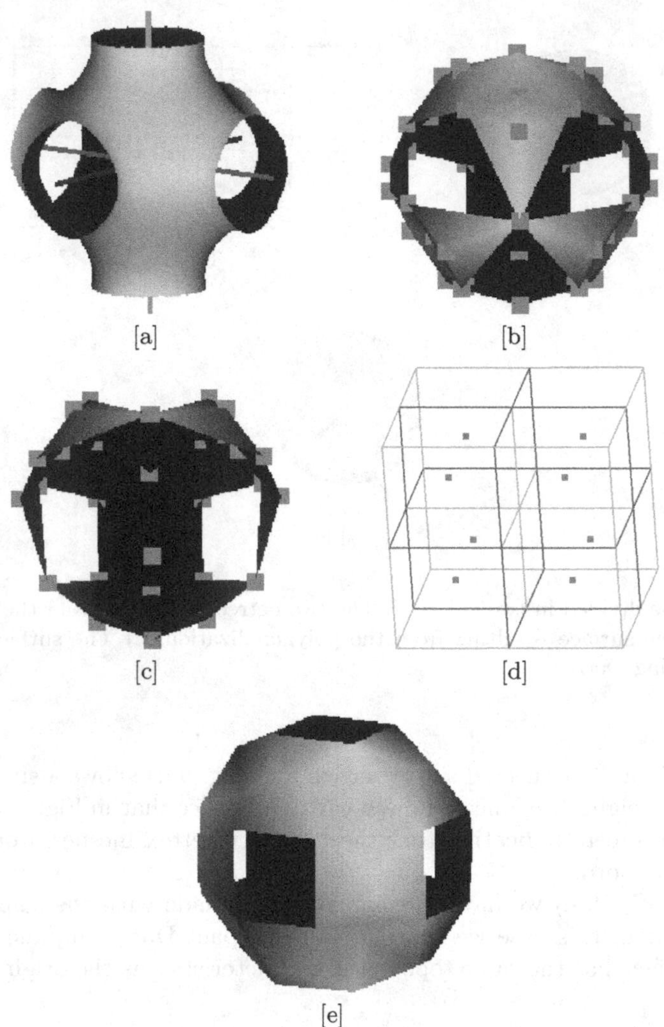

Fig. 9. Simplification. a) Original surface. b) Simplification with a portion of the surface degenerating in one vertex. c) Simplification generating a degeneration of the surface with a vertex disconnected from the rest of the surface. d) Simplification resulting in eight disconnected vertices. e) Simplification with the Non-Compact Dual Simplification, where the topological characteristics of the surface are maintained.

5.3 Example from Non-Compact Dual Simplification Coupled to Sewing Octree

To finish this section, we show a surface where the Sewing Octree must be used with the Non-Compact Dual Simplification to preserve the topology of a simplified surface.

In Fig. 12 we have a half sphere polygonalized with two octrees. Figure a) shows the two disconnected portions of the surface generated by the two octrees.

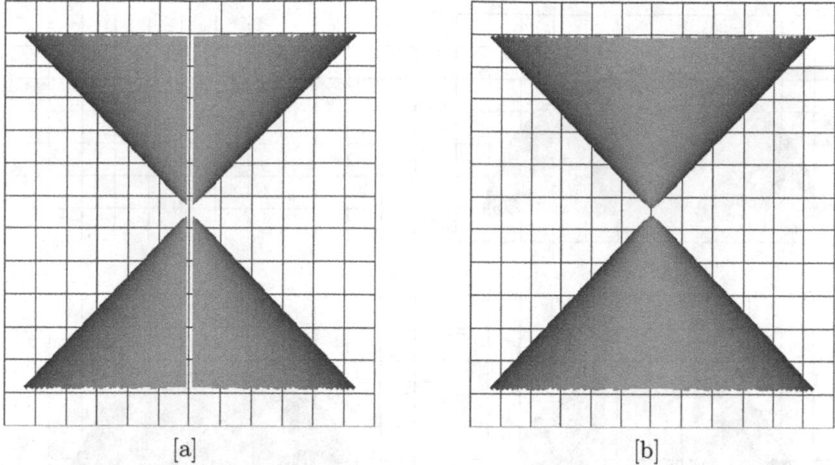

Fig. 10. Hyperboloid with Sewing Octree a) The two regions resulting from the two octrees. b) The final surface, glued with Sewing Octree.

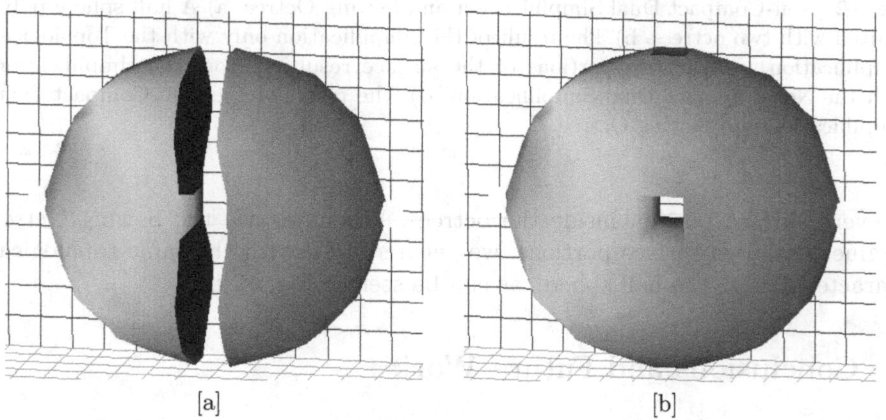

Fig. 11. Torus with Sewing Octree. a) The result of the polygonalization of the two octrees. b) The entire surface generated using the Sewing Octree.

Figure 12 b) shows the result obtained only with the Topological Simplification, which leads to two vertices, shown as the green points. In this case, every vertex is the result of an octree simplification, which points to a change in the surface topology inside each octree. With only two mesh vertices, even using the Sewing Octree we can not generate a portion of the surface.

In Fig. 12 c) we see the result of the simplification of the two octrees now with the Non-Compact Dual Simplification. Notice that in this case the result obtained corresponds to two portions of a surface, thus preserving the topology

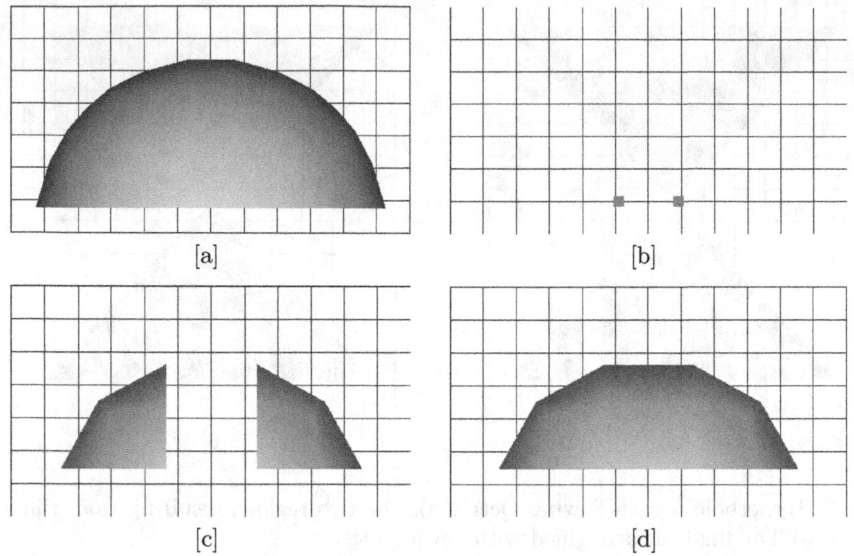

Fig. 12. Non-Compact Dual Simplification and Sewing Octree. a) A half sphere polygonized with two octrees. b) The result of the simplification only with the Topological Simplification. c) The two portions of the surface resulting from the simplification with the Non-Compact Dual Simplification. d) The result of the Non-Compact Dual Simplification and Sewing Octree.

of every surface portion inside the octrees. When we use the Sewing Octree to glue these two surface portions, we get a surface with the same topological characteristics as the half sphere, as can be seen in Fig. 12 d).

6 Conclusions and Future Work

In this paper we present two algorithms: The Non-Compact Dual Simplification and the Sewing Octree.

The Non-Compact Dual Simplification (NDS) algorithm can be coupled to the Topological Simplification to preserve the topology of non-compact surfaces and surfaces which are not totally contained in the original cube. On the other side, the Sewing Octree is used to glue two or more distinct octrees that meet on a face or edge and are used to polygonalize implicit surfaces.

With these two algorithms, we can split the original cube in small cubes, make the polygonalization and the simplification separately in each octree, using the Non-Compact Dual Simplification, and in the end glue these octrees with the Sewing Octree. This way, the resulting and the original surfaces bear the same topology.

Possible applications for this algorithm are to handle large data, where there is need to split the original volume. Even with this strategy a simplification may be necessary. Among our future works we consider to study parallel formulations for the presented algorithms for specific applications, in particular implemented with GPU environments.

References

[1987] Lorensen, W.E., Cline, H.E.: Marching Cubes: A High Resolution 3D Surface Construction Algorithm. In: Proceedings of the 14th Annual Conference on Computer Graphics and Interactive Techniques, SIGGRAPH 1987, pp. 163–169 (1987)

[1997] Bloomenthal, J., Wyvill, B.: Introduction to Implicit Surfaces. Morgan Kaufmann Publishers Inc. (1997) ISBN=155860233X

[1998a] Gibson, S.F.F.: Using Distance Maps for Accurate Surface Representation in Sampled Volumes. In: Proceedings of the 1998 IEEE Symposium on Volume Visualization, VVS 1998, pp. 23–30 (1998)

[1998b] Velho, L., Figueiredo, L.H., de, G.J.A.: Implicit Objects in Computer Graphics. Springer-Verlag New York, Inc. (1998) ISBN=0387984240

[2001] Kobbelt, L.P., Botsch, M., Schwanecke, U., Seidel, H.-P.: Feature Sensitive Surface Extraction from Volume Data. In: Proceedings of the 28th Annual Conference on Computer Graphics and Interactive Techniques, SIGGRAPH 2001 (2001)

[2002] Ju, T., Losasso, F., Schaefer, S., Warren, J.: Dual Contouring of Hermite Data. In: Proceedings of the 29th Annual Conference on Computer Graphics and Interactive Techniques, SIGGRAPH 2002, pp. 339–346 (2002)

[2004] Zhang, N., Hong, W., Kaufman, A.E.: Dual Contouring with Topology-Preserving Simplification Using Enhanced Cell Representation. In: 15th IEEE Visualization, IEEE_VIS 2004, pp. 505–512 (2004)

[2007a] Schaefer, S., Ju, T., Warren, J.: Manifold dual contouring. Proceedings of IEEE Transactions on Visualization and Computer Graphics, 610–619 (2007)

[2007b] Nguyen, H.: GPU Gems 3, 1st edn. Addison-Wesley Professional (2007)

[2009] Zomorodian, A.J.: Topology for Computing. Cambridge University Press (2009) ISBN=9780521136099

[2010] Atallah, M.J., Blanton, M.: Algorithms and Theory of Computation Handbook: Special Topics and Techniques. Chapman & Hall/CRC (2010) ISBN=9781584888208

[2013] Wenger, R.: Isosurfaces: geometry, topology and algorithms. CRC Press (2013) ISBN = 9781466570979

Closest-Point Queries for Complex Objects

Eugene Greene and Asish Mukhopadhyay

School of Computer Science, University of Windsor, Canada
egreene08@hotmail.com, asishm@uwindsor.ca

Abstract. In this paper we report on the implementation of a heuristic for preprocessing a set of n points in the plane to find one that is closest to the boundary of an object that is geometrically more complex than a simple point, to wit, a line, a half-line, a line-segment, a rectangle, a convex polygon with a fixed number of sides, a circle, an ellipse, or any other convex object whose boundary has a constant-sized description. Our experimental results show that the heuristic is more effective than a naive brute-force approach for query objects with small perimeter relative to the given point set, and large values of n.

Keywords: geometric optimization, proximity problems, Voronoi diagrams.

1 Introduction

There is a vast literature on the problem of preprocessing a set of n sites $S = \{p_1, p_2, p_3, \ldots, p_n\}$ to find the site that is closest to a query point p (see Clarkson [4], Liu et al.[9] for example). A problem that has been less studied is that of preprocessing the set of sites to find one that is closest to the boundary of an object that is geometrically more complex than a simple point, to wit, a line, a half-line, a line-segment, a rectangle, a convex polygon with a fixed number of sides, a circle, an ellipse, or any other convex object whose boundary has a constant-sized description. Thus the query objects that we shall consider are either curves that are convex in one-dimension or curves with constant-sized description that bound convex regions. We can also handle a non-convex curve if it can be decomposed into a constant number of convex pieces (a polyline, for example).

Assuming that the distance from a point to a query curve can be determined in constant time, then a query can be answered in $O(n)$ time, using $O(n)$ space to store the points in a list. The problem becomes harder if we want sublinear query time. In this paper, we address this problem.

Motivated by applications in pattern classification and data clustering, Mitra and Chaudhuri [11] have proposed an algorithm for this problem when the query objects are lines. Other applications include those in geographic information systems. We can also find other applications of this problem by interpreting the query object as the locus traced out by another object.

B. Murgante et al. (Eds.): ICCSA 2014, Part II, LNCS 8580, pp. 368–380, 2014.

A theoretical motivation is to show the connection of this problem with the very well-studied range-query problem in Computational Geometry. However, the data structures that one would borrow from range-query problems are rather complicated, and this brings us to the other goal of this paper - is there a simple way of doing this that can be easily implemented in practice? Indeed there is, and we describe a simple solution that can answer proximity queries for a number of different geometrically complex query objects in a uniform way.

The paper is organized as follows. In the next section, we discuss some prior work. In the following section, we briefly introduce the partition tree data structure, and show a general connection between range queries and closest-point queries in the section after that. In the fifth section, we outline our heuristic technique, and we conclude in the sixth section.

2 Prior Work

Cole and Yap [5] were the first to address this problem when the query object is a line. They reported a solution with both preprocessing time and space in $O(n^2)$ and query time in $O(\log n)$. Lee and Ching [8] obtained the same result using geometric duality. Mitra [12] reported an algorithm with both preprocessing time and space in $O(n \log n)$ and query time in $O(n^{0.695})$. In a subsequent paper, Mitra and Chaudhuri [11] improved the space complexity to $O(n)$. Mukhopadhyay [14] used the simplicial partition technique of Matousek [10] to improve the query time to $O(n^{1/2+\epsilon})$ for arbitrary $\epsilon > 0$, with preprocessing time and space in $O(n^{1+\epsilon})$ and $O(n \log n)$ respectively. Nandy et al. [15] reported an algorithm for the k-nearest neighbours of a query line with preprocessing time and space in $O(n^2)$ and $O(n^2/\log n)$ respectively, and query time in $O(k + \log n)$.

When the query object is a line segment, Goswami et al. [7] reported an algorithm for computing the k-nearest points with both preprocessing time and space in $O(n^2)$ and query time in $O(k + \log^2 n)$. The solution is reduced to two triangle range queries and to two point-locations in a k-th order Voronoi diagram and involves some complicated data structures. No implementation of this algorithm has been reported. We might point out that it is straightforward to extend the algorithm of this paper to the case when the query object is a half-line.

When the query object is a circle, Mitra et al. [13] reported two different schemes: one of these has both preprocessing time and space in $O(n^3)$ and query time in $O(\log^2 n)$; the other has preprocessing time and space in $O(n^{1+\epsilon})$ and $O(n \log n)$ respectively and query time in $O(n^{2/3+\epsilon})$. No implementation has been reported for the algorithms in this paper either. The latter scheme uses a lifting transformation that can be dispensed with if we make use of the partition trees of the next section. These trees reduce the complexity of a query to $O(\sqrt{n}\ polylog(n))$, but increase the preprocessing time to a higher degree polynomial in the input size.

A related problem is that of continuous nearest neighbor: given some query curve, partition the curve into intervals so that each point in a given interval has the same closest site, and report each such interval and closest-site pair. Tao et al [16] used R-trees for the continuous nearest neighbor problem for line segments, while De Almeida [6] used R*-trees for another implementation.

3 Partition Trees

Welzl [18] introduced a partition tree data structure to solve range query problems. His solution relied on ideas introduced by Vapnik and Chervonenkis [17]. This solution has an intimate connection with our problem, and to bring this out, we briefly discuss the range-query problem in a slightly formal setting. A range space is a pair (X, R), where X is a non-empty set and R is a set of subsets of X, called ranges. For example, let X be the two-dimensional Euclidean space and R the set of open half-planes. The range query problem is to determine the set of points of a finite subset S of X that lie in a given query range $r \in R$.

For the example range space above, if S is a set of three points in the plane then every subset of S is the result of intersecting S with some half-plane. We say that S is shattered by (X, R). However, if S is any set of four points in the plane then we can always find a subset B for which there is no half-plane H such that $H \cap S = B$. A range space (X, R) has finite VC dimension d, if d is the cardinality of the largest set S that can be shattered by (X, R). If no such d exists, then (X, R) is said to have infinite VC dimension.

In an important paper, Welzl [18] showed that if a range space has a finite VC dimension then the points of S can be organized into a partition tree (see Fig. 1) so that each range query visits few (sublinear in $n = |S|$) nodes of the partition tree. For a given range r, at each node of the partition tree, we have to resolve if the set of points (of S) at this node is *contained* in the range, is *disjoint* from it, or *crosses* it. In the first case we include the points for counting or reporting; in the second case we exclude the points completely. But for both cases the search stops at this node. In the third case, the query continues with the children of the node.

To come to our problem, consider cases in which the closest-point query object happens to be the boundary of a range (for example, the boundary of a half-plane is a line, that of a disk is a circle) from a range space of finite VC dimension. Once we have determined that a set of points at a node is on either side of this boundary, then we can structure these points so that the closest-point query can be answered efficiently. Thus Mitra and Chaudhuri [11] and Mukhopadhyay [14] organized the points into a convex hull (in fact, this has to be done for range queries also). Mitra et al.'s [13] algorithm, examining query circles, maintains the nearest and furthest point Voronoi diagrams at each node.

This correspondence between range and closest point queries is helpful whenever it is beneficial to group points on either side of the query curve. The only problem then becomes finding the closest points to the query on either side of it, which might be a nontrivial task. This happens when, for example, the query object is an ellipse, a rectangle, or a convex polygon with a fixed number of sides, even though these are boundaries of ranges in range spaces of finite VC dimension. Let us elaborate on this by taking the the ellipse as an example.

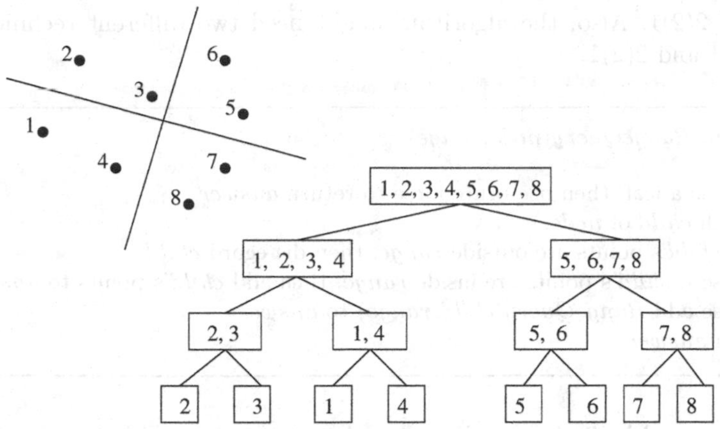

Fig. 1. An example partition tree

Here the first issue is to define how we measure the distance from a point p to the boundary of the ellipse. We could define it in terms of a point x on the ellipse such that $\overline{px}$ is normal to the ellipse at x. This, however, doesn't use any of the fixed points of the ellipse like its center or foci. Instead, we can define the distance as the difference $|2a - dist(p, f_1) - dist(p, f_2)|$, where $2a$ is the length of the major axis of the query ellipse, while $dist(p, f_1)$ and $dist(p, f_2)$ are the distances from p to the foci f_1 and f_2 of the query ellipse. This essentially reduces our problem to answering the following question for a query segment: find a point of S outside (respectively inside) the ellipse such that the sum of its distances from the end-points of the query segment is a minimum (maximum). Thus no matter how the distance is defined, identifying a set of points that lies entirely inside or entirely outside the ellipse does not help us in choosing a point from this set that is closest to the boundary in an obvious manner.

4 General Tree-Based Queries

Welzl's partition trees cannot be used when the query object is not the boundary of a range from a range space of finite VC dimension. Consider for example a line segment or a half-line. These have to be dealt with in a manner as shown, for example, by Goswami et al [7]. Depending on the type of query objects, it could be beneficial to use other tree structures, such as Matousek's simplicial partition trees [10], that put some bound on the number of nodes visited.

Consider a range query algorithm, based on any partition tree, of the form shown in Figure 2. Algorithm *RangeQuery* will determine the intersection of *range* with the points associated with some *node* of a partition tree. Then, after possibly adding some information to each node of the tree, this algorithm can be modified to a closest-point query algorithm (see Figure 3). The queries for Algorithm *ClosestPointQuery* are the boundaries of the ranges in Algorithm *RangeQuery*. The extra information at each node would be required for steps

2(1)1 and 2(2)1. Also, the algorithm might need two different techniques for steps 2(1)1 and 2(2)1.

Algorithm. *RangeQuery*(*node*, *range*)

1. If *node* is a leaf, then use brute force to return *answer*
2. For each *child* of *node*
   1. If *child*'s points are outside *range*, then disregard *child*
   2. Else if *child*'s points are inside *range*, then add *child*'s points to *answer*
   3. Else add *RangeQuery*(*child*, *range*) to *answer*
3. Return *answer*

Fig. 2. A general tree-based range query algorithm

Algorithm. *ClosestPointQuery*(*node*, *curve*)

– *curve* divides the plane $\mathbb{R}^2$ into *range* and $\mathbb{R}^2 - range - curve$

1. If *node* is a leaf, then use brute force to return *answer*
2. For each *child* of *node*
   1. If *child*'s points are outside *range*, then
      1. Find the point p of *child* that is closest to *curve*
      2. If p is closer to *curve* than *answer* is, then update *answer*
   2. Else if *child*'s points are inside *range*, then
      1. Find the point p of *child* that is closest to *curve*
      2. If p is closer to *curve* than *answer* is, then update *answer*
   3. Else
      1. $p = ClosestPointQuery(child, curve)$
      2. If p is closer to *curve* than *answer* is, then update *answer*
3. Return *answer*

Fig. 3. A general tree-based closest point query algorithm

5 A Heuristic Approach

Though the above data structures proposed for range queries are interesting in principle they are rather complicated. The construction of the partition trees relies on a variety of techniques, and each node of the tree could require three or more other data structures on the points the node contains. To the best of our knowledge, we are not aware of any implementations of these data structures. CGAL's [2] (Computational Geometry Algorithms Library) range query data structure, for example, is a Delaunay triangulation that uses depth-first search to answer queries.

As an alternative to such complex data structures, below we propose a practical heuristic for answering closest point queries, for query objects that are either curves that are convex in one-dimension or curves with constant-sized description that bound convex regions. It is easy to implement and works uniformly for many objects of this type. This method also works for the continuous nearest neighbor problem.

We construct a Voronoi diagram of the input point set and restrict this diagram to a large (implicit) bounding box that encloses all the points and all potential query objects. In addition, we require a point-location structure on this diagram. Given a query curve, we pick any point on the curve and locate the Voronoi cell containing this point. Now we traverse the diagram along the query curve. We can easily prove that the the query curve will cross the Voronoi cell of the site that is closest to the query object.

One of the main advantages of this approach is that the Voronoi diagram is a very common data structure, and implementations already exist in many libraries and/or languages. A system (a GIS for example) may even already be using a Voronoi diagram to store sites, in which case this technique requires little or no additional initialization time and requires little additional space.

To implement the traversal scheme we assume that the bounding curve of a query object permits the following operations:

1. Determine an initial point on the query curve from which to begin the traversal.
2. Compute the distance from a point to the curve. This assumes that the distance function is well-defined and computable.
3. We can determine the edge of the current Voronoi cell from which the curve exits into a neighboring cell in the direction of traversal or that it doesn't exit the cell at all.
4. If the curve exits through a vertex of the current Voronoi cell, we can compute the tangent to the curve at this point or in a small neighborhood of it. The latter condition is included to cover the possibility that the exit point on the curve may not have a well-defined tangent, as for example when the bounding curve is a convex polygon and the exit point is one of its vertices.
5. We can compute the intersection of the curve and a bounding edge of a Voronoi cell.

5.1 Implementation Details

This walk-based idea has been implemented using CGAL [2]. CGAL's Voronoi diagram data structure provides both point location and the Voronoi cells through which query curve passes. The main computational problem is this: given that a part of the query curve is in some cell in the diagram (after possibly having entered the cell through some edge), through which cell edge does this part of the query curve exit the cell? Four approaches to this problem were tested.

Counter-Clockwise Edge Traversal: Starting at some (possibly arbitrary) edge of the current Voronoi cell, we traverse its boundary counter-clockwise until we hit an edge that intersects the query curve.

Binary Search: This technique works only when the query curve is a straight segment or can be decomposed into a sequence of such segments. We assume that the query curve has entered the current Voronoi cell through some initial edge. (The implementation of this method actually uses the previous method to exit the very first cell we encounter.) This is not absolutely necessary, it is just simpler. We then perform a binary search on the edges of the cell to find the one that intersects the query curve. If a Voronoi diagram implementation does not support random access to edges of a cell (which is the case in CGAL's structure), then additional preprocessing and space is required to allow a binary search.

Radial Triangulation Traversal: Every Voronoi cell is triangulated, using the cell's site as a central vertex, to create one triangle for each edge. The motivation for introducing triangulations is that, given triangular cells, it is easier to determine the edge through which a query curve exits a cell. Note that this triangulation need not be explicit. This triangulation depends only on individual Voronoi cells, and so it can be used implicitly.

Random Triangulation Traversal: Every Voronoi cell is randomly triangulated without adding any vertices. Our implementation converted Atkinson and Sack's binary trees [3] into triangulations for this purpose. This triangulation has fewer triangles than the previous one. On the other hand, it might be that a query curve will end up intersecting more edges in this triangulation. Extra preprocessing time is required to add triangulation edges to a Voronoi diagram.

Whenever we encounter an edge intersecting the query curve, we jump over the edge into the next cell, and repeat the process of exiting the cell. For closed curve queries, we need to keep track of the starting point of the walk so that we stop when reaching it again. The implementation stores the first edge that the query intersects, and also the corresponding intersection point.

It could be that the query curve passes through a vertex of the diagram. In this case, we use the tangent of the curve (at that vertex) to determine the cell that the curve enters. If no more than 3 sites are ever on the same circle, then each Voronoi vertex will have exactly 3 adjacent edges, and it is straightforward to determine how the curve exits the vertex. If it is possible that Voronoi vertices can have many adjacent edges, then we can perform a binary search on the adjacent edges to find the exit.

5.2 Experimental Details

The implementation was tested against a brute-force algorithm (looking at each site and checking for the minimum distance). The tests covered circle, triangle, and rectangle queries. In addition to query times, the initialization times were also recorded (see Fig. 5). For comparison, the initialization times of the Voronoi diagram are listed with those of CGAL's range query structure. The range query structure is based on a Delaunay triangulation, and so the Voronoi diagram structure is actually constructed from the range query structure.

There are three potential variables for each query: the number of sites, the span of the sites, and the perimeter of the query object with respect to this span. The tests vary two parameters: the number of sites per set, and the perimeter of query objects with respect to the span of site sets; the span of the sites was kept at a constant 1000 units. If a query object has a relative perimeter of 1/2, then its perimeter is approximately half the vertical (and horizontal) distance spanned by the site set.

The test platform was as follows: 2 GHz AMD Athlon 64 X2 3800+ processor; 896 MB RAM; running under Windows XP Professional x64; compiled by Visual C++ 2005 Express. Site sets and query objects were randomly generated using a uniform distribution from the Boost library [1]. See Figure 5 for initialization times. This test generated a new point set for a number of different set sizes. Figures 6, 10, and 14 show how varying the number of sites affects the query time. Each of these tests kept a fixed query curve (boundary of a circle, triangle and rectangle respectively), and used the site sets generated in the previous test. Figures 7 to 9, 11 to 13, and 15 to 17 show how varying the perimeter of the query curve affects the query time. Each of these tests used ten different site sets, and generated ten different query objects for each set.

■ Brute Force
◆ Voronoi Cell Walk
▽ Voronoi Cell Bin S
▲ Radial Tri
▶ Random Tri
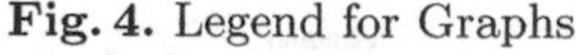
◀ Range Query Structure

Fig. 4. Legend for Graphs

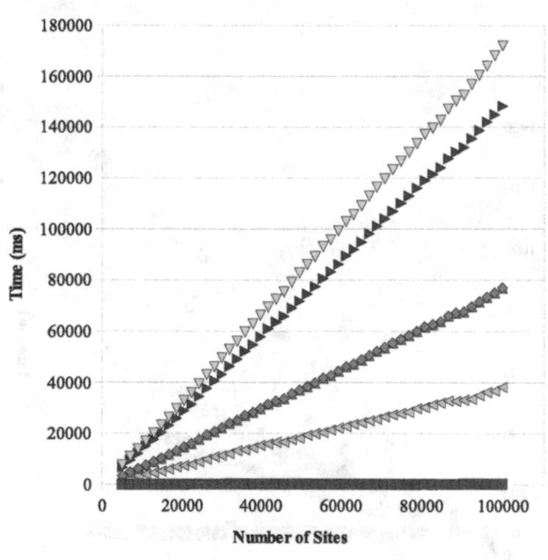

Fig. 5. Initialization Times

These graphs show that the Voronoi cell walk and the binary search techniques are advantageous over brute force when the number of sites is large. In addition, query objects with large perimeter are undesirable. Note that in the figures with circle queries, there appear to be two separate groups of data for the non-brute-force techniques. This is due to rounding errors that cause the algorithm

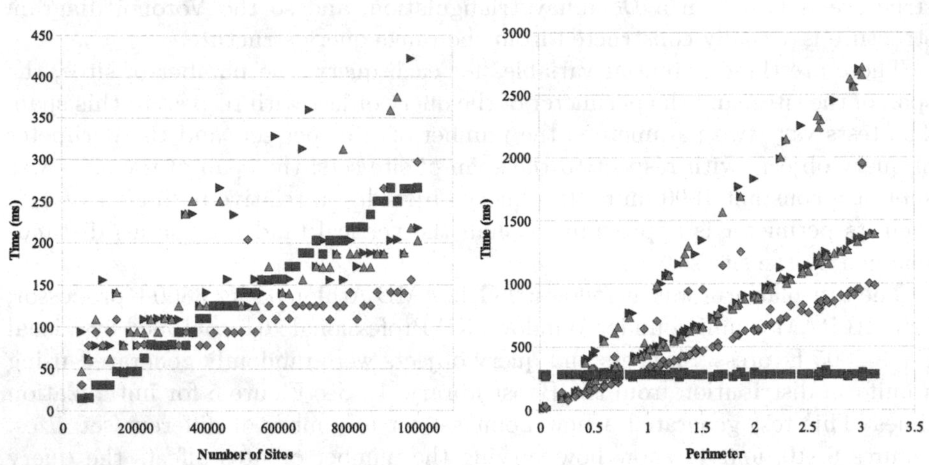

Fig. 6. Query Times for a 0.45 Perimeter Circle

Fig. 7. Query Times for Circles in 100 000 Sites

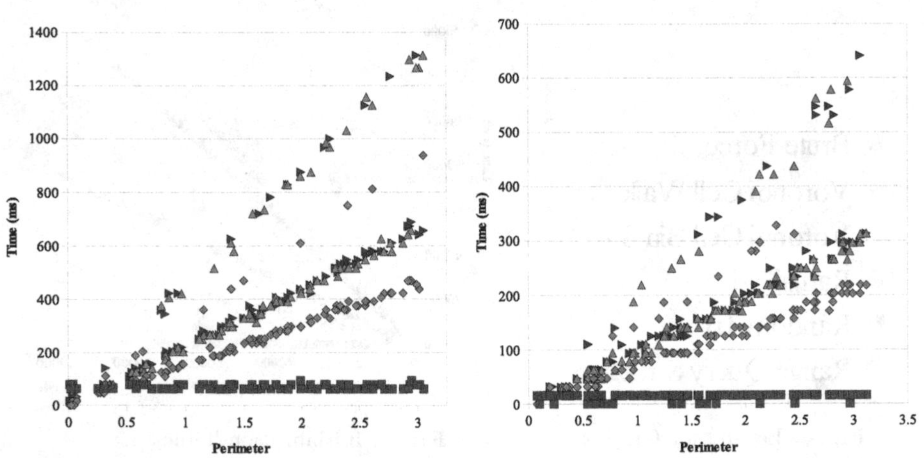

Fig. 8. Query Times for Circles in 25 000 Sites

Fig. 9. Query Times for Circles in 5 000 Sites

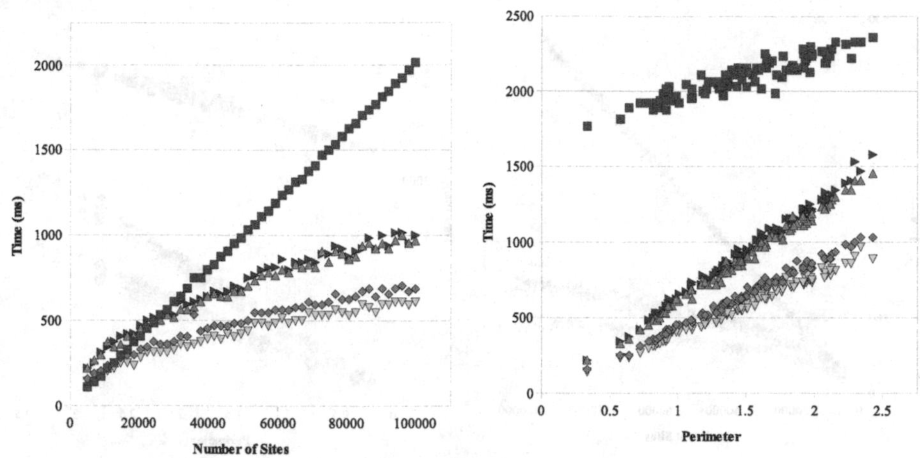

Fig. 10. Query Times for a 1.7 Perimeter Triangle

Fig. 11. Query Times for Triangles in 100 000 Sites

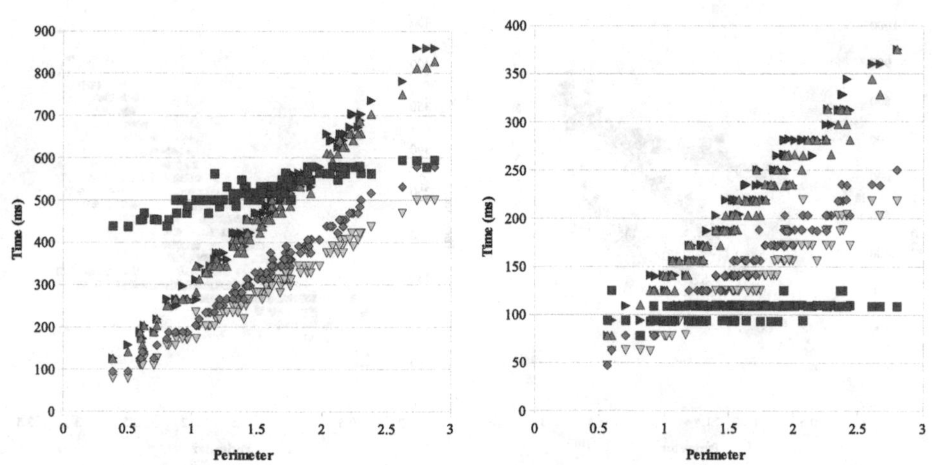

Fig. 12. Query Times for Triangles in 25 000 Sites

Fig. 13. Query Times for Triangles in 5 000 Sites

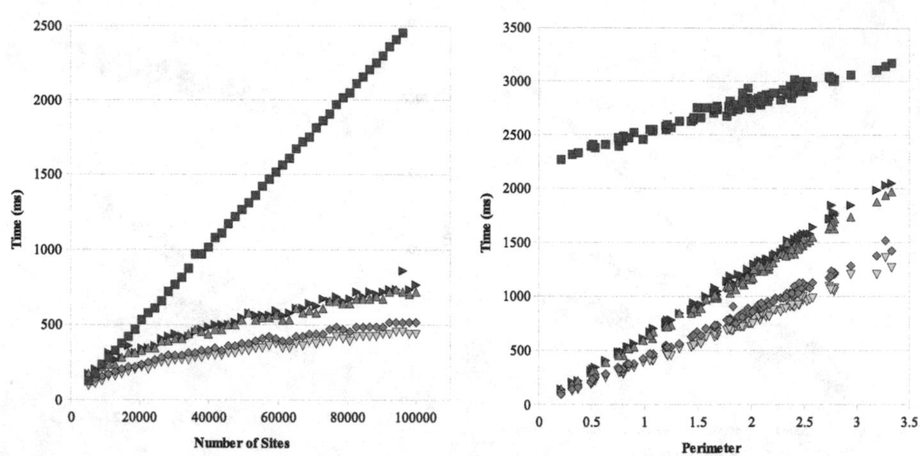

Fig. 14. Query Times for a 1.2 Perimeter Rectangle

Fig. 15. Query Times for Rectangles in 100 000 Sites

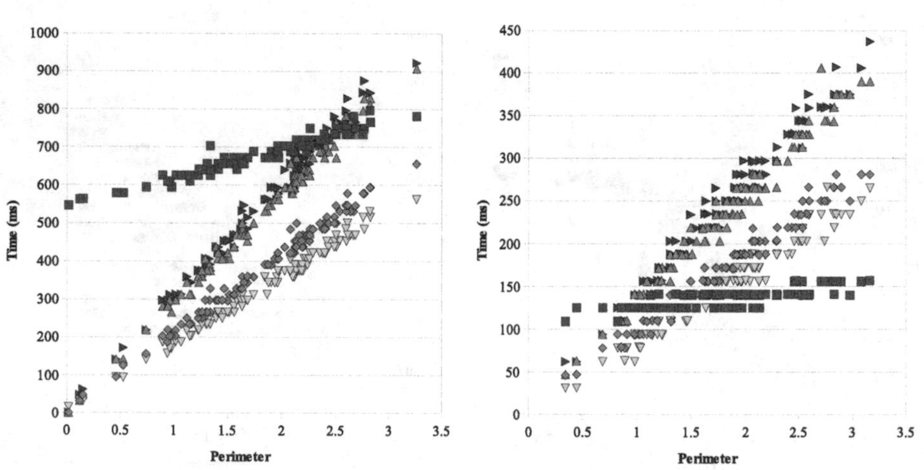

Fig. 16. Query Times for Rectangles in 25 000 Sites

Fig. 17. Query Times for Rectangles in 5 000 Sites

to miss the initial point of intersection and travel around the circle a second time. Circle queries do not do so well because the walk-based algorithms rely on (the relatively expensive task of) finding intersections, while the brute force algorithm only relies on (the relatively inexpensive task of) finding distances. Also, the initialization times for the radial triangulation method are the same as those for the Voronoi cell walk.

6 Conclusions

In this paper we have shown that there is an intimate connection between range-queries and closest-point queries for a variety of complex objects. We have also proposed a simple and implementable heuristic that works uniformly for a variety of complex objects. The method works well for large site sets and query objects with small perimeter relative to the site set. This technique can be used not only to answer closest point queries, but also continuous nearest neighbor queries. It is also conceivable to extend this approach for curves in three dimensions.

References

1. Boost C++ Libraries, `http://www.boost.org`
2. Cgal, Computational Geometry Algorithms Library, `http://www.cgal.org`
3. Atkinson, M.D., Sack, J.-R.: Generating binary trees at random. Inf. Process. Lett. 41(1), 21–23 (1992)
4. Clarkson, K.L.: A randomized algorithm for closest-point queries. SIAM J. Comput. 17, 830–847 (1988)
5. Cole, R., Yap, C.K.: Geometric retrieval problems. In: Proc. 24th Annu. IEEE Sympos. Found. Comput. Sci., pp. 112–121 (1983)
6. de Almeida, V.T.: Towards optimal continuous nearest neighbor queries in spatial databases. In: GIS 2006: Proceedings of the 14th Annual ACM International Symposium on Advances in Geographic Information Systems, pp. 227–234. ACM, New York (2006)
7. Goswami, P.P., Das, S., Nandy, S.C.: Triangle range counting query in 2d and its applications in finding k nearest neighbors of a line segment. Computational Geometry: Theory and Applications, 163–175 (2004)
8. Lee, D.T., Ching, Y.T.: The power of geometric duality revisited. Inform. Process. Lett. 21, 117–122 (1985)
9. Liu, T., Moore, A.W., Gray, A.G., Yang, K.: An investigation of practical approximate nearest neighbor algorithms. In: NIPS (2004)
10. Matousek, J.: Efficient partition trees. Discrete Comput. Geom. 8(3), 315–334 (1992)
11. Mitra, P.: Finding the closest point to a query line. In: Toussaint, G. (ed.) Snapshots in Computational Geometry, vol. II, pp. 53–63 (1992)
12. Mitra, P., Chaudhuri, B.B.: Efficiently computing the closest point to a query line. Pattern Recognition Letters 19, 1027–1035 (1998)
13. Mitra, P., Mukhopadhyay, A., Rao, S.V.: Computing the closest point to a circle. In: CCCG, pp. 132–135 (2003)

14. Mukhopadhyay, A.: Using simplicial partitions to determine a closest point to a query line. Pattern Recognition Letters 24, 1915–1920 (2003)
15. Nandy, S.C., Das, S., Goswami, P.P.: An efficient k nearest neighbors searching algorithm for a query line. TCS: Theoretical Computer Science 299(1), 273–288 (2003)
16. Tao, Y., Papadias, D., Shen, Q.: Continuous nearest neighbor search. In: VLDB 2002: Proceedings of the 28th International Conference on Very Large Data Bases, pp. 287–298. VLDB Endowment (2002)
17. Vapnik, V.N., Chervonenkis, A.Y.: On the uniform convergence of relative frequencies of events to their probabilities. Theory of Probability and its Applications 16(2), 264–280 (1971)
18. Welzl, E.: Partition trees for triangle counting and other range searching problems. In: SCG 1988: Proceedings of the Fourth Annual Symposium on Computational Geometry, pp. 23–33. ACM Press, New York (1988)

How Similar Are Quasi-, Regular, and Delaunay Triangulations in $\mathbb{R}^3$?

Donguk Kim[1], Youngsong Cho[2], Jae-Kwan Kim[2], Yuan-Shin Lee[3], and Deok-Soo Kim[4,*]

[1] Department of Industrial, Information, and Management Engineering Gangneung-Wonju National University, Wonju, South Korea
[2] Voronoi Diagram Research Center, Hanyang University, Seoul, South Korea
[3] Department of Industrial and Systems Engineering North Carolina State University, Raleigh, NC, USA
[4] Department of Mechanical Engineering, Hanyang University, Seoul, South Korea

Abstract. Voronoi diagrams and quasi-triangulations are powerful for solving spatial problems among spherical particles with different radii. However, a quasi-triangulation can be a non-simplicial complex due to anomaly conditions. While quasi-triangulation is straightforward to use when it is a simplicial complex, it may not seem so if it is not. In this paper, we report the experimental statistics of showing the phenomena related with two fundamental issues: i) How frequently anomalies occur in the quasi-triangulation of the arrangement of spherical atoms in $\mathbb{R}^3$ and ii) how much similar or dissimilar the three related structures (i.e., the quasi-triangulation, the regular triangulation, and the Delaunay triangulation of an atomic arrangements) are. The observations from the experiments are as follows: i) Anomalies occur extremely rarely in molecular structures and occur very rarely even in random sphere sets, and ii) the three dual structures of a given set of spheres are not similar.

Keywords: Voronoi diagram of spheres, quasi-triangulation, regular triangulation, Delaunay triangulation, beta-complex, anomaly, triangulation similarity.

1 Introduction

The Voronoi diagram of spheres, also called the additively-weighted Voronoi diagram, is useful for solving spatial problems among spherical particles [14]. In particular, in $\mathbb{R}^3$, the arrangement of spherical atoms has important applications in molecular worlds, both biomolecules and materials, because atoms are usually modeled by spherical balls with different radii. However, unlike the ordinary Voronoi diagram of points and the power diagram, the Voronoi diagram of atoms may at first seem a bit complicated for both computing it and using it to solve application problems. This is mainly because the size difference among atoms

* Corresponding author.

B. Murgante et al. (Eds.): ICCSA 2014, Part II, LNCS 8580, pp. 381–393, 2014.

may cause geometric and topological complications which may seem troublesome but are in fact not so.

The dual structures of the ordinary Voronoi diagram for points and the power diagram are the Delaunay triangulation and the regular triangulation, respectively, and they are known simplicial complexes which have nice topological properties [2,3,14]. However, the dual structure of the Voronoi diagram of spheres, known as the quasi-triangulation, is not necessarily a simplicial complex due to the phenomenon called *anomaly* [9,10,12]. It is called the *quasi*-triangulation because there are usually very few such anomaly cases in most sphere sets and the influence of the anomalies is local.

In this paper, we report the experimental statistics showing two related fundamental issues of the quasi-triangulation: First, how frequently anomalies occur in the quasi-triangulation of a set of spherical atoms in $\mathbb{R}^3$. Second, how much similar or dissimilar the three related dual structures of a given set of spheres (i.e., the quasi-triangulation, the regular triangulation, and the Delaunay triangulation) are. The experiments are based on 100 molecular structures from the protein data bank (PDB, [1]) and four random sphere sets where each set consists of 100 models of random spheres in $\mathbb{R}^3$. The observation made from the experiment is as follows:

– Anomalies occur extremely rarely in molecular structures and occur very rarely even in random sphere sets.
– The three dual structures of the arrangement of atoms are not similar.

Being a report of an experimental statistics, we avoid to attempt to explain the details of the Voronoi diagram of atoms, the quasi-triangulation, the beta-complex, and the anomaly. Readers are recommended to refer to the following articles: For the Voronoi diagram of spheres, see [6,7]; the quasi-triangulation, see [8,10,12,13]; for the beta-complexes, see [11]; for the anomaly, see [9,10,12].

We hope that the statistics reported in this paper are meaningful for designing algorithms and geometric library for sphere sets in $\mathbb{R}^3$. We maintain the data sets used in this paper in the VDRC web site (http://voronoi.hanyang.ac.kr/qtdb/testdataset/random_sphereset) and is available to public so that researchers can benchmark their experiments against the result reported in this paper. We will update this statistics on the web site when we have new information because we cannot rule out the possibility of the existence of a bug in our current code. This paper is an extension of the preliminary report presented in the International Symposium on Voronoi Diagrams in Science and Engineering 2013 [5]. In this paper, vertices in the Voronoi diagram and the quasi-triangulation are denoted as V-vertices and qt-vertices, respectively. Terms of V-edges, V-faces, V-cells, qt-edges, qt-faces, and qt-cells are similarly used.

This paper is organized as follows. Section 2 presents the definitions of anomaly types and the cases that actually occurred in the quasi-triangulations of the test data sets used in the experiment. Section 3 describes the test data set used in the experiment which consists of biomolecular structures available in the PDB and random spheres. Section 4 presents the statistics about the anomaly occurrences obtained from the experiment using the test data set. Section 5 presents the definition of similarity among the quasi-triangulation, the Delaunay triangulation, and the regular triangulation and the statistics obtained from the test data set. Then, the paper concludes.

2 Anomalies in Quasi-triangulation

Anomaly is a condition that makes a simplex set violate to be a simplicial complex. In the quasi-triangulation in $\mathbb{R}^3$, an anomaly can be defined between two tetrahedral qt-cells (an inter-cell anomaly) and between two worlds of qt-cells (an inter-world anomaly). There are three types of inter-cell anomaly: 2-adjacency anomaly, 3-adjacency anomaly, and 4-adjacency anomaly. There are also three types of inter-world anomaly: dangling-face anomaly, dangling-cell anomaly, and dangling-cluster anomaly. Note that a 4-adjacency anomaly and a dangling-cell anomaly are different interpretations of an identical anomaly phenomenon. Hence, there are only five distinct anomaly types in $\mathbb{R}^3$. Note that there is only one anomaly type in $\mathbb{R}^2$. For details, see [9,10,12].

An inter-cell anomaly occurs when a pair of cells has an abnormal face-connectivity. In a simplicial complex, two cells (or 3-dimensional simplexes) in the complex intersect at most one facet if they intersect at all. In the quasi-triangulation, however, two cells may intersect at more than one face and thus the simplicial complex condition is violated. If two cells share two, three, or four faces in the quasi-triangulation, it is called a 2-adjacency, a 3-adjacency, or a 4-adjacency anomaly, respectively.

Fig. 1 shows examples of the three types of inter-cell anomaly visualized in their Voronoi diagram counterparts. In this paper, we illustrate quasi-triangulations in their Voronoi diagram counterparts because two qt-cells involved in an inter-cell anomaly are identically displayed in the Euclidean space and their visualization does not help. This is because each V-vertex in the Voronoi diagram dual-maps to a qt-cell and each V-edge is dual-mapped to a qt-face shared by two qt-cells in the quasi-triangulation. Hence, in the three figures of Fig. 1, the two V-vertices are defined by the same set of four spheres and therefore the dual-mapped two qt-cells completely overlap in their realization in the Euclidean space from geometry point of view. However, in the topology space, the two qt-cells are distinct and intersect only at qt-faces. For example, the two qt-cells in Fig. 1(a) share two qt-faces which are dual-mapped from the two V-edges connecting the two V-vertices. In the figures, both V-vertices are defined by the same set of four spheres and therefore the two dual-mapped qt-cells have an identical set of four qt-vertices. Fig. 1 shows the cases that two qt-cells sharing two, three, and four qt-faces which are called the 2-adjacency, 3-adjacency, and 4-adjacency anomaly, respectively. Recall that two V-vertices are connected by a single V-edge in a Voronoi diagram if its dual structure is a simplicial complex. For details on the anomalies, please refer to the references [9,10,12].

Fig. 2 illustrates these anomaly cases which actually occurred in the random data set used in the experiment. Fig. 2(a) shows a 2-adjacency anomaly where

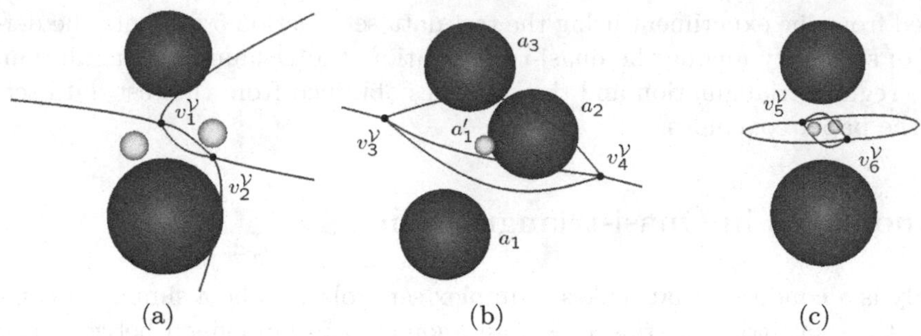

Fig. 1. The examples of three types of inter-cell anomaly shown in their Voronoi diagram counterparts (Figures quoted from [10]). (a) 2-adjacency anomaly, (b) 3-adjacency anomaly, and (c) 4-adjacency anomaly.

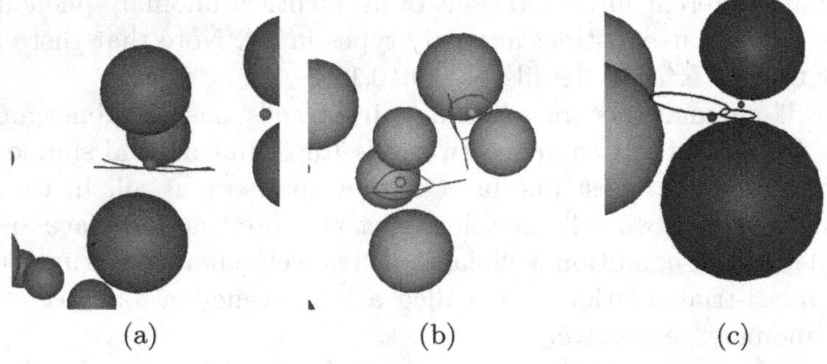

Fig. 2. The examples of the inter-cell anomaly occurred in random models. (a) 2-adjacency anomaly, (b) 3-adjacency anomaly, and (c) 4-adjacency anomaly.

the two V-vertices are defined by the same set of four spheres and these two V-vertices are connected by two V-edges. Fig. 2(b) shows two instances of a 3-adjacency anomaly where each one tiny sphere is contained in the convex hull of the three nearby big spheres. In each instance, a set of the four spheres define two V-vertices connected by three V-edges. Fig. 2(c) shows a 4-adjacency anomaly. In this case, two tiny spheres are contained in the convex hull of the two big spheres and the two V-vertices are connected by four V-edges. Hence, the two qt-cells (mapped from the two V-vertices) share all four qt-faces in the quasi-triangulation.

In the quasi-triangulation theory, a face-connected set of qt-cells is called a world and there can be a hierarchy of worlds in a quasi-triangulation. This definition implies the existence of an inter-world anomaly which occurs when two worlds have no face-connectivity. In the Voronoi diagram counterpart, the two sets altogether correspond to the case that the corresponding Voronoi graph, V-graph (a graph consisting of V-vertices and V-edges), is disconnected. The Voronoi diagram is of course connected from the V-face point of view. This case occurs when the following two conditions are met: i) one or more small sphere is contained in the convex hull of two big spheres, and ii) the V-graph G_{small} defined by all the small spheres plus the two big spheres in the neighborhood form a connected component but is disconnected from the V-graph G_{big} of the other part of the entire Voronoi diagram. We call G_{small} a small-world defined between the two big spheres and G_{big} a big-world. The biggest world is called the root world. Fig. 3 shows two examples. Fig. 3(a) shows a dangling-face anomaly and Fig. 3(b) shows a dangling-cells anomaly. The two big spheres define a qt-edge in the quasi-triangulation which is called a gate for both worlds because a small-world is connected with a big-world through the gate-edge; they are, however, not connected from face-connectivity point of view. In the inter-world anomaly, one or more worlds can be incident to a gate edge.

The simplest form of inter-world anomaly is a dangling-face anomaly as shown in Fig. 3(a). In the figure, one small sphere is contained in the convex hull of the two relatively big spheres and one elliptic V-edge exists between the big spheres. This elliptic V-edge is disconnected from the other part of the entire V-graph and itself forms a single small-world. In the quasi-triangulation, this elliptic edge corresponds to an isolated triangle dangling to a gate edge defined by the two centers of the two big spheres. Fig. 3(b) shows a dangling-cell anomaly where two qt-cells are face-connected through all four qt-faces. In the figure, the two V-vertices are connected through all four V-edges. Recall that the dangling-cell anomaly and the 4-adjacency anomaly are from an identical phenomenon. Note that a single cell cannot form a small-world unless it is an isolated triangular face. Hence, the two qt-cells in Fig. 3(b) form a single world and are attached to the gate edge defined by the two big spheres. When a small-world consists of more than two face-connected qt-cells, it is called a dangling-cluster anomaly. While it is possible in theory, we could not observe such a case in both PDB models and random models.

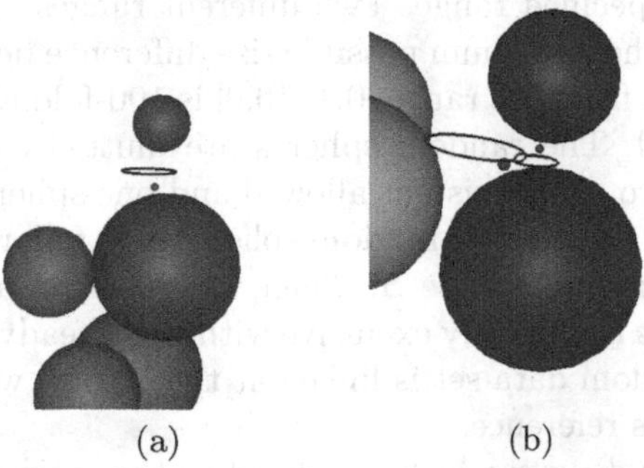

Fig. 3. Inter-world anomaly occurred in random models. (a) Dangling-face anomaly (black), and (b) dangling-cell anomaly.

3 Test Data Set

The test data set consists of two types of models: biomolecular structure models in the Protein Data Bank (PDB) and random sphere models. For biomolecules, we selected 100 PDB models (Accession codes listed in Table 1) where each is determined by x-ray crystallography and thus without any hydrogen atom. For each PDB model, we removed all water molecules and ligands before the experiment: Molecular size is in-between 268 and 5072 atoms, on average 2510.5 atoms. We used the Bondi radii [4] for the van der Waals atoms: C: 1.70; H: 1.20; O: 1.52; N: 1.55; S: 1.80.

Table 1. The 100 PDB models used in the experiment. All the protein models are generated via x-ray diffraction technique and contain no hydrogen atom.

1C26	1D2K	1D4T	1DC9	1DQZ	1EAI	1EDQ	1EQP	1EZG	1F2V
1F60	1FA8	1FHL	1I8K	1IZ9	1J27	1JYH	1K1B	1L7A	1LBW
1LF1	1LHP	1LZ1	1M0Z	1MHN	1MN6	1ORJ	1QB5	1QKD	1QQ1
1QXH	1R2T	1RAV	1RH9	1SWH	1SYQ	1T45	1T4Q	1T6F	1T7N
1TP6	1UGQ	1VDQ	1WLG	1WU3	1X7F	1XBA	1XG2	1XH3	1XIX
1XQO	1XWG	1Y0M	1Y2T	1Y9U	1YCK	1YM5	1YPF	1ZLM	1ZPW
1ZRS	1ZVT	1ZX6	2A8F	2AB0	2CAR	2CWC	2CWL	2EKC	2ERW
2ESK	2ET6	2F6L	2F82	2FN9	2FP8	2FTS	2G7O	2G85	2GAS
2GE7	2GGV	2GOI	2GPO	2GUV	2H2R	2I3F	2I49	2IGD	2NLS
2O37	2O7H	2OBI	2OL7	2OP6	2P19	2YZ1	3B7H	3BXY	4EUG

We also generated four sets of random sphere models, Random A, Random B, Random C, and Random D, where each set consists of 100 random models. Each random model consists of 50 to 5000 atoms with an increment of 50 atoms. In other words, the smallest model in Random A consists of 50 spheres and the second smallest one has 100 spheres, and so on. The largest model has 5000 atoms. The radii of the spheres are assigned at random with a uniform distribution within a specified range. Two different ranges are used: [0.1, 10.0] and [1.0, 10.0]. Hence, the maximum possible size difference between the smallest and the biggest spheres from the range [0.1, 10.0] is 100-fold and, from the range [1.0, 10.0], it is 10-fold. The random spheres are mutually exclusive: i.e., the intersection between two spheres is not allowed and one sphere does not include another. The algorithm to generate random spheres is as follows: We first assume a sufficiently large bounding sphere S. Then, we generate a random sphere s located within S while s is mutually exclusive with the already-generated spheres in the model. This random data set is linked at the VDRC web site and can be downloaded for reader's reference.

Let ρ be the *packing density* which is defined as the ratio of the union of the volume of the random spheres in a model to the volume of S. Then, it seems reasonable to believe that ρ is a parameter closely related with the frequency of anomaly in a model. Hence, we first set a target value of ρ and determined the radius of S from the radius distribution range so that ρ is maintained. Because we have a priori defined number of spheres in each model, the target value of ρ is approximately met. Hence, each random set has three associated parameters: the packing density, the minimum radius, and the maximum radius. The properties of the four sets are as follows:

- Random A: target ρ: 0.10, radius range: [0.1, 10.0],
- Random B: target ρ: 0.10, radius range: [1.0, 10.0],
- Random C: target ρ: 0.15, radius range: [0.1, 10.0],
- Random D: target ρ: 0.15, radius range: [1.0, 10.0].

4 Experimental Result for Anomaly Occurrence

Table 2 summarizes the statistics of the five sets (one PDB model set and four random sphere model sets) to count the number of qt-vertices, qt-edges, qt-faces, qt-cells, and the five anomaly types. The column (vi) shows the average of the four random sphere model sets. V, E, F, and C denote the numbers of qt-vertices, qt-edges, qt-faces, and qt-cells in the quasi-triangulation, respectively. Note that the number of vertices is identical to the number of atoms in the input model.

Table 2. (PDB + RANDOM) Average numbers of simplexes, anomalies, and worlds for the five data sets

		PDB (i)	Random A (ii)	Random B (iii)	Random C (iv)	Random D (v)	Avg(ii ~ v) (vi)
V		2510.50	2525.00	2525.00	2525.00	2525.00	2525.00
E		19276.84	17423.47	17829.17	17018.36	17524.28	17448.82
F		33462.15	29610.90	30418.75	28792.80	29806.79	29657.31
C		16694.81	14711.44	15113.58	14298.46	14806.51	14732.50
average number of atoms		2510.5	2525	2525	2525	2525	2525
expected packing density		–	0.10	0.10	0.15	0.15	0.125
min. radius		1.52	0.10	1.00	0.10	1.00	0.55
max. radius		1.80	10.00	10.00	10.00	10.00	10.00
average number of anomalies	2-adjacency	1.84	38.53	25.70	40.54	27.68	33.11
	3-adjacency	0	13.13	3.59	20.79	4.36	10.47
	4-adjacency	0	0.01	0	0.02	0	0.01
	dangling-face	0	5.62	0.58	8.10	0.68	3.75
	dangling-cluster	0	0	0	0	0	0
	total	1.84	57.29	29.87	69.45	32.72	47.33
average number of worlds		1	6.62	1.58	9.12	1.68	4.75
# gate edges		–	5.62	0.58	8.11	0.68	3.75
max. depth of small-world		–	1	1	1	1	1

4.1 Statistics of the PDB Models

Let V, E, F, and C denote the numbers of qt-vertices, qt-edges, qt-faces, and qt-cells in the quasi-triangulation, respectively. Then, the following is the summary of the experiment of the PDB models from the column (i) in the Table 2:

- $E \approx 7.62\,V$
- $F \approx 13.21\,V$
- $C \approx 6.58\,V$

From Table 2, we can make the following observations on the anomaly cases for the PDB models.

1. There are very few 2-adjacency anomaly cases (1.84 cases/model).
2. There is no 3-adjacency or 4-adjacency anomaly case.
3. There is no small-world (and therefore, no inter-world anomaly, either).

4.2 Statistics of Random Models

From the column (vi) of the Table 2, we obtain the following statistics of the quasi-triangulation for the random data sets:

- $E \approx 6.81\,V$
- $F \approx 11.52\,V$
- $C \approx 5.71\,V$

From the column (vi) of the Table 2, we find the occurrence of the five anomaly types as follows (the averages for the four random sets):

1. The average frequency of the 2-adjacency anomaly is 33.11, and 0.45% of total cells are associated with the anomaly.
2. The average frequency of the 3-adjacency anomaly is 10.47, and 0.14% of total cells are associated with the anomaly.
3. The average frequency of the 4-adjacency anomaly is 0.01, and 0.0001% of total cells are associated with the anomaly.
4. The average frequency of the dangling-face anomaly is 3.75.
5. There is no dangling-cluster anomaly.
6. The maximum depth of all small-worlds is 1.

Note that a single inter-cell anomaly corresponds to a pair of qt-cells. Unlike the PDB models, various anomaly types actually occur in random models.

From the experiment, we observed the following tendency of anomaly occurrence:

1. As the range of radii distribution becomes wider, the more anomalies occur, and
2. as the spheres are densely packed, the more anomalies occur.

5 Similarity and Dissimilarty among QT, DT, and RT: Definition and Experimental Statistics

The experimental statistics about anomalies above may be mistakenly interpreted so that the quasi-triangulation is not very different from the Delaunay and the regular triangulations of a set of spheres. If this is the case, it might be justified for some applications that the Delaunay and the regular triangulations can be used to approximate the solution related with the spatial properties among spherical atoms. However, the fact is that the opposite is true as shown below.

5.1 Distance between Two Triangulations

Let v^X, e^X, f^X, and c^X be vertex, edge, face, and cell in the triangulation X, respectively, where $X \in \{\text{QT, DT, RT}\}$.

Definition 1. *Two cells c^{QT} and c^{DT} are identical if their vertices are from identical set of generating spheres. Two faces f^{QT} and f^{DT} are identical if the qt-cells incident to f^{QT} and the dt-cells incident to f^{DT} are all identical. Two edges e^{QT} and e^{DT} are identical if the qt-cells incident to e^{QT} and the dt-cells incident to e^{DT} are all identical. Two vertices v^{QT} and v^{DT} are identical if the qt-cells incident to v^{QT} and the dt-cells incident to v^{DT} are all identical.*

Definition 2. *Let N_V^{Q2D} be the number of identical vertices between QT and DT. Let*

$$Sim_V^{Q2D} = \frac{2N_V^{Q2D}}{N_V^{QT} + N_V^{DT}} \tag{1}$$

be the measure of the similarity between QT and DT.

The similarity measures Sim_E^{Q2D}, Sim_F^{Q2D}, and Sim_C^{Q2D} for the edges, faces, and cells are defined similar to Eq. (1), respectively. Note that $Q2D$ is used for notational convenience and does not imply any directionality of the measure. In fact, it is directionless.

Note that $0 \leq Sim_V^{Q2D}$, Sim_E^{Q2D}, Sim_F^{Q2D}, $Sim_C^{Q2D} \leq 1$. If QT and DT are perfectly identical from topology point of view, all these four similarity measures are an identity. Fig. 4(a) shows a set of eight disks and its quasi-triangulation. Fig. 4(b) shows the Delaunay triangulation when each disk shrinks to a point. In the example, the two triangulations have different topological structure and we are interested to measure how much they are different. In Fig. 4(c) and (d), the topological structures of v_1 in both triangulations are identical and in Fig. 4(e) and (f), the topological structures of v_2 are different. There are four vertices with identical structures in both triangulations and thus $Sim_V^{Q2D} = \frac{2\times4}{8+8} = 0.5$ in this example.

Definition 2 can also be adapted between QT and RT and between RT and DT. These similarity measures calculated from the five test sets: the PDB model set and the four sets of random models. Fig. 5 shows the computed values of the similarity measures of the PDB models for the four simplexes: vertices V, edges E, faces F, and cells C. The numbers on a link in each figure denotes the average of the computed similarity value. Note that all the similarities are far from one. Fig. 6, 7, 8, and 9 are the counterpart for the Random A, B, C, and D data sets, respectively. From these computed similarity values, we conclude the following.

Observation 1. *The quasi-triangulation, the Delaunay triangulation, and the regular triangulation of a set of spheres are very different.*

Let $Sim(X,Y)$ be the *total similarity* defined by the arithmetic mean of the similarity measures of the vertices, edges, faces, and cells between X and Y, where each of X and Y is either QT, RT, or DT. Fig. 10 shows pairwise total similarities among QT, RT, and DT. From this experiment, we make the following observations:

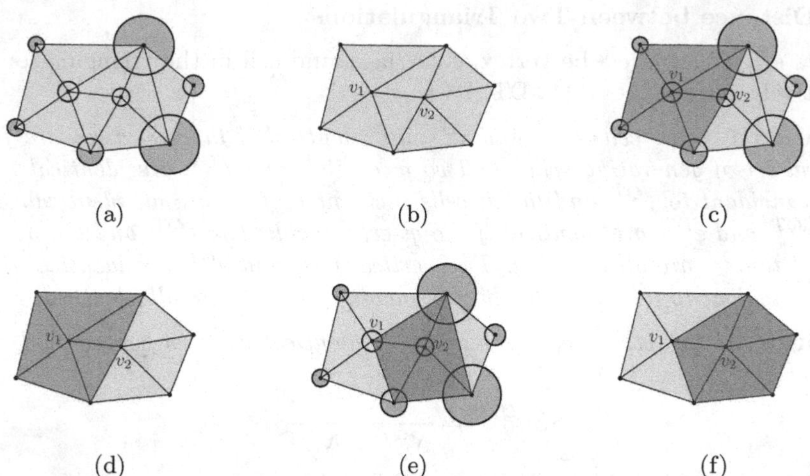

Fig. 4. Similarity measure: QT, DT, and RT. (a) QT, (b) DT, (c) incident cells of v_1 in QT, (d) incident cells of v_1 in DT, (e) incident cells of v_2 in QT, and (f) incident cells of v_2 in DT.

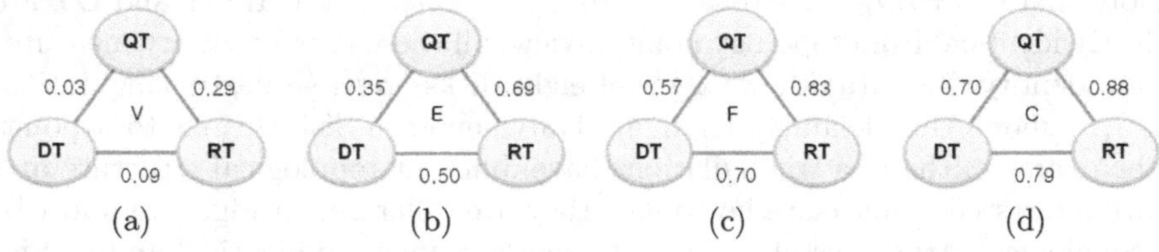

Fig. 5. (PDB) Similarity among QT, DT, and RT. (a) similarity of vertex, (b) similarity of edge, (c) similarity of face, and (d) similarity of cell.

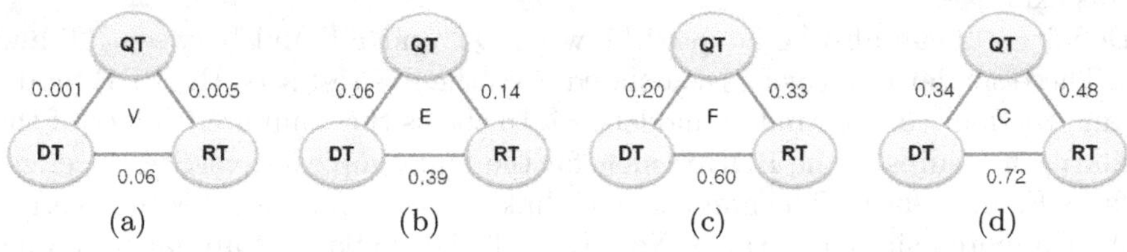

Fig. 6. (Random A) Similarity among QT, DT, and RT. (a) similarity of vertex, (b) similarity of edge, (c) similarity of face, and (d) similarity of cell.

1. QT is significantly different from both RT and DT.
2. For all models, $Sim(\text{QT}, \text{DT})$ is the lowest.
3. For PDB model set, $Sim(\text{QT}, \text{RT}) > Sim(\text{RT}, \text{DT})$.
4. For all random model sets, $Sim(\text{RT}, \text{DT}) > Sim(\text{QT}, \text{RT})$.

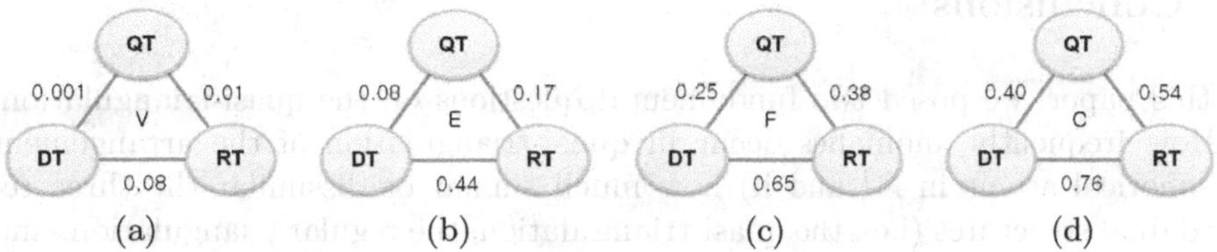

Fig. 7. (Random B) Similarity among QT, DT, and RT. (a) similarity of vertex, (b) similarity of edge, (c) similarity of face, and (d) similarity of cell.

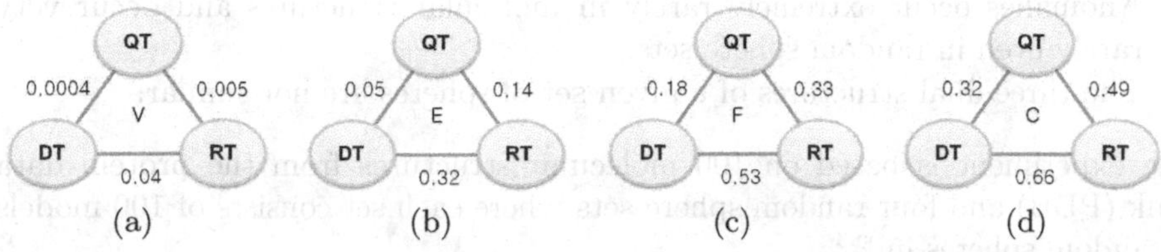

Fig. 8. (Random C) Similarity among QT, DT, and RT. (a) similarity of vertex, (b) similarity of edge, (c) similarity of face, and (d) similarity of cell.

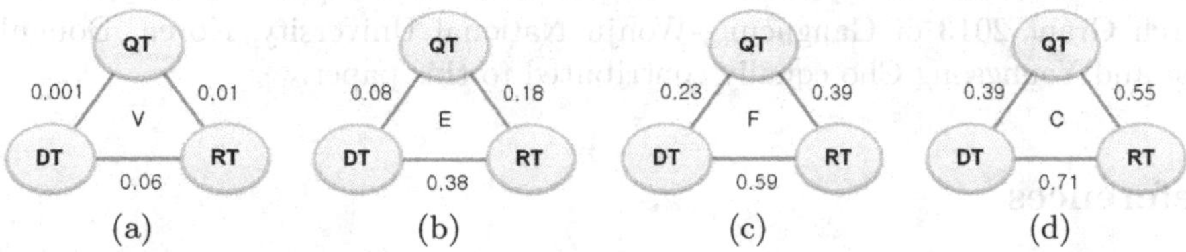

Fig. 9. (Random D) Similarity among QT, DT, and RT. (a) similarity of vertex, (b) similarity of edge, (c) similarity of face, and (d) similarity of cell.

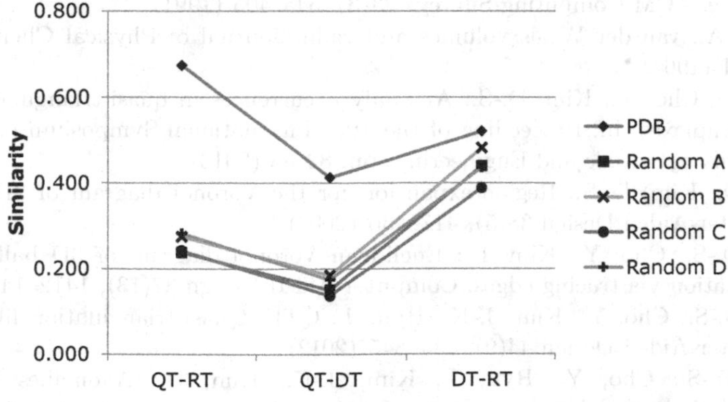

Fig. 10. Total similarity among QT, DT, and RT

6 Conclusions

In this paper, we posed two fundamental questions on the quasi-triangulation: i) How frequently anomalies occur in quasi-triangulation of the arrangement of spherical atoms in $\mathbb{R}^3$ and ii) how much similar or dissimilar the three related dual structures (i.e., the quasi-triangulation, the regular triangulation, and the Delaunay triangulation of an atomic arrangements) are. To answer to these questions, we performed an experiment and provided statistics supporting the following observations:

- Anomalies occur extremely rarely in molecular structures and occur very rarely even in random sphere sets.
- The three dual structures of a given set of spheres are not similar.

The experiment is based on 100 molecular structures from the protein data bank (PDB) and four random sphere sets where each set consists of 100 models of random spheres in $\mathbb{R}^3$.

Acknowledgements. This research was supported by NRF (No. 2012R1A2A1A05026395), Korea. Donguk Kim was supported by the Research Grant 2013 of Gangneung-Wonju National University, Korea. Donguk Kim and Youngsong Cho equally contributed to this paper.

References

1. RCSB Protein Data Bank, `http://www.rcsb.org/pdb/`
2. Aurenhammer, F.: Power diagrams: Properties, algorithms and applications. SIAM Journal on Computing 16, 78–96 (1987)
3. Aurenhammer, F.: Voronoi diagrams – a survey of a fundamental geometric data structure. ACM Computing Surveys 23(3), 345–405 (1991)
4. Bondi, A.: van der Waals volumes and radii. Journal of Physical Chemistry 68, 441–451 (1964)
5. Kim, D., Cho, Y., Kim, D.-S.: Anomaly occurrences in quasi-triangulations and beta-complexes. In: Proceeding of the 10th International Symposium on Voronoi Diagrams in Science and Engineering, pp. 82–88 (2013)
6. Kim, D., Kim, D.-S.: Region-expansion for the Voronoi diagram of 3D spheres. Computer-Aided Design 38(5), 417–430 (2006)
7. Kim, D.-S., Cho, Y., Kim, D.: Euclidean Voronoi diagram of 3D balls and its computation via tracing edges. Computer-Aided Design 37(13), 1412–1424 (2005)
8. Kim, D.-S., Cho, Y., Kim, J.-K., Ryu, J.: QTF: Quasi-triangulation file format. Computer-Aided Design 44(9), 835–845 (2012)
9. Kim, D.-S., Cho, Y., Ryu, J., Kim, J.-K., Kim, D.: Anomalies in quasi-triangulations and beta-complexes of spherical atoms in molecules. Computer-Aided Design 45(1), 35–52 (2013)
10. Kim, D.-S., Cho, Y., Sugihara, K.: Quasi-worlds and quasi-operators on quasi-triangulations. Computer-Aided Design 42(10), 874–888 (2010)

11. Kim, D.-S., Cho, Y., Sugihara, K., Ryu, J., Kim, D.: Three-dimensional beta-shapes and beta-complexes via quasi-triangulation. Computer-Aided Design 42(10), 911–929 (2010)
12. Kim, D.-S., Kim, D., Cho, Y., Sugihara, K.: Quasi-triangulation and interworld data structure in three dimensions. Computer-Aided Design 38(7), 808–819 (2006)
13. Kim, D.-S., Kim, J.-K., Cho, Y., Kim, C.-M.: Querying simplexes in quasi-triangulation. Computer-Aided Design 44(2), 85–98 (2012)
14. Okabe, A., Boots, B., Sugihara, K., Chiu, S.N.: Spatial Tessellations: Concepts and Applications of Voronoi Diagrams, 2nd edn. John Wiley & Sons, Chichester (1999)

WebGIS Solution for Crisis Management Support – Case Study of Olomouc Municipality

Rostislav Netek and Marek Balun

Dept. of Geoinformatics, Palacký University in Olomouc,
17. listopadu 26, Olomouc, 77146, Czech Republic
rostislav.netek@upol.cz

Abstract. Olomouc city, located in the east part of Czech Republic, is historically affected by floods from Morava River. This paper describes designing of web application for crisis management support. Especially it is focused on the support of inhabitants' evacuation from flooded objects and areas. The application deals with Rich Internet Application concept and Service-oriented architecture. It has been developed by Apache Flex technology. It allows implementation of number modern features and tools into the fully web application. Only web browser is needed for running; no installation process is required. Application itself does not contain any data because of following Service-oriented architecture. All spatial layers are connected as a web mapping services according to their character: Representational State Transfer, Web Map Service or Web Feature Service. This approach ensures that data are updated constantly. The main purpose of the application is visualization of flood areas of Morava River in Olomouc city. It contains two kinds of data. Real data was captured during floods in 1997 and 2006. On the other hand, data of statistical prediction flow for hundred-year (Q100), twenty-year (Q20) and five-year (Q5) flood periods. The flooded buildings and streets are displayed as affected depending on selected layer. Both spatial and tabular tools are enabled for selected features. User can display flooded cadastral area, streets and address points with a number of inhabitants, buildings and many other characteristics useful for evacuation process. This application has been developed in cooperation with the Department of Protection of Olomouc Municipality that provided the data for the application.

Keywords: floods, Olomouc city, Rich Internet Application.

1 Introduction

Historically, the city of Olomouc is one of the most important cities in the Czech Republic. Located on the vast plain called Hana, not far from Jeseníky Mountains, it is no stranger to floods from the Morava River, which flows directly through the city center. For example, in 1997 there were disastrous floods which affected a third of the city. About 50 people died and more than 26 000 inhabitants were evacuated from the area. Given these floods, crisis management plays a very important role in the protection of the population.

B. Murgante et al. (Eds.): ICCSA 2014, Part II, LNCS 8580, pp. 394–403, 2014.

Nowadays, Geographical Information Systems (GIS) are not only used for the presentation of spatial data [12]. Crisis management can also benefit from it. It serves as a great tool for decision-making in critical situations when lives and/or property are at stake. Right or wrong decisions have a big impact in these types of situations. That is the reason why the Operational Centre of Rescue systems are equipped with GIS that allows tracking and navigation of rescue vehicles in real-time, determine and calculate the threatened area and the number of population in those areas, among others. In addition, GIS tools are used for long-term strategic planning, such as the detection of flood areas according to the statistical prediction.

This paper describes the design of an application for crisis management support with a focus on the support of evacuation efforts. The application deals with the latest technologies such as Rich Internet Application and Service-Oriented Architecture.

2 Olomouc and Floods

The city of Olomouc is one of the biggest cities in the Czech Republic. From a historical point of view, Olomouc has always been the metropolitan and historical center of the region of Moravia. With a population of about 100 000 inhabitants, the city is the residence of the Archbishop, the second oldest university – Palacký University – and the world's biggest baroque sculpture – the Holy Trinity Column, which was declared a UNESCO World Heritage Site. Olomouc is located on the Morava River. Morava is the second biggest river in the Czech Republic which flows from Jeseníky Mountains, about 60 km away from Olomouc. Furthermore, Olomouc lies directly in the center of the vast plain called Hana, which is quite flat but a very fertile area. The center of the city (49 ° 35 'north latitude and 17 ° 15' E) is 219 m above sea level [2].

With these geographical characteristic, the area is prone to flooding. Where the nearby Jeseníky Mountains are located, there are almost no hills or geographical barriers around Olomouc. Historically, Olomouc was affected by floods from Morava River many times. As was mentioned before, the last great flood took place in the summer of 1997. The main reason behind the flooding was the ten-day continuous rain. Unfortunately, another flood took place in 2006 caused by the melting of the snow. These two incidents prompted the design and application of new flood protection measures. The municipality of Olomouc implemented some technical (modifications of channels, anti-flood barriers), organizational (early warnings, sensors, applications, decision-making support systems), economic and other measures.

3 Methods and Technology

The application combines the latest technologies and approaches such as Rich Internet Application concept and Service-oriented architecture. It has been developed by ArcGIS API for Flex technology; the map content utilizes the following mapping services: Representational State Transfer (REST), Web Map Service (WMS) and Web Feature Service (WFS).

Fig. 1. Disastrous floods in Olomouc city in 1997; source [5]

3.1 Data

Original data were provided by the Municipality of Olomouc. Few other layers were generated during the composition process of the geoprocessing tools. Individual datasets were available in shapefile (*.shp) or file geodatabase (*.gdb). The application deals with the following layers:

- Flood area for theoretical flow (statistical prediction) for hundred-year (Q100), twenty-year (Q20) and five-year (Q5) flood periods of Morava River (Polygon layers)
- Flood area from real floods in 1997 and 2006 (Polygon layers)
- Flood area for Bystřice River (Polygon layer)
- Cadastral areas (Polygon layer)
- Rivers, watercourses and water areas (Polygon layer)
- Buildings (polygon layer)
- Road network (Line layer)
- Streets (Line layer)
- Address points (Point layer)
- Labels

3.2 Service-Oriented Architecture

Currently, there is no precise definition of Service-Oriented Architecture (SOA). It is a general approach to the composition of services independent of the platform.

According to [11] it is a paradigm. From a Geoinformatics perspective, the most common example of SOA is Web services. The fundamental characteristic is the fact that a user does not deal with data in "raw" form and does not work with data files as was typical in desktop solutions. Service-oriented architecture operates with "remote" services. The data are not stored in the user's local computer.

In the case of SOA, service means a sequence of steps, an activity that leads to profit. The basic SOA model is composed of two parts — the Service provider and the Service consumer. The point where both sides will meet is the Registry services. There are descriptions and metadata of services. A consumer searches for a particular service in the register, then the communication with the selected service will be established.

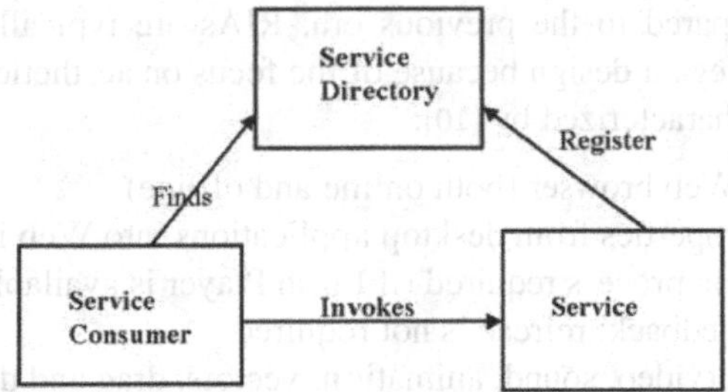

Fig. 2. Basic elements of SOA according to [11]

There are three basic types of service collaboration. The first one is called cooperation: service A works in conjunction with service B with the aim of satisfying consumer needs. The second type is called aggregation: service A connects to service B, and service C is established as a completely new service. The third type is called choreography: it is a collaboration of many services throughout the system in order to achieve business processes as a sequence of actions. The main characteristics according to [11] are:

- Users can work with data from more sources
- It is not required to save any data on the local computer or server
- Data are saved and managed in one place
- Data are still updated
- Centralized data management provides higher efficiency
- Lower financial costs
- Platform independence
- Only a web browser is required to access

There is a number of Web services used in the field of Geoinformatics: Web Map Service (WMS), Web Feature Service (WFS), Web Processing Service (WPS), Web Catalog Service (CSW), Geography Markup Language (GML), Keyhole Markup Language (KML), etc. [1], [9]

3.3 Rich Internet Application

The concept of Rich Internet Application entails trying to provide the same environment and opportunities for developing web applications as for desktop applications. RIA enables features for the web that were typically used in a desktop platform, such as complex graphical interface, user friendly behavior, drag & drop, keyboard shortcuts, etc. From a technical point of view, it is not strictly dependent on a request – response process; independent processing is allowed [4]. RIA is not just one technology. In fact, there are a number of different technologies, such as Apache Flex, Silverlight, AJAX, OpenLaszlo or HTML5.

The main advantage of applications built on the RIA concept is the web as a platform. It means that only a web browser is required for application startup. Recently, RIA has been extended to mobile platforms such as Android, iOS and Windows Phone as well. Compared to the previous era, RIAs are typically characterized by smart interface and elegant design because of the focus on aesthetic features [7], [8].

RIA is generally characterized by [10]:

- Runs in the Web browser (both online and offline)
- Combines properties from desktop applications into Web interface
- No installation process required (if Flash Player is available)
- Immediate feedback; refresh is not required
- Rich content (video, sound, animation, vectors, drag and drop, etc.)
- Platform independent
- Faster processing of requests
- Rich user interface
- Focus on aesthetics
- Easy distribution and start-up
- Mobile applications are fully supported and available
- Open source
- Huge possibility of customization

4 Case Study

The application is developed under the Apache Flex 4.9 technology in combination with ArcGIS API for Flex 3.1, in the environment Adobe Flash Builder 4.

4.1 Graphical User Interface

The layout of the application consists of a map field and a horizontal bar. There are icons and tools for the control of the map in the bar: home button, zoom in, zoom out, list of layers, flood zones, parking feature, info and numerical scale. An additional functionality is located in the pop-up. An interactive legend is located near the right edge; the legend could be turned on or off by users. Layers which are turned off are not displayed in the legend. If a user makes any changes in the symbology (color, transparency, border), it is reflected in the legend automatically. The graphical scale is located at the left bottom.

The first icon is used for "home" functionality; it returns the map back on the default position and scale of 1: 80 000. The second icon is used for zoom in and the third one for zoom out. These icons replaced default element zoomSlider. The fourth icon displays a pop-up window with a list of all available layers. Users can turn layers on or off by using the checkbox. There is a possibility to set transparency for polygon layers.

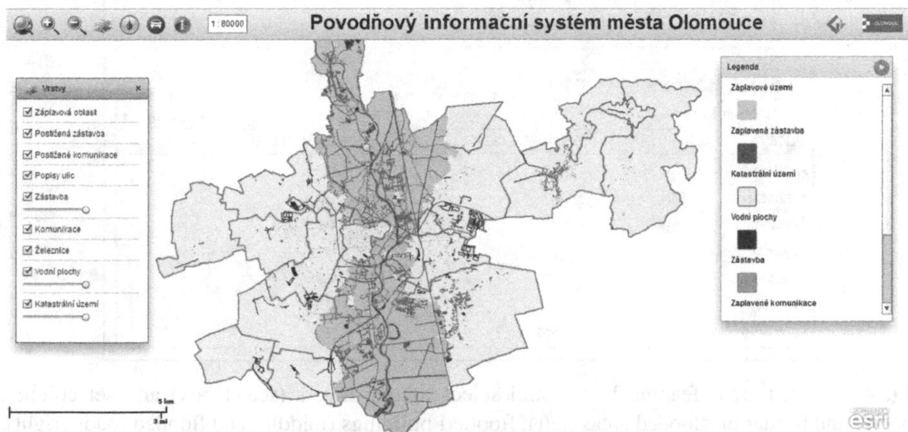

Fig. 3. Application interface; legend description: light blue=flooded area; red=threatened buildings; yellow=cadastral areas; dark blue=water areas; grey=buildings

The most important is the icon of flooded areas. In the pop-up window, users can choose which flood layer is displayed. Users can choose among real data captured during the destructive floods in 1997 and 2006 and the statistical prediction flow for the hundred-year (Q100), twenty-year (Q20) and five-year (Q5) flood periods for Morava River and the flood area for Bystřička River. The flood area is automatically drawn in the map by a blue color as well as the corresponding threatened streets and buildings. The window contains tools for the editing of visual parameters and symbology for individual customization. Finally, there is a table that has attribute data about threatened streets, address points and people that need to be evacuated.

All attributes are calculated dynamically depending on the selected layer, map scale and/or map extent. For example, when only one street is visualized, only attribute information about this street is shown. While buildings and streets outside flooded area are grey, buildings and streets inside are red. If a user clicks on any street, its basic characteristics are shown: street, number of address points, number of population. If a user clicks on any building, other characteristics are shown: number of population, number of apartments, and kind of object.

The red icon with a car is used for the parking feature described above. A user selects a cadastral area from the menu and then a suitable route from the selected area into the Olomouc Airport is shown. There is a possibility to change the color and width of the route line. The last icon is used for help that describes each icon and tools.

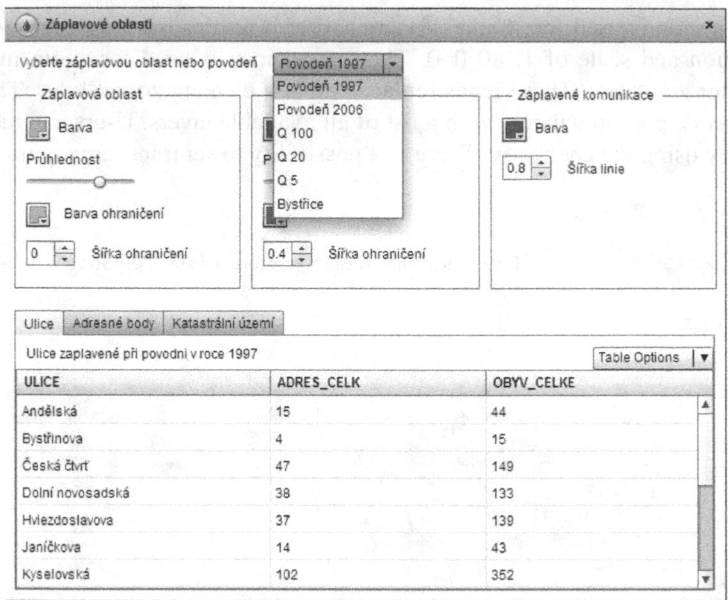

Fig. 4. Flooded areas feature. User could select among layers (select box) and set colours, opacity and border of flooded areas (left), flooded buildings (middle) and flooded roads (right). The table shows number of inhabitans (right), number of buildings (middle) for each street (left) in threatened area.

4.2 Output

The application has been developed through the cooperation of the Department of Geoinformatics, Palacký University in Olomouc with the Department of Protection, Olomouc municipality. The live version is available for public presentation at http://gislib.upol.cz/app/netek13/povodne/. The secured version for crisis purposes will be available after authentication on the intranet of Olomouc municipality.

It is an interactive web application that fully runs in a web browser environment. Based on selected technologies, no installation process is required, only Adobe Flash Player as a plugin for the web browser is required. All requirements from Olomouc municipality have been accomplished. Firstly, a user selects a flood layer (Q100, Q20, Q5, 1997, 2006, Bystřička River) and immediately flooded areas are highlighted by a blue color. The map provides a clear overview of the flooded and non-flooded areas in Olomouc city. According to the selected layer, cadastral areas, streets, buildings are highlighted in red color. Users can investigate on-screen how many people will be evacuated or how many streets will be under water during floods. The attribute table interactively calculates all parameters and enables export to CSV file. That is a great benefit for decision-making support during crisis events. It can be used for strategic planning and flood protection as well.

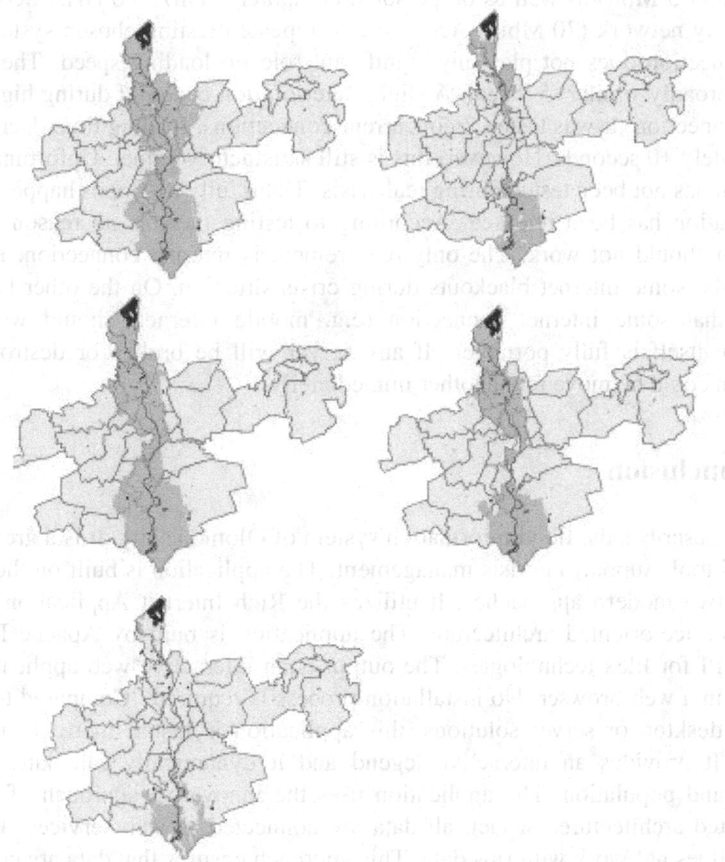

Fig. 5. User can choose among five flooded areas for Morava River. Legend description: light blue=flooded area; yellow=cadastral areas; dark blue=water areas

There is no background under the map. Typically, satellite images could be used for background. There are two reasons why only white color is used. First of all, there was an unsolvable problem with combination layers with different coordinate systems. While all the data in the application used Czech local coordinate system S-JTSK Krovak EastNorth (EPSG 5514), the Esri basemaps or common WMS with satellite images did not support S-JTSK. Unfortunately, when the transformation is made there is about a 20-meter difference between both layers. The second reason is it works faster without a background.

4.3 Testing

The application has been tested during out of crisis situation. The main disadvantage of designed solution is dependency on internet connection. The testing was performed on two different systems: notebook (Win7, 2.6 GHz, 4GB RAM) with slow internet connection (1.5 Mbit) as well as on personal computer (Win7, 3.6 GHz, 8GB RAM) on university network (70 Mbit). According to repeated testing chosen system or internet connection does not play any significant role on loading speed. The loading time was broadly similar (5-7 sec). A slight deterioration occurred during higher concurrent connection. It was tested 7 concurrent connections, loading time decreased to approximately 10 seconds. However, this is still satisfactory value. Unfortunately the application has not been tested during real crisis. Thankfully no floods happened since the application has been released. According to testing there is no reason why the application should not work. The only requirement is internet connection. Probably there will be some internet blackouts during crisis situation. On the other hand it is expected that some internet connection (e.g. mobile internet) should work. The application itself is fully portable. If any server will be broken or destroyed, the application could be move into another immediatelly.

5 Conclusion

The paper describes the flood information system of Olomouc city. It is a great example of GIS tools support in crisis management. The application is built on the combination of two modern approaches. It utilizes the Rich Internet Application concept and the Service-oriented architecture. The application is built by Apache Flex and ArcGIS API for Flex technologies. The output is an interactive web application that fully runs in a web browser. No installation process is required. Compared to the era of robust desktop or server solutions, this application is faster, more flexible, and effective. It provides an interactive legend and it dynamically calculates flooded buildings and population. The application uses the innovative approach of the Service-oriented architecture. In fact, all data are connected as web services (REST or WMS); it does not work with raw data. This approach ensures that data are constantly updated. It contains two kinds of data. Real data was captured during the destructive floods in 1997 and 2006. Statistical data predict flow for hundred-year (Q100), twenty-year (Q20) and five-year (Q5) flood periods. The main purpose of the output application is the visualization of flooded areas in Olomouc city. Depending on the selected layers and map extent, flooded buildings and streets are highlighted in both graphical and tabular form. Users can investigate on-screen which streets and/or population will be evacuated during floods. The application has been developed by the Department of Geoinformatics under the request of the Olomouc municipality.

References

1. An Introduction to Service-Oriented Architecture from a Java Developer Perspective, `http://www.onjava.com/pub/a/onjava/2005/01/26/soa-intro.html`
2. Burian, J., Brus, J., Voženílek, V.: Development of Olomouc city in 1930-2009: based on analysis of functional areas. Journal of Maps 9(1), 64–67 (2013) ISSN: 1744-5647, doi:10.1080/17445647.2013.778800

3. Cutter, S.L.: GI Science, Disasters, and Emergency Management. Transactions in GIS 7(4), 439–446 (2003)
4. Johansson, H.: Rich Web Map Applications, 68 p. Chalmers University of Technology, Sweden (2010)
5. Hrdličková, M.: Možnosti minimalizace povodňových rizik v Olomouci. Magistrát města Olomouce, Odbor ochrany, 43 p. (2012)
6. Konečný, M., et al.: Dynamická geovizualizace v krizovém managementu. Muni Press, Brno (2011)
7. Macromedia Flash MX - A next-generation rich client, `http://www.adobe.com/devnet/flash/whitepapers/richclient.pdf`
8. Meier, J.D.: Rich Internet Application Architecture Guide, 145 p. (2008)
9. Pinde, F., Jiulin, S.: Web GIS: Principles and Applications, 312 p. Esri Press (2011) ISBN:158948245X 9781589482456
10. Netek, R., Dobesova, Z., Vavra, A.: Innovation of Botany Education by Cloud-based Geoinformatics System. In: Wang, Q. (ed.) Innovative Use of Online Platforms for Learning Support and Management. Int. J. Information Technology and Management, 7 p. (2013)
11. Schreiner, V.: Implementace SOA pomocí moderních ICT principů. Masarykova Univerzita, Brno (2007)
12. Voženílek, V.: Artificial intelligence and GIS: mutual meeting and passing. In: 2009 International Conference on Intelligent Networking And Collaborative Systems (INCOS 2009), pp. 279–284 (2009) ISBN 978-1-4244-5165-4

Sensing World Heritage

An Exploratory Study of Twitter as a Tool for Assessing Reputation

Vasco Monteiro[1], Roberto Henriques[1], Marco Painho[1], and Eric Vaz[2]

[1] Instituto Superior de Estatística e Gestão de Informação, Universidade Nova, Lisboa, Portugal
{vmonteiro,roberto,painho}@isegi.unl.pt

[2] Department of Geography, Ryerson University, Toronto, Canada
eric@geography.ryerson.ca

Abstract. Social media services play an important role in today's society and are increasingly used by the scientific community to understand the human landscape, with regard to local sensitivities and the broadcasting of opinion. Collection of social media feeds has become a new data source for understanding and modelling phenomena. This article explores the possibility of understanding the relationship of people with world heritage using the information collected from Twitter. The information collected from late December 2013 to the end of January 2014, submitted to temporal, spatial and text mining analysis, shows that Twitter messages convey important meaning about local and global sensitivities regarding world heritage. Examples include the buzz regarding the possible delisting by the Australian government of the Tasmanian Wilderness World Heritage; the destruction of the heritage in Syria due to armed conflict, and the petition to include the Leuser Ecosystem in the world heritage list. It concludes that using data from social media services, such as Twitter, could improve our understanding of how people relate to world heritage, in a local and global perspective, and be a valuable tool in world heritage studies and management.

Keywords: web 2.0, spatio-temporal analysis, text mining, heritage reputation, Twitter, world heritage.

1 Introduction

Today, anyone holding a device with an internet connection has the ability to broadcast their opinions and everyday life experiences [1], functioning as a big network of citizen sensors [2]. Web 2.0 technologies and social media services, such as Twitter, are stimulating and shaping society [3] due to their growing popularity; and they are also shrinking the distance between people [4]. These services are facilitating the production and distribution of geospatial information by citizens providing unprecedented opportunities to understand users' behaviour regarding (i) local sensitivities and needs, (ii) how the public spaces are used, and (iii) regarding significant or unexpected events (e.g. wars, natural disasters, among others) [1], [5]. Due to

B. Murgante et al. (Eds.): ICCSA 2014, Part II, LNCS 8580, pp. 404–419, 2014.

location-based user-generated content, users are contributing to large quantities of geospatial data that provide unparalleled access to the public's "patterns of behaviour, opinions and preferences, sensitivities and specific needs" [5].

While there has been an increasing interest in using Twitter to track emerging trends, common interest in specific events, and daily activity patterns, a focus on the spatio-temporal aspects of such trends is relatively recent [1], [6,7]. Twitter provides a suitable platform for spatio-temporal data mining and information retrieval and, spatio-temporal trends can be mined and analysed by classifying tweet text content into topic-based groups [6]. Spatial information included in the tweet can be within the tweet text, when users talk about a geographic place; in a descriptive form in the tweet metadata, as a city name, for example [8]; or since 2009, in the form of coordinates with latitude and longitude, that contain the location of the GPS-enabled device every time a user tweets [9]. Twitter is of particular interest to researchers due to a number of factors: (i) the profiles are public [10]; (ii) tweets can be filtered and downloaded using the Twitter APIs [10] and (iii) more than 90% of the profiles and communication history is publicly visible, and therefore available to collect and process [5].

The objective of this research is to explore the potential of the social media feeds, such as the one shared by Twitter users, to understand both local and global sensitivities regarding heritage sites, specifically, those that are considered as world heritage by the United Nations Educational, Scientific and Cultural Organization (UNESCO); and to assess the heritage reputation using the framework in Figure 1 using the citizen sensing from the social media service Twitter [11].

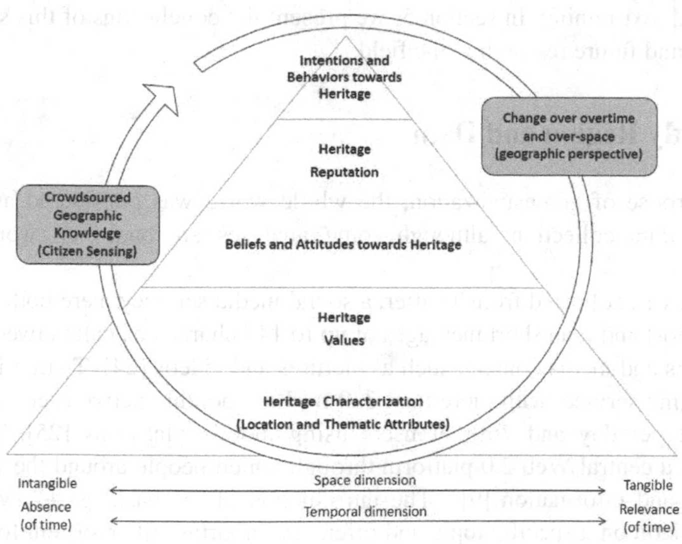

Fig. 1. Heritage Reputation Framework

The proposed framework entails that citizen sensing can help to measure heritage reputation, and its change over time and over place of a specific heritage, by assessing the heritage values - specifically 'whose' values (in this case the general public that uses the web 2.0/social media), 'where' values (the geographic perspective), and 'when' values (over time) [12] - knowing how these values impact in beliefs and attitudes and therefore in its intentions and behaviours towards heritage.

In recent years, accompanying the popularity of the World Heritage List, an extensive scientific literature has been produced regarding world heritage. These publications concern general issues [13]; advantages and disadvantages of being listed and/or economic value [14, 15]; the use of the world heritage sites (WHS) brand name [16]; the outstanding universal value of world heritage sites and tensions between national and global interests [17–19]; tourism development in WHS [20]; climate change [21]; case studies of specific WHS [22]; and the use of spatial technologies - Geographic Information Systems (GIS), remote sensing and Global Positioning System (GPS) – to inventory, manage and monitor the WHS [23].

It is proposed that studying the world heritage list, using citizen sensing, is a novel means to understanding how the general public relates with world heritage. This exploratory analysis look into concepts studied within the field of heritage studies field in a different perspective, more related to the general public view. For example, the advantages of being listed, the brand of world heritage, the universal value of world heritage and the tensions between local, national and global heritage.

The article is organized as follows. In section 2, we present the study region and the data collected from Twitter. In section 3 the methodology. In section 4, we present the results and discussion of the exploratory analysis, spatial and spatial-temporal analysis and text mining. In section 5, we present the conclusions of this study, some limitations and future research in the field.

2 Study Region and Data

For the purpose of geovisualization, the whole world was considered as the study region for data collection; although some analyses are based on world country boundaries.

The data was collected from Twitter, a social media service where both mobile and web users post and read short messages of up to 140 characters, called tweets, that can contain links and media content, such as pictures and videos [24]. Twitter is a popular microblogging service, with more than 230 million monthly active users, 500 million tweets sent per day and 76% of users using mobile platforms [25]. Twitter has emerged as a central Web 2.0 platform through which people around the world share knowledge and information [9]. The introduction of the hashtag "#" syntax facilitates discussion on a specific topic and offers an important filter system for a specific subject [26]. Tweets can have both primary and secondary status, the first when the user posts a message and the second when the tweet is a reply to another user or a re-tweet broadcasting of a prior message written by another user [27].

Using the ScraperWiki online tool [28] a total of 11,839 tweets in the period of Dec 20, 2013 and Jan 26, 2014 were extracted from Twitter.

3 Methods

This section presents an overview on the methodology used to analyse the spatial, temporal and content patterns of the tweets relating to world heritage.

3.1 Data Collection

In order to extract the tweets the ScraperWiki [28] was used. The ScraperWiki is an online tool that that uses both the Twitter REST and the Twitter Streaming API [29], to collect tweets from the preceding week and then automatically refresh for future tweets. ScraperWiki has the advantage of not needing coding, of working entirely online and of being dependent on an infrastructure and connection that has virtually a zero downtime.

Data collection was performed via the words *world* and *heritage* and *worldheritage*. The harvesting process began in Dec 27, 2013 and hence the first collected tweet using ScraperWiki was from one week forehand, it had the timestamp 2013-12-20 06:01:59+00:00. The data was downloaded as a spreadsheet for Microsoft Excel™ on Jan 26, 2014, and the final tweet has the timestamp: 2014-01-26 22:12:39+00:00. The following fields were collected, in a total of 15: (i) the unique identifier of the tweet, as *id_str*; (ii) the tweet link in the form of https://twitter.com/screen_name/status/id_str, as *tweet_url*; (iii) the timestamp of the creation date in the form of YYYY-MM-DD HH:MM:SS+00:00 for the GMT time-zone, as *created_at*; (iv) the message convey on the tweet, as text; (v) the language code, using the two letter code of the ISO 639-1, as *lang*; (vi) the count of retweets for the specific tweet, as *retweet_count*; (vii) the user profile name that has posted the tweet, as *screen_name*; (viii) the first link encountered in the tweet text, as *url*; (ix) the first user profile name encountered in the tweet text, as *user_mention*; (x) the first hashtag encountered in the tweet text, as *hashtag*; (xi) the latitude of the device when posting the tweet, as *lat*; (xii) the longitude of the device when posting the tweet, as *lng*; (xiii) the link of the image shared with the tweet text, as *media*; (xiv) the user profile name, if the tweet is a reply to a specific user, as *in_reply_to_screen_name*; and (xv) the tweet unique identifier, if the tweet is a reply to a specific status made by a user, as *in_reply_to_status_id*.

3.2 Data Processing for Geospatial Analysis

After the tweets were collected two pre-processing stages were performed: (i) filtering of the tweets that had latitude and longitude with the Microsoft Excel™ and (ii) then data input into Quantum GIS 2.0.1 Dufour, using the import CSV layer tool. The analysis of tweets per country was obtained through a spatial join of the tweets with the polygons of the countries. The coordinate system used for the spatial analysis was the World Geographic System (WGS) WGS84.

The heat map used in the spatial analysis uses the kernel density estimation; and a 1,000 km distance was used in order to show a regional distribution. A world map showing coastlines was used as a base at the 1:10m scale available from EarthData (http://www.naturalearthdata.com/downloads/10m-cultural-vectors/). In the country maps a world map with country boundaries at the 1:10m scale from the same source was used.

3.3 Data Processing for Text Mining

Text mining analysis was executed using the SAS Enterprise Miner WorkStation 12.1. The analysis was performed on a data source created using the spreadsheet downloaded from ScraperWiki. Since the tweets in Japanese were not correctly loaded to the data source, they were deleted from the spreadsheet. In order to perform the text mining analysis, only the text field of the database was used.

The following tools were used to perform the text mining analysis: (i) text parsing, which parse the tweet text in order to quantify the terms, showing for example the frequency of words as nouns, group nouns, verbs, among others; and the specific list of terms within the tweets and its frequency; (ii) text filter, which can be used to remove terms or to select only tweets that have a specific issue; and (iii) text topic, which explores the tweets collection by automatically associating terms and tweets according to topics describing main ideas or themes, both discovered and/or user-defined (e.g. you might be interested in mining tweets that discuss a "online petition") [30].

Text topic approach assigns a score for each term and tweet to each topic, based on quantification of terms (frequency of occurrence) and the derived weights of the terms for each tweet using statistical methods as singular value decomposition (SVD). After thresholds area used to determine the strength of the association in order to considerer if a tweet or term belongs to the topic. Using the method, tweets and terms can belong to more than one topic or to none at all that can belong to more than one topic or to none at all, differentiating from the clustering that only assign a tweet to a unique cluster [30]. A topics list example in shown in table 1 in section 4.3.

Figure 2 shows the methodology used to allocate a place to a tweet without coordinates available from Twitter; this place is the one mentioned in the tweet text and not the location of the device where the tweet was generated. After the parsing of the tweets database to discover the content terms, 25 topics were selected for this exploratory analysis. No user-defined topics were defined, because no prior knowledge of the tweets content existed. For each topic the terms were analysed to discover possible places. For those topics where some place was found, that place was directly allocated as the topic location. For those without a place term, the concept linking tool was used to find terms with some spatial connotation that are highly associated with the

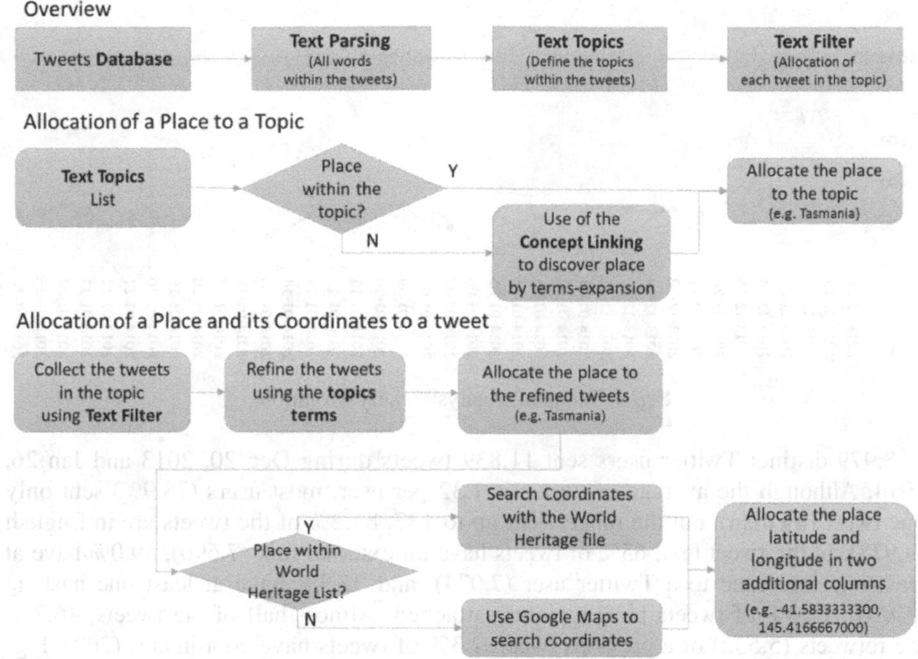

Fig. 2. Methodology to assess the tweet location using Text Mining

selected term. Since some tweets were allocated to the topics incorrectly, a refining of the terms was performed [30]. Finally to assign the place coordinates the documentation of UNESCO for WHS and the Google maps were used.

4 Results and Discussion

This section presents the results and discussion of the exploratory analysis conducted in the various fields collected with the tweets and temporal analysis; the geovisualization and geospatial analysis; and the text mining and discovery analysis of the tweets, for both topics and terms, and allocation of a place for non-geolocated tweets.

4.1 Exploratory Analysis

To better understand the tweets an exploratory analysis was conducted to extract knowledge and information from the fields collected other than the tweet text.

Figure 3 shows the tweets temporal evolution per day during the period of Dec 20, 2013 and Jan 26, 2014, with an average of 312 tweets per day. The period with the highest volume of tweets is Dec 21 - 27, 2013, with peaks in Dec 23, Dec 24 and Dec 27 with respectively 1,011; 935 and 826 tweets.

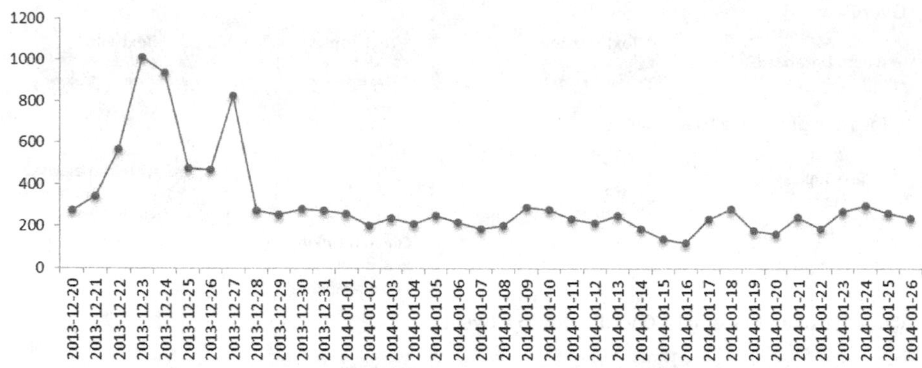

Fig. 3. Temporal Analysis: daily evolution

8,979 distinct Twitter users sent 11,839 tweets during Dec 20, 2013 and Jan 26, 2014. Although the average of tweets is 1.32 per user, most users (78.9%) sent only one tweet (median), but the range went up to 137. 84.2% of the tweets are in English (9,973). In the tweet text, 65% of tweets have an external link (7,690), 59.0% have at least one reference to a Twitter user (7,074), and 32.3% have at least one hashtag (3,820). 11.0% of tweets have a picture attached. Almost half of the tweets, 46.7%, are retweets (5,532) of a previous tweet. 2.3% of tweets have coordinates (262). Figure 4 summarizes the data mentioned above.

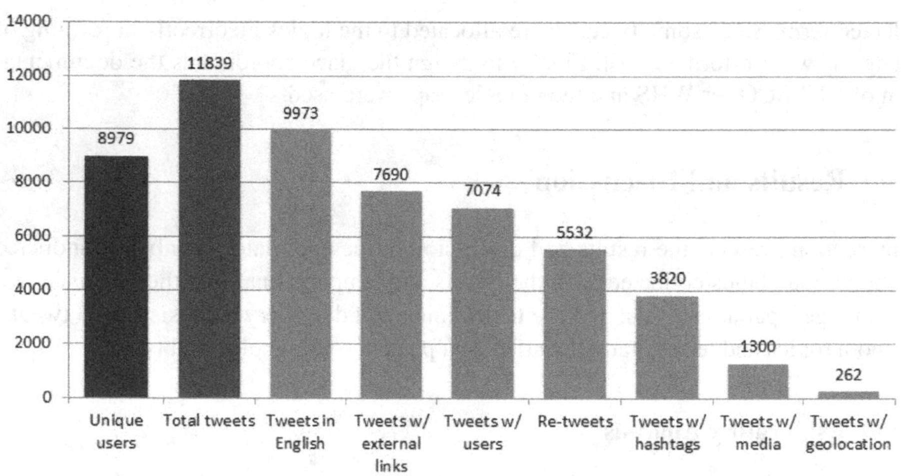

Fig. 4. Tweets Dashboard

As shown in Figure 4, 84.2% of tweets were written in English (9,973), with the remaining top 5 languages being Indian languages with 885 tweets (7.48%); Japanese with 343 tweets (2.9%); German with 260 tweets (2.2%) and the Slovak with 89 tweets (0.75%). All the non-English language tweets had the words *world* and *heritage* or *worldheritage* in the tweet text in English.

Considering the user_mention field, the most mentioned user was Thomas Keneally, an Australian novelist and historian. His tweet regarding the delisting of the Tasmania World Heritage, was retweeted by 435 users, representing 6.2% of all tweets that mention users: "RT @ThomasKeneally: Makes one proud we're the only country's that's ever asked to de-register one of our World Heritage sites". It is interesting that the UNESCO profile on Twitter is the fourth most mentioned, with 199 tweets, representing 2.2% of total tweets. Other mentioned users in the top 5 are FaktaGoogle with 252 tweets (3.6%), WOWFakta with 207 tweets (2.9%) and NatGeoTravel with 155 tweets (2.2%). 79.1% of the tweets mentioned one user within the text and 15.5% two users.

The most mentioned hashtag is #8D with 260 tweets representing 6.8% of all tweets with hashtags. The hashtag of the organization responsible for the listing of world heritage sites, #UNESCO, is in the third position with 171 tweets (4%) and the official hashtag for the world heritage #WorldHeritage is in seventh position with 81 tweets (2.1%); the sum of these two hashtags is 232 tweets, 6.1% of all tweets with hashtags. 83 tweets with the association of the hashtag #Travel with the world heritage sites, represents 2.17%. There are two hashtags referring to countries, namely #UK and #Syria, in the eighth and tenth positions, representing 3.5%. In second place, the hashtag #auspol, related to Australian politics, is mentioned in tweets related to the delisting of the Tasmanian Wilderness and Great Barrier Reef world heritages sites for the purpose of economic exploration; and in fourth place, the hashtag #SaveAceh, related to a campaign to include the Leuser Ecosystem, in the Aceh region in Indonesia, as a world heritage site, because there is an economic plan to explore the region. The remaining top 10 hashtags are #7d and #9d with 81 tweets (2.1%) and 78 tweets (2%), respectively. 55% of the tweets have one hashtag and 22% have two hashtags within the tweet text.

Fig. 5. Screenshots of the Top 10 external links shared in the tweets

Figure 5 shows the screenshots of the top 10 external links mentioned in the tweets. The most shared external link with 201 tweets (3%) is related to a picture of young students of the Indonesian province of Papua, using the Noken[1], a handcraft

---

[1] http://www.unesco.org/culture/ich/index.php?lg=en&pg=00011&USL=00619

bag that is an intangible cultural heritage from UNESCO since 2012, as shown in the top left of Figure 5. Information related to the Tasmanian wilderness world heritage and the plans to delist from world heritage were mentioned as the third and fourth most shared external links, accounting for 256 tweets. These were in the form of online news and a petition to keep the place as a world heritage site, respectively. The Leuser ecosystem is mentioned too in this field with an external link in the fifth position with 99 tweets (1%), in the form of a petition to declare the location as a world heritage site, for its protection.

Figure 6 shows the top 10 media included with the tweet text. The first image represents the Crater Lake in the Mount Ruapehu located in the Tongariro World Heritage National Park in New Zealand, with 364 tweets, representing 28% of all tweets with media included. The second image shared is related to the Leuser ecosystem, with 26 tweets (2%) and the fifth image, with 23 tweets (2%), is related to a campaign to list the Forth Bridge in Scotland as a world heritage site.

Fig. 6. Images shared with the tweets – Top 10

4.2 Geovisualization and Geospatial Analysis of Geolocated Tweets

Regarding the 262 tweets with latitude and longitude coordinates disclosed as metadata, Figure 7 shows their distribution in the world map, in points, and its heat map, in a grey scale that shows that most of the tweets are located in Europe and in the Far East (Asia).

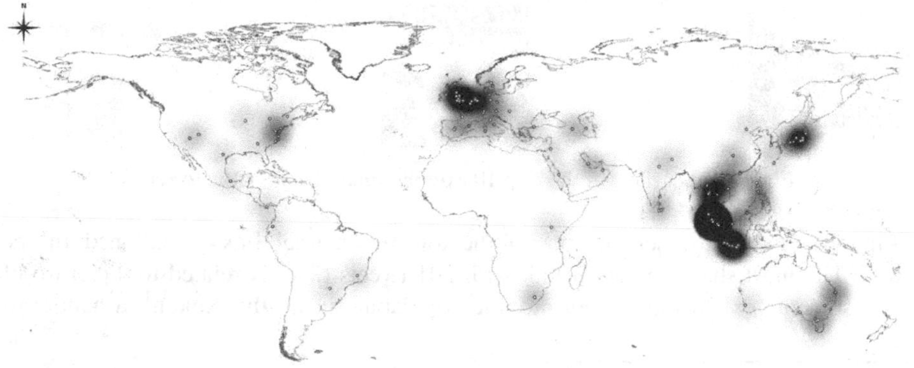

Fig. 7. Geovisualization with heatmap of the Geolocated Tweets

Figure 8 shows the tweets by country, where the dot is located in the centroid of the polygon that represents the country. The biggest density of tweets is in Malaysia with a total of 103 tweets, representing 39% of all tweets with geolocation. The remaining countries are Indonesia with 18 tweets (7%); Australia, Japan and tthe United Kingdom, each with 12 tweets (5%); United States, with 11 tweets (4%); Thailand, with 8 tweets, (3%); Netherlands and Vietnam, with 6 tweets (2%); and finally, Philippines, with 5 tweets (2%).

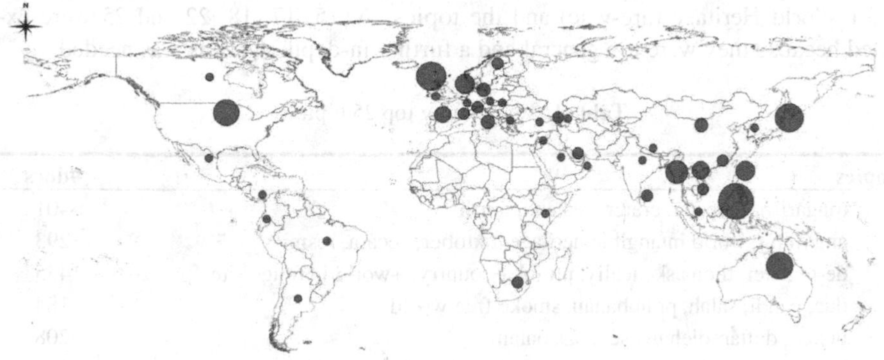

Fig. 8. Geovisualization of geolocated tweets per country

65.3% of the tweets with coordinates are related to the location where they were posted. Examples are *"I'm at George Town World Heritage Inc. (George Town, Penang)"*, *"I'm at Melaka World Heritage City (Melaka)"* or *"Finally in the world heritage city Quedlinburg @ Historische Altstadt Quedlinburg"*.

4.3 Geovisualization and Geospatial Analysis of Non-Geolocated Tweets

In order to assess the place of non-geolocated tweets, text mining analysis were performed. Table 1 shows the main 25 tweets topics, as in the SAS output. The topics 1, 4, 11, 12, 13, 14, 16, 19 and 20 are related to eight different places. The places that are included in the world heritage list are the Tongarino National Park[2] in New Zealand (topic 1); the Prambann Temple Compounds[3] in Indonesia (topic 4); the Tasmanian Wilderness[4] located in Australia (topic 12 and 16); Edinburgh[5] (old and new town) in Scotland – United Kingdom (topic 14); and the Great Barrier Reef[6] in Australia (topic 20). The remaining places mentioned in th top 25 topics are two places that are not in the world heritage list: the Leuser Ecosystem[7] in Indonesia (topic 11)

[2] http://whc.unesco.org/en/list/421
[3] http://whc.unesco.org/en/list/642
[4] http://whc.unesco.org/en/list/181
[5] http://whc.unesco.org/en/list/728
[6] http://whc.unesco.org/en/list/154
[7] http://leuserecosystem.org/

and the Waterberg Biosphere[8] in South Africa, a UNESCO Biosphere property (topic 19); and finally the England's Lake District[9] in the United Kingdom (topic 13), a place that is in the tentative list of the World Heritage (an inventory of sites that a country intends to consider for posterior nomination). Topics 2, 5 and 10 are about the UNESCO Intangible Cultural Heritage List[10], the Angklung and Noken from Indonesia, already inscribed elements in the list; and the German Brewers, which has the intention to be included. These topics were excluded from this analysis. Topic 24 was excluded because it is about the newly appointed director for the Arab Regional Centre for World Heritage (arc-whc) and the topics 6, 15, 17, 18, 22 and 25 were excluded because they were too general and a further in-depth analysis was needed.

Table 1. Text mining top 25 topics

Topics		#docs
1	tongariro, ruapehu, crater, zealand, north	401
2	angklung, world intangible heritage, oktober, secara, resmi	293
3	de-register, thomaskeneally, proud, +country, +world heritage site	435
4	dua, candi, salah, prambanan, smoke free world	184
5	noken, daftar, oleh unesco, tas, dalam	208
6	beautiful world heritage, beautiful, sites, luxury, website	578
7	+confirm, +plan, +world heritage area, +delist, federal	551
8	gb, +joke, world heritage de-listing, bloody, lnp	123
9	+forest heritage list, +threat, +fore, coalition, +listing	281
10	+brewer, un heritage status, +video, +seek, status	353
11	leuser, ecosystem, +declare, +protect, changeorg_id	347
12	growth,+old,forests, keep, tasmania	413
13	status, +district, world heritage status, lake, +nominate	969
14	scotland, uk, edinburgh, http://t.co/gwhxfzg0b7, travel	348
15	+site, +world heritage site, unesco, unesco world heritage site, world	2071
16	+forest, +move, tasmanian, +tasmanian forest, +world heritage list	563
17	+loss, universal value,+carry, universal, value	122
18	heritage, world, sites, unesco, travel	1232
19	entabeni safari conservancy, entabeni, safari, conservancy, Waterberg	76
20	tassie, barrier, amp, greghuntmp, reef	231
21	abbott, ha, auspol, +area, +push	711
22	list, unesco world heritage, rdal, unesco, +town	854
23	+destroy, regional, +regional strife destroy historical arab, +treasure, strife	276
24	mounir, bouchenaki, director, arc-wh, arab	169
25	+world heritage, +site, +world, +declare, unesco	987

[8] http://www.unesco.org/mabdb/br/brdir/directory/biores.asp?code=SAF+03&mode=all

[9] http://whc.unesco.org/en/tentativelists/5673/

[10] http://www.unesco.org/culture/ich/index.php?lg=en&pg=00011

For the remaining topics - 3, 7, 8, 9, 21 and 23 – the concept linking function available in the text analysis was used to understand what place the tweets referred to.

Figure 9 presents an example the concept-linking for the "word" *gb*, in topic 8, that shows that the tweets on this topic are mainly about the Great Barrier (GB) Reef, located in Australia. Using this method the topics 3, 7, 9 and 21 were allocated to the Tasmanian Wilderness world heritage (e.g "*Taking 170,000 ha of Tasmanian forest off the World Heritage List is not good for the climate @TonyAbbottMHR @GregHuntMP #auspol*" – topic 21); and the topic 23 was allocated to Syria, where all the world heritages sites are in the danger list, due to the armed conflicted[11] that the country lives (e.g. "*Tragic. Arab world is losing historical treasures. 6 @UNESCO sites in #Syria damaged or destroyed*").

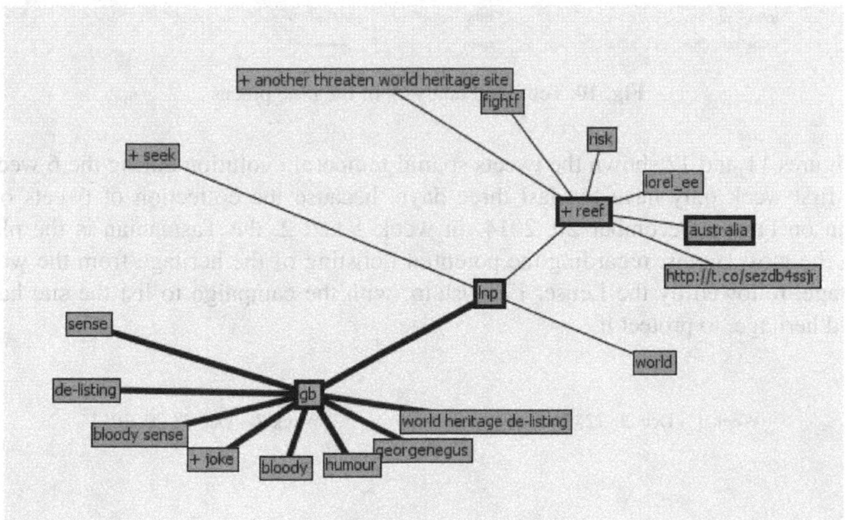

Fig. 9. Concept linking for the topic 8
(gb, +joke, world heritage de-listing, bloody, lnp)

In short, the following nine places were considered for spatial and temporal analysis below: (i) Tongarino National Park (topic 1) in New Zealand; (ii) Tasmania Wilderness (topics 3, 7, 9, 12, 16 and 21) and the Great Barrier Reef (topic 8 and 20), in Australia; (iii) Prambann Temple Compounds (topic 4) and the Leuser Ecosystem (topic 11), in Indonesia; (iv) the Lake District (topic 13) and Edinburgh (topic 14), in the United Kingdom; (v) the Waterberg Biosphere (topic 19) in South Africa; and (vi) the Syrian world heritage (topic 23) as a whole.

Figure 10 shows the temporal evolution of tweets for the nine places mentioned above, the left axis is for the Tasmania and the right axis for the remaining locations; it was used a secondary axis because the Tasmanian heritage was an outlier. The remaining eight places are presented in a stacked area chart with the cumulated totals of tweets for that locations.

---

[11] `http://whc.unesco.org/en/statesparties/sy`

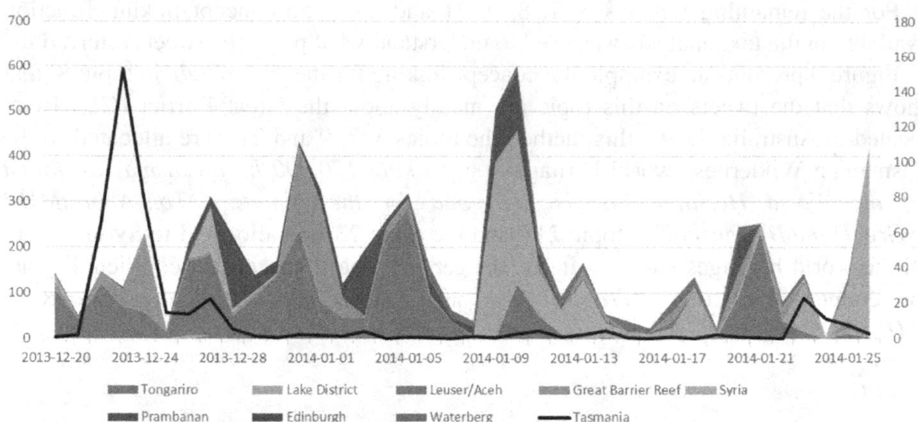

Fig. 10. Temporal analysis of the nine places

Figures 11 and 12 shown the tweets spatial temporal evolution during the 6 weeks. The first week only have the last three days, because the collection of tweets only began on Friday December 20, 2014. In week 1 and 2, the Tasmanian is the place with the most tweets, regarding the potential delisting of the heritage from the world heritage, followed by the Leuser Ecosystem, with the campaign to list the site has a world heritage, to protect it.

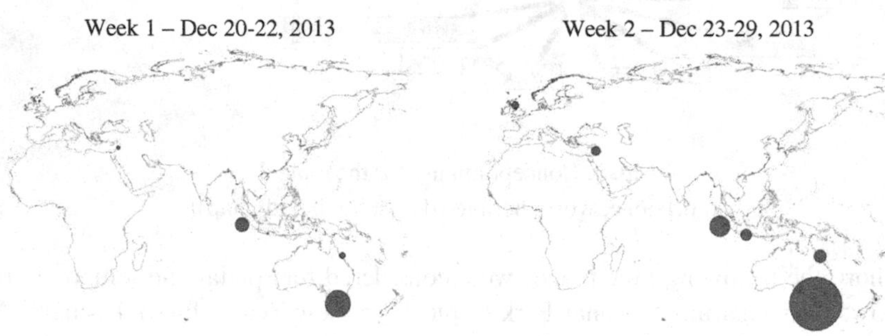

Fig. 11. Spatial-temporal geovisualization of tweets per place (Dec 20 – 29, 2013)

In week 3, the Tongarino is the place with the most tweets, followed by the Great Barrier Reef. In week 4, Lake District in England, with the nomination for world heritage, is the place with the most tweets, followed by Tongarino. In week 5, Lake District continues to lead the attention of the Twitter users, followed by Tasmania. In the last week, Tasmania regain the first place in the Twitter users attention regarding world heritage, followed by the Syria heritage sites, all in danger due to the armed conflict that the country lives.

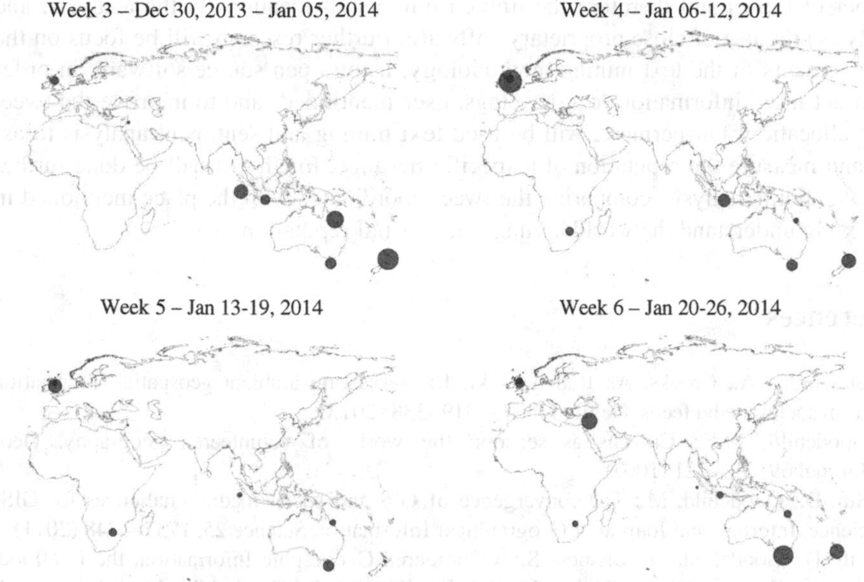

Fig. 12. Spatial-temporal geovisualization of tweets per place (Dec 30 2013 – Jan 26, 2014)

5 Conclusions

The analysis made in the paper, provided a different perspective for the heritage studies and for those who manage heritage sites, showing the sensitivities and valuing regarding a specific place, and how the value of the World Heritage brand and list is seen. The use of social media feeds, more specifically from Twitter, with text mining methods can help to understand better the relations between the general public and the world heritage, providing an opportunity to do spatial, temporal analysis; and therefore understand the reputation of places and the values, attitudes, intentions and behaviours towards a specific place. The use of temporal, spatial and spatio-temporal analysis helped to visualize in a simple way where or about which place and when this messages were conveyed.

In this paper, we detected a reaction of general public to the possible delisting of a world heritage site (e.g. Tasmania); the use of the inscription on the world heritage list, as a protection tool (e.g. Leuser Ecosystem); the concerns regarding the destruction of the sites due to armed conflicts (e.g. Syrian world heritage sites) or regarding places already threatened (e.g. Great Barrier Reef); the diffusion of the message that a site will be nominated as a world heritage site (e.g. Lake District); that one place has become a smoke-free site (e.g. Prambanan); or simply convey a message regarding tourism, travel or beauty (e.g. Edinburgh, Tongarino, Waterberg).

Limitations of this exploratory research were (i) the proportion of the tweets with coordinates that only represents 2.4% of total tweets; (ii) the use of the pre-fields of the collection platform; (iii) the temporal analysis time-frame; (iv) the use of a limited number of topics and therefore the limited number of locations in the analysis; and finally (v) the use of close proprietary software. Further research will be focus on the improvements of the text mining methodology, using open source software, in order to extract more information (e.g. hashtags, user mentioned) and to improve the tweet place allocation. Furthermore, will be used text mining and sentiment analysis to assess and measure the reputation of a specific heritage. Finally, it will be done further spatiotemporal analysis, comparing the tweet coordinates with the place mentioned in the text, to understand the world heritage site spatial reputation.

References

1. Stefanidis, A., Crooks, A., Radzikowski, J.: Harvesting ambient geospatial information from social media feeds. GeoJournal 78, 319–338 (2013)
2. Goodchild, M.F.: Citizens as sensors: the world of volunteered geography. GeoJournal 69, 211–221 (2007)
3. Sui, D., Goodchild, M.: The convergence of GIS and social media: challenges for GIScience. International Journal of Geographical Information Science 25, 1737–1748 (2011)
4. Sui, D., Goodchild, M., Elwood, S.: Volunteered Geographic Information, the Exaflood, and the Growing Digital Divide. In: Sui, D., Elwood, S., Good-child, M. (eds.) Crowdsourcing Geographic Knowledge, pp. 1–12. Springer Netherlands (2013)
5. Fischer, F.: VGI as Big Data: A new but delicate geographic data-source. GeoInformatics 15, 46–47 (2012)
6. Demirbas, M., Bayir, M.A., Akcora, C.G., Yilmaz, Y.S., Ferhatosmanoglu, H.: Crowdsourced sensing and collaboration using twitter. In: 2010 IEEE International Symposium on a World of Wireless Mobile and Multimedia Networks (WoWMoM), pp. 1–9. IEEE (2010)
7. Lee, C.-H.: Mining spatio-temporal information on microblogging streams using a density-based online clustering method. Expert Systems with Applications 39, 9623–9641 (2012)
8. Crooks, A., Croitoru, A., Stefanidis, A., Radzikowski, J.: # Earthquake: Twitter as a distributed sensor system. Transactions in GIS (2012)
9. Hale, S., Gaffney, D., Graham, M.: Where in the world are you? Geolocation and language identification in Twitter. Technical report (2012)
10. Liu, Y., Piyawongwisal, P., Handa, S., Yu, L., Xu, Y., Samuel, A.: Going be-yond citizen data collection with mapster: a mobile+cloud real-time citizen science experiment. In: 2011 IEEE Seventh International Conference on e-Science Workshops (eScienceW), pp. 1–6. IEEE (2011)
11. Monteiro, V., Painho, M., de Noronha Vaz, E.: Is the heritage really important? A theoretical framework for heritage reputation using citizen sensing. Habitat International (accepted not yet published)
12. Van der Aa, B.J.M.: Preserving the heritage of humanity? Obtaining world heritage status and the impacts of listing (2005)
13. Frey, B., Pamini, P., Steiner, L.: Explaining the World Heritage List: an empiri-cal study. International Review of Economics 60, 1–19 (2013)
14. Ashworth, G.J., van der Aa, B.J.: Strategy and policy for the world heritage convention: goals, practices and future solutions. In: Managing World Heritage Sites, pp. 147–158 (2006)

15. Frey, B.S., Steiner, L.: World Heritage List: does it make sense? International Journal of Cultural Policy 17, 555–573 (2011)
16. Poria, Y., Reichel, A., Cohen, R.: World Heritage Site—Is It an Effective Brand Name? A Case Study of a Religious Heritage Site. Journal of Travel Research 50, 482–495 (2011)
17. Labadi, S.: Representations of the nation and cultural diversity in discourses on World Heritage. Journal of Social Archaeology 7, 147–170 (2007)
18. Rakic, T., Chambers, D.: World heritage: Exploring the tension between the national and the "universal". Journal of Heritage Tourism 2, 145–155 (2008)
19. Von Droste, B.: The concept of outstanding universal value and its application:"From the seven wonders of the ancient world to the 1,000 world heritage places today. Journal of Cultural Heritage Management and Sustainable Development 1, 26–41 (2011)
20. Li, M., Wu, B., Cai, L.: Tourism development of World Heritage Sites in China: A geographic perspective. Tourism Management 29, 308–319 (2008)
21. Huggins, A.: Protecting World Heritage Sites from the Adverse Impacts of Climate Change: Obligations for States Parties to the World Heritage Convention. Austl. Int'l LJ 14, 121 (2007)
22. Leask, A., Fyall, A.: Managing world heritage sites. Routledge (2006)
23. Vileikis, O., Serruys, E., Dumont, B., Van Balen, K., Santana Quintero, M., De Maeyer, P., Tigny, V.: Information Management Systems for Monitoring and Documenting World Heritage-The Silk Roads CHRIS. International Archives of the Photogrammetry, Remote Sensing and Spatial Information Sciences 39, 203–208 (2012)
24. Reips, U.-D., Garaizar, P.: Mining twitter: A source for psychological wisdom of the crowds. Behavior Research Methods 43, 635–642 (2011)
25. Twitter: About Twitter, https://about.twitter.com/company
26. Longueville, B.D., Smith, R.S., Luraschi, G.: OMG, from here, I can see the flames!: a use case of mining location based social networks to acquire spatio-temporal data on forest fires. In: Proceedings of the 2009 International Workshop on Location Based Social Networks, pp. 73–80. ACM, Seattle (2009)
27. Schneider, J., Passant, A., Groza, T., Breslin, J.G.: Argumentation 3.0: how Semantic Web technologies can improve argumentation modeling in Web 2.0 environments (2010)
28. ScraperWiki: ScraperWiki, https://scraperwiki.com/
29. Twitter: Twitter Developers, https://dev.twitter.com/
30. SAS Institute Inc.: Getting Started with SAS® Text Miner 12.1SAS (2012), https://support.sas.com/documentation/onlinedoc/txtminer/12.1/tmgs.pdf

Municipal Building Regulations for Energy Efficiency in Southern Italy

Eleonora Riva Sanseverino[1], Raffaella Riva Sanseverino[2], Gianluca Scaccianoce[1], and ValentinaVaccaro[1]

[1] University of Palermo, Department of Energy, Information Enginering and Mathematical Models, DEIM, Palermo, Italy
{eleonora.rivasanseverino,gianluca.scaccianoce}@unipa.it, valentina.vaccaro.pa@gmail.com

[2] University of Palermo, Department of Architecture, DARCH, Palermo, Italy
raffaella.rivasanseverino@gmail.com

Abstract. The building sector is still one of the most energy consuming sectors in Italy, like developed countries in Europe. At European level, the main policy driver related to the energy use in buildings is the Energy Performance of Buildings Directive (EPBD, 2002/91/EC) and its recast. Through the EPBD introduction, requirements for certification, inspections, training or renovation are now imposed in Member States. In order to fulfill the expected changes, local regulations are a key factor aiming at sustainable territorial planning. It is thus required support the issue of local rules at municipal level in order to guide local administrators and technicians and to limit discretional power of bureaucracy. In this paper, a review of the most common practices for building regulations in Europe and in Italy is proposed, then the role and the framework of a municipal building regulation for the Southern European area accounting for sustainability features is discussed.

Keywords: Sustainable urban planning, Energy efficiency in buildings, Municipal building regulation.

1 Energy Efficiency in Buildings and Relevant Regulations in Italy

In Italy, the building sector is responsible for about 45% of primary energy consumption due to building materials and to the final use of energy from tertiary and residential buildings.

The data published from ENEA, in the annual report on energy efficiency (RAEE 2011) [1], show the incidence, in terms of primary energy consumption at national level, of the residential and no residential sector.

From the analysis of the use of energy in the residential sector, which is the one that mostly influence consumptions, emerges that heating (essentially based on natural gas) in 2010 covers more than two-thirds of the overall consumption (it was 68% of overall consumption in 2010), while hot water production and cooking

B. Murgante et al. (Eds.): ICCSA 2014, Part II, LNCS 8580, pp. 420–436, 2014.

respectively cover 9% and 6%, while 17% is devoted to the electrical energy consumptions like lighting, appliances and air conditioning.

In Italy the total energy consumption per household is decreased of 8,3% in 2010 as compared to 2000 (as an effect of the economic crisis in Italy), although the reduction is however lower as compared to the overall reduction across EU, since for EU-27 the average reduction is of 15,5%, as confirmed by Figure1.

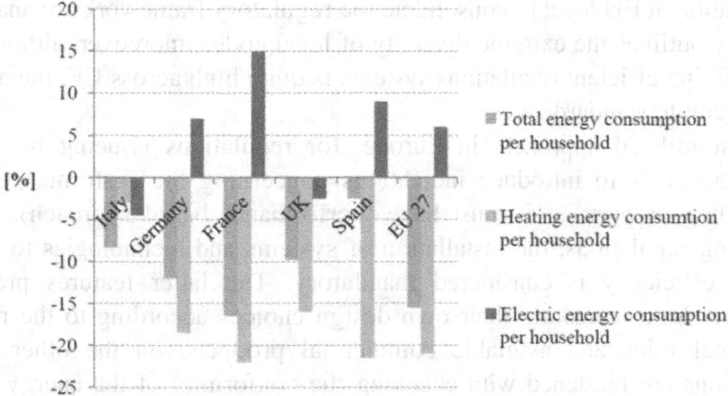

Fig. 1. Percentage variation of total electric and thermal energy consumption per household between 2000 and 2010 (Source: [1])

Essentially the electrical and thermal energy consumption determine the final level of energy efficiency in the residential sector, and, as evidenced from figure 1, the reduction of energy consumption in Italy is not large in the last 10 years and is mostly due to the economic crisis. Such element confirms that the energy efficiency actions in the buildings sector are not widespread in Italy. It is well-known that the policy orientations of EU strongly address towards sustainability in cities, buildings and technical infrastructures. At the European level, the main policy driver connected to energy use in buildings is the Energy Performance of Buildings Directive (EPBD, 2002/91/EC). Implemented in 2002, the EPBD Directive has been recast in 2010 (EPBD recast, 2010/31/EU) with more ambitious provisions.

Reaching objectives like the energy consuming reduction in buildings (Directive 2010/31/UE 'Energy efficiency in buildings', introduces the concept of *near zero energy buildings*), the rational use of energy, the integration of Renewable Energy Sources based production in buildings, etc. are the policy drivers in the member states and have a natural transposition in national laws and local regulations, with a different degree of detail according to the geographical scale to which they refer (provincial, town, etc.).

The European Commission considers local building regulations as a strategical element for the implementation of energy efficiency policies. A study commissioned by the EU in 2011 to the PRC Bouwcentrum International (Bodegraven, The Netherlands) in cooperation with Delft University of Technology (Delft, The

Netherlands) [2], reports a screening of national building regulations[1] in Europe. It analyzes whether and how the EU-27 member states currently regulate sustainable building. The report furthermore assesses whether and how these regulations are enforced on day to day basis. And finally assesses whether and how non-public market initiatives and public-private initiatives complement the formal building regulatory systems in addressing sustainability goals. Finally, the coherence and efficiency of the building regulatory systems are considered, and the possible needs for coordination at EU level to consolidate the regulatory framework are analyzed.

The study outlines the extreme diversity of local codes, moreover, although the interest in defining efficient regulations systems is quite high across UE, the approach it is very difficult to standardize .

The most utilized approach, in Europe, for regulations issueing in sustainable building practice is to introduce indications concerning the performances that the whole building or parts of it must have (performance-based approach). Rarely, in local building regulations, the installation of systems and technologies to improving the energy efficiency is considered mandatory. The latter features preserve the designers freedom of making their own design choices according to the most up to date technical rules and available commercial products, on the other hand, the administrations are burdened with checking the conformity of the energy efficiency measures with the requirements of the municipal building regulations. The latter is the main reason for which, some European countries, from a performance-based approach adopted a mixed performance-based/prescriptive approach, in which numerical and quantifiable parameters also expressing performance are included.

At European level, a larger consensus is reached about the different areas covered by the Municipal Building Regulation. As indicated in [2], the following four main areas are present in most cases, with a different degree of in-depth analysis: *ecological quality; economic quality; social quality and functional quality*.

The area to which greater attention is paid is *ecological quality* with particular reference to the *energy* related parameters. All countries introduce indications about the minimum energy performance of buildings, about the use of renewable energy sources and in general the performances of the building envelopes (e.g., minimum thermal transmittance and air sealing of the building values). This is due to the fact that these features else than being clearly ruled by the European Directives, are also easy to be quantified and thus can be easily checked by local administrations for issueing permits or giving incentives.

Also the attention towards the use of low environmental impact building materials, such as recycled, is a shared issue.

---

[1] The term 'building regulations' is used differently in the member states. For some it refers to the technical building regulations on construction works or construction products, as laid down in many countries in Building Codes. For others it has a more broad meaning, also including local government planning and zoning regulations, environmental regulations, regulations for safe working conditions etc. [2]. In this work we refer to all the legally municipal binding provision imposing mandatory or semi-mandatory requirements on the planning, the design, the execution, the maintenance and the use of construction works.

The aspects dealt with very diversified approaches are those that are not backed up by European Directives and that are ruled at local level. These concern indications for the economic and environmental optimization of the building productive process, as well as guidelines for the minimization of the use of resources in the building yard (energy and water), for the minimization of waste production and valorization of local economy. Other issues accounted for are the quality of the designed environment, thermal and acoustic comfort, lighting and access to green spaces.

Most countries have issued Building Regulation guidelines at national level, leaving the implementation details and and enforcement to the local decision level [2].

Notwithstanding the fact that Italy has transposed the European Directives on energy efficiency, a big effort has then been asked to support the implementation at local level. The distributed transposition on the Italian territory of such regulatory framework, at local level, is characterized by a delay, as compared to the european average [1], for what concerns the real implementation of *best practices* for energy efficiency in buildings.

In Italy, the regulation of urban-historical aspects is worked out at Municipal level. Such ruling system is traditionally connected to two complementary planning tools: the Technical Implementation Norms of the Municipal General Urban Planning and the Municipal Building Regulation. It is right at this level that a substantial lack of updating of the regulatory and planning tools can be found, especially in the energy sector [3].

Such condition, especially in small urban centers, is caused by the lack of sensitivity towards the issue of energy efficiency from municipal administrations. Such condition is even worsened by the absence of specific competences in the public administration staff and by the necessity to face the emergencies concerning the primary services to citizens such as the waste collection, mobility and vehicular congestion, hydrogeological risk, etc. This context leaves the update of the building techniques according to energy efficiency rules to the sensitiviy and the knowledge of the single operators in the different areas, giving rise to large violations even of the national regulatory framework. This happens mostly in small centers and in southern Italy.

Therefore it is clear that in Italy, like in other European countries that show a retardation on the issues of energy efficiency in buildings, it is mandatory to create tools to fruitfully address local administrators choices and policies towards the urban regeneration, trying to compensate the gap between Euroepan policies and local choices and implementations.

The new urban quality requires the integration of planning and design levels in the aim of attaining a real change even at local level.

The Municipal Building Regulation, MBR, referring to the building practice, seems to be one of the most effective technical-operational tools to embrace the integration of the two levels. It is indeed a fundamental turning point of the building process, since in it technical and procedural aspects converge. The MBR is strongly connected to the territory and is often defined through a 'bottom-up' approach, accounting for the critical aspects and the features of the ruled context, but at the same time it adapts to the rush towards sustainable development of cities that EU and then Member States are strongly supporting.

2 Sustainable Building Regulations in Italy

In Italy, below the national legislation level, the regional level law regulates some specific subjects. However, in some topics, national and regional levels are concurrent, such as in the case of urban planning, for which both national and regional policy levels can issue regulations. Within the planning process, the MBR integrates building aspects with energy efficiency measures. Cross competences of different actors in the fields of urban planning, building and energy sectors must be accounted for in the definition of MBR. Participation and sharing in the frame of the national framework are crucial aspects of the process of definition of the MBR.

The resulting document must be shared among stakeholders in the building sector (designers and companies) and must be applicable both at administrative and at technical levels.

In Italy, in the last years, a number of regional laws centered on energy efficiency have been issued in the aim of filling the regulatory gap that for years has characterised the italian national legislation in the field of energy efficiency. Many regions have indeed autonomusly defined methodologies, limits, energy efficiency criteria to be applied to new buildings or restructurings (e.g., Lombardia Region).

In most regions of Italy, the urban planning and building regulations are driven by 'regional guidelines for sustainable Municipal Building Regulations', in this way the municipal policy level can integrate shared rules for the revision of existing MBR.

In some cases the municipal administrations have formulated an Energy Annex to the existing MBR, entirely following the performance parameters indicated in the guidelines. The cited annex is an efficient tool to rule the building transformations following environmental compatibility and energy efficiency criteria, since the formulation of an 'Annex' is more flexible, allowing an easy and continuous updating of the performance and technological parameters, giving rise to sensible regulatory and market changes in continuous evolution.

On the other hand, the Energy Annex allows to keep, inside the main MBR document, all the features concerning general urban planning (like minimum/maximum distance requirements, interventions classification, disabled accessibility, etc), that are not strictly part of the energy issues.

A Legambiente[2] report [3], about the contents of the major Italian MBR, shows that a mixed performance based/prescriptive approach is the preferred choice in Italy. The most common structure includes operational rules concerning performances or specific requirements, that can be either mandatory or voluntary and that are distinguished by category of building intervention and by subject area concern the main fields of action of sustainable building practice:

[2] Legambiente is the most widespread italian environmental organization in the country. It is recognized by the Ministry for the Environment, Land and Sea as an association of environmental interest and it is part of the European Office of the Environment and of the International Union for Conservation of Nature.

- *reduction of energy consumption*. Interventions that reduce the energy consumption in dwelling, increasing the thermal insulation of buildings and enhancing the passive solar gains and efficiency in end uses;
- *renewable sources of energy*. Use and integrate in buildings the production of electrical energy, heating and cooling by renewable systems;
- *the water cycle*. Reduce water consumption needs in dwellings through recovery, purification, reuse for compatible uses also increasing the permeability of the soil.

In the most complete type of Energy Annex, each article is complemented by a qualitative description of the benefits that the provided actions are causing and the actions of verification and control that the City Council will put in place to verify the compliance of the project to the requirements. These contents, are better specified, for mandatory items.

Only in a few cases, the energy aspects of buildings are included in a more complex regulatory document. Infact, with the enactment of specific regional laws, in some italian regions (e.g., Emilia Romagna and Tuscany), the Building Regulations has taken on a new role, governing all possible interventions on the existing city, connecting the issues of sustainable building with ones of urban spaces and extending the view from a single building to public space and the neighborhood. The document then becomes a full-fledged Urban Building Regulations, with rules and performance limits, schedules and tables indicative of the organization of open spaces and uses, as well as the design criteria of the built-up area, the city's historic and rural areas, then going to define general guidelines for designers (e.g., the MBR of Bologna [4], in force since May 2009).

In these cases, the document assumes a more generalized performance-based character, characterized by the identification of the goals to be achieved without a pre-set and pre-figured solution to obtain the desired result. The Administration on its side must acquire the evaluation of interventions, with the assumption of responsibility to judge what is coherent with the MBR and what is not.

This performance based approach is applied in Italy only in municipalities that showed, for a long time, an increased focus on energy planning and in which there is a greater awareness from the stakeholders to the overall aspects of urban planning.

In other regions, however, especially those in southern Italy, we are witnessing a deadlock in terms of regulatory efficiency, therefore the contents of the MBR shall continue to be confined to regulate with binding rules and prescriptive technical aspects of buildings and their appurtenances according to a logic that does not refer to the performance of what is the object of regulation, but simply put numerical parameters over the technical-aesthetic, hygienic and health-related features, safety and livability of the property, without taking into account factors such as living comfort and energy consumption.

Below (Fig. 2) some data from the ONRE Report 2013 (National Observatory Building Regulations) [5], led by Cresme[3] and Legambiente, are reported. They relate to the distribution of Italian municipalities that have put in place to date a revision of

[3] Italian Centre for Social and Economic Research of the market for the building and the land.

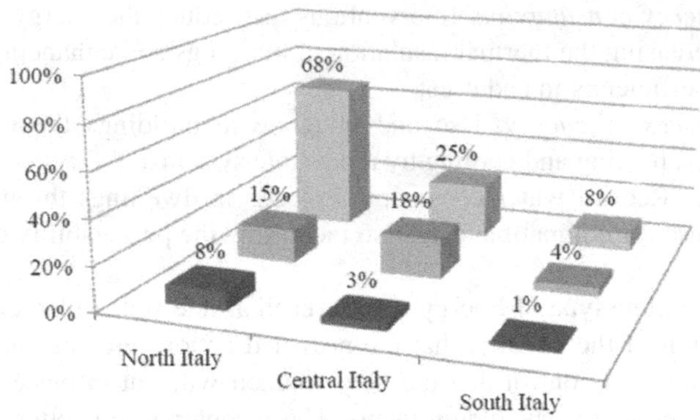

Fig. 2. Percentage of number of municipalities per geographical zone, that have revised the MBR until 2013 in Italy, respect to: overall Italian municipalities, Italian municipalities per geographical zone and overall Italian municipalities that have revised the MBR until 2013 (Source: [5])

its MBR. The municipalities that were engaged in the review and reformulation of the BR following energy-environmental issues, in Italy were about 1003 (12% of the overall Italian municipalities) in 2013, and were located in big cities and small towns. It is interesting to note that this change in Italy is recent (only 14% of the total of the Municipal Building Regulations is prior to 2006) and how, despite the geographical distribution shows the presence of at least one sustainable MBR in all regions of the country, the largest concentration is in the Centre-North. The latter are also the regions that have, first than others, in Italy enacted specific laws for energy efficiency and energy, thus defining a clear policy direction that allowed municipal administrations to start their journey towards sustainability in construction.

A significant element that may largely explain the high concentration of these innovative tools in the north of Italy is also linked with the climatic conditions of these regions and the resulting know-how in this field.

Italian territory from a climatic point of view is very different. A complete overview of this is given by a Italian Presidential Decree [6] that splits the country into Climate Zones[4] which are distinguished based on Degree-Day values (decreasing from F to A zone). The value of Degree-Days of a location, depends on the mean daily external temperature, this is the reason for which locations that are located in the same region can have a different number of Degree-Day. Infact, the daily average

---

[4] The Climate Zone is expressed in degree days that express the sum, extended to all days of an annual period conventional heating, of only the positive differences between the daily temperature, conventionally fixed at 20° C, and the average outside temperature daily.

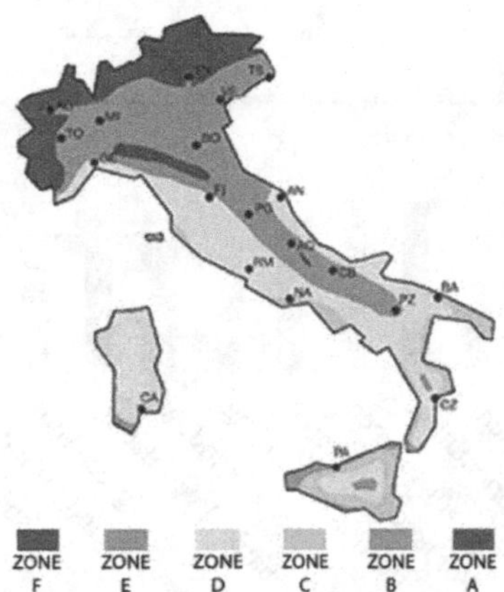

Fig. 3. Classification of the Italian country in Climatic Zones, according to the Italian Presidential Decree n.412/1993 (Source: [6])

outdoor temperature is influenced by geographical factors, primarily the height above sea level but also protection from the prevailing winds, proximity to the sea or lakes, etc.). Figure 3 illustrates this classification.

The regions located on the centre-north of Italy (F, E, D Zones) have a higher value of degree days, and then have an external temperature lower during the winter period, the opposite occurs, however, in the major part of the municipalities in the southern regions (C, B, A Zones).

If the Sustainable Italian MBR are analyzed [5], it can be seen how the key issues addressed are almost the same in all regions and are based on the one hand on the indications of the national legislation of the sector, very often expanding the contents and including improved performance indications, on the other hand on the most well-known parameters of sustainable building.

As previously mentioned, the largest energy consumption in Italy, in the field of residential building, are due to heating. The most common solutions placed within the Municipal Building Regulations in Italy, aim to achieve a reduction in consumption in heating and generally refer to consolidated knowledge about the thermo-physical behavior of buildings. The behavior of buildings in cold climates, and the subsequent identification of the most suitable solutions to be implemented to achieve high performance buildings, is a subject which, for many years, has been treated in the literature and which has been widely discussed by the most well known certification bodies in the energy field in Italy and abroad, i.e., the Agency for Energy Klima House - Alto Adige (Bolzano, Italy) and the Passivhaus Institut (Darmstadt, Germany), which are joined from operating in cold climates.

The following figure (Fig.4) is an analysis of the Cresme and Legambiente [5] concerning the parameters considered versus the number of MBR in which they are inserted. As it can be noted, the thermal insulation together with the obligation to

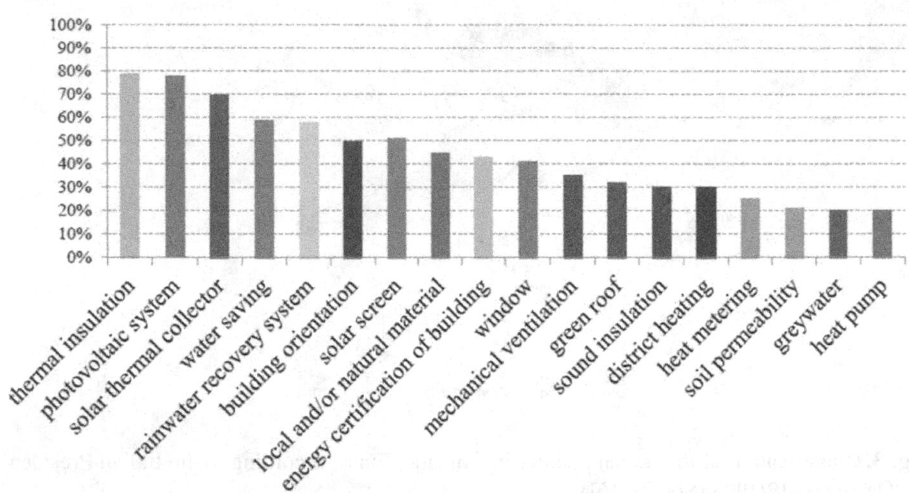

Fig. 4. Percentage values of the most considered parameters entered in the revised MBR in Italy until 2013 (Source: [5])

install photovoltaic panels and solar thermal systems are the most commonly adopted measures, indeed these measures are dictated as mandatory by Decree n. 28/2011[5] .

The fundamental difference between the various regulations is how these parameters are introduced into the regulatory document, this is essentially function of the climatic situation and territory, present in the municipal context. The indications are varied based on the place to which they relate. The obligations may then take several minimum requirements, however, acknowledging at least the minimum required by the regulations in force, in reference both to the use of renewable energy systems, as well as to the transmittance of the elements of the coating or the performance of heat generators.

Despite, for some years, the European Community shall initiate research projects with the aim of studying and promoting the most suitable methods to limit the summer thermal loads in buildings, mainly through passive solutions (e.g., Keep Cool project[6], [7] and Passive-On project[7] [8,9]). Infact demand for cooling energy is expected to rise dramatically in Europe in the coming years [7] and more in some countries in the Mediterranean area.

[5] The Decree n. 28 of May 3rd, 2011 transposes into the Italian legislation the Directive 2009/28/EC provisions on the promotion of the use of energy from renewable sources.

[6] The aim of this project (2005-2007) was to propose intelligent ways of getting passive cooling to penetrate the market and to establish a new definition of sustainable summer comfort.

[7] Passive-On is a completed research and dissemination project (2005-2007) which was funded within the Intelligent Energy for Europe SAVE programme. The project worked to promote Passive Houses and the *Passivhaus Standard* in warm climates.

For this reason, with the support of the most innovative MBR of Italy and of the general indications of the studies mentioned above, it was decided to present, in paragraph 3, the structure of a municipal building regulation for the municipalities in Sicily as an example of the Southern European area.

3 Contents of a Sample Energy Annex

As mentioned above, the most common way to revisit the MBR in Italy in order to include energy efficiency measures, is to structure the content relating to sustainable building within topic areas containing mixed performance/prescriptive based requirements, both mandatory or optional. The thematic areas identified are designed to group the main systems that can contribute to reducing the energy consumption of existing building and of new building.

Even if the proposed structure, in its general content, is applicable to all the regions in Italy and their climatic zones, it was decided, due to more scarce presence of sustainable MBR in southern Italy, to propose in the following section an application referred to mild climates. The example is that of the municipalities in the region of Sicily.

The Region of Sicily is, from a climatic point of view, the most varied region of Italy. As shown in Figure 3, it present almost all the Climatic Zones of Italy (although the area are mainly the B, C and D Zones); for this reason, the development of regional guidelines for the drafting of sustainable MBR is difficult to standardize. This is probably also the reason why the sustainable MBR in Sicily are very few. Moreover from the analysis of the energy consumption in buildings of the Sicilian region, not surprising that the peak comes during the summer season in the hours of major irradiation. Such data can be essentially connected to the climate features of Sicily that comprises different climatic zones, but is characterized by very hot summers in all the region. So in climatic conditions like this, parameters such as solar screens, massive construction of building envelopes, green or ventilated roofs, passive cooling systems, solar thermal system, permeability of the soil, acquire more importance than others that are more commonly addressed in building regulations in climates of central and northern Italy (Fig.4).

In Table.1 a draft of Energy Annex to the MBR is reported. For sake of completeness, this contains the major topics related to the definition of an Energy Annex, including those, more general, that are related to parameters shared by the municipalities belonging to all regions of Italy without distinction of climate zone. The areas covered are:

Area 1. Environmental sustainability and enhancement of the context
Area 2. Energy performance of building envelope
Area 3: Energy performance of technical systems
Area 4. Renewable energy sources

Table 1. Areas and contents of the energy annex to the municipal building regulation for a mild climate city

Thematic areas	Articles	Requirements
Area 1	Art.1 Building orientation	The layout solutions for new building and open spaces will be designed taking into account the apparent path of the sun and prevailing winds, favouring compact forms and conditions of exposure and orientation of buildings such as to maximize heat gains in winter and reduce them during the summer period. Arrange the windows of larger surface area and the rooms in which the majority of the inhabitants lives, in the south-east, south and south-west directions to bring in natural light, taking care to set the required screening elements at least 70% of the glass surface so as to allow the control of solar radiation in the summer period. If these measures are not adopted, design solutions that will reduce by at least 20% of the energy needs of primary energy will have to be implemented. [9,10,11]
	Art.2 Natural lighting	In order to ensure the optical-visual wellbeing of the occupants, the natural light into the building should be used at its best. Artificial lighting should be integrated by balancing the visual needs with energy savings. If the planimetric shape of new buildings does not have sufficient land to direct uptake of natural light, this can be ensured by means of zenithal lighting systems, this rule must be in accordance with the hygienic rules of the City [11].
	Art.3 External microclimate control	In order to limit the rise of local temperatures in densely built areas (urban heat island) for the benefit of the natural cooling in the summer, the trees around the building will be designed like protective elements of the fronts exposed to the winter wind and direct radiation in the summer. For external paving will be used materials with high reflectivity (cool material) such as turf, light stone, wood, or possibly draining pavement [11].
	Art.4 Usage of arboreal types	In new building constructions and existing buildings, if it is possible, to limit the thermal load in summer and preserve the soil permeability, at least 20% [11] of the area of the project is to be used as green space, with the planting of trees with large crown or suitable to create shading in the external walls of buildings exposed to the east, west and south. For the same reason is to prefer the achievement of green roofs. It is advisable to design the most suitable composition of vegetation, preferring the use of native plant species.
	Art.5 Water saving	In order to reduce the consumption of safe water, in all cases of new building and building renovations, is important to provide the mandatory installation of suitable devices to limit the use of it. Install individual meters of safe water per each housing, as well as provide filtration and storage systems of rainwater from the roofs and possibly even wastewater to be sent for recovery for compatible uses inside or outside of the building [12,13].
	Art.6 Usage of eco-friendly and recycled materials	To build is recommended to use natural or recycled materials, which require low embodied energy and low environmental impact throughout their entire life cycle. The use of sustainable materials, however, must ensure compliance with the regulations regarding energy efficiency, the acoustic quality and safety of buildings. The use of locally produced materials must to be duly certified and also incentivize by the municipality [11,12].
Area 2	Art.7 Acoustic pollution control	In new building constructions or major building renovations, the exposure to noise sources (e.g., external noise, noise from other properties, from the shared areas, the sounds of footsteps and plants) must be minimized, consistently with the context and constraints. Consider the possible use of specific architectural solutions that limit the exposure of the receptors to noise pollution, such as, articulation of the massing of the building, so as to achieve effective shielding against sound sources; use of large balconies or loggias with parapets and full use of sound-absorbing materials on ceilings; use of glazed elements with acoustic insulation properties and use of trees [10,11]

Table 1. (*continued*)

	Art.8 Ratio between dispersion surface and controlled temperature volume	For all categories of buildings for new construction and major renovations are preferred limited ratios of S/V (ratio of surface dispersants, flat and vertical to the outside or to unheated spaces, with respect to the heated volume), obtainable through the use of regular and compact shapes. The increasing of the surface area exposed to the outside, causes that the energy losses in the winter and heat gains in summer, increase too. It must be considered, however, that a more compact coating can make it difficult lighting and ventilation of internal spaces, this element should be taken into account in the design jointly with the necessity of limiting losses or heat gains [9], [14,15].
	Art. 9[8] Thermal insulation of buildings (building features of coatings–winter)	In order to enable a reduction in consumption of fossil fuels for heating in winter, for new construction and major building renovations, it is appropriate to reduce heat loss considering levels of thermal transmittance lower, if possible, for different parts of the building coating, both opaque (limited to walls and roof [9]) and transparent. Improvements with respect to the parameters imposed by the law will be rewarded by the administration (in Italy the decree [16] fixed, among other things, the requirements of national minimum transmittance of the envelope components and also the maximum primary energy demand for winter heating).
	Art.10[9] Thermal insulation of buildings (building features of coatings–summer)	In order to ensure the passive control of thermal hygrometric comfort in interior spaces, especially in the summer time, it is necessary that the structures of the building are able to attenuate and phase shift of the peak heat load due to the impact of external sun. Furthermore it is necessary that they are able to modulate the internal heat loads due, for example, to the presence of people and to the effects induced by lighting and electrical equipment as well as the infiltration of hot external air due to the opening of doors, as well as to diffuse solar radiation through transparent closures. For this purpose, in new constructions and major renovations, it is necessary to adopt building systems that give to the external closures and walls, appropriate behaviour in terms of thermal inertia, phase shift and attenuation of the thermal wave [17] or periodic thermal transmittance complying at least with the minimum limits of the Italian law [15]. In order to facilitate the night ventilation and the consequent

8 A possible improvement of the transmittance of envelope components allow to reduce the heat winter losses also thanks to a highly air-sealed building envelope and to the elimination of thermal bridges. An increased of the air-sealing in the building envelope on the other hand, reduces the air changes of the building envelope, therefore, it may be required an additional ventilation system (such as for example a forced ventilation system with heat recovery from exhaust air). The energy required for winter heating is also reduced if the design includes the possibility to exploit the solar heat gains through the glass shutters or glass walls installed on the south, east, west oriented walls, this notation must be addressed, including appropriate screening systems possibly modular, that during the summer, which must be designed to minimize solar gains.

9 The energy needs of the building in summer, are reduced if the solar earnings of the external surfaces are minimized thanks to a high thermal capacity (in Mediterranean climates especially if characterized by a high thermal capacity daily, about 15°C, and then by an effective night ventilation, this allows to reduce the thermal load by summer cooling by 10-40% compared to the case of enclosures with light same insulating performance [6], [18]) and also tank to the sun shadings (notations at art.11) or by suitable trees plantings (notations at art.4). A well-sealed and heavy structure is the ideal situation for the exploitation of the summer night cooling of the thermal mass of the building. During nighttime the air circulates through the building, using winds or natural air density gradients, or using a mechanical ventilation system (notations at art.17), setting partly free the building from the heat held by its mass. The greater or smaller efficiency of the system must however be verified with reference to the local climatic features of the site where the Municipality resides.

Table 1. *(continued)*

		cooling of the structure with high thermal capacity (especially in climates characterized by high daily temperature range), in order to remove part of the heat accumulated during the daylight hours, it is also recommended to evaluate the use of devices for automatic opening of windows, or natural ventilation, as well as the design of the interior spaces encouraging the movement of air, this in order to keep the internal temperatures significantly below the external environmental temperatures during the summer period [8,9]. In new buildings and renovations, the thermal energy for cooling in summer must be between 40-30 kWh/m2[10] per year, this according to the directions of the law [15]
	Art.11 Shutters performances and shadings control	It is necessary to adopt integrated solutions that simultaneously monitor the summer sunshine, favouring the winter sunshine and optimize the performance of passive buildings. For this purpose, in new constructions, renovation or repairs which affect even the replacement of windows, at least the limit value of transmittance of the windows indicated by Italian law [16] must be ensured (values improvements can be awarded by the municipality). It is also mandatory to use screening systems that will be congruent with the orientation of the front on which they are installed (they may be vertical or horizontal, fixed or mobile), so that in summer the sunny side of the transparent surfaces located on the south, east and west, is equal to, or less than, 30% [9,10]. Where this is not possible for reasons of architectural significance the use of innovative technologies is required, such as the use of selective glazing [19].
	Art.12 Roofs performances	For new buildings and those subject to demolition and reconstruction in total restructuring, for the extraordinary maintenance of continuous and discontinuous roofs with complete revision of the mantle, in the case where the roof is in direct contact with rooms that are accessible and the roof is a pitch roof (for example under the roof, attic, etc.), ventilated solutions or equivalent will be preferred, to reduce the surface temperatures of the summer covers and the amount of energy transferred inside the building, in order to reduce the 'heat island effect', it will be preferable to use light coloured surfaces, coatings with high reflectivity or construction of green roofs [9], [11].
	Art.13 Use of passive cooling measures	In new constructions and major renovations in order to promote energy conservation, ensuring the air natural conditioning in summer and improving wellbeing in the interior spaces, where possible, solutions that enable to take advantage of the temperature difference between opposite sides creating ventilation loop must be implemented (such as the construction of windows on the north and south) as well as to provide an organization of interior spaces which does not create an obstacle to the flow of air directed through the openings from the windward side to the leeward ones. Ventilation must be highest during the day, during the summer season, in the areas most frequently used by the inhabitants and the airflow should be up to the walls, those that store the most heat. If the intervention makes it possible the use of water reservoirs near the openings of the building, the cooling water potential in the evaporation must be exploited [8,9].
Area 3	Art.14 High performance heat generation systems, thermal centralized plants and metering systems	In new construction and major renovations to the entire heating plant is necessary to ensure the highest possible level of average seasonal efficiency of the heating system (levels greater than those imposed by law [15] will be promoted by the municipality) using centralized systems, in buildings with multiple dwellings, it is required the use of heating plants with high efficiency (condensing boilers, micro CHP plants, heat pumps, compression and absorption) that at least meets the performance requirements provided by law; materials, devices and products for the reduction of energy losses of the piping of heating systems or for a better performance of the final spread of heat. Where technically possible, it is mandatory the construction of facilities which permit the accounting and thermoregulation of heat per housing unit and possibly also for individual rooms or thermal zones [9].

[10] It refers to the walkable area.

Table 1. (*continued*)

	Art.15 Thermal regulation systems	For the purposes of thermo hygrometric comfort and reducing energy consumption, adequate levels of air and surface temperature in the interior spaces of the buildings should be guaranteed. Subject to any exceptions provided by law, during the period of operation of the heating the indoor air temperature (Ti) for the closed spaces for principal and secondary activities should be 18°C<Ti<22°C; it must also not show, if measured along the vertical axis (ideal straight floor ceiling), differences of more than 2°C. The operating temperature (Top) for residential uses should be around 20°C in winter and below 26°C in summer [9], [15].
	Art.16 Low temperature systems	In order to optimize the use of high efficiency generators for heating during winter is recommended the use of distribution systems at low temperature (like radiant panels, embedded in the floors or in walls, or fan coils). The radiating systems can also be used as air conditioning terminals provided that devices, for the control of relative humidity, are added in the plant[11]. The installation of radiant floor or ceiling panels in existing buildings must not compromise the achievement of the minimum heights of the premises laid down by the legislation in force [11], [21].
	Art.17 Mechanical ventilation systems	In order to ensure satisfactory conditions of hygiene, health and environmental well being, ventilation of internal spaces is an essential condition. Where it is not effective natural ventilation system, and mandatory in buildings for public use, it is necessary to install mechanical ventilation devices with heat recovery. That allow to control the relative humidity, mitigate the effects of steam condensation and prevent the formation of microbial colonies, this allows to limit the energy needs [9], [15], [20].
Area 4	Art.18 Solar thermal systems	For new buildings, and existing buildings subjected to demolition and reconstruction in maintenance, and for those subject to full re-structuring of the elements making up the building envelope it is mandatory to cover, through the use of energy produced from renewable sources, at least 50 % of the demand for domestic hot water [22], this requirement has to be met on a priority basis through solar thermal systems. It is preferable that the solar collectors are installed on roofs and façades exposed to the South, Southeast, Southwest, East and West, without prejudice to morphological elements, urban planning, and landscape protection. In the case of flat roofs the panels can be installed with the inclination considered optimal, although it is preferable that they are not visible from the road below, and it is mandatory that the system be achieved by ensuring compliance with the formal consistency of the building and minimizing the visual impact on buildings. It is allowed to install panels on facades and coating of the building if and only if conceived as elements that are functionally integrated into the facade. Where conditions allow, the integration of this solar thermal system with the heating system of the building will be prized by the administration. To connect the solar thermal system to the individual users appropriate rooms must be accounted for in the building. If the location of the building makes neither it nor possible the use of solar thermal panels an alternative renewable source will have to be found for the complete satisfaction of the obligation [14], [22]. In new and existing buildings (public and private) in which it is expected that the complete replacement of the cooling system, it is possible to consider the design and installation of solar cooling (associated with solar thermal cooling absorption machine) in order to achieve renewable energy based cooling systems [23].

[11] This is to limit the risk of condensation on the cold surface and to obtain a condition of comfort. As recommended by the UNI EN 7730, the relative humidity should not exceed 60 to 65% to ensure a feeling of comfort.

Table 1. (*continued*)

	Art.19 Renewable energy sources for electrical energy generation	In the case of new buildings or buildings undergoing major renovation it is required to install an renewable electric power system, at least equal to that required by law [22] (solutions that allow an almost total coverage of the needs of the building through these facilities will be rewarded by the local administration). Given the favorable conditions of average solar radiation are preferred photovoltaic systems, for the use of photovoltaic systems apply the same indications of architectural integration mentioned in the previous article. The projects of new buildings and major renovations of existing buildings ensuring a coverage of the consumption of heat and electricity for cooling in excess of at least 30 % compared with the minimum requirements set out in [22], will be prized, upon issuance of the building licence, with a bonus of 5% volume. It is permitted the installation of systems using geothermal, hydroelectric or wind power as long as they conform to the standards in force within urban planning and landscape of the town.
	Art.20 Electrical infrastructures for electric vehicles recharging	It is mandatory for new residential buildings or for residential buildings going under major renovation with surface area not greater than 500 m2 the installation of or predisposition for a charging infrastructure for electric vehicles likely to allow the connection of a car in each covered or uncovered parking space, and in each parking lot within a closed area, whether related to the building or not. Infrastructure, including private, intended for charging vehicles powered by electricity are primary infrastructure works realized in the entire municipality are exempted from construction taxes[12].

The table above is indicative of the contents that can be inserted in the energy Annex to the MBR for a municipal authority in the regions of central and southern Italy.

The more detailed practical and operational indications will provide a tool which will become as more effective, as more the indications can be easily translated into effective actions. It is the task of the Administration, on the basis of climatic parameters and urban context, to rule and specify more technical specifications.

4 Conclusions

In this paper the issue of energy efficiency in buildings is treated considering one of the most important implementation tools: the Municipal Building Regulation. After a comprehensive state of the art on the existing best practices in Europe a focus in Italy is proposed and the structure of the energy annex for sustainable Building Regulations in proposed. The application is devoted to southern Italy, proposing measures that are well designed for hot and mild climates.

[12] Notation introduced in accordance with the instructions contained in the Italian National Plan of Infrastructure for recharging vehicles powered by electricity, in relation to the voice of the Town Planning Review: incentives and obligations.

References

1. Unità Tecnica Efficienza Energetica dell'ENEA: Rapporto Annuale Efficienza Energetica. In: RAEE 2011. Executive Summary, Rome, Italy (2012)
2. PRC Bouwcentrum International, Delft University of Technology Bodegraven: The lead market initiative and sustainable construction: lot 1, screening of national building regulations. Final report, Delft, Netherlands (2011)
3. Legambiente: Il Regolamento edilizio d'Italia (2012)
4. Comune di Bologna: Regolamento Urbanistico Edilizio della città di Bologna. Bologna, Italy (2009)
5. Cresme Ricerche, Legambiente: L'Innovazione Energetica in Edilizia - I regolamenti edilizi comunali e lo scenario dell'innovazione energetica e ambientale in Italia. Rapporto ONRE (2013)
6. Italian Presidential Decree n.412/1993: Regolamento recante norme per la progettazione, l'installazione, l'esercizio e la manutenzione degli impianti termici degli edifici ai fini del contenimento dei consumi di energia (1993)
7. Varga, M., Bangens, L., Cavelius, R., et al.: Service Buildings Keep Cool: Promotion of sustainable cooling in the service building sector. Final report, Austrian Energy Agency, Vienna (2007)
8. Ford, B., Schiano-Phan, R., Zhongcheng, D.: The Passivhaus Standard in European warm climates: design guidelines for comfortable low energy homes - Part 3: comfort, climate and passive strategies. Part 3 report of Passive-On Project (2007)
9. Pagliano, L., Carlucci, S., Toppi, T., Zangheri, P.: Passivhaus per il sud dell'Europa-Linee guida per la progettazione. Rockwool guide, 'Passive-On' IEE Project, Milan, Italy (2009)
10. Comune di Bari: Nuovo Regolamento edilizio della città di Bari. Bari, Italy (2012)
11. Infoenergia, Dipartimento BEST Politecnico di Milano: Definizione di regole per strumenti urbanistici orientati alla valorizzazione energetica ed ambientale - Linee Guida, provincia di Monza Brianza. Italy (2012)
12. Sicilian Decree n. 18/Gab (Decreto attuativo, R.L n.6 del 23 marzo 2010): Caratteristiche tecnico costruttive per gli interventi di bioedilizia nei casi di demolizione e ricostruzione
13. Comune di Monterotondo: Norme per la sostenibilità ambientale allegate al regolamento edilizio. Monterotondo, Province of Rome, Italy (2009)
14. Italian Legislative Decree n.192/2005: Attuazione della direttiva 2002/91/CE relativa al rendimento energetico nell'edilizia (2005)
15. Italian Presidential Decree n.59/2009: Regolamento di attuazione dell'articolo 4, comma 1, lettere a) e b), del decreto legislativo 19 agosto 2005, n. 192, concernente attuazione della direttiva 2002/91/CE sul rendimento energetico in edilizia (2009)
16. Italian Legislative Decree n.311/2006 Disposizioni correttive ed integrative al decreto legislativo 19 agosto 2005, n. 192, recante attuazione della direttiva 2002/91/CE, relativa al rendimento energetico nell'edilizia (2006)
17. UNI EN ISO 13786:2008: Prestazione termica dei componenti per edilizia - Caratteristiche termiche dinamiche (2008)
18. Zangheri, P., Pagliano, L., Carlucci, S.: Passive house optimization for Southern Italy based on the "New Passivhaus Standard". In: ECEEE Summer Study: act! Innovate! Deliver! Reducing Energy Demand Sustainably, pp. 1643–1648. La Colle sur Loup, France (2009)
19. Dama, A., Pagliano, L.: Vetri ad alte prestazioni energetiche. In: Il Progetto Sostenibile, Italy, vol. 2005, pp. 60–65 (2005)

20. UNI EN 15251: Criteri per la progettazione dell'ambiente interno e per la valutazione della prestazione energetica degli edifici, in relazione alla qualità dell aria interna, all'ambiente termico, all'illuminazione e all'acustica (2008)
21. Initiative for Low Energy Training in Europe (ILETE): Labelling and Certification Guide. ILETE Report (2010)
22. Italian Legislative Decree n.28/2011: Attuazione della direttiva 2009/28/CE sulla promozione dell'uso dell'energia da fonti rinnovabili, recante modifica e successive abrogazione delle direttive 2001/77/CE e 2003/30/CE (2011)
23. Vaccaro, V.: The urban and environmental building code as implementation tool. In: Riva Sanseverino, E., Zizzo, G., Riva Sanseverino, R., Vaccaro, V. (eds.) Smart Rules for Smart Cities-Managing efficient cities in Euro-Mediterranean Countries, vol. 12, pp. 75–106. Springer for Innovation (2014)

On Fractal Complexity of Built and Natural Landscapes

Andrei Bourchtein[1], Ludmila Bourchtein[1], and Natalia Naoumova[2]

[1] Institute of Physics and Mathematics, Pelotas State University, Brazil
{bourchtein,ludmila.bourchtein}@gmail.com
[2] Faculty of Architecture, Pelotas State University, Brazil
naoumova@gmail.com

Abstract. In this study, some problematic points associated with the application of the box-counting method for the evaluation of the visual complexity of the chosen historic buildings and their surrounding environments are analyzed. The reliability of the results is measured in terms of the variation of the complexity level over different range of scales available in an image. Some rules for the choice of the box sizes used in the box counts are considered and tested in the case of classical fractals with the known theoretical fractal dimension. The proposed algorithm is applied to the evaluations of the fractal dimensions for the historic part of the city of Amasya and its surrounding environment. More accurate computations based on the traditional and new factual material show that there is a strong similarity between the fractal measures of the built and natural landscapes.

Keywords: computational fractal analysis, fractal dimension, box-counting method, natural landscape, historic buildings.

1 Introduction

The relationship between built settings and their surrounding environment is an important issue that has been the subject of numerous studies over the past decades. One important point of such investigations is the verification of the correspondence between the levels of the visual complexity of the two environments - artificial and natural. It can be expected that such a correspondence exists, at least for historic sites that did not suffer to a large extent from the centralized planned urban development and thus are much more related to inhabitant preferences and traditions. In these sites, the relationship can manifest through the local materials available for construction, the imitation of organic forms in buildings, the ways buildings are used, the main occupation of inhabitants, etc., with all of these depending (to a certain degree) on the surrounding areas and available natural sources.

It appears that all the studies on this relationship presented thus far can be divided into two groups. The traditional empirical approach consists generally of the following stages. Artificially created scenes of the urban and natural landscapes are presented to a chosen group of respondents who are asked to choose the scenes that match each other better or worse. The results of the answers are measured on a chosen scale for an analyzed feature (e.g., roughness, complexity, pleasure, etc.); in this way,

B. Murgante et al. (Eds.): ICCSA 2014, Part II, LNCS 8580, pp. 437–452, 2014.

all the results are transformed into numerical form and are subsequently subjected to statistical evaluation, including calculation of the correlation coefficient between urban and natural scenes, the standard deviation and other appropriate statistical characteristics. If the pool of respondents is sufficiently large and the standard deviation is sufficiently small, then in some statistics, one can derive conclusions on the validity and stability of the obtained results. Such an approach applied to the study of the visual complexity of natural and artificial sites is used in a number of publications (e.g., [19], [21], [35], [37] and references therein). The critical points of the empirical method include the subjectivity and artificiality in the creation of the proposed scenes, the subjectivity in the choice of the numerical scale for measuring the qualitative responses (which can strongly influence statistical evaluations), the necessity of having large and representative pool of respondents, and the absence of a mathematical theory on the validity of the results because the obtained distributions are usually non-normal.

The second approach was proposed about two decades ago based on a relatively new geometry theory. Geometry and architecture have close ties over the centuries [8], [20]. In the 20th century, a more complex geometry theory, fractal theory, was created and developed and, at the end of the century, became a strong mathematical theory with attractive opportunities for applications [26], [27]. First natural and then artificial structures in different fields of science have begun to be analyzed from the perspective of fractal geometry [26], [27], [32]. Studies on the fractal analysis of the built environment started in the 90s, first involving evaluations of the complexity of urban planes at the entire city and street scales [1], [2], [10], [30], [34] and then reaching a smaller scale of buildings and groups of buildings, the first results of which were presented in [3], [5]. Fractal theory provides a method for the quantitative evaluation of the visual complexity of a setting through computation of its fractal dimension. Usually, the box-counting method is applied to evaluate the fractal dimension of the characteristics chosen for comparison [3], [5], [25], [36]. By comparing the fractal dimensions of the built and natural landscapes, one can draw a conclusion about their relationship (or absence thereof) in different cases. First, such an approach was proposed by Bovill, who also applied it to the analysis of the historic part of the city of Amasya, Turkey [3], [5]. It seems that this approach provides more reliable and objective information than empirical studies because it is based on the use of factual material (photographs, digital images, plans and maps of real objects), which means that artificial elements in the scene compositions are largely eliminated and the preferences of significant groups of inhabitants are manifested in more objective ways. The method has attracted many researchers due to the relative simplicity of the algorithm and the availability of general-purpose and specific software for calculation of the box-counting dimension. Different built landscapes have been studied using the fractal dimensions, but sometimes the results obtained are inconclusive and controversial [3], [25], [35], [36], which can be partially attributed to the limitations of the box counting method, some intricate parts of its implementation, and misleading interpretations of the results of the computation.

The purpose of our study is to examine some problematic points of the box-counting method as applied to finding the fractal dimensions of real structure images. There are different studies on this subject and important results found in many areas of applications: Forountan-pour et al. [17] discuss the issues of the image resolution, the usable range of box sizes, the grid positioning and uniformizing the thickness of curves; Huang et al. [22] test the effect of resolution and box ranges on the computation of the box-counting dimension in order to find the version of the algorithm that matches better the known dimensions of some mathematical fractals; Chen et al. [11] and also Bisoi and Mishra [4] provide a theoretical discussion on the definition of the lower and upper bounds for the usable box sizes, supporting the arguments by computations with the artificial fractals; Feng et al. [16] point out that undercounting occurs at smaller scales, while overcounting - at larger scales, resulting in a smaller value of the dimension; Buczkowski et al. [6] propose some rules for the choice of the scaling coefficient and grid positioning that improve the performance of the method in their experiments with mathematical fractals; DaSilva et al. [14] consider such factors as the grid orientation relative to the image, the choice of scaling coefficient and the importance of the statistical error evaluation. Of course, this is just a small, albeit representative, fraction of contributions to the field, which are relevant for our study.

To the best of our knowledge, these issues were not systematically considered thus far in architectural and urban analysis, although some elements of the above studies were adopted and employed by researchers (e.g. [12], [13], [25], [31], [36]). More systematic application of the tuning parameters of the box-counting method and careful evaluation of the validity of the fractal evaluations allow us to adjust some of the results of the previous studies and to suggest a more reliable implementation of the box-counting method in the architectural and environmental studies. This report is structured as follows. In Section 2, the box-counting method is briefly reviewed, and a preliminary theoretical discussion of its basic properties is provided. The corresponding algorithm is used for calculation of the fractal dimensions of the known artificial geometric forms in Section 3. The obtained results of the simple simulations reveal the important properties of the box-counting method and lead to some propositions on optimization of the evaluation of fractal complexity. In the next two sections, the optimized version of the algorithm is applied to finding the fractal dimensions of the historic sites in Amasya, both by using the sources considered by the previous researchers ([5], [25], [36]), and by analyzing additional photographic materials of the same place. The concluding remarks are presented in the final section.

2 Box-Counting Method: A Theoretical Discussion

The well-known Mandelbrot definition of a fractal made in his fundamental essay states that "a fractal is ... a set for which the Hausdorff–Besicovitch dimension strictly exceeds the topological dimension" [26]. However, later, Mandelbrot regretted that " the definition is very general, which is desirable in mathematics. But in science its generality was to prove excessive: no only awkward, but genuinely inappropriate" [27]. Besides, the Hausdorff dimension itself is a complex mathematical concept and cannot be introduced here. Therefore, like many other authors (e.g., [1], [5], [32]), we appeal to a non-exact general description of a fractal as a geometric structure with the following properties:

1) it is irregular at any scale;
2) it cannot be described with the required precision using traditional geometry;

3) its non-traditional (fractal) dimension, defined in a proper way, is usually greater than its topological dimension.

Some fractals are self-similar, which means that any part of the entire figure is its smaller replica, and in this case, the entire structure can be obtained starting from a simple geometric form by an iterative process of making reduced copies and gluing them together (at each stage, maintaining the same reduction factor and number of copies). It can be shown that in such a case, there is a relation between the reduction factor r and the number of copies N in the form $N=(1/r)^{D_s}$ or, equivalently, $D_s=-\log N/\log r$, where D_s is called the similarity dimension. (A more detailed description and exact formulation of self-similarity can be found in [15], [32]. This is the case of classical fractals, such as the Cantor set, the Koch snowflake, the Sierpinski carpet, etc.

For example, the Koch curve can be obtained starting from a line segment that, in the first stage, is divided into three equal parts, and its middle part is replaced by two equal segments that would form an equilateral triangle with the removed piece. The same procedure should be applied to each side of the obtained figure, and so on. In this way, at each iteration, the Koch curve is made up of four copies of itself ($N=4$) scaled by a factor of 1/3 ($r=1/3$); therefore, its similarity dimension is $D_s=\log 4/\log 3\approx 1.2619$.

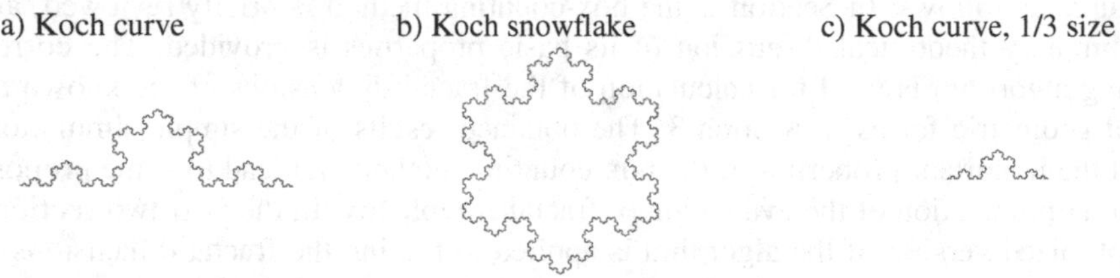

Fig. 1. The Koch curve and snowflake

Fig.1 shows an approximate Koch curve obtained after six iterations of the above procedure (Fig.1a), and also the Koch snowflake, which is composed of three Koch curves and has the same fractal dimension (Fig.1b), and a six-iteration Koch curve with its linear size reduced by a factor of 3 (Fig.1c). It is well-known that for self-similar fractals without (or with sufficiently small) overlapping, the similarity dimension coincides with the Hausdorff dimension (for the exact formulation of this result see [15]).

Generalization of this definition to the case of non-self-similar fractals can lead to a more general box-counting fractal dimension, which is one of the most popular measures of geometric complexity in the applied sciences due to its easy mathematical formulation and simple approximation through numerical algorithms for practically arbitrary shapes. To avoid unnecessary generality and simplify our considerations, let us consider a specific situation relevant for our study: a bounded figure located on a two-dimensional plane. Instead of counting exactly how many times the number of pieces increases under the chosen scale reduction, as is done in the similarity dimension, now we count the number of auxiliary simple figures that cover the given structure and evaluate the logarithmic dynamics of the change of this number under decreasing scales. More precisely, let N_r be the smallest number of squares of the side length r, which cover our figure; then the box-counting dimension is the following limit (if it exists): $D_b = -\lim_{r \to 0} \log N_r / \log r$. In the case of self-similar fractals, the two dimensions coincide: $D_b = D_s$ ([15], [32]).

In practice, one uses an approximation to D_b, evaluating N_r for a different fixed value of r and extrapolating the obtained results to the limit as r approaches 0. The most popular version of numerical approximation consists of the following steps: first, a figure is covered by a sequence of rectangular meshes with square boxes (cells) by reducing the linear size r of cells by a factor of 2; second, the number N_r of the boxes, which contain at least one point of the figure, is counted for each of the used meshes; further, the values of $\log N_r$ are plotted against $-\log r$; finally, the slope of the obtained curve is evaluated and it is considered to be an approximation to the exact value of the limit. This is a general algorithm used in many box-counting programs, including two programs we have applied in our study: the program Boxcount created by Moisy and available free from the Matlab Exchange File site [29], and the Fractalyse software available free from the web site given in [18].

Although very popular and simple in implementation, the box-counting method has some drawbacks related to the features of a given image. First, any real image has a finite resolution; therefore, according to the exact definition, the image cannot be considered a fractal but just its approximation. Actually, the same is true for any real (including natural and built) object, so even if images and computers had an infinite resolution, in reality, the non-artificial objects have a finite range of subdivisions and scales (at least we should stop before reaching the atom scale) and do not satisfy the exact definition of a fractal. Second, the original figure or image is not usually of a square or rectangular form and, almost certainly, does not contain the exact number of square boxes used for counts. This fact means some adjustments between the figure size and coverings on the figure/image boundaries are necessary, which can compromise the box counts at the largest sizes. Another issue is possible imperfections and even deficiencies of images, which are frequently photographs of real objects and include the imperfections of the camera, shooting conditions and the ways in which images are digitally saved. Additionally, the thickness of curves can be non-uniform over an image and larger than the smallest box sizes, which may contaminate the results of fractal evaluation and affect the reliability. Finally, the problem of a low confidence may occur if, for whatever reason, the obtained slopes at different space scales have very different values (for example, the standard deviation from their mean value is large). In such cases, even different methods of evaluations of the limit value (for example, the use of the mean value or the application of the least-square method) can provide quite different results. We will address this situation in our study in the following sections.

All these considerations lead to the following features of the box-counting method applied to define the fractal dimension of the real forms (both natural and built):
1) the final evaluation of the limit in the definition of D_b may have a very restricted (if any) relation with the characterization of the complexity we are looking for because the real objects frequently show different approximated fractal dimensions at

different space scales (multifractality) and their range of zooming is finite, that is, below some specific scale, they do not show any new details mainly due to their nature, the image resolution, or still computer restrictions;
2) the primary interest is shifted to the local slopes (local fractal dimensions) connected with rugged forms at the chosen range of scales (this range is frequently related to one of the approximated fractal dimensions presented in a multifractal);
3) the range of the spatial scales to be used depends on the specific aims of the analysis, but even when all spatial scales available in an image are of interest, the smallest and largest scales should usually be ignored (the former due to photography imperfections and the finiteness of the scales and the latter due to the boundary processing in the box-counting algorithms);
4) the dynamics of the local slopes as a function of the scale variable can reveal the distribution of the fractal characteristics for a specific class of objects, which may be a feature that distinguishes one class from another;
5) if the local slopes have large variations due to multifractality or a low image quality, then their joined evaluation can result in meaningless numbers, and even when these numbers can (to a certain degree) characterize the complexity of the form, their reliability is usually low due to the large values of the deviations.

Accordingly, the box-counting algorithm should be modified in such a way that it takes into account the above considerations to produce a sound characterization of complexity of the chosen real structures. In the first place, the focus should be on local slopes and local means instead of global values. To accomplish this objective, instead of obtaining only the final evaluation for the box-counting dimension, we will keep a sequence of the measurements obtained over different scales in the form of the local slopes, defined as $(\log N_{r_2} - \log N_{r_1})/(\log r_2 - \log r_1)$ for two consecutive linear box sizes r_1 and r_2. This measure can provide important information on the geometric complexity within the chosen scale window and on the reliability of the results.

All of these features of application of the box-counting method to evaluating the fractal dimension of two-dimensional geometric forms will be illustrated in the next section using examples of classical fractals with known theoretical fractal dimensions.

3 Tests with the Classical Fractals

In this section, we present some examples of the computation of the fractal dimension for the six-iteration Koch curve and its variations. We think these examples are instructive in light of the theoretical discussion in the previous section. The Boxcount and Fractalys programs practically provide the same results, which is one of the confirmations of their validity.

To simulate the subsequent treatment of the images obtained with a digital camera or prepared digitally using software, the Koch curve (Fig.1a) was generated by a Matlab program and then written as a jpeg file using two resolutions - 100 and 200 dpi, which produced files of 800x600 and 1600x1200 pixels, respectively (K100 and K200 notations in Table 1 below). Besides, the same curve but at 1/9 of its original size, placed on the same page (see Fig.1c), was saved in jpeg format with the two specified above resolutions (K100s and K200s notations in Table 1 below). Then, the color-formatted files of the saved images were processed by the box-counting program, which transformed the color format to a grayscale image, binarized it and counted the number of square boxes covering the obtained silhouette (black/white simplification) of the original image.

The results of the computations for the four images, shown in Table 1, include sequences of box numbers corresponding to different coverings with the linear box sizes equal to a power of 2 and also the final evaluation of the fractional dimension together with the standard deviation. The following notations are used. In the first line of Table 1, s is a power of 2, such that 2^{-s} is the linear size of the corresponding boxes in the fractions of the entire image. Therefore, the presented results are ordered with respect to increasing box size. In the first column, N stands for the number of boxes, and *fdloc* stands for the local slopes. In the last two columns, *fd1*, *fd2* and *sd1*, *sd2* stand for the fractal dimensions and corresponding standard deviations obtained over the different range of scales. Finally, K100, K200 - the Koch curve with resolution 100 and 200 dpi, respectively, K100s, K200s - the small size Koch curve with resolution 100 and 200 dpi, respectively.

Table 1. The local slopes and fractal dimensions for different versions of the Koch curve

	curve	scale characteristic s											
		10	9	8	7	6	5	4	3	2	1	fd1/sd1	fd2/sd2
N	K100	-	1429	639	293	122	47	21	8	4	2		
	K200	4739	1820	698	305	125	48	21	8	4	2		
	K100s	-	145	66	27	15	8	6	4	2	1		
	K200s	476	178	70	28	15	8	6	4	2	1		
fdloc	K100	-	1.16	1.12	1.26	1.38	1.16	1.39	1.00	1.00		1.19/0.15	1.30/0.11
	K200	1.38	1.38	1.19	1.19	1.38	1.19	1.39	1.00	1.00		1.25/0.16	1.26/0.08
	K100s	-	1.14	1.29	0.85	0.91	0.42	0.59	1.00	1.00		0.73/0.25	1.22/0.11
	K200s	1.42	1.35	1.32	0.90	0.91	0.42	0.59	1.00	1.00		0.99/0.34	1.36/0.05

As was expected based on the theoretical discussion in the previous section, the results in Table 1 show that the covering values of the smallest and largest boxes have weak (if any) relations to the theoretical fractal dimension. Indeed, if all the nine local slopes of the K200 computation are used to calculate the final approximation to the fractal dimension, then formally the result (1.25) is close to the theoretical value (1.2619), but it happens only due to accidental compensation of large and small wrong values in this specific case. The high standard deviation (0.16) indicates that, roughly speaking, the probable approximations to the theoretical result have large chance of being found in the entire interval of 1.09-1.41, which represents a very low level of precision. However, if we eliminate some problematic counts at both ends of the space spectrum, then the results are significantly improved. For example, using the scale window 8-5 (the boxes with s equal to 5 through 8), we just slightly improve the final evaluation (1.26), but the confidence level of this result is better because the standard deviation in this case is 0.08 (see *fd2* and *sd2* in the last column). The experiments with K100 and especially with K100s and K200s reveal the same properties, except that the overall mean value of the local slopes (penultimate column) does not show good compensation of the erroneous local results as was found for the K200 curve.

The comparison between the K100 and K200 curves and also between the K100s and K200s curves shows that an adequate resolution of the digital image can be an important factor for a more precise evaluation of the fractal dimension. As is natural to expect, in the case of the smaller curves, the subsequent decreasing of the resolution to 400x300 pixels leads to an image that has no information on the fractal structure of the Koch curve (the results are not shown in Table 1), although visually this image is rather similar to those with a better resolution. Therefore, the conclusion to draw is that the digital image should have sufficient resolution to contain information for the scales to be analyzed.

Let us discuss in greater details the results for the K100s and K200s curves. Because these curves occupy only 1/9 of the entire space on a blank page (different from the normal-sized K100 and K200 curves), it is clear that the relevant counts for K100s and K200s will be moved about three degrees toward smaller scales (three columns to the left in Table 1). Indeed, the first meaningful count for the K100s and K200s curves appears only at $s=8$, for the box size 2^{-8} of the original size (for K100s, this is also the only meaningful count). If one computes the approximation to the fractal dimension of K100s and K200s based on all the space scales, then the result is meaningful (0.73 and 0.99, respectively). However, the situation is not so disastrous if we take into account the scale of the K100s and K200s curves and use only the largest boxes for their evaluation, for example, using the scale window 9-8 for K100s and 10-8 for K200s (these windows were used for *fd2* and *sd2* in the last column in Table 1). In this case, a reasonable approximation can still be obtained that corresponds to the theoretical result stating that the fractal dimension does not change with modification of the curve scale.

The experiments with the smaller curves can also be interpreted in the terms of multifractality. Indeed, we have two different parts of a page: the first, which occupies about 8/9 of the space, is blank and has the fractal dimension 0, and the second, which occupies 1/9 of the space, is the six-iteration Koch curve with the fractal dimension 1.26. The theory states that the limiting box-counting dimension should coincide with the fractal dimension of a more complex figure. However, this result holds only when the limit value is considered and therefore larger scales have no influence on the final result. To reproduce a similar property in computational evaluations, an appropriate scale window should be chosen to obtain a reasonable approximation. In the case of the K100s and K200s curves, it can be the window 9-8 and 10-8, respectively.

Finally, it is worth to note that the presented results are quite accurate, which confirms good performance of both programs. The increased sensitivity of the provided results to local deviations is a consequence of the chosen form of evaluation by means of the local slopes, which allow clearer separation of the problematic parts of the computations. The more common form of representation of the results and evaluation of the final fractal dimension is based on the least squares linear fitting of $\log N_r$ versus $-\log r$. Using the last approach for the K200 curve, we obtain the values of the fractal dimensions 1.25 and 1.27, and the corresponding standard deviations 0.16 and 0.03 for the full and reduced space windows, respectively. Additionally, the correlation coefficients between the original and fitted values are 0.9988 and 0.9996. These results are in line with those reported in other studies (e.g., [17], [23], [28], [31]). In accordance with the theory ([15], [32]), similar results were obtained for the Koch snowflake (Fig.1b).

Hence, the presented results confirm the conclusions derived in a theoretical discussion about the properties of the box-counting method in the previous section. In particular, this simple simulation shows that the approximation of the theoretical fractal dimension depends strongly on the scale window (the box sizes) used for the calculation of the mean value. Therefore, to optimize the evaluation of the fractal dimension, specific filtration of the box counts should be applied. Apparently, the use of both extremely small and large boxes just contaminates the results in the middle part of the space spectrum. Therefore, it is reasonable to suggest that the one or two smallest sizes and at least three largest sizes should be eliminated from the final evaluation of the fractal dimension. In the remaining scale window, the choice of the box sizes should be based on a priori information about space scales of the chosen structures, especially when a multifractal system is studied. It means that special consideration should be given to the local slopes and bands of these slopes within the chosen scale window. Finally, one of the reasonable measures for the verification of the choice for the scale windows and the reliability of the evaluations is the standard deviation and similar parameters.

4 Analysis of the Amasya Site Using the Bovill Data

Using the box-counting fractal dimension, Bovill [5] tested the hypothesis on the relationship between the visual complexity of the built and natural settings in the case of the historic area of the city of Amasya, Turkey, which was founded over 2000 years ago. In particular, he compared the fractal dimensions of the images, representing the dominant hill in the historic part and a typical row of the traditional houses situated along a river at the foot of the hill (reproduced in Fig.2a and 2b) (see Bovill [5]). Later, the obtained results were reviewed independently by Lorenz [25] and Vaughan and Ostwald [36].

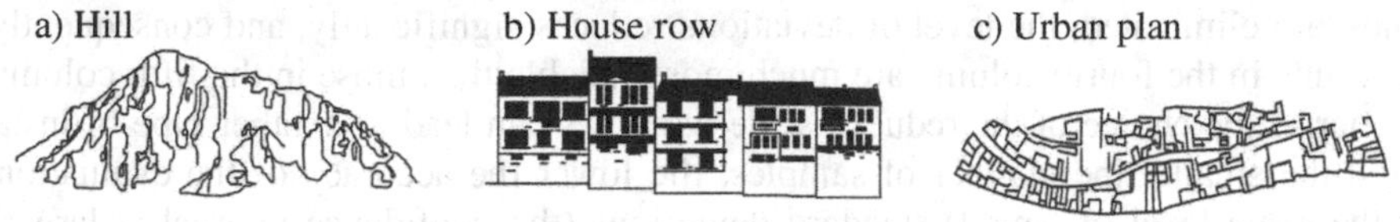

Fig. 2. The Amasya images, Bovill data

The fractal dimensions of the three images, which were found by applying the box-counting method, are presented in Table 2. The first three columns reproduce the results obtained by Bovill [5], Lorenz [25], and Vaughan and Ostwald [36], respectively (in a similar way as they are assembled in the last article), whereas the fourth column contains the results obtained in our study. Our computations are based on a version of the above figures with a resolution of about 600x200 pixels. Of course, formally, the digital resolution can be increased, but it will not provide more details on the factual forms. The first three columns contain only the final evaluations of the fractal dimensions, which were obtained apparently by averaging the results over all (or nearly all) space scales, whereas the fourth column contains the evaluation for the fractal dimension (*fd*) and standard deviation (*sd*) for the boxes remaining after filtration of the two smallest and three largest sizes.

Table 2. Fractal dimension of the Amasya images according to different sources

Element\Source	Bovill	Lorenz	Vaughan and Ostwald	this research *fd* / *sd*	all box sizes *fd* / *sd*	curve fitting *fd* / *sd*
D_{hill}	1.57	1.36	1.50	1.59 / 0.14	1.45 / 0.38	1.51 / 0.28
$D_{elevation}$	1.72	1.55	1.51	1.71 / 0.07	1.58 / 0.44	1.61 / 0.35
$D_{urban\ plan}$	1.43	1.49	1.59	1.62 / 0.09	1.42 / 0.36	1.49 / 0.32

It is seen that for each of the three evaluated elements, the results of the computations from the four independent studies are different. It was suggested by Vaughan and Ostwald [36] that the differences among the three first reports are due to the inconsistency of computational implementations of the box-counting method: Bovill [5] made his calculations manually, Lorenz [25] used an early version of the software Benoit, while Vaughan and Ostwald [36] applied a refined version of Benoit and Archimage software and presented the results averaged over the two programs. Another possible cause mentioned was that an appropriate treatment of thick lines in the images was not made in [5] and [25]. Indeed, these may be the sources of the discrepancy in the results, along with slight differences in the quality and computer form of photographs used by the authors and variations in the used box-counting algorithm. However, we would like to draw attention to another feature that seems to be the main cause of the differences among the results: the presence of significant deviations in the distribution of the box counts with respect to the box size (shown as *sd* in the fourth column of Table 2) even for the algorithm with filtration of the extremal sizes. In the fifth column, we also present the results averaged over the entire range of the used box sizes (without filtering the smallest and largest scales). It is seen that when the most problematic coverings, which correspond to a few smallest and greatest scales, are eliminated, the level of deviations reduces significantly, and consequently, the results in the fourth column are much more reliable than those in the fifth column. Of course, the choice of the reduced scale windows can lead to another type of instability: the smaller the number of samples, the lower the accuracy of the evaluations for the same level of (small) standard deviations (the confidence interval is larger). However, we cannot avoid such selective evaluations in the case of multifractal forms. Hence, just like in the case of the classical fractals, the range of the box coverings involved and the method of processing these counts can strongly influence the final result. This is the case of the considered Amasya images, as is exemplified in the fourth and fifth columns of Table 2.

Although details on the calculations are not presented in [36], the mere fact that the authors show only the final evaluation of the fractal dimensions and has no discussion on the validity of the obtained numbers suggests that the questions of the distribution of local slopes and the reliability of the obtained results were not considered in [36]. In additions, because the results in the last column considerably more closely match those in the first three columns than the results in the fourth column, it is very likely that the box coverings over all (or almost all) space sizes were used to obtain the fractal dimensions in [36]. Moreover, when we applied another common algorithm to find the value of the fractal dimension - the least square method, which defines the unique slope of a straight line based on the calculated set of all box coverings - the obtained results were even closer to those of [36] (see the sixth column in Table 2). This method (usually under the name of curve fitting) is one of the standard ways to evaluate the final fractal dimension and is used, in particular, in the software Benoit. Another observation is that the results attributed by Vaughan and Ostwald [36] to Bovill and Lorenz actually represent only mean values of the local slopes for different space scales given in [5] and [25]. Both Bovill and Lorenz take care of local slopes and provide analysis of the differences in these slopes, including explanations in terms of the building details and structures. On the contrary, in [36], all the considerations are based on the final numbers of the overall evaluation provided by the software, without concerning the variability and reliability of the results. In the course of their analysis of the Amasya case, Vaughan and Ostwald [36] arrive to the conclusion that the found gap of 0.09 (defined as the difference between the highest and lowest fractal dimensions, i.e., the gap=1.59-1.50) "suggests a significant difference in visual character". As one can see from Table 2, this statement is not applicable in the Amasya case because the standard deviations shown in the fifth column are so high (about four-five times greater than 0.09) that, based only on these numbers, one cannot draw a conclusion about the real value of the fractal dimensions with any reasonable precision. Even for the filtered results in the fourth column, such a conclusion is still impossible.

Of course, the approach used in [36] is just one of the examples of the application of the box-counting method without verification of the reliability of the obtained results, and this article was chosen only due to its relevance for the study of Amasya. As a matter of fact, we did not find any discussion on the implications of the dispersion of the local slopes in [5] and [25] either, but both authors pay due attention to the significance of the local slopes. Actually, as far as we know, basic information related to evaluation of the reliability of the results of the box-counting computations is presented only in a small fraction of articles concerned with architectural and urban studies (e.g. [7], [9], [19], [33]), and even when such information is provided, no discussion on the importance and implications of such evaluations exists. It seems that there is too much reliance on "ready to use" software, which certainly bring a great deal of automation and simplification to the analysis of complex structures, but these programs may also unintentionally hide some critical points of the employed algorithms, and the task of finding these points is frequently left to the users of the software. Of course, there are many articles and reports, especially in the area of image processing, that deal with the different issues of the practical implementation of the box-counting method (e.g., [4], [6], [11], [14], [17], [22], [24] and references therein); however, it appears that the results reported there were not brought to the attention of scholars in more distant areas, such as architecture and urban planning. This report is an attempt to fill this gap and highlight some problems inherent to the box-counting methods and possible solutions.

As for the relationship between the built and natural settings is concerned, it is still difficult to arrive at a definitive conclusion with the base on the tests presented above for Amasya, even in the case of the filtered results shown in the fourth column. There are too many indeterminations in all stages of the evaluation, beginning with the imperfections of the images and the question of the reliability of the hill image, and ending with relatively large deviations for the fractal dimension of the hill. To remedy this situation, in the next section, we present additional results for the Amasya case based on sources with better resolution.

5 Analysis of the Amasya Site Using the Improved Data

In this section, we continue the study of the Amasya case by analyzing the two images with better resolution related to the same hill and house row (see Fig.3a and 4). The first image has a resolution of 800x400 pixels and the second has a resolution of 1000x300 pixels, representing more detailed data on the studied structures compared to those in the previous section.

a) The Amasya hill ($Hill_1$)

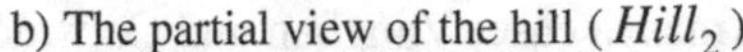

b) The partial view of the hill ($Hill_2$)

Fig. 3. The Amasya hill: general and partial view

Fig. 4. The house row at the foot of the hill

The obtained results are shown in the first two lines of Table 3 for the local slopes at the scales with *s*=8-4 (the boxes remaining after filtration of the two smallest and three largest sizes) and also for the evaluations of the fractal dimensions and corresponding standard deviations in the scale windows 8-4, 7-4 and 6-4 (in the last three columns). Comparing the results with those of the previous section, it can be noted that the found fractal dimensions have higher values, which can be expected because the new images contain more information and details on the forms. The additional evaluations for reduced scale windows (provided in the last two columns) show that within this range of scales, the variations of local slopes are small for both the hill and house row. For the "standard" window 8-4, the level of deviations from the mean value is acceptable for the hill but rather high for the houses. It is seen that the results in the first two rows are sufficiently close, especially in the reduced scale windows.

However, one should take into account that the real physical scales are different for the two images, and, consequently, the equally numbered scale windows in Table 3 correspond to different geographical distances. The hill frontal extent is about 1000 m, so the two windows with the greatest values of the local slopes are related to the real scale ranges of 50-100 m (window 4) and 12-25 m (window 6). At the same time, the length of the presented part of the house row is about 60 m, meaning that windows 4 and 6 (also among the highest values) correspond to the scale ranges of 4-8 m and 1-2 m, respectively. (Recall that local slopes are basically the differences between two consecutive box counts and therefore involve two different scales.) In general, the geometric parts of the hill image in all the space range between 6 m and 100 m have high visual complexity. The traditional house facades have greater complexity in the space range from 1 m to 8 m, with the maximum in the interval of 4-8 m, which corresponds approximately to the size (length and height) of an individual house. The high values for 1-2 m can be attributed to the salient details within each facade, such as windows and roofs. The image resolution does not allow us to make a reliable evaluation for the hill at smaller physical scales to compare with the house row. However, for this purpose we can use another image (Fig.3b), representing a local view of the part of the hill right above the houses. This part has a horizontal extent of about 120 m and the local slopes of the image, which correspond to the physical scale ranges of the house row, are shown in the corresponding scale windows in the third line of Table 3.

Table 3. Fractal dimensions: the hill versus the house row

image	scale characteristic *s*, local slopes					window 8-4	window 7-4	window 6-4
	8	7	6	5	4	*fd* / *sd*	*fd* / *sd*	*fd* / *sd*
Hill (Fig.3a)	1.72	1.87	1.89	1.86	1.89	1.84 / 0.07	1.88 / 0.02	1.89 / 0.02
House(Fig4)	1.40	1.68	1.91	1.92	1.96	1.77 / 0.24	1.87 / 0.13	1.93 / 0.03
Hill(Fig.3b)	1.42	1.75	1.89	1.88	1.94	1.78 / 0.21	1.87 / 0.08	1.90 / 0.03

It is seen that strong similarities between the values of the local slopes in the last two rows as well as between their spatial dynamics exist. Therefore, the presented results suggest that in the studied case of the Amasya historic site, a strong relationship between the visual complexity of the traditional historic buildings and the dominant parts of the natural landscape exists.

6 Conclusions

The box-counting method of the calculation of the fractal dimension was considered. A theoretical discussion of possible issues in the application of this method to determine the visual complexity of digital images was provided. The respective problematic points were practically analyzed through the computation of the fractal dimension of different versions of the Koch curve. The recommendations based on these simulations have led to an optimized version of the box-counting algorithm with filtration of the largest and smallest boxes, evaluation of the local slopes and local means and monitoring of the standard deviations. Application of this version to the well-known case of Amasya has shown that for the Bovill data set it is not possible to arrive at a definitive conclusion about the relationship between the traditional buildings and the dominant hill due to the low level of reliability of the obtained results over the entire range of scales. However, the computations based on the additional data with a better resolution have revealed that in the studied case, there is a strong relationship between the fractal characteristics of the traditional buildings and the natural landscape. The found correlations between the local slopes show the high level of visual agreement according to the fractal measures applied in architecture and urban planning ([5], [13], [31], [36]).

Concluding our analysis we can state that there are qualitative and quantitative similarities between the visual complexities of the built and natural sites in the historic part of Amasya at different space scales. Since the analyzed data present the images of real landscapes and constructions, this approach minimizes the presence of artificial elements and make possible a more objective evaluation of the preferences of large groups of inhabitants. Application of a similar method of the fractal analysis to chosen sites can reveal the underlying structures of complex natural and built environments and their interconnections, which are reflected in the local fractal dimensions. In this way, the architectural planning can be supported with the objective measures of pleasantness and harmony expressed by local residents. Moreover, one of the criteria in determination if the building project will be successful or not can be the assessment of the similarity between the fractal dimensions of the projected forms and those of the surrounding geographic settings.

Acknowledgements. This research was supported by the Brazilian science foundation FAPERGS.

References

1. Batty, M., Longley, P.: Fractal Cities: A Geometry of Form and Function. Academic Press, San Diego (1994)
2. Batty, M.: Cities and Complexity: Understanding Cities with Cellular Automata, Agent-Based Models, and Fractals. MIT Press, Cambridge (2005)
3. Bechhoefer, W., Bovill, C.: Fractal analysis of traditional housing in Amasya, Turkey. In: Changing Methodologies in the Field of Traditional Environment Research, pp. 1–21. University of California, Berkeley (1994)
4. Bisoi, A.K., Mishra, J.: On calculation of fractal dimension of images. Pattern Recognition Letters 2, 631–637 (2001)
5. Bovill, C.: Fractal Geometry in Architecture and Design. Birkhauser, Boston (1996)
6. Buczkowski, S., Kyriacos, S., Nekka, F., Cartilier, L.: The modified box-counting method analysis of some characteristic parameters. Pattern Recognition 31, 411–418 (1998)
7. Burkle-Elizondo, G., Valdez-Cepeda, R.D.: Fractal analysis of Mesoamerican pyramids. Nonlinear Dynamics, Psychology and Life Sciences 10, 105–122 (2006)
8. Calter, P.A.: Squaring the Circle: Geometry in Art and Architecture. Wiley, Chichester (2008)
9. Capo, D.: The fractal nature of the architectural orders. Nexus Network Journal 6, 30–40 (2004)
10. Cardillo, A., Scellato, S., Latora, S., Porta, S.: Structural properties of planar graphs of urban street patterns. Physical Review E 73, 066107-1–066107-8 (2006)

11. Chen, S.S., Keller, J.M., Crownover, R.M.: On the calculation of fractal features from images. IEEE Transactions on Pattern Analysis and Machine Intelligence 15, 1087–1090 (1993)
12. Cooper, J., Oskrochi, R.: Fractal analysis of street vistas: a potential tool for assessing levels of visual variety in everyday street scenes. Environment and Planning B: Planning and Design 35, 349–363 (2008)
13. Cooper, J., Watkinson, D., Oskrochi, R.: Fractal analysis and perception of visual quality in everyday street vistas. Environment and Planning B: Planning and Design 37, 808–822 (2010)
14. Da Silva, D., Boudon, F., Godin, C., Puech, O., Smith, C., Sinoquet, H.: A critical appraisal of the box counting method to assess the fractal dimension of tree crowns. In: Bebis, G., et al. (eds.) ISVC 2006. LNCS, vol. 4291, pp. 751–760. Springer, Heidelberg (2006)
15. Falconer, K.J.: Fractal Geometry: Mathematical Foundations and Applications. Wiley, Chichester (2003)
16. Feng, J., Lin, W.-C., Chen, C.-T.: Fractional box-counting approach to fractal dimension estimation. In: Proceedings of ICPR 1996, pp. 854–858 (1996)
17. Foroutan-pour, K., Dutilleul, P., Smith, D.L.: Advances in the implementation of the box-counting method of fractal dimension. Applied Mathematics and Computation 105, 195–210 (1999)
18. Frankhauser, P., Tannier, C.: Fractalyse (2012), `http://www.fractalyse.org/en-home.html`
19. Hagerhall, C.M., Purcell, T., Taylor, R.: Fractal dimension of landscape silhouette outlines as a predictor of landscape preference. Journal of Environmental Psychology 24, 247–255 (2004)
20. Hahn, A.J.: Mathematical Excursions to the World's Great Buildings. Princeton University Press, Princeton (2012)
21. Heath, T., Smith, S.G., Lim, B.: Tall buildings and the urban skyline: the effect of visual complexity on preferences. Environment and Behavior 32, 541–556 (2000)
22. Huang, Q., Lorch, J.R., Bubes, R.C.: Can the fractal dimension of images be measured? Pattern Recognition 27, 339–349 (1994)
23. Jelinek, H.F., Fernandez, E.: Neurons and fractals: how reliable and useful are calculations of fractal dimensions? Journal of Neuroscience Methods 81, 9–18 (1998)
24. Li, J., Du, Q., Sun, C.: An improved box-counting method for image fractal dimension estimation. Pattern Recognition 42, 2460–2469 (2009)
25. Lorenz, W.E.: Fractals and fractal architecture. Vienna University of Technology, Vienna (2003)
26. Mandelbrot, B.: The Fractal Geometry of Nature. Freeman, San Francisco (1982)
27. Mandelbrot, B.: Fractals and the rebirth of iteration theory. In: Peitgen, H.-O., Richter, P.H. (eds.) The Beauty of Fractals: Images of Complex Dynamical Systems, pp. 151–160. Springer, Berlin (1986)
28. Milosevic, N.T., Ristanovic, D.: Fractal and non-fractal properties of triadic Koch curve. Chaos, Solitons and Fractals 34, 1050–1059 (2007)
29. Moisy, F.: Boxcount (2008), `http://www.mathworks.com/matlabcentral/fileexchange/13063-boxcount`
30. Ostwald, M.J.: Fractal architecture: late twentieth century connections between architecture and fractal geometry. Nexus Network Journal 3, 73–83 (2001)

31. Ostwald, M.J.: The fractal analysis of architecture: calibrating the box-counting method using scaling coefficient and grid disposition variables. Environment and Planning B: Planning and Design 40, 644–663 (2013)
32. Peitgen, H.-O., Jürgens, H., Saupe, D.: Chaos and Fractals: New Frontiers of Science Springer, New York (2004)
33. Perry, S.G., Reeves, R.W., Sim, J.C.: Landscape design and the language of nature. Landscape Review 12, 3–18 (2008)
34. Rodin, V., Rodina, E.: The fractal dimension of Tokyo's streets. Fractals 8, 413–418 (2000)
35. Stamps, A.E.: Fractals, skylines, nature and beauty. Landscape and Urban Planning 60, 163–184 (2002)
36. Vaughan, J., Ostwald, M.J.: Using fractal analysis to compare the characteristic complexity of nature and architecture: re-examining the evidence. Architectural Science Review 53, 323–332 (2010)
37. Zacharias, J.: Preferences for view corridors through the urban environment. Landscape and Urban Planning 43, 217–225 (1999)

Using Ontologies to Support Land-Use Spatial Data Interoperability

Falk Würriehausen, Ashish Karmacharya, and Hartmut Müller

FH Mainz, University of Applied Sciences,
Lucy-Hillebrand-Str. 2, 55128 Mainz, Germany
{wuerriehausen,karmacharya,mueller}@geoinform.fh-mainz.de
www.i3mainz.fh-mainz.de

Abstract. This paper presents an ontology-based concept to support the land-use spatial data interoperability within the Infrastructure for Spatial Information in Europe (INSPIRE). Land-use spatial data specifications are available for different levels, which define the semantic requirements for the exchange of spatial planning data at local and the European levels. A large number of systems in the European Member States operate on various data schemas developed at the national and local levels to fulfil the statutory requirements of the state and local governments. Consequently, conceptual, structural, and formatting differences hamper information exchange between the European and the local level. This paper describes a spatial planning schema mapping approach to support interoperability in the context of INSPIRE implementation. The proposed solution relies on established standards like the GML Geographic Markup Language, the OWL Web Ontology Language, the RIF Rule Interchange Format, and others. Starting from an investigation of INSPIRE interoperability requirements, we outline the concept of a standard-based general platform which allows for the construction of runtime transformation services among different systems. We illustrate the feasibility of the concept by discussing land-use planning application cases within the area of the European Community.

Keywords: Interoperability, Planned Land-Use, Ontology, INSPIRE.

1 Introduction

To achieve interoperability is one of the greatest challenges not only for the Infrastructure for Spatial Information in Europe (INSPIRE), but also for the local communities and authorities of the Member States [1]. In many cases, there are storages of secondary data sets to connect different data models, even if the same area of land use, utilities and alike is represented. In urban land-use planning and data management, the use of ontologies has been underrepresented and only sporadically considered in planning data management. With a view that provides an implementation at the local environment, a comprehensive management and deployment of spatial information exchange is necessary to create interoperability.

B. Murgante et al. (Eds.): ICCSA 2014, Part II, LNCS 8580, pp. 453–468, 2014.

The national administrative structures with their own specific features and datasets have led to heterogeneous information infrastructures at the local level. This heterogeneity in data management is a reason for more cooperation at the operational level. A semantic and organizational concept in a heterogeneous organization and data structure landscape is, therefore, imperative. Without binding concepts for local government, interoperability between land-use planning data in municipalities and INSPIRE cannot be generally realized.

In Europe, different institutions like European Environmental Agency (EEA) and Joint Research Centre (JRC) and ongoing initiatives like GMES [2] and GEOSS [3] are acting in the field of Earth observation. Several projects aim at detecting changes in existing land-cover and land-use vector data. The German project DeCOVER2 [4] develops algorithms for updating the digital landscape data model of German biotopes and land use. Remote sensing data is often used to do the change detection. Because the input and output datasets use different classifications, knowledge about the mapping among classifications through the use of ontologies is required for interoperability [5].

In this study we will analyze the current situation in the context of planned land-use (plu) spatial datasets and the usage of an ontology to solve heterogeneity problems in data management. In many application fields like land use, detection of equivalent objects, data modeling and data mining processes, ontologies are most beneficial. With standard-based rules for local government data, we will demonstrate how extensive knowledge which is already available for local government data can be used at the European level. The German projects DeCOVER2 [4], XPlanung [6] and other relevant European initiatives like Plan4All [7], SENSOR [8], Harmony [9], SEIS-BASIS [10], HUMBOLDT [11], GENESIS [12], etc. also might benefit from the developed solution for improved knowledge management in the semantic web. The activities to establish the appropriate rules are guided by the objectives of searchability, usability, reusability, composability and interoperability of digital data in the context of the legal framework of the INSPIRE directive [1].

2 The European Spatial Data Infrastructure (INSPIRE)

The goal of the European Spatial Data Infrastructure (INSPIRE) is the cross-border use of interoperable spatial data. To combine data from different sources and use them in a seamless way, data from different countries and sources should be provided [13]. The requirements for the development of a Spatial Data Infrastructure (SDI) confront the European Commission with the previously identified data exchange problems of Regional SDI level [14]. These obstacles include differences in the availability, quality, organization and accessibility of spatial data between the various Member States, which cover nearly all levels of government. National spatial data regularly are restricted exclusively to a selected area and are rarely cross-border. Through the establishment of INSPIRE, seamless cross-border access to spatial data should be facilitated in the future. In the INSPIRE directive it is determined by the EU that the exchange of spatial data can best be done at the European level.

2.1 Requirements of the INSPIRE Conceptual Model

Parts of the European spatial data infrastructure with spatial data, metadata, network services and arrangements of data and services are already available. INSPIRE underpins the principle of decentralized data storage within centrally aligned structures. The specifications are determined by different stakeholders in the European Community, including the thematic working groups and the users of the data. Member States especially are responsible for that and should give public access to the data, share the data and make them accessible to many users. Users should be able to do a simple search for spatial data, and the usability and usage conditions should be quickly recognizable [15].

The provinces and municipalities in Member States in Europe are required to implement the legal framework and rules of the European INSPIRE directive. The Commission Regulation (EU) No 1089/2010 of 23/11/2010 as implementing Directive 2007/2/EC of the European Parliament and of the Council as regards to interoperability of spatial data sets and services is also applicable [16]. The purpose is to compile the technical and organizational requirements for local governments. It can already be seen from the discussion above that interoperability can often be viewed in different contexts, firstly, from the perspective of data, metadata and services and, secondly, from the interoperability of systems and applications. Therefore, a description of components is necessary, to fulfil the requirements of interoperability in Europe.

Table 1. Data transformation (J) and corresponding components (M, P, S) as a part of the INSPIRE Data interoperability, adapted from [15]

Data interoperability components		
(A) INSPIRE Principles	(B) Terminology	(C) Reference model
(D) Rules for application Schemas and feature catalogues	(E) Spatial and temporal aspects	(F) Multi-lingual text and cultural adaptability
(G) Coordinate referencing and units of measurement model	(H) Object referencing modeling	(I) Identifier Management
(J) Data transformation	(K) Portrayal model	(L) Registers and registries
(M) Metadata	(N) Maintenance	(O) Data & information quality
(P) Data transfer	(Q) Consistency between data	(R) Multiple representation
(S) Data capturing rules	(T) Conformance	

The components of the INSPIRE data interoperability, *Data transformation (J)* with *Data transfer (P)*, plus *Metadata (M)* and *Data capturing rules (S)* for description (see Table 1), require the data to be restructured and supplemented to correspond to the semantics of a different data model (target data) on the semantic level. *Metadata (M)* should be reported for individual spatial objects (spatial object-level metadata), for complete datasets and for dataset series (dataset-level metadata). In a standards based environment, spatial object-level metadata are fully described in the application schema. Particularly needed are agreed *Data capturing rules (S)* for the provision of *Data transfer (P)* based on network services, as well as for services which are able to process comprehensive spatial data based on different, even local, spatial data themes. Such data concepts and data models can help to guarantee the data interoperability in a heterogeneous environment of organization units. In-depth analysis of data structures, therefore, is imperative. By the *Data transfer (P)* it is possible to exchange heterogeneous data sources by INSPIRE Transformation Network Services. Based on the INSPIRE requirements, aspects of data interoperability will be discussed in the next section. The goal will be to define all information of a local dataset at the highest possible level as interoperable dataset in context of INSPIRE.

2.2 Requirements of INSPIRE Data Interoperability

Interoperability of spatial data sets is an overall goal to be achieved in the legal context of INSPIRE. It is necessary to establish a common understanding of the interoperability goals and concepts. Following the INSPIRE Directive [1], the term *'interoperability' means the possibility for spatial data sets to be combined, and for services to interact, without repetitive manual intervention, in such a way that the result is coherent and the added value of the data sets and services is enhanced;*

According to Article 8, paragraph 1 of the INSPIRE Directive 2007/2/EC, spatial data sets correspond to one or more themes, with the requirements of the implementing provisions of Article 7 paragraph 1 being met at the same time [16]. The benefit of implementation is reflected in a re-use that can be counted. Results of the cost-benefit analysis for municipal e-government with INSPIRE show demonstrable advantages in terms of availability of information, the interoperability and the quality [17]. The advantage of transparency and an open data policy with INSPIRE can be achieved for municipal e-government, a significant added value. For its realization, a technical and substantive evaluation of regulations is necessary. In the implementation rules the objectives are described [16]:

- In order to achieve interoperability and benefit from the endeavours of users' and producers' communities, when appropriate, *international standards* are integrated into the concepts and definitions of the elements of spatial data themes listed in the Directive 2007/2/EC Annex I, II or III.
- In order to ensure interoperability and harmonisation across spatial data themes, the Member States should meet *requirements* for common data types, the identification of spatial objects, metadata for interoperability, generic network model and other concepts and rules that apply to all spatial data themes.
- In order to ensure the interoperability and harmonisation within one spatial data theme, the Member States should use the *classifications* and definitions of spatial objects, their key attributes and association roles, data types, value domains and specific rules that apply to individual spatial data theme.

It has to be ensured that *'interoperability and harmonisation'* is not only applied in the context of the original system modeling, but also as a general framework for the "interoperability of systems" or, more accurately, be used by computer systems. The systems have to be conceptually designed, including metadata, processes and conditions of use. From the machine-readable information, concrete attributes and relations of spatial objects are produced. With the overall goal of firstly, understanding different data sources and, secondly, its use in decision support assistance, the tools of Semantic Web Technology [18] can be used to solve semantic problems of interoperability.

3 Methodology

To measure interoperability, a concept to describe interoperability is needed. Wang et al [19] present a Levels of Conceptual Interoperability Model (LCIM), which they originally design as a general framework for conceptual modeling. The LCIM can be used as a descriptive as well as a prescriptive model. In that way the model can be used to describe the interoperability of systems not only in the originally intended modeling and simulation context, but also as a general purpose framework to describe the interoperability of machine systems in general or, more specifically, of computer systems. The categorization of interoperability in technical, syntactic, semantic and organizational parameters is necessary.

The study is founded upon the methodology, that the semantic interoperability in the LCIM [19] can be supported by an ontology. Ontologies are typically specified in languages that allow abstraction apart from data structures and implementation strategies; in practice, the languages of ontologies are closer to first-order logic than languages used to model databases. For this reason, ontologies are said to be at the "semantic" level, whereas database schema are models of data at the "logical" or "physical" level [20]. The used technologies would include the Standards Ontology Web Language (OWL), the Resource Description Framework (RDF), and the transformation of Extended Markup Language (XML) with XML Schema definition (XSD) and their representation in Semantic Rule Web Language (SWRL) or Rule Interchange Format (RIF). The following sections, present a discussion on how semantic interoperability between local governments and the European level could be realized with semantic rule definition in SWRL/RIF.

The novelty here is not to develop a new methodology for the transformation, but rather to define heterogeneous local datasets as Semantic Web Resources. When ontology mapping and transformations must be applied on heterogeneous models, we have to reduce the complexity of the original capabilities of Semantic Modeling. Therefore, we define the following approach:

1. Available land-use datasets and services are to be documented on the *syntactic level* when a database schema resulting in a XML file or another encoding file has been recognized as model output data
2. If spatial data are presented in *semantic models*, with the heterogenic data models represented as an RDF resource, then the knowledge of datasets and their relations in OWL can be mapped with RIF rules. Execution Services are to be developed on the foundation of Semantic Web technology.

3. If only documentation, data sets at raster level or non-referenced data are in use, the INSPIRE specifications have to be applied directly at the *organizational level*. In that case at least codelists representing all aspects of local level semantics as a "knowledge about source" should be available.

To illustrate case 2 in particular, an application case on the INSPIRE theme of "land-use" and its compatibility with one of the national e-government initiatives will be discussed. An example for case 3 that will be investigated is the topic of codelists and knowledge about the sources in urban land-use planning.

4 The Standard-Based Semantic Web Technology

4.1 Knowledge Management and the Semantic Web

Knowledge Management can be described as the process of identifying, creating, and distributing the experiences, expertise and insights possessed within an individual, group or even an organization. Knowledge is commonly distinguished from data and information [21]. Data are representation of observations or any singular fact which are kept out of context. Data are meaningless until they are put in the context of space or an event. Information understands the nature of the data, but does not provide the reasons behind the existence of data and is relatively static and linear by nature. Information is a relationship between data and has great dependence on context for its meaning and with little implication for the future. However, Knowledge Management is about the capture and reuse of knowledge at different knowledge levels. Beyond every relationship, a pattern arises which has the capacity to embody the completeness and consistency of the relations to an extent of creating its own context [22]. Such patterns represent knowledge on the information, and consequently, on data.

Knowledge Management re-evolved with the rise in the *Semantic Web*. The explosion of information on the World Wide Web (WWW) has led to the problem of managing it. It is generally perceived that this vast information cannot be managed through a human effort only. In some form there should be interference from machines to assist humans in information management. In order to have machines interfere and assist humans in managing information, it should understand the information first. This would require knowledge formalized from the information. In their paper [23], Berners-Lee et al. have envisaged the next generation of the Web which they call "the Semantic Web". In this Web, the information is given with well-defined meaning, better enabling computers and people to work in cooperation. Moreover, the Semantic Web aims at machine-readable information, enabling intelligent services such as information brokers, search agents and information filters to offer greater functionality and interoperability [24]. The association of knowledge with the Semantic Web has provided a scope for information management through knowledge management. Since both the technologies use ontologies to conceptualize the scenarios, Semantic Web technology could provide a platform for developments of knowledge management systems [25].

The ontologies are core to both the technologies in whichever methods they are defined. The Semantic Web defines ontologies through XML based languages and advancements in these languages.

4.2 Ontology and the Ontology Languages

The term *Ontology* has been used for centuries to define an object philosophically. The core theme of the term remains the same in the domain of computer science; however the approach in defining it has been modified to adjust the domain. Within the computer science domain, an ontology is a formal representation of knowledge through the hierarchy of concepts and the relationships among those concepts. In theory, an ontology is a *formal, explicit specification of shared conceptualization* [20]. In any case, an ontology can be considered as a formalization of knowledge representation and *Description Logics* (DLs) [26], [27] provide logical formalization to the Ontologies [28].

The World Wide Web Consortium (W3C) has standardized the Web Ontology Language to model ontologies. OWL is actually a family of three language variants of increasing expressive power: OWL Lite, OWL DL, and OWL Full. The standardization of OWL has sparked off the development and/or adaption of a number of reasoners and ontology editors, including Protégé [29]. OWL 2 is a new version of OWL, the ontology language which considerably improves the datatype [30].

4.3 Query and Rule Languages

SPARQL is a query language for the Resource Description Framework RDF triplets with which OWL is syntactically aligned [31]. In this manner, SPARQL queries the knowledge within OWL. As a query language, SPARQL is "data-oriented" in that it only queries the information held in the models; there is no inference in the query language itself. SPARQL is able to query OWL ontologies which use RDF graphs to structure it. SPARQL uses filters to limit the solutions to only those which are returned true with the expression.

The Semantic Web Rule Language SWRL is a logic program which infers the knowledge base to derive a conclusion based on the observations and hypothesis [32]. The SWRL as the form, *antecedent* --> *consequent*, where both *antecedent* and *consequent* are conjunctions of atoms written *a1* ^ ... ^ *an*. Variables are indicated by using the standard convention of prefixing them with a question mark (e.g., "*?x*"). URI references ("URIrefs") are used to identify ontology elements such as classes, individual-valued properties and data-valued properties. For instance, the following rule asserts that one's parents' brothers are one's uncles, where parent, brother and uncle are all individual-valued properties (see equation).

hasParent(?x,?p)^hasBrother(?p,?u)→ hasUncle(?x,?u) (1)

The Rule Interchange Format RIF is a format which enables different rule languages within the Semantic Web framework to exchange rules among them.

Currently, many rule languages exist which include SWRL as primary rule language. The primary objective of an RIF is to address the problems of integrating rule sets from different rule languages and to make a rule engine work when executing these diverse sets of rules. It is a W3C standard [33].

5 Ontology-Based Approach to Support Interoperability

5.1 Relation of Semantic Web and Spatial Data Objects

The potential of the *Semantic Web technologies* in INSPIRE directives is one of the upcoming areas of research. Querying heterogeneous data through a common entry point like in the initiative Linked Data within the Semantic Web framework attracts the increasing attention of interested researchers. An approach presented in [34] translates sophisticated Geographic Markup Language (GML) [35] into the Semantic Web OWL ontologies [30] as Linked Data Resources. GeoSPARQL [36], a spatially modified SPARQL, can wrap the INSPIRE Data and Web services and provides a semantic platform for querying data from different sources.

A set of rules in RIF [33] were used to transform queries into GeoSPARQL, both by attributes and by the geometry. For the description of, among others, topological relations, the vocabularies of simple features are taken from the OGC [37]. The most important classes in GeoSPARQL, the "geo:SpatialObject" with subclasses "geo:Feature" and "geo:Geometry", are linked by the relationship "geo:hasGeometry".

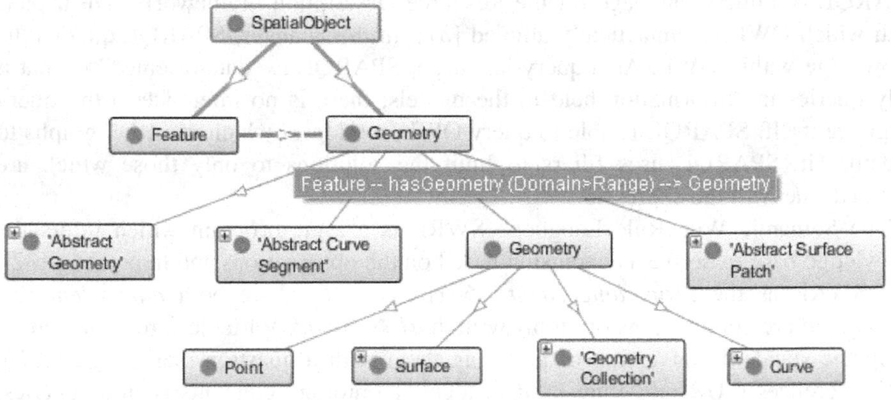

Fig. 1. Representation of Spatial Object, Feature and Geometry and their relationships as Ontology Graph

A conceptual Spatial Object has a geometry, defined by different types of geometry, represented in Geographic Markup Language (GML) or Well Known Text (WKT). Classes for many different types of geometry are provided, "geo:Geometry" for WKT, "geo:AbstractGeometry", "geo:AbstractCurveSegment" and "geo:AbstractSurfacePatch" for GML. The properties "geo:asWKT" and "geo:asGML" link the entities as so-called literal representations of geometry. Objects that have one of both properties, use "geo:wktLiteral" and "geo:gmlLiteral", the specific GML Simple Feature or data types, by separating the actual relationships (entities) and their locations. In that way several geometries may be linked to a single feature for different purposes. With those relations of geometry and feature, spatial objects, including their data source and target requests for the ontology execution, can be represented (see section 5.2).

5.2 The Concept of Ontology Use for Data Execution in INSPIRE

In the previous sections we defined how and with which conceptual models INSPIRE Data interoperability can be described. The needs at different levels were defined, which are necessary for the full *implementation of interoperability* from the local level to INSPIRE. Essentially, four main categories can be found in all models, which have to be taken into account in spatial data management. All categories in terms of a description of the data corresponding categorizations are required, to be provided as INSPIRE profile. Ontology and Semantic Web Technology can be set to apply the developed interoperability model to an urban land-use spatial data source and target data model in INSPIRE.

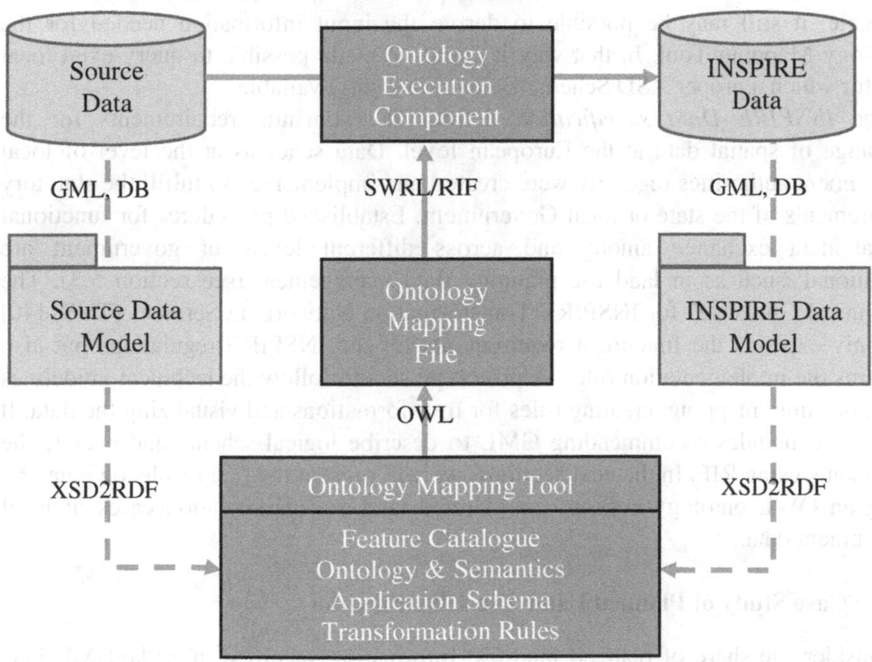

Fig. 2. Implementation Concept of Ontology-based Spatial Data Integration in INSPIRE

Fig. 2. shows which requirements are to be met by the *Ontology Mapping Tool*. Both the land-use planning data (Source Data) and target data (INSPIRE Data Model) are described by application schemas, but are not necessarily present in the RDF Language. However, this is necessary to create a common class and property model, especially in the local data sets with significant heterogeneity. State of the art work in the field of semantic transformation is available. Althoff [38] already present and describe conceptual model driven approaches, however without ontologies. The Ontology Mapping Tool contains both the "Feature Catalogue" and the "Application Schema" in RDF notation, edit functions for the "Ontology & Semantics" and the "Transformation rules" in Semantic Web Language.

In our ontology based concept, the description of two different sources in RDF is a precondition for the use of the OWL. The Local Database (DB) source is given by an XML-Schema Definition and its encoding in Geographic Markup Language. The step of XSD to RDF "XSD2RDF" in Fig. 2 can be performed by using XML tools. In general computer science there are approaches, such as the transformation from XSD and RDF in ReDeFer project [39]. In the concept of the Ontology Mapping Tool, this step is defined as a general requirement. By querying the Ontology Mapping Tool (see also section 5.2) Ontology Mapping Files can be created which serve as an SWRL/RIF input for an Ontology Execution Service (see central arrow in Fig. 2). As can be seen from Fig. 2, the initial requirement for an Ontology Execution Service is the availability of a formal data model description as it exists in some cases (see section 5.3). However, if a formal description of the Local Data Model is not available, it still may be possible to derive the input information needed for the Ontology Mapping Tool. In that way it could be made possible to query even local data for which a proper XSD Schema Definition is not available.

The *INSPIRE Data specification* defines the semantic requirements for the exchange of spatial data at the European level. Data schemas at the level of local government authorities regularly were created and implemented to fulfill the statutory requirements of the state or local Government. Established procedures for functional spatial data exchange among and across different levels of government are operational, such as in land use planning data management (see section 5.3). The "Technical Guidance" for INSPIRE Transformation Networking Services (TNS) [40] not only explains the functional requirements for the INSPIRE regulations but also explains the implementation rules. A prototype should follow the technical guidelines for publishing, mapping, creating rules for transformations and visualizing the data. It however concludes recommending GML to describe logical schema and encode the documents using RIF. In the next sections, we will present the first results obtained by using an OWL ontology concept for planned land-use classification cases of local government data.

5.3 Case Study of Planned Land-Use Classification

A considerable share of planned land-use information specified in XPlanGML [41] addressed by the INSPIRE Annex III theme land-use [5] is maintained at the level of local government. In addition to the work in [42] we will discuss the use of ontologies in the context of German land-use planning and the Hierarchical INSPIRE Land-Use

Classification System (HILUCS). An ontology mapping step is assumed as a prerequisite towards data integration as a separate problem in data management [43].

Table 2. Comparison of local data model (left) and equivalent classes (≡) in the Hierarchical INSPIRE Land-Use Classification System (HILUCS) (right)

BP_GemeinbedarfsFlaeche	**≡**	**3_3_CommunityServices**
BildungForschung	≡	3_3_2_EducationalServices
Kirche	≡	3_3_4_ReligiousServices
Sozial	≡	*3_3_3_HealthAndSocialServices*
Gesundheit	≡	*3_3_3_HealthAndSocialServices*
Kultur	≡	3_4_1_CulturalServices
Sport	≡	3_4_3_SportsInfrastructure
OffentlicheVerwaltung	≡	3_3_1_PublicAdministrationDefense AndSocialSecurityService
SicherheitOrdnung	≡	3_3_1_PublicAdministrationDefense AndSocialSecurityServices
Infrastruktur	≡	*4_3_Utilities*
Sonstiges	≡	3_5_OtherServices

The needed *mapping and transformation* rules are those between any local SDI source data model and the common INSPIRE target data model, here defined in the Annex III specification Land use [5]. The feature types of the local level source data model had to be mapped to the INSPIRE target data model. A short extract of the large set of mapping rules will be described in ontology schema description language. Table 2 shows an extract of transformation rules for planned zoning elements in 'Community Services' with attributes of the land-use in local and INSPIRE systems. We can represent the corresponding mapping rules in an ontology graph, created with our prototype of the ontology mapping tool. Ontology mapping can be described in accordance with Tschirner et al. [34] as a step to integrate GML-/XSD- Source into the RDF/OWL- language. Rules are presented in two distinct forms by using the ontology mapping platform detailed in Fig. 2: through restriction axioms and through SWRL/RIF rules for the execution component. Fig.3 demonstrates the graphical overview of local and INSPIRE vocabulary and how they are mapped through OWL ontology.

The schema in the Fig. 3 presents some *specialized classes* which illustrate the structure of the INSPIRE specification and their relationships with the local data structure. Mapping is done by relating similar concepts through "Equivalent Class" or "sameAs" axioms (cf. Table 2).

As per example

$$\text{INSPIRE} \sqsubseteq \text{4\_3\_Utilities} \tag{2}$$

$$\text{Local} \sqsubseteq \text{Infrastruktur} \tag{3}$$

And

$$\text{Infrastruktur} \equiv \text{4\_3\_Utilities} \tag{4}$$

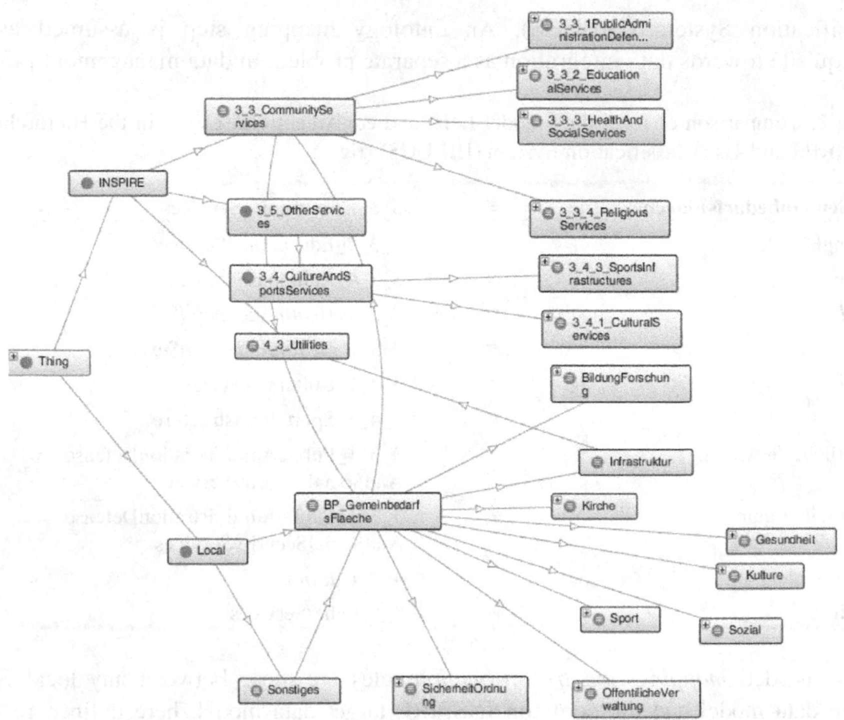

Fig. 3. Graphical Overview of Mapping Schema in OWL ontology

Equation 2 and 3 denote that "4_3_Utilities" is a sub class (⊑) of "INSPIRE" and "Infrastruktur" is a sub class of "Local" (Local datasets modeled in the OWL ontology). Equation 4 illustrates that "4_3_Utilities" is an equivalent class (≡) of "Infrastruktur", inferring that any instance of "Infrastruktur" is also an instance of "4_3_Utilities" in the HILUCS INSPIRE codelist. This provides a schema level mapping of local data structures to INSPIRE directives though an ontology. In a similar fashion, rules could be formulated that map INSPIRE specifications at the general INSPIRE schema level to local data schema. An example of such a rule is presented in equation 5, which infers that all instances of "Infrastruktur" are instances of "4_3_Utiltities".

$$\text{Infrastruktur}(?x) \rightarrow \text{4\_3\_Utilities}(?x) \tag{5}$$

In summary these few examples may give evidence that the logical inference capabilities of an OWL ontology provide a strong argument for a possible framework of schema mappings at different data levels with a description of ontology data relationships. This especially also applies for information which is not localized. Codelists of statistical data or other topics (e.g. population) can be present. With this methodology for a national codelist (see case 3 in section 3) it is possible to create a mapping definition for existing codelists and knowledge in owl classes (see OWL step in Fig. 1), presented in a *.owl file.

```
<!-- http://www.semanticweb.org/i3/ontologies/2014/01/
untitled-ontology-39#3_3_3_HealthAndSocialServices -->

<owl:Class rdf:about="http://www.semanticweb.org/
i3/ontologies/2014/01/untitled-ontology-39#
3_3_3_HealthAndSocialServices">

<owl:equivalentClass
rdf:resource="http://www.semanticweb.org/i3/ontologies/20
14/01/untitled-ontology-39#Gesundheit"/>

<owl:equivalentClass
rdf:resource="http://www.semanticweb.org/i3/ontologies/20
14/01/untitled-ontology-39#Sozial"/>
</owl:Class>
```

This OWL description shows the equivalence of planned land-use "HealthAndSocialServices" with referenced 1:n local classes "Gesundheit" and "Soziales". The local sources with their own semantics in codelists are made available through the semantic web. The concept given in Fig. 1 in conjunction with INSPIRE specifications can help to further develop the existing solutions, but is it sufficient to maintain land planning law or rules which are created in *.owl Ontology Web Language. The ontology description of codelists and languages is most beneficial, especially when the rules are defined at a more general level rather than using feature manipulation in most detailed 1:1 ratios.

6 Conclusions and Future Work

Urban land-use planning is the most important planning tool for controlling and organizing the urban development of a local authority district. The principal instrument is the preparatory and the urban land-use plan in Germany, which represents the planned spatial development of the municipal area. The Ontology-based approach can be used to develop a concept of a standard based general platform which allows for the construction of knowledge between different systems.

In this paper we demonstrate that such a concept can support interoperability in the context of the European INSPIRE implementation. With ontologies the step from a data based application to a linked open data environment can be taken. It can be expected that the requirement of easier usage of different data models can be met if the study of the rules can significantly reduce the cost of a survey. This can be achieved by semi- automated creation of the class descriptions in the data collection process. It can be expected that due to existing legal frameworks and other national restrictions, INSPIRE implementation in the foreseeable future will not lead to the replacement of existing data models. The presented concept has a potential to considerably facilitate the direct data transfer from local data sources into the target INSPIRE data model.

Future work in this context is the creation of a flexible Ontology Service Toolbox as can be realized on the Internet. The implementation of redundancy-free data management can only be realized from a service level. A central ontology-based network service needs a unique definition of underlying mapping rules. The tradeoffs between using Ontology and Semantic Web based rule definition in SWRL / RIF and GML versus OWL / RDF requirements have to be further investigated.

The implementation of knowledge management in the INSPIRE application field can help to support interoperability for the benefits of the local level. The benefits of the approach can be found in the unique flexibility of an ontology instead of some feature manipulation, in the standard-based application and the versatility of use. In that way ontologies can be used to model extensive knowledge about the data, for rule definition, and other objectives in order to reduce the increased complexity caused by INSPIRE for the local government.

References

1. European Union: Directive 2007/2/EC of the European Parliament and of the Council of 14 March 2007 establishing an Infrastructure for Spatial Information in the European Community (INSPIRE) (2007)
2. GMES: Global Monitoring for Environment and Security. In: Copernicus, The European Earth Observation Program (2012), http://www.copernicus.eu/
3. GEOSS: Global Earth Observation System of Systems. In: GEO, Group on Earth Obervations (2013), http://www.earthobservations.org/
4. DeCover2: Space-based service for German land cover (2012), http://www.de-cover.de/
5. INSPIRE Thematic Working Group Land use: D2.8.III.4 INSPIRE Data Specification on Land use – Draft Technical Guidelines (2013)
6. XPlanung: German eGovernment project of exchanges among IT systems of zoning plans, regional plans and land use plans (2013), http://www.xplanung.de
7. Plan4All: Geoportal for spatial planning (2012), http://www.plan4all.eu
8. SENSOR: Sustainability Impact Assessment: Tools for Environmental, Social and Economic Effects of Multifunctional Land Use in European Regions. In: Peer, Partnership for European Environmental Research (2009), http://www.peer.eu/projects/peer_flagship_projects/sensor/
9. Harmony: Development and demonstration of Marine Strategy Framework Directive tools for harmonization of the initial assessment in the eastern parts of the Greater North Sea sub-region (2012), http://harmony.dmu.dk/
10. SEIS-BASIS: Shared Environmental Information System Baseline and Evolution Study (2011), http://ies.jrc.ec.europa.eu/seis-basis-project
11. HUMBOLDT: Towards the Harmonisation of Spatial Information in Europe (2013), http://www.esdi-humboldt.eu/
12. GENESIS: Generic European Sustainable Information Space for the Environment (2011), http://ies.jrc.ec.europa.eu/genesis-project
13. Drafting Team "Data Specifications": D2.3: Definition of Annex Themes and Scope V3.0. European Union (2008)
14. Rajabifard, A., Feeney, M., Williamson, I.: The Cultural Aspects of Sharing and Dynamic Partnerships within an SDI Hierarchy (2002)

15. Drafting Team "Data Specifications": D2.5: Generic Conceptual Model, Version 3.4rc3., European Union (2013) INSPIRE Generic Conceptual Model
16. European Union: Commission Regulation (EU) No 1089/2010 of 23 November 2010 implementing Directive 2007/2/EC of the European Parliament and of the Council as regards interoperability of spatial data sets and services (2010)
17. Schilcher, M., Fichtinger, A., Jaenicke, K., Kraut, V., Stahl, J., Straub, F.: INSPIRE - Fundamentals, Examples, Test Results, Munich, Germany (2009)
18. Hitzler, P., Krötzscher, M., Rudolph, S.: Foundations of Semantic Web Technologies. Chapman & Hall/CRC Taylor & Francis Group, US (2009)
19. Wang, W., Tolk, A., Wang, W.: The Levels of Conceptual Interoperability Model: Applying Systems Engineering Principles to M&S, San Diego, CA, USA (2009)
20. Gruber, T.: Ontology. In: Liu, L., Özsu, M. (eds.) Encyclopedia of Database Systems, pp. 1963–1965. Springer Science+Business Media, LLC, USA (2009)
21. Zack, M.: Managing Codified Knowledge. Sloan Management Review 40(4), 45–58 (1999)
22. Bellinger, G.: Knowledge Management - Emerging Perspectives (2004), http://www.systems-thinking.org/kmgmt/kmgmt.htm
23. Berners-Lee, T., Hendler, J., Lassila, O.: The Semantic Web. Scientific American, 34–43 (2001)
24. Decker, S., Melnik, S., Harmelen, F., Fensel, D., Klein, M., Broekstra, J., Erdmann, M., Horrocks, I.: The semantic web: The roles of XML and RDF. IEEE Internet Computing, 63–74 (2000)
25. Stojanovic, N., Stojanovic, L., Handschuh, S.: A framework for knowledge management on the Semantic Web. In: The Eleventh International World Wide Web Conference, Hawaii (2002)
26. Calvanese, D., Giacomo, G., Lenzerini, M., Nardi, D.: Reasoning in expressive description logics. In: Robinson, A., Voronkov, A. (eds.) Handbook of Automated Reasoning II, pp. 1581–1634. Elsevier Science Publishers (2001)
27. Baader, F., Sattler, U.: An overview of tableau algorithms for description logics. Studia Logica, 5–40 (2001)
28. Baader, F., Calvanese, D., McGuinness, D., Nardi, D., Patel-Schneider, P.: The Description Logic Handbook: Theory, Implementation and Applications. Cambridge University Press, Cambridge (2003)
29. Stanford University: Protégé (2013), http://protege.stanford.edu
30. W3C: OWL 2 Web Ontology Language Profiles. In: W3C Recommendation (2009), http://www.w3.org/TR/2009/REC-owl2-profiles-20091027/
31. W3C: SPARQL Query Language for RDF (2008), http://www.w3.org/TR/rdf-sparql-query/
32. W3C: SWRL: A Semantic Web Rule Language Combining OWL and RuleML (2004), http://www.w3.org/Submission/SWRL/
33. W3C: RIF Overview, 2nd edn. (2013), http://www.w3.org/TR/rif-overview/
34. Tschirner, S., Scherp, A., Staab, S.: Semantic access to INSPIRE: How to publish and query advanced GML data. In: Terra Cognita: Foundations, Technologies and Applications of the Geospatial Web, Bonn (2011)
35. OGC: Geography Markup Language (2010), http://www.opengeospatial.org/standards/gml
36. OGC: GeoSPARQL - A Geographic Query Language for RDF Data (2012), http://www.opengeospatial.org/standards/geosparql

37. Herring, J.: OpenGIS® Implementation Standard for Geographic information - Simple feature access - Part 1: Common architecture. OpenGIS Implementation Standard (2011)
38. Althoff, J.: Model-driven tools to support conceptual geospatial modelling. Diss. Nr. 19918, ETH Zürich (2011)
39. ReDeFer: A compendium of RDF-aware utilities organised in a set of packages. In Rhizomik (2009), http://rhizomik.net/html/redefer/
40. Howard, M., Payne, S., Sunderland, R.: Technical Guidance for the INSPIRE Schema Transformation Network Service. Version:3.0, EC JRC Contract Notice 2009/S 107-153973 (2010)
41. Karlsruhe Institute of Technology In: XPlanung Specification (2010), http://www.iai.fzk.de/www-extern/index.php?id=680&L=1
42. Müller, H., Würriehausen, F.: Semantic Interoperability of German and European Land-Use Information. In: Murgante, B., Misra, S., Carlini, M., Torre, C.M., Nguyen, H.-Q., Taniar, D., Apduhan, B.O., Gervasi, O. (eds.) ICCSA 2013, Part III. LNCS, vol. 7973, pp. 309–323. Springer, Heidelberg (2013)
43. Doan, A., Halevy, A., Ives, Z.: Principles of Data Integration. Morgan Kaufmann Publisher, Elsevier Inc., Waltham, USA (2012)

Crowdsourced Monitoring, Citizen Empowerment and Data Credibility

The Case of Observations.be

Jihad Farah

University of Liege
Sart Tilman Campus, Chemin des chevreuils 1, B52, 4000, Liege, Belgium
jihadfarah@gmail.com
http://www.lema.ulg.ac.be/

Abstract. Crowdsourcing is today a revolutionary phenomenon changing profoundly our ways of communicating and producing. This article is interested in two issues that are crucial to its development and impacts. On one hand, it investigates the forms and limits of crowdsourcing-related citizen empowerment. It is less concerned by the now recognized fact that crowdsourcing is empowering, but rather focuses on the ways it does so and the architecture of the relations between citizens, scientists and institutions in this new context. On the other, it discusses the question of credibility of data produced through crowdsourcing. This question represents, in fact, the Achilles' heel that destabilizes the rise of citizen power in the face of experts and institutions. In its discussion of these two issues, the article relies on a particularly interesting case study: the online platform of participatory monitoring of biodiversity in Belgium Observations.be. The creation of databases is an occasion here for reflexivity, learning and mobilisation. It is also an occasion for the liberation of the lay citizen, as an individual, from the straightjackets delimiting the institutional, scientific and associative spaces where he remains a subject, a collaborator or a member – always in a subordinate position. He becomes a peer producer, partner and discussant. More important, learning and action networks that develop in the platform cut transversally through the three spheres. We find unexpected and new cooperations between citizen, scientists and civil servants. Likewise, actions developed through the platform, mainly reporting, counting campaigns and early alert systems attest new modes of action that transgress the functional and ontological division of the three spheres.

Keywords: Crowdsourcing, Empowerment, Credibility, Monitoring, Brussels, Observations.be.

Crowdsourcing is today a revolutionary phenomenon changing profoundly our ways of communicating and producing. In the last few years, Crowdsourcing has known important media coverage, especially regarding its role in recent international events like the post-earthquake Haiti rescue efforts in 2010 (Heizelman & Waters, 2010; Gao et al, 2011) and the "Arab Spring" in 2011 (Allagui & Kuebler, 2011; Castells, 2012).

B. Murgante et al. (Eds.): ICCSA 2014, Part II, LNCS 8580, pp. 469–485, 2014.

However, crowdsourcing is becoming more ubiquitous and is mobilized in a large number of fields. This includes new practices of design based on ideas from the "crowd"[1], new modes of reporting, live and from the scene of action, through social media, the development of a new more qualitative cartographic tradition based on data produced individually and voluntarily by large numbers of people, and a better management of catastrophic situations by putting at the disposition of citizen and institutions bottom-up produced information that could have crucial importance during and after the crisis[2].

Crowdsourcing is clearly empowering. It is especially so for lay citizens. It allows them to have considerably larger impact in two arenas where they have long been marginalized: science and politics. Despite the growth of participatory practices, these arenas remain ontologically built as spaces of expertise. For some decades now, practices of citizen science and citizen participation have brought citizens into the "sacred" spaces of the laboratory and the government council. However, he remains in a subordinate position in no way in control of the process. With crowdsourcing we can see a restructuration of the equilibrium set between scientists', public authorities' and citizens' powers at the founding moments of the modern project. Crowdsourcing allows citizens to "construct" the confines of Nature and Society out of laboratories and government councils and without waiting the initiative – or even the consent – of scientists and politicians.

Regarding this profound undergoing revolution this article is interested in two issues. On one hand, it investigates the forms and limits of crowdsourcing-related citizen empowerment. It is less concerned by the now recognized fact that crowdsourcing is empowering, but rather focuses on the ways it does so and the architecture of the relations between citizens, scientists and institutions in this new context. On the other, it discusses the question of credibility of data produced through crowdsourcing. This question represents, in fact, the Achilles' heel that destabilizes the rise of citizen power in the face of scientific and political institutions. In its discussion of these two issues, the article relies on a particularly interesting case study: the online platform of participatory monitoring of biodiversity in Belgium Observations.be.

Crowdsourcing: What Kind of Power?

Crowdsourcing is a word that has been applied to quite different kinds of activities that have in common a call to the "crowd" to contribute in the development of a product, an idea or an activity. Howe (2006) coined this particular word to describe practices by private sector companies that amount to a business model in countercurrent to outsourcing strategies to third world countries by relying on "crowds'" contribution. However, the word has since long traveled and is used to talk of many different things. This includes "*collective intelligence/crowd wisdom, crowd creation, crowd*

[1] This is the case of the emblematic Threadless.com and its crowdsourced T-shirt designs.

[2] This is the case with post-Katrina. Data, describing the situation of neighbourhoods and properties, produced by volunteers and mashed-up on a site was crucial for households' decision to return (Hudson-Smith et al, 2008).

voting, crowd funding" (Seltzer and Mahmoudi, 2012). Definitions of crowdsourcing diverge if not conflict. Zhao and Zhu (2012) define crowdsourcing as any collective intelligence system comprising three elements, an organization that benefits from the crowd's work, the crowd itself and a platform that links all together. Seltzer and Mahmoudi (2012) set a more restrictive definition insisting that the following conditions should be met: a diverse crowd, a well-defined problem, ideation, internet and solution selection. Hudson-Smith et al (2008) insist on different characteristics, mainly a decentralized format and the possibility for all users of the platform to share and use the data. To that we can add the distinction made by some authors between crowdcasting and crowdsourcing. The former would be more restrictive limiting the contribution of the crowd to answering a particular question, in a precise format, while the latter suggests a larger capacity of proposition and initiative on the part of the user of the platform.

All this to say that, in these conditions, crowdsourcing will have different relations to power depending on what we are talking about. In their study of the contribution of crowdsourcing to citizen participation in planning, Brabham et al (2009), Brabham (2009) and Seltzer and Mahmoudi (2012) defend the great potential of crowdsoucing in improving such a process and giving more voice to the crowd. However, as pointed out by Seltzer and Mahmoudi (2012) the real contribution of crowdsourcing to the planning process is mainly in the development of ideas once the strategic choices are set. In fact, as stressed by Brabham et al (2009), crowdsourcing works best in aggregating the crowd's individual contributions rather than averaging[3] them: "*individuals in the crowd incorporate discussion and exchange as they develop potentially a series of individual solutions to contribute to a commons*".

Of course, in the case of private companies capturing the added value of the crowdsourcing exercise, speaking of commons would be clearly misplaced. As some authors have stressed concerning the economics of the crowdsourcing business model, it clearly favors the organization to the detriment of the laborer who does all the work and gets to suffice of a moral consolation (having fun, engaging in a community…). In this perspective, crowdsourcing gives recognition to the citizen but could hardly be called empowering. In the case of a public body being the sponsoring organization, crowdsourcing could enhance the quality of the expected product but cannot replace the need for traditional democratic arenas for setting strategic goals. Power is still confined in other decision-making arenas. Crowdsourcing might even contribute in this case to maintaining a lure of participation to the benefit of dominant actors. It is however in the case where groups and communities build and manage their own crowdsourcing platform and capture its value that words like "empowerment" and "commons" get to have solid anchorage and meaning.

Rappoport (1981) defines empowerment as a construct that links individual strengths and competencies, natural helping systems and proactive behavior to social policy and social change. This involves also important elements of group psychology as caring and maintaining mutual respect (Cornell empowerment group, 1989) as well as the construction of a collective intelligence through a critical understanding of the

[3] Understand making compromises to reach consensus.

group's environment (Zimmermann et al, 1992). Empowerment processes are diverse and may take different turns. The building of commons is one of the most radical paths to empowerment.

Commons means an environmental or cultural resource as well as the stewardship system that allows maintaining it and collectively benefitting from it (Ostrom, 1990, Ostrom et al, 1999). Commons is usually opposed to the private ownership and state systems of managing collective goods. In this process the community is responsible has to improvise sustainable social and economic organizations that maintains the resource and ensure a long-term profit of it. The very existence of commons is seen by its promoters in research and policy as somehow destabilizing for traditional hierarchal power systems and empowering to communities.

The rise of Web 2.0 has brought a new wind to the commons theory. Speaking of the nebula of groups and communities dedicated to the open source culture in the digital world, Boiller (2008) says that the "*viral spiral*" of these "*commoners*" has "*democratized creativity on the global scale challenging the legitimacy and power of all sorts of centralized hierarchal institutions*". Crowdsourcing, when it is mobilized by civil society's communities and groups, contributes to this power shift by building commons.

The case of Observations.be describes a mode of empowerment through crowdsourcing that is closer to this perspective of power relations than that held in deliberative democracy theory and described by Brabham (2009) and Selzer and Mahmoudi (2012). By building on crowdsourced monitoring of biodiversity, Observations.be is representative of this new mode of management and production of collective goods (biodiversity and data) and the power shifts it implies.

The Story of a Disruption: Biodiversity, IBGE and Observations.be

In this part we will present the case of the online platform for the monitoring of biodiversity in Belgium Observations.be. The choice of this platform comes from the fact that it shows the on-going transition in power equilibriums, but also the complexity of relations between institutional, scientific and citizen spheres. It illustrates this complexity by staging collaborations, oppositions and interpenetrations where some actors belong to the different spheres at the same time (scientist-activist, civil servant-activist). It does so also by stressing the role of the credibility controversy as a point of rupture and divergence between these three spheres.

Biodiversity, A Destabilizing Theme

Far from being just a technical exercise, the management of biodiversity is also a complex political question. It touches a number of controversial issues: common heritage of humanity, state sovereignty on their resources, rights of peasants, modes of development, precaution principle, genetic manipulation, right to pollute (Aubertin et al, 1998). These debates and controversies on issues of biodiversity are very present for some time now in scientific circles. However, since the rise of environmental and ecological movements since the 1960s and later the Brundtland report's (1987) position on the protection of biodiversity as a major component of sustainable development, biodiversity started to be more present in public debates.

Hence, data on habitats and species became central to the capacity of public institutions in charge of biodiversity management to elaborate action strategies and justify them, as well as to the capacity of the civil society to follow up and mobilize. From international to scientific organizations, to public institutions and civil society actors, we can see as of the 1990s an increasing interest in the development of tools allowing monitoring and database building on biodiversity (Salem, 2003 ; Scholes et al, 2008 ; Buchanan et al, 2008 ; Coops et al, 2009).

In the field of nature observation and monitoring, the participation of citizens is not something new. On the contrary, as the history of science teaches us, the citizen – as "amateur" – has long been at the centre of the development of certain natural sciences all through the 19th and 20th century. This is mainly the case in ornithology, botany and mammology (Charvolin et al, 2007). These amateurs have been pioneers, partners and competitors to scientists. The latter have indeed appropriated their legacy and integrated it in different ways in their practice. However, with the formalization of sciences, especially the development of elaborate practices and research protocols depending on complex procedures and materials inaccessible to the large public, the amateur started to lose his place and ended up marginalized, even caricaturized.

The return of the amateur happens paradoxically with the development of national and international injunctions imposing on public institutions the provision of databases on the state of biodiversity. Be it on the proper initiative of the associative sector or on the demand of public institutions totally submerged by this task, we can see the establishment of biodiversity monitoring groups in a lot of developed (Bell et al, 2008; Whitelaw et al, 2003) and developing countries (Danielsen et al, 2000). In fact, these institutions do not have the capacity, in terms of human and financial resources – to engage by themselves in this exercise (Levrel et al, 2010). Hence, they try to rely on existing associative groups, or create themselves new participatory monitoring networks through top-down networking and structuring.

The development of citizen science (Charvolin et al, 2007; Bonney et al, 2009; Cohn, 2008) reinforces this dynamic. For twenty years now, laboratories in natural sciences are reviving the tradition of the 19th century pioneers by increasingly calling on citizens' collaboration. Hence, if some scientists have faith in technological performance and expert institutions networking to answer the lack of data on biodiversity challenge (Scholes et al, 2008; Buchanan et al, 2008), others defend an approach mostly based on the input of local communities and amateurs of nature (Cohn, 2008). This citizen science produced through cooperation between citizens and scientists has a double asset. On one hand, it allows bringing a wealth of details inaccessible or too expensive to insure by solely scientists. On the other, it brings to the local communities and amateurs networks a certain "pedagogy" for better uses and protection devices of nature.

Environmental themes, by their complexity and multi-scale nature, destabilize power structures based, since the beginning of modernity, on an understanding between the representatives of the political sphere and those of the scientific one (Latour, 1991, 1999). Monitoring of biodiversity, by the extent of the space to cover, adds to that by putting these representatives in the position where they have little choice but to call on the input of, long marginalized, lay citizens. New relations emerge where the breaking line between scientist, institution and citizen spheres becomes increasingly blurred and relations of power redefined.

The IBGE : An Administration That Wants to Open Up

The *Institut Bruxellois de Gestion de l'Environnement* (IBGE)[4] is the Brussels-Capital Region's public institution in charge, since its creation in 1989, of the administration and management of issues related to energy and environment. With the regionalisation of environmental competences in Belgium, it becomes in charge of research, public information and management of issues as diverse as green spaces, quality of air and water, waste, the use of energy and ground pollution. IBGE regroups a large number of departments and employs 850 persons. The department of nature management administers the question of biodiversity and its protection in the Brussels-Capital Region. This includes, on one hand, the production of data on the situation of biodiversity, on the other, the management of the natural reserves in Brussels.

With the development of the European legislation on issues of environmental protection, member states are in the obligation to proceed into a regular and systematic monitoring of the state of biodiversity. With directives on birds (1979), habitats (1992) and more recently the directive NATURA 2000, IBGE elaborates, since 1993, strategies to build databases on biodiversity in Brussels. Each six years it must submit an exhaustive report on biodiversity to the European Commission. This process has been strengthened with Brussels-Capital Region's own legislation that extended the list of species to monitor in order to include species with regional importance.

Despite the relatively large size of the institution, it was in no way capable to proceed alone in the development of such databases. This would have necessitated human and financial resources that are not available. Between 1993 and 1998, IBGE chose to subcontract monitoring and base its reports on specialized centres, mainly universities. It even gave the coordination of the process and the writing of the report to a university. Since 1998, IBGE has decided to take back the coordination responsibility and open more the monitoring subcontracts to specialized NGOs.

This opening up to the associative sector is due to different logics. First, there is the practical need to cover a monitoring field that keep enlarging with newer directives and that only NGOs, with their human resources, seem capable of covering. Second, this occurs at a moment when a new generation of employees at the IBGE and elected officials become to defend a larger constructive partnership with the

[4] Brussels Institute for Environment Management.

historical opposition that the associative sector represents in Brussels. These employees believe that a sustainable partnership is possible and it will serve a better long-term management of the environment in the Region. Finally, the orientation of some ministers' cabinets close to the environmentalist associative world contributed to this dynamic.

This opening up to citizens is an important change that should be however relativized. First, it must be said that "citizens" here are a somehow particular form of associations that gravitates at the intersections of the institutional, scientific and citizen spheres. In fact, if we find certainly amateurs in these NGOs dedicated to the study and protection of particular species, they are mostly under the supervision of specialized scientists members. In other NGOs, open to larger environmental themes, solid relations can exist with political parties and universities – some of the members of the latter often being founding members of these NGOs. Second, the citizen-amateur has here a very restrictive executive role strongly framed by scientists and is in no way master of the process that, paradoxically, depends essentially on his contribution.

It is mainly at this level that the arrival of Observations.be represents a major and destabilizing change. Two figures summarize the situation. In 2010, the sum of all the databases accumulated since 1993 by the IBGE reaches around 30000 entries. For the same year alone, Observations.be has more than 65000 observation entries.

Observations.be : An Earthquake in Biodiversity Monitoring in Belgium

Observations.be is the name of the French version of the online participatory monitoring platform of biodiversity in Belgium. It is a joint venture, created in 2008, by two nature protection NGOs in Belgium: Natuurpunt (Dutch-speaking) and Natagora (French-Speaking). The platform is based on the format of Waarneming.be created in 2005 by a network of nature protection NGOs in the Netherlands. In 2013, the Dutch site has around 22 million entries and more than 10000 registered persons, and the Belgian site has around 8 million entries and more than 5000 registered persons.

Based on a convention between the Dutch and Belgian NGOs, the sites have identical interfaces. The functions of the site allow the fast creation of statistics and maps by specie. The sites cover hundreds of thousands of species divided in around twenty categories (birds, mammals, reptiles, insects…). Amateurs of ornithology are the spearhead of the movement and around 65% of the entries relate to birds. Observations.be can present data in twenty languages. It has also many versions – or sub-sites – for different regions, cities or zones (e.g. parks), like bru.observations.be for Brussels.

The functioning of the site is based on crowdsourcing. It is the individual contribution of the user, in its different forms, that nurtures the site and makes it work. The user, in his entry, identifies a specie, geo-localize it on online maps. He can add photos, sound files and commentaries (see Figure 1). He can also give his opinion and propose action. Two forums exist: one in Dutch administrated by Natuurpunt and covers a large number of subjects divided in types of discussion, the other Brussels Birding Forum is focalised on birds and administrated by some Brussels volunteers.

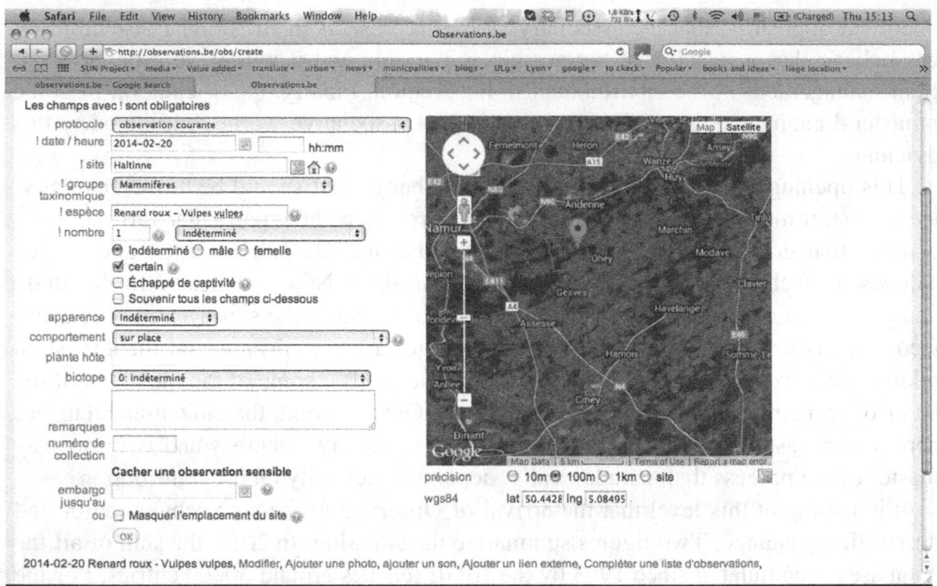

Fig. 1. Example of an entry in Observations.be

However, Observations.be is structured also to meet organized collective action. The existence of working groups is central to its functioning. These groups are divided by categories and sub-categories of species or by place (e.g. cities and natural reserves). Each group has its administrator who is in charge of validating each entry for it to be integrated in the database. A Projects section serves also for counting campaigns organized from time to time by some NGOs for certain species in different areas in Belgium. Finally, the existing of regional sub-sites allows local groups to have their own homepage where they can have local information and news.

Observations.be brings a reconfiguration to the relations between citizens, scientists and public institutions even more profound than that imposed by the biodiversity theme. On one hand, in the citizen sphere, the individual citizen has his own existence vis-à-vis organized NGOs. He can even use the platform beyond the basic functionalities expected from an amateur in monitoring and develop his own networks of exchange and learning. Hence, though the platform is created and administered by specific NGOs, it gives a large space for the individual citizen. It allows every person, with no link to these NGOs, to contribute. It facilitates links between citizens that do not go through any institutional framework, mainly through the forums. More importantly, it offers a space of learning on the monitoring of different species in a practical and accessible way to any person interested. This learning that benefits from the exchanges in the forums with, among others, veterans and scientists is one of the major contributions of the platform to the citizen. The breaking line between citizen knowledge and scientist knowledge, that represented historically the fracture line between their respective powers, has been greatly moved. The perspectives of action and mobilization based the monitored data and its interpretations have been considerably widened.

On the other hand, Observations.be represents an important change in the relation between the associative sector and the public institution in charge of the management of the environment. In the last two decades we have seen a shift from a logic of marginalization of associative structures considered as incompetent to an opening up and partnership dynamic that could be explained by necessity or the injunctions of an increasingly growing participatory and democratic ethos, to, lately, a certain form of dependence where the IBGE database is largely tributary to that of associations. In fact, nowadays, the IBGE subsidizes the platform and integrates its database.

In this new landscape of monitoring based on crowdsourcing, the citizen conquest seems formidable. However, this conquest is still very fragile. Despite the acute dependency of IBGE on Observations.be for data regarding biodiversity, it publishes its new report on nature in Brussels excluding data coming from the platform. The displayed reason: the "doubtful" credibility of this data.

The Loch Ness Monster of « Data Credibility »

In an interview with the persons in charge of biodiversity issues at the IBGE, we can see, on their part, a strong enthusiasm for Observations.be. They are themselves regular users of the platform. For them, the platform is a considerable levier for a better monitoring of biodiversity and they give it the gratifying grade of 9/10. However, we see them hesitate to adopt its data as a reference to the studies and reports they publish. Beyond their personal adherence to the Observations.be movement, they stress that they officially represent a public institution that cannot provide reports that do not rely on solid credible data produced in the rules of science. For them, the mode of production of the platform's data do not offer sufficient guarantee on this level. "*Maybe in the future*" they say.

The question of the credibility of data produced by volunteers, especially through crowdsourcing, is today a heavy debate in number of disciplines and fields that rely on observation and monitoring. Literature on citizen science and neogeography are particularly concerned by this debate. It is also in these fields that we find interesting answers to this challenge.

Data Credibility in Citizen Science and Neogeography Literature

We must first say that the question of the credibility of data is today less clear than it ever was. Inexactly evaluating credibility of data could have important scientific, social, personal, educational and even political consequences (Flanagin and Mitzger, 2008). The credibility of information is traditionally given to the information produced by a person that has a history of providing reliable data, or based on the recognized – scientific or professional – status of that person. However, these basic indicators of credibility become obsolete in cyberspace (Callister, 2000).

In fact, producers of this information, especially in crowdsourcing, do not necessarily do it in the perspective of professional and scientific production. Many other reasons lead them to participate in crowdsourcing activities. These include simple fun, an enthusiasm for new technologies, emulation in a social phenomenon, the search of auto-representation, activism for a particular cause and the will to participate in the creation of a reliable product that is not a multinational copyright (Neis et al, 2012, Hudson-Smith et al, 2008, Tulloch, 2007, Boulos et al, 2011, Seltzer and Mahmoudi, 2012; Brabham, 2009). All these reasons lead some to think that this mode of data production could be subject to serious biases and lack of necessary seriousness, undermining its scientific credibility.

On the opposite side, and in the face of this scepticism some authors show, by comparing data produced by experts and those relying on crowdsourcing, that this is largely a prejudiced position. In neogeography, Neis et al (2012) proceed to a comparative study between OpenStreetMap (OSM) and more classic commercial cartographic systems. They show that for Germany for example, OSM tends to have the same coverage in terms of surface as the classic systems, but at the same time it brings 27% more qualitative and descriptive data than they do. Likewise, Levrel et al (2010), concerning participatory monitoring of biodiversity – by relying on different studies done by a number of researchers – show that the quality of the produced data does not correlate with the profile and experience of the observer but rather to other parameters. These parameters are principally linked to the existence or absence of guidebooks, the quality of observation tools and their methodologies, the modes of verification of protocol and the animation of the observers' networks. To decrease bias and augment the quality of data, other authors advance different technical and social devices.

Increasing the quality of Metadata can be an efficient way to reduce possibilities of confusion and errors at the level of entries (Brando and Bucher, 2010). This includes, for example, the continuous improvement of Metadata through – automatic – adaptation of classification categories, rendering them the most simple and heuristic possible. Likewise, programmes allowing the display of the IP address of an entry's author's could contribute to the credibility of data – especially for local data (Flanagin and Mitzger, 2008). Other programmes built on powerful algorithms capable of filtering "noise" and "suspect" entries in data built through crowdsourcing do exist (Boulos et al, 2011, Bonner et Cooper, 2012)[5].

Beside these technical devices, the principal mean to reinforce data credibility remains peers' control. The more there is positive commentaries and reviews the more an entry has a chance of being credible. Likewise, edition wars in Wiki formats, where peers undertake corrections of one another's entries in the same document, is an indication of refinement in the quality of produced data. Hence, the presence of functions allowing to see the history of corrections and follow their evolution – like in Wikipedia and Google Documents – can serve to better understand the entry and corroborate or not its credibility. Besides, this mode of verification and validation by peers is capable of questioning the credibility of established sources coming from

[5] This is the case for example of WSARE and SwiftRiver that treat data built through the "firehose" of the social Web (Boulos et al, 2011).

scientific expertise. This is what some authors call the "*wisdom of the crowd*" (Surowiecki, 2004) or "*of the masses*" (Madden and Fox, 2006; Boulos et al, 2011).

Another type of social devices to increase data credibility is the improvement of the technical knowledge of users (Flanagin and Mitzger, 2008; Cohn, 2008; Bonner and Cooper, 2012). Training users through online tutorials and webinars is becoming a widespread practice in number of nature observation projects in citizen sciences (Toomey, 2014).

The notion of credibility itself is also questioned reviewed and relativized. For Flanagin and Mitzger (2008) credibility is divided into two components that have different roles depending on the situation: veracity and expertise. In communication sciences, credibility is treated through its objective dimension: degree of precision in respect to norms that are considered as acceptable by specialists in the field. Contrariwise, in psychology, credibility is dealt with through its subjective dimension, mainly the degree of acceptance by the receiver of the information. In this perspective, there is a differentiation between credibility and precision. It is not the fact that it is based on norms considered as references, from the point of view of peers that makes the information credible but the fact that the receiver believes that this is the case.

This opens the way to a reflexion on the means to reinforce credibility of data produced through crowdsourcing. This includes, for example. The consideration of heuristic elements that would increase the perceived credibility of a site and its data: the professionalism in the design and display of a webpage, the easy navigation through the site, the absence of displayed commercial intentions (.org and .edu vs .com)[6]. Likewise, the longevity of a site and the number of its users are also source of credibility. Finally, on the individual level, the more a person uses a site the more he tends to believe it credible. This perspective puts non-experts credibility under a new light. This concerns mostly those that have a local knowledge of a phenomenon that hence might seem more "true" and source of trust than accredited experts (Flanagin and Mitzger, 2008).

These different resources to understand and deal with the issue of credibility, based on citizen science and neogeography literatures, are of course relevant to the issue of crowdsourcing in the field of monitoring. In fact, as we will see in the next section, the actors of Observations.be are already mobilizing some of them.

The Strategy of Observations.be for Crediblizing Its Databases

Confronted with the position of IBGE representatives, Observations.be's administrators defend their platform by advancing arguments justifying the credibility of their databases and, beyond, the unquestionable input of the platform for a better understanding and management of biodiversity.

[6] The perceived motivation of the authors of the entries is important in the credibility of an information material: political and commercial agenda vs. neutrality and altruism.

Justification Through technicity: protocols and « ruses »

Observations.be's administrators stress the existence of protocols and "ruses" that allow the validation of data and the bypass of certain issues and problematic situations. The validation touches here two distinct issues: the right identification of the reported specie and the credibility of the aggregated data – statistics and maps.

Concerning the first issue, for the administrators, the primary and most basic filter is the user himself. In case of doubt concerning the right identity of the observed fauna or flora (its specie), the user can proceed in two ways. He can put the observation in different categories among which he believes the specie would be most accurately identified. He can also check the category case with (?) to say he does not know exactly what it is. In these cases, the specie(s) coordinator will contact him to help him identify it. A second filter is an automatic message that pops up when the user signals rare species. The message ask him to answer a series of questions that help confirm his declaration, it asks him also – if not already done – to upload a photo backing his claim. The third filter is the specie(s) coordinator. He is usually a veteran of nature observation, sometimes even a specialist. This person has an experienced knowledge of the category of specie(s). He is in charge of validating entries. He contacts the user when he has doubts. Photos are usually a valuable mean for confirming claims, however, it is not always the case, as with some mushroom species for example. In cases of remaining doubt, the coordinator can also refer to the forums where through discussions with other users, veterans and specialists, he might be capable of taking a decision: accepting or discarding the entry.

As for the credibility of maps and statistics built on the sites' databases, there is a double challenge. On one hand, observations are not equitably distributed across territory. On the other, we can end up with a situation of multiple entries reporting the same observation.

In fact, clearly, some areas and places are better covered than others. This is the case for example between Flanders and Wallonia. In the latter, the nature monitoring movement is not as developed as in Flanders. Consequently, if a map shows a larger presence of a specie in the North of the country than in the South, that might be related to the number of observers than to the true density of the specie in these areas (see Figure 2). Likewise, certain areas in Brussels are more covered than others. This is the case for example of a park near the IBGE building, where a lot of the institution employees, users of the platform, take their snack in sunny days.

There is also the possibility of having multiple entries for the same plant/animal at a certain time and place. This is particularly the case when a rare and exotic animal is spotted. Social networks of nature amateurs will get us in a situation where a lot would want to take him in photos.

These questions are not particular to biodiversity monitoring and are found in neogeopgraphy. Hence, density of population or use of a place contributes to its overrepresentation. Likewise, we can find an interest for large open spaces among users of some neogeography platform and Observations.be. In fact, as stressed by Neis et al (2012) concerning OSM, users are remarkably interested in developing the covered area and reducing "Terra Incognita". The latter usually corresponds to large

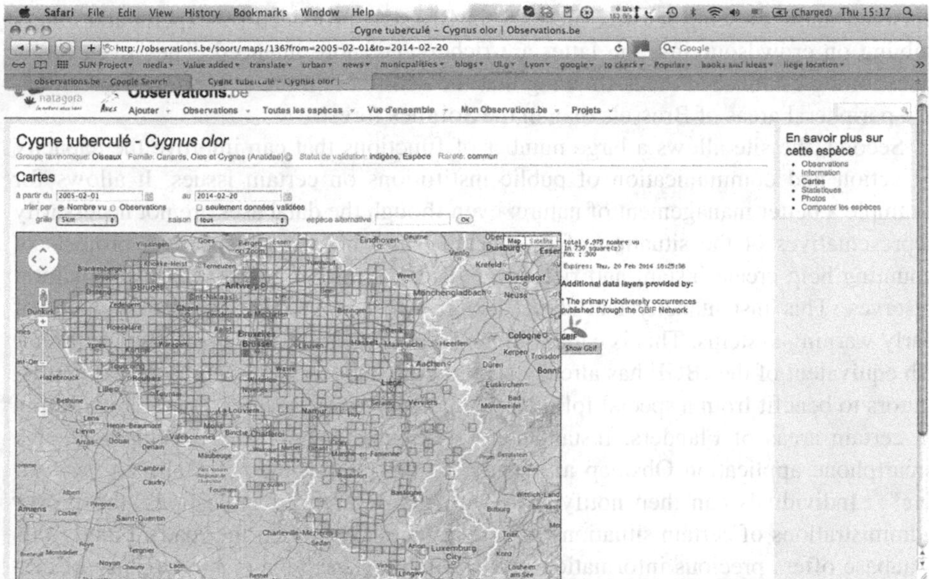

Fig. 2. An example of disparities between Flanders and Wallonia in Observations.be coverage

uninhabited areas. As for Observations.be users, we see a particular coverage of these spaces visited by a large number of users involved in tracking and nature activities[7].

In the face of these situations, the platform's administrators try to introduce elements of rebalancing. One way is through counting campaigns. Dividing a territory in zones, and based on users and local affiliated NGOs and actors, a counting of specific species is organized. The user participating to the campaign can signal for a certain zone he is covering the number of observed specimen, or he can say "I saw nothing to report". This allows a global cartography of the specie for a certain territory. On the other, the coordinator has the possibility to hide some observations or take them off if he believes there is some excess.

Justification through Efficacy : A « Super-Site »

For the administrators, the platform does not pretend to provide a detailed global cartography of biodiversity but to show tendencies. In any case, the potentialities of the platform and its databases render the question of the scientific validity of the data marginal: "*it does so much more than the scientists' tools have done that it compensates its possible weaknesses*". The large number of functions that it insures answers multiple objectives and different issues that are of interest for various actors.

First, the size of databases is by itself a considerable asset. As shown by one IBGE employee in charge of biodiversity, in one of his presentations, through a comparison

[7] We point that we did not engage in our research in a survey of users' backgrounds and practices and we base these claims on the sayings of the platform's administrators – with all the limitations that this implicates.

between maps showing red fox presence in Brussels produced by experts and others relying on crowdsourcing, the latter are richer and more nuanced. They even allow questioning established ideas showing that the red fox is more present in the residential peripheral areas of Brussels than in the Soignes forest.

Second, the site allows a large number of functions that can improve the capacity of action and communication of public institutions on certain issues. It allows for example a better management of nature: even though the databases are not necessarily representatives of the situation of biodiversity in a holistic perspective, projects of counting help create instant and rich inventories of certain places, especially natural reserves. This instantaneity can serve also in other ways, like the development of early warning systems. This is capital in the fight against intrusive species. The Flemish equivalent of the IBGE has already signed conventions with the platform administrators to benefit from a special follow up of certain exotic species and their evolution in certain areas of Flanders. Instantaneity – augmented with the development of a smartphone application Obsmap articulated to the platform – helps also in *"reporting"*. Individuals can then notify, live with geo-referenced indications, the proper administrations of certain situations (e.g. dead animal body on the road). Finally this database offers precious information on the situation in terms of biodiversity of certain construction lots. This is in the interest of construction and real-estate professionals, especially architects. They can then anticipate, through conception and impact studies, potential regulatory oppositions to their construction permit by competent authorities.

Third, the site has undeniable impacts in terms of empowerment for activists and NGOs of nature protection. They can use the site's databases to back their arguments and contest institutional projects. This is the case for example in Ghent where, relying on these databases, NGOs brought strong arguments for their campaign against intensive agriculture, showing its impacts on the drastic reduction of the population of nesting birds. This is the case, paradoxically, also in Brussels, where the databases were used to counter in justice an IBGE project to set a children playground on a fringe of a terrain that the associations consider as a natural area to protect.

Conclusions

Crowdsourcing could have major destabilizing consequences in terms of power capture. The case of Observations.be helps better understand and ground this power shift. In fact, Crowdsourcing moves the centre of gravity of power equilibrium from the alliance of scientific and political institutions to the "*masses*" [8]. It does so by turning around two principal limitations of the citizen sphere: its lack of credibility on scientific issues and its weak efficacy in the organization of large-scale action.

By building a system of crediblization based mainly on peer review and learning, crowdsourcing deconstructs credibility and authority. In the situation of data scarcity, institutional and scientific elites were the only parties capable of validating and using

[8] We use here the words of Hudson-Smith et al (2008) and Boulos et al (2011) that speak of "mapping for the masses" when talking about neogeography.

effectively this information for large-scale action. This elite centralized power drew top-down government modes of action that we know now the limits. With crowdsourcing, information is produced and diffused on large scale, and put at the disposition of "*masses*". In these "*masses*", profiting from abundant and omnipresent information as well as of the proximity between individuals and groups geographically scattered, thanks to Internet and Web 2.0, multiple networks of action are emerging around specific issues and logics of learning.

In the face of the scientific model of production of veracity based on the empiricism of laboratories, crowdsourcing advances a procedural model built through monitoring and the contribution of "*masses*". The first marks the norm and the limit, allowing the validation of what is "true" and acceptable in a certain context. The second announces a continuous process of development that adapts continually the norm and the limit. As said by Hudson-Smith et al (2008) « *OSM like Wikipedia is a process of evolving a good product not a product in itself because there is no end goal in sight as to what constitutes the best map (or the best entry in the case of Wikipedia)* ». This reflexivity and the refusal to limit oneself to one meaning and one finality is the true power of crowdsourcing and its transformative potential.

Observations.be shows this well. The creation of databases is an occasion here for reflexivity, learning and mobilisation. It is also an occasion for the liberation of the citizen, as an individual, from the straightjackets delimiting the institutional, scientific and associative spaces where he is just a subject, a collaborator or a member – always in a subordinate position. He becomes a peer producer, partner and discussant. More important, learning and action networks that develop in the platform cut transversally through the three spheres. We find unexpected and new kinds of cooperation between citizen, scientists and civil servants. Likewise, actions developed through the platform, mainly reporting, counting campaigns and early alert systems attest new modes of action that transgress the functional and ontological division of the three spheres.

As said by Latour (1999, 2005) the redefinition of the world – nature and society – through participatory description is one of the best ways to democratically transform our institutions. Crowdsourced monitoring could well serve as a plural and democratic mode of power reconstruction.

References

1. Allagui, I., Kuebler, J.: The Arab Spring and the Role of ICTs: Editorial Introduction. International Journal of Communication 5, 1435–1442 (2011)
2. Aubertin, C., Boisvert, V., Vivien, F.-D.: La construction sociale de la question de la biodiversité, In. Natures Sciences et Sociétés 6(1), 7–19 (1998)
3. Bell, S., Marzano, M., Cent, J., Kobierska, H., Podjed, D., Vandzinskaite, D., Reinert, H., Armaitiene, A., GrodziŃSka-Jurczak, M., MuršIč, R.: What counts? Volunteers and their organisations in the recording and monitoring of biodiversity. Biodiversity and Conservation 17, 3443–3454 (2008)
4. Boiller, D.: Viral Spiral – How the Commoners Built a Digital Republic on their Own, copy[at]viralspiral.cc (2008)

5. Bonner, D., Cooper, C.: Data validation in citizen science: a case study from Project Feeder Watch. Frontiers in Ecology and the Environment 10, 305–307 (2012), http://dx.doi.org/10.1890/110273
6. Boulos, K.M.N., Resch, B., Crowley, D.N., Breslin, J.G., Sohn, G., Burtner, R., Pike, W., Jezierski, E., Chuang, K.-Y.S.: Crowdsourcing, citizen sensing and sensor web technologies for public and environmental health surveillance and crisis magement : trends, OGC standards and application examples. International Journal of Health Geographics 10(67) (2011), http://www.ij-healthgeographics.com/content/10/1/67
7. Brabham, D.: Crowdsourcing the Public Participation Process for Planning Projects. Planning Theory 8, 242–262 (2009)
8. Brabham, D., Sanchez, T., Bathelomew, K.: Crowdsourcing Public Participation in Transit Planning: Preliminary Results from the Next Stop Design Case. Paper presented at TRB (August 1, 2009)
9. Brando, C., Bucher, B.: Quality in User-Generated Spatial Content: A Matter of Specifications. In: Proceedings of the 13th AGILE International Conference on Geographic Information Science, Guimarães, Portugal, May 10-14 (2010)
10. Bruntland Report. Our Common Future. Oxford University Press, Oxford (1987)
11. Buchanan, G., Nelson, A., Mayaux, P., Hartley, A., Donald, P.: Delivering a Global, Terrestrial, Biodiversity Observation System through Remote Sensing. Conservation Biology 23(2), 499–502 (2008)
12. Callister, T.A.: Media literacy: On-ramp to the literacy of the 21st century or cul-de-sac on the information superhighway. Advances in Reading/Language Research 7, 403–420 (2000)
13. Castells, M.: Networks of Outrage and Hope: Social Movements in the Internet Age. Polity Press, Cambridge et Malden (2012)
14. Charvolin, F., Micoud, A., Coord, N.L.: Des sciences citoyennes ? La question de l'amateur dans les sciences naturalists. Éditions de l'Aube, Aigues (2007)
15. Cohn, J.: Citizen Science: Can Volunteers Do Real Research ? BioScience 58(3), 192–197 (2008)
16. Coops, N., Walder, M., Iwanicka, D.: An environmental domain classification of Canada using earth observation data for biodiversity assessment. Ecological Informatics 4, 8–22 (2009)
17. Cox, L.-P.: Truth in Crowdsourcing. Security & Privacy IEEE 9(5), 74–76 (2011)
18. Danielsen, F., Balete, D., Poulsen, M., Enghoff, M., Nozawa, C., Jensen, A.: A simple system for monitoring biodiversity in protected areas of a developing country. Biodiversity and Conservation 9, 1671–1705 (2000)
19. Flanigin, A., Metzger, M.: The credibility of volunteered geographic information. GeoJournal 72, 137–148 (2008)
20. Gao, H., Barbier, G., Goolsby, R.: Harnessing the Crowdsourcing Power of Social Media for Disaster Relief. In: Zeng, D. (ed.) Cyber-Physical-Social Systems. IEEE (2011)
21. Heinzelman, J., Waters, C.: Crowdsourcing Crisis Information in Disaster-Affected Haïti, special report. United States Institute of Peace, Washington (2010)
22. Hudson-Smith, A., Batty, M., Crooks, R., Milton, R.: Mapping for the Masses: Accessing Web 2.0 through Crowdsourcing. UCL Working Papers Series, Paper 143. University College of London, London (2008)
23. Latour, B.: From realpolitik to dingpolitik or how to make things public. In: Latour, B., Weibel, P. (eds.) Making Things Public: Atmospheres of Democracy. ZKM/MIT Press, Karlsruhe (2005)

24. Latour, B.: Politiques de la nature: comment faire entrer les sciences en démocratie. La Découverte, Paris (1999)
25. Latour, B.: Nous n'avonsjamaisétémodernes:essaid'anthropologiesymétrique. La Découverte, Paris (1991)
26. Levrel, H., Fontaine, B., Henry, P.-Y., Jiguet, F., Julliard, R., Kerbiriou, C., Couvet, D.: Balancing state and volunteer investment in biodiversity monitoring for the implementation of CBD indicators: A French example. Ecological Economics 69(7), 1580–1586 (2010)
27. Madden, M., Fox, S.: Riding the Waves of Web 2.0: More than a Buzz Word but Still not Easily Defined. Pew Internet and American Life Project (2006), http://clearwww.co-bw.com/Ethics/ethics_Web_2.pdf
28. Neis, P., Zielstra, D., Zipf, A.: The Street Network Evolution of Crowdsourced Maps: OpenStreetMap in Germany 2007–2011. Future Internet 4, 1–21 (2012)
29. Ostrom, E.: Governing the commons. Cambridge University Press (1990)
30. Ostrom, E., Burger, J., Field, C., Norgaard, R., Policansky, D.: Revisiting the commons: local lessons, global challenges. Science 284, 278–282 (1999)
31. Rappaport, R.: In praise of paradox: A social policy of empowerment over prevention. American Journal of Community Psychology 9, 1–25 (1981)
32. Salem, B.B.: Application of GIS to biodiversity monitoring. Journal of Arid Environments 54, 91–114 (2003)
33. Scholes, R.J., Mace, M., Turner, W., Geller, G.N., Jürgens, N., Larugauderie, A., Muchoney, D., Walther, B.A., Mooney, H.A.: Towards a Global Biodiversity Observation System. Science 321, 1044–1045 (2008)
34. Seltzer, E., Mahmoudi, D.: Citizen Participation, Open Innovation and Crowdsourcing: Challenges and Opportunities for Planning. Journal of Planning Literature 28(1), 3–18 (2012)
35. Surowiecki, J.: The Wisdom of the Crowds. Anchor Books, New York (2005)
36. Toomey, D.: How Rise of Citizen Science is Democratizing Research. Environment 360 (2014), http://e360.yale.edu/
37. Tulloch, D.: Many, many maps: Empowerment and online participatory mapping. First Monday 12(2) (2007), http://www.firstmonday.dk/ojs/index.php/fm/article/view/1620
38. Whitelaw, G., Vaughan, H., Craig, B., Atkinson, B.: Establishing The Canadian Community Monitoring Network. Environmental Monitoring and Assessment 88, 409–418 (2003)
39. Zimmermann, M.A., Israel, B.A., Schulz, A., Checkoway, B.: Further explorations in empowerment theory: An empirical analysis of psychological empowerment. American Journal of Community Psychology 20, 707–727 (1992)

City as Commons: Study of Shared Visions by Communities on Facebook

Maria Célia Furtado Rocha[1], Pablo Vieira Florentino[2], and Gilberto Corso Pereira[3]

[1] Federal University of Bahia, CAPES (proc. n. 11527/13-7), Salvador, Brazil
mariacelia.rocha@prodeb.ba.gov.br
[2] Federal Institute of Bahia, Salvador, Brazil
pablovf@ifba.edu.br
[3] Federal University of Bahia, Salvador, Brazil
corso@ufba.br

Abstract. The present article is an exploratory exercise of analysis of texts present in digital social networks organized by communities that operates on urban public spaces. This text present some ideas about civic and political participation to support our point of view of "expanded participation" provided by social interactions throughout social networks digital platforms. We intend to improve our knowledge about how groups are sharing visions by analyzing a words network created from those interactions. We try to characterize relationship between words used in posts in Facebook pages from two groups, one in Italy and other in Brazil. This work analyzes networks and relations, from this point, it is possible identifying in most important word pairs, words with higher centrality metrics and the main idea expressed in the content of posts and discussions. Finally we discuss possibilities and difficulties found in the usage of this kind of tool.

Keywords: Digital Social Networks, Semantic Networks, Civic Participation.

1 Introduction

Dealing with time and space in liquid modernity, Bauman [4] states it is necessary the existence of available spaces for exercising of public *personas*, thus learning the skills involved in civility. Thereby, the city appears as a common good for its residents.

However, public spaces in contemporary cities may segregate or incorporate people, but only as agents of consumer acts. Therefore, these would not constitute public spaces but civilian spaces [4].

Amin [2] reminds that, nowadays, places of civic and political education are varied and distributed, involving civic practices that take place in flows and associations and go beyond the urban, materializing through books, magazines, television, music circuits transnational associations. Unlike Bauman [4], Amin argues that even through consumption and leisure, experience with public spaces remain related to sociability and social recognition and acceptance of codes of civil conduct.

B. Murgante et al. (Eds.): ICCSA 2014, Part II, LNCS 8580, pp. 486–501, 2014.

Much of western political is linked to the idea that penetrating the sphere of politics implies the existence of something we can define as the common good [24]. Arenilla [3] says that both models - the Greek democracy and the different republic models - coincide when focus on the need for citizens to establish the common good ahead of private interests to preserve the political community and defend freedom [3]. For Vitale[24], even though, the modern individualistic philosophy has its starting point in the methodological and ontological precedence of the individual over the collective. He does not consider the notion of the common good and public space as residual.

Rodotà [21] considers that public goods are those which can not be subjected to pure economic logic, such as water, clean air, healthy environment, knowledge, food, health. These assets are fundamental to human existence. The Internet, for example, would be a key resource for understanding what happens near or far and creating new possibilities of participation and dialogues with others. The novelty here was the fact that the common good, in this case, is not just something that can serve the individual, but a good produced by each and therefore should remain common.

Common goods, according to the author, would speak of the connection between people, not something that we manage or use solely in our sole interest. They would speak of social connections because when goods is common and we use it together with others, we must defend it in conjunction with others [21].

The demands currently placed by movements that stir crowds, as the case of Brazil in mid-2013, are generally based on speeches in favor of social justice to finally claim what they consider to be investments in the common to all – the right to transport, health, education and so on. Utilizing social networks, groups seek, for example, to reclaim areas of the city to be enjoyed by all or to participate in choices about the destination to be given to such areas. The rights do exist, but the parsimonious investment or the weakness of stakeholders to act in the field of decision-makers tend to override the right, degrading the common good.

In other cases, online social networks have been used intensively for participation in local projects. This is the case of the movement in favor of the deployment of a large park in Basento (Potenza, Italy) in 2012, which thereby enabled the expansion of the discussion and the acceptance of many proposals that have arisen in these digital spaces [15] and even the amendment of a law by which the municipality resumes its planning capacity for the area.

The present article is an exploratory exercise of how the concern with the common good is being primarily expressed through digital social networks (DSN) that organize and are organized by social movements that aim to improve urban public degraded area or waste urban spaces. We can speculate that democratic values subsidize present visions of the common good, understood here as the right to the city [11].

In this text we firstly visit some ideas about civic and political participation to support our vision about the main subject and some authors that are concerned with social interactions provided by social networks in digital platforms. We intend to improve our knowledge about how groups are sharing visions by analyzing semantic networks created from those interactions. Two cases are analyzed in different countries: Italy and Brazil. Finally we discuss the possibilities and difficulties found in developed process analyzing complex semantic network.

2 Networked Conversations: An Augmented Vision of Participation

Carpentier [7] points out that theoretical expansions, sometimes incorporated by democratic theory, grew from a diversity of political practices originated from actors who were often (strictly speaking) located outside the realm of institutionalized politics. They may be called interest groups, social movements, civil society or activists and these actors would have expanded the scope of political participation and would become more heterogeneous and multidirectional [7]. He emphasizes that power is a decisive element for participation.But he considers politics as a dimension of social, allowing participation in several different social fields, macro and micro, respecting diversity and referring to civil society, economy and family as places of political practice [7].

Referring to the scope of Political Science, Cantijoch and Gibson [9] argue that the understanding of the need to include non-institutionalized forms of political behavior appears to have become widespread in the area. Although, debates keep on questioning how far this extension can and should arrive. In particular, they question if less active and instrumental types of political engagement would qualify as forms of participation. The authors cite recent studies that refer to political activity on blogs and social networking sites as new forms of participation.

Noting the extension of the sense of participation beyond the political system, Dahlgren [8] discusses different trajectories of participation, either through "consumers", civil society or political participation itself. This author affirms that a first step towards civic participation offers easy access to symbolic communities and to a reality beyond individuality. Debates related to public issues can emerge anywhere and any time in the social space by discursive interaction that occurs, for example, through discussions, as well as experience and reflection [8]. The process of discursive production of a political dimension can be considered as a form of political participation [8].

Bakardjieva (cited in [8]) uses the term "*subactivism*" to describe the form of civic participation that takes place between people in their everyday life. In this case, the rules are put in discussion, contested and negotiated, the social world is evaluated according to types of moral perspectives, and questions are raised from what is fair and what is wrong – all that issues become a political dimension or create links with the policies.

Maia, Mendonça and Marques [13] state the process of public discussion would not be confined to formal arenas such as parliaments and ministerial meetings. In [12], Maia proposes thinking of communicative interaction modalities observed on the Internet as a comprehensive cultural dimension, where the formation of preferences and beliefs is a cumulative process.

Papacharissi [16] realizes that the interactions enabled by web 2.0 would reinforce contemporary values of self-expression and the proliferation of online personal/private spaces, such as blogs and social networks, but she considers it has little in common with Habermas's public sphere. However, she recognizes that, regardless of motivations, blogs and forums like virtual environments encourage the proliferation of voices that would expand the public sphere.

As Benkler [5] says, through blogs, wikis, online social networks, but also mobile phones, people started to register and publicize their impressions, like digital traces, and to establish conversations. This amplifies the urban experience and can provide input to public debate, in a distinct mode of what was practiced before, when strictly privilege of some organizations.

By verifying practices of sharing images via mobile phones integrated with social interactions in Japanese everyday life Ito [10] suggested the emergence of a form of visual sharing centered on personal, intimate and pervasive nature of social connections via handheld devices. This trend certainly indicates a change in the perception of context, which in this shared environment, beyond rational argument, is likely to be able to approach and give opportunity to the convergence of views.

As stated in [8], the network environment provides a great civic potential, makes available a wide range of participatory forms, which we can define as "civic practices". The online media is part of a wider social, political and cultural world. The commons, as we saw, are broadened and fight for rights is renewed, either for inclusion of the Internet in this "catalog" or by updating it in view of its shortage, case of the water. All this seems to lead to the need for a more complex agreement, allowing the emergence of uncovered themes. This certainly enables creation of new shared social values.

In this work, we assume "expanded participation" as a way to include new social arrangements, present in cultures that use networked communication as a means to share visions, values and produce meanings and actions collectively. We observe how democratic values are present in assumed notions about the common good, whatever are the results and the actions performed through the support of digital social networks. The ease of interaction allowed by web 2.0 is not only virtual, but real, and opens up many opportunities for participation through self-expression. It's an enlarged participation that tends to fall under the biases of more maximalist political participation, in accordance with the vision presented in [7].

3 Discovering Shared Views through Digital Social Networks Interactions

Recent works have studied behaviors based on interactions in digital social networks: in [22] is performed an analysis about local politicians engagement on Facebook considering a municipality in Norway; in [17] social networks of about 100000 Facebook users and their interactions on pages of presidential candidates in Finland are examined. The digital social networks seem to act in the public sphere, helping to support and share opinions by groups, just like their political actions and manifestations. Such processes happen as prescribed by those adept to the democratic deliberative perspective: certainly sustained in values, habits, ideals that spread and provide the bases for contemporary urban culture, avoiding rational forms.

Interactivity is a relevant characteristic in the constitution of texts on Internet. According to Hilgert (quoted in [14]), the interlocutors want to interact, therefore, creating a kind of “spontaneous text”. This way of writing is more flexible and informal. Unlike the slow and planned process of writing, typed language – or as we call *digital talk* – appears on the screen in real time, similar to face to face conversation [14].

According to [18], the terms interaction and conversation are not always synonymous. In her book *The Conversation Network*, she observes social interactions, those constructed by social actors with the purpose to negotiate, construct and share meanings. These interactions occur through talks. Orecchioni (also cited in [14]), explains that conversation would require interlocutors engagement in a pragmatic exchange – not only the alternation of speech.

Thus, this paper is not concerned with digital conversations. It verifies through the networked interactions, relevant themes and terms in two digital social networks (using their respective Facebook groups). Both are used to support movements, whose characteristics in these socio-cultural practices are designed to amend sharing and new meanings to physical, social and political contexts in favor of using the public space as a common good, in the interest, if not all, of many.

4 Semantic Networks

The networked interaction was analyzed using complex semantic networks. We try to characterize relationship between words used in the posts on Facebook pages from both groups. This work analyzes words and their relations forming semantic networks. From this point, by considering pairs of elements, it is possible identifying words with higher centrality metrics – an indication to reach main ideas expressed in the content of virtual posts and discussions.

In both actions of speech and writing there is a previous stage for planning in which words are evoked in a associatively mode. This characteristic permits us representing association of words as a semantic network of terms [1]. Some recent works about this kind of networks have been developed analyzing specifically the speech [23] or different kinds of literature produced by men and women in different languages ([6],[1]). All of them are based on the ideas of complex network analysis, which permit non traditional statistic methods and data analysis [25]. Within this technique, metrics considering existing relations, such as centrality and density help us to understand and identify the kind of structure of network itself.

In all referenced examples, the research process established a sentence as the unit of analysis inside each text studied. In this approach, a relation among all possible pairs of words forming one single sentence is created. This process creates a network of all words present in the texts analyzed, taking into account every single sentence. This semantic network can be investigated using the theory of complex networks, as it consists of hundreds or thousands of nodes: the words. Let´s take as example the two following sentences: *"Some beers in Brazil are produced with corn. This fact has transformed corn in an important item of leisure, but not for feeding"*, from which will rise a semantic network of words.

We shall firstly convert all verbs to their canonical format (infinitive) and eliminate grammatical words (articles, pronouns, conjunctions, prepositions), following premisses and heuristics defined in previous works cited above. This would give us the following items as elements for a semantic network and the network represented in Figure 1: beer – Brazil – produce – corn – fact –have – transform – important – item – leisure – feeding.

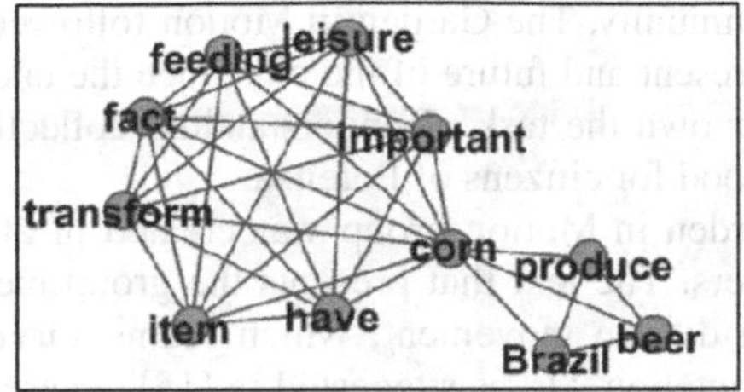

Fig. 1. Example of a network of words generated from two different sentences

The use of techniques of social network analysis may identify terms of higher relevancy in developing conversation and interactions expressed in DSN as identify complex network topologies.

Given any two existing words A and B in the text analyzed, metrics considered are: a) the frequency of each pair of words AB (*FreqPar*): number of occurrences of each pair, b) *Force*: the division of *FreqPar* by total number of sentences in text analyzed and c) *Fidelity*: the division of *FreqPar* by the sum of the total number of occurrences of words A or B, together or not [23]. Therefor, this research does not aim working with the most relevant words, but considering the most relevant pairs of words and the measures generated from such relations, observing interactions of texts analyzed.

5 Studying Groups Interactions

In this section we will firstly present a brief resume of studied cases in focus. Secondly we will explain retrieval process of data and its format, as well as method of analysis. Finally we will present results.

5.1 Cases in Focus

Both cases discussed own a forum group and a *fanpage* in Facebook, where actions of gardening and art intervention are massively organized, debated and disseminated. In the group area of each project, the respective participants have autonomy to publish and interact with each other trough posts, likes and comments, or even sharing the content for beyond the group. Public actions are organized and invite local community and unknown people interested in participating of initiatives in a hyper-connected way using Facebook as main platform.

The Garden in Motion

“Il Giardino in Movimento” (The Garden in Motion) started after the relatively successful experience in a project enriched from discussions with stakeholders and mobilized many citizens of Potenza (Italy) involving a degraded area and the initiative to create a City Park (Parco di Basento) as an opportunity to think about the city. This experience permitted participating in the planning of urban space and experiencing new forms of public participation. As a result of this process, a group of people continues to discuss the use of urban space through Facebook and meetings.

Through *fanpage* or community, The Garden in Motion followers and participants are sharing visions on the present and future of the city since the end of 2012. In this way, they experience on their own the task of understanding collective interest – in short the currently common good for citizens of Potenza.

The page of The Garden in Motion group was created in 2012. In December 19, 2013, it had 562 members. The text that presents the group mentions the movement itself as self-organizing diffuse movement, which seems survive at the expense of their own process of adaptation. Or, as suggested in [15], a case of *wikiplanning*. The *fanpage* was created after the group formation, on June 26, 2013. Until December, 2013, the *fanpage* reached 826 followers.

Already in its presentation, the page is defined geographically as hyperlocal. The area near Musmeci Bridge has great symbolic meaning for the group. The bridge and the whole area adjacent would become valued with the desired realization of Parco di Basento project.

Collective Yards

"Canteiros Coletivos" (Collective Yards) group and *fanpage* were born from discussions on Facebook forums related to the city of Salvador (Bahia, Brazil), in February 2012, amid demonstrations from other movements that used to pressure the municipality to restore infrastructure and ensure democratic use of city public spaces. The actions of "planting, maintenance, painting and cultural occupation of spaces" have succeeded in various parts of the city. Through actions that take place with anyone interested in collaborative efforts and voluntary participation, the group strengths the educational aspect "proposing a new relationship, exchange, learning the day-to-day, collective work for the common good". On December 19, 2013, the group was composed by 1,502 members and the *fanpage* reached 2,704 followers.

Both cases work through the occupation of public space: urban residual spaces in Brazil and in Italian case a space whose quality has been degraded by previous usage. We can say that these movements change the public space in a sense they try to improve its quality to become usable by citizens.

Similarities and differences in terms of goals, forms of speech and action can be observed. The Garden in Motion comes to ensure the development and selection of participatory projects with involvement of voluntary associations, university professors, designers and others professionals to be realized by public administration.

Collective Yards, on its side, wants to provoke changes in daily behavior of local communities, utilizing direct actions on urban space to promote environmental awareness. In this case, there aren't references or mentions to public administration or even relationships with it. There is involvement of distinct and distant neighborhood associations as well as university projects.

5.2 Tools, Process and Data Used for Analysis

This paper focuses on textual content published in Facebook groups forums. Considering theory presented in previous section, both sets of texts analyzed in this paper, representatives of each project, have naturally no linearity: differently of other works considering semantic networks, these are short and fragmented posts written by different users of digital social networks. The start and end points in sets are never clear. All these aspects bring a higher probability of diversity, but no idea about standards, even considering relevant part of interactions based on image sharing, which, often, produce textual comments. Thus, regarding these characteristics, we present the steps utilized in our research process.

The first part of our process focus on extracting posts from Facebook groups and *fanpages* via Netvizz application [19], which permits access to different kinds of data and statistics related to projects analyzed. At first we retrieved all textual content of posts in each group and respective comments, just as the statistics and aggregated values of comments and "likes". Then we analyze the statistics of 20% of the posts with the greatest commitment, computed here as the amount of comments and "likes" addressed directly in original post ("likes" on comments are not considered). Statistics and characteristics relating to data extracted are the following[1]:

- Garden in Motion: group posts collected between October 1st, 2012 until September 2nd, 2013. During 337 days, there were 80 posts concerned to: informative and/or calling for mobilization or participation in open meetings, opinion and knowledge expression. Between July 15 and September 3, 2013 (50 days), there were 52 posts, on average 1.04 post per day on the *fanpage*;
- Collective Yards: group data collected between June 4 and July 29, 2013. In 56 days there were 107 messages posts, most often calling to participate in any event, but also sharing knowledge, opining on issues concerned to the group or reporting initiatives that reinforce the image of the group, through adherence or news about their activities in traditional media. In relation to *fanpage*, the period of posts collected refers to interactions between May 17, 2012 and September 26, 2013. Six hundred and eighty (680) posts were published in 497 days. On average 1.37 posts per day.

This allowed us an initial statistical analysis, observing content extracted, which permitted future correlations with semantic networks obtained. The second step in the process concerns to unifying and performing human intervention on textual content of all posts and comments extracted from each group in two different files, representing each project. This includes cleaning and adjusting texts by removing or manually converting unrecognized characters, meaningless words (most of them originated from colloquial way of writing in virtual applications and special combination of characters, not processable by tool utilized), signs of avoidance as the apostrophe (widely used in Italian) and eliminating auxiliary verbs as HAVE – BE – GO – COME.

The third step concerns to applying an open source set of tools (most of then modules of UNITEX2, modified or not) for separating all sentences and respective words in each file of both projects. These tools generate a semantic network and metrics of occurrences of words into text. The last step permits generating metrics of produced network in previous step and classifying them.

---

[1] All periods considered in this work were defined by publishing dates from posts collected via Netvizz [19]

[2] UNITEX is an open source software that comprises "a set of programs that makes possible the processing of a text in human language using linguistic features such as lexicon-grammar tables, electronic dictionaries, grammars" [23].

5.3 Statistics, Analysis and Discussions

The Garden in Motion Group

Observing the 3 top posts with higher engagement, it is evidenced that the first two posts own images with link to photo albums on the intervention group. The posts with largest number of comments (18) and likes (8) refer to photos of an event under the Musmeci Bridge. The comments are however a private conversation between three friends. The next post is another photographic record (9 comments, 10 likes) of an event. Third-placed post brings images made at workshop for participatory design to the area under the bridge, one of which to be chosen by vote on *fanpage*. The post got one of the highest amounts of likes (14) for the analyzed period, but only 2 comments which do not address the park designs.

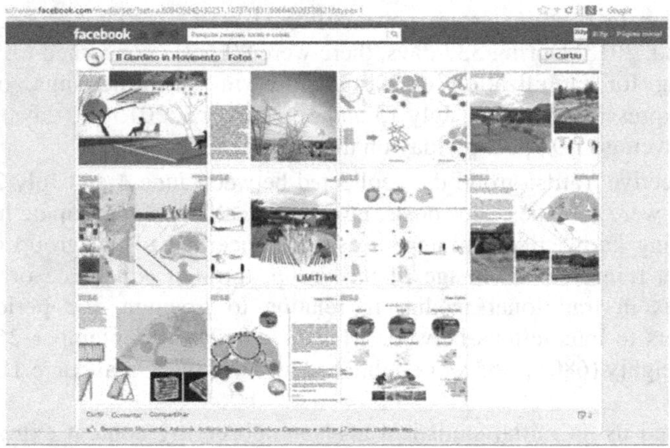

Fig. 2. Image of the third-placed post demonstrates the group projetual concerns

The network in Figure 3 represents the size of each node (words) as a function of its centrality degree value. This metric represents the number of connections and is highly influenced by the size of the sentence where the word appears, either by the quantity of occurrences in different sentences.

The most emphasized words in the second largest cluster for the interactions of the Garden in Motion group are: *piante* (plants), *laboratorio* (laboratory), Roma, Pompei, *soprintendenza*. The word *piante* appears 8 times, followed by *Pompei* that occurs 5 times (2 regarding the name of the institution and 3 regarding a place), *laboratorio* (4), *Roma* (3) and *soprintendenza* twice. All of these words own a very high degree centrality, remaining in a selected subgroup of nodes with largest number of connections.

The largest cluster (Figure 4) stands out the words: *antica* (ancient) and *Hortus*. *Antica* occurs 10 times – 6 of which refers to a road (Via *Appia Antica*) and 4 employed as an adjective – and *Hortus* occurs 9 times – 4 of which related to the Hortus Urbis project and 5 to some links. These words have respectively the third and fourth highest degree centrality in the network.

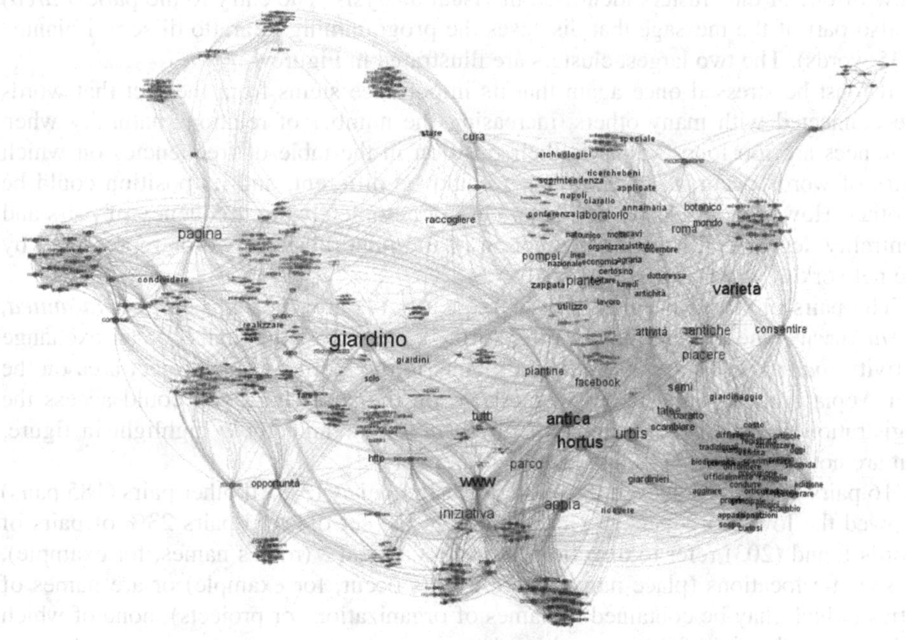

Fig. 3. Garden in Motion group interactions – the networked words

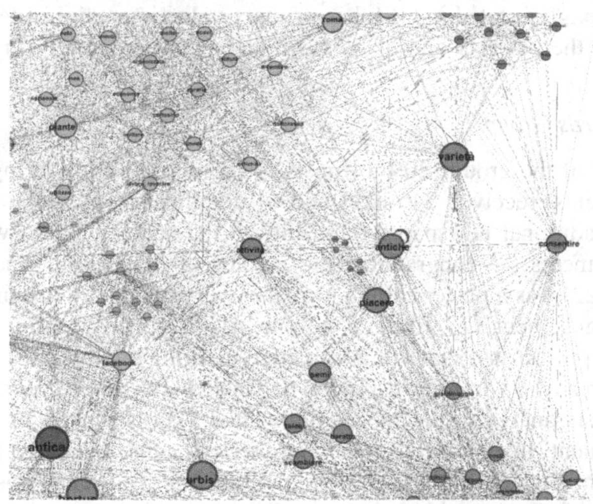

Fig. 4. Garden in Motion group interactions – main cluster view in the networked words

The Parco dell'Appia Antica participates in this project and hosted a lecture given by a biologist at the Laboratorio Ricerche Applicate Soprintendenza Speciale Beni Archeologici Napoli Pompei dealing with the reconstruction of the ancient Vesuvian landscape. It is precisely the call for this conference (204 words) that is shown in the view of one of the clusters identified in visual analysis. The entry to the park (*Parco*) is also part of the message that discloses the programming "Baratto di semi i piante" (435 words). The two largest clusters are illustrated in Figure 4.

It must be stressed once again that its importance stems from the fact that words are connected with many others, increasing the number of relations naturally when sentences are too long. Vis-a-vis their position in the table of frequencies on which pairs of words occur, it appears that it could be different, and its position could be another. However, at least in this case, but not completely, the frequency of pairs and centrality degree confirms the perception of the importance of words represented by the network of words (Figure 3).

The pairs of words occurring most frequently (4) are: *baratto-semi, appia-antica, hortus-www*. These pairs mainly refer to the largest post that calls for an exchange activity (*baratto*) of seeds (*semi*) and plants in the Hortus Urbis project area on the Via Appia Antica, which conveys message on the site where you could access the registration to attend the event. *Antica*, *Hortus*, *www* and *Appia* highlight in figure, but are not so clear on visual inspection of network.

16 pairs of words achieved the subsequent frequency (3). All other pairs (185 pairs) showed the lowest frequency (2). Considering the set of word pairs 23% of pairs of words found (203) refer to directions to access a local (routes names, for example), to specific locations (place name where events occur, for example) or are names of cities (which may be contained in names of organizations or projects), none of which is located in the city of Potenza, where the group acts.

This data may be pointing to the importance of considering references to geographic locations when it comes to composing the contexts of interactions. Recovering contexts, as stated in [18], has primary value in studies of talks, including digital talks, and the place is one of its main elements particularly in initiatives on urban space.

The Collective Yards Group

The top 22 posts of the group (20% of posts in the period) considering engagement (likes and comments) received 246 likes and 61 comments. The most liked (28) was the post about traditional newspaper reporting on the state of Bahia with statewide circulation. Comments (4) congratulate the creator of movement. The second post with most likes (22) was a poster calling to make everyday the Environment Day. The third highest engagement (18 likes, 5 comments) is related to the disclosure of Collective Yards project in a class of a university and the expectations of students to occupy an area on a heavy traffic road in the expansion area of the city of Salvador.

Among 22 posts analyzed, the more commented (11) is ranked in sixth place in terms of engagement. Its comments focus on different subjects. The 3 posts that had higher engagement contains pictures: newspaper *timeline* photos and posters pictures.

The following figure provides the overall vision of the words network from Collective Yards group. Performing a visual inspection, we can identify large areas shaped by a kind of cloud of words that suggest three major groups with recursive references to the following terms: *movimento / pauta / estado / bem / poder / social* (movement/agenda/state/good/power/social); *cidade / evento / dia / grande* (city/event/day/large) and *dar / http / www / br* (to give/http/www/br).

The primer group is formed by top 6 words in number of connections with others words (centrality degree, varying from 240 to 278 connections). The following groups are also formed by words of high degree centrality, but in a lower dimension, varying from 196 to 234 connections.

The word *movimento* (movement) occurred 22 times in the set of interactions, and 20 times in a particularly post. This post, the greatest of all, it has 1114 words and puts on debate propositions to give "organicity" to the movements that exploded on the streets of several Brazilian cities in June 2013. The author of the post was concerned with the supposed "absence of a common utopia" and proposed that groups and more decentralized movements, like Collective Yards, constitute spaces for the articulation and synthesis of claims raised by those movements.

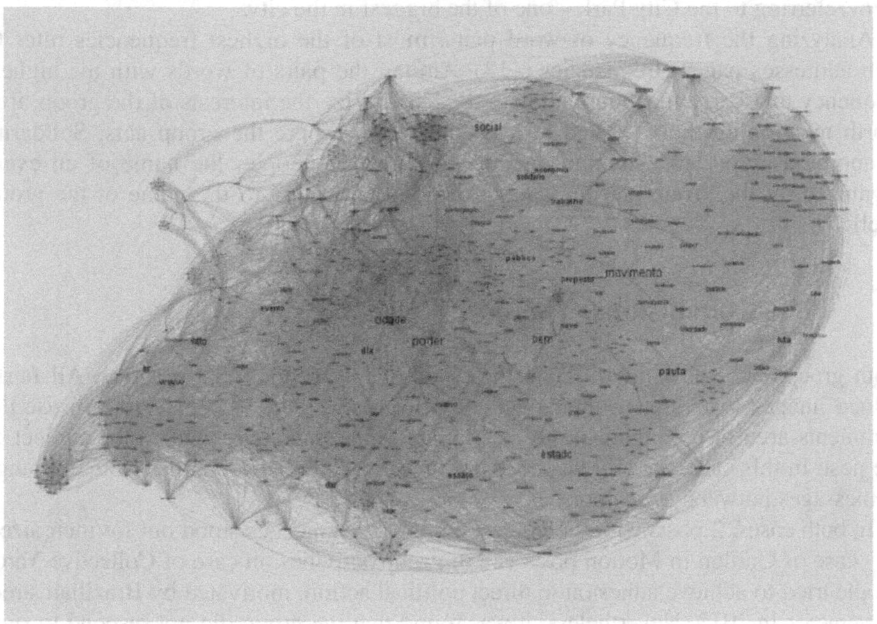

Fig. 5. Collective Yards group interactions – the networked words

Curiously, this post seems to have been ignored by the group that apparently reacted with silence, since it got no likes, no share or comments. The word "movimento" (movement), highlighted in the image of the network in function of such post, can not characterize the most relevant issues of interest in the group.

The second largest post found, composed of 552 words, it suggests "10 reasons for joining the street demonstrations and movements". Although 4 people have enjoyed this post, there were no comments or share. Like the previous, this does not seem to be interesting to the group, as it can not be noticed by the reactions of participants, but by the very different contexts of words usage in postings.

An interesting observation points when considering the context of use of the word "agenda". This occurs 18 times in the set of interactions: 16 times in the first post mentioned but never in the second largest post. The second largest post, albeit not cite the term ("agenda"), brings a cast of political demands, ranging from the fight against corruption to issues of urban mobility.

The word "state" on the other hand is mentioned 9 times in the set of interactions analyzed, 5 of them referring to the territory. The remaining 4 concern the state as institution, and 3 of them are concentrated in the post that proposed "10 reasons to go to the streets".

The words "good" and "power" had the following occurrences, in their respective contexts: the primer occurred 8 times, but 5 of which as adverb; the second occurred 5 times, but only once was used in a political sense. Both are examples of how difficult is interpreting based on the amount of occurrences exclusively.

As for the remaining clusters, it is not important for the present work the references to the Web itself. The word "town" on the other hand points with 22 occurrences, 2 of them referring to the City Park – one of the biggest in the city.

Analyzing the frequency of word pairs, most of the highest frequencies refer to web addresses pages (frequencies 6-13). Among the pairs of words with the highest frequency that carry meaning and help to characterize the interests of the group, it is worth mentioning those related to a neighborhood where the group acts; Solidarity Economy as types of economic practices that value people; the name of an event organized by the group and the pair the words that refers to the name of the group itself.

6 First Conclusions

Both groups showed similarities in the types of interactions that practice. All fairly valued images that refer to their own activities, using "like". People in turn use the comments area to communicate to each other, sometimes unrelated to the subject of the post. In this case, the tool has been used more as a bulletin board for the exchange of messages between acquaintances.

In both cases, 2 posts with the highest number of sentences stood out for their sizes. In case of Garden in Motion posts call to group activities, in case of Collective Yards people tried to achieve adhesion to direct political action, motivated by Brazilian street movements in 2013. Nevertheless, it was found that the group did not respond to such clamor.

While immersed in civic practices for improvement and recovery of public spaces, interactions of Garden in Motion group do not mention the "common good". On the other side, the term appears only 2 times in Collective Yards in the most political oriented post, which seems to have no impact in the group interactions.

The fact that posts of greater adherence (engagement) in both groups presented little or no degree of conversation lead us to wonder if, at that moment, digital social networks are contributing for expanding public sphere or if they just become vehicles for amplifying urban experience and diffuse information. If there was no exchange of views, can we say there is a real debate? So, could we say themes and terms emerge and bring new meaning to the social and political context?

However, it is important to note this work as a snapshot of a given period. Other dynamics may be established in the life of the group to establish and define more precisely their identity through their practices.

On the other hand, online and offline practices observed in the groups show a large engagement in civic actions, which aim primarily to consolidate new behaviors in social life, also tuned with the idea of city as a common good – case of Collective Yards – or directly influencing decisions of the government, adopting practices that reinforce ideas of participatory urban planning – case of Garden in Motion. In our view, such findings converge with points of view of [8] in defense of different places where one develops the sense of civility required by urban life and with the idea of producing the city as commons through the notion of "enlarged participation".

Focusing on studies about microcosms constituted by group interactions with and/or through social networks, Rodotà [20] notes the daily actions of many individuals. Such actions bring an uninterrupted declaration of rights, that draws its strength not from a deep conviction but of men and women so they can gain recognition and respect for their dignity and their humanity. They are not the "historical subjects" of the great modern transformation: the bourgeoisie and the working class [20]. In fact, they are a plurality of individuals already interconnected in a global network. Not a collective intelligence, or an indeterminate crowd, but a plethora of hard-working women and men who find and create political occasions for not succumbing into passivity and subordination.

A resource that can assist in the composition of such urban framework is the use of Social Network Analysis [25], allowing identification of relation structures in groups interactions. Even though this is not the object of the research in the paper, showing the network of interactions among group members (Figure 6) sets out the major players of these talks networks.

The following figure demonstrates the large concentration of interactions in a single individual in the case of Collective Yards, which have a bit more distributed at the Garden in Motion group. This finding suggests further investigation of the role played by some actors in the constitution of the group's agenda.

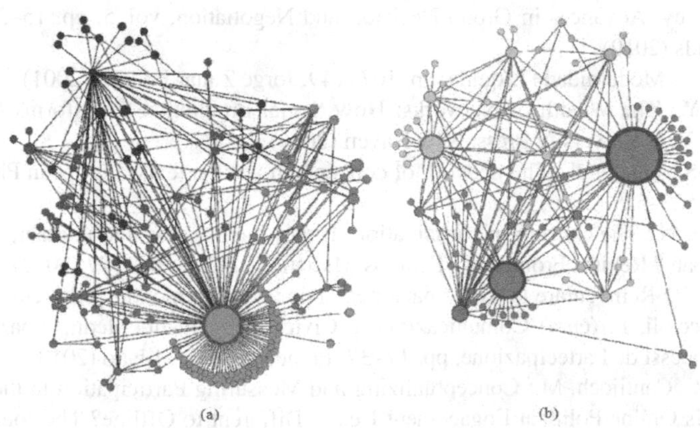

Fig. 6. Interactions networks - Collective Yards (a) and Garden in Motion (b)

Reviewing the process utilized, the experience gives us the opportunity to conclude for now that:

a) first of all, the networks of words can deceive or mislead, since it is based on sentences – the highlighted words tend to be those at larger sentences. This possibility of misconception was greatly demonstrated in the case of Collective Yards;

b) in some cases, words may not appear at a given sentence, but the concept related to them could be transmitted trough others pairs of words ;

c) it is necessary observing carefully the context of ambiguous words or composed terms with an auxiliary role in the construction of meaning, when processed by software (eg. composed expressions with verbs such as "may," "should" and "could");

d) there is difficulty to identify the geographic information (a lateral concern to this paper) in interaction texts. However, we believe there is a need to expand the analysis to include features that clarify the context, especially in studies where place has symbolic value;

e) it is fundamental to know group´s targets for studying interactions, considering different kinds of metrics that represent levels of interactions, such as number of "likes", comments, shares and so on. This will permit cutting out textual postings that are not concerned with the groups interests.

These research let us conclude that it will be a good starting point to know the average size of group interactions to identify cases distant from patterns and then examine individually. We may consider that complex network analysis can be used in a recursive way to guide researchers on selecting posts to be analyzed in detail, including images and web links.

References

1. Aguiar, M.: Words Networks in Written texts: An Analyzes of verbal language using complex networks. Mastering dissertation, 116 p. UFBa, Salvador, Brazil (2009)
2. Amin, A.: Collective culture and urban public space. City 12(1), 5–24 (2008)
3. Arenilla, M.: Concepts in Democratic Theory. In: Insua, D.R., French, S. (eds.) e-Democracy. Advances in Group Decision and Negotiation, vol. 5, pp. 15–30. Springer, Netherlands (2010)
4. Bauman, Z.: Modernidade Líquida, pp. 107–149. Jorge Zahar Ed., RJ (2001)
5. Benkler, Y.: The Wealth of Networks: How Social Production Transforms Markets and Freedom. Yale University Press, New Haven (2006)
6. Caldeira, S.M.G., et al.: The network of concepts in written texts. European Physical Journal B 49, 523–529 (2006)
7. Carpentier, N.: The concept of participation. If they have access and interact, do they really participate? Revista Fronteiras - Estudos Midiáticos 14(2), 164–177 (2012)
8. Dahlgren, P.: Reinventare la partecipazione. Civic agency e mondo della rete. In: Bartoletti, R., Faccioli, F. (cura) Comunicazione e Civic Engagement. Media, Spazi Pubblici e Nuovi Processi de Partecipazione, pp. 17–37. Franco Angeli, Milano (2013)
9. Gibson, R., Cantijoch, M.: Conceptualizing and Measuring Participation in the Age of the Internet: Is Online Political Engagement Really Different to Offline? The Journal of Politics 75, 1–16 (2013)
10. Ito, M.: Intimate Visual Co-Presence. In: International Conference on Ubiquitous Computing, UbiComp 2005, Tokyo, Japan (2005), `http://www.itofisher.com/mito/archives/ito.ubicomp05.pdf` (accessed on December 3, 2012)

11. Lefebvre, H.: The right to the city (1968)
12. Maia, R.C.M.: Democracia e a Internet como esfera pública virtual: aproximação às condições de deliberação. In: Gomes, W., Maia, R.C.M. (eds.) Comunicação e Democracia: Problemas & Perspectivas. Coleção Comunicação, pp. 277–292. Paulus, São Paulo (2008)
13. Maia, R.C.M., Marques, A.C.S., Mendonça, R.F.: Interações mediadas e deliberação pública: a articulação entre diferentes arenas discursivas. In: Primo, et al. (orgs.) Comunicação e interações. Livro da COMPÓS 2008, pp. 93–110. Sulina, Porto Alegre (2008)
14. Modesto, A.T.T.: Processos interacionais na Internet: Análise da Conversação Digital. Tese de Doutorado, Faculdade de Filosofia, Letras e Ciencias Humanas da Universidade de Sao Paulo, Sao Paulo (2011)
15. Murgante, B.: Wiki-Planning: The Experience of Basento Park In Potenza (Italy). In: Borruso, G., et al. (ed.) Geographic Information Analysis for Sustainable Development and Economic Planning: New Technologies, pp. 345–359 (2013)
16. Papacharissi, Z.: The Virtual Sphere 2.0: The internet, the public sphere, and beyond. In: Chadwick, A., Howard, P. (eds.) Handbook of Internet Politics, pp. 230–245. Routledge, Taylor & Francis Group, London, New York (2009)
17. Parviainen, O., Poutanen, P., Salla-Maaria, L., Rekola, M.: Measuring the effect of social connections on political activity on Facebook. Mimeo (2012)
18. Recuero, R.: A Conversação em Rede. Ed. Sulina, Porto Alegre (2012)
19. Rieder, B.: Studying Facebook via Data Extraction: The Netvizz Application. In: ACM WebSci., pp. 346–355 (2013)
20. Rodotà, S.: Il diritto di avere diritti. Ed. Laterza, Roma-Bari (2012)
21. Rodotà, S.: La democrazia e il bene comune. In: Gallina, E. (a cura) Vivere la democrazia, pp. 83–96. Edzione Gruppo Abele, Torino (2013)
22. Rustad, E., Sæbø, Ø.: How, Why and with Whom Do Local Politicians Engage on Facebook? In: Wimmer, M.A., Tambouris, E., Macintosh, A. (eds.) ePart 2013. LNCS, vol. 8075, pp. 69–79. Springer, Heidelberg (2013)
23. Teixeira, G.M., et al.: Complex Semantic Networks. Int. J. Mod. Phys. C 21, 333 (2010), doi:10.1142/S0129183110015142
24. Vitale, E.: Contro i Beni Comuni. Una critica illuminista. Editori Laterza, Roma-Bari (2013)
25. Wasserman, S., Faust, K.: Social Network Analysis: Methods and Applications. Cambridge University Press, Cambridge (1994)

Evaluating Urban Development Plans

A New Method in the Toolbox

Valerio Cutini

Department of Energy, Systems, Territory and Buildings,
University of Pisa,
Pisa, Italy
valerio.cutini@ing.unipi.it

Abstract. The configurational approach to the analysis of urban settlements, first introduced by Bill Hillier in the mid '80s and since then developed and applied by large groups of researchers all over the world, has been appreciated as a reliable tool for enhancing urban interpretation and understanding as well as supporting town planning.

The configurational techniques allow to simulate and predict the effects of urban plans and projects on a wide range of variables, thus orienting the decision making process in evaluating and choosing among different options. Any urban project, in that it modifies the consistency of the urban grid, involves a transformation of its configurational state and therefore the inner geography it can reproduce, as well as all the variables it is able to predict. The issue with the present paper is to pinpoint the configurational parameters that result suitable for characterizing any urban planned transformation; what is expected to concretely facilitate its evaluation and to allow a deeper comprehension of its likely consequences.

Keywords: urban development plan, grid configuration, evaluation.

1 Introduction

The configurational approach, first introduced by Bill Hillier in the mid '80s [1] and since then developed and applied by large teams of researchers all over the world, has been proved and highly appreciated as a useful and reliable tool for supporting urban interpretation and understanding and town planning.

On the one hand, the configurational techniques can strongly enhance the comprehension of the inner urban geography, either at the present (or even at a past or future) date or at several, successive temporal phases. Such applications are respectively said synchronic and diachronic, and both are aimed at reproducing, by means of configurational parameters, the distribution of several urban aspects or phenomena: pedestrian or vehicular movement, attractiveness towards activities, urban rent values, economic or social segregation, urban safety and so on. The

B. Murgante et al. (Eds.): ICCSA 2014, Part II, LNCS 8580, pp. 502–519, 2014.

outcome of the synchronic configurational analysis has been demonstrated and used as a precious reference and support tool for several planning operations, such as the working out of transport plans, urban retail plans and projects, accessibility plans and so on. Obviously the synchronic analysis is suitable for enhancing the comprehension of such phenomena and, more in general, the inner urban geography at a selected date, while the diachronic analysis was proved a unique tool for understanding and following out its genesis and historic evolution, as well as to interpreter its likely spatial causes.

On the other hand, the configurational techniques are currently used to support town planning, in that they allow to simulate and somehow predict the effects of urban plans and projects on a wide range of variables, thus supporting the decision making process in evaluating plans and projects as well as choosing among different options. Any urban plan or project, in that it modifies the present consistency of the urban grid, involves a transformation of its configurational state and therefore the inner geography it can reproduce, as well as all the variables and the aspects and phenomena it is able to predict.

The issue with the present paper is to pinpoint the configurational parameters that result as characterizing any urban planned transformation, so as to propose an objective method for analyzing it, what is expected to concretely facilitate its evaluation (or its comparison with others) and to actually favour a deeper comprehension of its likely consequences.

2 Backgrounds

The configurational approach to the analysis of urban settlements, introduced in the mid ‘80s by Bill Hillier [1] and since then applied, improved and developed by large teams of researchers all over the world, is based on the role of the urban grid as the primary element in the phenomena occurring along its paths, in particular movement distribution and activities location. The hypothesis is the existence of a portion of movement, which is called 'natural movement', that does not depend on the presence and the position of the located activities but it’s a mere function of the spatial relationships between its elements, that is the grid configuration [2]. Analyzing the grid configuration means to appraise the specific spatial value that each of its element is provided with, due to the spatial relationships between it and all the other elements of the system. The measure of such relationships is what is properly defined as the configuration of the urban grid [3].

This assumption does not involve the denial of movement depending on the interaction between activities, or attracted movement, which can even greater than the natural one; still, as depending on the grid configuration, the natural movement is assigned a primary role, derived from the primary role of the urban grid. It's then the grid configuration what indicates and suggests the likely distribution of movement flows an hence the pre-condition for the use of urban land [4]. On such basis, several operational techniques have been so far introduced and developed: the first one, introduced by Hillier in 1984 [1], is the axial analysis, which transforms the grid into a system, called axial map, composed of the fewest and longest lines that cover all the grid, connecting all its convex spaces. Other techniques, including visibility graph analysis [5], angular analysis [6], segment analysis [7], mark point parameter analysis [8], road-centre lines analysis [9] sharing the same conceptual basis, differ from one another for the way of reducing the grid into a system, and actually present different advantages and limits, so as to coexist, suitable to be used in different cases and with different purposes [10].

All those operational methods provide each element of the system with a full set of configurational variables; among them, connectivity (in mathematics usually known as degree, that is the number of elements directly connected to the observed one), integration (in mathematics usually known as closeness, that is the mean depth of the observed element with respect to all the others) and choice (in mathematics usually known as betweenness, that is the frequency of the presence of an element in the shortest paths mutually connecting all the couples of other elements) deserve a specific consideration. Integration, probably the most significant configurational index, is usually used in a normalized expression with regard to the number n of the elements of the system, in order to make it independent of the dimension of the grid; what is necessary to make it suitable for comparing spatial grids of different size. Among them all, the normalization expression below, one of the first to be introduced [1], will be used in the following:

$$I = 2\,(D_M - 1)/(n-2)$$

This normalized value of integration resulting from the expression above (where D_M is the mean depth of the observed line and n the number of lines of the system) will vary from 0 (minimum value, correspondent to a spatial element that is directly connected to all the others, as each of the lines of a star-shaped grid) to 1 (maximum value, correspondent to the terminal element of a tenial grid).

In particular, the relevant significance of integration derives from several studies, which in these last decades have tested and proved it suitable for narrowly reproducing the actual distribution of urban centrality, meant as the measure of the attractiveness towards activities [11]. Such notion of centrality can be said pure, in that it does not depend on the located activities, but is merely determined by the spatial relationships between the elements, that is by the grid configuration. In other words, the distribution of integration appears to narrowly reproduce the distribution of the positional appeal of the spatial elements of the grid; what allows overcoming the notion of centrality as a morphologic or functional state, rather appraising it as a process, depending on the spatial relationships between the elements of the grid [12]. Integration value can be concretely computed either in a global view (taking into account all the elements of the whole grid, so as to obtain the so-called global integration – or radius n integration - value) or in a local view (computing only the elements lying in a topologic circle with pre-established radius R around the observed one; usually R= 3 and the resulting index is usually called local integration, or radius 3 integration). In order to highlight the distribution of the most integrated parts of the urban area, the lines whose integration value results over an established threshold (usually 90° or 95° percentile) form the so-called integration core, whose position and extent is useful to somehow materialize the centrality of the settlement as well as to follow its possible shifting.

On such basis, the assumption of the grid as the primary element of many urban phenomena actually provides the space of the settlement with a foreground role, and allows configurational analysis to face questions concerning the morphology of urban texture: in fact, the configurational techniques assume the urban space with its material consistency (from the pattern of streets up to the presence of squares and to shape of blocks and buildings) as their input variable, so as to provide results that regard several aspects of the working of the settlement. In that they relate their findings to the physical features of a settlement, the configurational techniques appear an outstanding tool for bridging the gap between urban design and territorial quantitative analysis [3]. More in concrete, such feature makes them suitable for predicting the effects of any material transformation of the grid, which normally concerns the geometry, shape and position of streets, squares, blocks and buildings, on a wide range of material and immaterial variables of the urban system (distribution of movement flows, distribution of rent values, attractiveness, accessibility, and so on) [13].

3 Methodology

The idea this research is based on is that any urban transformation (either actual or just planned, of course) involves some transformation of its grid configuration, and therefore some change in the inner urban geography that can hence be captured by means of a configurational approach. In order to allow an informed and well-aware evaluation of an urban plan or project as well as to facilitate an objective comparison of different options, a rigorous analysis procedure is strongly advised. What will be shown and discussed in the following.

On the basis of what was shown above, several parameters can be assumed in order to describe the configurational state of the system as a whole. A first one, probably the roughest, is the mean value C_M of connectivity; since connectivity represents the number of connected elements (that is intersected lines) to the observed one, its value is included between 1 (any terminal line, directly connected to only another line) and n-1 (a line connected to all the other lines of the system). Obviously also C_M will then vary from 1 to n-1, hence approaching 1 in the case of a tenial grid and k-1 in a star-shaped one, where each line is connected to all the others; figure 1 provides both a geometrical representation of those paradigms (on the left) and a relational one (on the right). It was shown, discussed and verified how this simple configurational parameter can contribute to reproduce the level of resilience of an urban system [14] [15]: since the redundancy in connections between couple of elements is an essential element of resilience, in that any interrupted link can be easily be substituted by another, the mean connectivity of a spatial system is suitable for reproducing its level, that is the degree to which any interrupted path can be substituted by an alternative one without significant change in the mutual spatial impedance. In graph theory this property is usually reproduced (to a certain extent) by the so-called, defined as the smallest number of edges we can delete in order to disconnect the graph [16]. Yet, the edge-connectivity reproduces the resilience of systems with respect to their possible cracks for disconnection, while the mean value of connectivity can more suitably represent their steadiness in front of any perturbation and transformation, as it can indicate the richness in alternative paths; moreover, a grid with many terminal lines (that is a graph with many bridges, in graph theory [16]) not necessarily lacks in resilience, since it may be provided with a redundant number of internal connections.

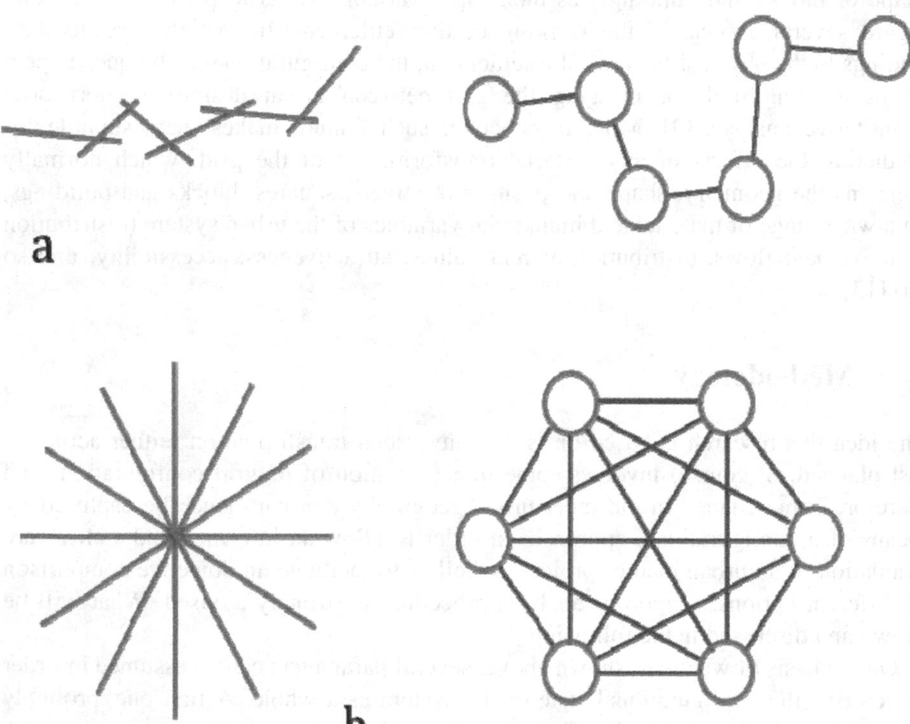

Fig. 1. Two paradigms of configuration: a tenial grid (above) and a star-shaped one (below)

Another configurational parameter suitable for describing the properties of a whole spatial system is the mean value I_M of global integration, which is obviously included between 0 and 1. This parameter is suitable for reproducing the mean depth of a system, that is the mean accessibility of its elements. Such index increases as the system gets deeper and decreases, approaching 0, as it gets shallower. Once more, the tenial grid and the star-shaped grid respectively represent the extreme and paradigmatic references: in a tenial grid the mean integration value approaches 1 as the number of its elements tends to infinity, while in a star-shaped grid (which in graph theory is usually defined a complete graph [16]), the shallowest one could imagine, being each line directly connected to all the others, $I_M = 0$. Roughly speaking, this parameter somehow reproduces the relational compactness of an urban settlement, that is how far its single elements are meanly from all the others. Furthermore, since integration was proved a reliable indicator of accessibility, low values of I_M are expected standing for a uniform distribution of attractiveness and activities from the inner core to the edge of the settlement, and arguably also for a uniform distribution of rent values; on the contrary, low values of I_M will likely stand for a strong gradient centre/periphery with regard to attractiveness, appeal and rent values; what the above mentioned references (the tenial grid and the orthogonal grid) can actually exemplify in real urban cases.

A further global configurational parameter of a system is the gradient in the distribution of choice value, which can be represented by the degree the line with maximum choice value intercepts all the possible connections within the system . Since choice represents the frequency of the presence of an element in the shortest paths that connect all the couples of other elements, a high gradient stands for a pyramidal distribution of the connections, a large part of which (or even all them) passes through few elements (or even just one of them, as in a bottleneck). If the tenial grid is once more the ideal reference of high choice gradient, the paradigm for its low values is the orthogonal grid, where all the lines are equally present on the shortest paths between the others. On a computational regard, this feature can be suitably reproduced by the ratio of the highest choice value and the maximum frequency a line could present, what would occur if it were located on all the shortest paths between any couple of the other lines of the system. If we consider an axial map of n lines, it can be shown that the total amount of the shortest paths between any pair of its elements (excluding the observed line, to be assumed just as a possible intermediate element) is

$$k = \tfrac{1}{2}\, n\, (n - 1) - (n - 1) = n^2/2 - 3/2\, n + 1$$

Then the proposed index, to be named frequency index, would be

$$f = \text{choice}_{max}/k$$

Of course the proposed parameter will be included between 0 to 1, increasing with the gradient of choice so as to reproduce the degree to which the distribution of movement can be said uniform (low frequency values) o pyramidal (high frequency values). Obviously, also this parameter can be assumed as a reliable indicator of resilience [14]; in fact, should a single spatial element result along all the shortest paths that connect all the others, its interruption would be expected to radically upset the relationships all over the grid, and hence the whole configuration of the system and the inner geography of the urban settlement. In such case (which is obviously characterized by f = 1), the system would result vulnerable to its highest degree, in that each of its paths, no matter the source and the destination, would depend on that single line.

Beside those presented so far, two further parameters can usefully be obtained by means of the correlation of configurational indices; as they depend on the values of other indices, they can be considered second order parameters. In particular, two parameters appear suitable for describing important global features of the grid, that is the R^2 coefficient of the correlation connectivity versus radius n integration and the same R^2 coefficient of the correlation radius 3 versus radius n integration. Both those parameters, obviously varying from 0 (representing total spread and hence absence of correlation) to 1 (perfect correlation), are worth briefly discussing in that they respectively account for different relevant aspects.

The R^2 coefficient of the correlation connectivity versus radius n integration (generally exponential) is often called 'intelligibility index' and will here be reported as X, as the fact that the most accessible spaces give access to a larger number of other spaces accounts for a clearer spatial perception, so as to facilitate the comprehension of the whole urban consistency and to better orient in its experience. In concrete terms, the intelligibility value describes how the global configuration of an urban settlement can be appraised from its single parts, that is, in other words, how any observer in the urban grid can be informed about his position in the spatial system from every location he may actually occupy. Bill Hillier does incisively define it as "the degree to which what can be seen and experienced locally in the system allows the large–scale system to be learnt without conscious efforts" [3].

For what concerns the correlation of local versus global integration (usually radius 3 versus radius n integration, and generally linear), which is often called 'synergy index' and will here be reported as S, the idea is that when the local perception actually corresponds to the global geography of the settlements, it will support and orient the path finding all over the grid. In case of narrow correlation, local integrators are also prominent integrators at a global scale, thus creating a stronger and perceivable interface between the whole settlement and its single parts. This synergy index can therefore be usefully assumed as an indicator of the degree to which the different scales of the settlement are actually correlated, so as to concur in synergy to the global working of the city, which is commonly acknowledged a vital property in urban areas [17].

Summing all up, so far five parameters (including the second order ones, intelligibility and synergy) have been presented above, suitable for reproducing significant global features of urban geography. And obviously each of those parameters are likely to change, either slightly or radically, as a consequence of any - little or great - transformation the observed settlement is subject to. The direct comparison of their respective values ante and post the planned transformation can therefore point out its likely effect on urban geography and hence significantly support its evaluation. Yet the purpose of this research goes further, aiming at objectively appraising, and even at numerically quantifying, the actual impact the observed transformation is likely to determine on the inner geography of the settlement, so as to allow comparing different planning options. This purpose will be pursued using a further parameter, suitable for synthesizing the comparison of the ex-post and the ext-ante configurational state and thus measuring the impact of any urban transformation, as normalized with reference to its actual extent. This parameter, to be named 'impact coefficient', can be written as follows

$$K = |I_1 - I_0 / I_1|_M$$

being I_1 the global integration value in the transformed grid and I_0 its value in the present state. This parameter can also be normalized with reference to the number of added (or varied) lines, thus making it indifferent to the extent of the new development and allowing to compare different plans; nonetheless, the expression above appears to more effectively and immediately reproduce the global impact the development plan is likely to determine. As it can be easily seen, such parameter varies from 0 (no change to the integration values) to 1 (maximum change to all the integration values). Furthermore, it's worth noticing that the value of this coefficient cannot be indifferent to the actual dimension of the whole grid, in that smaller systems will receive from a similar size transformation a stronger global impact than bigger ones; or, in other words, the same transformation could be likely to cause a relevant impact into a small system and a poor impact into a large one. What actually occurs in cities, being large urban settlements generally steadier and more stable than small cities when subject to any transformation.

A basic example, an elementary spatial system that goes under several possible transformations, can be usefully discussed for a better comprehension of the method. Let us consider the system here represented by the graph of figure 1, which can be assumed as corresponding to the spatial grid of a very small settlement, composed by four spatial elements; the values of the three primary configurational parameters discussed above are reported in the table on the right. Obviously, due to the minimum size of the example, in this case the second order parameters are not considered, as not significant.

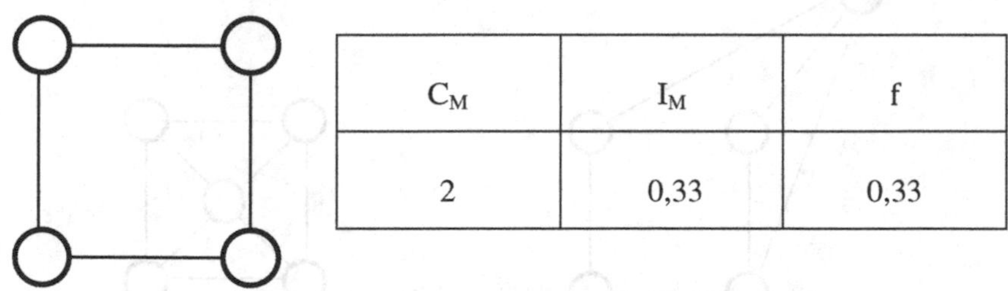

C_M	I_M	f
2	0,33	0,33

Fig. 2. An example: graph representation and correspondent configurational parameters

Let us then hypothesize a slight transformation of the system, consisting of the simple addition of a single element (for instance, consider just one more street in the urban grid), which obviously is represented by an additional node in the graph. From a relational (that is also configurational, as it was shown above) point of view, this addition can modify the graph of figure 2 in only four different ways, depending on the possible relationships of the added node with the existing ones; and these different possibilities are here respectively represented in figure 3 as a, b, c and d. Obviously, as the number of the added elements grows, the variety of their possible relationships and hence the number of the resulting configurational states does exponentially increase. The three parameters introduced above and reported in figure 2 change as a consequence of those different transformations, and their respective values are here reported in table 1, which includes also the value the impact coefficient K assumes as respectively referred to the four different circumstances.

The values in table 1 allow a prompt visualization and understanding of the effects of the four hypothesized elementary transformations. Proceeding from a to d, we can notice the progressive increase of mean connectivity as well as the progressive decrease of mean integration and frequency. In the whole, those values attest that the system gets more and more resilient (high values of C_M, low values of f) as the additional node is more and more connected; in other words, the system will be provided with a bigger inertia with respect to any perturbation. Moreover, as the number of its connections increases (from a to d), the system gets globally shallower, as characterized by a more uniform (that is less pyramidal) distribution of accessibility and hence a wider integration core.

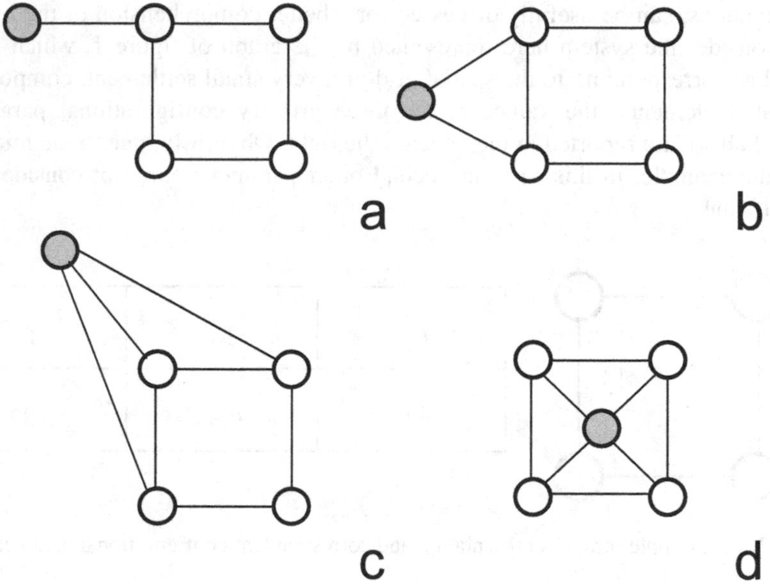

Fig. 3. The addition of a node (in grey) to the graph of fig. 2. The four relational possibilities

Table 1. Configurational parameters of the transformations of fig. 3

Parameters	C_M	I_M	f	K
state 0	2	0.33	0.33	-
transformation a	2	0.40	0.67	0.25
transformation b	2.40	0.27	0.33	0.25
transformation c	2.80	0.20	0.17	0.375
transformation d	3.20	0.17	0.33	0.50

For what concerns the degree to which the different transformations are likely to actually affect the geography of the settlement, that is the quantification of their impact on the initial state of the system, the K value clearly shows a progressive increase of that effect as the added node gets more connected to the existing ones; the connection to all the other elements radically appears to upset the whole grid configuration, as it is attested by K = 0.50.

4 Case Studies

In order to implement and test the method sketched above, five significant real transformations, undoubtedly the last relevant development plans worked out in the urban settlement of Pisa in the last fifty years, have been selected and analyzed. Pisa, as it is well known, has a long lasting urban history, which over the centuries has gone determining significant successive transformations of the settlement from its ancient origins up to the present times; yet, in this research no attention will be paid to the historic genesis of the grid configuration of Pisa, which was already analyzed and discussed elsewhere. With reference to each of the five selected transformation, the research will rather assume the ex-ante consistency of the settlement as its starting point, since the only purpose of the experimentation of the method above is to investigate around the planned interventions, so as to show how different the parameters that were introduced for measuring their respective effects on urban geography can actually result. The selected cases appear quite different one from another, and here briefly listed.

The first one is the neighbourhood called 'I Passi', which was planned and then worked out in the late '50s around the extreme southern edge of Pisa. The second development plan is the Saint-Gobain village, a company neighbourhood worked out in the same period for the glass factory employers between CEP and the historic city. The third case, approximately contemporary, was called CEP e was (and presently still is) located at the extreme western edge of the city. The fourth one is Cisanello, a great development plan of the mid '60s whose declared purpose was to represent a new centrality area along the eastern end of the settlement, aimed at relieving the crowded historic core of Pisa. The fifth case is a recent relevant intervention within the ancient town walls, hence focused on a urbanized area and presented and acknowledged as internationally highly significant. The area presently corresponds to the Santa Chiara Hospital, which has been occupying it, with dozen remarkable buildings, since the Middle Age. The area is just close to the monumental site of Piazza dei Miracoli, as the historic hospital is part of the natural cycle of life (birth, life, illness, death), here respectively symbolized and materialized by the Baptistery, the Cathedral and its Leaning Tower, the Hospital and the Cemetery. According to the prescriptions of the Master Plan of Pisa of 1998, it was decided to shift the great hospital into the peripheral area of Cisanello, along the eastern edge of the settlement, thus opening and returning the whole inner area (over 110,000 square meters) to the community civil use. In order to rearrange the whole area with new buildings, urban spaces and activities, in 2007 the municipality announced an international contest, open to well known architects from all over the world; the project presented by the group headed by David Chipperfield was declared winner and is here synthesised in the table reported in figure 4. Presently, the shifting of the hospital activities and services is underway, and, despite the definitive conclusion of the international contest in 2007, the actual assignment of this wide urban area is still to be defined and concretely decided. In this paper, the consistency of the grid as it results from the Chipperfield winning project will be assumed as our second case study, here named Santa Chiara project. As it will be easily understood, this further case is the very target of the research, since it is the widest and most significant urban plan in Pisa in the last decades and at present is subject to evaluation and discussion, in that it involves such a relevant part of the historic core; the comparison with other relevant urban development plans in the recent history of Pisa is then expected to enrich that evaluation and to reliably test the proposed method as well.

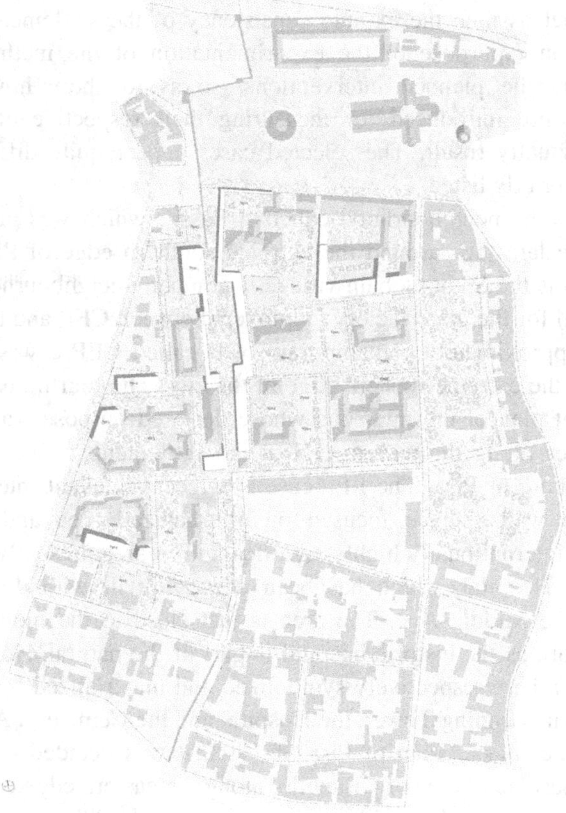

Fig. 4. The Chipperfield project for the area of Santa Chiara hospital in Pisa – Table 1

Fig. 5. The distribution of global integration in the axial map of Pisa – urban development 'I Passi'

Fig. 6. The distribution of global integration in the axial map of Pisa – urban development 'Villaggio Saint Gobain'

Fig. 7. The distribution of global integration in the axial map of Pisa – urban development C.E.P

Fig. 8. The distribution of global integration in the axial map of Pisa – urban development Cisanello

Fig. 9. The distribution of global integration in the axial map of Pisa – Santa Chiara development project

In fact, each of the five selected case have been subject to the same method, analyzing the configuration of both ex-ante and ex-post grid consistency and then computing the values of the six configurational parameters introduced above. The distribution of global integration values with reference to each of the five development plans, as it results from the respective grid configuration analysis, is here reported in figures 5, 6, 7, 8 and 9; the chromatic representation uses warm colours (up to red) for integrated lines and cold tones (up to violet) for integrated ones.

The tables here reported in the figures above clearly show the successive changes in the distribution of integration, which synthesise the transformation in the whole configurational state of the system that each of the selected development plans has actually caused (or is likely to cause, in the case of Santa Chiara development plan); more in detail, it can be noticed how (and where) the integration core of the settlement actually shifts as a consequence of the development plan, thus providing a significant qualitative view of such transformation of urban geography; what has been proved to affect several urban variables and phenomena, from the distribution of movement to the land values. For instance, for what specifically concerns the Santa Chiara development plan (fig. 9), the representation highlights the shifting westbound of the integration core, previously steadily anchored to the rough cross composed by the north-south axis and the embankments along the Arno, which pivot in the Ponte di Mezzo, the main urban bridge that appears and actually remains the very earth of the

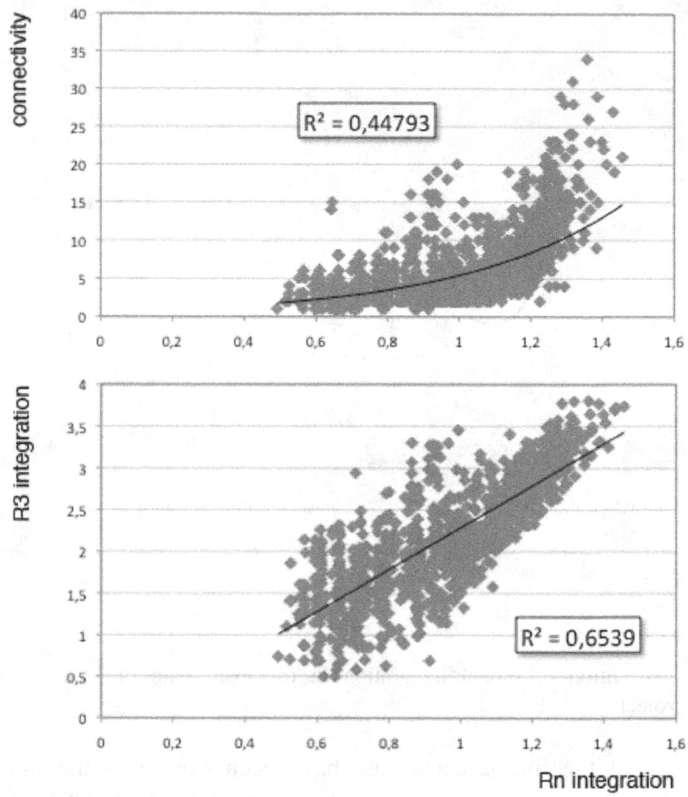

Fig. 10. Correlation of global integration versus connectivity (above) and local integration (below) in the axial map of Pisa - Santa Chiara project consistency. The intelligibility and synergy indices are highlighted.

settlement. As an example of the second order parameters, again referred to the recent Santa Chiara development plan, the analysis of the correlation of global integration versus local (radius 3) integration and connectivity is here represented in figure 10, so as to show the finding of the so-called intelligibility and synergy values.

So far, the present study has not achieved any really original result, since it represents a mere application of the configurational techniques to some planned case studies, which essentially describe the qualitative shifting of integration core due to two different urban projects; what appears quite significant but scarcely innovative if only we consider the amount of previous researches on similar issues. Yet, the very purpose of the research is rather to implement the method sketched above in order to measure the actual impact the selected transformations have determined, or are likely to determine (in the case of the Santa Chiara development plan), thus allowing their direct and objective comparison as well as their shared and unequivocal evaluation. For this aim, the values of the six configurational parameters previously introduced have been computed with reference to the respective ex-ante actual consistency of the whole settlement and to the grids that result from the five selected urban development plans; the results are here summarized in table 2.

Table 2. Configurational parameters in the selected case studies

case studies	parameters	C_M	I_M	f	X	S	K
'I Passi'	ex-ante grid consistency	7,44	1.11	0.28	0.47	0.65	-
	ex-post grid consistency	7,37	1.10	0.29	0.47	0.64	0.05
Saint Gobain	ex-ante grid consistency	7.37	1.10	0.29	0.47	0.64	-
	ex-post grid consistency	7.26	1.08	0.28	0.47	0.64	0.03
C.E.P.	ex-ante grid consistency	7.26	1.08	0.28	0.47	0.64	-
	ex-post grid consistency	7.14	1.07	0.28	0.47	0.64	0.09
Cisanello	ex-ante grid consistency	7.04	1.04	0.28	0.47	0.64	-
	ex-post grid consistency	6,45	0.96	0.26	0.46	0.63	0.18
Santa Chiara	ex-ante grid consistency	6.45	0.95	0.26	0.45	0.63	-
	ex-post grid consistency	6.49	0.97	0.26	0.45	0.65	0.04

Several interesting aspects can be drawn out of the interpretation and mutual comparison of the results above summarized. First, and most obvious: it is easy to observe that each of the selected development plans involve some (slight or prominent) change in the global proposed parameters, thus materializing some transformation of the ex-ante configurational state; what appears confirming the distribution of integration observed in figures 5, 6, 7, 8 and 9. Less obvious, such effect can be actually measured, by means of the impact coefficient K, which record poor effects (with the Saint Gobain development plan) as well as strong and upsetting impacts (CEP and Cisanello). Furthermore, and not obvious in the least, the Santa Chiara development plan materializes a radical change in the previous trend: all the four realized developments had in fact determined a decrease in mean connectivity and integration and the weakening of intelligibility and synergy, as it was usual in most twentieth-century developments [18]; on the contrary, the recently planned intervention in Santa Chiara hospital area appears to involve a (slight but constant) increase in those values, as well as a slight decrease in the frequency coefficient. As it was discussed above, this increase can be interpreted as a reinforcement of the global resilience of the system, an improvement of its intelligibility and mutual integration. In few words, such effect can be summarized saying the resulting system appears shallower and more integrated than the pre-existing; what, as compared with the other results, appears materializing a clear inversion with respect to the trend so far, characterized by the weakening of integration, resilience and intelligibility of the whole settlement.

5 Conclusions

The results so far can be regarded as a useful concrete contribution, as they provide the present discussion and debate around the crucial transformation of the Santa Chiara hospital area in Pisa with some objective knowledge elements on the actual impact the development plan is likely to determine on the whole settlement, suitable for supporting the decision making on this area. More in detail, the findings allow acknowledging it as the first development plan in the last decades that considerably strengthens internal connections and integration, increasing urban resilience and intelligibility; what in Pisa actually represents a clear inversion with respect to the trend determined by the twentieth-century sprawling growth.

Even more, and far beyond our case study, such results can be considered as interesting and highly significant on a methodological regard, in that the proposed method appears broadening the outlook for evaluating urban plans and projects, thus concurring in bridging the gap between territorial analysis and urban design. Not only, in fact, the distribution of configurational values can reproduce the features of the urban inner geography, hence allowing simulating and predicting several urban aspects and phenomena. Even more, each single development plan can be objectively analyzed with reference to the actual effect it is likely to produce, and provided with objective indices and measures. From a quantitative point of view, in fact, those parameters can appraise the relevance of the impact on the whole urban settlement, that is the extent to which its configuration gets actually modified, thus understanding if the development plan is likely to confirm or upset its existing inner geography. Furthermore, the values the global parameters assume clearly point out the specific sense of such impact, that is how (and how much) the development plan is going to modify the resilience, the integration, the intelligibility of the settlement, either increasing or reducing them. What suggests the proposed method as a new, useful tool for interpreting, evaluating and comparing urban plans and projects.

Some limitations in its use appear worth mentioning. First, it ought to be noticed that a configurational approach, as focused on the spatial relationships between the elements of the grid, is indifferent to the functional state of the system, and hence cannot appraise any merely functional transformation, nor obviously predict its likely effect on urban geography: a change in urban land use (for instance from industrial to commercial or residential) won't therefore modify the grid configuration. Furthermore, the analysis is indifferent to the change of buildings height or density, if not involving some change in the consistency of the grid (streets, squares, blocks, etc.). Nonetheless, such limitations are somehow physiological and don't actually contradict the utility and reliability of the proposed method; they rather specify and encompass its field of use, that is the spatial features of the urban grid: when it comes to space and some physical transformation are concerned, then configurational analysis can usefully enlighten town planning. To summarize, the configurational approach to the evaluation of urban plans is anything but an exhaustive and all-around method; rather, just one more powerful instrument in the toolbox.

References

1. Hillier, B., Hanson, J.: The Social Logic of Space. Cambridge University Press, Cambridge (1984)
2. Hillier, B., Penn, A., Hanson, J., Grajevski, T., Xu, J.: Natural Movement: or, Configuration and Attraction in Urban Pedestrian Movement. Environment and Planning B, Planning and Design 20, 67–81 (1993)
3. Hillier, B.: Space is the Machine: a Configurational Theory of Architecture. Cambridge University Press, Cambridge (1996)

4. Hillier, B.: Cities as Movement Economies. Urban Design International 1(1), 41–60 (1996)
5. Turner, A., Doxa, M., O'Sullivan, D., Penn, A.: From Isovists to Visibility Graphs: a Methodology for the Analysis of Architectural Space. In: Environment and Planning B: Planning and Design, vol. 28, pp. 103–121 (2001)
6. Turner, A.: Angular Analysis. In: Proceedings of the 3rd Space Syntax Symposium, Atlanta (GA), May 7-11, pp. 30.1–30.11 University of Michigan, Alfred Tauban College of Architecture (2001)
7. Hillier, B., Iida, S.: Network and Psychological Effects in Urban Movement. In: Cohn, A.G., Mark, D.M. (eds.) COSIT 2005. LNCS, vol. 3693, pp. 475–490. Springer, Heidelberg (2005)
8. Cutini, V., Petri, M., Santucci, A.: From Axial Maps to Mark Point Parameter Analysis (Ma.P.P.A.) – A GIS Implemented Method to Automate Configurational Analysis. In: Laganá, A., Gavrilova, M.L., Kumar, V., Mun, Y., Tan, C.J.K., Gervasi, O. (eds.) ICCSA 2004. LNCS, vol. 3044, pp. 1107–1116. Springer, Heidelberg (2004)
9. Turner, A.: From axial to road-centre lines: a new representation for Space Syntax and a new model of route choice for transport network analysis. In: Environment and Planning B: Planning and Design, vol. 34, pp. 539–555 (2007)
10. Cutini, V.: Grilling the Grid: a Non-Ultimate (Nor Objective) Report on the Configurational Approach to Urban Phenomena. In: The Dynamics of Complex Urban Systems, pp. 163–183. Physica-Verlag, Heidelberg (2007)
11. Cutini, V.: Centrality and Land Use: Three Case Studies on the Configurational Hypothesis. Cybergeo, Revue Européenne de Geographie (188) (March 26, 2001)
12. Hillier, B.: Centrality as a Process: Accounting for Attraction Inequalities in the Deformed Grids. Urban Design International (3/4), 107–127 (2000)
13. Hillier, B., Vaughan, L.: The City as One Thing. Progress in Planning 67(3), 205–230 (2006)
14. Cutini, V.: The City, when It Trembles. Earthquake Destructions, Post-earthquake Reconstructions and Grid Configuration. In: Proceedings of the 9th Space Syntax Symposium, Seoul, South Korea, October 31-November 3, pp. 102.1–102.17 (2013)
15. Cutini, V., Rabino, G.: Searching for Ariadne's Thread. Some Remarks on Urban Resilience and Orientation. TEMA 5, 7–22 (2012)
16. Wilson, R.J.: Introduction to Graph Theory. Longman Scientific and Technical, Harlow (1994)
17. Hillier, B.: The Art of Place and the Science of Space. World Architecture 185, 96–102 (2005)
18. Cutini, V.: Working on the Edge of Town: The Periphery as a Spatial Pattern. In: Gervasi, O., Murgante, B., Laganà, A., Taniar, D., Mun, Y., Gavrilova, M.L., et al. (eds.) ICCSA 2008, Part I. LNCS, vol. 5072, pp. 26–41. Springer, Heidelberg (2008)

Characteristics of Sprawl in the Naples Metropolitan Area. Indications for Controlling and Monitoring Urban Transformations

Rocco Papa[1] and Giuseppe Mazzeo[2]

[1] University of Naples Federico II, Department of Civil, Architectural and Environmental Engineering (DICEA), P.le V. Tecchio 80, 80125 Napoli, Italy
rpapa@unina.it

[2] National Research Council (CNR), Institute of Studies for the Mediterranean Societies (ISSM); University of Naples Federico II, Department of Civil, Architectural and Environmental Engineering (DICEA), P.le V. Tecchio 80, 80125 Napoli, Italy
gimazzeo@unina.it

Abstract. Naples metropolitan area is one of the largest in Italy, located in the South of the Country with a population of about 4 million inhabitants. Several studies have proposed for this area different spatial metropolitan boundaries but, despite the presence of national regulations, the area is not still defined at administrative level. One of the main characteristics of the area is the high level of unplanned urban expansion, with the consequence of a wide level of fragmentation, an extensive illegal urban development and an incorrect use of the agricultural land. The paper analyze the urbanization process in the study area and proposes a monitoring tool of the sprawl phenomenon, with a smart spatial planning approach.

Keywords: Naples metropolitan area, Sprawl, Planning tools for spatial monitoring.

1 Introduction

The analysis of the phenomena related to the expansion of the Naples conurbation is the starting point to suggest actions for the control of urban transformations, also using social networks technological innovations as a mean to create smarter cities.

When talking about large conurbations, we use specific concepts, such as "metropolitan area" [1, 2, 3] and "urban sprawl" [4, 5]; besides, the analysis of urban conurbations are often related also to the concept of "urban hierarchy" [6].

With "metropolitan area" we refer to the urban expansion that quickly pour the area surrounding cities. Some planning models have been defined with the goal of controlling and governing the urbanization processes in a systematic and efficient way. Nerveless, in the long term, these models have revealed to be ineffective and have left the field to strategic designs, characterized by disconnected actions.

B. Murgante et al. (Eds.): ICCSA 2014, Part II, LNCS 8580, pp. 520–531, 2014.

The processes of uncontrolled urbanization, named as "urban sprawl", derive mainly by the self-referential nature of the market economic system and by its indifference towards the environment: the city becomes a testing site of the economic theories based on unlimited appropriation of the space.

The two concepts are timely sequential, but not alternative. The concept of "sprawl" overlaps with the notion of "metropolitan area" between the 1970s and 1980s, in the last phase of the urban crisis. Both concepts are products of urban developments addressed to a new urban renaissance characterized by economic, cultural, and political preponderance of the city on the rest of the land. Moreover, while the metropolitan expansion between 1950s and 1960s is additive and based on industrial activities expansion, the 1990s city growth is characterized by a more complex phenomenon. In addition to the dismissal of some activities from the city, the re-use of urban areas blowup in an uncontrolled way: generally, their transformation into high quality districts, as well as the activity of promotion and marketing develops, as one of the "creeds" of the recent urban policies.

The paper analyses the transformation processes in the Naples conurbation underlining how much they are closer to the concept of "sprawl" than of "metropolitan area"; this have certainly influenced the poor urban quality of the area and the rearguard position of the city in international urban hierarchies [7].

According to this, the paper proposes an interpretation of the changes in the study area according to two time-series data: the settled population and the rates of urbanized land.

Starting from the above considerations, the paper stresses that it is a key priority to implement effective systems of governance in a largely damaged territory. The challenge is to stop the processes of expansion (if not reverse it) identifying in the already urbanized land and the areas in which to locate – in a smart way – the new developments, with a view to creating a different future for Naples and its conurbation.

The results of the paper should be considered as a base for further research developments, which would also use socio-economic data and building dataset analysis for the definition of a system for spatial transformations control and management. The development of such control techniques, is furthermore supported by the political pressure for the creation of metropolitan cities as administrative subjects, foreseen in recent Italian regulations [8].

2 Methodological Structure of the Process

The paper is organized into two sections: a first analytical part, where the changes of the territory are described, and into a second section, where some actions for a monitoring Naples metropolitan area urbanization are proposed. The two sections are strictly connected.

The analysis is structured into the following steps:

- choice of the study area and splitting it into sub-areas;
- collection and analysis of the time-series population data from 1861 to 2011;

- analysis of the urbanization process within the Province of Naples administrative boundary;
- clarification of the first results;
- identification of the next steps, i.e. support actions for the monitoring of the land use.

The paper is directed to delineate intervention's policies grounded on clear analyses that highlight the underway processes on the basis of effectively observed trends.

3 Naples Metropolitan Area: Characteristics and Boundaries

One of the key features of the urbanization process of Naples metropolitan area is its confused growth. If we analyze the planning tools that have been set up in the past 50 years, it is possible to note that they have almost never turned into effective planning tools [9]. Nevertheless, they contained hypotheses of spatial organization and identified trends that, in an indirect way deeply impacted the operative policies implemented at the regional, provincial and local administrative scales, even if they were not able to guide the urbanization of the metropolitan area in a rational way.

Such situation has influenced the construction of large scale infrastructures, the acceptance of illegal urban developments and uncontrolled urbanization processes. This has been reflected both in the indiscriminate multiplication of volumes and in the transformation of rural land into different uses, not suited to agricultural production.

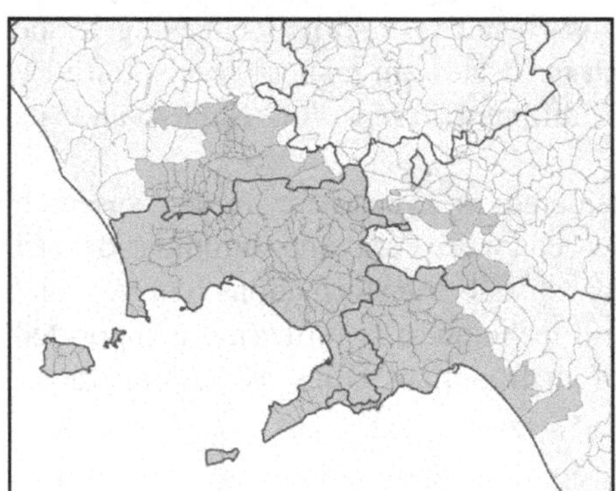
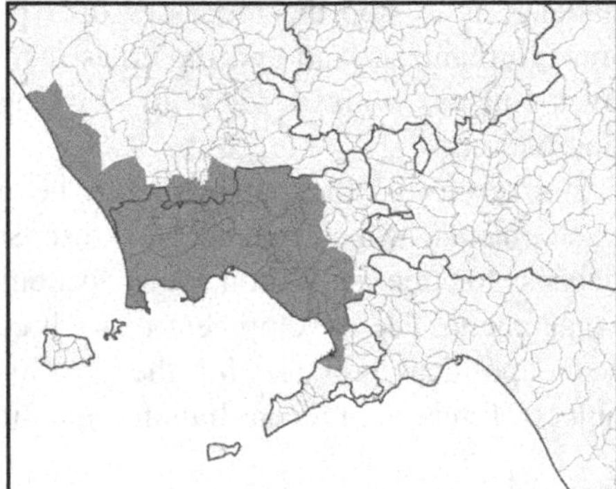

Fig. 1. Proposes of Naples metropolitan area delimitation by SVIMEZ [10] and by the Regional Development Plan [11]

The Naples metropolitan area has undergone profound changes over the last 50 years. It changed from an area characterized by a significant rural matrix in which well-defined urban centres were built, into a deeply urbanized and infrastructured land where you can glimpse large shreds of what was the Campania Plain. Urban centres, in the meantime, have spread linking up in an indefinite conurbation without quality.

A second element to consider is the delimitation of the metropolitan area. The Naples metropolitan area is characterized by a mono centric shape, centered in the city of Naples as main pole of the spatial system. Various attempts have been done in the recent past to define its delimitation. In particular, two hypotheses can be cited.

The first is proposed by a SVIMEZ research [10] that assumes a wide metropolitan area, made by 169 Municipalities belonging to the Provinces of Naples, Caserta, Salerno and Avellino and with an extension equal to the 17% of the Campania Region. In the boundary were also included the towns of Caserta, Avellino and Salerno.

The second boundary is proposed by the Regional Development Plan [11] and supposes a metropolitan area made by some of Municipalities of the Province of Naples and of the South-West Province of Caserta, for a total of 63 Municipalities.

Even today the delimitation of the metropolitan area is not definite from the administrative point of view, even if, with great probability, it will correspond to the boundary of the Province of Naples, as stated in the recent Law nr. 56, issued on April 7, 2014.

4 Population and Urbanized Areas as Key Factors

4.1 Choice of the Study Area and Its Splitting

The proposed delimitations of the Naples metropolitan area reported in Figure 1 are the starting point to define the study area.

The paper assumes an area that is in between the two cited proposals. This choice is not another attempt to define a boundary, but the response to the need of defining a spatial basis in which to deeply analyze the urbanization process. Moreover, a too wide extension includes areas with highly specialized characters and with poor functional relations with Naples. For the opposite reason we think that a boundary that matches with the Province of Naples is highly reductive.

For these reasons the spatial analysis reported in the study starts by choosing a study area consistent with the physical and functional characteristics of the Neapolitan conurbation but without any actual relapse in administrative terms. It is a space that includes 142 Municipalities belonging to the Province of Naples, Caserta and Salerno (Figure 2). These Municipalities were assigned to 6 belts roughly concentric, with Naples as the core pole; on them was carried out an analysis of the changes of the resident population from 1861 to 2011 [12, 13, 14, 15, 16].

The correspondence of the 142 Municipalities to the 6 belts is made according to the criterion of the geographical proximity. Starting from Naples (belt 0), belt 1 contains the closest Municipalities, and so on up to the fifth belt formed by the Gulf of Naples islands and by the more far localities of the Sorrentina Peninsula.

4.2 Evolutionary Trend of the Settled Population

The data underline that the main growth of population occurs from 1936 to 1981, when it almost doubled in absolute value (from 2,122,243 inh to 3,741,661 inh), while in the following decades (1981-2011) the growth is characterized by lower gradients (3,999,784 inh in 2011), with a substantial stasis starting from 2001 (Table 1). In the same period, in Naples (belt 0) the population grew until 1981; afterwards, the trend changes and the population decreases. The belts from 2 to 4, for their part, present a constant population growth, with a stabilization in the more recent periods.

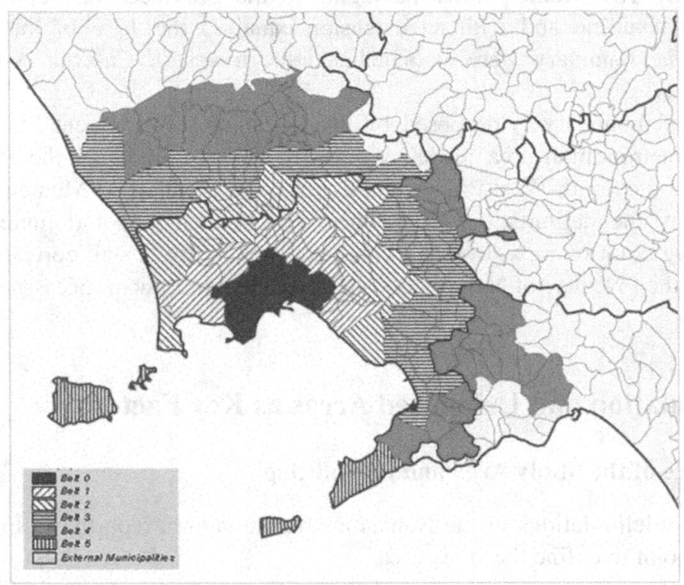

Fig. 2. Naples metropolitan area. Study area for the population analysis

Table 1. Population from 1921 to 2011 in the belts of the study area. The population derives from official Census data.

Belt	1921	1936	1961	1981	2001	2011
0	770,611	865,913	1,182,815	1,212,387	1,004,000	962,003
1	222,246	285,187	449,310	765,106	907,445	905,943
2	210,127	247,677	375,249	526,763	646,692	690,896
3	253,142	295,140	455,515	543,186	605,993	628,762
4	320,273	344,417	474,803	573,742	641,289	666,917
5	86,319	83,909	97,089	120,477	137,973	145,263
Abolished	90,906	0	0	0	0	0
TOTAL	1,953,624	2,122,243	3,034,781	3,741,661	3,943,392	3,999,784

Figure 3 shows the population change in terms of percentage values. The percentage of population living in Naples grows until 1936 (40.80% of the population of the area is resident in the city) decreasing, concurrently, in the external Municipalities. Since 1936 the trend changes and in 2011 the residents rate in Naples is only the 24.05% of the total.

The belt 1 has a similar trend to the 0, although with a temporal delay of about 40 years. In 1901 the percentage of resident in the belt 1 is 11.85%, increased up to 23.01% in 2001. Approximately from this year the percentage begins to descend and in 2011 it falls to 22.65%. It is interesting to underline that this is almost similar to percentage of the Naples inhabitants (24.05%).

Belt 1 includes the Municipalities around Naples, characterized by oldest urbanization and by a growth occurred in less recent times, in the same time of the population growth in the main center. These Municipalities, besides, have absorbed most of the migration flows coming from the inland areas that have been attracted by the coastal areas with a consequent great increase of population. Finally, the belt is characterized by the presence of a series of Municipalities where in the last period the population decreases, comparable to those of Naples.

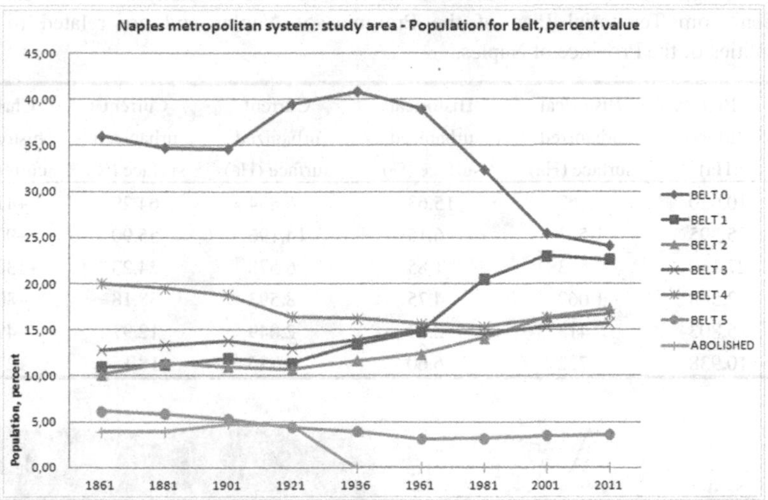

Fig. 3. The population time-series analysis in the Naples metropolitan area. Changes of the population in the six belts. The belt 0 is made by the Municipality of Naples.

It is important to underline that in the period 2001-2011, only the belts 0 and 1 present declining percentages of population; this is valid only for the belt 0 if we consider the absolute values.

Regarding the other belts, the Municipalities belonging to the belt 2 present a steady growth that leads them to reach the value of 17.27% of residents in 2011. The belt 3 contains the 15.72% of residents, while the Municipalities of the belt 4, after a decline rate that has been going until 1981, began slowly to grow getting 16.67%; in a stationary state the resident's percentage in the belt 5 (3.63%).

In the period 1961-2011, the percentage change of resident population in the Municipalities ranging from a maximum positive equal to 604.50% (Castelvolturno, CE, belt 3) to a maximum negative equal to 50.46% (Boscotrecase, NA, belt 3, for change of administrative boundaries); in this period, furthermore, 39 Municipalities have experienced a growth exceeding 100%, and 47 a growth between 50 and 100%.

Therefore, the population decrease of the city of Naples is accompanied by a parallel growing concentrating, in particular, in the Municipalities of the first, second, and third belts.

4.3 Trend of the Urbanized Areas

The process of population dispersal in the study area is confirmed by the data of the urbanized land. In particular for this analysis, the data of the surveys of the Environment Report of the Territorial Plan of the Province of Naples [17] have been used and those were assigned to the 92 Municipalities of the Province and then to the belts (Table 2).

Table 2. Changes of the urbanized areas in the five belts of the study area. The data are an elaboration from Territorial Plan of the Province of Naples and are related to the 92 Municipalities of the Province of Naples.

Belt	Belt surface (Ha)	Historical urbanized surface (Ha)	Historical urbanized surface (%)	Current urbanized surface (Ha)	Current urbanized surface (%)	Changes historical / current (%)
0	10,630	1,662	15.63	6,834	64.29	+411.19
1	25,395	1,563	6.15	14,196	55.90	+908.25
2	27,150	503	1.85	6,578	24.23	+1307.75
3	22,508	1,069	4.75	8,593	38.18	+803.84
4	15,803	415	2.63	2,049	12.97	+493.74
5	10,938	722	6.60	1,985	18.15	+274.93

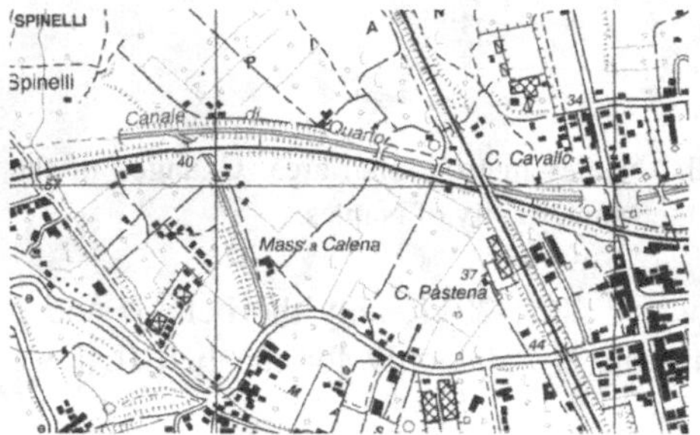

Fig. 4. Example of spatial changes. Municipality of Quarto, Province of Naples. Maps from the site Natura 2000 of the Ministry of the Environment [18]. Original scale 1:5,000

The historical urbanized area represents the urban tissue until 1945. The early 2000s urbanized area includes the residential and productive urbanized areas, the environmental restoration areas and the infrastructural area (ports, railway stations, interchanges poles, etc.). From the analysis, it is evident that the expansion of the urbanized areas in the belts 1, 2, and 3, those closer to the city of Naples, have double or triple values with respect to the central city, representing, therefore, the basin in which the urban sprawl's phenomenon occurred in a more massive way.

According to the same data, some areas located mainly in the North of Naples, in the Phlegraean Fields and in Vesuvius area show percentages of urbanized land over 50%, reaching up to 90% and more.

Figure 3 shows an excerpt of example of the Municipality of Quarto, in the Province of Naples. As it is evident the change from the 1960s to now, it is clear also that it affected residential, commercial and the infrastructural systems.

5 First Results

From the proposed analysis it is noticeable that the diffusion of the population in the study area has caused a redistribution of population's weight from the central pole (Naples) to the Municipalities of the first, second and third belt. This had an impact on the expansion of the urbanized space.

This urbanization, however, did not take place in an orderly way, but with disorder and decay. In particular:

- the diffusion of the urbanized spaces in outer belts with lower density than the ones in central urban areas;
- the widespread lack of planning tools with a consequent fragmentation, heterogeneity and randomness of the distribution of volumes and activities;
- the persistence of illegal buildings construction, as a response to real or supposed delays of land use rules and as a witness of the strong presence of organized crime;
- the mobility networks secondary role, built to respond to an imposed transport demand and not as an occasion for integrated transport and spatial planning;
- the lack of urban and regional facilities;
- the overall lack of building quality.

The impact of these factors on the Naples metropolitan area was devastating. Furthermore the size of the urbanized areas is likely to jeopardize a whole range of environmental activities related to the use of agricultural land.

This makes clear the need for a change in the land use planning in the study area.

6 Governing Spatial Transformations

In the complex and degraded reality of the Naples metropolitan area it is necessary to create innovative planning tools for spatial phenomena smart management.

The proposed process starts by a basic premise, namely the presence of modes and effective operational systems – both on the urban design and planning ground (urban plan) and on the administrative and technical ground – able to act with quick and effective times and reliable results. A proper use of urban planning processes is the way to control the land use changes, because the plan is based on measurable data (the size of the urbanized areas sorted by categories), and on this data it builds forecasts (the amount of new areas to be urbanized). It is obtained, therefore, a reliable datum relating to the possible land consumption. This represents a clear valuable monitoring indicator.

This part of the process is critical because any activity of control and monitoring cannot work without effective plans and without operative structures directly addressed to urban expansions control and to existing tissues regeneration.

Starting by this substantial premise it is possible to argue on the identification of systems, high-tech devices and communication networks that can support a real-time control of the ongoing transformations (Figure 5). This section of the process is based on a clear definition of the data and information systems to gather and to process; it is based also on an advanced architecture (sensors for information collection, transmission networks, data storage, software for the research and characterization) explicitly directed towards a smart control of the spatial transformations [19].

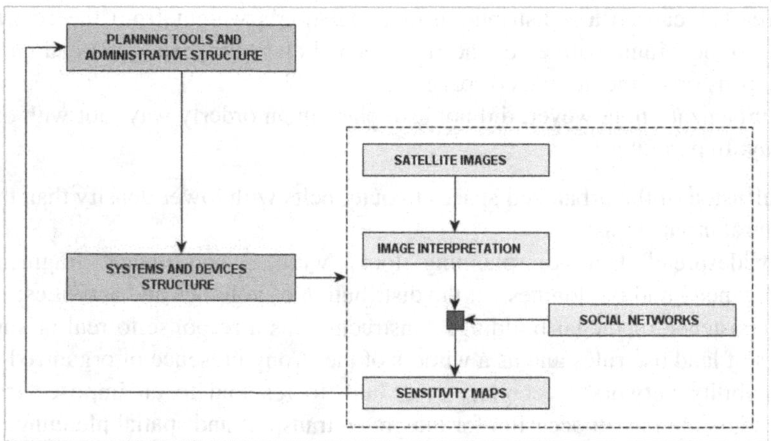

Fig. 5. Structure of the management system for spatial transformations monitoring

To do this, it is possible to define hardware and software systems made by four basic modules that can generate the row data, and the subsequent structured information, concerning the spatial changes.

The first module refers to the collection of the images from satellite systems that can be clear or spectral images, with a view to associating the scale of colors to specific materials [20].

The second module allows the interpretation of the data from satellite. For this purpose mathematical algorithms are applicable for the analysis and control of detectable changes in the same territory in different times.

The third module is based on the dissemination in order to involve the users in reporting the changes taking place. In relation to this module it is necessary to make some considerations.

If the territory has undergone profound degenerative processes, the causes are to be found also in the social acceptance of these phenomena. Such aspect cannot be omitted, as well as cannot be unspoken the phenomena of participation that are present in the area and that seem to have found in the social networks the ideal tools through which fulfill their role. It follows that an intelligent management of the role of the social networks can be an added value in the process of land management, especially if it is possible to structure defined and recognized modes for the reporting of the information, such as to prevent incorrect or prejudicial uses that can affect the rights of the other citizens.

Satellite data, ground processing, integrations of the results with the reports of the social networks lead to the final module in which hardware and software systems allow the construction of maps showing areas with different sensitivity depending on the particular incidence of transformative processes. These maps identify the areas particularly affected by the sprawl phenomena and address in a more direct and effective way the land control and monitoring. Among the sensitive areas are to include those potentially affected by catastrophic events, such as the Vesuvius area, the Phlegraean Fields, and the protected area, relevant from the landscape and environmental points of view.

The modular structure has in the third and fourth modules the points of greater innovativeness, while the two initial modules are more consolidated. The use of the social networks can be a real experiment of participation of citizens who may be interested in participating to programs of land management in which the consequences of social control can be clearly detected.

Therefore, the opportunities inherent in the involvement of social networks in the control of the territory are evident, as well as the danger that such involvement implies, given the importance and sensitivity of the data included into the management system.

The same sensitivity maps can be built with strong dynamic characteristics, with a clear readability and the chance to be easily accessible via Internet, so as to be perceived as a technical product and as a direct result of the participation of the citizens through social networks.

7 Conclusions

The paper considers in detail the spatial changes of the Naples metropolitan area. Discussion has been based on the underlying assumption that in order to deeply understand the phenomena, it has to be analysed using two main time-series data: population and urbanized land.

This has been done for identifying characteristics and ongoing processes. Beyond this, the process has underlined the necessity to identify and discuss other spatial factors, in particular those connected with a continuous control of the spatial transformations.

The Italian spatial planning was marked by a series of failings that can be summarized in the lack of effectiveness of the land use plans. Among the negativities it is to highlight the lack of the definition of the metropolitan areas, both because the importance of the main cities in the economic and social Italian context, and for the relevance of the conurbation's phenomena that is in act around them.

The paper has shown that in the Naples metropolitan area the phenomena can be defined as "expansive" with a lack of metropolisation characters; this have been affected by the population trends and on its distribution between central pole and crown, with clear consequences on the spatial urbanization, on the landscape reduction and on the loss of agriculture land.

A more effective control of these phenomena can achieve a turnaround. The paper argues that this is possible adopting advanced technologies for the spatial control, able to process a large amount of information and to interface with a monitoring function that the citizens can take using with an intelligent and controlled use of the social networks.

In this way it would be possible to have an impact both on the short term and on the longer term processes connected with the metropolitan area spatial planning. It is clear that in both cases the presence of an up to date and complex data system represents a quality leap forward in the implementation of their tasks.

The possible conclusions are:

1. The spatial transformation planning and control of the Naples metropolitan area is a complex process with many potential factors contributing to make it possible.
2. A number of elements (population and urbanized land) have been identified as significant variables for understanding the phenomena.
3. A number of matters require further research and analysis.
4. The quality of the control and management of the urban transformations is fundamental to reach a higher level of spatial quality.

Note

The paper is one of the analytical results of the National Operational Programme (PON) denominated "*SINERGREEN - RES NOVAE - Smart Energy Master per il Governo energetico del territorio*" (PON04a2_E) funded by MIUR (Ministry of Education, University and Scientific Research).

Although the work is unitary, Rocco Papa has developed the sections 1 and 6, Giuseppe Mazzeo the sections 2, 3, 4, 5. Section 7 is to assign the two authors.

References

1. Berry, B.J.L., Goheen, P.G., Goldstein, M.: Metropolitan Area Definition. A Reevaluation of Concepts and Statistical Practice. Working Paper, U.S. Bureau of the Census, U.S. Government Printing Office, Washington D.C. (1968)
2. Marchese, U.: Aree metropolitane in Italia – anni '80 economia e fattori di centralità, trasporti e movimenti pendolari. CEDAM, Padova (1989)
3. Marchese, U.: Aree metropolitane in Italia alle soglie del Duemila. ECIG, Genova (1997)
4. Altshuler, A.: Review of the costs of sprawl. J. of American Inst. of Planners 43, 207–209 (1997)
5. European Environment Agency: Urban sprawl in Europe. The ignored challenge. Technical report, 10 (2006)
6. Kirby, A.: Cities. A Research Agenda for the Close of Century. Cities 12(1), 5–11 (1995)
7. Mazzeo, G.: Impact of the High Speed Train on the European Cities Hierarchy. TeMA Journal of Land Use, Mobility and Environment 3(SP), 7–14 (2010)

8. Mone, D.: Città metropolitana. Area, procedure, organizzazione del potere, distribuzione delle funzioni (April 9, 2014), http://www.federalismi.it/
9. Regione Campania: Piano Territoriale Regionale, http://www.sito.regione.campania.it/PTR2006/PTRindex.htm
10. Svimez: Rapporto, sull'Economia del Mezzogiorno. ESI, Napoli (1981)
11. Regione Campania. Proposta per il Piano Regionale di Sviluppo (Marzo 1, 1990)
12. ISTAT: Comuni e loro popolazione ai Censimenti dal 1861 al 1951. Roma (1960)
13. ISTAT: 10° Censimento generale delle popolazione. Dati sommari per comune. Fascicoli 61-Caserta, 62-Napoli, 65-Salerno. Roma (1965)
14. ISTAT: 12° Censimento generale della popolazione. Dati sulle caratteristiche strutturali della popolazione e delle abitazioni. Tomo 1. Fascicoli 61-Caserta, 62-Napoli, 65-Salerno. Roma (1985)
15. ISTAT: 14° Censimento generale della popolazione e delle abitazioni (2001), http://dawinci.istat.it
16. ISTAT: 15° Censimento generale della popolazione e delle abitazioni (2011), http://dawinci.istat.it
17. Provincia di Napoli: Piano Territoriale di Coordinamento Provinciale. Rapporto Ambientale. Tecnical report (2009)
18. Ministero dell'Ambiente e della Tulela del Territorio e del Mare, Rete Natura (2000), http://www.minambiente.it/home_it/menu.html?mp=/menu/menu_attivita/&m=Rete_Natura_2000.html
19. Papa, R., Gargiulo, C., Galderisi, A.: Towards an Urban Planners' Perspective on Smart City. TeMA Journal of Land Use, Mobility and Environment 6(1), 5–17 (2013)
20. Fiumi, L., Landolfi, M.: Analisi su aree urbanizzate mediante tecniche MIVIS. Applicazione a Pomezia (RM). TeMA Journal of Land Use, Mobility and Environment. 5(1), 49–62 (2012)

ClickOnMap: A Framework to Develop Volunteered Geographic Information Systems with Dynamic Metadata

Wagner Dias de Souza, Jugurta Lisboa-Filho, Jean Henrique de Sousa Câmara, Jarbas Nunes Vidal Filho, and Alcione de Paiva Oliveira

Departamento de Informática, Universidade Federal de Viçosa (UFV)
Viçosa, MG, Brasil
wagnerdiasdesouza@gmail.com, jugurta@ufv.br, jean.camara@ufv.br, jarbasfito@hotmail.com, alcione@dpi.ufv.br

Abstract. Volunteered Geographic Information (VGI) features a specific type of "user-generated content" that involves spatial data. Geobrowsers are environments that present spatial data dynamically and can be accessed through a compatible browser. A Geobrowser can own a collaborative Web system to collect VGI. This paper presents the ClickOnMap, a framework to develop collaborative environments in Geobrowsers in a quick and standardized way. The framework has an architecture to document VGI and supports the generation of metadata following the rules of a dynamic metadata template for VGI called DM4VGI. Thus, the data can be interoperable with other systems that do not use ClickOnMap. CidadãoViçosaMG is a system to collect VGI about Viçosa city that was developed from ClickOnMap. The VGI collected in this system was documented using the DM4VGI template. The paper presents a quantitative analysis of VGI collected and of dynamic metadata generated to assess the usability of ClickOnMap along with DM4VGI.

Keywords: VGI, Geobrowser, framework, dynamic metadata.

1 Introduction

Topographic mapping reached its peak as a government-sponsored activity in the 1950s and 1960s, but has been declining over the years because it is a costly and time consuming activity [7]. Governments have been gradually reducing use and investment in mapping to reduce costs. Researchers and government agencies from many countries seek alternatives to reduce costs and increase production of spatial data [2], [4].

The evolution of Web 1.0 to Web 2.0 allowed the development of Web Collaborative Systems, including systems that use spatial data. A Web phenomenon known as user-generated content [9] is increasing and diversifying the creation of data provided

B. Murgante et al. (Eds.): ICCSA 2014, Part II, LNCS 8580, pp. 532–546, 2014.

via collaborative environments [13], [19]. The free encyclopedia Wikipedia[1], the global volunteer mapping OpenStreetMap[2], Wikimapia[3] and even customer reviews about the quality of a product (e.g. Ebay[4]) are examples of this phenomenon.

Collaboration involving geographic data, so-called Volunteered Geographic Information (VGI) [7], characterizes a specific type of user-generated content. VGI combines three key elements, the Web 2.0 [11], the concept of Collective Intelligence [10] and what has been defined as Neogeography [16].

Geobrowser is a system that allows the user to access, in a simple and intuitive way, a data set and dynamic geospatial information [12]. It also can provide a collaborative module that allows search, access, visualization and integration of geospatial data interactively. However, the lack of standards for collecting and documenting VGI has been a huge barrier to recovery and interoperability of volunteers' data between different Geobrowsers.

By using volunteers interoperable data is possible, for example, to cross data from a system that collects information about infrastructure problems of a city with a system that maps diseases in that city. Based on VGI, the public authority can act using information provided by the population or the population may carry out actions independent of the government. Souza et al. [14] proposed the DM4VGI template for documenting VGI via dynamic metadata, i.e., metadata automatically captured during a voluntary contribution.

This paper presents the ClickOnMap framework designed to enable the rapid development of VGI collection systems, with support for documentation of dynamic metadata using the template DM4VGI. It also presents quantitative analyses of a set of VGI collected in the collaborative Web system CidadãoViçosaMG[5] that was implemented from the ClickOnMap framework.

The rest of the paper is structured as follows: Section 2 lists the top Web map services that currently exist, a review on VGI and discusses some related works; Section 3 introduces the framework ClickOnMap, its functionalities and structure; Section 4 describes a case study developed from ClickOnMap; Section 5 presents the conclusions of the study and proposes future research.

2 Volunteered Geographic Information

VGI, according to Elwood et al. [4], represents a dramatic shift in content, feature and mode of creation of geographic information, as well as sharing, dissemination and use of these data. In a VGI environment, the user can constantly collaborate to increase the database, and help in the collection, validation and analysis of data quality, reducing costs of production and data management.

[1] https://pt.wikipedia.org/

[2] http://www.openstreetmap.org/

[3] http://wikimapia.org/

[4] http://www.ebay.com/

[5] http://www.ide.ufv.br/cidadaovicosa

Citizen participation as a 'human sensor' to collect, edit and update data in a Geo-browser has been discussed and analyzed in the scientific community [1], [7]. These users are also seen as 'helpers for validation of information', as they may provide ratings or reviews of the collaborations, besides describing the experience of VGI post-use [6].

VGI can be used to approximate citizens to political and administrative sectors, for instance, helping to map infrastructure problems or crime occurrences in a city, which is called e-Government [5].

2.1 Web Map Services

There are a number of web map services; Table 1 lists some of them. Google Maps API was chosen as the basis of the framework of this work because of the following factors: (1) greater compatibility with most browsers[6], thus covering a larger number of users; (2) Street View assists in identifying sites at the moment you collaborate; (3) access speed as fast or faster than competitors. Nevertheless, several other Web Map Services on Table 1 can be used in the replication of this work without impairing the results.

2.2 Related Works

Davis Junior et al. [3] proposed a framework for developing collaborative environments to collect and make VGI available in the Web and mobile device applications. The structure includes all components required to implement various VGI applications. The objective of this framework is to reduce the effort of developing and publishing new VGI topics. Therefore, it is possible the creation and the quick customization of VGI environments to collect data from current events.

Another environment available is the Ushahidi Platform [17]. It originated from the site Ushahidi that means 'testimony' in Swahili (official language of Kenya, Tanzania and Uganda). This site used VGI to map, expose and process information of crime, violence and peace efforts during the post-election period in the early 2008 in Kenya, based on reports submitted via the Internet and mobile phones. Later, the site turned into the computing platform Ushahidi Platform that can be used by people worldwide for collection of information, visualization and interactive mapping. It is based on dynamic maps that receive voluntary cooperation through markers associated with information balloons, where the user can click a point on the map and make a contribution, characterizing, in this way, a framework for developing collaborative systems.

These two systems do not provide functions for documentation (metadata) and validation of VGI quality. There are no tools with extra functionality for analyzing volunteer data. They do not provide methods to improve VGI quality and do not allow Wiki revisions on collaborations such as is proposed by this work.

---

[6] http://en.wikipedia.org/wiki/Comparison_of_web_map_services

Table 1. Web Map Services and their APIs

Service Name	Official Website	API and Libraries
Apple Maps	http://www.apple.com/br/ios/maps/	https://developer.apple.com/library/ios/documentation/userexperience/conceptual/LocationAwarenessPG/MapKit/MapKit.html
Bing Maps	http://br.bing.com/maps/	http://www.microsoft.com/maps/choose-your-bing-maps-API.aspx
CloudMade	http://maps.cloudmade.com/	http://blog.cloudmade.com/category/api/
Géoportail	http://www.geoportail.gouv.fr/	http://www.programmableweb.com/api/geoportail
Google Maps	https://maps.google.com.br/	https://developers.google.com/maps/
Map.Geo.Admin.ch	http://map.geo.admin.ch/	http://api.geo.admin.ch/main/wsgi/doc/build/
Mappy	http://en.mappy.com/	http://www.programmableweb.com/api/mappy
MapQuest	http://www.mapquest.com/	http://developer.mapquest.com/
Nokia Here	http://here.com/	http://developer.here.com/
OpenLayers	http://openlayers.org/	http://dev.openlayers.org/apidocs/files/OpenLayers-js.html
OpenStreetMap	http://www.openstreetmap.org/	http://wiki.openstreetmap.org/wiki/Develop
Yahoo! Maps	http://maps.yahoo.com/	http://developer.yahoo.com/maps/

3 The ClickOnMap Framework

The ClickOnMap framework was designed to develop collaborative modules in Geobrowsers based on the Google Maps API. This API provides functions, methods, maps and satellite imagery to assist in the development of a Geobrowser. It has support to the capture of user actions, e.g., to capture latitudes and longitudes in geographic coordinates of points selected by mouse clicks. It has classes of markers, info windows and geometry of point, line and polygon. Thus, it enables the development of the collaborative module.

The ClickOnMap environment is customizable according to the categories and types of data one wants to collect, for example, data on urban issues such as infrastructure, security and entertainment, or environmental issues such as natural disasters, forest fires and floods. The system can accept anonymous or identified collaboration depending on the system policy. For instance, in an environment about crime and anonymous crime tips, the user may want not to be identified, but in an environment that collects data on the positive quality of shops the users can even benefit from their collaboration. Figure 1 shows the structure of the ClickOnMap database.

All VGI captured via ClickOnMap is dynamically documented by the DM4VGI template [15]. Thus, the data collected can be interoperable among different Geobrowsers. For example, every spatial collaboration has its limits documented by a bounding box element present in the DM4VGI. Then, it is possible to perform fast searches based on spatial comparisons. It is also possible to perform temporal or thematic data searches from the DM4VGI elements that allow it, e.g., date and time of the collaboration (or instance) to carry out a temporal comparison over time, and category and type of collaboration to perform a thematic comparison. In addition, from VGI metadata collected by systems that use DM4VGI one can achieve interoperability of VGI.

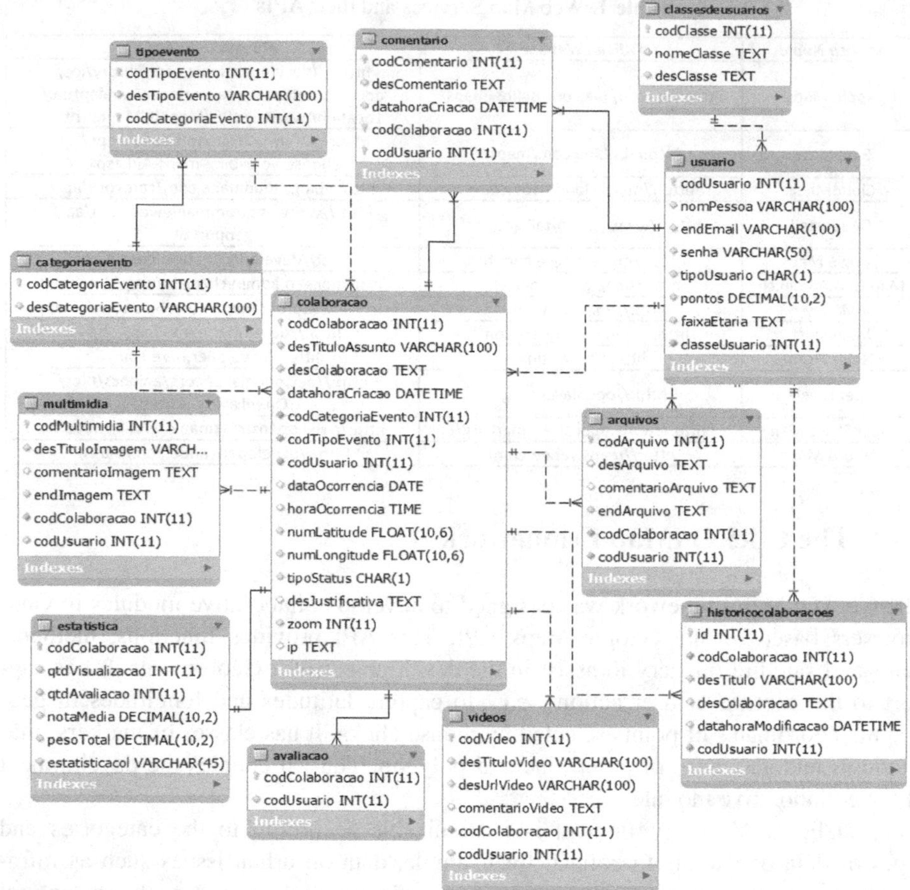

Fig. 1. Structure of the ClickOnMap database

The framework has full support for the automatic capture of all elements of the DM4VGI template. The template values are built from: data present in VGI, e.g., title, date of collaboration, category and type; data provided by Geobrowser functions, e.g., zoom level, user's IP, name of the collected site; data generated from the interaction between the user and the Geobrowser functions, e.g., the level of user ranking at the time the collaboration is performed and the final score of the collaboration.

The main features of the ClickOnMap system are described below.

3.1 User Types

ClickOnMap allows four distinct types of users. Each type of user has a set of privileges and acts differently on the system.

- Unidentified user - a user who does not perform any type of identification and cannot work or evaluate data in the system; user can only view and download the volunteer data.
- Anonymous user - when a user wants to collaborate without directly being identified. The system automatically creates a record of anonymous user with generic name and login: 'anonymousX', where 'X' is an incremental integer (e.g. 'anonymous 77'). Each time a user requests this type of login another record of anonymous user is created. After a logout, an anonymous user (e.g. 'anonymous 77') cannot be reused by another user. For security and political issues, the system captures the Internet Protocol (IP) and time of the user's collaboration, because if the user breaks the law using the system he can be identified by the authorities.
- Registered User – the user who has registered on the website or used a Facebook account or Google to identify her/himself. Only registered users can be notified about the evolution and acceptance of their collaboration.
- Administrator - user responsible for managing the system and collaborations.

ClickOnMap has a 'User Ranking' mechanism in which the more points they have the higher their level will be. Every user has a score and is linked to a reliability class. These classes can provide special privileges to users. Table 2 illustrates an example of how users can earn points. This scoring can be configured in each system.

Anonymous users only accumulate points during the current login session. If a user logged in as anonymous logs out, there will be no point accumulation for him because each time a user makes a login as anonymous, a new anonymous user will be created with an initial default score. Therefore, for a user to level up in the hierarchy she/he must have a registration. Users at different hierarchical levels help other users identify the degree of collaborative reliability because the DM4VGI template has elements that store the reliability class and points of users who created and updated the VGI.

Table 2. Users' points system

User's Action	Points
Registered for the website	+5
Collaboration performed	+10
Comment performed	+5
Evaluated collaboration	+1
Extra files sent	+2 per file
Performed a Wiki review in a VGI	+1
Had a collaboration evaluated	+ (Average or weighted average via user ranking of scores 0-5 of all their collaborations) * (number of reviews)
Made a false collaboration	-15
Made a malicious collaboration	-50
Made a criminal collaboration	Baned

3.2 Wiki Review Functions, Queries and Statistical Analyses

The system allows the collaborations are reviewed and edited by any user who has logged in. This type of review can be called WikiVGI. This method may be one way to improve VGI quality. All changes made in VGI are stored and displayed in a historical format. Thus, if necessary, you can return the collaboration to a previous state. This security procedure is also used on Wikipedia.

When an anonymous user provides a VGI, before this collaboration is made available to the general public, the data are moderated by word filter algorithms and also by administrator users, which becomes a type of control to prevent criminal or malicious collaborations from being disclosed. Registered users do not have their collaborations moderated, i.e., their collaborations are available to the public immediately.

The system has several tools to assist in data analysis and decision making, including spatial data analysis, such as kernel map and visualization with marker clustering, both to identify what are the geographical areas that receive the most specific type of collaboration. There is a function that allows historical data search and also follow the Wiki evolution of a VGI. The environment provides pie charts with data on the percentage of the amount of each type and category of contribution in a particular region of the map, thus enabling to identify the incidence of 'what' is occurring in a region of the map . For example, using this method it is possible to identify the main problems of a neighborhood or a street. There is a visualization based on filters of categories and types, so it is possible to view on the map only contributions from one or more types or categories. The system also features a search on metadata, which may run text, temporal, spatial and thematic searches. Metadata can be downloaded in XML format as shown in Figure 2.


```
▼<marker>
 ▼<identification>
    <title>Semaforo para pedestres nao funciona</title>
   ▼<abstract>
      O semaforo para pedestres nao funciona, forcando quem deseja atravessar a contar com a sorte.
    </abstract>
    <category>Infraestrutura</category>
    <type>Semaforo</type>
    <software>ClickOnMap</software>
    <website>http://www.ide.ufv.br/cidadaovicosa</website>
  </identification>
 ▼<time_record>
    <date_and_time_of_the_contribution>10/09/2013 - 12:48:06</date_and_time_of_the_contribution>
    <date_and_time_of_occurrence>10/09/2013 - 00:00:00</date_and_time_of_occurrence>
    <date_and_time_of_updates>10/09/2013 - 15:48:32</date_and_time_of_updates>
  </time_record>
 ▼<geopositioning>
   ▼<bounding_rectangle>
      <north>-20.757017</north>
      <south>-20.757017</south>
      <east>-42.877377</east>
      <west>-42.877377</west>
    </bounding_rectangle>
    <geometry>Point</geometry>
  </geopositioning>
 ▶<vgi_quality>...</vgi_quality>
 ▶<auditing_and_distribution>...</auditing_and_distribution>
 ▶<multimedia>...</multimedia>
 ▼<metametadado>
    <identifier>cidadao-vicosa-mg-200.235.131.170-12</identifier>
    <date>08-01-2014 - 16:24:58</date>
    <language>portugues</language>
    <charset>ISO-8859-1</charset>
    <standard>DM4VGI</standard>
    <profile>nenhum</profile>
  </metametadado>
</marker>
```


Fig. 2. Example of metadata downloaded

3.3 Validation of VGI Quality

ClickOnMap validates the VGI quality using four methods: (1) VGI scores; (2) users' scores; (3) moderators; and (4) administrators.

Each user can give a score from 0 to 5 for each collaboration, aiming to evaluate the quality or reliability of VGI. The final score of the collaboration is calculated by the weighted average on the users ranking. Thus, each user also has a score.

The more points a user has, the higher her/his hierarchical level in the ClickOnMap users ranking. Users at higher levels in the ranking have greater weight in the final collaboration score than users at lower levels. A user who sends quality collaborations may have them rated with higher grades. Thus, this user earns more points than a user who performs many collaborations rated with low grades.

3.4 Administrative Module

ClickOnMap has an administrative module enabling users who have administrative permissions to configure a new collaborative system, as well as to manage collaborations and dynamically customize the application. Using this module the user can quickly define a new application based on ClickOnMap. It is necessary to define a set of configurations and establish usage policies for each new system. Table 3 shows the elements that need to be configured.

Table 3. Elements of initial application configuration

Information	Item	Element	
Site	1.1	Name	
	1.2	*Email*	*Login*
			Password
	1.3	*Link* to homepage	
Map	1.4	Latitude	
	1.5	Longitude	
	1.6	*Zoom* level	
	1.7	Type of map	
Login	1.8	*Facebook*	
	1.9	*Google+*	
	1.10	Anonymous	

Item 1.1 in Table 3 registers the name of the system that appears as a title on every page of the system. Item 1.2 registers the contact email of the application. This email can be used to notify collaborators about changes in the validation status of their collaborations. Item 1.3 registers the link of the system home page; this link is used to configure the website navigation. Items 1.4 and 1.5 must be filled according to the location where the application is set up, as the map starts with the center at this point. Item 1.6 registers information on the level of the initial map zoom. Item 1.7 registers the type of map, for example, a street map, a satellite image or hybrid. Items 1.8 to 1.10 register the types of optional login the system will have: Facebook, Google or Anonymous. The login from a diary kept in the system will always be available.

When developing a new application, it is also possible to establish the categories and types of collaborations, which are the classes of users and their range of points, and what will be the score policy of the application, i.e., according to each user's action on the system, what will be the amount of points that the user will earn or lose.

The administrative module provides tools to create new users with special permissions and a complete environment for managing collaborations, categories and types. One can also change some rule of the system, for example, modify the range of points in user classes or change the score of each user's action on the system.

The collaboration management enables the validation of a collaboration and write a justification linked to the collaboration that is made persistent in the database. After the positive or negative validation one can send a message to the user who created the collaboration. Figure 3 shows a visualization interface of collaboration to be evaluated and validated by the administrator. Any collaboration can be deleted by an administrator user.

Fig. 3. Assessment and validation environment of collaboration

4 The Collaborative Web System CidadãoViçosaMG

The ClickOnMap framework has been used to develop three VGI applications. The system VGI-Pantanal [15] is an environment to gather information related to the environmental issues of the Pantanal floodplain, located in central-western Brazil. The second system that used ClickOnMap was the system MossoróCrimes [18], aiming to collect voluntary information in the area of public safety of the municipality of Natal, Rio Grande do Norte. The third system, described in detail in this section, was CidadãoViçosaMG, aiming to collect more comprehensive voluntary information on various topics related to Viçosa city, state of Minas Gerais.

Data are collected and stored in a database hosted on its own server, that is, volunteers data are not sent to Google. Figure 4 shows the home page of CidadãoViçosaMG. Figure 5 shows the collaboration page.

The system offers three login options to ClickOnMap: login from the registration made in the system itself; login from Google and Facebook accounts; and login as anonymous user. Despite in this last option the user does not identify herself or himself, her/his Internet Protocol (IP) is captured automatically for safety, but only the system administrators will have access to this information.

Fig. 4. Homepage of CidadãoViçosaMG

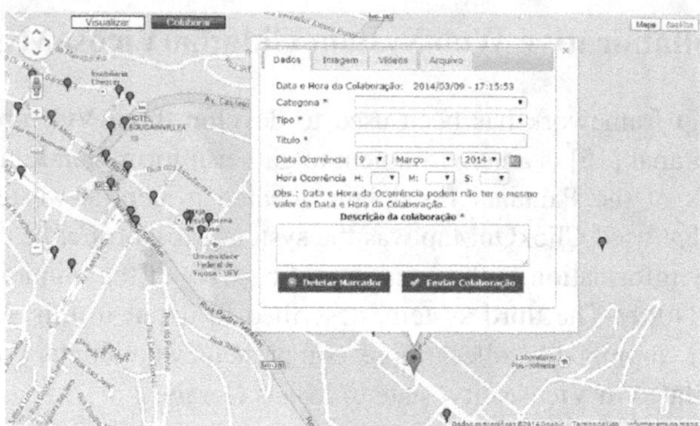

Fig. 5. Collaboration page

When performing a login, the user is allowed to enter a new contribution or edit, review, update and evaluate any collaboration in the system. Each collaboration has a forum. So users can comment and discuss about the veracity or quality of a VGI. All contributions can also be Wiki revised.

The system has filters for categories and types to better visualize the information of interest on the map. The user also has access to a news system to see the most viewed, recent, revised or evaluated collaborations.

Some examples of classes defined in the collaborative environment CidadãoViçosaMG are: Entertainment, Infrastructure, Environment, Health, Safety, Service or Product. Each category can be sorted by a set of more specific types. For example, the category Infrastructure has the types garbage collection, water distribution, crosswalk, street lighting, traffic lights, vacant lots. The category Security has the types burglary, anonymous crime tip, theft, policing, location of drug activity, illegal trafficking, domestic violence, others.

CidadãoViçosaMG provides tools for the user to analyze VGI, for example, it has charts with the percentage of each category and type of collaboration . These graphs can also be generated for a specific area on the map, i.e., one can easily see which categories or types of collaborations that occur in a more specific neighborhood.

In the context of validating VGI quality, the system keeps a ranking of users, VGI evaluation through rating scores provided by users and approval via administrators. Administrators analyze the amount of ratings assigned to the collaboration and its final rating. Then, the collaboration can be validated positively or negatively. To do this, the administrators change the status of the collaboration from 'in review' to 'approved' or 'disapproved'.

Some users at higher hierarchical levels have permissions to approve or disapprove a collaboration. CidadãoViçosaMG has six reliability classes that can provide special permissions to users. Table 4 shows the reliability classes of this system. The scores can be changed according to the evolution of the system and collaborations.

All VGIs are automatically documented by the VGI documentation module based on the template DM4VGI. Therefore, metadata are dynamic, i.e., as VGI is collected or changed, the corresponding metadata are created or updated.

Table 4. Reliability classes of CidadãoViçosaMG

Name of the reliability class	Scores required to belong to the class
Special Collaborator	Over 800
Experienced Collaborator	Between 500 e 799
Master Collaborator	Between 250 e 499
Legal Collaborator	Between 100 e 249
Basic Collaborator	Between 0 e 99
Malicious Collaborator	Less than 0

4.1 Quantitative Analysis of VGI Collected in the System CidadãoViçosaMG

To analyze quantitatively the collaborations to the system CidadãoViçosaMG, a period of three months (from 01/08/2013 to 31/10/2013) was established. A simple disclosure was made through Facebook and in a list of students of the Federal University of Viçosa (UFV). The data for this period are presented next.

According to Google Analytics, 809 visits were performed by 405 different users, i.e., with different IP. The system had 45 collaborations and 100 registered users, with 68 registrations created by 'login as anonymous user' and the remaining 32 users are registered in the system or identified by Facebook and Google accounts. The increased number of 'login as anonymous user' occurred because: (1) users are afraid of being identified, even the system making it clear that the author of the VGI will not be identified by other users; (2) users do not want to waste time performing registration; and (3) users do not want to log in using their Facebook or Google accounts, even the system making it clear that they will not be identified by other users.

The low number of collaborations in relation to the number of accesses may be due to a number of factors: (1) the users do not know exactly with 'what' to collaborate to the system; (2) users had difficulty locating a region of interest; (3) people are afraid of being identified, even the website stating that is anonymous collaboration; (4) users may have the feeling that their collaboration will not be judged and analyzed by authorities, hence the problems will not be resolved from a VGI system; and (5) the user has no time to perform a collaboration.

The Wiki-VGI system was also little used, having only 6 collaborations of the Wiki type in just 3 different collaborations, i.e., 6.7% of contributions had Wiki revision. The reasons may be due to the following factors: (1) the collaborations may be correct in the view of users, hence no need for revision; (2) users did not realize that they could review the collaboration; (3) users identified errors but did not want to inform or did not know how to correct; and (4) users had to provide more information about collaborations but they would not cooperate.

All 45 VGIs were viewed by other users. The total number of VGI visualizations was 1998, that is, on average each of the 45 collaborations were visualized approximately 44 times. Thus, it is possible to verify that users are seeking collaborations more than collaborating.

Regarding the amount of metadata collected by this system, all required elements were captured automatically. All optional elements that are generated from the Geobrowser were captured automatically. All optional elements that have been provided by users were captured automatically. Thus, provided that the information exists in the datum or in the Geobrowser, it is captured automatically. For example, the date of the collaboration occurrence can only be captured if the user has provided this information in the voluntary datum, or, the final VGI score can only be captured if any user has provided the score and if the Geobrowser has a module of assessment to calculate this final score. Therefore, the documentation was carried out from the VGI documentation module and succeeded in all collaborations.

4.2 Visualization and Statistics of VGI

Figure 6 shows three types of maps provided by ClickOnMap. The first presents VGI clusters; the second displays all VGI markers; and the third shows the Kernel Map (radius value = 20 of Google Maps API) collaborations. The highest number of collaborations was performed in the city center. Only 3 collaborations were performed over the campus area of the Federal University of Viçosa. Collaboration only occurred in the urban area of the municipality of Viçosa.

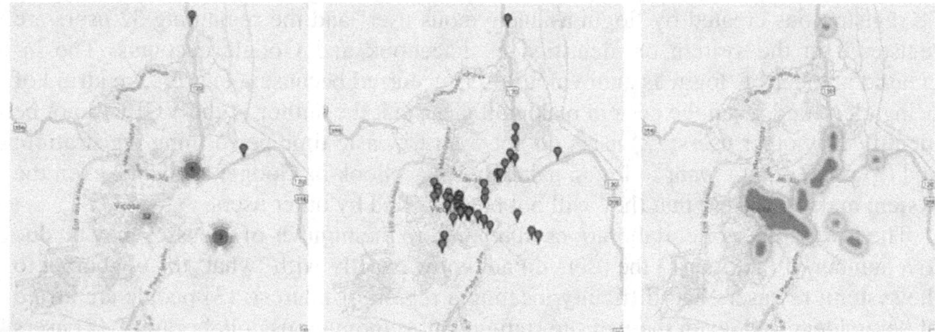

Fig. 6. VGI visualization

Table 5 presents a quantitative summary of collaborations according to the categories with the highest number of collaboration. It shows the total amount of collaborations each category received and the percentage of collaboration of each category relative to the total collaborations. Most collaborations are related to the categories infrastructure and security.

Table 5. Quantitative analysis of data in relation to categories

Category	Amount of collaboration	% in relation to total collaboration
Missing	0	0
Entertainment	4	8.89
Infrastructure	19	42.22
Environment	1	2.22
Health	3	6.67
Security	7	15.56
Service or Product	8	17.78
Other Categories	3	6.67

5 Conclusions

This paper presents the ClickOnMap framework designed to enable quick and easy development of applications in VGI Geobrowsers. ClickOnMap allows quick customization of a collaborative environment defining a region of interest, user roles, collaboration permissions and the categories and types the system will offer. The system is useful, for example, to create a VGI application for supporting disaster management or to make a survey of geographical information on an environmental event which has to be fast responded (e.g., are there or were there fires this week in the region X ?).

The framework has full support to automatic capture of elements of the dynamic metadata template DM4VGI. All data collected by systems using this framework will be interoperable data.

The use of ClickOnMap with DM4VGI in the project CidadãoViçosaMG showed that it is really possible to collect metadata dynamically in collaborative Geobrowsers. All collaborations were properly documented. The VGIs are interoperable because the metadata have a standardized XML structure, which make data both human and machine-readable.

One realizes how difficult it is to obtain a high number of collaborations if the system is not properly disclosed. If the collaborative Web system has an interface with the possibility of several login options, one can increase the number of collaborations. Tools for both data analysis and user feedback help attract more collaborative users.

The participation of public sectors in managing the Geobrowser is a theme to be explored in future research. For example, collaborations regarding problems or lack of infrastructure or complaints regarding public safety need response from these agents, which could be one more attraction for more users to collaborate positively in the system.

Acknowledgements. Project partially funded by FAPEMIG, CNPq and CAPES and the companies Sydle and Gapso.

References

1. Cooper, A.K., Coetzee, S., Kourie, D.: Perceptions of virtual globes, volunteered geographical information and spatial data infrastructures. Geomatica 64(1), 73–88 (2010)
2. Cooper, A.K., et al.: Challenges for quality in volunteered geographical information. In: Proceedings of Africa GEO 2011, Cape Town, South Africa (2011)
3. Davis Junior, C.A., Vellozo, H.S., Pinheiro, M.B.: A Framework for Web and Mobile Volunteered Geographic Information Applications. In: Brazilian Symposium on Geoinformatics. Campos do Jordão (SP). Proceedings, vol. 14. Instituto Nacional de Pesquisas Espaciais (INPE), São José dos Campos, SP (2013)
4. Elwood, S., Goodchild, M.F., Sui, D.Z.: Researching Volunteered Geographic Information: Spatial Data, Geographic Research, and New Social Practice. Annals of the Association of American Geographers 102(3), 571–590 (2012)

5. Furtado, V., et al.: Collective intelligence in law enforcement–The WikiCrimes sys-tem. Information Sciences 180(1), 4–17 (2010)
6. Georgiadou, Y., Bana, B., Becht, R., Hoppe, R., Ikingura, J., Kraak, M.J., Lance, K., Lemmens, R., Lungo, J., Mccall, M., Miscione, G., Verplanke, J.: Sensors, empower-ment, and accountability: a Digital Earth view from East Africa. International Journal of Digital Earth 4(4), 285–304 (2011)
7. Goodchild, M.F.: Citizens as Voluntary Sensors: Spatial Data Infrastructure in the World of Web 2.0. International Journal of Spatial Data Infrastructures Research 2, 24–32 (2007)
8. Goodchild, M., Linna, F.: Assuring the quality of volunteered geographic information. Spatial Statistics 1, 110–120 (2012)
9. Hollenstein, L., Purves, R.: Exploring place through user-generated content: Using Flickr tags to describe city cores. Journal of Spatial Information Science (1), 21–48 (2013)
10. Lévy, P., Bonomo, R.: Collective intelligence: Mankind's emerging world in cyberspace. Perseus Publishing (1999)
11. O'Reilly, T.: What is Web 2.0: Design Patterns and Business Models for the Next Generation of Software, Disponível (2005), http://www.oreillynet.com/pub/a/oreilly/tim/news/2005/09/30/what-is-web-20.html
12. Schrader-Patton, C., Ager, A., Bunzel, K.: GeoBrowser deployment in the USDA forest service: a case study. In: International Conference and Exhibition on Computing for Geo-Spatial Research & Application, Washington, Anais, vol. 1, ACM Digital Library, New York (2010)
13. Shriver, S.K., Nair, H.S., Hofstetter, R.: Social ties and user-generated content: Evidence from an online social network. Management Science 59(6), 1425–1443 (2013)
14. Souza, W.D., Vidal Filho, J.N., Ribeiro, C.A.A.S., Lisboa Filho, J., Oliveira, D.F.: Informação Geográfica Voluntária no Pantanal: um sistema Web colaborativo utilizando a API Google Maps. In: Simpósio de Geotecnologias no Pantanal, Bonito, MS. Proceedings of Simpósio de Geotecnologias no Pantanal (GeoPantanal), vol. 4, pp. 763–772 (2012)
15. Souza, W.D., LisboaFilho, J., Vidal Filho, J.N., Câmara, J.H.S.: DM4VGI: A template with dynamic metadata for documenting and validating the quality of Volunteered Geographic Information. In: Simpósio Brasileiro de Geoinformática (GEOINFO). Campos do Jordão-SP. Proceedingsof São José dos Campos, vol. 14, pp. 1–12. MCT/INPE (2013)
16. Turner, A.: Introduction to Neogeography. O'Reilly Media, Inc. (2006)
17. Ushahidi: Open source software for information collection, visualization and interactive mapping, http://ushahidi.com/
18. Vidal-Filho, J.N., Lisboa-Filho, J., de Souza, W.D., dos Santos, G.R.: Qualitative Analysis of Volunteered Geographic Information in a Spatially Enabled Society Project. In: Murgante, B., Misra, S., Carlini, M., Torre, C.M., Nguyen, H.-Q., Taniar, D., Apduhan, B.O., Gervasi, O. (eds.) ICCSA 2013, Part III. LNCS, vol. 7973, pp. 378–393. Springer, Heidelberg (2013)
19. Yildirim, T.P., Gal-Or, E., Geylani, T.: User-Generated Content in News Media. Management Science (2013)

Government Tools for Urban Regeneration: The Cities Plan in Italy. A Critical Analysis of the Results and the Proposed Alternative*

Antonio Nesticò and Gianluigi De Mare

University of Salerno, Department of Civil Engineering
Via Giovanni Paolo II, 132 - 84084 Fisciano (SA) - Italy
{anestico,gdemare}@unisa.it

Abstract. The study discusses the application of a national law in the period 2012/13 aimed at stimulating the aggregation of entities – public and private – involved in the regeneration of deprived urban areas. With limited funding, little more than €300 million, the Italian State managed to entice local governments into the coagulation of initiatives to improve the quality of urban life. More than 400 proposals were received and evaluated by the institutional Control Room, resulting in a classification that awarded 28 municipalities. This paper analyses the results and proposes a multi-criteria methodology and alternative rationale to prioritise the projects, highlighting the consistencies and weaknesses of the management approach.

Keywords: economic evaluation of projects, multi-criteria techniques, urban regeneration.

1 Introduction

The *National Cities Plan*, pursuant to art. 12 of l.d. 83/2012 issued by the Italian government and re-proposed in September 2013, has the objective of coordinating interventions in urban areas related to new infrastructure, car parks, housing, schools and, more generally, to the redevelopment and enhancement of the inhabited fabric, with particular reference to degraded areas, through the use of economic resources, from both economies or the revocation of building programmes that are no longer feasible, as well as the activation of strong public-private synergies.

For the implementation of the interventions, the decree provides for the setting up of a fund, called *Fund for the implementation of the National Cities Plan*, with €224 million in the first edition in 2012. Approximately €95 million has been made available from other resources.

The process put in place by the government is, therefore, based on an informal bidding procedure through which to select proposals with the objectives of the Plan and

* This paper is to be attributed in equal parts to the two authors.

B. Murgante et al. (Eds.): ICCSA 2014, Part II, LNCS 8580, pp. 547–562, 2014.

the aim of identifying projects that will attract additional sources of funding in order to respond to the real needs of the area as well as improve the urban quality.

457 proposals were submitted by the municipal authorities to the ANCI (National Association of Italian Municipalities), for the year 2012, with 28 being selected for funding: a relatively low number, but compatible with the modest government resources made available to the Plan.

On the basis of a large dataset, created through a study of the summary reports of all the 457 proposals (thanks to eng. M. Marra for making the data available), the initiatives submitted to the Ministry are commented upon in this paper and a multiparametric model for the economic evaluation of the proposals developed [2-3-5-6]. It is based on the formalization of an objective function that allows to assign a score to each project initiative between 0 and 1, which is useful in drawing up a temporal priority list. For the elaboration of the model, it is essential to make use of coefficients that can interpret the selection criteria mentioned in the decree. Each criterion is assigned a weight p.

In the selection phase, the institutional Control Room favoured the proposals from the municipalities that were either a regional or provincial capital, especially those with a high number of inhabitants. In order to consider this additional factor, it is necessary to implement an additional parameter to the objective function of the model.

The two models highlight the similarities and contradictions with the selective logic adopted by the institutions.

2 Selection Criteria for the Selection of the Projects

The Control Room at the MIT (Ministry of Infrastructures and Transport) made determinations on the classification to be drawn up for the projects presented by a simple majority, *on the basis of the following criteria*:

a) immediate possibility to realise the interventions;
b) ability and way to involve public and private entities and funding and implementation of a multiplier effect of public funding towards private investment;
c) reduction of the phenomena of housing pressure, marginalization and social deprivation;
d) improvement of infrastructural facilities also with reference to the efficiency of the urban transport systems;
e) improving the quality of urban life, the social and environmental fabric and containment of new undeveloped land (Article 12 of Decree 83/2012).

The subjects of the redevelopment interventions were predominantly brownfield sites and degraded areas/neighbourhoods. The interventions also included the realisation of complementary and integrated structures of different types and intended use, consisting mainly of the construction of residential areas (public and private); improvement of infrastructural facilities (intermodal traffic, car parks, public lighting, sewers, parks and public furniture); upgrading of school facilities (construction of schools as well as functional and energetic requalification), consolidation of economic, cultural and social activities (development/ requalification of commercial areas; areas of historical/architectural importance; sports and social centres). In addition, in several cases, the interventions proposed by the municipalities were not located in a single circumscribed area but were integrated into a larger redevelopment plan with specific goals and/or in support of certain social categories. For example, there was the redevelopment of existing public schools or residential buildings, which meet the requirements set out in the comprehensive plans of support to children, young couples, families and immigrants. Finally, there were interventions located in distinct urban areas but connected to a single plan for the improvement and enhancing of a strategic axis of urban roads (www.mit.gov.it).

3 The Projects Presented

The ANCI, which was assigned the role of first examiner of the *Cities Plan* projects, concluded the drawing up the classification on November 19, 2012 subsequently presenting all the documentation to the MIT.

On the deadline of October 5, 2012, 457 proposals were submitted by municipalities to the ANCI, more than expected. Figure 1 shows the number of proposals submitted by each Italian region.

Fig. 1. Number of proposals presented by each individual region (www.mit.gov.it)

Only 5% of the total number of the municipalities presented proposals. However, they represent more than 22 million people (more than a third of the Italian population) and of these 22 million, the majority reside in the South (27%), with the Centre and the North-West also having a significant percentage (25% and 22% respectively). Furthermore, 44 % of the projects came from towns in the South, with this reaching 55% if taking into account the proposals from the islands.

Of the 457 proposals, 15 included metropolitan cities (Turin, Milan, Venice, Trieste, Genoa, Bologna, Florence, Rome, Naples, Bari, Reggio Calabria, Palermo, Messina, Catania, Cagliari), constituting 3.5% of the all the municipalities interested and almost 17% of the planned investment. In addition, 92 proposals came from cities with more than 50 thousand inhabitants (21.2%), with a proposed investment of € 8.3 billion, with the two categories therefore accounting for 61.8% of the planned investment.

The concentration of potential investments in municipalities with a population between 15,000 and 50,000 people (25.1%) is also significant.

The metropolitan cities (42.3%) and municipalities with more than 50,000 residents (37.2%) represent the majority of the population affected by the available interventions, for a total value of 79.5%.

More than half of the proposals submitted relate to the re-use of existing buildings (50.9%), 30.2% of the total investments. There are also those that provide for the construction of new facilities (36.5%), based on most of the investments that can be activated (44.6 %). Only a small part of the proposals received are projects related to both these types of interventions (11.6%), with investments amounting to 24.6%. Table 1 lists the types of interventions in more specific detail for each type/intended use/structure, the number of proposed interventions and investments. The highest percentage of potential investments is for projects relating to multiple functions/intended uses (53.5%). However, it also worth noting those that relate to the construction of covered public offices and/or public use (11.4%) and systems for accessibility and viability (10.8%), followed by social housing (housing with rent control, mostly through private investment with a substantial public contribution) with 6.5%.

Finally, the interventions are principally located in historic centres (39.1%) and consolidated cities (30.2%), highlighting how the urban centres have significant potential to become a major engine for development and economic growth, at a time of profound crisis. The suburbs (13.6%), neighbourhoods (8.7%) and areas of expansion (3.8%) make up the minorities.

As previously stated, the projects involved investments worth approximately €18.5 billion. These investments were already covered by resources for approximately €8 billion, while it will necessary to raise €10.5 billion from public and private resources.

Table 2, below, highlights how the intervention categories with the highest proportion of resources to be found in the total relate to the following sectors: commercial (80%), tourism and hospitality (76%), private directional (76%) and covered public office and/or public use (73%).

Table 1. Number of proposed interventions and investments that can be activated for every type of intervention/intended use/structure (ANCI data)

Type function/intended use/structure	*N. Proposals*	*% of the total*	*Total investments*	*% investments*
Free residences	63	1,9%	€ 469.096.205	2,5%
Erp (only subsidised)	40	1,2%	€ 272.632.477	1,4%
Social housing	183	5,6%	€ 1.229.871.611	6,5%
Commercial	53	1,6%	€ 358.837.171	1,9%
Tourism/hospitality	81	2,5%	€ 282.026.227	1,5%
Private directional	10	0,3%	€ 39.860.291	0,2%
Offices and/or covered public structures and/or public use	556	17,1%	€ 2.140.136.195	11,4%
Accessibility and viability systems	612	18,9%	€ 2.039.983.665	10,8%
Squares, street furniture and public facilities discoveries, urban network and sub-services	375	11,6%	€ 662.839.452	3,5%
Sport centres	82	2,5%	€ 163.106.997	0,9%
Structures and installations aimed at improving energy and environmental efficiency	104	3,2%	€ 391.468.890	2,1%
Other	153	4,7%	€ 616.896.788	3,3%
Multiple interventions (with more functions/intended use/structures)	904	27,9%	€ 10.085.601.039	53,5%
Non specified	28	0,9%	€ 94.480.864	0,5%
Overall total	3.244	100,0%	€ 18.846.837.869	100,0%

Table 2. Resources already found and to be found for each type of intervention, with % of resources to be found on the total (ANCI data)

Type function/intended use/structure	*Resources already found*	*Resources to find*	*% resources to be found on investments*
Free residences	€ 314.905.847	€ 154.190.358	33%
Erp (only subsidised)	€ 106.421.350	€ 166.211.126	61%
Social housing	€ 669.188.630	€ 560.682.981	46%
Commercial	€ 71.047.042	€ 287.790.129	80%
Tourism/hospitality	€ 67.723.795	€ 214.302.433	76%
Private directional	€ 9.474.048	€ 30.386.243	76%
Offices and/or covered public structures and/or public use	€ 569.347.234	€ 1.570.788.961	73%
Accessibility and viability systems	€ 694.672.238	€ 1.345.311.427	66%

Table 2. (*continued*)

Squares, street furniture and public facilities discoveries, urban network and sub-services	€ 256.331.652	€ 406.507.800	61%
Sport centres	€ 65.085.121	€ 98.021.876	60%
Structures and installations aimed at improving energy and environmental efficiency	€ 140.575.653	€ 250.893.237	64%
Other	€ 117.565.303	€ 499.331.484	81%
Multiple interventions (with more functions/intended use/structures)	€ 5.043.229.132	€ 5.042.371.907	50%
Non specified	€ 29.706.042	€ 64.774.822	69%
Overall total	€ 8.155.273.086	€ 10.691.564.784	57%

4 The Government's Evaluation of the Proposals

After the preparatory stage, all the proposals were graded by the Control Room according to three degrees of priority (high, medium, low), based on compliance with the objectives of the Plan. At this point, a definitive list was not drawn up with a score for each proposal being assigned, but rather a qualitative comparison between the proposals themselves was carried out, dividing them using the aforementioned degrees of priority.

To divide the eligible proposals into three separate priority orders, considering the large number of projects received, "it was necessary to adopt a criterion for determining the order of presentation of the projects and then proceed to the stage of qualitative assessment provided by the law. On the basis of this objective necessity, the MIT proposed the criterion resulting from the funnel application of the following selection parameters, which were used to decide the order of presentation of the projects to the Control Room. The path involved the following steps:

- identification of the proposals for which the ratio between existing resources (both public and private) and the total investment is more than 15%;
- identification of the proposals for which the ratio between public resources to be found (from the MIT or other public entities) and the total investment is less than 60% (overall average found in relation to all the proposals submitted);
- verification of the municipalities included in the CIPE deliberation dated 13/11/2003, by means of which the regions have formulated lists of densely populated cities.

In order to avoid discarding some interesting and particularly significant proposals, the municipalities that had the following parameters simultaneously were also considered:

- existing financial resources above 15% of the total investment;
- Municipalities listed in the CIPE resolution of 2003;
- provincial capitals (www.anci.it).

Secondary criteria used for selection were also the immediate possibility to realiase in relation to increased investments that can be activated, the contrast to the social degradation, improving the quality of urban infrastructure and the relevance of the economic impact generated in the territories of the interventions and their environmental sustainability.

At the beginning of 2013, the National Cities Plan definitively entered into full operation. 28 proposals were selected for funding in the last meeting of the Control Room, held in mid-January of 2013: a relatively low number, but compatible with the modest government resources made available for the Plan. The cities selected were: Milan, Genoa, Rome, Taranto, L'Aquila, Trieste, Venice, Verona, Turin, Settimo Torinese, Pieve Emanuele, Pavia, Reggio Emilia, Bologna, Rimini, Florence, Ancona, Naples, Foligno, Eboli, Bari, Lecce, Potenza, Matera, Lamezia Terme, Erice, Catania and Cagliari. Additionally, Naples, Lamezia Terme, Catania, Erice and Taranto, were also be assigned integrative funds.

The Control Room selected the proposals favouring those able to generate a greater investment volume due quickly realisable interventions. More precisely, of the 457 proposals received, only 124 met the minimum requirements, which were:

- percentage ratio of existing resources (both public and private) and the total investment in excess of 15 %;
- ratio between public resources to find and the total investment less than 60% ;
- the presence of the municipalities in the densely-populated lists included in the CIPE deliberation dated 13/11/2003.

Of the 124 proposals mentioned above, which the Control Room assigned a high priority, it was decided to finance only 28, due to the limited resources available (just over €300 million), keeping in mind that, of these resources, approximately €95 million can be used only for the proposals relating to the integration funds.

Part of the 28 winning projects, 9 involved property owned by the State, the State Property Management Agency, which, in conjunction with local authorities, will start the processes of development and recovery of the same and will endeavor to implement policies for access to European cohesion funds for the redevelopment of urban areas. These state owned assets, involved in the projects eligible for funding are located in the cities of Ancona, Bari, Cagliari, Florence, Lecce, Naples, Rimini, Trieste and Venice (www.italiaoggi.it).

Table 3 shows the number of proposals selected for each region and the value of the investments planned, specifying the contribution of the Control Room.

Table 3. The proposals selected for each Region, the planned investments and the Control Room contribution (www.mit.gov.it)

Region	*Number of proposals to be funded*	*Planned Investments (in millions €)*	*Control Room contribution (in millions €)*
Abruzzo	1	37,1	15
Basilicata	2	67,9	21,2
Calabria	1	51,9	30
Campania	2	144,1	25,2
E. Romagna	3	348,2	29,4
F. Venezia G.	1	10,8	4
Lazio	1	113	12,9
Liguria	1	221	25
Lombardia	3	98,8	19,7
Marche	1	66,4	8,8
Piemonte	2	557,8	16,9
Puglia	3	419,6	40,5
Sardegna	1	111,1	11
Sicilia	2	138,6	20
Toscana	1	467	14,7
Umbria	1	90,3	6,6
Veneto	2	1.428,4	17,7
Total	*28*	*4.372*	*318,6*

5 A Multi-criteria Model for the Economic Evaluation of the Proposals

The multidimensional evaluations rationalise the choices driven by conflicting objectives. These assessments are conducted from the public point of view, for which the objectives to be achieved are numerous, heterogeneous and, consequently, conflicting. With the multi-dimensional assessments "the heterogeneous aspects of the different alternatives or different implementation strategies in predictions and opinions on the level of well-being of a community are transformed, on the basis of logic diagrams that highlight the differential advantages and disadvantages compared to predetermined objectives" [4-8].

The multidimensional evaluations, due to providing the results through a quali-quantitative profile , allow for a complex assessment of the projects, namely an evaluation that includes both the commercial components as well as those extra-market of the overall value, which takes into account the different categories of users – direct, indirect, potential, present and future – and the overall capacity of the projects to meet the diverse needs of man, in relation to the different dimensions in which they occur, economic, but also social, cultural and environmental.

On the basis of a large dataset constructed through a study of the summary reports of all the 457 proposals, a multi-parametric model for the economic evaluation of the same was developed, based on the formalisation of an objective function [9] that allows to assign each project initiative a score between 0 and 1, indicative of their ability to pursue various criteria. Thus, it is possible to easily perform comparisons and rank the proposals in descending order, starting with the most deserving, according to a classification of temporal priority for funding.

5.1 Construction of the Objective Function and Application to Case Studies

For the elaboration of the model, it was decided to follow the same selection criteria cited by l.d. 83/2012.

Thus, the parameters used for the interpretation of the criteria of the l.d. 83/2012 are:

1. immediate possibility to realise;
2. capacity to involve public and private funding;
3. reduction of the phenomena of housing pressure, marginalization and social degradation;
4. improvement of infrastructures;
5. improvement of the urban quality of the social and environmental fabric.

Furthermore, an additional parameter, not present in the reference l.d., was used, making it possible to consider the design, documentation and synthesis scheme deficiencies of the proposals. In fact, the documentary quality of the proposals influenced the choice of the projects to be funded.

Each criterion is assigned the same weight p, corresponding to the ratio of 1/6 = 0.167 (where 6 is the number of criteria considered), with the exception of criteria 3 and 5, represented by a single parameter with a double weight equal to 0,167×2 = 0,334.

The first criterion is represented by a coefficient C1 equal to 0 in the case of not immediate realisable and 1 on the contrary:

C1 = 0 Proposal not immediately realisable;
C1 = 1 Proposal immediately realisable.

The second criterion is represented by a coefficient C2, between 0 and 1, equal to the ratio between existing financial resources and total resources, the latter being the sum of the resources already existing and those to be found, both in the Cities Plan as well as from public or private sources:

$$C2 = \text{Existing financial resources/Total resources}; \quad 0 \leq C2 \leq 1.$$

As previously stated, criteria 3 and 5 are considered through a single coefficient C (3-5), with a double weight, the ratio between the m^2 of the total area of the intervention provided for each proposal and the number of inhabitants of the proposing municipality (according to data from Istat updated to 01/01/2013). This parameter is always between 0 and 1 because, through a normalization procedure, it is proportionate to the maximum value of the ratio, which is set equal to 1 (essentially, each value of the ratio is divided by the maximum value):

$$C(3\text{-}5)_{\text{not normalized}} = \text{m}^2 \text{ of the total area of intervention/ number of inhabitants of the Municipality}$$

$$C(3\text{-}5) = C(3\text{-}5)_{\text{not normalized}}/C(3\text{-}5)_{\max} \quad \text{with } 0 < C(3\text{-}5) \leq 1.$$

The fourth criterion is represented by a coefficient C4, the ratio between surface m^2 for the construction work or re-use of the infrastructure/car park/services related to the transport sector and the number of inhabitants of the proposing municipality (according to Istat data updated to 01/01/2013). This coefficient is always between 0 and 1 because, through a normalization procedure, it is proportionate to the maximum value of the ratio, which is set equal to 1 (each value of the ratio is divided by the maximum value):

$$C4_{\text{not normalized}} = \text{m}^2 \text{ for interventions on the infrastructures-car parks/number of inhabitants of the Municipality}$$

$$C4 = C4_{\text{not normalized}}/C4_{\max} \quad \text{with } 0 \leq C4 \leq 1.$$

Criterion 6 is represented by a coefficient C6, between 0 and 1, which simultaneously takes into account the quality of the proposal (lack of design, documentation and diagrams in the summary report) and the presence or absence of a feasibility study/financial analysis/cost benefit analysis of the economic impacts:

$$0 \leq C6 \leq 1.$$

In detail, the value of C6 has been assumed, according to the different cases, as specified in Table 4.

Table 4. Values assumed by criterion C6

lack of documentation and diagrams in the summary report	*feasibility study/financial analysis/cost benefit analysis of the economic impacts*	*C6*
no	yes	1
no	no	0,9
in part	yes	0,6
in part	no	0,5
yes	yes	0,1
yes	no	0

It should be noted that, with the same degree of lack of project documentation, the difference between the number of C6 presence or absence of a feasibility study/Cost Benefit Analysis/analysis of economic impacts is always equal to 0.1. This is an indication that it is intended to confer a modest weight to this type of analysis because, in almost all of the proposals in which it is present, it is not a rigorous examination of the economic and financial aspects but provides limited information, with little significance.

Thus, the objective function is defined as follows :

$$\text{function value} = p\times C1 + p\times C2 + 2\times p\times C(3\text{-}5) + p\times C4 + p\times C6.$$

Table 5 shows the results obtained from the application of the model, specifying the name of each proposing municipality, its position in the final standings, here revised, and the indicative numerical value attributed to the technical and economic evaluation of the proposal. Understandably, the 28 proposals selected by the Government are shown.

Table 5. The reclassification of the proposals funded by the Government

n°	*Municipality*	*Position*	*Proposal Evalutation*
1	Erice	10	0,4750
2	Pieve Emanuele	20	0,4453
3	Eboli	24	0,4365
4	Matera	44	0,4088
5	Trieste	60	0,3857
6	Lamezia Terme	67	0,3822
7	Roma	84	0,3598
8	Pavia	165	0,3118
9	Verona	168	0,3067
10	Bari	174	0,2973
11	Settimo Torinese	175	0,2961
12	L'Aquila	181	0,2844
13	Rimini	185	0,2797
14	Potenza	198	0,2714
15	Catania	203	0,2636
16	Foligno	209	0,2616
17	Napoli	235	0,2365
18	Genova	241	0,2314
19	Firenze	250	0,2272
20	Ancona	256	0,2232
21	Lecce	274	0,2089
22	Taranto	280	0,2055
23	Cagliari	284	0,2034
24	Milano	308	0,1861
25	Reggio nell'Emilia	312	0,1854
26	Bologna	324	0,1781
27	Torino	340	0,1647
28	Venezia	392	0,0598

In the analysis of the results obtained, it appears that 16 proposals funded by the government are in the first half of the rankings resulting from the application of the objective function to 5 parameters .

Considering the top 100 proposals of the classification, it can be observed that 10% of them come from cities with a population greater than 100,000 inhabitants, while 15% of the municipalities with a population between 50,000 and 100,000 inhabitants, 27% of municipalities with a population between 50,000 and 15,000 inhabitants, 26% of municipalities with a population between 5,000 and 15,000 inhabitants and 22% from small towns (a population less than 5,000). In addition, 22% of them are projects presented by a provincial capital (Matera, Trieste, Rome, Chieti, Brindisi, Trani, Andria, Caserta, Latina, Verbania, Nuoro, Novara, Lucca, Cuneo, Benevento, Vicenza, Frosinone, Parma, Alexandria, Sassari, Bergamo and Siracusa).

Almost half of these 100 proposals (49%) come from the South, significantly higher than the projects from other national geographical areas. In fact, the most representative cities from this point of view are those of Campania (18 proposals out of 100) and those of Puglia (18 proposals out of 100).

5.2 Introduction of an Additional Parameter to the Model

In the selection of projects for funding, it should be noted that the Control Room wanted to fairly distribute throughout the whole country the financial resources made available by the MIT. In fact, there is a maximum of 3 funded proposals per Region (this is the case of Emilia-Romagna, Lombardy and Puglia). This equitable distribution among the various regions was rightly wanted in order to revive all the national geographical areas in the same way, trying not to create more economically disadvantaged areas than others.

Moreover, proposals were selected from the municipalities that are the either a regional or provincial capital, especially in the case in which they had a large population. Only in a few cases, were proposals preferred, although not from provincial or regional capitals, that involved substantial investments capable of stimulating the local economy and subsequently the national one.

In order to consider this additional factor, it was necessary to create a second model, adding an additional coefficient C7 in the objective function. This new parameter is assumed to be equal to 1 in the case of a provincial capital and 0 on the contrary:

$$C7 = 0 \text{ Municipality not a Provincial Capital;}$$
$$C7 = 1 \text{ Municipality a Provincial Capital.}$$

Ultimately, the objective function is set as follows:

$$\text{value function} = p\times C1 + p\times C2 + 2\times p\times C(3\text{–}5) + p\times C4 + p\times C6 + p\times C7,$$

where p is the ratio 1/7 = 0.143 (this time 7 is the number of criteria considered).

In summary, Table 6 shows only the municipalities financed, their position in the final classification and the indicative numerical value of the technical and economic evaluation of their proposals.

Table 6. The reclassification of the proposals financed by the Government according to the model with parameters

n°	*Municipality*	*Position*	*Proposal Evaluation*
1	Matera	9	0,4932
2	Trieste	13	0,4735
3	Roma	23	0,4512
4	Pavia	37	0,4101
5	Erice	39	0,4071
6	Verona	40	0,4058
7	Bari	45	0,3977
8	L'Aquila	49	0,3867
9	Rimini	50	0,3826
10	Pieve Emanuele	52	0,3817
11	Potenza	59	0,3755
12	Eboli	61	0,3742
13	Catania	65	0,3688
14	Napoli	87	0,3456
15	Genova	91	0,3412
16	Firenze	97	0,3376
17	Ancona	99	0,3342
18	Lamezia Terme	107	0,3276
19	Lecce	111	0,3219
20	Taranto	116	0,3190
21	Cagliari	120	0,3172
22	Milano	133	0,3023
23	Reggio nell'Emilia	134	0,3018
24	Bologna	149	0,2955
25	Torino	173	0,2840
26	Settimo Torinese	222	0,2538
27	Foligno	250	0,2242
28	Venezia	283	0,1941

The results obtained from the application of the second model show how the "criterion of the capital" has a significant influence on the final classification. In fact, with the addition of this parameter, almost all the funded projects (24 out of 28) are located in the first third of the list (some examples are Rome, Trieste, Milan, Matera, Pavia, Verona, Bari, L'Aquila, Potenza, Catania, Naples, Genoa, Florence, Ancona, Taranto, Cagliari and Bologna). This indicates a greater logical affinity and consistency between the results obtained from the model and the criteria used by the Control Room.

6 Considerations on the Results Obtained

Comparing the results obtained from the models with the decisions taken in the Control Room, there are discrepancies due to the nature of the technical and economic parameters upon which the objective functions are implemented. Evidently, the economic analysis models exclude political reasons, even in important decision-making purposes. It is worth considering, for example, the logic regarding the needs and specific requirements of a given territory, which vary according to specific and local issues (including the cases of the winning projects, aimed at enhancing the cities that have suffered substantial inconvenience: Taranto, which proposed the redevelopment of the district Tamburi, affected by a high rate of environmental pollution due to the absolute contiguity with the industry ILVA; Genoa, with the safety measures for the tributary Rio Fereggiano; L'Aquila, which proposed the redevelopment of Piazza d'Armi, acquired by the city before the 2009 earthquake). Nevertheless, the use of political criteria alone could lead to questionable results if not supported by scientific evidence. The *optimum* decision making condition could be obtained by overlaying technical, economic and political-institutional criteria.

An analysis [1] of the projects and summary reports highlights how the decision-makers have focused on the proposals of the capital cities and, more generally, highly populated municipalities (and therefore with a greater number of people affected by the intervention), while also paying attention to those not satisfying these requirements but very valid from the point of view of the potential investment. Especially in these times of economic crisis, they are mostly large urban centres that may become important engines for economic growth, since they are able to produce new revenue through the structures that potential interventions can generate.

Moreover, during the selection, the Control Room focussed on projects from urban free zones (for the model, the best are Lamezia Terme and Erice) as well as those that had large intervention surfaces in order to reduce social and urban decay.

The reasons described above were decisive in the choice of the best projects and have highlighted the differences between the results obtained from the model and the selections made by the Control Room. In fact, the first funded proposal that is classified by the model is in 9th place (Matera), the second in 13th (Trieste), the third 23rd (Rome), and so on. One of the cases where there are major differences is the Puglia region, where the proposals of Lecce (111th in the ranking of the model) and Taranto (116th) are preferred over those of Brindisi (3rd) and Andria (5th). However, it should be noted that Taranto was chosen due to it being more populated than the two excluded municipalities as well as affected by ILVA, which has attracted national attention. Lecce was also chosen because it includes an larger intervention and redevelopment area, equal to the other parameters. Another case is the Campania region, where the proposals of Naples (87th) and Eboli (61th) were preferred over those of Caserta (6th), Benevento (16th) and Salerno (35th). Again, Naples was chosen due to it being considerably more populated, with the intervention areas being much larger as well as urban free zones. Eboli was chosen due to the considerable resources already available, and the meager funds requested from the Cities Plan. Among the excluded proposals, there is also Atrani (1st), due to the modest impact that it could generate (Atrani is a town of 887 inhabitants and the second smallest in Italy, with an area of 0.2 km^2). Another example is Tuscany, where the proposal from Florence (97th) was chosen at the expense of Lucca (14th), because of the number of inhabitants potentially affected by the impacts and more consistent available resources.

Ultimately, the selection of the most deserving implemented by the Control Room appears, in general terms, acceptable and consistent with what is the primary objective of the Cities Plan.

7 Conclusions

In light of the analyses carried out in this study, the Control Room can be considered a successful experiment, since it not only satisfies the needs of the various Ministries, Regions and Organizations involved but also combines the sometimes conflicting selection parameters in the decision-making phase.

The criteria used by the Control Room for the selection of the most deserving proposals, in general, seem to be shareable. The analysis of the projects and summary reports highlights how the decision-makers have focused on the proposals from capital cities and, more generally, highly populated municipalities, however also paying attention to those that seem valid from the point of view of the potential of the investment.

Especially in these times of crisis, they are mostly large urban centres that can qualify as a major engine for economic growth, since they are able to produce new revenue through the structures that potential interventions can generate. Moreover, during the selection phase, the Control Room focussed on projects from Urban Free Zones as well as those that involved high intervention surfaces in order to reduce social and urban decay.

Ultimately, the criteria applied are consistent with what is the primary objective of the *Cities Plan*, i.e. the redevelopment of urban areas, with particular attention to degraded ones, assuming that the construction and valorisation of large urban areas can contribute to triggering the revival of the national economy.

The second year of this research, however, will borrow from this present study as well as other analytical experiences transparency and rationalisation tools that may be useful in making joint decisions.

References

1. Curto, R.: Un approccio economico alla pianificazione. In: Mantini, P., Oliva, F. (eds.) La Riforma Urbanistica in Italia, Pirola, Milano (1996)
2. De Mare, G., Nesticò, A., Tajani, F.: Building Investments for the Revitalization of the Territory: A Multisectoral Model of Economic Analysis. In: Murgante, B., Misra, S., Carlini, M., Torre, C.M., Nguyen, H.-Q., Taniar, D., Apduhan, B.O., Gervasi, O. (eds.) ICCSA 2013, Part III. LNCS, vol. 7973, pp. 493–508. Springer, Heidelberg (2013)

3. De Mare, G., Nesticò, A., Tajani, F.: The rational quantification of social housing. In: Murgante, B., Gervasi, O., Misra, S., Nedjah, N., Rocha, A.M.A.C., Taniar, D., Apduhan, B.O. (eds.) ICCSA 2012, Part II. LNCS, vol. 7334, pp. 27–43. Springer, Heidelberg (2012)
4. Keeney, R.L., Raiffa, H.: Decision with multiple objectives: preferences and value trade off. Weely, NY (1976)
5. Girard, L.F.: La valutazione multidimensionale nella pianificazione territoriale paesistica. Genio Rurale (3) (1992)
6. Girard, L.F., Nijkamp, P.: Le valutazioni per lo sviluppo sostenibile della città e del territorio, Franco Angeli, Milano (1997)
7. van Delft, A., Nijkamp, P.: Multi-Criteria Analysis and Regional Decision-Making, H. E. Stenfert Kroese B. V., Leiden (1977)
8. De Mare, G., Nesticò, A.: An expeditious model for equalization of municipal tax on real estate: an Italian case. Advanced Materials Research 931-932, 551–554 (2014) ISSN: 1022-6680, doi:10.4028/www.scientific.net/AMR.931-932.551
9. Munda, G.: Multicriteria evaluation in a fuzzy environment. Physica-Verlag, Berlin (1995)

A Model for the Economic Evaluation of Energetic Requalification Projects in Buildings. A Real Case Application*

Antonio Nesticò, Gianluigi De Mare, Pierfrancesco Fiore, and Ornella Pipolo

University of Salerno, Department of Civil Engineering
Via Giovanni Paolo II, 132 - 84084 Fisciano (SA) - Italy
{anestico,gdemare,pfiore}@unisa.it, ornellapipolo@virgilio.it

Abstract. The study is part of the current debate on the selection of cost-effective interventions to ensure the energy efficiency of buildings. It is generally considered a major issue for the conscious exploitation of environmental resources and, in particular, the economic sustainability of the management processes of buildings.

In accordance with the provisions of the relevant legislation, a model for the quantification of energy audits and the verification of the cost effectiveness of energetic requalification projects in buildings is proposed. There is also the aim to provide professionals, agencies and operators within the sector a standardised methodology for the selection of useful and practicable steps for the reliable estimation of the relevant costs and revenues. The model is structured on a logical sequence of operational steps that range from the simulation of the thermodynamic behaviour of the building up to the implementation of a Cost-Benefit Analysis, a technique traditionally used to express judgements of convenience of the implementation of investments. The reliability of the model is tested through the application to a real case.

Keywords: economic evaluation of projects, energy efficiency requalification, economic models.

1 European Union Guidelines on Bioclimatic Construction

The current political and economic debate, as well as the social and cultural one, cannot be separated from energy and environmental issues. High energy costs, pollution and environmental damage, urban congestion require a reinterpretation of the concept of development, which must necessarily meet the criteria of sustainability [1, 2]. It therefore follows that the lifestyles of production and consumption must look to the conservation of resources in addition to the reduction of pollution sources [3].

The cornerstones of eco-sustainability are saving and efficiency, with saving being the non-consumption of energy and efficiency the intelligent exploitation of the energy resources available.

---

* This paper is to be attributed in equal parts to the four authors.

B. Murgante et al. (Eds.): ICCSA 2014, Part II, LNCS 8580, pp. 563–578, 2014.

The objectives of eco-sustainability should be pursued with particular commitment in the field of construction, that covers at a European level about 40% of energy demand.

With the intent of the fulfilling the commitments made by signing the Kyoto Protocol, Europe has approved the first European Union Regulation n. 761 of 2001, which sets out the methodology of voluntary membership to an EU eco-management and audit scheme (EMAS). This has been followed by Directive n. 91 of 2002, which led to the environmental policy establishing guidelines with which the different Member States must comply by enacting appropriate legislation or adapting the existing one.

The new Directive 2010/31/EU, which entered into force on 9 July 2010 replaces and repeals 2002/91/EU and promotes the improvement of the energy performance of buildings within the EU, taking into account local external climatic conditions as well as the provisions relating to indoor climate and effectiveness under the terms of costs. It also provides provisions on: the methodology for calculating the integrated energy performance of buildings and housing units; energy certification and the minimum energy performance of buildings; independent control systems for energy performance certificates and inspection reports; national plans for increasing the number of buildings with “nearly zero energy”; regular inspections of heating and air-conditioning systems in buildings.

In addition, to containing the energy requirements, Directive 2010/31/EU requires the Member States to set out requirements for the construction of technical systems in relation to the overall energy performance, the correct installation and size as well as regulation and control. By 31 December 2020, it is expected that all new buildings will be “nearly zero energy”, thus obtaining a building with a very high energy performance and whose energy needs are almost entirely covered by renewable energy sources. In fact, Directive 2010/31/EU goes exactly in the direction of environmental protection as set out by the climate and energy package, better known as 20-20-20, approved on December 17, 2008 and with which Europe is committed to reducing by 20% the emissions of greenhouse gases, increasing up to 20% energy saving and bringing up to 20% the consumption of renewable energy, all by 2020.

2 The Situation in Italy

In Italy, the law gives the companies distributing electricity and gas the responsibility to realise increases in energy efficiencies in the end-users, according to a precise time schedule. The savings are certified by “white certificates”, negotiable on the market, that the distribution companies must procure in proportion to the amount of energy delivered, either with their own interventions or by buying certificates from third parties. Thus, as with the European directives, the Italian regulations intervene decisively in the building sector. The list of initiatives to be proposed for the issue of energy efficiency certificates is, in fact, mainly aimed at residential and tertiary buildings, although inventions in the industrial sector and, albeit marginally, in the transport sector are also included [4].

Presidential Decree n. 59 of 2 April 2009 contains the regulations that define the calculation methods and minimum requirements for the energy performance of buildings and heating systems. The regulation is aimed especially at those regions that have yet to draw up any form of legislation on energy certification. In contrast, in the presence of regional regulations, this shall prevail over the national one, albeit with the commitment of the regions as much as possible to adapting this legislation to the national guidelines.

Ministerial Decree dated 26 June 2009, containing the "National Guidelines for the energy certification of buildings", sets out the parameters for the certification services and specific connection, coordination and cooperation tools between the State and Regions. In particular, there are the details on the Certified Energy Certificate (ACE), which all buildings being sold must possess as well as new buildings and those being restored.

The Ministerial Decree dated 19 February 2007 (later modified by the decrees of 26 October 2007, 7 April 2008 and 6 August 2009) regarding the New Energy Bill is also important. The regulation, with the purpose of reaching a photovoltaic capacity of 3,000 MW by 2016, provides:

- more advantageous tariffs for small photovoltaic systems integrated into buildings, typically subservient to residential users;
- tariffs increased by 5% to reward installations in public buildings (schools, hospitals, local authorities of small towns) and those replacing roofing containing asbestos;
- even higher tariffs are possible if the installation of the system (up to 20 kW) is accompanied by a requalification certificate and/or energy certification;
- simplified administrative procedures to obtain the incentives.

Legislative Decree 63/2013, which implements Directive 2010/31/EU, introduces the Energy Certificate APE, replacing the ACE. Compared to the previous certificate, the APE proposes cost effective energetic requalification interventions.

3 The Economic Evaluation of Energy Efficiency Requalification Projects in Buildings

Most of the energetic requalification interventions mainly arise from a legal obligation which must be adhered to rather than from the growing awareness of environmental or economic problems. In other cases, the measures aim to achieve set standards in order to qualify for financial incentives.

In relation to the legislative obligations, it is worth noting that the technical and economic evaluations that are part of energy efficiency requalification projects are carried out by considering: the national legislation concerning the containment of the energy requirements for buildings and facilities for the evaluation of the technical requirements imposed on the interventions; the national legislation transposing the UNI regulations for the calculation of the energy demand of buildings; the local regulations.

In strictly economic terms, energetic requalification aims to achieve profitable operations capable of returning the extra cost of the design choice, or in other words the most technologically advanced intervention. It is only through a cost-benefit analysis that these types of intervention can be justified. In fact, they can be carried out in terms of cost-effectiveness either due to the depreciation deriving from the lower cost of gas for heating or electricity, with it being thanks to government incentives that pay off a part of the investment or maintenance work that leads to carrying out work that improves the energy class of the building and thus can, even in part, be tax deductible. Thus, quantifying the "change" of energy class is decisive, through calculations that are able to take into account the different variables that contribute to the energy characterisation of the building [5, 6].

Following the approach described, it is possible to estimate not only the financial convenience of the intervention but also the reduction in the consumption of fossil fuels as well as the reduction in pollution.

4 The Structure of the Protocol for the Evaluation Study

This paper proposes a model to study the energy audits of buildings as well as verify the economic convenience of energy efficiency requalification interventions of the building and its systems.

The study of energy audits consists of a systematic procedure aimed at the knowledge of the end-use of energy as well as the identification and analysis of any potential inefficiencies and problems. Through the cost-benefit analysis, an assessment of the economic convenience allows to ensure the financial sustainability of the project both during the carrying out of the works as well as in the management process [7, 8].

The model consists of the following seven steps.

4.1 Collecting and Analysing Data for the Measurement of Energy Audits of the Building in Its Current State

The first stage consists of collecting and studying climatic and general data on the property, the geometric-dimensional data of the building, the thermo-physical data of the casing components of the building, the performance data of the energy conversion and distribution systems under operating conditions.

Climatic and general data on the property:
- degrees days DD (°C);
- climatic zone;
- intended use.

Geometric-dimensional data:
- gross surface area in typical floor area (m^2) and heights (m);
- net heated surface area S (m^2) and net heated volume V (m^3);
- S/V coefficient (m^2/m^3).

Thermophysical data of the casing:
- thermal resistance R (K/W);
- transmittance of simple components U=1/R and stratigraphic $U=1/\sum R_i$ (W/m²K).

Performance data of the systems:
- type (heating, electrical, hot water);
- fuel used (propane, methane, oil);
- delivery systems (radiators, heaters, etc.);
- distribution systems (vertical, horizontal).

4.2 Processing of the Collected Data and Simulation of the Thermodynamic Behaviour of the Building

An energy profile that summarises the type of user, the installed energies, the type, the use profiles and operation hours of the plants should be drawn up. The preparation of a statement that describes the development of the characteristic energy flows of the building make it possible to evaluate the specific fuel consumption. It makes use of algorithms implemented in a computer structure through a specific software for thermal design (e.g. in Italy: *Docet*, developed by ITC - Institute for Construction Technologies and ENEA - Agency for New National Technologies, Energy and Sustainable Development; *TermiPlan* by Analist Group; *TerMus* by ACCA; *Blumatica Energy*; *MasterClima*).

According to the methodology specified in the UNI TS 11300, the indices of primary energy are calculated:

$$Ep_i = \frac{\left(\frac{Q_h}{A_{pav}}\right)}{\eta_g} = \frac{\left[\frac{(Q_{h,tr} + Q_{h,ve}) - \eta_s \times (Q_{int} + Q_{sol})}{A_{pav}}\right]}{\eta_g},$$

where

Ep_i = primary energy for winter heating [kWh/m²K],
Q_h = thermal energy demand of the building [kWh],
A_{pav} = useful floor area [m²],
η_g = average global seasonal performance coefficient,
$Q_{h,tr}$ = transmission losses [W/K],
$Q_{h,ve}$ = dispersions due to ventilation [W/K],
η_s = coefficient of use of free inputs, generally assumed to be equal to 0.95,
Q_{int} = free internal inputs [MJ],
Q_{sol} = solar inputs [MJ];

$$Ep_{acs} = \frac{\left(\frac{Q_w}{A_{pav}}\right)}{\eta_r} = \frac{\frac{\rho_w \cdot c_w \cdot [V_w \cdot (\theta_{er} - \theta_o)] \cdot G}{A_{pav}}}{\eta_r},$$

with

Ep_{acs} = primary energy for hot water [kWh/m²K],
Q_w = energy demand for domestic hot water [kWh],

ρ_w = volumetric mass density of water [1000 kg/m^3],
c_w = specific heat of water [1.162 * 10^{-3} kWh/(kg · K)],
V_w = daily volume of water required by activity or service [m^3/day],
θ_{er} = water supply temperature [40° C],
θ_0 = entry temperature of cold water [15° C],
G = number of days in the calculation period,
η_g = global seasonal average performance coefficient;

$$Ep_{e,inv} = \frac{(Q_{int} + Q_{sol}) - \eta_{c,ls} \times (Q_{h,tr} + Q_{h,ve})}{A_{pav}},$$

where

$Ep_{e,inv}$ = energy performance index for casing summer cooling [kWh/m^{2}K],
$\eta_{c,ls}$ = thermal loss coefficient.

4.3 Verification of the Energy Requirements Resulting from the Simulations by Comparison with the Actual Consumption of the Building. Calibration of the Model Parameters

It should be ascertained that the data on the thermodynamic calculations relating to the thermal characteristics of the casing and performance of the systems are representative of the actual energy consumption of the building, as inferred from the bills.

The difference between the two output values is considered acceptable if less than 15%, otherwise it is necessary to vary the input data of the model. In formal terms, the verification of consistency is summarised in the relation:

$$\frac{Q_{i,th} - Q_{i,real}}{Q_{i,th}} \times 100 < 15\%,$$

where

$Q_{i,th}$ = theoretical energy requirement for heating or acs [kWh];
$Q_{i,real}$ = real energy demand for heating or acs [kWh].

4.4 Identification of Critical Points of the Building in Terms of the Energy-Performance Profile

After the three aforementioned steps, it is possible to identify the components of the building casing and/or system equipment that are the reason for the low energy class of the building. The identification of the problems is where certain elements of the building have evident poor thermal qualities. In contrast, it is possible to proceed with innovative support techniques and tools, such as thermography, useful in detecting moisture, thermal bridges and dispersions.

In particular, a *thermographic survey* allows for the rapid and reliable recognition of differences in material or construction defects using radiant energy. This is a function of the surface temperature of the materials, in turn dependent on the thermal conductivity and specific heat of the same. These physical quantities express in quantitative terms the ability of the material to transmit or retain heat and thus its speed in the heating up or cool downing. Due to the different values of the thermal conductivity and specific heat, the various components of a structure, such as a building casing, assume different behaviours and thus different temperatures when subjected to thermal stresses such as sunlight or when they are heated by the systems. In the electromagnetic spectrum, radiation emitted by a heated body is indicated with a wavelength λ , which is roughly between 0.1 and 100 μm.

4.5 Selection of Possible Energy Efficiency Interventions

These include:

- works on the building casing (insulation of masonry walls, roofs and floors, replacement of fixtures, solar shading);
- partial or total replacement of systems (heating, domestic hot water, lighting);
- use of renewable energy sources (solar thermal, photovoltaic, geothermal).

Naturally, the investment is aimed at reducing the demand for primary global energy Ep_{gl} [kWh/m^{2}K or kWh/m^{3}K]:

$$Ep_{gl,post\ investment} < Ep_{gl,ante\ investment} \, .$$

4.6 Definition of Management Processes. Search for Financial Support for the Types of Intervention Proposed

Several European funds (European Energy Efficiency Fund EEEF, ELENA) and national ones (in Italy, there are the resources from the Energy Bill or Law 90/2013) support energetic requalification interventions. In order to obtain incentives or special concessions, a licensed professional must issue the Energy Certificate (APE) of the building. The management of the intervention can be:

- *direct*, if carried out by the user or company that intends to improve the energy performance of the building;
- *indirect*, if carried out through the *Energy Service Company* (ESCo). The ESCo is "a living or legal person that delivers energy services or other energy efficiency improvement measures to the user's facility or premises and, in doing so, accepts some degree of financial risk. The payment for the services delivered is based wholly or partly on energy efficiency improvements and on the meeting of the other agreed performance criteria" [Legislative Decree 115/2008].

For the construction and property sectors, the ESCo can develop diagnostic and energy certification activities, management and maintenance programmes of the buildings and systems, design of buildings and its systems focussing on energy efficiency, including the use of renewable sources. The ESCo can draw up a contract with either a public administration or private company, at no cost to the institution, which allows them to pay only through the results of the intervention, in terms of energy savings, incentives as well as through national or European incentives. Basically, the owner of the system continues to pay the same energy bill that he paid prior to the intervention, reimbursing the ESCo with the difference.

4.7 Economic Evaluation of Possible Energetic Requalification Projects and Optimal Investment Choice

The protocol is summarised in the economic evaluation of selected interventions, which consists of an estimate in monetary terms of the construction and operation costs of the works as well as revenues resulting from the savings on primary energy. It is therefore possible to express a judgement on the economic viability of the single intervention or their combinations, preferring the one which corresponds to the highest Net Present Value (NPV) and Internal Rate of Return (IRR) or the lesser Payback period (PB).

If the criterion is that of the NPV, the following is estimated for each design solution:

$$NPV = CF_0 + \frac{CF_1}{(1+r)} + \frac{CF_2}{(1+r)^2} + \cdots + \frac{CF_i}{(1+r)^i} - I_0 = \sum_{i=0}^{n} \frac{Cf_i}{(1+r)^i} - I_0 ,$$

with CF_i cash flow to the i-th year; r discount rate, I_0 the initial investment cost. For the IRR and Payback period, the following relations are applied, where A_f indicates any tax breaks and R the annual recovery:

$$IRR = \sum_{i=0}^{n} \frac{CF_i}{(1+r)^i} - I_0 = 0 ;$$

$$PB = \frac{I_0 - A_f}{R} .$$

It is worth noting that the benefits resulting from the energy efficiency interventions can be:

- *direct, in the form of saving on energy consumption and therefore increase in the property value*. It generates a measurable revenue through the economy of management, with lower energy bills, which produces an increase in the market value of the asset, whose best energy class is certified;
- *indirect, in terms of environmental improvement*. This is because the reduction in consumption of fossil fuels results in lower emissions of pollutants.

In the light of the analysis and the calculations carried out, it is possible to select the hypothesis of optimal intervention in function of the ability to pursue technical and functional, economic, environmental, aesthetic and architectural integration objectives [9, 10, 11]. Figure 1 summarises the phases of the model.

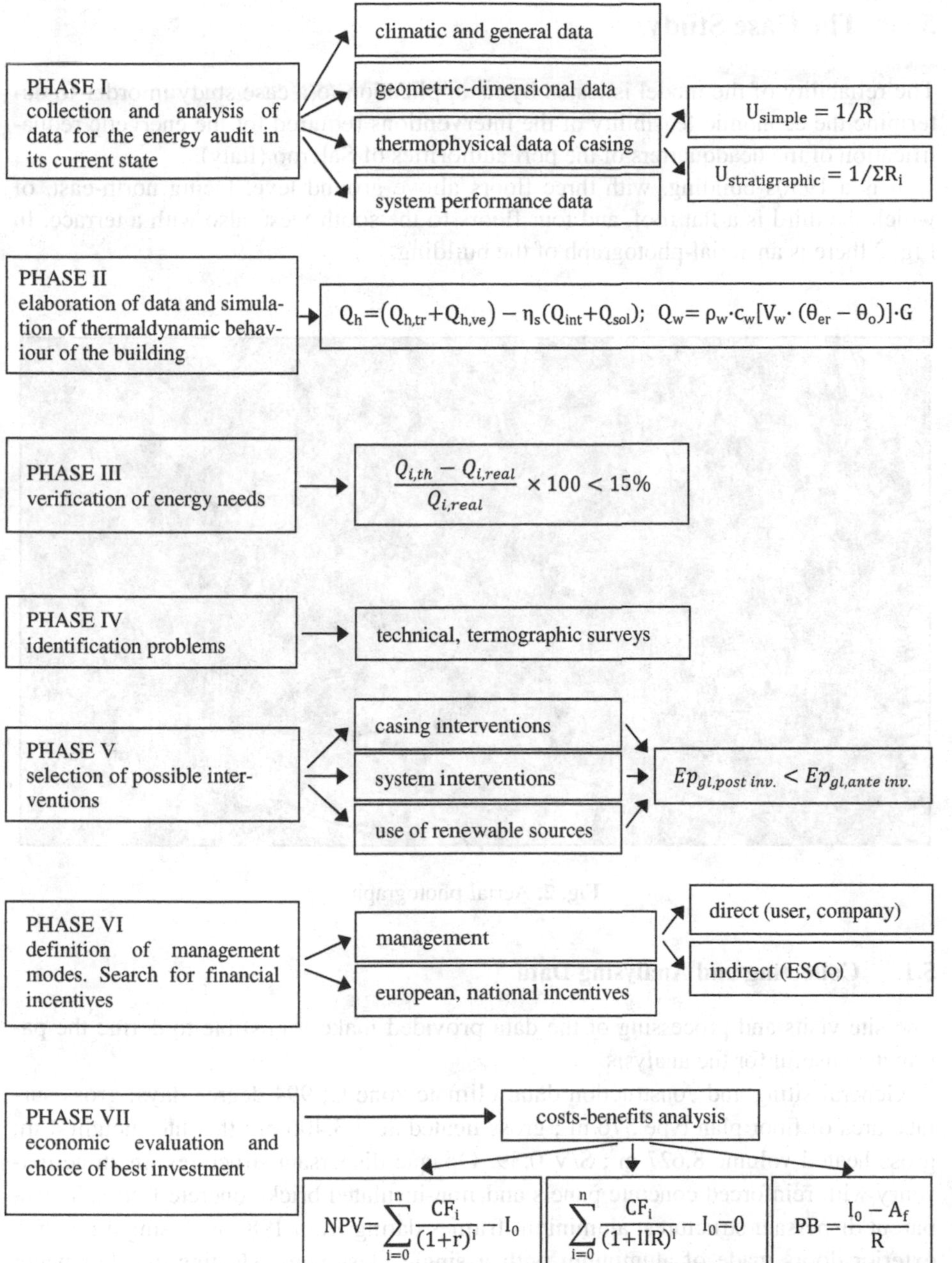

Fig. 1. Logical phases of the model

5 The Case Study

The reliability of the model is tested by its application to a case study in order to determine the economic feasibility of the interventions required for the energetic requalification of the headquarters of the port authorities of Salerno (Italy).

It is a 1950s building, with three floors above ground level facing north-east, of which the third is a flat roof, and four floors to the south-west, also with a terrace. In Fig. 2 there is an aerial-photograph of the building.

Fig. 2. Aerial photograph

5.1 Collecting and Analysing Data

The site visits and processing of the data provided make it possible to define the parameters useful for the analysis.

General siting and construction data: climate zone C; 994 degree days; gross surface area of floor plan type 576 m^2; gross heated area 3,406 m^2; the inter-height 4 m; gross heated volume 8,627 m^3; S/V 0.39. Opaque dispersant structures: in stone masonry with reinforced concrete panels and non-insulated brick-concrete floors. Transparent dispersant structures: aluminium frames dating from 1987 with single glazing, exterior doors made of aluminium with a single glass pane. Heating and hot water systems: 4 heating pumps with a power of 80-100-140-160 kW; methane gas boiler to heat corridors and bathrooms, heat distribution through cast iron radiators and convectors. The values are taken from the tables of thermal transmittance U for dispersive surfaces.

5.2 Simulation of the Thermodynamic Behaviour of the Building

The data processing allows to draw up the end-use energy budget for winter heating and the production of hot water. The *TermiPlan* software was used to quantify the consumption of the building in compliance with the technical standards UNI/TS11300. The simulation of the thermodynamic behaviour of the building provides the requirements listed below, which place the building in the energy class E:

- primary energy for winter heating $Ep_{i,b}$ (boiler) = 15.16 kWh/m^3;
- primary energy for winter heating $Ep_{i,hp}$ (heating pumps) = 9.20 kWh/m^3;
- primary energy for hot water Ep_{acs} = 0.38 kWh/m^3;
- energy performance index for casing summer cooling $Ep_{e,inv}$ = 11.21 kWh/m^3;
- global primary energy $Ep_{gl,b}$ (boiler) = 15.53 kWh/m^3 (class E);
- global primary energy $Ep_{gl,hp}$ (heating pumps) = 9.53 kWh/m^3 (class D).

5.3 Calibration of the Model

The next step is a process to validate the model by comparing actual and theoretical consumption, so as to verify the accuracy of the input data. Where necessary, adjustments should be made so as to make the two needs congruent.

To obtain the energy values corresponding to the actual use of the building, the bills over the last two years are analysed. The information is collected in a database. With reference to 2013, electricity, gas and water are recorded with the following consumption and real costs respectively: 93,813 kWh and € 23,419 (with 6.21 tons of CO_2 produced); 6,268 m^3 and € 3,537 (10.23 tons CO_2), 723 m^3 and € 1,860. The cost of electricity is 0.25 €/kWh as stipulated in the supply contract.

For the comparison between actual and theoretical consumption, it is necessary to convert the amount of natural gas consumed into thermal energy. 1 m^3 of methane generates 10.35 kWh, for which 6,268 m^3 produce 64,873.8 kWh. It is worth considering that 10% of the electricity is used for lighting and electrical equipment, the real consumption $Q_{h,real}$ for winter heating and for summer cooling is 84,432 kWh. The theoretical energy requirements for electricity $Q_{h,th(el)}$ and for methane gas $Q_{h,th(gas)}$ are:

$$Ep_i + Ep_{e,inv} = \frac{Q_{h,th(el)}}{V_{gross}} \rightarrow Q_{h,th(el)} = 20.41\ \text{KWh/m}^3 \times 4{,}422\ \text{m}^3 = 90{,}258\ \text{KWh};$$

$$Ep_{gl} = \frac{Q_{h,th(gas)}}{V_{gross}} \rightarrow Q_{h,th(gas)} = 15.53\ \text{KWh/m}^3 \times 4{,}205\ \text{m}^3 = 65{,}307\ \text{KWh}.$$

Since the difference between $Q_{h,real}$ and $Q_{h,th}$ is less than 15%, it can be concluded that the model correctly simulates the thermodynamic behaviour of the building:

$$\frac{Q_{h,th(el)} - Q_{h,real(el)}}{Q_{h,th(el)}} \times 100 = 6.90\% < 15\% \rightarrow \text{model validated;}$$

$$\frac{Q_{h,th(gas)} - Q_{h,real(gas)}}{Q_{h,th(gas)}} \times 100 = 0.66\% < 15\% \rightarrow \text{model validated.}$$

5.4 Identification of Problems

In the light of what emerges from the analysis of the building-plant system, it is evident that the energy performance can be improved. The greatest needs are related to the consumption of electricity for heating, cooling and lighting. The critical points are found in the excessive power used by the neon lights and the dispersion of the building. In particular, the transparent dispersant structures are not thermally adequate, with it being necessary to intervene in order to improve the transmittance of the building through the replacement of the fixtures.

5.5 Selection of Possible Interventions

These include replacing any fixtures, installing a photovoltaic system as well as substituting the neon lights with LEDs.

Replacing Fixtures. The intervention reduces transmittance values of the entire glass-frame component. PVC/aluminium fixtures can be used. The windows have a double low-emissivity. The 7 room structure of the profile ensures excellent thermal insulation ($U = 1.1\ W/m^2K$). The middle seal guarantees the best permeability values (class 4 according to DIN EN 12207) and hard rain resistance (class 9, DIN EN 12208).

The simulation of the thermodynamic behaviour of the building with the new transmittance parameters of the windows gives a decrease in energy demand, with Ep_{gl} going down from 15.53 to 14.17 kWh/m^3 for the boiler and from 9.53 to 7.39 kWh/m^3 for the heating pumps. The energy class remains unchanged for the boiler and improves for the heating pumps (from D to C).

Installation of a Photovoltaic System. The capability needed to meet the energy needs depends on solar radiation, in Salerno approximately 1,350 kWh/kWp. Thus, knowing the final consumption of electricity (93,813 kWh), the system – with a fixed, permanent one being chosen – gives power of 70 kWp.

Noting the latitude of the site, it is possible to calculate the optimum angle of solar collectors. The study of the available surface area and the corresponding effects of shading identify 1,738 m^2 of the car park for the installation of the photovoltaic canopies. These not only give a good impression of eco-sustainability but also allow cars to be protected against direct sunlight during the summer months.

The system allows the building to pass from energy class D to class A^+, with $Ep_{gl,hp}$ decreasing from 9.53 kWh/m^3 prior to the intervention to 0.00 kWh/m^3 after the intervention. It should be noted that following the replacement of the windows, the photovoltaic system produces energy not consumed in amounts equal to the lower thermal energy requirement for heating. In the current state, $Ep_{i,hp}$ is equal to 9.20 kWh/m^3. In the project with the replacement of the fixtures, $Ep_{i,hp}$ is 7.06 kWh/m^3, with a reduction in consumption of 2,14 kWh/m^3 that, multiplied by the entire volume of the building, entails a saving of 9,464 kWh. Subsequently, the energy fed into the grid increases and improves the energy class of the building.

Replacing Fluorescent Lights with LED Lights. Table 1 shows the number of lamps for each floor of the building and the corresponding watts of consumption for both the fluorescent lighting and the replacement LEDs.

Table 1. Number of lights and consumption in watts of neon lighting and LEDs

	Neon			Led		
floor	type	n°	watt	type	n°	watt
basement	2 x 36 w	35	2.520	2 x 18 w	35	1.260
	1 x 18 w (EM)	17	306	1 X 9 w (EM)	17	153
	2 x 58 w	5	580	2 x 22 w	5	220
	4 x 18 w	12	864	4 x 9 w	12	432
mezzanine	2 x 36 w	4	288	2 x 18 w	4	144
	1 x 18 w (EM)	19	342	1 X 9 w (EM)	19	171
	2 x 18 w	7	252	2 x 9 w	7	126
	2 x 58 w	1	116	2 x 22 w	1	44
	4 x 18 w	65	4.680	4 x 9 w	65	2.340
first	2 x 36 w	2	144	2 x 18 w	2	72
	1 x 18 w (EM)	19	342	1 X 9 w (EM)	19	171
	2 x 18 w	6	216	2 x 9 w	6	108
	4 x 18 w	73	5.256	4 x 9 w	73	2.628
second	2 x 36 w	4	288	2 x 18 w	4	144
	1 x 18 w (EM)	12	216	1 X 9 w (EM)	12	108
	4 x 18 w	29	2.088	4 x 9 w	29	1.044
		W tot.	18.498		W tot.	9.165

5.6 Management Procedures and the Search for Financial Incentives

The management is direct. The replacement of the fixtures respects the legislation constraints that provide access to tax deductions equal to 65% of the cost of work carried out by June 30, 2014, for a maximum of € 60,000.

In contrast, the photovoltaic system cannot take advantage of the incentives of the V Energy Bill, which has now expired, and not even the tax deduction of 50% provided only for households below 20 kWp. However, the investment will pay for itself through the non-purchase of energy at 0.25 €/kWh and with the feed-in of 0.10 €/kWh of energy produced and not consumed.

5.7 Financial Evaluation of the Investment

The cost-benefit analysis is implemented starting from the estimate of the costs of investment and management as well as revenue, with the latter being represented by the savings resulting from lower energy consumption. This is followed by the timing of the data. The period of analysis is assumed to be 25 years. The start-up takes place immediately after the first year needed for the execution of the works. The evaluation is carried out at constant prices. The cash flows are discounted at the discount rate of 5%, already adjusted for inflation (European Commission, 2007-2013). It is assumed that the interventions are financed 50% with equity and the other half with loan capital. A ten-year amortization schedule of the loan at a constant rate is drawn up, at the current rate of interest of 7%.

The detailed economic study conducted for each of the three interventions considered as individually is followed by the financial plan of Table 2, which gives an account of the financial sustainability of the entire investment by virtue of the positive financial results:

NPV = € 141,780, IIR = 8.80%, Payback period = 15.12 years.

Replacing Fixtures. The cost of the intervention is derived from the estimate, which returns an amount of € 86,544 for all the work done, gross 10% VAT. Due to the use life of 50 years, there is a residual value of € 43,272. Replacing the existing fixtures, starting from the second year, energy consumption falls from € 58,265 to € 51,094, with an annual saving of € 7,162. The interest rate on a ten-year loan goes from € 1,666 in the first year to € 222 in year 10. The indicators give:

NPV = € 59.760, IRR = 12,08%, Payback period = 9,81 years.

Installation of a Photovoltaic System. The photovoltaic system has an average life of 25-30 years. The only maintenance that is needed is periodic cleaning of the modules and the replacement of the inverter board (€ 24,886) every 10 years. The working costs of an on-site exchange system are related to the stipulation of the agreement with the GSE (Energy Services Supplier) at € 45 per year. Polycrystalline modules with a nominal power of 240 Wp are used for the canopies, which cost € 235.20. Thus, the total cost of 292 modules is € 68,678, which becomes € 83,101 VAT included. The cost of the inverter, 75 kWp and use life of 50 years, is € 20,530 (€ 24,841 VAT included). The cost of a steel canopy is € 89,600 (VAT included), calculated according to the Public Works Pricelist of the Campania Region, with a residual value of 75% at 25 years. Therefore, the total investment cost is € 197,542.

The energy produced in the first year is worth 94,608 kWh, which for subsequent years assumes a 1% decline in performance. It can be assumed that 84,000 kWh are consumed, resulting in a saving in the bill of € 21,000/year. 10,000 KWh are collected annually at the rate of exchange on site of 0.14 €/kWh, with an exchange contribution of 1,400 €/year. The remainder of the energy produced is fed into the grid at a rate of 0.10 €/kWh. In 10 years, the incidence of the mortgage goes from € 6,914 to € 920, so

NPV = € 75.326, IRR = 8,19%, Payback period = 17,07 years.

Replacing Fluorescent Lights with LED Lights. Having evaluated the lighting equivalence between LED and fluorescent lighting and identified the number of lights in the building, this is followed by the assessment of the cost of purchasing – from the Philips catalogue – of both LED lamps (€ 47,650 VAT included) and fluorescent lights (€ 6,959 VAT included), the estimation of the amount of time the lights are on (2184 hours/year) in order to quantify the power consumption in KWh. The LED lamps are to be replaced every sixteen years, while the neons every four. From the calculations, it follows that the replacement of neon lamps with LED ones produces a reduction in the cost of energy for lighting from € 7,070 to € 4,003/year, with an annual saving of € 3,067. In a projection horizon of 25 years, with a ten-year loan at 50% of the total cost of investment I_0 = € 245,141, the NPV of the intervention is negative € –25,824. Nevertheless, the intervention, which still pays for itself over a longer period of time, has the possibility of being realised due to the very positive impact on the environment.

Table 2. Total cash flows of the investment

year	savings from fixtures subst. [€]	fixtures subst. costs [€]	revenues photovolt. [€]	costs photovolt. [€]	savings led [€]	costs led [€]	mortgage interest rate [€]	cash flow [€]	discounted and accumulated cash flow [€]
1	0	86.544	0	197,542	3.067	47,650	10,248	-340,281	-318,720
2	7,172	0	23,461	45	3.067	0	9,506	22,884	-292,861
3	7,172	0	23,366	45	10.026	0	8,712	30,647	-261,527
4	7,172	0	23,273	45	3.067	0	7,863	24,557	-236,696
5	7,172	0	23,180	45	3.067	0	6,955	25,494	-212,314
6	7,172	0	23,088	45	3.067	0	5,982	26,503	-188,339
7	7,172	0	22,997	45	10.026	0	4,942	34,550	-159,787
8	7,172	0	22,907	45	3.067	0	3,829	28,762	-136,512
9	7,172	0	22,818	45	3.067	0	2,638	30,023	-113,533
10	7,172	0	22,730	24,886	3.067	0	1,364	6,537	-97,128
11	7,172	0	22,643	45	10.026	0		39,795	-73,860
12	7,172	0	22,556	45	3.067	0		32,750	-55,624
13	7,172	0	22,471	45	3.067	0		32,664	-38,302
14	7,172	0	22,400	45	3.067	0		32,593	-21,840
15	7,172	0	22,400	45	10.026	0		39,552	-2,814
16	7,172	0	22,400	45	3.067	0		32,593	12,117
17	7,172	0	22,400	45	3.067	47,650		-15,057	5,548
18	7,172	0	22,400	45	3.067	0		32,593	19,091
19	7,172	0	22,400	45	10.026	0		39,552	34,743
20	7,172	0	22,400	24,886	3.067	0		7,752	43,153
21	7,172	0	22,400	45	3.067	0		32,593	54,852
22	7,172	0	22,400	45	3.067	0		32,593	65,994
23	7,172	0	22,400	45	10.026	0		39,552	78,871
24	7,172	0	22,400	45	3.067	0		32,593	88,977
25	7,172	0	22,400	45	3.067	0		32,593	141,780

6 Conclusions

The application of the model to the case study gives important both technical and financial information. The overall project, that the analysis of the critical issues of the building-plant systems leads to the replacement of the fixtures, the installation of a photovoltaic system and the substitution of the lighting, reduces the global primary energy rate $Ep_{gl,hp}$ from 9.53 kWh/m^3 in its current state to 7.39 kWh/m^3 after the works are carried out, and $Ep_{gl,b}$ from 15.53 kWh/m^3 to 14.17 kWh/m^3. The energy class of the building improves, going from D to A$^+$ for the electric plant. The CO_2 emissions into the atmosphere are almost halved, falling from 16.44 to 9.31 kg/m^3. In addition, the values of economic convenience indicators are positive, with NPV = € 141,780, IRR = 8.80%, payback period of about 15 years.

The model is easy to apply and can be adapted to different cases. The calculation procedures are accompanied by tables and graphs, with the aim of facilitating the work of the analyst who will perform energy audits as well as the economic evaluation of the investment. References to current regulations ensure transparency and reproducibility of the results.

On a methodological level, the protocol covers the entire logical sequence of operational steps ranging from the collection and analysis of data for the measurement of energy audits of a building in its current state up to the processing of datasets for the simulation of the thermodynamic behaviour of the building. This is then followed by the verification of the energy needs resulting from the simulations through the comparison with the actual consumption of the building, so as to calibrate the parameters of the model. The identification of the critical points of the building in terms of the energy-performance profile, including a thermographic survey, is followed by the selection of possible energy efficiency interventions and, then, the search for the optimal form of management and financial incentives for possible initiatives. The last phase is the economic evaluation of any possible energetic requalification interventions, with the best investment being chosen. Thus, providing professionals, agencies and operators in the sector a tool that, by implementing the cost-benefit analysis, can serve as a guide for the selection of practicable initiatives.

References

1. Redclift, M.: Sustainable Development: Exploring the Contradictions. Routledge, London (2004)
2. Organisation for Economic Co-operation and Development (OECD): Sustainable Development: Critical Issues, Paris (2001)
3. Adams, W.M.: Green Development: environment and sustainability in a developing world, 3rd edn. Routledge, London (2008)
4. ENEA: Rapporto Annuale sull'Efficienza Energetica, Roma (2011)
5. Hanley, N., Spash, C.L.: Cost-Benefit Analysis and the Environment. Edward Elgar, Cheltenhm, Northamptom (1993)
6. De Mare, G., Manganelli, B., Nesticò, A.: The Economic Evaluation of Investments in the Energy Sector: A Model for the Optimization of the Scenario Analyses. In: Murgante, B., Misra, S., Carlini, M., Torre, C.M., Nguyen, H.-Q., Taniar, D., Apduhan, B.O., Gervasi, O. (eds.) ICCSA 2013, Part II. LNCS, vol. 7972, pp. 359–374. Springer, Heidelberg (2013)
7. Dasgupta, A.K., Pearce, D.W.: Cost-benefit analysis: theory and practice. Macmillan (1972)
8. Pennisi, G.: Tecniche di valutazione degli investimenti pubblici. Istituto Poligrafico e Zecca dello Stato, Roma (1991)
9. De Mare, G., Nesticò, A., Tajani, F.: Building investments for the revitalization of the territory: A multisectoral model of economic analysis. In: Murgante, B., Misra, S., Carlini, M., Torre, C.M., Nguyen, H.-Q., Taniar, D., Apduhan, B.O., Gervasi, O. (eds.) ICCSA 2013, Part III. LNCS, vol. 7973, pp. 493–508. Springer, Heidelberg (2013)
10. De Mare, G., Nesticò, A.: Efficiency analysis for sustainable mobility. The design of a mechanical vector in Amalfi Coast (Italy). Advanced Materials Research 931-932, 808–812 (2014) ISSN: 1022-6680, doi:10.4028/www.scientific.net/AMR.931-932.808
11. Morano, P., Tajani, F.: Break Even Analysis for the financial verification of urban regene-ra-tion projects. Applied Mechanics and Materials 438-439, 1830–1835 (2013) ISSN: 1660-9336, doi:10.4028/www.scientific.net/AMM.438-439.1830

The Paradigm of the Modern City: *SMART and SENSEable Cities* for Smart, Inclusive and Sustainable Growth

Ilaria Greco[1] and Massimiliano Bencardino[2]

[1] Department of Law, Economic, Management and Quantitative Methods, University of Sannio, Via delle Puglie, 1, – 82100 Benevento, Italy
ilagreco@unisannio.it

[2] Department of Political, Social and Communication Sciences, University of Salerno, Via Giovanni Paolo II, 132 - 84084 - Fisciano (SA)
mbencardino@unisa.it

Abstract. The concept of the "smart city" has recently been introduced as a strategic tool to encompass the modern functioning processes of urban development and, in particular, to highlight the importance of Information and Communication Technologies (ICTs) for developing competitive and sustainable of a city. The present paper aims to shed light on the often still elusive definition of the concept of the "smart city" and tries to define a new reading of the "smartness" of a city that includes the size of the equity as a parameter to (re)definition of the International and European rankings of smart cities. In detail, the classifications made by Boyd Cohen's of *Top Ten Global Smart Cities* and *Top Ten Smartest European Cities* based on the metric of "Smart Cities Wheel" of Vienna Polytechnic will be reinterpreted according to the values of the *Equity City Index*, compiled by the UN-Habitat 2012. Finally, comparing the two dimensions (*smartness* + *equity*) will be presented different possible models of *Smart-Equitable Cities* and policies.

Keywords: smart city, equity, urban development, human capital, ICTs.

1 Introduction[1]

The discussion around the cities and their development has never stopped. Moving from the "industrial city" to the "city of services" and, finally, to contemporary urban conurbations, the city has indeed experimented issues of a different nature related to its growth. In these years of "urban reconcentration" social problems have emerged, related to the provision of essential services such as housing and education at an affordable

[1] The paper is the result of a common reflection of the authors; however, the single sections can thus be attributed to: Ilaria Greco paragraphs 1, 4, 5 and Massimiliano Bencardino paragraphs 2 and 3. The reflections presented are the result of participation in the session of the World Urban Forum 6 "The Urban Future", organized at the University of Sannio in collaboration with UN-Habitat, 1-7 September 2012.

B. Murgante et al. (Eds.): ICCSA 2014, Part II, LNCS 8580, pp. 579–597, 2014.

cost, a social environment without crime, safe and inclusive, but also innovative services, in addition to the problems of congestion, pollution and physical degradation.

In response to these challenges, after the models of urban sustainability represented by the "green city" and the "creative city", taking shape new paradigm for the modern city, the "*smart city*" as *"intelligent city"*.

"A Smart City is a city well performing in six characteristics, built on the 'smart' combination of endowments and activities of self-decisive, independent and aware citizens: mobility, environment, people, living, governance, economy": this is the most complete definition dictated by the Report of the European Smart Cities, in line with the new European vision for the future development of global cities (Horizon 2020 Urban Forum, Digital Agenda - Strategy 20.20.20, Decree Digitalia, etc.).

But if the "*smart city*" is a city where investments in human and social capital, in the participation processes and in the technology infrastructure, are directed to sustainable and competitive economic development, ensuring a high quality of life and providing for the responsible management of natural and social resources from a shared governance, "the *SENSEable city*" is the which encourages dialogue between the different elements that make up the urban life, and that encourages more informed and fair decisions about their urban environment, with a new approach to the urban planning and an efficient use of networks.

The new dimension that completes the process of smartness ensuring the development of a city in terms of *SENSEable City* is the *equity*: a city should not just be smart, but its smartness must cover all the inhabitants!

Starting from the concept and the debate about smart cities, the contribution represents an attempt at a new interpretation of the "smartness" referring to a city that includes, in addition to the size that they have acquired the best-known research and studies on the subject (Boyd Cohen, European Smart Cities, City Protocol, Smart City in Europe, MIT Senseable City Lab, The European House-Ambrosetti, iCity Lab PA Forum, etc.), the size of the *equity* as a parameter to (re)definition of international and European rankings of smart cities.

The study examines two rankings: the ranking of the *Top Ten Global Smart Cities* by Boyd Cohen published in the journal "Co.Exist" and a further ranking by the same Cohen of the *Top Ten Smartest European Cities* based on the metric of "Smart Cities Wheel" of Vienna Polytechnic. The rankings, compiled by Cohen, are reinterpreted according to the values of the *Equity City Index,* compiled by the UN -Habitat 2012. On the basis of performance achieved with respect to the dimensions considered (*smartness + equity*) will be presented four different models of *Smart-Equity Cities*: *i) Potential Smart/Equity City, ii) Smart City; iii) Senseable City, iv) Equitable City.*

2 Smart Cities: Paradigm of the Modern City

The recent and rapid changes that have transformed the economy, especially in the last century, inevitably have started processes of urban transformation which, however, have succeeded and sometimes superimposed in a convulsed and often unregulated drawing of the cities, both in developed countries and in the other ones. In a few years, we have seen the transformation of the industrial cities in cities of services up to the current forms of conurbation and urban agglomeration, that have had as a single common factor to extend and subtract more and more space to a more rarefied rurban space.

During the nineteenth century, the economic production function of the city takes on a prominent role on all other and the city became central to the development of national economies. This is the moment in which cities are beginning to grow and to expand in the suburbs. Thus, the economic and industrial development, the increase of population and the increase in urban population are linked, throughout the century, with a double thread in a cumulative growth. This was much more evident in the cities of the northern hemisphere (especially in North American and European ones) where the possibility of work related to the localization of mineral deposits formed a precise condition for regional development and industrial cities, which have designed the space, according to precise geographic forms, opposing workers' quarters to bourgeois ones, in an urban pattern of well distinguished areas.

But the industrial city has its beginning and its end, leaving its inheritance. The twentieth century was especially the century of the city and the suburbs. Far from growing and multiplying according to an idealized model of rationalist "modern city" rethinking existing cities, in most cases the suburbs have come to be arranged "like wildfire" around antique and nineteenth centers, laying the foundations for the development of the current shapeless metropolis and megalopolis [12].

Then, at the end of the twentieth century, the crisis of the economic model based on Fordism and the development of the tertiary sector impress a further acceleration to the city's transformation from an industrial city to a city of services, resisting the loss of its functions redesigning itself, individually and as part of the overall urban system.

The current urban transformations require us a further reflection. In fact, since 2009 for the first time in human history, the urban population has been surpassed the one established in the rural area and, according to a report from the Worldwatch Institute, by 2050 the 70% of the Earth's population will live in the city, a percentage than in industrialized countries will rise up to 84%, by the middle of the twenty-first century [29].

The uniqueness of this stage is that these changes have not resulted in the reduction or containment of urban development of the city, which continue to grow, mutate and diversify its forms. The cities that have developed and grown in the industrialization have dilated and have become territorial systems of various types in which the flows of the industry are only one element of the complexity and, through agglomeration and conurbation, they have developed over more and more vast areas.

This was due to phenomena of various kinds: the relocation of production and services, a more and more intensive use of the car and finally to the phenomena of peri-urbanization that have pushed people to migrate to places where a greater *well being* is guaranteed, or to places where the "capability set" of goods and services and access to them is retained by the citizens and, at the same time, living conditions in terms of environmental quality are better [31].

Then, what is a city today? Contemporary cities share a number of distinguishing characteristics: high density of population, a population size significantly different from that of rural area, a complexity of cultural, social and economic functions, being centroid of power and services, being creative and dynamic environments and places of contradictions and conflicts [21]. In urban areas also remains the need to provide services such as housing and education at an affordable cost, clean water, good air quality, social environment without crime, an efficient transport system, energetic sustainability and rational use of resources.

Therefore, it becomes necessary to solve these problems in a defined border of the city, and put a solution to the current uncontrolled spatial development of the same.

In particular, environmental issues have become central to economic development and their solution must start from urban areas. In fact, the cities are responsible for the majority of pollution, producing up to 70% of the total emissions of carbon dioxide, even occupying a residual portion of the earth surface. These problems appear more acute in developing countries, where the increase of the urban population are higher than in industrialized countries and it is expected that in the coming decades there will be 95% of the world's urban population growth[2].

Migrations from rural area combined with population growth cause the formation of megacities, ranging to be consolidated with a sprawl. Today, these cities are inform agglomerations, a collection of non-places that do not encourage the meeting, communication, without which the groups retire into their shells.

In Europe, the threats to sustainable urban development are taken into particular consideration and are addressed in an integrated approach, taking into account both of environmental issues and of those social, economic, cultural and political ones.

In fact, since 1999, through ESDP European Spatial Development Perspective (ESDP)[3], the EU has been beginning to lead to programs focused to an integrated territorial development, primarily oriented to territorial balance and cooperation between the cities of the local territorial systems.

Today, the end of a period of steady economic growth causes stagnation and economic decline in many cities, particularly in those ones that are not European capitals and in the old industrial cities of Western Europe. That is leading to the gradual withdrawal of the welfare state in most European countries. So, in the framework program for research and innovation "Horizon 2020" the aging population, the low-density urban sprawl, that threatens sustainable development and that makes the service more expensive difficult to secure, the over-exploitation of resources, the lack of public transport networks, the risks to biodiversity and, finally, the issues related to the protection and maintenance of the land, threatened by the widespread hydrogeological, are all considered central problems [16].

However, for some years first in the world and then in Europe, the researchers have been beginning to analyze the modern city through the paradigm of the *smart city*. The main feature of the smart city seemed to be on the role of ICT infrastructure, although much research has also been carried out on the role of human capital, the social and relational capital and the environmental quality as important drivers of urban growth [5].

Also various institutions and organizations have long devoted constant efforts to devising a strategy for achieving urban growth in a 'smart' sense for its metropolitan

---

[2] In Africa and Asia the most significant change will be, because is estimated that in 2050 a total of 86% of the increase in urban population will record [11].

[3] Approved by the Informal Council of Ministers of Spatial Planning of European Commission in Potsdam in 1999.

areas. So, we can find in the Oslo Manual (2005), developed jointly by Eurostat and the OECD, the importance of role of innovation in ICT sectors but we also can detect that a method is provided to identify various consistent indicators, that form a sound framework of analysis on urban innovation. In particular, we observe renewed attention for the role of "soft infrastructure" (governance, innovation forums and network and community organizations) in determining economic performance.

As well as Caraglio and Del Bo (2009) have written, the availability and quality of the ICT infrastructure is not the only definition of a smart or intelligent city. Other definitions stress the role of human capital and education in urban development. Berry and Glaeser (2005) and Glaeser and Berry (2006) show, for example, that the most rapid urban growth rates have been achieved in cities where a high share of educated labour force is available. In particular Berry and Glaeser (2005) model the relation between human capital and urban development by assuming that innovation is driven by entrepreneurs who innovate in industries and products which require an increasingly more skilled labour force [8]; [4].

Without going into details of the various attempts to arrive at an univocal definition of a smart city, we can summarize the different ways in which it has been interpreted the concept of smart city into three types of approaches: (1) a *techno-centered approach* characterized by a strong emphasis on "hardware", new technologies and infrastructure that ITC would be the key to the smart city, (2) a *human-centered approach* where there is a large weight of social and human capital in defining the smart city; (3) an *integrated approach* that defines a smart city from the possession of both the foregoing qualities, because the intelligent city has to ensure integration between technology and human and social capital to create the suitable condition for a continuous and ongoing process of growth and innovation.

But even this interpretation seems still limited. In fact, if a smart city is a city that knows how to exploit their human capital so that there is a creative and qualified context for economic development, other factors that are not exclusively linked to economic growth seem very important.

We know that in 2030, five billion people will live in cities, and it is clear that the "human overhead" can not be supported by the urban centers as well as designed until now and the need for smart cities and smart communities can not remain much longer concepts that are prerogative of a restricted array of experts, but will instead become a shared concept for improving the quality of life and for adapting it to the future needs of urban housing. Therefore, the development of tangible and intangible infrastructure can not be addressed only to the economic and political efficiency but rather they should promote social inclusion, quality food and good life. Neither a smart city can be identified simply as a green city. A smart city not only encourages the use of bicycles, urban green or waste recycling, but invest wisely on more items such as mobility, environment, people, liveability, governance and economy.

So, a smart city is not a project but the beginning of an overall process of *sensing* and *actuating* for the transformation of the city, where there are particular needs of citizens, active and passive actors in the process.

In conclusion, a smart city requires a combination of "smart people" and "smart governance". Where, smart people refers to citizens aware of the importance of participation in public life, capable of peaceful coexistence, responsible for their choices in life. But a smart city is also a city that considers the population one of its most important resources for the future and who knows how to direct the development policies of the questions of the community in its various phases (for example services for the elderly or for children). While, smart governance means an administration with a strategic vision of sustainable development, investing in communications and technologies for environmental sustainability and that is able to promote awareness-raising around the common good. A smart city is a city that can support the establishment of public-private partnerships, able to involve citizens in decision-making in public policy, focusing more and more on participatory processes, such as online consultations and deliberations, as well as through the activation of participated creativity workshops [30].

3 Towards a New Definition of the Concept of Smartness: The Size of the *Equity*

Although it is difficult to arrive at a single definition of the concept of "smart city", because of the variety of studies on the subject, and although the concept of smart city is difficult to measure in a uniform way, we can find an increasing number of plans, policies, actions carried out under the label "smart" in many cities around the world.

However, as pointed out by Hollands (2008), this terminological vagueness could not be just a problem of defining a uniform framework for benchmarking but, behind a deliberate choice and an artificial generality, all the contradictions that characterize the new urban forms may be hidden [23].

Hollands shows clearly that, today, there are no studies that correlate the smart city projects with the most critical aspects of the city and its transformations, as instead it had been when the *entrepreneurial city* was born (Harvey, 1989), or when the dominance of the activities and neo-liberal spaces was increasing (Peck and Tickell, 2002), and he emphasizes the risk that the smart city can be only a high-tech variation of the entrepreneurial city [22]; [28].

In fact, the growing assertion of the concept of territorial competitiveness that has had great influence on the way of understanding cities and development (through industrial clusters in Porter, 2000, the innovative milieux in Scott, 2000, and Nevarez, 2003, or creative cities in Landry, 2000, and Florida, 2002), generating a process of enterprising of urban policies, was supported in the time by the guiding principle of sustainable development at the urban scale (Gibbs, 2002; Gibbs and Krueger, 2007). This has led to the development of other paradigmatic interpretations such as: the "ecological city" (Platt, 2004), the "compact city" (Breheny, 1995), the "green urbanism" (Beatley, 2000), up to the measurement of the "ecological footprint" (Wackernagel and Rees, 1996) [17].

But the city can not be strictly sustainable, because it is an area of high concentration of fuel consumption and environmental impact; it is a "predator" organism of external resources, that stresses the energetic and environmental imbalance. The city depends on the outside (the rural area in peripheral systems or the entire world in the case of "global cities") and consumes resources for its own development.

Therefore, it is necessary to identify the criteria that make development aspects comparable to sustainability issues as well as issues of social justice in an urban scale.

In fact, in the new "smart urbanization", processes of inclusion and exclusion can be born, that are worth to be observed and analyzed in a more consistent way [19].

These transformations can redevelop parts of the city and portions of the population on the one hand, instead they may increase the gap in access to information networks of other portions of the city on the other hand. So, it is useful to refer also to the studies of the spatial justice (Soja, 2009) to analyze the processes of segmentation and social fragmentation that the introduction of new technologies can cause. [32].

To arrive at a definition that brings together different criteria of analysis and the previously mentioned aspects, a few years ago the Vienna University of Technology - in collaboration with the University of Ljubljana and the Delft University of Technology - gave birth to a research on European medium-sized cities (with population less than 500,000 inhabitants). Later, this research became the ranking instrument of approximately 1600 city of EU27, plus Iceland, Liechtenstein, Norway and Switzerland.

This project, called "European smart cities", was born as part of a wider project ESPON 2013 (ESPON Project 1.1.1) and showed not only a final ranking of 70 cities, but it has remained a reference model to identify factors that make cities "smart".

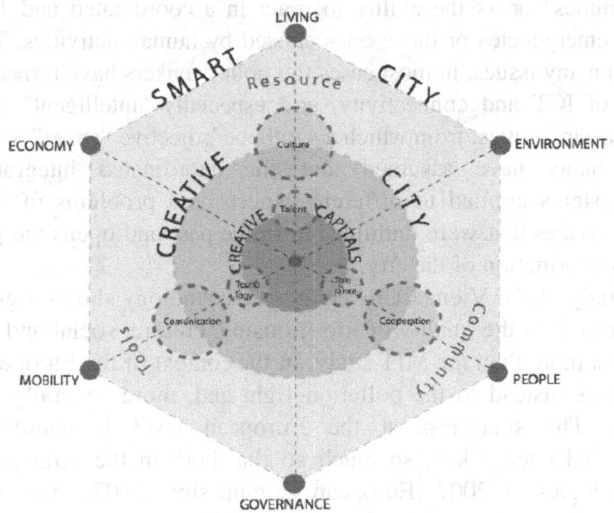

Fig. 1. Six main axes of smartness and three C of the creative city (Carta M., 2013)

In this context, smart cities can be identified and ranked along six main axes or dimensions, that are: *a smart economy; smart mobility; a smart environment; smart people; smart living; and, finally, smart governance* (Fig.1). These six axes connect the traditional theories of urban growth and development, with the modern aspect of sustainable development of a city. Then, a middle city can be defined as "smart" when investments in human and social capital and traditional (transport) and modern (ICT) communication infrastructure fuel sustainable economic development and a high quality of life, with a wise management of natural resources, through participatory governance [8].

In the last years, the scientific debate has developed at the same rate with political investments, to rethink the cities in a "smart" way. In fact, according to the report GreenItaly of 2011, these investments will bring a worldwide turnover of about $ 39 billion in 2016, compared to 8 in 2010, and in the same five-year period, the city will spend $ 116 billion to become "smart structures".

But whole design of smart cities, in terms of policies, plans and actions was oriented to predominantly engineering and selective interventions in comparison wit the urban areas and portions of the population affected by them. The implemented measures concerned essentially the "high impact" sector (as the energy, the transport of goods, the mobility, the waste management, etc.), based mainly on high-tech solutions [19].

Although sustainability has also been seen so far strictly in energy and environment key, through choices and technologies that save energy, or from a functional point of view, through integration of e-participation techniques such as online consultation and deliberation over proposed service changes to support the participation of users as citizens in the democratisation of decisions taken about future levels of provision [18].

This has stimulated the development of new paradigms of sustainability, which is today understood as quality of life given by social participation, a key element of the "smart communities" or as the ability to react in a coordinated and flexible way to environmental emergencies or those ones caused by human activities. To achieve this "glue" between many issues, in most cases the policy makers have turned to a large and advanced use of ICT and connectivity, and especially "intelligent" technology and systemic design capabilities, from which a synthetic 'adjective "smart" was born. [3]

Therefore, many have assumed that the coordinated integration of more "intelligent" systems applied to different aspects and problems of the city makes possible new services that were unthinkable in the past and opens the possibility of a progressive transformation of the city.

But if the study of the Vienna University of Technology shows a greater attention to the issues related to the quality of life (housing, culture, social and environmental conditions, etc.), more than the MIT study, in the context of the European institutions, the concept binds instead to the pollution fight and, more generally, to the Europe 2020 strategy. The smartness at the European level is mainly read in the environmental and energy key, so much so that both in the Strategic Plan for the Energy Technologies of 2007 (European Commission, 2007), and in the resulting Technology Roadmap (European Commission, 2009), there is precise and explicit reference to the smart city and a specific budget dedicated to this axis [19].

And also, in 2012, the European Commission launched a specific initiative for the development of smart cities of the Old Continent: "Smart Cities and Communities European Innovation Partnership". This program has provided € 365 million for innovative ideas and demonstration projects within the energy, transport and ICT in urban areas [9]. These policy (initiatives) are then witness to a European commitment to the sustainability of our cities, especially viewed in terms of technological innovation, in order to reduce the load of greenhouse gases and to improve the quality of the life of the citizens.

Though, all this is directed at a fair balance in the development of a city, it is still insufficient and we need to continue to question the meaning we want to give to smart option in the city and, therefore, to understand the implications.

If the "Smart City" is a city where investments in human and social capital , participation in the processes and technology infrastructure fuel sustainable economic and competitive development, ensuring a high quality of life and providing for the responsible management of natural resources and from a social participatory governance , the "SENSEable city" is (the) one that encourages dialogue between the different actors of the urban reality and that promotes more informed decisions for the development of the city in all its parts and components, with a new participatory approach to urban development and with a more efficient and equitable use of resources and networks. The transition from a Smart to a SENSEable city is, therefore, allowed by a new dimension of analysis, the "equity". The equity is, in fact, the dimension that completes the process of "smartness" of a city in terms of ensuring the development SENSEable City.

In this regard, it is interesting to take the warning issued by the sociologist and economist Sassen (2011), who believes that the new challenge is the attempt to "urbanize the technologies", that make them actually useful to new urban needs [33]. Therefore SENSEable city is a city that is also smart and fair.

Following, the size of the equity will be used to reread the ranking of the *Top Ten Global Smart Cities* and the *Ten Smartest European Cities* by Boyd Cohen.

4 The Reinterpretation of the Concept of Smartness According to the Values of the *Equity City Index*

In the debate on smart cities, next to the attempts of definition and declination of the concept of smartness, have been conducted a series of studies to formulate the rankings, through which to read the performance of the most smart cities at global, regional and local level. At the base of these classifications, the multiple dimensions that in time have been used to measure the ability of a city to be smart and, therefore, the search continues for a parameter, a variable, an indicator that better than others to be able to explain and measure this ability.

The objective of this work is not a discussion of the benchmarking existing or definition of a new system of dimensions and set of indicators, but will try an analytic experiment much more limited: to compare the traditional dimensions of *smartness* (competitiveness and sustainability), use in different classifications and taxonomies quantitative (ranking), with the concept of *equity* as well as declined by UN-Habitat, in order to re-read these dimensions according to the values of the *Equity City Index*.

Of course, from a methodological perspective, it is evident that the quantitative indicators used in the analysis constitute a partial representations of reality and an extremely simplification of complex and multidimensional phenomena: in other words, there are many ways to be "smart" and many ways to be "equity". In this sense, it is a comparison between representations, namely between discourses (which use a technical-quantity language) relating to the ideas of *smartness* and *equity* previously presented.

Accordingly, the present analysis is developed starting from the study and comparison of three urban classifications produced by different actors and that presuppose logical and analytical perspectives different: two related to the concept of *Smart city* and one of *Equity city*. For the dimension "smart", the study examines two rankings: the ranking of the *Top Ten Global Smart Cities* by Boyd Cohen[4] published in the journal Co.Exist and a further ranking by the by the same Cohen of the *Top Ten Smartest European Cities*. The rankings compiled by Cohen adopt a scientific approach to the topic of smart cities, the result of a measurement system of "intelligence" based on the metric of "Smart Cities Wheel" of the Vienna Polytechnic, declined through six "key components" and three "factors key" for each component.

The six key components identified by Cohen are: "smart economy" (intelligent econo-my), "smart enviro" (intelligent environment), "smart gov" (government policies), "smart living" (liveability), "smart mobility" (mobility), "smart people" (people). Each component has, in turn, three "key factors" that determine the positioning in the final ranking[5].

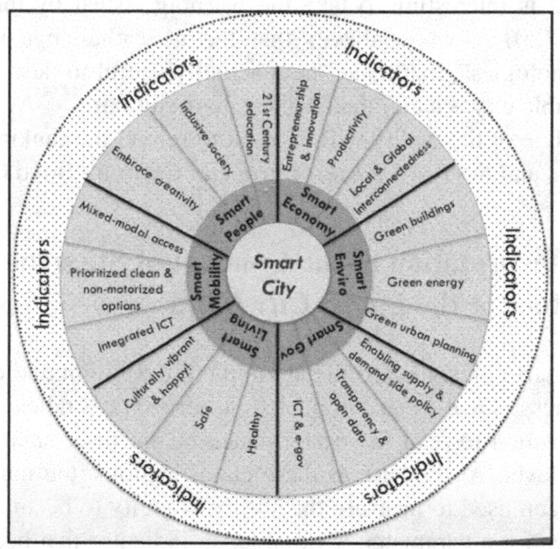

Fig. 2. The Smart Cities Wheel

Cohen's vision of smart city is inclusive of all of these dimensions, even if the quality of life is the central element: "*Smart cities use ICT to become more intelligent and efficient in the use of available resources, with the effect of reducing costs and energy consumption and at the same time, improve the delivery of services and quality of life citizens, reducing the ecological footprint and developing innovative and sustainable economy*" [14].

---

[4] Boyd Cohen Ph.D., LEED AP, is an, researcher, professor, and consultant expert in planning strategies in the context of the sustainable issues related to climate, to urban development and to urban sustainability. He is author of numerous publications on the subject.

[5] For example, in "smart economy", Cohen has put "entrepreneurship & innovation", "productivity", and "local & global interconnectedness".

At the base of the study, several regional and global rankings theme of innovation, environmental sustainability, quality of life and digital governance[6].

Table 1. The Top Ten Smart Cities on the Planet[7]
Source: www.factorexist.com

Ranking	Innovation	Environment	Quality of Life	Digital City
1) Vienna	5	4th in Europe	1	8
2) Toronto	10	9th in North America	17	10
3) Paris	3	10th in Europe (RC: 6)	30	11
4) New York	4	3rd in North America (RC: 8)	47	4
5) London	11	11th in Europe (RC: 9)	38	13
6) Tokyo	22	Above Average in Asia (RC: 10)	46	15
7) Berlin	14	8th in Europe	17	32
8) Copenhagen	9	1st in Europe (RC: 1)	9	39
9) Hong Kong	15	Above Average in Asia	70	3
10) Barcelona	19	NR in Siemens (RC: 3)	40	2 in IDC
Reference values	**125**	**22 Asia, 30 Eu, 27 NA, 10 RC**	**221**	**87, 44 IDC**

Vienna leads the overall standings of the Top Ten Smart Cities on the Planet and is the only city to rank among the top ten in all categories, with ambitious goals and programs like Vision Energy Smart 2050, Roadmap 2020 and Action Plan 2012-2015[8]. Toronto follows, active member of the network of cities Clinton 40 (C40) for the

---

[6] In detail, the sources used by the author for each dimensions: The Brookings Institute and Europe's biggest Startup cities (*1. Smart Economy*);. Siemens Green City Index, List of membership of Cities Climate Leadership Group (C40) and Local Governments for Sustainability (ICLEI), the Carbon Disclosure Project for Cities. For some cities, this dimension has been integrated with the Resilient Cities ranking (RC) developed by the same author on the basis of efforts by the city in this field (*2. Smart Environment*); The Worldwide E-Governance Report 2011, The City Data Sets-Data Catalogs, The City Protocol. For some cities, this dimension has been integrated with the IDC Ranking (*3. Smart Governance*); Mercer Quality of Living Report (*4. Smart Living*); The Eurotest - Quality Safety Mobility, Siemens Transit Ranking, The Emerging Markets Trade Association (*5. Smart Mobility*); The Economist Global Competitiveness Rankings. (*6. Smart People*).

[7] The size of each Ranking:

- *Innovation*: Innovation Cities Top 100 Index: 162 indicators grouped into 31 sectors, studying a total of 331 cities in 2011, of which 125 were classified;
- *Environment*: Green City Index: three areas studied (30 cities in Europe, 27 in North America and 22 in Asia);
- *Quality of Life*: Mercer Quality of Living 2012: observing more than 460 cities, 221 of them were classified on the basis of ten categories with 39 factors;
- *Digital Cities*: Digital Governance in Municipalities Worldwide: we have studied the 100 most wired nations in the world according to data from 2009 Telecommunications Union (ITU) of the UN.

[8] The three programs, respectively of long, medium and short term, within the project "Smart City Wien" proposed for the call "Smart Energy-Demo - Fit for SET" dall'Austrian Climate and Energy.

reduction of greenhouse gas emissions, and promoter of the project "Smart Commute Toronto", launched in 2001 to reduce the number of cars and facilitate urban mobility through a service of carpooling, currently used by 12,000 commuters and that has reduced from 25 to 30,000 transfers in the car in a month. The other North American city is New York present in fourth place, but penalized in terms of quality of life.

In the top ten proposed by Cohen there are, together Vienna, five other European cities: Paris (third) appreciated in different categories, London (fifth) awarded for actions in the field of sustainability as the "London Congestion Charge" and the digital, Berlin (seventh) with good results in innovation and the environment (eighth in Europe), Copenhagen (eighth) evaluated as number one in the ranking of green cities in Europe proposed by Siemens with a leadership role for sustainable innovation and, finally, Barce-lona to close the ranking as one of the first cities to be interested in smart city planning.

Two, finally, the Asian cities: Tokyo and Hong Kong, respectively, in the sixth and ninth position. Tokyo has very high values in innovation and in digital, thanks to a series of plans for the creation of a smart grid in the periphery of the city, Hong Kong, instead, achieves good performance in key areas including e-governance, but is penalized in the quality of life. The city is also leader in the adoption of smart cards (Smart Identity Card and the Octopus Card), used by millions of residents for services such as public transportation, access to libraries, payments in shops and car parks, etc.

The European, Asian and North American cities in the top of ranking are different in many aspects, starting from culture and history of urban planning. Many North American cities have been built when the car have started to spread. So, big roads and suburbs have had priority over the public transport. The European cities depart, instead, advantaged in the transition to the smart city as more oriented to public transport and denser. This increa-sed density makes possible, for example, a more efficient applications of ICT solutions.

One area in which, however, the United States is in the lead is to support entrepreneurial ecosystems. In many American cities, in fact, research institutions are partners of EU initiatives and this helps to spread innovation and local economic development.

The six European cities included in the *Top Ten of Smart Cities on the Planet* also includes in the ranking of the *Top Ten Smartest European Cities* written by the same author in 2013[9]. Into the *Top Ten Smartest European Cities*, Copenhagen is the city with the best performance, getting the first place in Environment and People according to the City group/Economist Intelligence Unit Global City Hotspots Report but, however, with high scores in all components of the Wheel.

---

[9] The dimensions of analysis are integrated, compared to those seen with "The Brookings Institute and EU Startups" per la *Smart Economy*; with participation in the C40, ICLEI and Carbon Disclosure Project for Cities for the Smart Environment; with the Worldwide report E-Governance 2011, City Data Sets-Data Catalogs, and with participation in the City Protocol for *Smart Governance*; with Eurotest del 2010 on public transport, the Siemens Transit Ranking, and the Barometer of EMTA 2009 for *Smart Mobility*; with ranking of Economist Intelligence Unit for *Smart people*. Cfr. [15].

Appear, instead, for the first time the city of Stockholm, Amsterdam, Monaco and Frankfurt, with the first two excluded for very few positions from the list of *Top Ten Smart Cities on the Planet*. Specifically, Stockholm has a good position in the economic and environmental dimension, while Amsterdam is the undisputed leader in Europe in non-motorized transport: almost 70% of all displacements takes place on foot or by bicycle. The city is in the top five in Smart People, Smart Environment and Smart Living.

Table 2. The Top Ten Smartest European Cities
Source: www.factorexist.com

Rankig	Economy	Environment	Governance	Living	Mobility	People
1) Copenhagen	7	1	8	4	3	1
2) Stockholm	2	2	5	7	5	4
3) Amsterdam	6	3	7	5	1	5
4) Vienna	10	8	1	1	2	7
5) Paris	4	5	4	8	7	2
6) Berlin	1	4	6	6	6	9
7) London	5	6	3	9	10	3
8) Barcelona	8	7	2	10	4	6
9) Munich	3	9	9	2	8	10
10)Frankfurt	8	10	10	3	9	8

In this classification Vienna gets the fourth place, positioning first or second in three components (Mobility, Governance and Living). Paris, with a fifth place in the ranking, is among the top five in four dimensions (Economy, People, Governence and Environment). Berlin, elected as the first city in Europe to start up business by EUStartups is, instead, first in the Smart Economy and fourth in Smart Environment.

London occupies the highest places for Governance and People, but it is placed at the end of the rankings for Quality of life and Mobility. Barcelona, finally, distinguishes itself for governance, with the highest volume of open data and a management commitment to make Barcelona a smart city model for aspiring smart cities around the world.

Let us now to re-read the rankings established by Cohen according to the values of the Equity City Index compiled by UN-Habitat 2012[10].

In the Report "State Of The World's Cities 2012/2013, prosperity of Cities" is explained the concept of equity in the composition of the index. It is measured, in particular, taking into account the distribution and redistribution of the benefits of prosperity of a city, with a consequent reduction of poverty, the supply of adequate housing, the protection of the rights of minorities and vulnerable groups, of gender equality and the public participation in political and cultural life of society.

From the comparison of rankings, the new classification according to equity leads to different distributions. In ranking order according to equity, in fact, are the great

[10] UN-HABITAT (2012), "State Of The World's Cities 2012/2013, prosperity of Cities", World Urban Forum Edition.
`http://www.unhabitat.org/pmss/listItemDetails.aspx?publicationID=3387`

metropolises of North America such as New York and Toronto and the mega-city of Hong Kong[11] to present the lowest values, as characterized by congestion, real estate prices out of control and growing disparities in income. The only major city that improves its positioning is Tokyo, despite its weakness in terms of quality of life.

Among European cities, Vienna confirms its discreet positioning and London his fifth position, while Copenhagen recovers seven positions relative to the global ranking of Cohen, with the best performance according to the City Equity Index. Barcelona scale two positions, Amsterdam and Stockholm achieve a good positioning, while Paris lost three positions, with a value of *Equity City Index* less than 0,8.

Table 3. Comparison between Rankings
Source: www.factorexist.com.; State Of The World's Cities 2012/2013 www.unhabitat.org

Ranking	Innov.	Environment	Q.L.	D.C.	Equity	Reclassification
1) Vienna	5	4th in Europe	1	8	***0,883***	= (2)
2) Toronto	10	9th in North America	17	10	***0,733***	(9)
3) Paris	3	10th in Europe (RC: 6)	30	11	***0,788***	(6)
4) New York	4	3rd in North America (RC: 8)	47	4	***0,502***	(10)
5) London	11	11th in Europe (RC: 9)	38	13	***0,793***	= (5)
6) Tokyo	22	Above Average in Asia (RC: 10)	46	15	***0,828***	(3)
7) Berlin	14	8th in Europe	17	32	***n.d.***	= -
8) Copenhagen	9	1st in Europe (RC: 1)	9	39	***0,922***	(1)
9) Hong Kong	15	Above Average in Asia	70	3	***Unequal***	(11)
10) Barcelona	19	NR in Siemens (RC: 3)	40	2(IDC)	***0,755***	(8)
11) Amsterdam	6	5th in Europe	12	43	***0,818***	(4)
12) Stockholm	21	2nd in Europe	19	34	***0,767***	(7)
Reference Values	**125**	**22 Asia, 30 Eu, 27 NA (10 RC)**	**221**	87 10 (IDC)	***0,217-0,967***	

The entity and nature of the different distributions, together with the assessments made, allow you to develop a series of reflections between the ideal of city "smartness" and of city "equity". Considering, in fact, the *SENSEable city* according to an inclusive and sustainable approach as a smart and fair city, in which the technology is aimed at increasing the participation and wellness of all citizens, the position of cities like New York, Toronto and Hong Kong can be questioned. These could leave the place to cities like Melbourne, Seattle, Stockholm, Vancouver, Amsterdam that, though smaller in size and weight internationally, are characterized by a strong focus not only on innovation and environment, but also on social and equity.

Other authors such as Joel Kotkin, while not establishing a ranking, include in the list of The World's smartest cities published in Forbes magazine in 2009, cities such as Amsterdam, Seattle, Curitiba, Monterrey (these cities are also included in the list Smart Cities in the world, edited by CITTALIA-foundation ANCI), while excluding cities like New York, Tokyo [26];[13]. Among these cities Seattle, in particular, with its energy policies based on innovative solutions has been able to achieve and exceed

[11]Hong Kong does not have a numeric value of the index, but the report of the UN-Habitat defines one of the most unequal in the world.

the goals of the Kyoto Protocol to which the United States have not acceded, and Mayor Greg Nickelst has spread its commitment to other U.S. cities, promoting the U.S. Conference of Mayor's Climate Protection Agreement, consisting of a federal network that includes 590 municipalities.

On the other hand, if on the one hand, the great metropolis, with a high concentration of economic activities relevant, tend to be the main object of urban research and, therefore, of smart practices, other numerous studies primarily within Europe, put in evidence as are the medium-sized cities to host the majority of the urban population, with a high potential for developing "intelligent" due to the fact that the average size enables their attitudes more smartness with a greater propensity to equity. In this regard, for example, in the *Ranking of European medium-sized cities* carried out by the Vienna Polytechnic in cooperation with the University of Ljubljana and the Delft Polytechnic, are the Scandinavian cities of the Benelux and of the Austria to dominate the overall ranking of smart cities, with Luxembourg in the first place [20].

5 A Possible Evolution of the *Smart City*: Four Different Models of *Smart-Equitable Cities* and Policies

In this paper, we have presented an overview of the concept of the "smart city", with a critical review of the traditional approaches to this concept.

The relationship between *smartness* and *equity* is far from linearity, as shown both by theoretical considerations proposed above, both from simple benchmarking of the rankings presented. in relation to the assessments exposed is, however, interesting to try to outline possible models of *Smart-Equity City* according to different combinations of values of *smart* and *equity*. To this end, we can consider a high perfomance in the *smartness* component when the city is ranked among the top four in at least three of the six dimensions considered (see Table 2), and a high performance in the *equity* component when the Equity City Index has values between 0.7 and 1.

Table 4. Matrix Smartness/Equity. Possible models of *Smart-Equitable City*
Source: Our elaboration

Smartness			
	High	**SMART CITY** Toronto; New York; Hong Kong; Paris; Amsterdam; London	**SENSEABLE CITY** Vienna; Copenhagen; Stockholm
	Low	**POTENTIAL SMART/ EQUITY CITY** Berlin; Barcelona	**EQUITY CITY** Tokyo
		Low	**High**
		Equity	

From the combination of chosen parameters we obtain four different models of Smart-Equitable Cities: *i) Potential Smart/Equity City, ii) Smart City; iii) Senseable City, iv) Equitable City.*

i) Potential Smart/Equity City. To low performance in both components (*smartness and equity*) corresponds a model of city still largely tied to a first view of the paradigm of smart cities, where the main feature of the smart city seemed to be the role of the ICT infrastructure and environmental sustainability as important factors of urban growth, with low investments in planning, in designing and creating social infrastructure, in the enhancement of social capital and in the processes of inclusion as widely as possible, according to equity. It is, therefore, a model of the city that has the potential to become a *smart or equity city*, but it is not yet fully. Among the cities considered we could define *Potential Smart/Equity City* cities like Berlin and Barcelona, with excellent performance in terms of the Smart Economy and Smart Environment and important examples of strategic planning oriented to the smartness, but not yet able to respond to a growing need for quality of life, social and good governance on the part of citizens and away from processes of inclusion. It is on these aspects that should point to future policies and future plans for the creation of the city really "smart" and "equitable".

ii) Smart City. They are cities that have high values in smartness, but not yet high in equity. It is a model of city that reaches, in fact, high performance in terms of economic competitiveness, creativity, in the field of technology and e-government, with particular attention to aspects green, but in many cases the urban development is focused on only economic efficiency, environmental and policy to achieve still mainly through technological innovation, with less emphasis on factors of social inclusion, quality of life and sociality. One example are cities such as New York, Hong Kong, Toronto, Amsterdam, London and Paris that fall in the rankings of the leading smart cities in the world for the considerable investments in digital, in green innovation, in sustainable urban mobility, in e-governance, etc. For these cities is, therefore, essential to continue along this process, but improving equity through policies and programs oriented to Social Cohesion, to the partecipation in decision-making, partecipation in public life, accessibility and usability of e-government, etc.

iii) Senseable City. With a high performance in both dimensions, the *Senseable city* can be defined the model that best responds to the concept of "Smart City-Equity". This is not simply a city "affirmed" in the process of smart city, but with a strong and constant attention to the needs of citizens. The *Senseable City* is, therefore, a liveable city, which has been able to seize the opportunities offered by the external environment in terms of competitiveness and technological and digital development, by implementing policies directed at city and citizens based on innovation but, at the same time, investing in the enhancement of territorial identity, creativity and human development, with improvements to education, employment, health care, social assistance and social and cultural integration. This is the model developed in recent years by cities such as Vienna, Copenhagen and Stockholm that represent a successful model to be imitated not only on a European scale, but global.

iv) Equitable City. With an average/low value in terms of smartness, but very high in the equity component, the *Equitable city* is a city model that interprets urban development as a result of processes of distribution and re-distribution of the economic benefits of a city, with consequent reduction of poverty, protection of rights of minorities and vulnerable groups, gender equality and participation of citizens in social, political and cultural life. Many the opportunities yet to be exploited, by increasing the use of inclusive technologies, of technological and digital capital, and the management of local governance.

The paradigm of smart city outlined by the studies of definition and measurement of its smartness shows a city that makes of the technologies a key tool to improve urban sustainability. The common vision is that of a wired city, which uses the network infrastructure such as business services and ICT to support the development of social, cultural and urban through connectivity [8].

It is clear, based on the considerations set out above, how this is now a reductive concept of smart city to overcome.

The future smart policies should, therefore, take into account some elements: it is necessary, first, put the citizen at the center of the interventions of smart city, considering ICT as a "human-oriented technologies", ie you have to orient technologies to improve the quality of life and the realization of the real needs of man. If on the one hand, technological innovations reduce the contribution of man, on the other hand they can play an important role when used to create a "sense of community", making it possible cooperation between public institutions, businesses and citizens in joint projects .

These policies must also incorporate the concept of *SENSEable City*. A smart city takes the form of *SENSEable City* when its citizens become active actors, informed and motivated to be involved in the modelling of the city on their own needs. In this sense, is fully shareable the vision of Carlo Ratti[12], who advises to the Administrators of cities of replace a smart approach that has as its focus the large firms that offer engineering solutions according to a "top down approach", with a "bottom up approach" which involves the citizens, whose good behaviors are replicate "in a kind of social contagion". This must be the basis of intelligent city, as well as the smart city is the backdrop to a more smart world.

Give attention to the citizens also implies a greater consideration to the periphery of city, not only to increase the participation, but also social inclusion. The programs and plans of action are, in fact, often limited to the city center, operating only in part as a tool for urban regeneration and rebalancing between the heart and other parts of the city.

Citizens are, in fact, the passive actors but above all active of what must be understood as a process rather than a project (*build a smart city*). To do this, you need to make a long-term policy agenda, with interventions in the short and medium term, to allocate resources to the world of work, of education, of professional formation, of sustainable transport, of communications and enabling technologies, in a continuous manner.

This is not only to cope with global pressures (from climate change to poverty latent), but also to reverse or at least mitigate an ongoing process, probably

[12] Carlo Ratti is founder and director of the MIT Senseable City Lab, a research laboratory founded at the Massachusetts Institute of Technology to study the changes in the technological sense in urban areas.
http://senseable.mit.edu/papers/pdf/2012_Ratti_WIRED.pdf

unstoppable, very dangerous [33]. The planning of interventions aimed at making cities intelligent must be, therefore, built as a “roadmap” including a well-defined overall plan, according to objectives and feasible projects every 2-3 years, free from innovations sectoral or generalists [3].

References

[1] AA.VV.: Smart Cities nel mondo, CITTALIA-fondazione ANCI Ricerche (2012)
[2] AA.VV.: Smart Cities in Italia: un’opportunità nello spirito del Rinascimento per una nuova qualità della vita. ABB e The European House-Ambrosetti (2012), http://www.abb.it/
[3] Annunziato, M.: La roadmap delle Smart Cities. Energia, Ambiente, Innovazione 4-5(1), 33–42 (2012)
[4] Berry, C.R., Glaeser, E.L.: The divergence of human capital levels across cities. Regional Science 84, 407–444 (2005)
[5] Berthon, B., Guittat, P.: Ascesa delle città intelligenti. Outlook 2 (2011), http://www.accenture.com
[6] Camagni, R.: On the concept of territorial competitiveness: sound or misleading? Urban Studies 39(13), 2395–2411 (2002)
[7] Campbell, T.: Beyond smart cities. How cities network, learn and innovate. Earthscan, Londra-New York (2012)
[8] Caragliu, A., Del Bo, C., Njkamp, P.: Smart cities in Europe. Paper presented at the Conference III Central European Conference in Regional Science, CERS (2009)
[9] Cardone, M.: La rivoluzione delle smart city è in corso. QualEnergia.it (2012), http://www.qualenergia.it
[10] Carta, M.: Città Creativa 3.0. Rigenerazione urbana e politiche di valorizzazione delle armature culturali. In: Cammelli, M., Valentino, P.A. (eds.) Citymorphosis, pp. 213–221. Politiche culturali per città che cambiano, Giunti, Firenze (2011)
[11] Centro Regionale di Informazione Nazioni Unite (Unric): Città, 18 giugno 2012 (2012), http://www.unric.org/it
[12] Ciorra, P.: La fine delle Periferie. Nascita e morte della periferia moderna (2010), http://www.treccani.it
[13] Cittalia: Smart cities nel mondo. Roma (2011)
[14] Cohen, B.: The Top 10 Smart Cities on the Planet (2012), http://www.fastcoexist.com/1679127/the-top-10-smart-cities-on-the-planet
[15] Cohen, B.: The Top 10 Smartest European Cities (2013), http://www.fastcoexist.com
[16] Comunicazione COM (2011) 808 della Commissione al Consiglio, al Parlamento Europeo. Programma quadro di ricerca e innovazione “Orizzonte 2020” (2011)
[17] Crivello, S.: Competitive city and sustainable city: some reflections on the relationship between the two concepts. Sociologia Urbana e Rurale 97, 52–67 (2012)
[18] Deakin, M.: From city of bits to e-topia: taking the thesis on digitally-inclusive regeneration full circle. Journal of Urban Technology 14(3), 131–143 (2007)
[19] De Luca, A.: Come (ri)pensare la smart city. EyesReg Giornale di Scienze Regionali 2(6), 143–146 (2012)

[20] Giffinger, R., Kraman, H., Fertner, C., Kalasek, R., Pichler-Milanovic, N., Meijers, E.: Smart Cities - Ranking of European medium-sized cities. Centre of Regional Science, Vienna (2007), `http://www.smart-cities.eu`

[21] Greiner, A., Dematteis, G.: Geografia umana. Un approccio visuale, Torino, UTET (2012)

[22] Harvey, D.: From Managerialism to Entrepreneurialism: the Transformation. Urban Governance in Late Capitalism, Geografiska Annaler 71(1), 3–17 (1989)

[23] Hollands, R.G.: Will the real smart city please stand up? Intelligent, progressive or entrepreneurial? City 12(3), 303–320 (2008)

[24] Komninos, N.: Intelligent cities: innovation, knowledge systems and digital spaces. Spon Press, London (2002)

[25] Kotkin, J.: The World's Smartest Cities (2009), `http://www.forbes.com`

[26] Komninos, N.: Smart Cities are more competitive, sustainable and inclusive. Cities. Brief (2) (2011)

[27] Papa, R.: Smart Cities: Researches, Projects and Good Practices for the City. TeMA Journal of Land Use Mobility and Environment 6(1) (2013)

[28] Peck, J., Tickell, A.: Neoliberalizing space. Antipode 34(3), 380–404 (2002)

[29] Potter, G.: Urbanizing the Developing World (2012), `http://www.vitalsigns.worldwatch.org/vs-trend/urbanizing-developing-world`

[30] Ratti, C.: Smart city, l'onda può partire da noi, Avoicomunicare (Telecom) Interview, `http://www.avoicomunicare.it/blogpost/ambiente/smart-city-l-onda-puo-partire-da-noi`

[31] Sen, A.K.: Commodities and Capabilities. Oxford University Press, Oxford (1985)

[32] Soja, E.W.: Seeking Spatial Justice. University of Minnesota Press, Minneapolis (2010)

[33] Sassen, S.: Who needs to become "smart" in tomorrow's cities. keynote speech at the LIFT Conference. The Future of Smart Cities (2011)

[34] Schaffers, H., Komninos, K., Pallot, M.: Smart Cities As Innovation Ecosystem Sustained by the Future. Internet, Fireball White Paper (2009), `http://www.fireball4smartcities.eu/`

[35] Shapiro Jesse, M.: Smart Cities: qualità della vita, produttività, e gli effetti di crescita del capitale umano 88(2), 324–335 (2006)

[36] Woods, E., Bloom, E.: Smart Cities, Intelligent Information and Communication Technology Infrastructure in the Government, Buildings, Transport, and Utility Domains, Executive Summary, Pike Research. Cleantech Market Intelligence (2011)

The Geographic Turn in Social Media: Opportunities for Spatial Planning and Geodesign

Michele Campagna

University of Cagliari, Cagliari, Italy
campagna@unica.it

Abstract. This paper introduces the concept of Social Media Geographic Information (SMGI) as an emergent pluralist source of information which – it is argued- may find valuable application in spatial planning and Geodesign. On the base of empirical research, the author proposes a tentative framework for SMGI Analytics in spatial planning. Among other methods, Spatial-Temporal Textual Analysis (STTx) is proposed as a tool to investigate people perceptions and interest in space and time. Possible implications and benefits of SMGI analytics for the planning practice emerge from the overall discussion.

Keywords: Social Media, Geographic Information, Spatial Planning, Geodesign, Spatial-Temporal Textual Analysis.

1 Introduction

The term Geodesign [19] has recently gained popularity among scholars in planning and related disciplines to refer to an approach to planning and design as an integrated process which includes project conceptualizations, knowledge building, design of alternative scenarios, evaluation of impacts, decision-making, collaboration and participation, wherein all activities are grounded on robust scientific geographic knowledge support. Unlike the term itself, many of the concepts that shape such an approach are not new and actually rooted in the environmental planning and landscape architecture tradition, as well as in the sustainability of development discourse. In fact, Geodesign in a nutshell refers to a process able to inform design by geography in its broader holistic sense, including its physical, biological, social, cultural facets. Rather the innovation in the term Geodesign springs from the awareness that current developments in Information and Communication Technology (ICT) and Geographic Information Systems (GIS) provide unprecedented power for more effective use of scientific and societal knowledge in planning, design and spatial decision-making [6].After two decades of academic research, paralleled by more limited experimentation in the practice, many scholars and industry experts argue technology is nowadays mature enough to overcame former barriers to a proper exploitation of ICT support in the planning and design practices, which until recently have been acknowledged to be related among other issues to lack of specific training and data availability [7]. While

B. Murgante et al. (Eds.): ICCSA 2014, Part II, LNCS 8580, pp. 598–610, 2014.

on the one hand the Geodesign movement is fostering the diffusion of Geodesign programs in several North American Schools of Planning such as Pen State University or Northern Arizona University, on the other hand the consolidation of new digital data sources may help to overcome the data barriers and even foster, in some cases, the ICT and GIS diffusion in the planning practice.

Complex problem-solving and decision-making workflows in the spatial planning and design processes may nowadays be supported by an avalanche of digital data which thanks to powerful geospatial analytics and computing functions can help us to achieve more informed – thus arguably more sustainable- decision-making. As a matter of facts, in the last decade we faced tremendous development and diffusion of both authoritative and user generated spatial data resources.

We can refer to at least two major categories of data resources which become widespread available in many developed and developing countries, namely i) Spatial Data Infrastructure [15] and ii) Volunteered Geographic Information (or VGI) [9] [20]. These two types of resources are very different in nature, but both of them are so relevant to planning and design which can eventually drive the evolution towards new practices such as Geodesign, which in turn, if we accept the assumption that more informed decision-making is valuable, should contribute to the achievement of more sustainable development processes [4].

The results of the study presented in this paper deals with the above issues focusing on a special type VGI, which recently gained widespread diffusion: Social Media Geographic Information (SMGI). SMGI may be defined as the georeferenced multimedia which every second is publicly shared by social network (SN) users. This paper discusses why SMGI is special and proposes a novel analytics which - is argued- may find valuable application in urban and regional planning.

In the light of the above premises in the next section some issues on the relevance of new GI resources for spatial planning are discussed. The third section presents the case study and the fourth one reports on the results of the research on SMGI analysis developed by the author, on the base of which a SMGI analytics framework is proposed. The last section briefly synthesizes the overall results and adds some concluding comments.

2 Innovation in Planning Data Sources

The first generation SDI paradigm dates back to the 1990s or earlier. In 1994 President Clinton's Executive Order 12906 defined the National SDI in the United States as "the technology, policies, standards, and human resources necessary to acquire, store, distribute, and improve utilization of geospatial data" to be created on the assumption that availability of geographic Information is critical in promoting sustainable development. Since the last decade SDIs are facing widespread diffusion and development in many countries worldwide. In Europe the implementation of the Directive 02/2007/EC requires Member States to contribute through the development of National SDIs to the construction of the Infrastructure for Spatial Information in Europe (INSPIRE) with the objective of easing the sharing and the integration of spatial data in order to improve decision-making for sustainability of development. With regards to 34 data themes, the INSPIRE Directive requires spatial data produced by the public authorities at all levels to be shared according to interoperability standards and made publicly accessible through geoportals at no costs.

An implication for spatial planning is that planning professionals can seamlessly access and use official spatial data for their work. Actually in practice several cases are found for which it is a mandatory requirement by the public administration for the practitioners to use official data available in the SDI in the plan-making process. Along with the process of preparation, adoption, and implementation of the INSPIRE Directive in facts, several regional governments across Europe in the last decade already started to build their Regional SDI which in many case can be considered already advanced in their development [5], and their impact is slowly bringing innovation into the planning practice: as an example in Italy, in some Regions such as Lombardy and Sardinia, the availability of spatial data and services of the regional SDIs through their geoportals are eventually enabling the implementation of policy measures which foster the use of GIS in plan-making [5] making it compulsory to use the Regional SDI data in the process. The evolution in the professional practices can be considered still in its very infancy, but the ground is becoming more fertile for a wider diffusion of Geodesign methods.

Beside institutional initiatives which are giving authoritative spatial data public access on the web, in the last decade or less the diffusion of the geobrowsers and GPS-enabled mobile devices fostered the widespread production, collection and sharing of georeferenced information by Internet users. Since few years the term of Volunteered Geographic Information (VGI) became popular to indicate the avalanche of directly or indirectly (i.e. through geocoding) georeferenced information which every second is shared on the web by citizens acting as sensors [8,9]. In broader terms VGI may include both geographic information collected by groups of people within crowdsourcing initiatives (e.g. OpenStreetMap, OSM) and geo-tagged multimedia (e.g. images on Flickr) collected for personal purposes by Internet users and publicly shared through archives in the cloud. VGI has been proven useful in many application contexts such as emergency response, environmental monitoring and spatial planning [17]. As a matter of facts, the concept of citizens observatories for environmental protection is an issue of interest in the EU Framework Program for Research and Innovation Horizon 2020 (i.e. Call SC5-17-2015).

More recently the convergence of GIS and social media [21] granted by interoperability of geo-web tools is further enriching the possibility of sharing the knowledge not only about measured facts on the Earth surface but also about all the biological, social and cultural phenomena there happening. Furthermore, the integration of geo-web and social media has also become a new media on its own supporting constructive dialogue about social issues in space. This is a double dimension not of minor relevance for spatial planning and Geodesign for SMGI can convey both people measurements (i.e. multimedia data) and at the same the observers perspective on given observations, fully expressing the pluralism potential of this new type of data source. Indeed, this feature solves a possible limit that was acknowledged with regards to early Geographic Information Systems which were often seen as elitist tools in the hands of restricted circles of power-holders [16].

In order to explore the potential of SMGI as a tool for eliciting community knowledge and fostering citizens dialogue on urban and regional planning, the author developed a number of experiments in order to collect data for testing these assumptions. The remainder of this paper reports the results of this project, on the base of which a framework is proposed as a contribution to the formalization of the concept of SMGI, and highlights opportunities for SMGI analytics in urban and regional planning and Geodesign.

3 SMGI: From Collection to Analytics

Social Network (SN) platforms can be used to support the community planning dialogue [2]. In facts, social network platforms thanks to web 2.0 technology and user-friendly interfaces may be used flexibly for different purposes. As an example, if we look at some of the most popular SN, Facebook is supposed to be used to manage the user personal social network, while other applications such as LinkedIn.com or Academia.edu or Researchgate.net have more specific purposes (i.e. supporting professional, academic, or scientific social networking respectively). Nonetheless, some argue Facebook is the "Inadvertent" Business Network for Generation Y [23], highlighting the flexibility of use of the platform for diverse purposes. Some scholars even investigate the case for the use of Facebook as medium in the college classroom [14] or in enterprises [18]. Accordingly a geographic social media platform can be used both for leisure or for more professional purposes ideally allowing for the integration and sharing of all the resulting information streams. Some users may create a project for documenting their georeferenced multimedia travel log and sharing it with friends, or the collection of any kind of thematic hotspots. Likewise, a local administration may ask the community to report local infrastructure maintenance issues, such as the case of the popular Fixmystreet.com, or to document forgotten cultural sites, or to assess local landscape values. Possible uses in education applications at any level are limited only by the creativity of teachers. Children may report issues of interest in the neighborhood, while university students may carry on scientific surveys and share the results.

However, when a social network starts working successfully a great deal of information starts accumulating and it may become difficult to extract useful knowledge from the big data avalanche. Computational social science is a new paradigm emerging not without difficulties related to data accessibility (for researchers) and interdisciplinary background training (i.e. social science, computer science, statistics, to name some) [11] with the aim of finding new methods and tools for knowledge discovery from big data sources. If we consider geospatial information we can include several of them such as point-of-sales (POS) data, location-aware technologies, geo-sensor networks, simulation and the Internet [13]: all of them produce terabytes of data and new knowledge discovery methods are required to extract useful information.

From an analytical perspective, the integration of GIS and Social network models allows to discover a lot of useful hints than either GIS or SN alone would. In other words, the underlying assumption is that the discourse developed within a SN-map may offer more support than a SN alone would. In a SN-map a post is in facts placemarked (i.e. purposely by the user) in its relevant location enabling the implementation of spatial analysis to extract knowledge.

Unlike in the case of traditional spatial data though, what is under the focus of analysis are not (only) spatial patterns of measured facts (i.e. authoritative spatial data), but rather the distribution in space of people perceptions of facts, say the geography of the people (i.e. VGI), which in turn may be anyway related to the geography of the place.

We can imagine a scenario where a city planner would aim to "listen" what the local community feels about civic issues or to interact with them to ask what alternative projects or development options would be welcome to the community. This would be a possible common use-case in the tradition of Public Participation/Participatory GIS (PPGIS) domain. However, until recently, PPGIS initiative required substantial endeavor in order to set-up the technology and manage an initiative. The availability of geo-enabled Social Network platforms may ease the process both in technology and social term. As a matter of facts, while on the one hand almost no technology set-up is needed, on the other one hand the social network functions require less commitment by the potential participants which growingly use the media in their daily lives. The participation to the initiative would blur with the everyday social networking activity. Accordingly we can use the platform to create a dynamic dialogue and monitor people interest about the place on a routine basis, or launch time-limited initiatives to ask the assessment of development alternatives.

The capability of extracting relevant knowledge for supporting spatial planning and Geodesign however requires a novel approach for SMGI analytics as explained in the remainder of this section.

4 Cagliari, I care! 2.0: From Data Collection to SMGI Analytics

A pilot project called Cagliari, I care! 2.0 (CIC 2.0) was developed by the author using the web application Place, I care! (PIC)(Campagna et al, 2012). PIC is a SMGI platform which allows users to register to a project and publish and interact with SMGI in a geo-browser wherein the working space is an interactive map (Figure1).

The CIC 2.0 pilot was set-up asking the participants to spot and report issues of appreciation and/or concern from an environmental, social, or cultural perspective in Cagliari, Italy. In PIC each post is a multimedia placemark to which the user can attach textual descriptions, pictures, video or audio clip. The model of the SMGI placemark is shown in Figure 2.

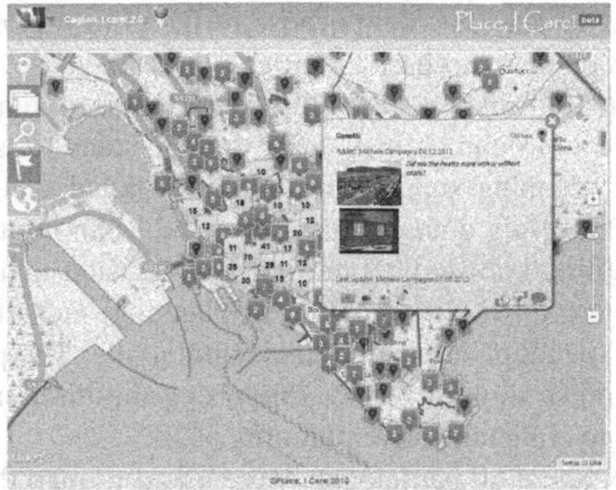

Fig. 1. Cagliari, I care! 2.0 project

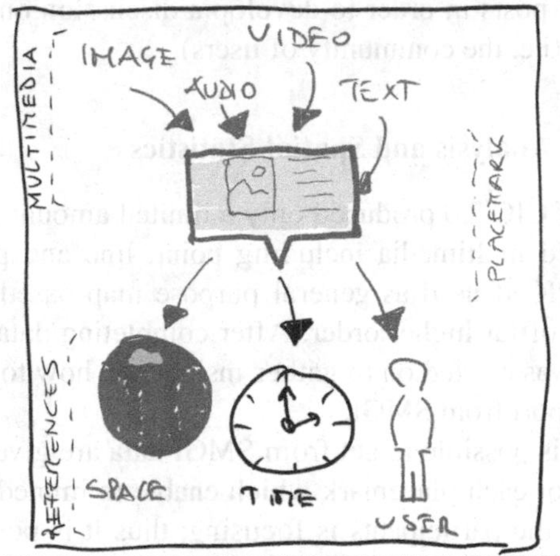

Fig. 2. A general SMGI placemark model

The general SMGI model in Figure 2 shows how SMGI follows a different structure from traditional vector dataset model (i.e. a point shapefile or a point feature class in a geographic database) causing two major implications. First, unlike traditional vector feature data models, the thematic attributes are not alphanumeric but multimedia clips (i.e. long-texts, images, video or audio clips); hence traditional query languages (i.e. SQL) cannot be used to extract knowledge from data and novel analytics should be integrated in spatial analysis methods. Secondly, not only spatial (i.e. placemark location) and temporal (i.e. placemark creation time) references are supported, but the user becomes a semantic reference itself enabling the extension of placemark spatial-temporal analysis with user behavioural analysis (i.e. activity of the user in space and time). To some the latter analysis may seems entailing sensitive privacy issues, but it should be noted that even in a blind user behavioural analysis – as will be explained later- extremely useful information may be extracted from a dataset in order to understand citizens participation dynamics in a SMGI initiative.

As a matter of facts in our case, the purpose of the pilot was not only to stimulate the participation to an online map-based dialogue, but also to develop a novel analytics to be applied to the information stemming from the discussion among the participants with other applications and tools in order to understand possible ways to extract useful knowledge for planning support. It should be noted that the term analytics, often related to analysis of big data, recently emerged as an integrated framework of different analysis enabling knowledge building and decision-making support. In this sense we may refer here to SMGI analytics in planning, meaning the extraction of useful knowledge for spatial planning and decision-making. The potential of SMGI is then to express a pluralist view of reality supporting the discourse according to the most recent paradigm in communicative planning [10].

In the CIC 2.0 pilot about 60 participants were asked to map through SMGI placemarks the location of environmental, cultural or social issues of interest (i.e. positive or negative) and to describe them. The users were also allowed to like/dislike and comments other users' posts in order to develop a discussion on the issues of greater interests for the group (i.e. the community of users).

4.1 SMGI: Spatial Analysis and Spatial Statistics

The implementation of CIC 2.0 produced only a limited amount of data (i.e. about ten hundreds georeferenced multimedia including point, line and polygon placemarks), but the potential for PIC if used as general purpose map-based social network is to produce data volumes of far higher order. After completing data collection an exploratory analysis phase was carried on to gather insights on how to extract useful knowledge for planning support from SMGI.

The first insights it is possible to get from SMGI data are given by spatial analyses thanks to the location of each placemark which enable to immediately detect anytime where the attention of the participants is focusing; thus it is possible to answer such questions as those related to what areas, places, or artefacts in the city attract the attention of the participants at any given time.

Summarizing the number of placemarks by census track it is then possible to detect hot-spots. In CIC 2.0 not surprisingly the Cagliari city center attracted the main attention of the participants, while the mostly residential Pirri district represents a cold spot. Actually Pirri developed in the past as independent small town in the rural periphery of the old Cagliari, and it was only later annexed administratively to the latter. This may be the reason why Pirri was possibly perceived by the inhabitants as an independent town (as physically it actually is) and not as part of the Cagliari municipality.

Generalizing, the most interesting question arising from this kind of analysis may be why certain areas or artefacts are not considered as part of a given place - Cagliari, in this case- by the participants and the answer may give useful hints to city planners and decision-makers for further large scale analysis. Such a kind of line of inquiry can be supported by the integration of SMGI data with other authoritative and official spatial data resources on demographic, land-use, transport or socio-economics accessible from local SDIs. One interesting research question to be further investigated would be whether traditional spatial statistic methods such as regression analysis can be used to understand whether the spatial interest of the participant is influenced by environmental or socio-cultural factors.

We can also focus on the content. If we want understand not only the where but the what people think, we need to find ways to analyze hundreds of multimedia posts or more: we can first select keywords, than we can explore keywords we are more interested in, where they are located, and then access and review related multimedia documentation. In the current PIC version, advanced post query functions are available in the web interface, which allow to select post by keyword, author (if permission is granted), date, and multimedia content to all the participants.

Further analyses we report in the remainder were carried after exporting CIC 2.0 data, making them anonymous and processing them with other desktop software. Data collected in CIC 2.0 feature the structure shown in Figure 3. The PIC's data structure available for CIC 2.0 enables several dimensions to be analyzed such as time, interest (other users' Like/Dislike for the posts), users' behavior, and multimedia content. In the reminder of this section the analytical framework for CIC 2.0 data is explored with the aim of developing an integrated SMGI analytics for spatial planning.

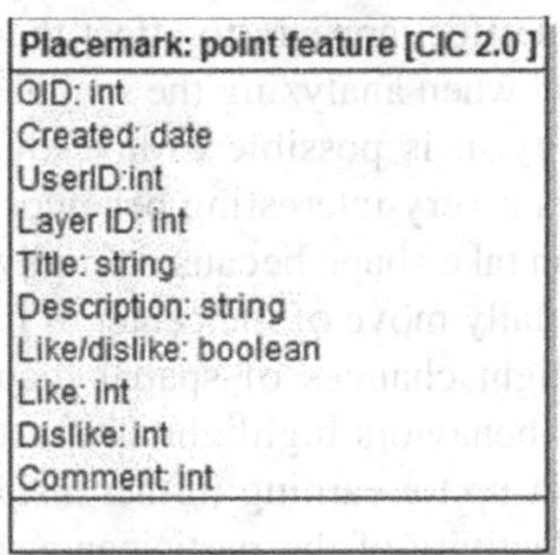

Fig. 3. Data model in CIC 2.0

Three public thematic layers were set-up for placemarks so that they could be analyzed independently. This may be considered a generic approach to a kind of map-based community listening and discussion in CIC 2.0, where citizens use the interactive map to propose issues of interest and discuss them. Hence, the main question is: what kind of knowledge can be elicited from the discussion?

A first order of questions or analysis dimension relates to the spatial distribution of people interests. The PIC data structure allows to analyze not only the overall distribution of people places of interest, but statistical function can be used to analyze patterns, mapping clusters, and data mining exporting data in a GIS application. As an example hot-spot analysis was used to detect spatial clusters of appreciation or areas concern in the city. Then spatial regression analysis may be used to investigate the why of such distributions. Moreover the possibility to analyze changes in participants' spatial focus in time may be very valuable to understand the reactions in participants' spatial interest especially if the platform is administrated in order to support a dynamic discussion. The latter issue relates to a second order of questions regarding the participants' behavior.

4.2 SMGI: User Behavioural Analysis

A very interesting dimension to investigate is the behavior of participants both in quantitative, temporal and spatial terms: how active are participants? Does the spatial center of interest of single or of group of participants move along the discussion? What/how big is the area of interest, and how it is perceived by the participants? Does the area of interest correspond to any administrative or physical or any other boundary?

The analysis of other participants' appreciation and comments on placemarks are very helpful in order to assess the most interesting issues for the overall community, or for given groups, along the discussion. Furthermore the possibility to analyze the appreciation in time may help to investigate the move in general spatial interest patterns along the discussion in order to detect possible imitation effect (i.e. the behavior of one participant affects the behavior of others by imitation) [22]. As a matter of facts, in CIC 2.0 non-random behavior were detected in participants' spatial center of interest in time. The imitation effect appears to affect the discussion and its influence should be properly understood when analyzing the spatial focus patterns.

Charting participant activity, it is possible to investigate which participants were more active than other: this is a very interesting behavior to analyze for it may affect how a participatory discussion take shape because of individual participants influence. It is also possible to map the daily move of the center of interest, which can be used as geo-visual indicator to highlight changes of spatial focus along the discussion. All these aspects on participants behaviors highlight stimulating research issues and tailored experiment should be set-up for earning further insights.

Monitoring the degree of activity of the participants showed the activity is higher before set (intermediate and final) deadlines. This kind of analysis may be very useful for the administrator of a participatory project (e.g. a manager of a local administration) to timely assess along the process the degree of interest or participation at any stage in an urban management or development process.

4.3 SMGI: Multimedia and Spatial-Temporal Textual Analysis

As mentioned in earlier in this section, each multimedia type supported in the placemark (i.e. long texts, images, videos, sounds) should be analyzed with relevant pattern recognition or mining techniques. In this study only simple textual mining techniques were explored with the aim of earning first insights on the potential of this kind of analytics, on the base of which, it was possible to define a novel method for Spatial-Temporal Textual Analytics (STTx) as described in the reminder.

Firstly, long text in the placemarks can be analyzed directly in PIC by querying the title and body of the posts. A post-query box allows to select and visualize on the map relevant posts not only by keyword but also by time, author, or presence of multimedia (i.e. images, video, audio) for retrieval: e.g. 'select all the posts related to "church" ["chiesa"] with images and video'. However, more sophisticated text analytics techniques [3] [1] can be applied for knowledge discovery exporting data in other applications (i.e. desktop GIS), especially when the amount of information rapidly grows.

Textual analysis enables summarising the number of posts by theme and time interval, and reporting the most used keywords in each category. Looking at the top keyword it is possible to note what issues are perceived as of major interests (i.e. n° of likes) or concerns (i.e. n° of dislikes) for what the participants considered environmental, cultural or social issues.

To give an example, in an earlier pilot, called CIC 1.0, the tag cloud was generated from the long text of all placemarks showing interesting hints. The global tag cloud clearly shows the focus of the majority of the words in the posts refers to spatial or physical aspects of the city, such as city, zone, church, or to some specific areas such as the Castello district in the historic city center where the majority of the spots were located. While it is interesting to note that the spatial and physical structure characterize the participants perspective, it should be noted that the social and cultural perspectives were somehow neglected. In other words in CIC 1.0 the City of Cagliari was mostly seen as a physical place by the participants. The situation was different in CIC 2.0 where the tag cloud includes keywords related to retails services, public and green spaces, or sport facilities referring to social activities in the city. It should be noted that both the participants and the technology used to collect SMGI was different in the two pilots, and this may have influenced the results. If it is not surprising that different group of participants express a different view of the city in different time and contextual situation, the possible influence of the supporting SNGI platform should be further investigated.

The most interesting results were obtained applying text analysis to local subsets of data obtained by selecting placemarks by location (i.e. spatial-textual analysis). Using three neighborhoods namely Castello (the innermost historic district in the center), Marina (the most commercial district linking the historic center to the harbor) and Is Mirrionis (a poor-housing district of the 1960s urban expansion at the hedge of the center) the overall perception resulting by the posts defined a clear image for each of them as shown in table 1.

Table 1. Perception of the neighborhood by the participants for CIC 2.0

Neighborhood	Posts	Likes	Dislikes	Comments	Keywords frequency
Castello	57	69	25	1	*Cagliari (11), Church (9), Century (8), Tower (7), It was/were (7)*
Marina	76	80	24	4	*Ladies-ware (14) Restaurant/bar (12), Shop (5), Cagliari (4)*
Is Mirrionis	41	47	7	2	*Football field (26), Park (5), University (4), Students (4)*

This is not the kind of information one usually find in land use planning documents, but its potential for design and decision-making support may be considered of highly valuable.

The above examples show early analyses carried-on thanks to the crowd-sourced data collected in CIC1.0 and 2.0 pilot studies. Beside spatial and temporal analysis of people interests and concerns, interesting hints on the perception of the place and the participation behavior of the group of participants the examples entail unusual data analysis perspectives which should further investigated. Thanks to the insights of this first experiences further pilots will be set up to more specifically address such analytical dimensions as those regarding user spatial, temporal, and thematic perception of the place, degree and of activity and behaviors of the participants. Moreover the analysis of statistical relationships between the perception, concerns, and interests of the community and the geography of the place expressed by available authoritative data from SDI (i.e. physical structure of the city as well as social, cultural and economic conditions in the city) is a promising domain to further investigate. Last but not least, another further promising analytic dimension is to explore data mining methods for images, audio, and video [12].

5 Results and Conclusions

This paper proposes a framework for Social Media Geographic Information analytics as a methodology support for spatial planning.

The peculiarities of the SMGI data model enable the analyst to perform the following analysis on multimedia data:

- Spatial analysis of user interests;
- Temporal analysis of user interests;
- Spatial Statistics of user preferences;
- Multimedia content analysis on texts, images, video, or audio;
- User behavioral analysis;

or

- a combination of two or more of the previous such as in Spatial-Temporal Textual Analysis (STTx).

Together in an integrated framework SMGI analytics may help to understand user observations, perspectives, interests, or needs in spatial planning. Granted the access to a Social Media spatial-temporal database it is possible to analyze data outside the social media platform (i.e. PIC in this case), by exporting the data and processing with GIS desktop or other applications. Either accessing data from a social media platform (i.e. Facebook, Twitter, Flicker or YouTube, to name some of the most popular) or from a project specific Social Media platform (i.e. PIC) in both cases it is possible to generate useful knowledge for urban and regional planning, which may foster citizens' dialogue about places and facilitate the development of computational social science methods to extract knowledge which may be of valuable support in decision-making. The latter challenge may require the planning team to make advanced use of computational frameworks for analysis (i.e. not only GIS, but also text data mining, image processing, and the like) as an opportunity of great potential to eventually enrich the social dimension in Geodesign. This is the way forward we may now foresee for the integration of experiential and pluralist spatial information with authoritative spatial data sources and sensor-web. Thus, the big data avalanche can be enriched with the social sphere of knowledge springing from the computable city.

Would it be in planning education or practice, or just for social or cultural interests the integration of GIS and social networks may be used to understand the community's image of the city and to support discussion about past, present and future development of places people care of. Moreover, if we focus on public participation in planning, we can now study how people participate, how active they are and how they behave, and to what extent they may eventually influence the overall discourse and the decision-making process about the present and the future of places.

Acknowledgements. The work presented in this paper was developed by the author within the research project "Efficacia ed efficienza della governance paesaggistica e territoriale in Sardegna: il ruolo della VAS e delle IDT" [Efficacy and efficiency of landscape and environmental management in Sardinia: the role of SEA and of SDI] CUP: J81J11001420007 funded by the Autonomous Region of Sardinia under the Regional Law n° 7/2007 "Promozione della ricerca scientifica e dell'innovazione tecnologica in Sardegna". The author wish to thanks Anastacia Girsheva and Konstantin Ivanov for their contribution to the development of Place, I care! and Roberta Falqui for her contribution to data analyses.

References

1. Adams, B., McKenzie, G.: Inferring Thematic Places from Spatially Referenced Natural Language Descriptions. In: Sui, D., Elwood, S., Goodchild, M. (eds.) Crowdsourcing Geographic Knowledge, pp. 201–221. Springer (2013)
2. Afzalan, N., Evans Cowley, J.: The role of online neighbourhood groups in creating self - organized and resilient communities. In: Proceedings from Using ICT, Social Media and Mobile Technologies to Foster Self - Organisation in Urban and Neighbourhood Governance. Delft University of Technology, Delft, Netherland (2013)
3. Berry, M.W., Kogan, J.: Text mining applications and theory. Wiley, Chichester (2010)
4. Campagna, M.: GIS for Sustainable Development. In: Campagna, M. (ed.) GIS for Sustainable Development, pp. 3–20. CRC Press, Taylor & Francis Group, Boca Raton, USA (2006)
5. Campagna, M., Craglia, M.: The socio-economic impact of the spatial data infrastructure of Lombardy. Environment and Planning B: Planning and Design 39(6), 1069–1083 (2012)
6. Ervin, S.: A system for Geodesign. In: Digital Landscape Architecture Conference. Dessau, Germany (2011), http://www.kolleg.loel.hs-analt.de/landschaftsinformatik/fileadmin/user_upload/_temp_/2011/Proceedings/305_ERVIN_2011May10.pdf (accessed May 5, 2013)

7. Göçmen, Z.A., Ventura, S.J.: Barriers to GIS Use in Planning. Journal of the American Planning Association 76, 172–183 (2010)
8. Goodchild, M.F.: Citizens as voluntary sensors: spatial data infrastructure in the world of Web 2.0. International Journal of Spatial Data Infrastructures Research 2, 24–32 (2007a)
9. Goodchild, M.F.: Citizens as sensors: the world of volunteered geography. GeoJournal 69, 211–221 (2007b)
10. Healey, P.: The communicative turn in planning theory and its implications for spatial strategy formations. Environment and Planning B: Planning and Design 23(2), 217–234 (1996)
11. Lazer, D., Pentland, A., Adamic, L., Aral, S., Barabasi, A.-L., Brewer, D., Christakis, N., Contractor, N., Fowler, J., Gutmann, M., Jebara, T., King, G., Macy, M., Roy, D., Van Alstyne, M.: Computational Social Science. Science 323, 721–723 (2009)
12. Manovich, L.: Trending: the promises and the challenges of big social data. In: Gold (ed.) Debates in the Digital Humanities, The University of Minnesota Press, Minneapolis (2012), http://www.manovich.net/DOCS/Manovich_trending_paper.pdf (accessed July 15, 2013)
13. Miller, H.J.: The data avalanche is here. Shouldn't we be digging? Journal of Regional Science 50, 181–201 (2010)
14. Munoz, C., Towner, T.: Opening Facebook: How to Use Facebook in the College Classroom. In: Gibson, I., Weber, R., McFerrin, K., Carlsen, R., Willis, D.A. (eds.) Proceedings of Society for Information Technology & Teacher Education International Conference 2009, pp. 2623–2627. AACE, Charleston (2009)
15. Nebert, D.: The SDI Cookbook, Version 2.0., Global Spatial Data Infrastructure Association Technical Working Group Report (2004), http://www.gsdi.org/docs2004/Cookbook/cookbookV2.0.pdf (accessed May 5, 2013)
16. Pickles, J. (ed.): Ground Truth: The Social Implications of Geographic Information Systems. Guilford, New York (1995)
17. Poser, K., Dransch, D.: Volunteered Geographic Information for disaster management with application to rapid flood damage estimation. Geomatica 64(1), 89–98 (2010)
18. Skeels, M.M., Grudin, J.: When social networks cross boundaries, p. 95. ACM Press (2009)
19. Steinitz, C.: A framework for Geodesign: changing geography by design. Esri, Redlands (2012)
20. Sui, D., Elwood, S., Goodchild, M.: Crowdsourcing geographic knowledge: Volunteered Geographic Information (VGI) in theory and practice. Springer, Dordrecht (2013)
21. Sui, D., Goodchild, M.: The convergence of GIS and social media: challenges for GIScience. International Journal of Geographical Information Science 25, 1737–1748 (2011)
22. Surowiecki, J.: The wisdom of crowds. Abacus, London (2005)
23. Zappe, J.: Facebook is the "Inadvertent" Business Network for Gen Y. ERE.net Recruiting Intelligence. Recruiting Community (2009), http://www.ere.net/2012/01/09/facebook-is-the-inadvertent-business-network-for-gen-y/ (accessed May 5, 2013)

An Agent-Based Model as a Tool of Planning at a Sub-regional Scale

Fernando Pereira da Fonseca[1], Rui A.R. Ramos[1], and Antônio Nélson Rodrigues da Silva[2]

[1] Department of Civil Engineering
University of Minho
Braga, Portugal
ffonseka@gmail.com, rui.ramos@civil.uminho.pt
[2] Department of Transportation Engineering
School of Engineering of São Carlos
University of São Paulo
São Carlos, Brazil
anelson@sc.usp.br

Abstract. This paper describes an agent-based model developed to simulate the impact that different planning policies may have in enhancing the attractiveness of the industrial estates located in a network of four municipalities located in the North of Portugal. The policies were simulated using three scenarios that can be distinguished by the municipal level of coordination they are implemented and by the type of action performed. In the model, enterprises are agents looking for a suitable location and the estates attractiveness is based on their level of facilities, amenities, accessibility and in the cost of soil. The coordinated qualification of the industrial estates is the most effective policy to strengthen their attractiveness. It was in this scenario that more industrial estates become attractive and more enterprises relocated. Results also indicate that the promotion of diffused and unqualified industrial estates is an inefficient policy to attract enterprises.

Keywords: Agent-based models, Industrial estates, Planning, Territory.

1 Introduction

One of the main obstacles faced by the territorial planning is the difficulty in conciliate the expectations and interests of a large number of actors with the limited nature of the own territory. To address these challenges, the use of more robust tools has been conceived as a way to deal with the increasing levels of uncertainty and complexity related with the territory. The agent-based models (ABM) have emerged as a tool with potential to deal with these phenomena. In fact, ABM have been used to study many issues, such as traffic, accessibility, urban growth, suburbanisation, gentrification, segregation, etc. When compared with other more conventional tools of spatial analysis, such as the geographic information systems (GIS), and even with

B. Murgante et al. (Eds.): ICCSA 2014, Part II, LNCS 8580, pp. 611–628, 2014.

more advanced tools, such as cellular automata, the ABM have a number of advantages. They can represent the heterogeneity of the territorial actors, because each agent can represent the behaviour of each actor. They are based on a bottom-up approach, that is, the macro trends result from a variety of interactions kept at an individual level. They are dynamic because the agents have the ability to move around the environment, simulating the actors and the objects behaviour and to simulate the evolution of a phenomenon over time.

In this context, the aim of the paper is to present an ABM developed to simulate the impact that different policies may have in enhancing the attractiveness of the industrial estates located in a network of four municipalities located in the north of Portugal (Quadrilátero). The main goal of this network is to improve the competitiveness and the internationalisation of the territory, by acting in some critical sectors. One of them is the adoption of common actions to strengthen the capacity of the industrial estates for attracting enterprises, mainly by implementing planning policies. Inspired by this agenda, policies were simulated in the model using the three following scenarios: 1) maintenance of an uncoordinated and unqualified policy by the municipalities; 2) adoption of a coordinated policy of qualification of all the industrial estates; 3) impact of the strategies predicted by the municipalities in a scenario of: uncoordinated and low qualification (3a); and of coordinated and advanced qualification (3b). The model was also developed to validate four hypotheses related with the implications of the policies in the estates' performance. These hypotheses are described with more detail in the Methodology Section.

Besides the introduction, the paper is composed by four main Sections. The first one comprises a theoretical framework of the subjects under analysis, with a focus on the ABM that use enterprises as the main topic of the study. The methodological approach and the main steps followed to construct the model are described in the second Section. The results and the main findings obtained with the model are discussed in the third Section, which is followed by the conclusions and some final remarks. Besides the findings obtained, this study is also an attempt to strength the literature in the industrial estates planning domain at a sub-regional scale, a topic scarcely exploited in the literature [1, 2, 3].

2 Modelling Enterprises: A Review

Agent-based modelling is a form of computational simulation [4]. These models can be used to study a great diversity of urban and territorial phenomena, being the location one of the topics more analysed. The location is a problem that involves many entities (inhabitants, enterprises, public administration, real estates, etc.) which try to reply to the basic question: where should I locate? In fact, the location decision arises from a complex process, where individual aspects are articulated with economic, social and political factors [5]. On the other hand, there is a strong competition between the territories to attract resources, mainly those that produce more returns. Enterprises always have been central in the location models, since they generate jobs, revenues, taxes, traffic and can influent the transformation of soil, the location of residential areas and the attraction of more investments and enterprises. The model conceived by Alfred Webber in 1909 is a reference on the industrial location theories. The main goal of his model was to determine the optimal location for an industry, assuming that transport costs are a function of the euclidean distances between the sources of raw materials and the markets. This theory knew a remarkable expansion and inspired other studies. August Lösch gave a new emphasis to the location theory through the principle of the profit maximisation, considering a larger scale (the economic regions). In the second half of the last century, the behavioural theories gained more importance, being the approaches of Edgar Hoover, Walter Isard and Melvin Greenhut good examples of this new conception. The location only based in economic purposes is replaced by the principle that the best location is one that satisfies the particular requirements of entrepreneurs in a more efficient way. The selection becomes a probabilistic process where the individual preferences and motivations have an important role. In the last decades of the 20th century, the integrated or relational approach emerged [6]. This new conception takes in consideration both the economic and the behavioural aspects in the industrial location. The theory advocates that there is a strong interaction between an entrepreneur and the possible locations through a negotiating process. In this case, the territory with more suitable conditions that respond to the requirements of the entrepreneur will be the selected.

Enterprises also played a significant importance in the integrated land use and transportation models developed in the second half of the last century. The gravitational models were the first generation of these integrated models [7]. The cities were conceived as a system composed by several interacting areas in a continuous equilibrium condition. The goal of these models is, through mathematical equations (linear regressions), to predict the movements between the different sectors of a city, namely between the residential areas and the zones where the economic activities are located [8]. There are several examples of these interacting models based in spatial aggregation such as METROPOLIS TOMM, PLUM, ITLUP and IRPUD [9]. The IRPUD was one of the most complex working as macroanalytic model of economic and demographic change, composed by interlinked submodels that deal with relocation of enterprises, change of job, change of residence, etc. In the 1980's emerged a new generation of integrated models: the utility-maximising multinomial logit based models [7]. The spatial interaction of the first generation was replaced by logistic regression techniques that allows the prediction of events based in a set of observations. In comparison with the gravitational, these models are more integrated (they combine modules of land use and transportation) and more dynamic (the transportation module predicts the demand and its suitability while the land use module analyses the spatial distribution of residential areas and those that influence the employment, namely the industrial and the commercial areas [10]. The interaction between the modules of land use/transportation determines how transportation affects land use, which influences the location of activities and the traffic generation. There are several examples of these multinomial logit based models, such as MUSSA (predicts the location of households and enterprises), URBANSIM (integrates a household and economic mobility models to simulate whether households and enterprises decide whether to move) and the IMREL (integrated model of residential and employment location). In the beginning of this century, plans have incorporated micro-simulation methods. This new generation have notable differences when compared with the previous generations: they are more dynamic in the spatial and temporal dimensions; they use individual data in the simulation (objects, vehicles, parcels, etc.); the spatial interaction between the different parts of a territory are replaced by the interactions that arise from the interactions between the individual elements; and they conceived the territory in a condition of disequilibrium, due to the continuous changes operated by the actors over time [8], [11, 12]. This third generation was expanded through the development of two main tools: the cellular automata and the ABM [8]. The agent-based modelling is considered more powerful than the cellular automata, because they are more flexible and suitable to represent and simulate the individual behaviour and the territorial dynamics [11], [13, 14].

The ABM have had a large utilisation in the planning domain in the last two decades [15]. The applications encompass topics such as the urban grow [16], suburbanisation [17], gentrification [18], residential segregation [19, 20], accessibility [21], the impact of urban policies [22], the land use change [22, 23], the evolution of urban networks [24], among many others.

In the context of the ABM, enterprises usually appear as one of the agents simulated. A first selection can be made between the ABM that encompass many typologies of agents (inhabitants, households, dwellers, enterprises, public administration, etc.) from the models that only integrate enterprises in the modelling. In the first case enterprises are usually one of the entities simulated due to their implication in urban development, soil demand, location of residential areas and in traffic generation [7]. These generic models are composed by a changeable number of modules, one for each agent simulated. The enterprises' module controls the way these agents interact among them and with the remaining agents. There are several examples of these models, such as the Ilute [22] and ILUMASS [25]. The second category integrates models that can be grouped into three main topics: (i) to analyse the relationships between enterprises; (ii) the firmographic models; (iii) and the location models. The models developed to analyse the relationships between enterprises, namely in territories with a high concentration of enterprises (industrial districts and clusters) are the most representative. The relationships comprise variable domains: the concentration benefits achieved by the enterprises [26, 27]; the multiple forms of cooperative and competitive relationships that can be found is these areas [28]; how social phenomena may affect clusters' dynamics [29]; and how innovation propagates among the enterprises [30]. In turn, the firmographic models are focused in identify scenarios related with the enterprises' evolution in a specific territory, not only in terms of birth, grow and death, but also their implications in terms of jobs, urban development, mobility and to anticipate the location and the relocation decisions of the enterprises [2], [31]. There are several firmographic models, such as the models developed by Khan [1], Wissen [32] and by De Bok [33]. The last category is the less exploited in the literature and comprises the approaches carry out to find a suitable location for a specific enterprise or industrial sector. The ABLOOM model [34] and the work of Manzato et al. [35] are two examples of simulations developed with this specific intent. In other type of applications, ABM have been used to improve the productive and the organizational processes at the individual scale of the enterprises [36].

In this global framework, we present in the next Sections the main steps undertaken to develop an ABM that use enterprises to assess the impact of some policies in the attractiveness of industrial estates. In the literature we hadn't found an identical approach, which highlights the innovator character of this study.

3 Methodology

The implementation of the model followed several sequential steps. The first one was the acquisition of data. As the initial purpose was the construction of a spatially explicit model, the first step was the acquisition of detailed data about the industrial estates and with the industrial enterprises. The industrial estates data was gathered through a survey addressed to the municipal services with the aim of collecting information about their location and their general characteristics (surfaces, number of plots, number of enterprises installed, price by m^2, etc.). Further, the survey intended to identify the coverage of 16 facilities and 14 amenities in each industrial estate. Furthermore, the survey included some questions to disclose the strategies predicted by the municipalities, namely to understand whether the existing supply would be enlarged in the future through the creation of new estates or through the expansion of the existing ones. As the data collected with the surveys had several gaps, all the industrial estates identified by the municipalities were visited in order to obtain the missing data and to check all the information provided. During this fieldwork, the business owners located in the industrial estates were also interviewed. The goal of this survey was to learn about the past and future locative behaviour of the entrepreneurs and their satisfaction level regarding the provided conditions. The data obtained with this survey was helpful to calibrate the model. In short, the survey showed that the majority of enterprises located in the industrial estates come from a diffuse situation (88%), having predominantly micro and small size (78%). Moreover, the survey indicated that 90% of the entrepreneurs are reasonably satisfied with the conditions provided and, for that reason, they were not interested in moving to another industrial estate.

In the case of the enterprises, the data used was the Integrated Business Accounts System (SP 2011). This is a data base compiled by the Portugal Statistics that contain disaggregated data about enterprises, including location, sectors of activity, number of employees, turnover, etc. Based on the location provided in the database, all the enterprises and their attributes where mapped through points using the ArcGis software.

Once the data collection was conclude, the next step was the definition of the rules that control how the model run. To accomplish the model intentions, the definition of the industrial estates attractiveness was the starting point of this task. Following the recommendations of Bodenmann and Axhausen [38], the attractiveness was based on the level of facilities, amenities, accessibility and on the land cost in each industrial estate. Based on the presence (1) or on the lack (0) of the different facilities and amenities, the respective coverage (in percentage) was estimated for each industrial estate according the formulas presented in the Table I. The criteria proposed by Ramos and Mendes [39] were used to estimate the accessibility of each industrial estates to five transportation infrastructures (national road, motorway intersection, railroad terminal, seaport and airport). Thus, the accessibility of an estate can be classified as excellent (≤ D (Dmax / 2), acceptable (Dmax ≤ D < (Dmax / 2) or bad (D > Dmax). Based on this classification different weights were assigned to each estate (1 = excellent; 0.5 = acceptable; 0 = bad) through which their final accessibility (in percentage) according to the formula presented in the Table I. Regarding costs, several variables were pondered, but in the end only the cost of land was included, considering the average price identified in each estate. In fact, the fieldwork showed that the labour and the utilities costs were identical due to the subregional scale of the study. Further, the land cost was the variable most difficult to identify due its strong variation from plot to plot and to the presence of several owners/promoters in each estate.

In the model, enterprises have the ability to move and relocate whenever an attractive estate is found. However, relocation is not a random process. Inspired on the conclusions obtained with the survey, relocation follows criteria related with the enterprises' size, location and distance (to an industrial estate). Since the supply is targeted for small enterprises, it was stipulated that only the small and the micro enterprises can relocate in an estate. From the locative point of view, the relocating enterprises are those with a diffuse location with bad (0) coverage in terms of accessibility and of facilities and amenities. In scenario 1, the relocating enterprises were those with bad coverage in the three locative criteria described (258), but in the scenarios 2 and 3, due to the larger supply simulated, the relocating enterprises encompass all that had a bad performance in two of those criteria (1145). Since the facilities and amenities coverage in the area of the diffuse enterprises was unknown, it was assumed that enterprises located in rural areas and predominantly rural spaces would have a worse coverage (0) than those located in urban areas (1). The distinction between the urban and the rural spaces was made through the Classification of Urban Areas published by the Portugal Statistics at the statistical subsection unity (the most disaggregated available). Finally, the maximum range of relocation of an enterprise was set at 20 km. This criterion was introduced because the survey showed that most relocations occur in the same municipality.

Table 1. Procedures used in the evaluation of the industrial estates attractiveness

Facilities	**Amenities**
$R_{Inf} = \left(\frac{\sum (Faci) \times 100}{16} \right)$	$R_{Ament} = \left(\frac{\sum (Ameni) \times 100}{14} \right)$
Accessibility	**Land cost**
$R_{Acess} = \left(\frac{\sum (Access) \times 100}{5} \right)$	$L_c = \bar{X}_{AAE}$

After concluding the theoretical procedures of the model, the next step was the selection of a suitable toolkit to simulate the described phenomenon. After the evaluation of several options, the NetLogo toolkit was chosen mostly because NetLogo has a GIS extension that supports vector data in the form of ArcGis shapefiles, where all the geographic information was initially processed. Due to the NetLogo characteristics, the patches (cells in which the environment is divided) were used to represent the industrial estates while the enterprises were the agents of the model. Each industrial estate was represented by a changeable number of patches, which contain information about the facilities, the amenities, the accessibility and the cost of soil. The model was developed with many buttons and commands to optimise the simulation process.

In the relocation process, enterprises interact with patches looking for a better location. The relocation only occurs if the new place represents an improvement of the conditions experienced by the diffused enterprises and if the estate has vacant plots. The locative conditions are determined by the combination of the four parameters mentioned above (facilities, amenities, accessibility and cost of soil), having as reference the average values obtained in the four municipalities. The attractiveness of an industrial estate is defined by minimum values that are changeable according to the scenario simulated (Table II). Each scenario correspond to a different policy. For instance, the qualification of the industrial estates was achieved through an improvement of the facilities, amenities and, occasionally, of the accessibility values. The improvement induced in those parameters arise not only from the inclusion of some basic facilities and amenities that were missing in some estates, but also from the inclusion of more advanced infrastructures and services. With the enhancement of the general conditions offered by the estates, it was assumed that the locative requirements of enterprises would also be higher. To avoid the exclusion of industrial estates without vacant plots from the simulation, in scenarios 2 and 3 a fictional number of vacant plots was attributed to those estates. In the model, enterprises do not have pre-defined preferences for any industrial estate or for any municipality and can move to any one of them according to the established criteria. In the relocation process, when an enterprise settles on an industrial estate, a vacant plot was deducted, ending the process when all the vacant plots are occupied by enterprises.

Table 2. Minimum locative requirements considered in the enterprises relocation

Scenarios		Facilities	Amenities	Accessibilities	Cost (€/m²)
1		≥54,6%	≥19,9%	≥56,8%	<733
2		≥73,2%	≥42,4%	≥58,9%	
3	3a	≥50,1%	≥18,6%	≥58,2%	
	3b	≥61,4%	≥30,1%	≥58,2%	

4 Case Study

The model was performed to the Quadrilátero, designation given to a network of four municipalities (Barcelos, Braga, Guimarães and Vila Nova de Famalicão) located in the North of Portugal. This territory has a strategic location due to the proximity of the Oporto Metropolitan Area and has a good accessibility and proximity to several transportation and logistics facilities located in the region (airport, seaport, railroad terminal, logistic platforms, etc.). According with the last census [40], the Quadrilátero has 593.841 inhabitant's, being 39% urban residents, mainly in Braga, the biggest urban centre in the region. Despite the bigger importance of services in jobs, 45% of the active people works in secondary activities mainly in the 4722 industries located in this territory [40]. In fact, the Quadrilátero is an old industrial area. In the second half of the 20th century, manufacturing knew a great development in the region, mainly due to the establishment of textile and clothing industries. However and since the 1990s, the region where the Quadrilátero is inserted experienced an industrial crisis with the exit of several enterprises. As Barbot [41] highlights, competition from East European and Asian countries may have been one of the leading factors of the crisis, together with a wage raise due to a shortage in the labour offer. Gross investments were spent and several studies and plans were made to rehabilitate the declining sectors. Despite the fall detected in some sectors and in the employment, the region experienced some recovery through the development of trade and services and to a small but increasing industrial diversification and modernisation of the traditional sectors. The last decade was distinguished by the emergence of new sectors, based on technology, innovation and knowledge, being very competitive in the international context. This include software, nanotechnologies, health, electronic and optic, among others. Even the traditional sectors have slowly move to a pattern based on quality, design and innovation. As a consequence, the Quadrilátero have installed enterprises that occupies the first position in the Portuguese ranking considering the turnover in many industrial sectors, namely in textile, clothing, leathery, rubber (tyres production) and electrical and optical components. The global turnover achieved by the industrial enterprises located in the Quadrilátero was 5500 M€ in 2010 [37]. The Quadrilátero trade balance is also very favourable (1800 M€) which attests the productive capacity of the industries located in this territory.

Recognising this potential, local entities decided to constitute a network among the four municipalities in order to exploit in a more efficient way their common potentialities. The goal is to improve the competitiveness and the internationalisation of the territory by acting in some critical points. One of them intends to turn the industrial estates more attractive to enterprises by adopting common planning practises, by the qualification of the spaces and by the adoption of a common management structure for all the estates. The model was developed to give some insights to the impact that the adoption of some policies in a coordinate manner by the four municipalities, namely in terms of the estates' qualification, can play in their attractiveness. In the Section VI we present the results of that simulations.

5 Main Features of the Industrial Estates

In this Section, the main features of the industrial estates located in the Quadrilátero are described according with the data collected with the surveys. These four municipalities have 79 industrial estates (Fig. 1) in two different categories: the planned areas (51 estates) and the spontaneous spaces which become of industrial use by the concentration of several enterprises in these areas (28 estates). The 79 industrial estates occupy a global surface of 1440 ha, being 20 ha areas of expansion (without occupation). Guimarães is the municipality with more industrial soil (569 ha) while Barcelos provides the smallest surface (107 ha). These spaces are strongly diffused by the municipal territories and are mostly of small and medium dimension (50% have less than 10 hectares of global surface and only one estate have more than 100 hectares).

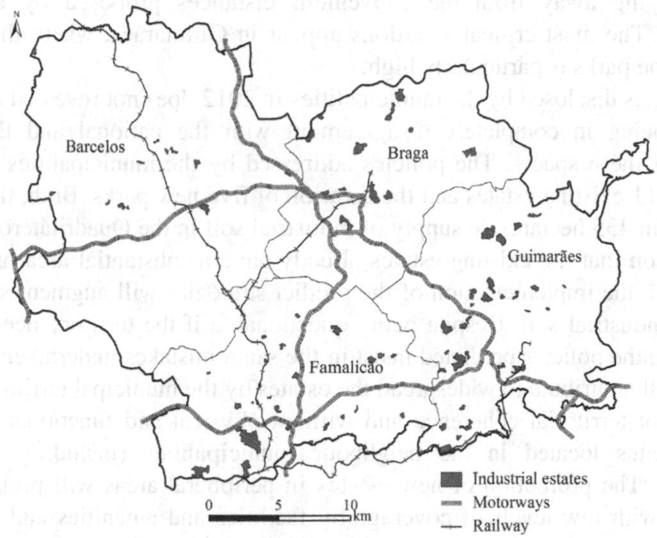

Fig. 1. The industrial estates location in the Quadrilátero

The 79 estates are subdivided in 2404 lots with widely varying dimensions. In 2012, when the survey was made, 639 of these lots were vacant. This problem affected particularly the estates located in Guimarães, where 36% of them were unoccupied. The land cost is also strongly variable from estate to estate, being the most difficult variable to estimate, due to the private nature of almost all the parks. In several situations, the land cost does not have correspondence with the facilities and with the amenities installed in some estates. The strong differences diagnosed in the prices arises from objective reasons related with the conditions offered by the estates (modern buildings, the location, advanced facilities, etc.), but also depends from the individual management made by each owner/promoter (financial and speculative purposes).

Regarding the facilities and the amenities, the industrial estates have a globally poor coverage. From the 16 facilities considered in the survey, only four were identified in the 79 estates (parking, internal roads, telecommunications and electricity supply). The most advanced facilities in a technological and in an environmental point of view (natural gas, optical fibre and wastewater treatment plant) are only present in a few numbers of parks. The coverage by amenities is even worst since many of the 14 services considered in the study were not installed in many estates, such as post office, service station, bank and multifunctional buildings. However some of these amenities can be found in the adjacent urban areas, performing a proximity service to the entrepreneurs located in the industrial estates. The diagnosis also demonstrated that the 28 spontaneous spaces have worst coverage by facilities and amenities than the planned areas.

In other hand, the accessibility is favourable. The industrial estates are located in a distance of 1.8 km from a main road and in a distance of 5.7 km from a motorway intersection. However, it was found that 14 industrial estates have particularly bad accesses, staying away from the convenient distances proposed by Ramos and Mendes [39]. The most critical situations appear in Guimarães, where the territorial diffusion of the parks is particularly high.

The strategies disclosed by the municipalities in 2012 does not revealed any level of integration, being in completely disagreement with the national and the regional guidelines for these spaces. The policies addressed by the municipalities include the expansion of 11 existing estates and the creation of five new parks. Both, these actions will increase in 435 hectares de supply of industrial soil in the Quadrilátero. If we take in consideration that the existing estates already have a substantial area of expansion (220 hectares), the implementation of the predict strategies will augment significantly the offer of industrial soil. Despite being questionable if the territory needs so many industrial soil, the policies predicted insist in the same mistakes undertaken in the past. Thus, they will contribute to widespread the estates by the municipal territories without any purpose of territorial coherence and without physical and functional articulation with the estates located in the neighbour municipalities (including those from Quadrilátero). The promotion of new estates in peripheral areas will probably create more spaces with low levels of coverage by facilities and amenities and can worsen some problems, namely the number of vacant lots and the declining of some parks (the less attractive). The simulation will confirm or not this suspicions.

In relation to the enterprises (Fig. 2), the Quadrilátero had 4722 industries located in his territory (INE, 2011). The greatest part of these enterprises were not located in an industrial estate (82%) and most of them (93%) were small or micro enterprises. The industries of the Quadrilátero generated a turnover of 5495 M€ and an international commerce (exportations) of 2772 M€ (INE, 2010). In the industrial estates were located 873 industries, being predominantly micro and small enterprises. The number of companies installed in these parks was very variable (between three and more than 40).

The turnover generated by each estate is also very different. For instance, the turnover produced in the industrial estate of Lousado (Famalicão) is bigger than the turnover generated by all the enterprises located in the 11 estates of Barcelos. This is related with the dimension of the enterprises and with the industrial sectors presented in each park. In the estates of Barcelos and Guimarães the most relevant sectors are the clothing and textile; in Braga is the metallurgic and the metalomecanic industries; while in Famalicão, despite the clothing, the rubber and the food sectors are also well represented.

The survey addressed to the entrepreneurs located in the industrial estates gave some relevant data, which was used to calibrate the model. The survey showed that 92% of the enterprises came from a diffuse location. The relocation was explained by several reasons, being the most relevant the following: the existence of an available area in the estates, the geographical location and the accessibility of the parks. In other hand, the majority of respondents (90%) does not intend to relocate to another estate. Indeed, entrepreneurs make a satisfactory evaluation of the conditions offered by the estates (3.3 on a scale of 5). The most criticised aspect was the quality and the low diversity of the amenities presented in the estates (this component received an evaluation of 2.3). In the other side, the most valued component was the good atmosphere and the good relations between the enterprises located in the estates (3.8).

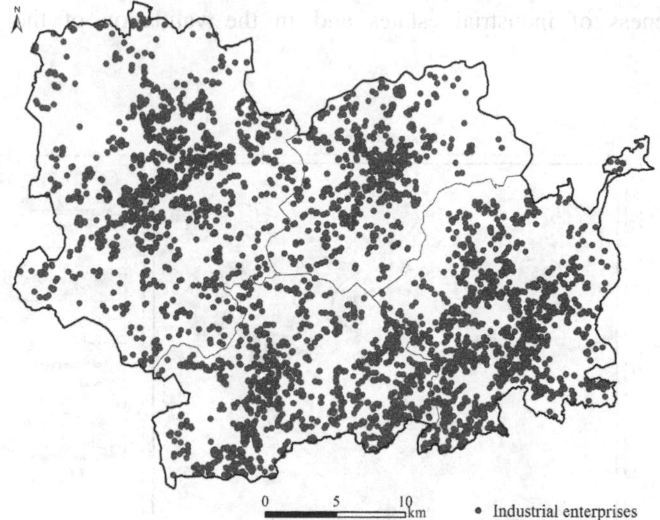

Fig. 2. Geographic distribution of the enterprises in the Quadrilátero

The findings obtained with the diagnosis and with the survey were useful to define the model and its rules. The main goal was to identify which policies may be useful to overcome the problems identified and to turn the industrial estates more competitive and attractive in a regional scale. The results of this simulation are described in the next Section.

6 Results and Discussion

This Section summarises the results obtained with the model. According with the methodology described in the Section III, the data related with the industrial estates and the enterprises was imported from the ArcGis to the NetLogo with several attributes necessary to the simulation. The simulation environment is represented in the Fig. 3.

As described in the Introduction, the model was performed to test the effect of the following three scenarios on the attractiveness of industrial estates: 1) maintenance of an uncoordinated and unqualified policy by the municipalities; 2) adoption of a coordinated policy of qualification of all the industrial estates; 3) impact of the strategies predicted by the municipalities in a scenario of: uncoordinated and low qualification (3a); and of coordinated and advanced qualification (3b). Further, the model was used to validate four hypotheses related with the impact of those policies on industrial estates. The envisaged hypotheses were the following: 1) the coordinated policies increase the attractiveness of all the industrial estates; 2) the coordinated policies leads to a common pattern of the estates occupancy in all municipalities; 3) a qualification policy improves the occupancy rate of all estates; 4) the promotion of diffused estates with low levels of qualification is an inefficient policy to attract enterprises. The following discussion will be focused on the impact of the scenarios in the attractiveness of industrial estates and in the validation of the hypotheses described.

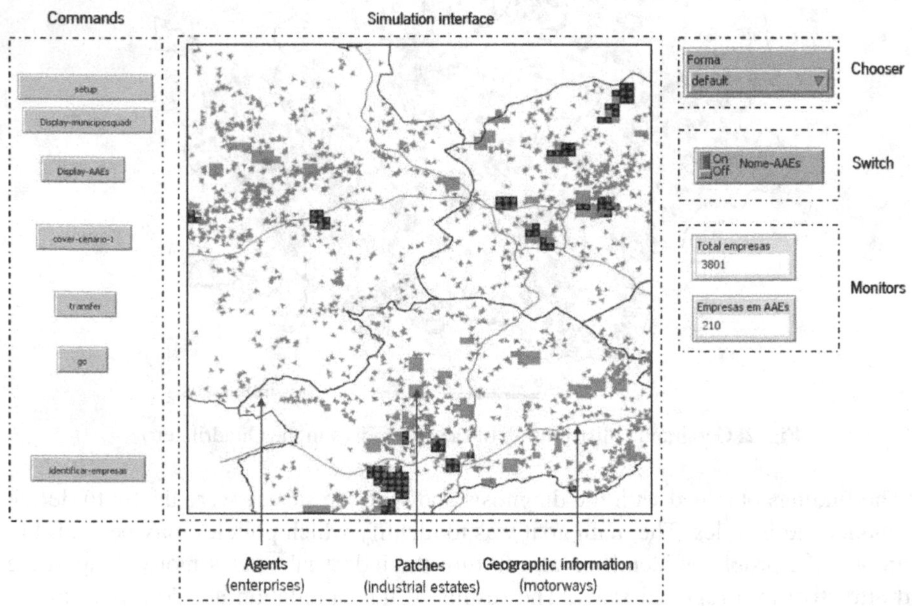

Fig. 3. Partial view of the model developed in the NetLogo toolkit

Despite the low level of locative requirements given to the enterprises in the scenario 1 (which represents the current situation and the starting point of the simulation), the results show that most of the industrial estates remain unattractive to enterprises (Table III). In fact, the lowest number of relocated enterprises was attained in this scenario (235) and 404 of the available plots remain vacant after the simulation (37% of the available supply). From the 79 industrial estates only 24 were attractive to enterprises. This bad performance is explained by the very low level of amenities found in many estates (19.9% on average) and by the lack of many basic facilities and by the poor accessibility diagnosed in many estates. During last years, municipal policies worsened this problem by encouraging the diffusion of a significant number of small estates with low levels of qualification by the municipal territories.

The impact of a coordinated qualification was tested in the scenarios 2 (existing estates) and 3b (where the predicted estates were included). The results show that this policy has a positive impact in the attractiveness of industrial estates. Regarding the existing estates, the comparison with the results obtained in the scenario 1 (Table III) indicates that there are more attractive industrial estates (+26), the number of relocated enterprises more than doubled and the number of vacant plots was substantially lower (-291). In the case of the scenarios that include the predicted strategies (3a and 3b), the simulation indicated that the most efficient policy is the provision of qualified estates. In fact, in scenario 3b the number of relocated enterprises is considerably higher (+223) and the number of vacant plots decreases proportionally (from 917 to 694). These results clearly demonstrate that qualification is an important policy to improve the capability of industrial estates attract enterprises. Nonetheless, these results only partially validate hypothesis 3. The qualification policy improves the occupancy rate of the estates, but many remain vacant. In fact, in the scenarios 2 and 3b many estates remain unattractive even if qualified. This is related with structural deficiencies that arise from the location defined for those industrial estates.

The implementation of coordinate policies between the municipalities can be more efficient in the locative performance of the industrial estates than if municipalities pursue with their individual initiatives. Results showed that both in the cases of the existing estates as well as in the case of the predicted sites, coordination improves the number of attractive estates (except in the scenario 3b) and the number of relocated enterprises (more 291 in the scenario 2 and more 223 in the case of the scenario 3b).These benefits are extensive to all municipalities, especially those where accessibility is more favourable. Thus, the results show that a coordinated policy has a global positive impact on the capability of industrial estates attract enterprises. However, these results only partially validate the hypothesis 1. The implementation of coordinated policies enhance the locative performance of many estates, but a significant number of them remain unattractive (Table III). On the other hand, results do not validate the hypothesis 2. Policies coordination does not produce an identical pattern of the industrial estates occupancy at the municipal level. In fact, at the Quadrilátero level and even at the internal scale of each municipality there are industrial estates with very different locative conditions. The most affected are the municipalities with more problems of accessibility and those which encourage the territorial diffusion of the estates. In this context, Guimarães emerge as the municipality where these problems are more visible and where the impact of these policies is less effective. On the other hand results indicate that municipalities with estates less diffused (which are usually the more attractive) are the most benefited by the adoption of coordinated policies. This is the case of Famalicão and Braga.

Considering the attractiveness of industrial estates, the model demonstrated that the adoption of a coordinated policy of qualification produces different municipal. The comparison of the scenarios 3a and 3b reveals a reduction in the number of the attractive estates (-3). This performance is explained by the loss of attractiveness suffered by some existing estates which were simulated as unqualified in scenario 3b. Thus, they become less competitive when compared with the advanced estates that will be provided in the future. Results also suggest that the new estates only make sense if inserted in a broader strategy of provision of qualified and competitive estates in order to respond to the requirements of contemporary enterprises. Besides, results also show that the existing estates should be framed in that larger strategy, otherwise these will become less attractive in the future.

Table 3. Main results obtained with the simulation

Indicators	Scenario 1	Scenario 2	Scenario 3	
			3a	3b
Number of attractive estates	24	50	33	30
Number of unattractive estates	55	29	53	56
Number of unoccupied plots	404	113	723	500
Number of relocated enterprises	235	526	694	917

The results obtained with the simulation of scenario 2 showed that 29 estates remain unattractive even after being qualified. This is related with the extremely low levels or, in some cases, with the total lack of basic facilities and amenities in those estates. The unattractive profile is often related with the questionable location defined for these estates, which restricts the accessibility and the facilities/amenities coverage. Thus, the industrial estates with minimum conditions become attractive when submitted to a qualification; the poorest remain unattractive even if upgraded. These unattractive estates will hardly play an important role in the Quadrilátero context, being more suitable for encourage the economic development at a local level. To turn these estates attractive, larger investments would be required to overcome the debilities related with their location. Further, municipalities with more industrial estates scattered around the territory exhibit low efficiency in the simulated policies, because dispersion causes a worse level of accessibility and a lower coverage of facilities and amenities. These findings highlight the importance of accurately planning the industrial estates in order to fully accomplish the economic, territorial and environmental requirements for which they were promoted and to avoid additional costs with their qualification in the future. Thus, the results validate the hypothesis 4, that is, the promotion of dispersed industrial estates with low levels of qualification is an inefficient policy to attract enterprises.

The verification of the four hypotheses described above and, particularly, the validation of the last one, means that the paradigm of arbitrarily supplying industrial estates, without criteria of territorial coherence and without advanced levels of qualification should be rethought by municipalities. Policies, to be sustainable and effective, should be properly planned and discussed within the municipality and in articulation with the neighbours. In an urban network, such the Quadrilátero, strengthening the municipal coordination of the policies is vital to achieve a global pattern of development.

7 Conclusions

Planning industrial estates accurately is an important issue since it determines the territorial and the entrepreneurial competitiveness. The use of appropriate tools is vital to ensure that industrial estates are planned in a more efficient way. This article describes an agent-based model that was developed as a tool for planning a set of industrial estates located in four municipalities. Agent-based simulation is a tool with growing application in the planning domain, because it allows dynamic and individual interactions among several territorial entities in a bottom up approach, highlighting macro patterns that arise from the individual interactions.

The model was geographically explicit and used the enterprises as agents able to assess the industrial estates' attractiveness. The attractiveness was defined by four parameters: facilities, amenities, accessibility and the cost of the land. The industrial estates were represented by patches (cells in which the environment is divided) that contain the data related with the attractiveness. Enterprises could move to another estate when they find a suitable location (the process involves an upgrade of their locative situation). The relocation process encompasses criteria related with the size, the territorial situation of the enterprises and how close or far they are from suitable estates. The impact of the policies in the estates' attractiveness was assessed by three scenarios and four hypotheses. Policies were distinguished by the level of municipal coordination under which they were implemented and by the type of intervention considered.

In the model, the diffused enterprises with bad locative position had the purpose of finding a new location in an industrial estate. The new location would find an upgrade in relation to their original place. The relocation process was concluded when an enterprise would find an industrial estate with vacant plots with a minimum level of attractiveness. In the model, the attractiveness of the industrial estates can be described as their capacity to attract new enterprises.

The adoption of a coordinated policy of qualification was found to be the most effective policy to strengthen the attractiveness of industrial estates. It was in this scenario that more industrial estates became attractive, more enterprises were relocated and more vacant plots were occupied. Despite the general positive effect, the adoption of a coordinated policy of qualification can produce different impacts in a local/municipal scale, as some industrial estates remain unattractive to enterprises. The simulation showed that the estates with the worst locations and with low coverage of facilities and amenities remain repulsive for enterprises even if they are subject to an upgrade. Furthermore, the results also indicated that the promotion of diffused and unqualified industrial estates is an inefficient policy to attract enterprises.

The authors believe that these conclusions can be an important contribution to guide the planning policies addressed to industrial estates for enhancing territorial and entrepreneurial competitiveness.

Acknowledgment. The authors would like to thank the Portuguese Foundation for Science and Technology for funding this research (SFRH/BD/48567/2008 grant).

References

[1] Khan, A.: A system for microsimulating business establishments: analysis, design and results (PhD Thesis), University of Calgary, Calgary (2002)

[2] Maoh, H., Kanaroglou, P.: Agent-based firmographic models: a simulation framework for the city of Hamilton. In: Proc. Second International Colloquium on the Behavioural Foundations of Integrated Land-use and Transportation Models: Frameworks, Models and Applications, Toronto (June 2005)

[3] Levy, S., Martens, K., Heijden, R.: An agent-based model of transport and land use policy coordination between municipalities. In: Proc. Bijdrage aan Het Colloquium Vervoersplanologisch Speurwerk, Antwerpen (November 2011)

[4] Gilbert, N.: Agent-based models, quantitative applications in the social sciences. Sage Publications (2008)

[5] Crooks, A.: Using geo-spatial agent-based models for studying cities. Working Paper Series (Paper 160). Centre for Advanced Spatial Analysis, UCL, London (2010)

[6] Witlox, F., Timmermans, H.: MATISSE: a knowledge-based system for industrial site selection and evaluation. Computers, Environment and Urban Systems 24(1), 23–43 (2000)

[7] Timmermans, H.: The saga of integrated land use-transport modelling: how many more dreams before we wake up? In: Axhausen, K. (ed.) Moving Through Nets: The Physical and Social Dimensions of Travel, pp. 219–248. Elsevier, Oxford (2003)

[8] Devisch, O., Timmermans, H., Arentze, T., Borgers, A.: Towards a generic multi-agent engine for the simulation of spatial behavioural processes. In: Van Leeuwen, J.P., Timmermans, H. (eds.) Recent Advantages in Design & Decision Support Systems in Architecture and Urban Planning, pp. 145–160. Kluwer Academic Publishers, Dordrecht (2004)

[9] Bowman, J.: A Review of the literature on the application and development of land use models. ARC Modelling Assistance and Support. Atlanta Regional Commission (2006)

[10] Zhao, F., Chung, S., Shaw, S., Xin, X.: Modelling the interactions between land use and transportation investments using spatiotemporal analysis tools, Lehman Center for Transportation Research, Miami (2003)

[11] Torrens, P.: Cellular automata and multi-agent systems as planning support tools. In: Geertman, S., Stillwell, J. (eds.) Planning Support Systems in Practise, pp. 205–222. Springer, London (2003)

[12] Robertson, D.: Agent-based models to manage the complex. In: Richardson, K. (ed.) Managing Organizational Complexity: Philosophy, Theory, and Application, vol. 24, pp. 417–430. Age Publishing (2005)

[13] Brown, D., Riolo, R., Robinson, D., North, M., Rand, W.: Spatial process and data models: toward integration of agent-based models and GIS. Journal of Geographical Systems 7, 25–47 (2005)

[14] Arentze, T., Timmermans, H.: Multi-agent models of spatial cognition, learning and complex choice behaviour in urban environments. In: Portugali, J. (ed.) Complex Artificial Environments, pp. 181–200. Springer, Heidelberg (2007)

[15] Matthews, R., Gilbert, N., Roach, A., Polhill, J., Gotts, N.: Agent-based land-use models: a review of applications. Landscape Ecology 22, 1447–1459 (2007)

[16] Kim, D., Batty, M.: Modelling urban growth: an agent-based microeconomic approach to urban dynamics and spatial policy simulation. Working Paper Series, vol. 165. Centre for Advanced Spatial Analysis (UCL), London (2011)

[17] Brown, D., Robinson, D.: Effects of heterogeneity in residential preferences on an agent based model of urban sprawl. Ecology and Society 11(1), 46 (2006)

[18] Diappi, L., Bolchi, P.: Smith's rent gap theory and local real estate dynamics: a multi-agent model. Computers, Environment and Urban Systems 32, 6–18 (2008)

[19] Crooks, A., Castle, C., Batty, M.: Key challenges in agent-based modelling for geo-spatial simulation. Computers, Environment and Urban Systems 32, 417–430 (2008)

[20] Singh, A., Vainchtein, D., Weiss, H.: Limit sets for natural extensions of Schelling's segregation model. Commun. Nonlinear Sci. Numer. Simulat. 16, 2822–2831 (2011)

[21] Campo, S.: Developing the land use and transportation integrated modelling framework for Lisbon Metropolitan Area (LUTIA-LX). In: Proc. 11th International Conference on Computers in Urban Planning and Urban Management (CUPUM), Hong-Kong (2009)

[22] Miller, E., Hunt, J., Abraham, J.: Microsimulating urban systems. Computers, Environment and Urban Systems (28), 9–44 (2004)

[23] Ettema, D., Kor, J., Timmermans, H., Bakema, A.: PUMA: Multi-agent modelling of urban systems. In: Proc. 45th Congress of the European Regional Science Association, Amsterdam (August 2005)

[24] Sanders, L.: Les modèles agent en géographie urbaine. In: Amblard, F., Phan, D. (eds.) Modélisation et Simulation Multi-Agents, Applications Pour les Sciences de L'homme et de La Société, pp. 151–168. Hermes Science Publications (2006)

[25] Moeckel, R., Spiekermann, K., Schürmann, C., Wegener, M.: Microsimulation of land use. International Journal of Urban Sciences 7(1), 14–31 (2003)

[26] Squazzoni, F., Boero, R.: Economic performance, inter-firm relations and local institutional engineering in a computational prototype of industrial districts. Journal of Artificial Societies and Social Simulation 5(1) (2002)

[27] Fioretti, G.: Agent-based models of industrial clusters and districts. In: Tavidze, A. (ed.) Progress in Economics Research, vol. IX, ch. VIII, pp. 125–142 (2006)

[28] Albino, V., Carbonara, N., Giannoccaro, I.: Coordination mechanisms based on cooperation and competition within Industrial Districts: an agent-based computational approach. Journal of Artificial Societies and Social Simulation 6(4) (2003)

[29] Giardini, F., Tosto, G., Conte, R.: A model for simulating reputation dynamics in industrial districts. Simulation Modelling Practice and Theory (16), 231–241 (2008)

[30] Albino, V., Carbonara, N., Giannoccaro, I.: Innovation in industrial districts: an agent based simulation model. International Journal of Production Economics (104), 30–45 (2006)

[31] Kumar, S., Kockelman, K.: Tracking the size, location and interactions of businesses: microsimulation of firm behaviour in Austin, Texas. In: Proc. 87th Annual Meeting of the Transportation Research Board, Washington (January 2008)

[32] Wissen, L.: A micro-simulation model of enterprises: applications of concepts of the demography of the firm. Papers in Regional Science 79(2), 111–134 (2000)

[33] De Bok, M.: Infrastructure and firm dynamics: a micro-simulation approach. PhD Thesis, Delft University of Technology, Delft (2007)

[34] Otter, H., Veen, A., Vriend, H.: ABLOoM: location behaviour, spatial patterns, and agent based modelling. Journal of Artificial Societies and Social Simulation 4(4) (2001)

[35] Manzato, G., Arentze, T., Timmermans, H., Ettema, D.: A support system that delineates location-choice sets for enterprises seeking office space. Applied GIS 6(1), 1–17 (2010)

[36] Leal, A.: Modelação do sistema rodoviário na perspectiva do conflito emergente. MSc Thesis, ISCTE, Lisboa (2009)
[37] SP – Statistics Portugal, Integrated Business Accounts System, Unpublished data base, INE (2011)
[38] Bodenmann, B., Axhausen, K.: Synthesis report on the state of the art on firmographics. Institute for Transport Planning and Systems, ETH, Zurich (2010)
[39] Ramos, R., Mendes, J.: Avaliação da aptidão do solo para localização industrial: o caso de Valença. Engenharia Civil (10), 7–29 (2001)
[40] SP – Statistics Portugal, Census 2011, INE, Lisbon (2012)
[41] Barbot, C.: Industrial determinants of entry and survival: the case of Ave. Working paper 111. FEP, Porto (2001)

Some Preliminary Remarks on the Recreational Business District in the City of Sassari: A Social Network Approach

Silvia Battino[1], Giuseppe Borruso[2], and Carlo Donato[1]

[1] DiSEA– Department of Economic and Business Sciences, University of Sassari, Via Muroni, 25 - 01700 Sassari, Italy
{sbattino,cadonato}@uniss.it
[2] DEAMS – Department of Economic, Business, Mathematic and Statistical Sciences, University of Trieste, Via A. Valerio, 4/1 – 34127 Trieste, Italy
giuseppe.borruso@econ.units.it

Abstract[1]. In geographical studies the Central Business District represents the 'central place' of the city, or its core, where the activities typical of a city, and different from those carried on in rural settlements, are carried on and realized. Several studies have been carried on in the past but a general characteristic is given by the concentration of the central activities in the city. Talking about specialized activities, authors, particularly studying tourism, identified the concept of Recreational Business District, as that part of a city mostly dedicated to free time and leisure, both frequented by locals and by pass-byers and tourists. Starting from previous experiences of research on Central Business District Activities, in this paper we present the first results of a research aimed at highlighting the Recreational Business District in urban areas, starting from the city of Sassari (Sardinia, Italy), with the aim of a first spatial delimitation of the district. Also, we analyzed the presence of recreational activities on the world of social networks ad media, in order to observe if and to what extent such 'virtual' connections hold a spatial component in tourist terms. Point pattern analysis is used for the analysis over the recreational activities and particularly a Kernel Density Estimation is performed over the different datasets.

Keywords: City, Urban Core, Density Estimation, Sassari, Sardinia, Tourism, Recreational Business District, Central Business District.

---

[1] The paper derives from the joint reflections of the three authors. Silvia Battino realized paragraphs 1 and 3, while Giuseppe Borruso wrote paragraphs 2 and 4. Carlo Donato wrote paragraph 5.
The geographical visualization and analysis, where not otherwise specified, have been realized using ESRI ArcGIS 10.2.

B. Murgante et al. (Eds.): ICCSA 2014, Part II, LNCS 8580, pp. 629–641, 2014.

1 Introduction

The city is the object of geographic research since the Twenties of the Twentieth century, when different authors ([1], [2], [3], [4], [5]) carried on theoretical analyses on the evolution of urban spaces. Such studies led to highlighting an area in cities of a certain dimension known as Central Business District (CBD), characterized, mostly in the US context, by a low population density and other secondary spaces around main road accesses dedicated to shopping and services, or other peripheral areas dedicated to commercial activities.

All of these areas are characterized by the presence of tertiary activities and functions, often different from each other, originating districts different according to the dominant activities being carried out, being them administrative, residential, commercial, industrial and dedicated to free time or entertainment. In this latter kind we find tourism that, particularly from the end of World War II, played an important role in the evolution of the city as a system: a new urban function that in some cases changed radically the economic structure of the city into a mono-functional space, in any case leading to new zoning division [6].

So an area called Recreational Business District (RBD) can be overlaid and added to the functional zoning of the city. Such an area is characterized by the presence of activities that can be enjoyed by tourists as well as locals. The RBD concept is studied initially by Stansfield and Rickert in the Seventies of the Twentieth century [6] and is defined as an area where services and goods used by visitors and tourists are concentrated around natural phenomena or historical, cultural and architectural attractions. As Zanini and Lando recalls [7], in time the willingness to consume of tourists made the RBD focusing also on retail activities, accommodation and leisure, activities highly specialized and often organized so to determine a "mood" with its own attracting capacity ([8]; [9]). Getz [10] stated that in most European cities the Tourism Business District was spatially consistent with the Central Business District, and we can reinforce such statement, saying that a high significance could be found in the overlapping of CBD and a tourist district or RBW. The metropolitan services serving the CBD are in fact dedicated to satisfy also the needs expressed by tourists and locals in their free time.

Here in this paper we tackle a first analysis to highlight the RBD of the city of Sassari (Sardinia Island, Italy) where, in a previous research, was studied in terms of the extension and characters of its CBD [11] [12] [13], highlighted in the historical center and neighboring districts.

The rest of the paper is organized as follows. In Paragraph 2 the methods adopted and the type of analysis carried out is presented, while Paragraph 3 is focused on a short description of the study area and on the data used. Results and discussion are presented in Paragraph 4 and the Conclusions are dedicated to some remarks concerning the recreational and tourist aspects of the city are presented, together with suggestions for future research activities and directions.

2 The Methods. Point Pattern Analysis and Social Networks

The research carried on in this paper is based on the analysis of recreational activities considering their geographical location in order to detect areas of clustering and therefore suggest some spatial definition of a recreational business district at urban level. We started with collecting urban activities and classifying them in categories referred to free time and recreation, moving then to georeference them. The starting point was a research carried on recently on the definition of CBD in Sassari ([11] [12] [13]) and from that we moved to the more in depth analysis of its recreational part. The work therefore involved the update of the list of activities at urban level particularly dedicated to recreation. Also we proceeded to enrich such list of features inserting other attributes than the categories as derived from the Yellow Pages and the geographical components as addresses and coordinates, therefore populating the list with the presence of the activities on the Internet and on social networks and media.

The analysis was done over the point pattern represented by the recreational activities at urban level. In particular a *Kernel Density Estimation* (KDE) was used to transform point events in space in a continuous density function over the study region, in order to visualize the phenomenon as a kind of 'heat map' or a pseudo - 3D surface that shows area of concentration of point features in a given area. The method is quite used in spatial statistics and analysis and widely used in several research areas, in case in which geographical elements can be presented by means of point patterns and we are interested in grouping them to highlight 'hot spots' or areas of major concentration of point events ([11] [12] [13] [14] [15] [16] [17] [18] [19] [20] [21])

Briefly recalling the formula, the kernel functions are three dimensional, characterized by a moving window visiting all events in a point pattern and weighting other events within a certain range according the their distance from the point where density is being estimated [22].

$$\hat{\lambda}(s) = \sum_{i=1}^{n} \frac{1}{\tau^2} k\left(\frac{s - s_i}{\tau}\right) \tag{1}$$

$\hat{\lambda}(s)$ is the density estimation of the point pattern measured at location s, while s_i represents the observed i^{th} event. $k(\ )$ is the kernel weighting function and the parameter τ is the radius of research of the function, or bandwidth, to be centered in location s, and searching for events s_i to be computed into the density function. The searching radius τ is the main arbitrary variable and a wider distance will produce a smoothed surface, good for visualizing hot spots over a wider area, while a shorter distance will produce mainly local peaks in the density distribution while wider values tend to dilute the phenomenon and over smooth the observed phenomenon [23]. The continuous density function is represented, in a GIS environment, by means of grid cells whose values represent either a density or a probability function. The variation of values between neighboring cells is smooth so that their distribution approximates a 3D distribution.

The KDE can be performed over the pure distribution of events, therefore considering just the geographical location of events. The function can also consider weights attributed to events belonging to the point pattern. In doing so, not only geographical proximity to the estimation point will provide denser values but also the presence of weights could change the shape of the density function over the study region. In this analysis we considered a weighted density function in which the presence of recreational activities on social networks and media was taken into consideration.

3 The Case Study. The Study Area and the Data Used

The city of Sassari is the capital of the homonymous Province and is the second city of Sardinia, after the regional capital Cagliari, in terms of inhabitants. It is located in North-Western Sardinia and counts to-date around 126,000 inhabitants [24]. The city hosts an old University dating back at 1617 although it does not seem holding the characteristics of national and international historical cities having an attractiveness as tourist centers.

After a period of prevalence of industrial activities the city gained a role in the tertiary functions, partly dedicated to free time both for resident and non-residents. The 'touristic' function has therefore widely grown and assumed stable characters thanks to the increase in cultural tourism that altered the urban framework and its economic structure.

From the tourism point of view in 2012 nearly 68,000 tourists visited Sassari in 2012 and 70 % of them were Italians, with an average presence of 1.9 days. It can be defined as a not proper and tourism, only partially increasing in the summer period when the city is visited by people spending their holidays in the neighboring and most renown coast locations.

The Yellow Pages service was used ([25] accessed February 2014) to collect the recreational activities of the Recreational Business District of the City of Sassari. These represent a subset of those used for the definition of the Central Business District, although updated at more recent times, as several recreational activities changed in their denomination and position in the years separating the two researches. Data were georeferenced at address point level using the GIS data provided by the Municipality of Sassari [26] and checked via on-line geocoding services [27].

The activities selected are those dedicated to recreation as "Art and Culture", "Retail", and "Free time". Nine sub-categories were also highlighted for a total of 321 activities as reported in Table 1.

Table 1. The activities determining the RBD in the city of Sassari. Source: Our elaboration from Yellow Pages [25] (2014)

Sub Category	Hotels and B&B	Bars and Coffee shops	Wineries	Take away food	Museums	Ice cream - cakes	Restaurants	Theaters	Other restaurants	Total
N.	16	91	15	3	5	37	134	10	10	321

After this reclassification of recreational activities we georeferenced and visualized them on top of a digital map of the Municipality of Sassari in order to understand how they distribute over the urban territory.

Over than analyzing the spatial distribution of the recreational activities our aim was also analyzing how they participate actively to the world of social networks and media. Such participation generally is represented by the presence of a website – although many activities do not hold a webpage – and by a profile on popular social networks or media, as Facebook, Twitter, Google+ or Instagram, just to cite a few among the most popular ones (Figure 1).

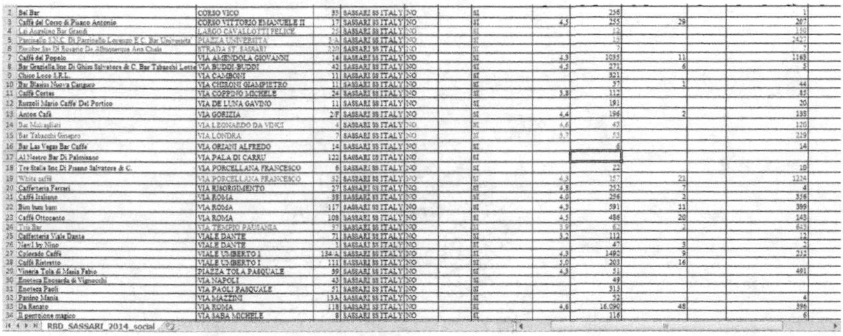

Fig. 1. Part of the recreational activities and their presence on the Web and Social Networks (2014)

Source: our elaboration on GIS data from Yellow Pages [25] and Social networks and media

Of the 321 activities considered, a 29.28 % holds a website, while 51.09 % have a social profile. Just 22.74 % of them have both (Table 2). Restaurants are among the categories with higher presence over the social networks and internet, with a 54.79% having a web site as well as a social profile. Hotels and Bed & Breakfast follow (10.96 %) and bars and coffee shops (8.22%).

Among the social networks considered, specific attributes were defined and referred to the most popular applications. Facebook, Twitter, Google+ and Instagram were considered as the social media and network to check for each economic activity. Of such activities, characteristics as the ‘Likes’, ‘Followers’, ‘Following’, etc. were considered and eventually used as weights in one of the elaborations.

Table 2. Recreational activities grouped by their presence on the web and on social networks

Source: Our elaboration from Yellow Pages [25] (2014)

	Activities		Activities (%)	
	Yes	No	Yes	No
Web site.	94	227	29.28	70.72
Social network / media	164	157	51.09	48.91
Web site + Social network / media	73	248	22.74	77.26
Total	321		100	

4 Results and Discussion

A first visual analysis can be done on the scatterplot of point features referred to recreational activities.

Fig. 2. Central activities in the city of Sassari
Source: our elaboration on GIS data from Yellow Pages [25]

As a reference we represent the distribution of activities used for highlighting the Central Business District (Figure 2) and those used in the present research for the (social network) analysis of the Recreational Business District (Figure 3).

The analysis carried on by Battino, Borruso and Donato [12] highlighted an area of concentration of the full dataset of the central activities in the compact city's urban districts and particularly in the central districts (Centro Storico, Piazza d'Italia, Viale Dante and Viale Amendola - Viale Italia) that alone covered more than 55 % of the total.

The observation of the point pattern given by the recreational activities shows obviously a less dense presence, as the recreational activities appear as a subset of the central activities' dataset (321 events versus 1,980 belonging to the central activities datasets). It can however be noticed that the areas of concentration are not very dissimilar between the two datasets, so more in depth analyses can be performed to highlight true hotspots in the study region.

A Kernel Density Estimation was therefore performed on the central activities and on the recreational ones to observe if the hot spots overlay or some other pattern arise. Also, the density estimation was performed over the activities that are actually active in social networking, or those demonstrating a higher openness to new media and new opportunities to accessing new customers.

Fig. 3. Recreational activities in the city of Sassari
Source: our elaboration on GIS data from Yellow Pages [25]

Different bandwidth were tested, while here we present the results related to a bandwidth of 355 m, that corresponding to the 150 nearest neighbor activities computed for the central activities. This implies the "computation of the average of intra-events distances of different orders [27], thus linking the control of the variable to a k-nearest neighbor choice instead of an arbitrarily chosen radius" [12]. We decided to maintain such a bandwidth also for the recreational activities and for the social recreational activities, as such a distance is compatible with an average 5 minutes walking distance, a good approximation of accessibility to services at urban scale.

The results from the density analysis on central activities are portrayed in Figure 4, where a suggested shape of the CBD is presented, following a North-west – South-east orientation, from the boundary of Centro storico district (in the centre map), crossing Piazza d'Italia and ending up in Viale Dante. Also a secondary hot spot can be observed on the left side of the map, West from the main elongated cluster, highlighting a shape similar to the Greek letter 'lambda' (λ).

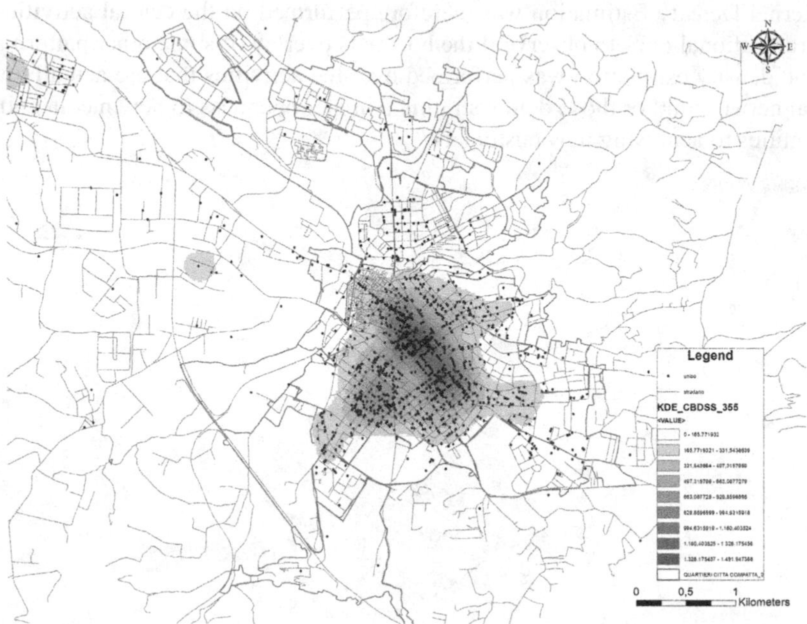

Fig. 4. Central business activities in the city of Sassari (355 bandwidth)
Source: our elaboration on GIS data from Yellow Pages [25]

However the main core of the CBD follows mainly an elongate shape, whose vertices are the areas of the two districts Centro Storico and Viale Dante, linked together by the two parallel streets Corso Vittorio Emanuele II and Via Roma.

The same density analysis was performed over the recreational activities, with the aim to identifying the area (s) of concentration of activities and therefore determine a shape for the Recreational Business District. Figure 5 shows the results of such an analysis. We can observe that a general trend comparable with that represented in figure 4 and related to the CBD can be noticed.

Some differences however arise, particularly with a major elongation and higher levels of density present at the two extreme points of the elongated shape, in the 'Centro Storico' and 'Viale Dante' districts, with a denser area in the 'Centro Storico' district. So two 'hot spots' seem to be visible following the same elongated shape of the CBD, without necessarily describing a 'lambda' shape. Also, the area of concentration of the hot spots of recreational activities is not completely overlapping with the CBD area showing some mismatch.

The two hotspots or clusters in the two areas mentioned above are even more visible and neatly identifiable when the analysis is performed over the subset of the recreational activities, as those characterized by being active in social networking.

Figure 6 reveals a neatly elongated shape of the RBD following the two main streets connecting the 'Centro Storico' and 'Viale Dante' districts, with nearly equal density values for the two areas.

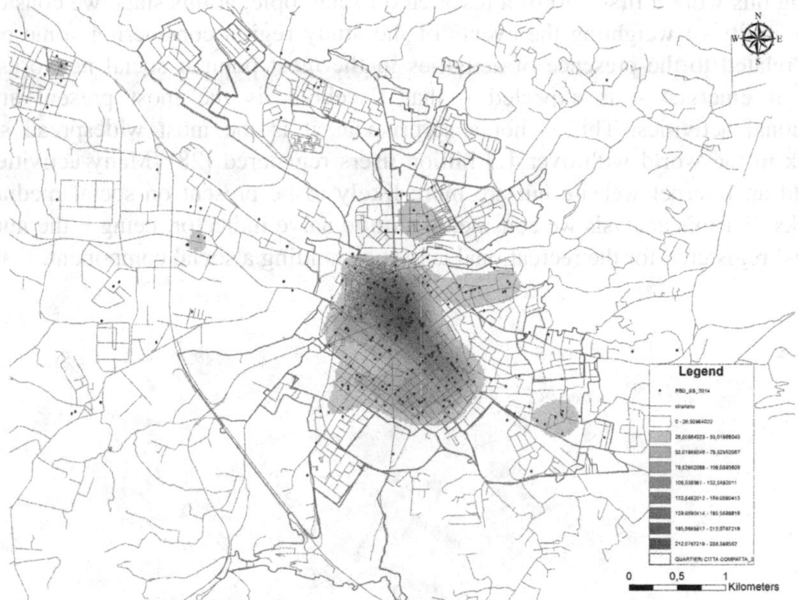

Fig. 5. Recreational activities in the city of Sassari

Fig. 6. Recreational activities in the city of Sassari on social networks

Being this work a first start of a research on such topic, at this stage we considered the possibility of weighting the events of the study region considering some of the values related to the presence of activities in the most popular social networks and media. It emerged - as expected – that Facebook is the most present among recreational activities. This is not a surprise as it is the most widespread social network in the world with over 1.2 billion users registered [28]. Many activities do not hold an Internet website but are more likely to be present on social media and networks. For this analysis we considered a quite naïve indicator, being it the number of 'likes' registered for the recreational activities holding a social component.

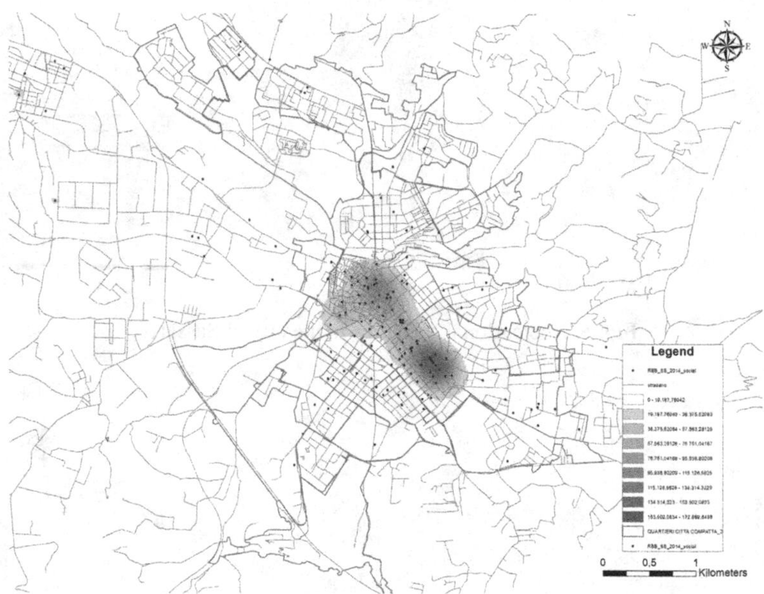

Fig. 7. Recreational activities in the city of Sassari on social networks (weighted)

In figure 7 the results are displayed. Here the clustered area is more neatly defined that in the previous cases along the two main high streets and with peaks in the same areas highlighted before. However an interesting element to be noticed is that, after weighting the density function, the main hot spot moves from the city center towards the South-Eastern part of the figure. Such an area is actually very close to the university and also hosting several recreational activities accessed mainly by students and university personnel working in the area. So it appears quite interesting to notice that possibly the activities more dedicated to a young public are also those more interesting in having an appealing profile on the usual social networks, so to establish a link between activities and customers that has both the characters of the virtual one (as many connections on social structures) but also between people and places, having the 'non social network component' of the advertisement a very strong and visible linkage to space – being such activities inserted in a physical, urban space.

Surely new recreational structures try to attract people and particularly, being mobile devices and social networks and media more and more widespread, it is an important way to be in touch with potential customers. Also, this represents an opportunity to test the response from clients, as social networks and media are open to comments and therefore the users have the potential to rise doubts on possible bad choices of the recreational activities and to monitor, to some extent, their performances.

5 Conclusions

Some final reflections can be done with reference to the methods adopted and the importance of delimiting central urban areas. The results were interesting as the methods allowed highlighting clusters of central activities in the urban areas, and also as their application, at different scales and using different parameters, can be easily repeated not only to other urban cases but also to highlight different characters and specialization of sub-areas in an urban environment, helping also in re-drawing, if needed, administrative subdivisions.

This initial research reveals as the recreational activities that characterized Sassari Recreational Business District are mostly located in proximity of the same Central Business District, although with some differences in the hot spots. In particular the hot spots concerning the activities more active in social networking and media. The presence on popular social networks and media enhances an activity's own attractive capacity and therefore allows playing a real touristic role.

However, the Recreational Business District seems to be dedicated mainly to satisfying the needs of free time of its local population and, only partly of tourists that concentrate during the summer months.

As a final comment, the perspective of Sassari as a 'historical and cultural city' must be supported by urban planning policies aimed at qualifying the urban spaces hosting the Recreational Business District and allow a sustainable tourism, trying to interpret the city as a place where functional activities different from the basic ones are located (i.e., those dedicated to residents), as well as the basic ones (or those dedicated to tourists).

The research is however at an initial stage for different reasons. On one side there is a need to debate and further develop a theoretical discussion on the activities dedicated explicitly to tourism and their relationship with the wider ones targeted on recreation – these latter including actually also locals as consumers and not just tourists. Another issue is dedicated to the weight and importance of the presence on the Internet of the activities and particularly on the social networks and media and how such elements can be effectively related to spatial elements.

A third element is dedicated to visualization issues and the perspectives of the correct choice of the cartographic representation for mapping phenomena like those observed. What used here was based mainly on a point pattern analysis and on methods to visualize them, but it must be considered that the elements considered hold a different and more complex nature in terms of their forms.

References

1. Burgess, E.W.: The growth of the city. In: Park, R.E., Burgess, E.W., Mc Kenzie, R.D. (eds.) The City, pp. 47–62. University of Chicago Press, Chicago (1925)
2. Christaller, W.: Die Zentralen Orte in Suddeutschland. Fischer, Jena (1933)
3. Hoyt, H.: The structure and Growth of Residential Neighborhoods in American Cities. U.S. Government Printing office, Washington D.C. (1939)
4. Harris, C.D., Ullman, E.L.: The nature of cities. Annals of the American Academy of Political and Social Science 242, 7–17 (1945)
5. Alonso, W.: Location and land use. Toward a general theory of land rent. Harvard University Press, Cambridge (1965)
6. Stansfield, C.A., Rickert, J.E.: The recretional business district. Journal of Leisure Reseach 2, 213–225 (1970)
7. Zanini, F., Lando, F.: Impatto del turismo sulla struttura terziaria urbana. Note di Lavoro. DSE, Università Ca' Foscari di Venezia 5, 1–25 (2008)
8. Timothy, D.J., Butler, R.W.: Cross-border shopping. A North American perspective. Annals of Tourism Research 22(1), 16–34 (1995)
9. Jasen-Verbeke, M.C.: Leisure shopping. A magic concept for the tourism industry. Annals of Tourism Research 12(1), 9–14 (1991)
10. Getz, D.: Planning for tourism business districts. Annals of Tourism Research 3, 583–600 (1993)
11. Battino, S.: Estensione e delimitazione dei core urbani della città di Sassari. Bollettino A.I.C. (143), 29–48 (2011)
12. Battino, S., Borruso, G., Donato, C.: Analyzing the Central Business District: The Case of Sassari in the Sardinia Island. In: Murgante, B., Gervasi, O., Misra, S., Nedjah, N., Rocha, A.M.A.C., Taniar, D., Apduhan, B.O. (eds.) ICCSA 2012, Part II. LNCS, vol. 7334, pp. 624–639. Springer, Heidelberg (2012)
13. Battino, S., Borruso, G.: Analisi GIS del Central Business District di Sassari. Visualizzazioni cartografiche. Atti 16° Conferenza ASITA (Vicenza 6-9 Novembre), pp. 183–190 (2012)
14. Thurstain-Goodwin, M., Unwin, D.J.: Defining and Delimiting the Central Areas of Towns for Statistical Modelling Using Continuous Surface Representations. Transactions in GIS 4, 305–317 (2000)
15. Borruso, G.: Il ruolo della cartografia nella definizione del Central Business District. Prime note per un approccio metodologico. Bollettino dell'Associazione Italiana di Cartografia 126-127-128, 255–269 (2006)
16. Borruso, G., Porceddu, A.: A Tale of Two Cities. Density Analysis of CBD on Two Midsize Urban Areas in Northeastern Italy. In: Borruso, G., Lapucci, A., Murgante, B. (eds.) Geocomputational Analysis for Urban Planning. SCI, vol. 176, pp. 37–56. Springer, Heidelberg (2009)
17. Borruso, G.: Network Density Estimation: a GIS Approach for Analysing Point Patterns in a Network Space. Transactions in GIS 12, 377–402 (2008)
18. Danese, M., Lazzari, M., Murgante, B.: Kernel Density estimation methods for a geostatistical approach in seismic analysis: the case study of Potenza Hilltop Town (Southern Italy). In: Gervasi, O., Murgante, B., Laganà, A., Taniar, D., Mun, Y., Gavrilova, M.L. (eds.) ICCSA 2008, Part I. LNCS, vol. 5072, pp. 415–429. Springer, Heidelberg (2008)
19. Danese, M., Lazzari, M., Murgante, B.: Geostatistics in historical macroseismic data analysis. Transactions on Computational Sciences 6(5730), 324–341 (2009)

20. Murgante, B., Danese, M.: Urban versus Rural: the decrease of agricultural areas and the development of urban zones analyzed with spatial statistics, Special Issue on Environmental and agricultural data processing for water and territory management. International Journal of Agricultural and Environmental Information Systems (IJAEIS) 2(2), 16–28 (2011)
21. Gatrell, A.: Density Estimation and the Visualisation of Point Patterns. In: Hearnshaw, H.M., Unwin, D.J. (eds.) Visualisation in Geographical Information Systems. Wiley, Chichester (1994)
22. Levine, N.: CrimeStat III: A Spatial Statistics Program for the Analysis of Crime Incident Locations. Ned Levine & Associates, Houston, TX, and The National Institute of Justice, Washington, DC (2004)
23. Municipality of Sassari, Comune di Sassari: Popolazione residente al 2013. Sassari (2014)
24. Italian Yellow Pages, `http://www.paginegialle.it`
25. Municipality of Sassari, `http://www.comune.sassari.it`
26. Chainey, S., Reid, S., Stuart, N.: When is a hotspot a hotspot? A procedure for creating statistically robust hotspot maps of crime. In: Kidner, D., Higgs, G., White, S. (eds.) Socio-Economic Applications of Geographic Information Science. Innovations in GIS, vol. 9, Taylor and Francis, London (2002)
27. GPS Visualizer, `http://www.gpsvisualizer.com/geocoding.html` (accessed April 10, 2014)
28. Cosenza, V.: Social Media Statistics, `http://vincos.it/social-media-statistics/` (accessed April 10, 2014)

Orienteering and Orienteering Yourself. User Centered Design Methodologies Applied to Geo-referenced Interactive Ecosystems

Letizia Bollini

Department of Psychology, University of Milano-Bicocca,
20126 Milano, Italy

Abstract. The research proposes a critical mapping of evaluation methods and user-centered approaches typical of the user experience design applicable to interactive geo-localized data systems – for example, usability user test – according to the emerging patter of mobile user experience and geo-interfaces usability in location-based services.

Keywords: Mobile interface design, geo-based user experience design, ambient findability, environmental data usability.

1 Introduction

The mobile revolution, the so-called *Web 3.0* – geo-referred, semantic and social – and the environmental pervasiveness of digital technology in its various forms (*pocketable* web, Internet of the Things, ubiquitous, Augmented Reality, intelligent environments, wearable devices etc.) are opening up new scenarios of research and development both in the field of geo-referenced information and in user-centered design.

The opportunity given to people to geo-locate themselves and to use information based on their presence in the space stratifies different layers of experience: firstly, the real territory and its cartographical representation mediated by digital technology; secondly, the level of spatial perception and the resulting mental model built by the people's perception; and the logical level of information, textual, multimodal, interactive of the Internet.

Access to these different story-telling levels can also occur from many different access points and through a variety of devices within what might be called a *digital communications ecosystem* in which *real* and *virtual* coexist in the experience and cognitive dynamics of the users.

This shift represents a turning point in the people's interaction model both with the space and the way they use Internet.

B. Murgante et al. (Eds.): ICCSA 2014, Part II, LNCS 8580, pp. 642–651, 2014.

1.1 From Places to Experiences in the Mobile Era

The first generation of devices – such as mono-function GPS – allowed the user to *find* themselves in a territory effectively replacing the analog mapping systems with a digital technology.

Smartphones introduce instead the generation of multi-function devices that *also* allow geo-localization and to interact or to find information related to physical territory: the cognitive model moves therefore from the task/function of being and find yourself in a place, to find information on that site, i.e. the pins and balloons of Google Maps. Until the third generation of devices that Husson [1] defines the *pocketable web*, thath allow th users to interact in a place with their own social networks – for example, the check-in to Foursquare – with the user geo-based generated content related to that place, with their own temporal tracks related to that place – the geo-tagging of images originally introduced by Flick'r and then picked up by Facebook, Twitter and iOS – with unknown people who are at the same time in that place etc.

Therefore the experience of geo-based data it is not longer the purpose of the activity, but the access point to other tasks and interactions mediated by the mobile devices.

1.2 Space as the Form of Knowledge Organization

This change implies – first of all – the need for a redefinition of the concept of space itself both from the user's and the designer's perspective.

On one hand the concept of space has much more to do with the idea of *perception* and *memory* as it emerges from studies conducted by Baddeley [2], Kosslyn [3] and Cornoldi [4]. But is the concept introduced by Tolman [5] of *mental models* or – better to say – *cognitive maps* as the way people acquire, organize and interpret information about the locations and attributes of phenomena in their everyday environment both spatial and metaphorical, i.e. as type of mental representation a way of building a mental image-based experience both physical and conceptual, thath seems to be the connecting link with the forms of spatial representations.

On the other hand the conceptualization of space as a metaphorical form of interaction in user experience and interface design is deeply embedded since the birth of one of the primary branch of user-centered design that means Information Architecture. The term firstly mentioned by Wurman [6] in the book *Information architects* – that became the foundation of IA discipline – explicit the use of the spatial metaphor as a reference both in theoretical and design work in wen design.

As noticed by Rester [7] there is therefor a difference of approach involved in the production of spatial metaphor in opposition to the real essence of the physical space: "an architect begins with an abstraction – a blueprint – and creates from that abstraction a concrete structure existing in physical space. The cartographer, on the other hand, starts with concrete structures existing in physical space and creates from that an abstraction: a map."

The map, as a form of conceptual *meta-representation* devoted to communicate an idea or as the final result of a *translation* activity from the 3D world to the 2D space of a page, become the elective *tool* to explicit and convey both abstract and physical issue and information – at least if its language is clear and familiar to users – and a visual *text* to tell about the *sémiotique plastique* of a space as underlined by Greimas [8] or the visual representation of the spatial meaning, i.e. the *semiologie graphique* as stated by Bertin [9].

1.3 The Social Construction of Space

If the space is first of all a physical dimension of our perception and its two-dimensional representation is a form of interpretation also of derived from the cognitive model it became – in a diffused and collective experience – a social construction. As well as shown by studies – carried out almost in the same period – by Lefebvre [10] and Lynch [11] the space is a complex form and produced by people who live, move, inteact in/with it.

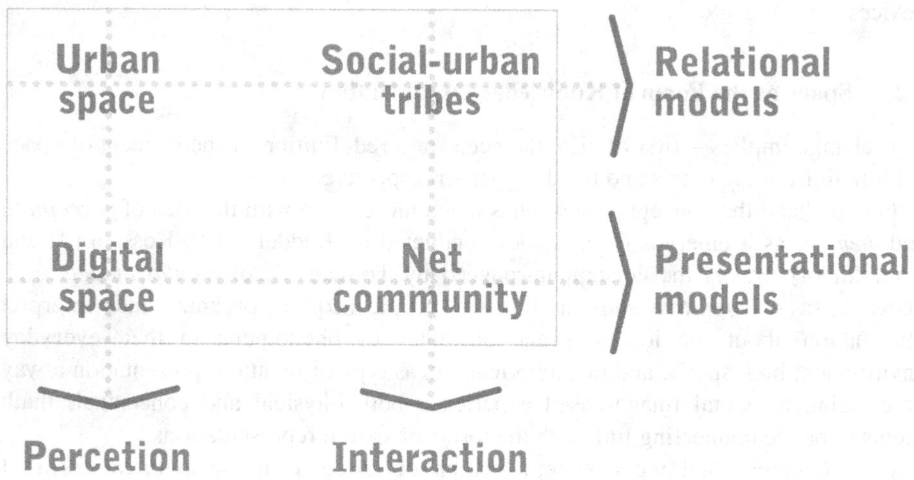

Fig. 1. The *matrix* of spatial/social/digital layer of interaction built and experienced in an physical environment mediated by mobile devices. The figure is part of the research project Bollini, L. (2012) "Hybrid methodology for social & digital space design." Presented at the IV International Design Forum of Design as a Process in Belo-Horizonte, 2012.

In the mobile era the social dimension become even more pre-domimant in such an interaction among space, people and digital sphere of experience as metaphor, "social space contributes a relational rather than an abstract dimension […] has received a large variety of attributes, interpretations, and metaphors" as noticed by Kellerman [12]. As long as social media - not just or no more network - become the experiential layer that overlays the physical territory that digital devices allow you to bring in a continuum of *here and now*, the relationship moves from the dual combination person/environment to the reticularity and multiplicity of the interaction between individual, networks, spaces – simultaneously and in copresence or in another time/place – both real and virtual. The cartographic or geo-referred data becomes an access point that opens up virtually limitless possibilities for communication and relationship.

2 Maps as Interaction Paradigm

The coexistence of these different layer of story-telling and information both spatial (territory and map) and logical (text and hypermedia) may converge in a sole communication ecosystem which unifies the user activities and tasks.

A visual and touch direct manipulation of the map – and consequently the data referred to the geo-spatial dimension – become in this context a paradigm of interaction in which mobile and GIS technologies are just framework where to build the user experience. On the other hand, also elements presented in a *logical* way [13], i.e. lists of items, text and paragraphs, documents etc. could be *converted* – or at least – presented linked to the spatial representation.

As proposed already in 1993 by Egenhofer & Richards [14] a map-overlay metaphor based on "direct manipulation and iconic visualization [...] a *geographer's desktop* by replacing the familiar notion of files and folders with the concepts of *map layers* and a *viewing platform* on which layers can be staked" could be consider the original core of the possible exploration of this pattern of interaction.

2.1 Geo-interfaces Usability

The research work of Cartwright, Crampton, Gartner, Miller, Mitchell, Siekierska and Wood [15] explores further the question of the representation of planar data or other "spatialized representations of non-geograhic phenomena" introduces both the problem of *aesthetics* – better to say in the user-centered design gergon *affordance* – according to the visual dimension of interaction and user interface, and *usability* focusing therefor on user experience and abilities.

That "implies that the user is able to both navigate the visualization itself and navigate the synthetic geography that it represents. Navigation can refer to navigation between "map" representations; navigation within "map" representations (e.g., pan, zoom, scale, generalization); navigation between and within datasets; navigation between spatial objects and related temporal and thematic attributes and related information; and navigation between spatial objects and metadata. Navigating/browsing a geospatial representation (whether it be a map, a 3D virtual environment or whatever) requires the user to be able to [...] changing the scale, map projection, level of generalization and field of view; pan, move, browse across the "map" extent/content [...]. Interface tools to support these tasks should be designed so that they correspond to cognitive usability principles."

According to this research we can now discuss about *geo-visualization* and usability standards and methods applied to the field coming both from human-computer interaction and cognitive sciences and from cartographic research to design usable *geo-interfaces*.

3 The Evaluation of Geo-Based Mobile Experience

The investigation methods to be adopted within the research field of *geo-interfaces usability* can not be that hybrids. A mashup of the experiences of (information) architecture, environmental psychology based on the experimental model of Lynch [11] Parr [16] Sonnenfeld [17] and Bagnara [18] and methods of user-centered design, with particular reference to usability test/tasks personas-based, according to the heuristics of Nielsen [19] and recently updated for the mobile version [20] and approaches quick&dirty such as of Krugg [21] and Morville ambient findability conceptualization [22] able to cover and taking in account both aspects of the user interaction: spatial and digital.

3.1 "In the Field" Experience

The majority of applications up to now, however, was designed for the desktop version producing a gap – de facto – between the real use of the geographic data and the possibility to access them outside of the natural context, mobile devices, restore the original vocation of a map as a tool for interaction "in the field" or *field-based tasks* and activities here-and-now and enriches the user experience with the continuous flow of digital information and social interactions.

Moving from the closed space of an office where interaction is mediated by mouse and keyboard on a wide screen and with a stable connection to the real world with a noisy background, a dynamic use of the multi-touch device or its vocal interface assistant, in a crowded and loud surround is definitely a different experience to design, to test and to assess.

3.2 The Paradox of Lab Mobile User Test Bias

The traditional methodologies to conduct a usability test on a mobile device risk to fail in their purpose or – at least – to return results correct in itself, but not real realted to the context of use of the device or application.

A *in person moderated test* [23] implies to discuss tasks and interactions in a laboratory more adequate to researchers and clients needs rather to an efficient evaluation of features and problems.

Firstly the lab guarantees a constant condition of illumination, weather, noise, temperature that means a stable and controlled environment that will not impact on test results and will allow researchers to have a good comparision of experimental data. Also the technical conditions and infrastructure will be controlled and uninfluential: same device, same setting, operating system and installed apps, bold and costant connection and performances. Finally the user will be put in confortable condition even if the device will be in a fixed position – to be recorded by the camera – to a costant distance and inclination and will be possible for the user to interact with both hand without the need to hold the device.

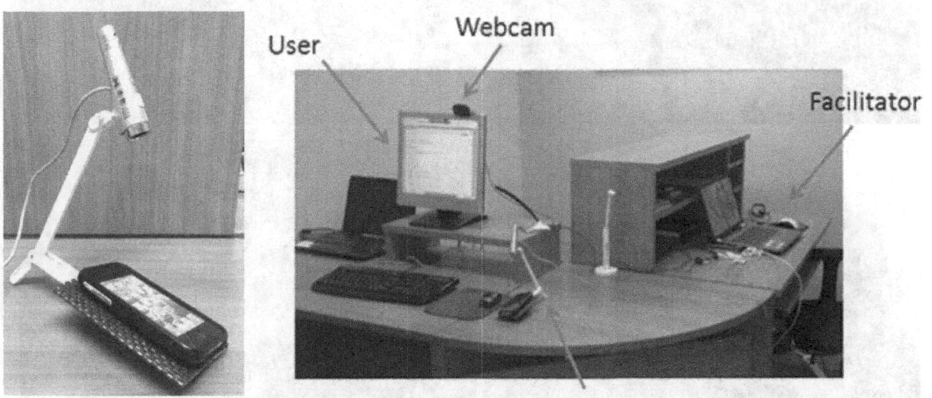

Fig. 2. Laboratory settings and tool to conduct usability user test on mobile devices mediated by a facilitator. The images area available at the web site *measuring usability* (URL: http://www.measuringusability.com/blog/mobile-usability-test.php)

Although these conditions allows you to limit the bias of research and obtain comparable and *clean* data it is not obvious how these results might be significant to design or improve an application with respect to its actual context of use.

Especially this type of methodological approach eliminates the fundamental aspect in case of interaction via mobile or the possibility of geo-based interaction with the surrounding environment.

3.3 *Open-Air* Mobile User Test

The conditions that in the context of a *normal* experiment ensure the accuracy of the results in the case of geo-referred applications and interfaces turn into a strong limit.

To assess the usability of those – we have referred to as – *geo-interfaces* the “in the field” research is essential. The experimental limits are part of the conditions to be tested and that will have a significant impact on the user experience: the complexity and variables increase by an order of magnitude, but at the same time feed-back will be much closer to the actual experience.

The disadvantages are in fact balanced by a series of positive aspects, especially if valued in terms of qualitative feedback and as guidelines for the design rather than as quantitative data:

Fig. 3. Usability user test on mobile devices open-air. The images are part of an on-going research project *MoBi - Mobile (Milano) Bicocca* focus on the opportunity given by geo-located services in the field of university choice as an alternative to printed student guide or services.

What You See Is What You Get. Put into the physical space the user does not have to simulate behaviors through the mobile device and geo-referred data or make abstract assumptions as in the case of the laboratory. The geo-interface become a trasparent tool to *naturally* interact with the surrounding allowing the acces to information and social interactions.

Personal Is Better. The more informal context of an open-air test, people can directly use their own devices. In this case knowing the hardware, the operating system and settings they will be directly focus on tasks without facing the difficulty of learning how to interact with an unfamiliar smartphone or tablet.

Weather Conditions. The *sterile* environment of the laboratory ensures costant physical conditions for the single test sessions and for the research as a whole. On the other hand whether condition – such as illumination, humidity, noise – could can cause great difficulties During the test tap and other touch gestures do not play well if fingers are wet, too cold or if the user wears gloves without capacitive patches.

Display *Blindness*. If the backlit display of the device is illuminated by an oblique light it turns *black* making it impossible to interact.

***Freestyle* Interaction.** Not only the user is free to move in space, but also to hold and to interact with the device without constraints or fixed conditions of use, position, distance etc. This makes the interaction more efficient and spontaneous and it could suggest to a good observer unconventional pattern in addition to those prefigured in the plot of the experiment.

Connection Hiccups. The conditions of access to Internet connetio – controlled and veirified in a lab project – become one of the most significant variables in an open-air evaluation. Evaluating the effects on the user experience of the worst-case scenario will make possible, for example, to opt for a standalone native app – not sensitive to the connection speed – or to develop a web based one highly optimized in terms of weight and content.

Authors such as Li & Longley [24] propose and verified a further approach for testing a digital product or app devoted to open-air use. It is an *immersive virtual reality*, created specifically for the test that users are going to accomplish a "urban VR models that allow individuals to 'walk around' at street level, a mobile device as information source which simulates LBS applications, and software for recording participant actions and reactions within the test environment". In this way, the moderator can even control those variables that *outdoor* would be uncontrollable and unpredictable.

Moreover the mobile app to be used outdoors and which provide a substantial use of geolocation, impilies that the test is carried *in the field* and, once again, the researcher must be careful to note any deviation from the typical behavior due to variable uncontrollable, and take them into account at the time of the discussion of the results.

4 Conclusions

The mobile revolution is already changing the way we interact with the world around us and with each other inserting multiple levels of information remedies available through digital technology.

Interactions and interfaces geolocated become more and more the gateway that allows us to exploit the potential of interfacing in a local, social and virtual environment, on the other hand they introduce a new paradigm of interaction that needs a conceptual change from *logical* form to the spatial shape of metaphors in user interface design. Therefore we also need new approaches and tools to determine the outcome, efficiency and usability of these new touch points of access to communication ecosystems both connected to reality and virtual.

References

1. Husson, T.: Another Year in Review: Revisiting 2013's Mobile Trends (2014), http://blogs.forrester.com/thomas_husson/14-01-09-
2. Baddeley, A.D.: Working memory. Claredon Press, Oxford (1986)
3. Kosslyn, S.M.: Image and mind. Harvard University Press, Cambridge (1980)
4. Cornoldi, C.: La memoria di lavorovisuo-spaziale (MLVS). In: Marucci, F.S. (ed.) Le Immaginimentali. Teorie e processi, pp. 145–181. NIS, Roma (1995)
5. Tolma, E.C.: Cognitive maps in rats and men. Psycohlogical Review 55(4), 189–208 (1948), doi:10.1037/h0061626
6. Wurman, R.S.: Information Architects. GraphisInc., New York (1996)
7. Rester, A.: Mapping Memory: Web Designer as Information Cartographer. A List Apart 266 (2008), http://alistapart.com/article/mappingmemory
8. Greimas, A.J.: Sémiotique figurative e sémiotiqueplastique. ActesSémiotiques. Documents. 60 (1984)
9. Bertin, J.: SemiologieGraphique: Les Diagrammes, Les R Seaux, Les Cartes. Walter de Grutyer, Berlin (1967)
10. Lefebvre, H.: La production de l'espace. Anthropos, Paris (1974), http://www.scribd.com/word/removal/57689655
11. Lynch, K.A.: The image of the city. MIT Press, Cambridge (1960)
12. Kellerman, A.: The Internet on Earth. A Geography of Information, pp. 31–32. Wiley (2002)
13. Bollini, L., Cerletti, V.: Kowledgesharing for local community: logical and visualgeoreferenced information access. In: International Conference on Enterprise Information Systems and Web Technologies (EISWT 2009), pp. 92–99. ISRST, Orlando (2009)
14. Egenhofer, M.J., Richards, J.R.: The geographer's desktop: A direct-manipulationuserinterface for mapoverlay. In: AutoCarto-Conference, pp. 63–71. ASPRS (1993)
15. Cartwright, W., Crampton, J., Gartner, G., Miller, S., Mitchell, K., Siekierska, E., Wood, J.: Geospatial Information Visualization User Interface Issue. Cartography and Geographic Information Science 28(1), 45–60 (2011)
16. Parr, A.E.: Mind and milie. Sological Inquiry 107, 273–288 (1963)
17. Sonnenfedl, J.: Variable values in space landscape: an inquiry into the nature of environmental necessity. Journal of Social Issues 22, 71–82 (1966)
18. Bagnara, S., Misiti, R. (eds.): Psicologiaambientale. Il Mulino, Bologna (1978)
19. Nielsen, J.: 10 Usability Heuristics for User Interface Design (1995), http://www.nngroup.com/articles/ten-usability-heuristics/
20. Budiu, R., Nielsen, J.: Mobile Usability. New Riders Press (2012)

21. Krugg, S.: Rocket Surgery Made Easy: The Do-It-Yourself Guide to Finding and Fixing Usability Problems. New Riders Press (2009)
22. Morville, P.: Ambient Findability: What We Find Changes Who We Become. O'Reilly Media (2005)
23. Sauro, J.: How to Conduct a Usability test on a Mobile Device (2012), `http://www.measuringusability.com/blog/mobile-usability-test.php`
24. Li, C., Longley, P.: A Test Environment for Location-Based Services Applications. Transactions in GIS 10(1), 43–61 (2006)

User Experience & Usability for Mobile Geo-referenced Apps. A Case Study Applied to Cultural Heritage Field

Letizia Bollini[1], Rinaldo De Palma[2], Rossella Nota[3], and Riccardo Pietra[1]

[1] Department of Psychology, University of Milano-Bicocca
Piazza dell'Ateneo Nuovo 1, 20126 Milano, Italy
letizia.bollini@unimib.it
[2] Bitmama SRL, Corso Francia, 110, 10143 Torino, Italy
[3] Avanade SpA, Via Lepetit, Milano, Italy

Abstract. The mobile revolution has become the main access gate to Internet in a diffused communication ecosystems. Consequently the apps become the interface mediation between the user and the real/virtual space of information. The opportunity of associating data to a physical position requires deeper thought: this means the interweaving between perception of the real-world, representation in digital systems and the mental model of users.

The research presented proposes an extension of user testing methodologies – of the user-centered design – applied to user interaction with mobile app in which the geo-referenced feature is the focus of the user experience. In particular, the case study *The Bethroted 2.0* becomes the experimental research field. Using a mixed methodology approach – qualitative interviews, in-field research and spatial user tests according to both ambient psychology and user-centered design – the prototype is tested. The experimental activities were addressed analyze both the environmental interactions and with the app trying to better understand the relations between real and digital space and the mental model of the users.

Keywords: Mobile app usability, geo-referenced app, mobile storytelling, spatial and ambient findability, geo-based User Experience evaluation.

1 *Mobile*: A Spatial Relational Tool

Although we already have had devices to connect ourselves with spatial information – that means NavSTAR GPS or other space-based satellite navigation devices – is only the introduction of smartphone – first of its kind the iPhone 3G in 2008– natively equipped with GPS that the mobile revolution begins to change our habits, behaviors and relation with the environment surrounding us.

As stated in the last *Forrester Research 2013* on mobile trends [1] the opportunity given by smartphones – and nowadays *phablet*[1], the hybridization between

[1] Sincratic neologism in use since 2008 and defined in 2013 in a *The Wall Street Journal*'s that defines them as "devices […] with a screen size of five to seven inches."

B. Murgante et al. (Eds.): ICCSA 2014, Part II, LNCS 8580, pp. 652–662, 2014.

(smart)phone and tablet – is to carry constantly the Internet with us. To say with Husson's words, to make the Internet *pocketable*.

Nevertheless the so-called *mobile revolution* is not just a matter of technology and devices innovation, but rather a changing point in the way we relate with the space unknown or of the our daily life, and the way we interact with others in the co-presence in the same space & time and/or remotely & deferred.

On one hand, ITC have made the web a constant and personal part of our life pushing this experience to the last frontiers of wearable devices, on the other hand the environment around us is further more a sort of digital signals "speaker". Phenomena like the *IoT-Internet of the Things*[2], *Ubiquitous Computing* [2][3] and *AmI-Ambient Intelligence* [3] are deeply changing the space itself and the way we interact with it.

In this double polarization the social dimension of the web 2.0 is inserting itself in the spatial dynamic of relation among individuals, places and social groups or community.

As a matter of fact the mobile revolution is dumping Internet from the monitor – the use of desktop or laptop computer, with keyboard and mouse and a *fixed* or at least moveable location the so-called *first screen* – to the *second screen* and shifting the web to its *third* generation. That is the *web 3.0* – as expected by many authors among them Berners Lee himself – largely driven by *geolocation* features, *social* dimension of interactions and *semantic* organization of knowledge.

1.1 Physical and Virtual *Ambient Findability*

In this sort of ideal triangulation – among *people*, *data*, and *space* – applications play a vital role, whether they are native-apps that give leverage directly to the potential of the device such as GPS or they are web (based) apps, which use a third parts software such as APIs to offer similar functionality.

Consequently the apps become the interface of interconnection between these actors and mediators between the user and the real/virtual space of information and other people. The mobile devices become the main access gate to touch-points of the cross-media and diffused communication ecosystems. The design role is, therefore, to face "the challenge of reconciling our dual citizenship in the physical and digital worlds" as declared by Hiroshi Ishii of *Tangible* Media *Group* at MIT Media Lab [5] to balance the *mirror* that allow us to find something and to be *findable* at the same time throughout the same *matching structural surface* – to mention the definition of *interface* proposed by Bonsiepe in 1993 [6] – the spatial interface of our geo-referred position.

In this meaning *findability* becomes the focus of the design process: as reminded by Morville "The quality of being locatable or navigable. The degree to which a particular object is easy to discover or locate. The degree to which a system or environment supports navigation and retrieval." [7]

---

[2] Invented by Kevin Ashton co-founder of *Auto-ID Center* MIT in 1999

[3] The concept was firstly introduced by Mark Weiser in 1998 and then defined in the Adam Greenfield's book: *Everyware: The Dawning Age of Ubiquitous Computing*

But we have to keep in mind the difference between the physical and the virtual world, between the real geography of a place – and consequently the conceptual and visual representation of it – and the mental model that its *figurability* build in our mind or – better to say – the *images* that we have and share as individual and collectively as found out by Lynch's research [8] in the '60s.

If the first one is naturally perceived by our sensorial systems and made by quality like size, shape, color etc. that allow us to create a cognitive map where things are clearly recognizable and distinct, in the second case we relate more often on a logical approach – as Morville underlines – "in the digital realm, we rely heavily on words. Words as labels. Words as links. Keywords."

1.2 For a *Spatial Information Literacy*

So if the mobile revolution is shaping a new emerging "world where we can find anyone or anything from anywhere at anytime" we probably need new conceptual tool to understand these phenomena that enable us to face both with the potentiality of the *always in and on* experience and the enormous information overload that it brings in our daily life. We already well know the *information anxiety* – described by Wurman [8] in the late '90s with the coming up of the Net – but now we can hardly figure out the impact that such a revolution – parallel and coexistent with the phenomenon of *big data* and the incredible acceleration that the production of information have had in recent time – will have on us and the environment we live in.

The access to geo-referred or -localized data of the digital experience fosters the transition "from the *textual/logical* dimension of interaction – the traditional search engines and index structure, based on syntactic logics and its potential semantic evolution – to a more visual, graphical and *spatial systems of interaction* with the data based on *topologies* rather than on *typologies* that emphasizes also the relational dimensionality of the given results" [10]. Then – paraphrasing Morville – we need a *spatial* information literacy that means a "set of abilities requiring individuals to recognize when information is needed and have the ability to locate, evaluate, and use effectively the needed information" in the mobile context also driven by the social interaction.

2 Real, Coded, Simulative, Augmented Space

While Ux and UI designers try to understand how to manage the new touch, small and geo-based interfaces to shape a mobile experience that gives users affordance, efficiency and a *good feeling*, software developers and other technology giants – e.g. Google, Apple – are setting standards and design patterns to face with. On one hand this standardization guarantees a sort of a priori and largely tested level of usability to maps and geo-referred information. But on the opposite it risks to flatten and overwhelm the opportunity to conceptualize and to design experience *out of the box* namely interactions, representations and behavioral model specific related to the challenges posed by both the devices and the context of use.

The transition from the real space to its coded representation – typical of traditional cartography and most of its digital *translation* – to the extreme possibilities given by the Augmented Reality which overlays a further data-layer mediated by the interaction with space and the device, opens more questions about usability and cognitive models.

In the following paragraphs are given a brief description of the services provided natively or via API on mobile devices focusing on the user experience and aspects of usability standards and design patterns of the GUI and touch gesture interaction now become a reference point in the sector.

2.1 Google Maps

GMaps is one of most important web mapping service application present on the web since 2005 powered by Google and it aims to provide detailed maps to users with information about famous places, points of interest, the traffic and for bigger cities, the stops of public transportation, both desk and mobile. In addition are provided social experiences about the place, like the user rating, comments and other details according to the category.

The maps look like common topography maps and users can navigate the space with the interface of the device which they using. Or they can choose to watch the satellite images of the map: although the images are not uploaded in real time, they are no more than 3 years old. GMaps provide tools with which the user can create content, enriching the user experience by making it interactive. In particular, the user can creates custom paths indicating routes and points of interest. The paths thus created can be shared on social networks or remain private. Lastly the feature *Street View* allows to explore the map with a particular view built on 360° photos. All these features are available also for the mobile version natively or through APIs that allow apps development companies to use them for their own purposes. GMaps has become the standard de facto for the development of geo-geolocalized services and applications.

2.2 Apple Maps

In 2012 Apple launched the first version of Maps based on iOS6. The event was a totally failure: app worked very badly and provided a non-competitive user experience with GMaps: often the maps were not updated and the navigator suggested obsolete paths that disoriented users. With the last release of iOS7 – one year later – the company of Palo Alto released a new version providing a high level service compared to Google, previous partner and now competitor in this market.

With this release Apple built new vector maps in line with the look&feel of iOS7 developed in collaboration with TomTom and others partners like DigitalGlobe, Waze and Yelp. The service provided is very powerful and detailed: traffic, restaurants, shops, hotels information are given and much more. The app is fully integrated with the hardware and the OS: in fact, thanks to the optical sensor that captures ambient light, the brightness change map showing a night version in case of low light. Furthermore with iCloud the user can share routes and points of interest created on the app with all devices he owns. Maps are available for Mac, iPhone and iPad so if the user creates a path on one of his devices, it is automatically synchronized to the others in a very efficient cross channel experience.

The most interesting features are twofold. The first is the *Navigator* that guides the user through voice commands showing his position in a 3d map with a perspective view of the space. The second is the *Flyover* that allows the exploration of a space through a 3d high fidelity rendering of the environment.

Apple provides a set of APIs too to integrate Maps into native applications developed by other companies. [11] [12]

2.3 Open Street Map

OSM is an open source web service that provides geographic data to websites, mobile apps and hardware devices. Based on a database built by a community of mappers who contribute and maintain data with aerial images and use low-tech GPS devices/maps to verify that OSM is accurate and updated. The service is free but is necessary to assign to OSM the copyright for maps and data. This project is supported by *The Open Street Map Foundation* [13] that supervises the growth, development and share of the platform.

OSM [14] has built a database composed of a series of different maps that can be used in different contexts. The integration within a site or an app is through the use of a large XML file which contains all the information about the map choice. Using data from OSM is for free. Otherwise if to produce a series of maps to be integrated into existing database to use it within own app is needed, it costs a yearly fee [15].

2.4 Augmented Reality

According to the last research by ComScore Inc. [16] about the app usage in the US, GMaps is the most used mobile app of mapping. This data is given only by the large spread of devices that have Android as the operating system. In fact over 52% of the mobile market (smartphones and tablets) is oversee from the OS made by Google.

With the last release of iOS (September 2013), Apple has regained the share of market that had lost with the 2012 version of Maps, thanks to an app in line with the new operating system in terms of look and feel and user experience providing a new 3d view and a better integration with iOS and other Apple devices. Apple users prefer the native maps rather than Gmaps or other competitors because the gap between the two major applications has elapsed. Furthermore the new innovations of Maps have prompted Apple users to set aside Google Maps.

3 A Multimodal Storytelling's Approach to Space Experience in Cultural Heritage Field

If digital technologies give us tool and framework where to develop services and experiences to enhance the life of users, nevertheless the meaning and the benefit that these will bring is a matter of design.

Storytelling is a narration. It is a discipline that uses principles drawn by rhetoric and narratology. It is included in every literary and audiovisual work, but it can be found in verbal world too.

It is an art and it's used as an instrument to represent real or fantastic events with words, images, sounds and places. It is a natural vehicle of effective communication: it involves contens, emotions, intentions and environments.

3.1 Physical and Augmented Space in Storytelling

In the case study *The Bethrothed 2.0* formerly presented and discussed in ICCSA2013 [17] there are several point of interest mapped in three different historical periods to discover the city of Milan respectively through the places visited by Renzo Tramaglino one of the main charactes of the novel – set in the XVII century – and through the places where Alessandro Manzoni lived in '800, and the contemporary city. In particular, in every historical layer the multimodal geo-referred storytelling is emphasized with the functionality named "look around". It can return an augmented experience thanks to the photo-camera: pointing it at the point of interest, the surrounding landscape will be shown on screen and some balloons or tips will compare over it: they show the presence of multimedia contents associated (audio, video, text, external links to the web) to the place experienced by the user.

An example of the carrying out of the augmented reality in historical and cultural contests in "MediaEvo". It can be ascribed in this experimental phase: the project proposes the making of a didactic device (a videogame) used to improve the knowledge about the history, the culture and the life of the medieval society through the rebuilding of the town of Otranto (Italy) in the age of XIII century.

Another very interesting project under this point of view is "TechCoolTour"s, created for visited tours in archeological areas. The application can frame the archeological heritage with a device and associate directly on the image multimedia contents such as texts, video or a personal guide in 3D (and typical dress) that describes the wonders of the tour!

There's a last example of augmented reality that can be included in this research: "ARmedia". It is an application that draw and recognize objects in 3D independently of their shape and size: one of the most interesting demonstrations of this app is the reconstruction of the Colosseum in Rome (Italy) in augmented reality.

3.2 Geo-localized Storytelling

An important cue in the approach of storytelling with geo-localization is Eli Horowitz's application "The Silence History". The landscape becomes a fundamental paradigm on which it can be developed a real web of conceptual and narrative connections. In "The Silence History" the Wikipedia app is enriched by geo-localization, that will allow users to consult items and uptake them via mobile. The user can decide to make a chronological or a geo-localized lecture. With the geo-localized one the users will participate to the writing of the item thanks to the Field reports: live reports published automatically on a map.

"Livehood" is an application that reveal how the people and places of a city come together to form the dynamic character of local urban areas. The app offer a new way to conceptualize the dynamics, structure, and character of a city by analyzing the social media its residents generate.

3.3 Timemachine

To explain the feature of *The Betrothed 2.0* called "Time machine" we carry the attention on *Time Machine* of the site of Faragola realized by the University of Foggia with the *Itinera project*. The traveler can use a panel control to interact with the web site moving from "how is today" to "how the archaeologists image the old site". "Time machine" forecast four different ages: the first period is contextualized in about 400 AC, the second around the 450 AC, the fourth in the Middle Ages and the last around the 700 AC.

Another best parctice regarding the time machine is the "Chichen Itza" project that is a good example of the iPad's interactive nature. TimeTours is not just a typical guidebook. Instead, this virtual trip through time brings the past back to life using modern 3D reconstructions. The Now & Then time windows let you experience what the city looked like when ball games and human sacrifices were performed here. The new "Now & Then & ..." bring the time of the rediscovery alive.

4 Mobile User Testing Open Air: *The Bethroted 3.0*

The prototype of the mobile app will allow users to discover the city of Milan through the places visited by Renzo Tramaglino in *The Betrothed*, and through the places dear to Alessandro Manzoni the aothor of the novel, and the nowadays city of Milano and it is addressed to make possible to perform in all the subjects interviewed, the test on their personal smartphone or mobile devices, regardless of the environment used.

The prototype is accessible via the web at the following address: http://men11q.axshare.com/ loading # p = & view = a.

4.1 Testing the Mobile User Experience in the Field

The special feature of User Test was their performance *in the field*. It was decided to undertake a non- simple, which is the Mobile User Testing Open Air, because it was considered appropriate to test the user experience under the real conditions of actual use. A digital product such as *The Betrothed 3.0* app is mainly used outdoors on certain territory of which offers information and inspiration from which it takes to present material multimodal action to tell the story of the place itself. For this reason, in the mobile industry, the activity of the Test User on the field is increasingly practiced and discussed [17]

In the past [18] have been proposed solutions such as immersive virtual reality (immersive virtual reality), created specifically for the test that users are going to play. With the wide spread of mobile app focused in the area, and with features such as Geolocation and Augmented Reality, the user experience of mobile app is increasingly focused on the area surrounding the user, and the device becomes a means to bring connecting the user with what surrounds him. For this reason, it becomes increasingly necessary to test the alpha versions and prototypes directly in the field. This fact implies, however, certain attentions, such as being able to identify the elements of disorder arising from the surrounding environment (eg noises that distract the user, weather conditions, strong light, low light, etc.) and check how they affect the tests carried out by testers. When you analyze the video documenting the User Test and write down the difficulties encountered by users, it is essential to recognize exactly what errors and bias resulting from the interaction due to the user interface and which are instead due to environmental conditions of the surrounding area. For both types of problems, you need to find the appropriate solutions: Redesign of interventions on the interface of the app for the first, and technical and functional improvements for the latter (for example, offer the opportunity to raise the volume more than is usually possible to do, if the elements are mostly multimodal Audio or Audio Visual).

See now how the prototype created: the user chooses which of the two available paths to follow, then select the time level he wants to receive the information. At this point the user begins to follow the suggested route when approaching near a point of interest, an alert warns. The user can now display the tab for the point of interest. Tab presents the basic information of the stage, referring to the level previously set time.

The user can, from the card, switch to look around, where, thanks to augmented reality, you can access a variety of multi-modal content.

Thanks to the temporal filtering in look around, the user can view the digital elements associated with the three levels proposed. Look around mode, the user can also access the list of contents.

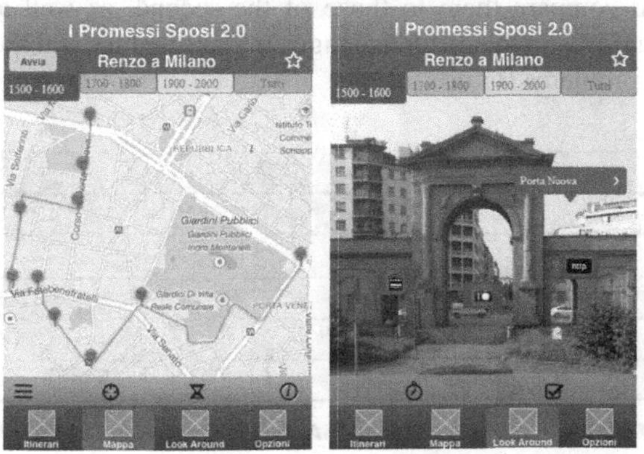

Fig. 1. Screenshot of the prototype

For preparing and executing User Test the steps were followed:

• *Defining the objectives of the User Test:* check the user experience and the usability of the prototype of the app *The Betrothed 3.0*.
• *Method of Selection of Sample:* Based Personas: 7 personas were identified, belonging to three main categories – residents, tourists and teachers. For each personas were selected two subjects, for a total of 14 experimental subjects (8 women 6 men, age: 38.8).
• *Defining scenarios:* scenarios have been described, namely the contexts of use that would fit well with the 7 personas.
• *Task Definition:* according to the scenario , the functions and content available, have been defined were tested, in detail the tasks to be executed users during the tests.
• *Data collection*: each user was filmed during the execution of the test. In addition, the subjects were asked to complete a pre-test questionnaire, which was intended to frame than the personas, and to participate in an interview at the rear of the trial, which investigated their impressions of the product. At the end of the test user, video were analized to record errors committed by users to be cathegoryzed and counted to define the total of feedback.

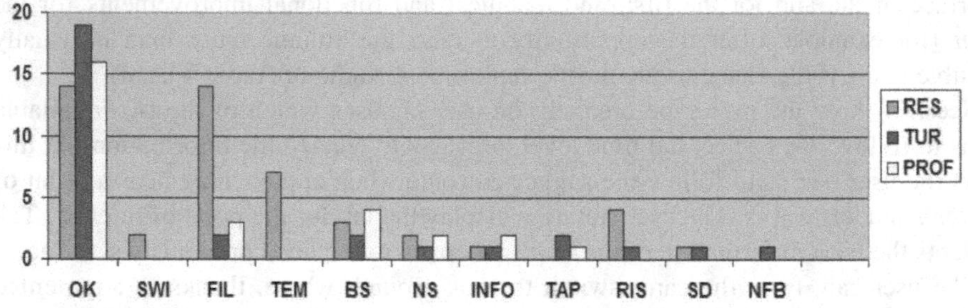

Fig. 2. Critical tasks: comparision by personas

At this point it was possible to compare the performance made by the various parties , belonging to the three major categories and verify the performance made on the first route and compare them to those of the second, as well as comparing the performance of subjects on the different stages in the itinerary and then see what tasks they caused more problems for users.

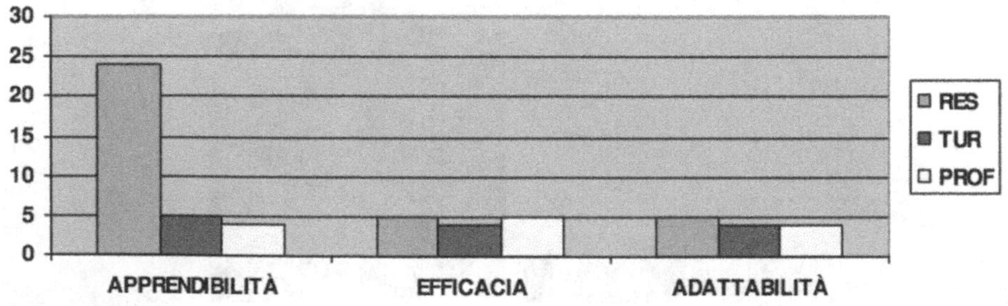

Fig. 3. Critical tasks: clusterization according to the Nielsen's eurystic

5 Conclusions

The results of the experimental experience prove interesting evidences. Although emerging critical issues of high priority, none of these concerns directly georeferenced data itself or its representation.

Because some features are only simulated in the prototype , such as the Reality increases and the Path Hint, it was not possible to gather ideas for technical improvements to improve the general use of these two functions. However, it was possible to note, some of the dynamics related to User Experience of two particular functions.

The user interacts with the user interface correctly on the map that shows the route, dividing it by stages. Users are properly related with the surrounding territory, being able to identify the pathways to reach the various stages of the route. They also successfully interacted with the map, moving within this and doing a zoom lens where it is necessary to understand how to proceed on the proposed route. Users also interacted well with the general user interface , being able to switch from tab stop to this page map to view the route to follow to reach the next stage.

Even compared to Augmented Reality, users are properly related with the surrounding territory, setting it through the camera of your smartphone. The simulation of augmented reality has been made through a still image of the territory, in which it was entered multimodal elements with which the user can interact. When users used the function, trying as much as possible to frame the surrounding area, than the image they saw on the screen of your device.

These indications suggest that, on average, users are ready and eager to interact with this type of digital products, focused on the surrounding area and designed for outdoor use – or better – in the real world.

Acknowledgments. Although the paper is a result of the joint work of all authors, Letizia Bollini is in particular author of parts 1, 2, 4, 3 and 5, Rinaldo De Palma is author of paragraphs from 2.1 to 2.5 included, Rossella Nota is author of paragraphs 3.1 to 3.2 and Riccardo Pietra of paragraphs from 4.1.

References

1. Husson, T.: Another Year in Review: Revisiting 2013's Mobile Trends (2014), http://blogs.forrester.com/thomas_husson/14-01-09-another_year_in_review_revisiting_2013s_mobile_trends
2. Greenfield, A.: Everyware: The Dawning Age of Ubiquitous Computing. New Riders (2006)
3. ISTAG, Ambient Intelligence: from vision to reality. For participation – in society & business (2003), ftp://ftp.cordis.europa.eu/pub/ist/docs/istag-ist2003_consolidated_report.pdf
4. Sterling, B.: Shaping Things. MIT Press, Boston (2005)

5. Ishii, H.: Tangible Bits: Beyond Pixels. In: Proceedings of the Second International Conference on Tangible and Embedded Interaction (TEI 2008), Bonn (2008), http://tactile-resources.wikispaces.asu.edu/file/view/Tangible+Bits+Beyond+Pixels.pdf/256894244/Tangible+Bits+Beyond+Pixels.pdf
6. Bonsiepe, G.: Il ruolo del design. In: Anceschi, G. (ed.) Il Progetto Delle Interfacce. Oggetti colloquiali e protesi visrtuali. Domus Academy Edizioni, Milano (1993)
7. Morville, P.: Ambient Findability. New Riders (2006)
8. Lynch, K.: The Image of the City. MIT Press, Cambridge (1960)
9. Wurman, R.S.: Information Anxiety 2. Hayden/Que (2000)
10. Bollini, L., Cerletti, V.: Knowledge sharing and management for local community: logical and visual georeferenced information access. In: International Conference on Enterprise Information Systems and Web Technologies (EISWT 2009), pp. 92–99. ISRST, Orlando (2009)
11. https://www.apple.com/it/ios/maps/
12. https://developer.apple.com/library/ios/documentation/MapKit/Reference/MKAnnotationView_Class/Reference/Reference.html#//apple_ref/doc/uid/TP40008207-CH1-DontLinkElementID_1
13. http://www.openstreetmap.org/
14. http://wiki.osmfoundation.org/wiki/Main_Page
15. http://www.theguardian.com/technology/2013/nov/11/apple-maps-google-iphone-users
16. Bollini, L., De Palma, R., Nota, R.: Walking into the Past: Design Mobile App for the Geo-referred and the Multimodal User Experience in the Context of Cultural Heritage. In: Murgante, B., Misra, S., Carlini, M., Torre, C.M., Nguyen, H.-Q., Taniar, D., Apduhan, B.O., Gervasi, O. (eds.) ICCSA 2013, Part III. LNCS, vol. 7973, pp. 481–492. Springer, Heidelberg (2013)
17. Buxton, B.: Sketching User Experiences: Getting the Design Right and the Right Design: Getting the Design Right and the right Design. Morgan Kaufmann (2010)
18. Ly & Longley (2006)

PALEOBAS: A Geo-application for Mobile Phones – A New Method of Knowledge and Public Protection of the Paleontological Heritage of Basilicata (Southern Italy)

Maurizio Lazzari[1,*], Agostino Lecci[1,2], and Nicola Lecci[3]

[1] CNR-IBAM C/da S. Loja Zona Industriale Tito Scalo (PZ) 85050-I
m.lazzari@ibam.cnr.it
[2] MIUR, Matera 75100-I
agostino.lecci@istruzione.it
[3] Freelance, Matera Italy

Abstract. In this paper a new mobile application, designed for Android, to create social awareness for a sensitive topic, such as cultural heritage of a specific region, is proposed. The theme is the knowledge of paleontological assets located in Basilicata (Southern Italy) for their subsequent enjoyment and enhancement. The proposal for the use of the paleontological heritage of Basilicata is based on the use of new technologies, currently widely spread in the population. This project aims to census of the published sites in paleontological topic, displaying their geographical distribution, the understanding of their relevance from the scientific point of view and from the educational point, to promote their enhancement and the use of the heritage of Basilicata through an application for Smartphone and tablets.

Keywords: geo-application, paleontology, mobile phone, heritage, Basilicata, Southern Italy.

1 Introduction

The recent few years have been characterized by a revolutionary change in the ways that geographic information can be utilized. The rapid growth of the World Wide Web, as a medium of acquisition and dissemination of geographic information, triggered a transition from stand-alone to distributed Geographic Information Systems. Concurrently, the evolution of wireless communications and positioning technologies together with the emerge of increasingly powerful and affordable mobile devices, such as the mobile phones and pocket PCs, led to the creation of the Mobile applications.

Mobile system, above all GIS-type, opened a whole new world of opportunities for the development of innovative and useful applications that can provide amongst others location-based information to mobile users. Vehicle and pedestrian navigation systems and mobile tourist guides are examples of mobile geo-applications, which often use interactive maps as interfaces.

* Corresponding author.

B. Murgante et al. (Eds.): ICCSA 2014, Part II, LNCS 8580, pp. 663–676, 2014.

Mobile geo-applications, mainly GIS mobile application, are able to bring to the mobile user geospatial information in the forms of text, image, voice, and video in the field. As mobile geo-applications are emerging fast, and there is a strong interest for implementing new applications that can promote high end-user acceptability, it is important to develop proper usability testing frameworks that can improve the process of design, testing and evaluation, such as proposed by Delikostidis [1].

Adaptation is a key concept in mobile geo-applications, as it can help overriding the limitations of the current mobile software and hardware platforms [2], [3], [4]. It adjusts a system by changing its characteristics in order to respond according to a particular context of use. Adaptation role is to make the system able to cover the specific user group requirements efficiently, by giving them the needed information.

In light of this "adaptation" concept, the present work extends the "Geo"-application concept including also those disciplines regarding the Earth Science, such as geology, geomorphology, paleontology, geoarchaeology and so on.

Starting from the awareness of low assistance of general population to the local cultural heritage, particularly the school-age population as several researches report (experience shows that more interest is generated when interactive activities are offered), this paper presents the development stages of a geo-guide to paleontological heritage of Basilicata (Southern Italy) using an integrated solution of mobile app and internet website.

The aim is to contribute towards spreading the rich cultural of this Italian region through the use of new technologies. The point focused in this early stage of development is the ability to read cultural material as well as browse specific paleontological information, anywhere and even without any data connection.

Considering the sensible increase in the usage of mobile phones and handheld devices such as tablets, there is an special growth in usage of mobile apps.

Mobile applications are generally used for entertainment, social network, for increasing productivity as well as education. In this project the use of mobile applications is proposed to create social awareness for a sensitive topic, such as cultural heritage of a specific region.

Several obstacles prevent cultural topics to reach the wider audience and cover a larger community. Moreover, some topics are supposed to be focused to specific age interval or specific gender, while on the contrary the diffusion of mobile applications is completely wide, overcomes any barriers, and can reach even people who simply do not expect to find such a knowledge into a mobile application.

The topic of this paper is to promote the knowledge and dissemination of the results of a research project carried out by CNR-IBAM aimed to the census and analysis of palaeontological sites of Basilicata region in order to a subsequent enjoyment and valorization of each interested geo-palaeontological site. In fact, the protection and enhancement of the wealth of the territory cannot be separated from its knowledge which is the goal to be aware for the protection of sites.

The main aims include the census of the paleontological sites known in literature (scientific and popular publications) and by field survey, displaying their geographical distribution, the understanding of their relevance from the scientific point of view and from the educational point, to promote their enhancement.

Basic research is linked to the census of sites characterized by the presence quantitatively and / or qualitatively significant of fossils, through a reasoned analysis of literature sources and field observations carried out over the past two years. This kind of work has never been realized in Basilicata and the app is a unique product, by typology, at national and partly international level.

2 Material and Methods

The first step of the work has been that of the bibliographic research of paleontological sources and the following localization and georeferencing of sites. The surveyed paleontological heritage of Basilicata region (Figs. 1, 2 and 3) becomes the main subject to insert in a special mobile application. We thought this technological solution because these devices [5], are widely disseminated, are portable and allow promoting a cultural image using an object of daily use.

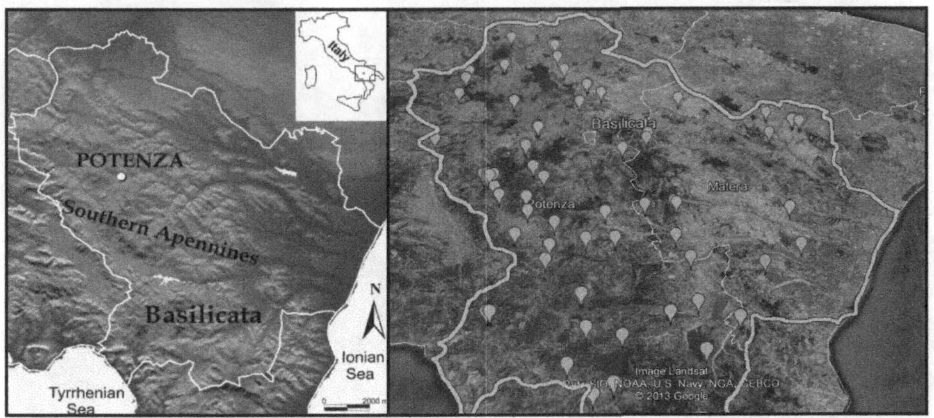

Fig. 1. Geographic location of the study area and georeferenced paleontological sites (on the right)

The bibliographical study permitted to identify about 800 fossil records (according to phylum, classes, order, families, and generates species, Fig. 2) in the region distributed into 50 municipalities, derived from 144 publications consulted (Fig. 3).

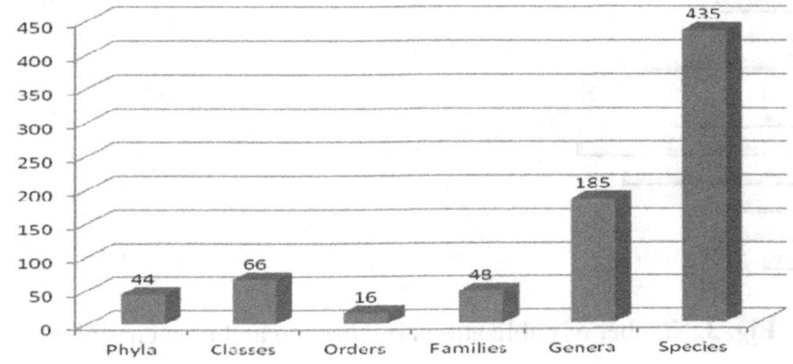

Fig. 2. Histogram showing the number of fossil for each hierarchical level

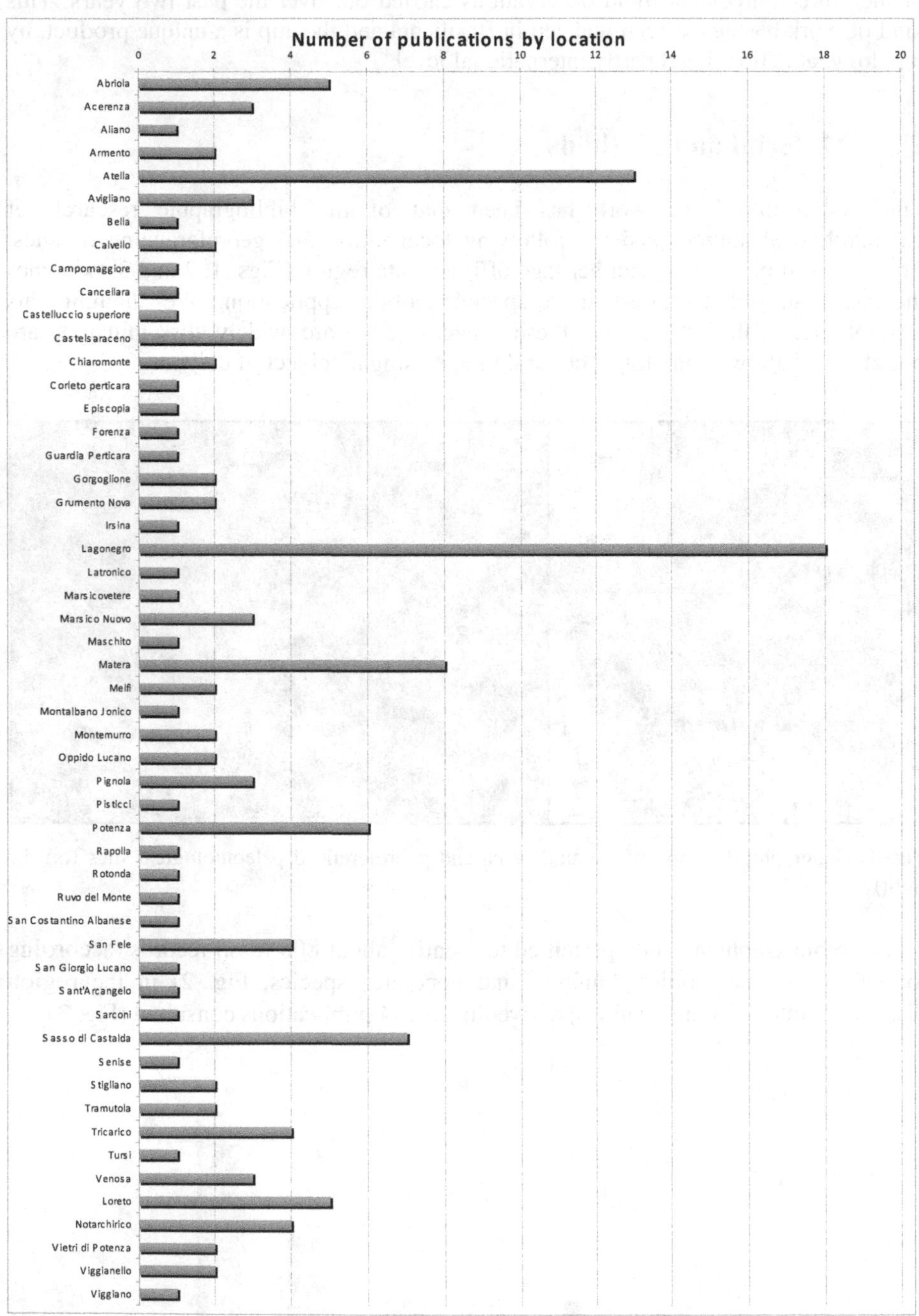

Fig. 3. Number of publications by location and municipalities

The application has been designed for Android. Market share for smart phones are reviewed to make sure that Android is the most suited platform, but plans are available within this project to cover other valuable platforms.

The Eclipse development environment has been used along with the software development kit (SDK) provided by Google to develop Android apps. The language used is Java, Java version 7 update 21 for 64-bit systems, to develop the interaction with the operating system of the device, while the presentation of the content has been developed by means of HTML language, used in conjunction with CSS for the graphics and Javascript for data management. The application uses the graphical user interface of the smartphone to flexibly display the proposed information.

In particular, the application displays data correctly with either horizontal or vertical screen orientation, allowing for flexible use.

The graphic presentation is based on HTML5 and CSS3 language adapted from "It Fits" by $hekh$r d-Ziner (@shekhardesigner) released on 4/26/2012 under Creative Common License, available on the web at
`http://www.cssjunction.com/freebies/html5-css3-responsive-web-template`

Interfacing between HTML5/CSS3 and the Android system has been created through the platform Apache Cordova version 2.9.0, licensed under the Apache License, Version 2.0.

The development system used is the Android SDK version 4.4, in the Eclipse ADT plugin plus version containing also Android SDK Tools, Android Platform-tools, the latest Android platform, the latest Android system image for the emulator.

Programming in Eclipse environment has been made using the Java programming language, and the contents have been developed in HTML5/CSS3. HTML5 is a markup language used for structuring and presenting content for the World Wide Web and a core technology of the Internet. It is the fifth revision of the standard HTML. Its core aims have been to improve the language with support for the latest multimedia while keeping it easily readable by humans and consistently understood by computers and devices.

With regards to CSS3, it is currently divided into several separate documents called "modules". Each module adds new capabilities or extends features defined in CSS 2, preserving backward compatibility. The earliest CSS 3 drafts were published in June 1999. Due to the modularization, different modules have different stability and statuses.

3 The State of Art of Similar Geo-apps

The proposed app has been developed for Android systems, using a Google platform.

In Google Play store, a few apps are related to palaeontological topic. The majority of them are readers specifically targeted for paleontology. An existing app is, for example, that of the "*Texas Paleontology*" developed by "Mattingly Apps", but last updated is of July 03 2011. The app allows the user to explore the paleontology of Texas through an interactive geologic timeline as well as maps of Texas Geological exposures. It also includes Google maps additions to help guide the user to interesting palaeontological sites.

The iPhone and iPad app store is richer in educational apps: besides Quizzes and Games about paleontology, that are not related to any specific region, it are possible to find a few apps developed by foundations and universities. It is worth reporting the app of the *Fondazione e Museo dei fossili del Monte San Giorgio – Meride*, even if it seems a mobile retargeted version of the foundation website.

The app of Houston Museum of Natural Science, *Morian Hall of Paleontology*, is interesting since, according to the description available in the app store, it helps to optimize the exploration through the hall which encompasses more than 4 billion years and hundreds of species.

Another app developed in Italy in the Apple store is the *ISIPU*, the official app of the *Italian Institute of Human Paleontology*. It allows the user to virtually visit the palaeontological sites of *Fontana Ranuccio* and *Coste San Giacomo* using augmented reality. The user can use maps or augmented reality to understand where the palaeontological sites are located. It is possible to receive information by means of a locations list. For each palaeontological site the user can find information, photos, illustrations and some 3D models of animals who lived in the area.

No apps about fossils and paleontology are available on Windows Phone app store.

4 The Geo-application: PALEOBAS

This project aims to promote the knowledge of the heritage of Basilicata through the PaleoBas geo-application for smartphones and tablets. It works offline and does not require any network connections, any data subscription and any telephone data traffic.

As regards content, the application offers information research about paleontological studies of Basilicata organized for location, epoch and fossil species (Fig. 4). Large navigation buttons are available that allow user to enable specific screens on fossil list and specific species (Figs. 5 and 6).

Clicking the first button opens a screen that displays the distribution of single references (Fig. 7). The "starts" button, found at the bottom of the screen navigation area, allows user to open a screen with detail map of the distribution of citations per municipality and its legend. User can notice how the municipalities having the largest number of citations are those described in darker color. To continue browsing the application, the user can use the back button that is visible at the bottom of each screen of the Android system, or any of the navigation menus in each application screen. To continue browsing for epoch user needs to activate the corresponding menu button.

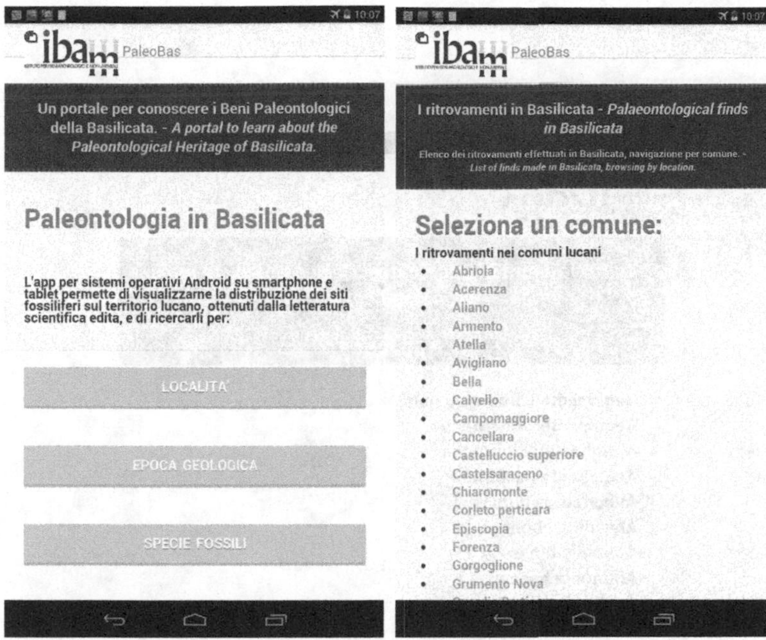

Fig. 4. Display screens and querying geo-PaleoBas app. (Paleontology of Basilicata)

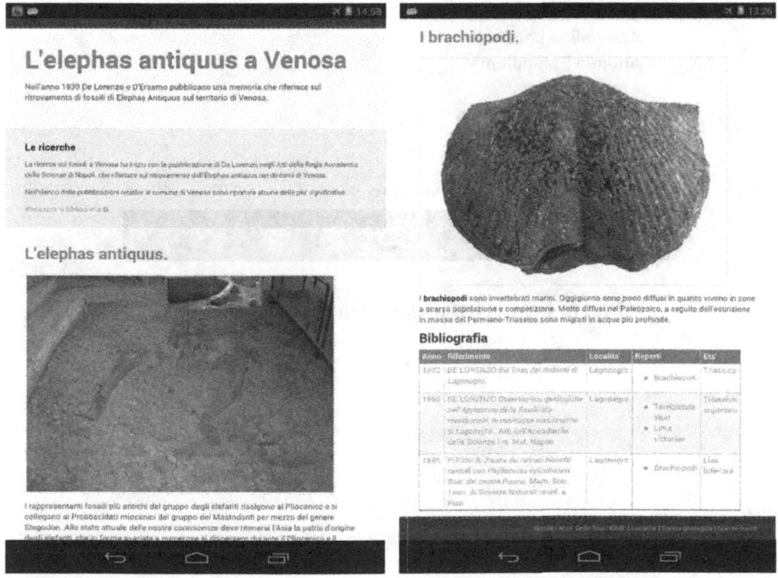

Fig. 5. Example of visual representation on Smartphone screen of each fossil specie linked to a specific geographic location, accompanied by references

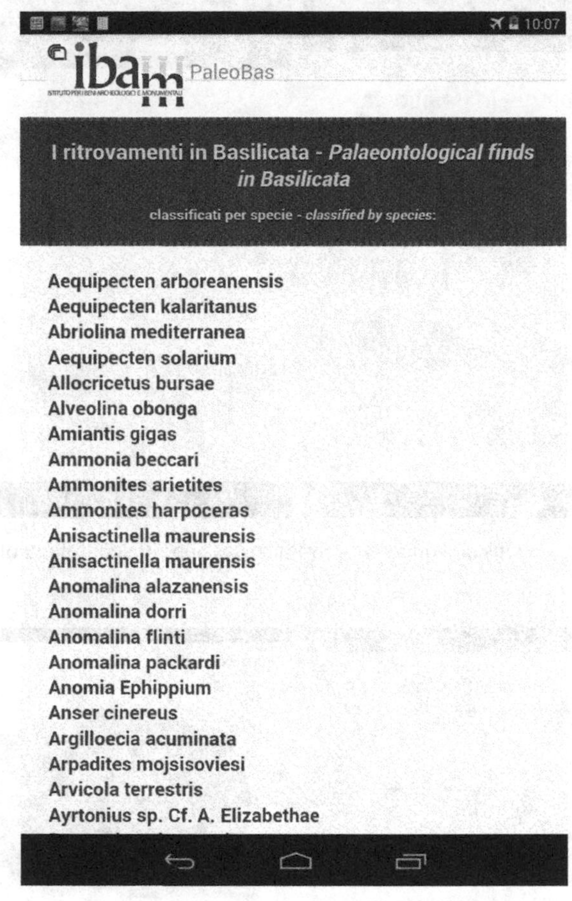

Fig. 6. Fossils list included in the PaleoBas app

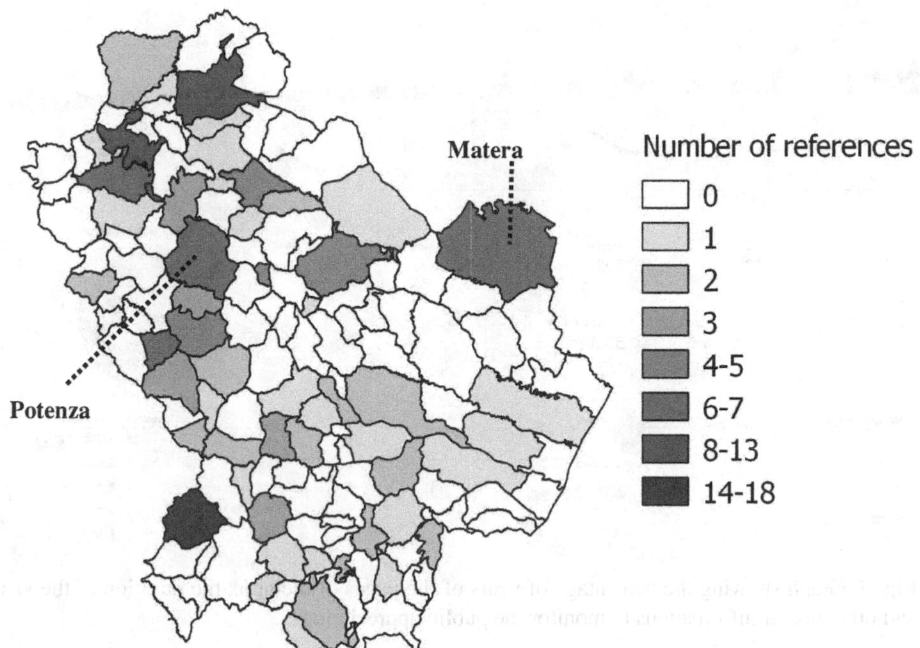

Fig. 7. The intensity map of the statistic distribution of publications relating to paleontological findings in Basilicata for each municipalities

The graph showing the statistic distribution of publication (Fig. 7) allowed to verify that the north-central and western sector of Basilicata is the most cited because of a greater variability in the number of geological formations (the age covered is 250 My), while the southern and eastern part is less cited for the lithological homogeneity with sands and conglomerates, normally sterile, having poor fossil content (Quaternary age, last 2My).

4.1 Monitoring Access

The Android publishing platform allows monitoring the number of installations classified for Android version, device, country, language, application version, telephone operator.

Furthermore, additional services of Google have been activated to monitor visits of the website *www.paleobas.it*, which are also classified according to different voices of analysis including language and locality (Figs. 8 and 9).

Application and web site monitoring is important during the development of the project because it allows to verify the impact of the introduction of new content and new features on the distribution of users.

In order to promote a global distribution of the palaeontological heritage of Basilicata and to translate the scientific contents of the website, linked to the PaleoBas Geo-app, in the preferred language, a Google Translator Toolkit has been also inserted (Fig. 10).

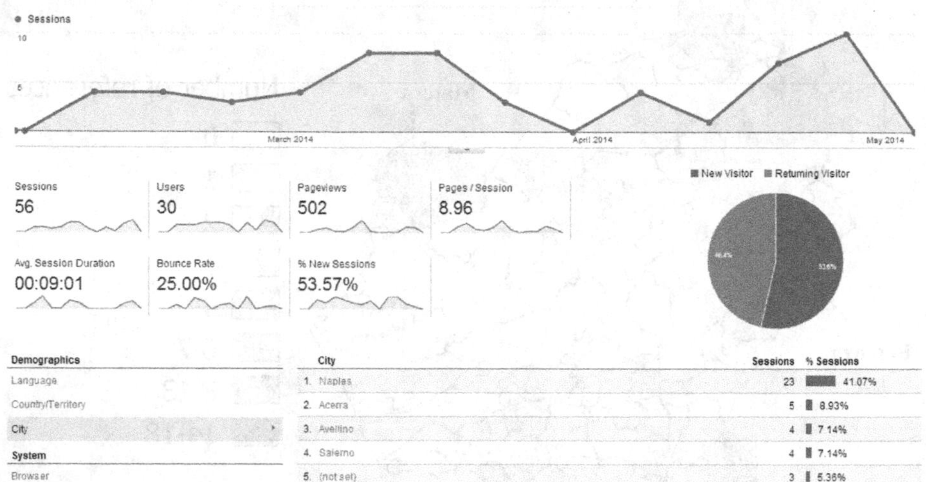

Fig. 8. Graph showing the percentage of visits of the pages of the app., the duration of the visit and other useful informations to monitor the public appreciation

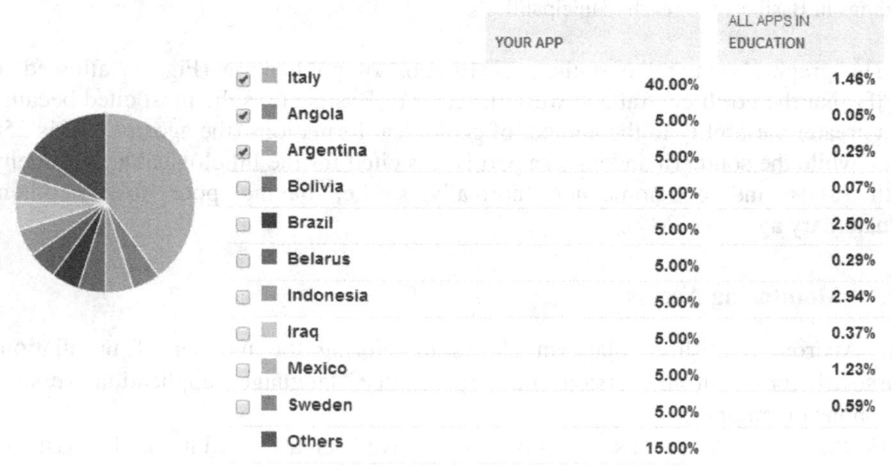

Fig. 9. Geographic distribution of visits in percentage of the PaleoBas app respect to other apps in educational topics

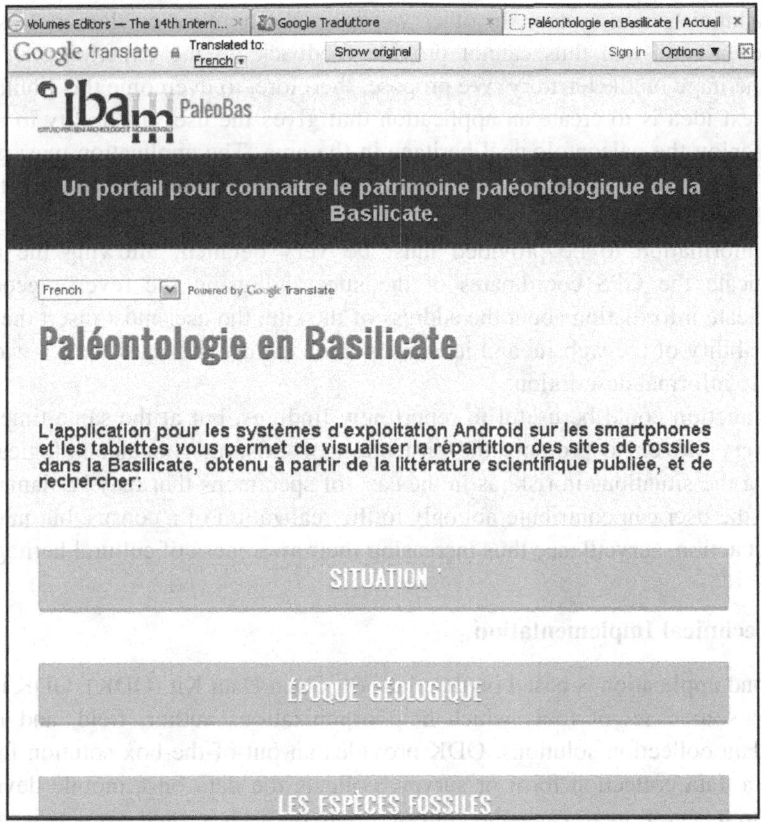

Fig. 10. The website *www.paleobas.it*, linked to the PaleoBas Geo-app translated in the French language by Google Translator

5 PALEOBAS 2.0

Starting from a thorough research work based on field survey and analysis of the bibliographic sources, it has been possible to reconstruct the state of the art of knowledge concerning the presence and location of fossils in Basilicata. Thanks to the spatial distribution of geo-referenced palaeontological sites, it has been also possible to give an interpretation of results, showing the sites most studied or, conversely, those less known today.

After successfully tested the first version of the application for smartphone and tablet devices, we propose to develop a new application with a more advanced set of features.

First, it must be held that the initial version the application has the advantage of working without data connection and operate in all conditions, online or offline. This however does not allow users to send information directly via the application to a central administration system. In other words, if the application does not use data, cannot even send and thus cannot provide feedback on the experience of use of cultural heritage on the territory. We propose, therefore, to overcome this limitation.

The next idea is to create an application that gives the user the ability to actively live and enjoy the paleontological heritage in the area. The application must give the user the ability to send alerts to an email concerning paleontological sites that are already known or novel.

The information to be provided must be very detailed, allowing the user to communicate the GPS coordinates of the site, and using the reverse geocoding, communicate information about the address of the site; the user must insert the degree of accessibility of the website and its conservation status, and must have a wide field devoted to informal description.

This function could be useful to report new findings, but at the same time, it can allow every citizen to promote the protection of existing geopaleontological sites, indicating the situations in risk, as in the case of specimens that may be damaged. In this way the user can contribute not only to the realization of a census, but may carry out direct action, surveillance thus increasing their awareness of cultural heritage.

5.1 Technical Implementation

The second application is based on the platform Open Data Kit (ODK). ODK is a free and open-source set of tools which help organizations author, field, and manage mobile data collection solutions. ODK provides an out-of-the-box solution for users to build a data collection form or survey, collects the data on a mobile device and sends it to a server, aggregates the collected data on a server and extracts it in useful formats.

In addition to socio-economic and health surveys with GPS locations and images, ODK is being used to create decision support for clinicians and for building multimedia-rich nature mapping tools.

Main menu of the application allows to the user to do several actions. The user can create a new report of paleontological discovery, can edit unsent reports (Fig. 10), send completed report to server and download report templates to fill in.

The basic form that has been used is based of three information that are divided into several screens to swipe through. The first screen records device GPS position, with the intent that the user should send feedback in the place of interest. The second screen of the report asks a confirmation for current date.

The third screen has a text field to be filled with information about the discovery of new findings or the feedback about existing findings.

After these screens are completed it is possible to send data to the server by means of a simple button. Data can be sent when a data connection is available; otherwise the reports are stored in the devices and can be sent successively.

Data are saved on the server available at paleobasilicata.appspot.com

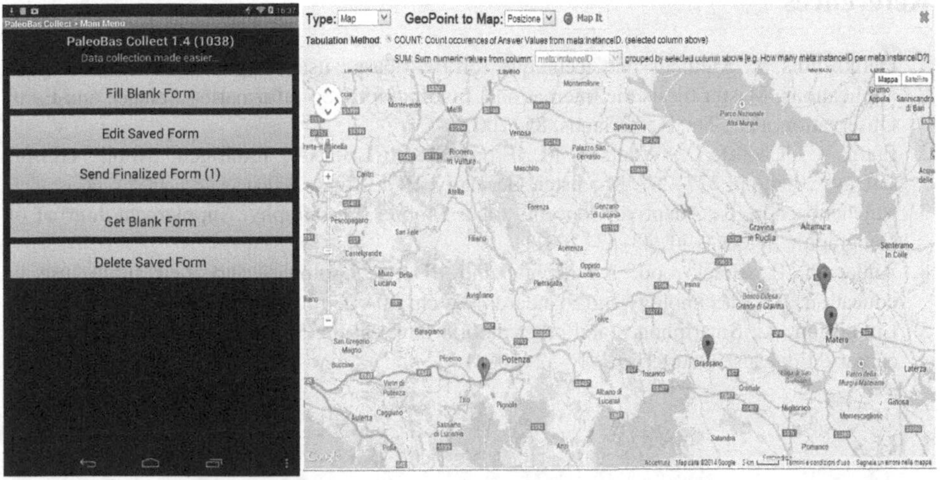

Fig. 11. Example of the technical implementation of the PaleoBas 2.0. On the left, the basic form and main menu of the application, that allows to the user to communicate with app., is showed. On the right, it is represented the screen with which the user can locate the palaeontological site, giving the geographic coordinates.

6 Final Remarks

The app presented in this paper confirms that smart phones are well suited to implement local cultural spreading all round the world, as several internal metrics disclose.

The work has helped to reconstruct the state of the art knowledge on the occurrence and geographical distribution of fossils in Basilicata. From the representation of census data through the places mentioned in the publications , it has been possible to check which sites are the most studied , or, conversely , to date less known.

In order to such a work would not remain an end in itself it was decided to build a solution that would facilitate the understanding and enjoyment of the sites through an easy-to -use smartphones and tablets .

As a further development, it will make a subsequent application (PaleoBas 2.0) that also gives the user the ability to send messages to other sites.

This feature can be useful for both new report finds that in order to protect existing paleontological sites, reporting any risk situations, as in the case of objects that could be damaged and / or exposed to human activities or natural factors, such as the degradation due to climatic weathering.

In this way, the external user can not only contribute actively to the realization of a census, but also play a direct action of surveillance and protection, thus increasing their awareness of cultural heritage preservation.

References

1. Delikostidis, I.: Methods and technique for field-based usability testing of mobile geo-applications. Master thesis the International Institute for Geo-information Science and Earth Observation, ITC, The Netherlands, 85 (2007)
2. Frank, C., Caduff, D., Wuersch, M.: From GIS to LBS: An Intelligent Mobile GIS.fu: GIDays 2004, pp. 277–287. Muenster, Germany, IfGI Prints (2004)
3. Reichenbacher, T.: Adaptive Concepts for a Mobile Cartography. Supplement Journal of Geographic Sciences 11, 43–53 (2001)
4. Anderson, P., Blackwood, A.: Mobile and PDA technologies and their future use in education, JISC Technology and Standards Watch: I 04-03 (2004)
5. Livingston, A.: Smartphones and other mobile devices: the Swiss army knives of the 21st century. Edueause Quarterly 2, 48–52 (2004)

URLs

HTML5 template:
http://www.cssjunction.com/freebies/html5-css3-responsive-web-template
HTML5 differences from HTML4
http://www.w3.org/TR/2011/WD-html5-diff-20110405/
Selector level 3: http://www.w3.org/TR/selectors/
Apache Cordova: https://cordova.apache.org
The Android SDK: https://developer.android.com/sdk/index.html

GIS Assessment and Planning of Conservation Priorities of Historical Centers through Quantitative Methods of Vulnerability Analysis: An Example from Southern Italy

Maurizio Lazzari[1,*], Maria Serena Patriziano[1,2], and Giovanna Alessia Aliano[3]

[1] CNR-IBAM C/da S. Loja Zona Industriale Tito Scalo (PZ) 85050-I
m.lazzari@ibam.cnr.it
[2] School of Engineering, University of Basilicata via dell'Ateneo Lucano Potenza, 85100-I
[3] Istituto Pilota s.r.l., via Sicilia Potenza, 85100-I

Abstract. This paper aims to propose a new approach to identify, through the GIS, the priorities for action and structural conservation in the historic center through the calculation of the decay index, the quality of the typological and the conservation status of each building and these elements were subsequently made in relation to each other for a more accurate analysis of the historical aspects of the urban historical center of Potenza (southern Italy).

Keywords: decay index, conservation, vulnerability, GIS, historic heritage, Potenza, Basilicata, southern Italy.

1 Introduction

The preservation of Cultural Heritage is very important as it represents the legacy of human beings on the planet as well as evidence of their activities in different living conditions and environments. However, cultural heritage is vanishing at a global scale, especially in developing countries and if this trend continues, many significant cultural features which represent the remaining foundations of humankind's history will be lost forever. According to the Global Heritage Fund (2009), available on www.globalheritagefund.org, the losses in the past decade include ancient monuments, buildings, archaeological sites, and even entire historic cities and townscapes, all of which survived for hundreds or even thousands of years.

The damage to cultural heritage sites appears widespread and accelerating and represents a permanent loss to the planet, comparable to endangered species loss. So, in this paper we propose a method, aimed to identify the priorities for action in a historical centre, that allows to better address economic budgets oriented to interventions of recovery of the ancient buildings, being based either on the data objectively surveyed with a census or on thematic derived cartography, such as the typological quality map, typological emergencies map and state of conservation map, all expressions of the real conditions of degradation and vulnerability of the site.

* Corresponding author.

B. Murgante et al. (Eds.): ICCSA 2014, Part II, LNCS 8580, pp. 677–692, 2014.

2 Methods and Case Study

In this paper we focus on a new integrated GIS-based approach to evaluate the state of conservation and decay of the architectural heritage of a historical centre.

We used two quantitative indexes defined on the basis of the total number of architectural elements and the number of constructive typologies for surface unit.

These two parameters, together with Decay Index (DIx) and state of conservation, can represent a useful and objective tool to address economic contribution to recovery historical buildings, above all for those that at present could be not included in a priority list of interventions by the Monuments and Fine Arts Office of Basilicata.

Concerning the first point inherent in the state of conservation, at present an objective criteria to assess it does not exist, but in most cases the method consists in a qualitative direct evaluation considering a range value varying from ruin to very good state [1,2,3]. Concerning the DIx, we used the method proposed by Lazzari et al. [4], in which a value range defined on the basis of the load bearing structures deterioration and degradation of facades, has been considered.

To apply the above-mentioned criteria in order to assess the *DIx* and state of conservation, we have chosen the case study of Potenza hill top town, located in Basilicata region (southern Italy) along the main axial zone of southern Apennines chain. Geographically, the site selected for this study is located in the northern sector of Basilicata region (40°38'43" N; 15°48'33.84") at an average altitude of 800 m. a.s.l., on a elongated narrow asymmetrical ridge geologically characterized by outcrops of gray-blue clays (substratum) overlapped by sandy-conglomerate Pliocene deposits, which varies in thickness along both west-east and north-south directions [5].

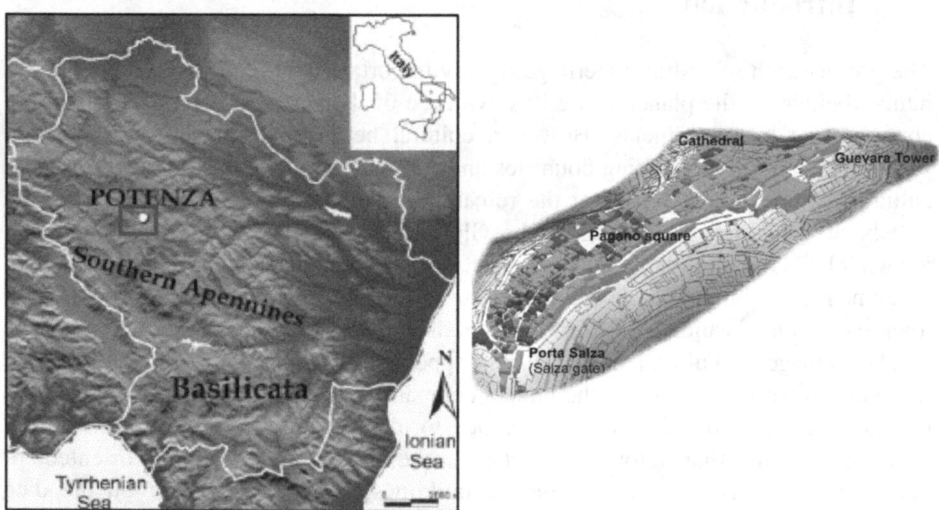

Fig. 1. Geographic location of the study area and detailed extension of the urban historical centre of Potenza town

2.1 Architectural Analysis and GIS Applications

The work has been focused on the analysis of the historical built heritage of Potenza, which has been carried out in the following steps:

i) census and survey of building elements characterized by significant artistic, historical, constructive values (aged from the 16th to the 19th century);
ii) classification of building elements in building types (such as loggia, balconies, chimneys, cornices, windows, portals, external scales, vaults, arcs and ligneous covers);
iii) Cataloguing and georeferencing of the building types in a geodatabase;
iv) Evolution of the state of conservation (low, medium, high o ruin, bad/mediocre, good/restructured);
v) calculation of quantitative indexes.

In particular, we adopted a specific form in which several information have been inserted, such as the age, state of conservation, metric data, materials, formal features, building characteristics. Afterwards, a qualitative and quantitative analysis has been carried out in a GIS, producing, respectively, typological architectural emergencies and architectural typology quality maps.

The field work has been based on a preliminary bibliographic analysis of sources [6,7,8,9,10] and the census of the historical-architectural heritage of the Potenza town (total surface of 0.23 km^2). For each building (518 buildings numbered progressively) the number of historic architectural elements and the total number of architectural typologies has been calculated.

Thirty-one buildings on 518 surveyed are represented by historical-architectural monuments (churches, palaces, gates, towers, theatre, walls), of which only a limited part (7) is subject to the listed ministerial. During the field survey several roman epigraphs, located on churches and traditional historical buildings, have been also signed.

The *typological emergencies map (*Figs. 2 and 4; Tab. 1) was built using surveyed information on each building of Potenza, considering all the typological elements characterized by constructive and aesthetic historical value. The total number of these elements ($\sum_{ntot}$) for each building has been divided for the external lateral surface *A* (m^2). In this way, it has been possible to calculate the number of constructive elements found for surface unit, *Ec*:

$$E_c = \frac{\sum n_{tot}}{A_{(m^2)}} \qquad (1)$$

The *typological quality map* (Figs. 3 and 4; Tab. 1) was built considering the total number of typologies surveyed ($\sum_{ntip}$), characterized by constructive and aesthetic historical value, for surface unit *A* (m^2), *Et* (Fig.):

$$E_t = \frac{\sum n_{tip}}{A(m^2)} \tag{2}$$

Both qualitative and quantitative analyses have been carried out through a survey of the outer architectonical elements, because in many cases it has not been possible to verify directly the conditions inner to historical buildings.

Table 1. The two tables show the percentage and number of buildings for each class used to map the emergency and quality values of the historical centre

EMERGENCY VALUE	N° BUILDINGS	%
0.01 - 0.04	251	48
0.05 - 0.06	144	28
0.07 - 0.10	98	19
0.11 - 0.25	25	5
TOTAL	518	100

QUALITY VALUE	N° BUILDINGS	%
0.01 - 0.04	321	62
0.05 - 0.06	109	21
0.07 - 0.10	74	14
0.11 - 0.25	14	3
TOTAL	518	100

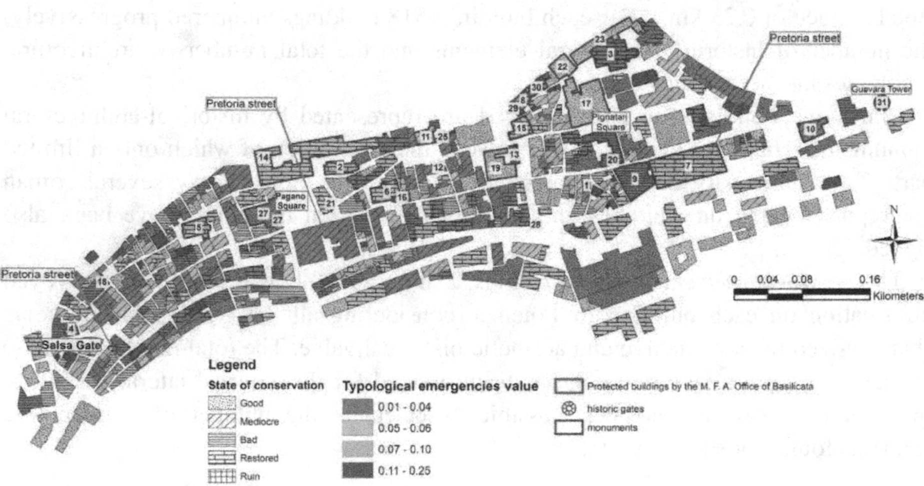

Fig. 2. The typological emergencies values compared with the state of conservation of each building

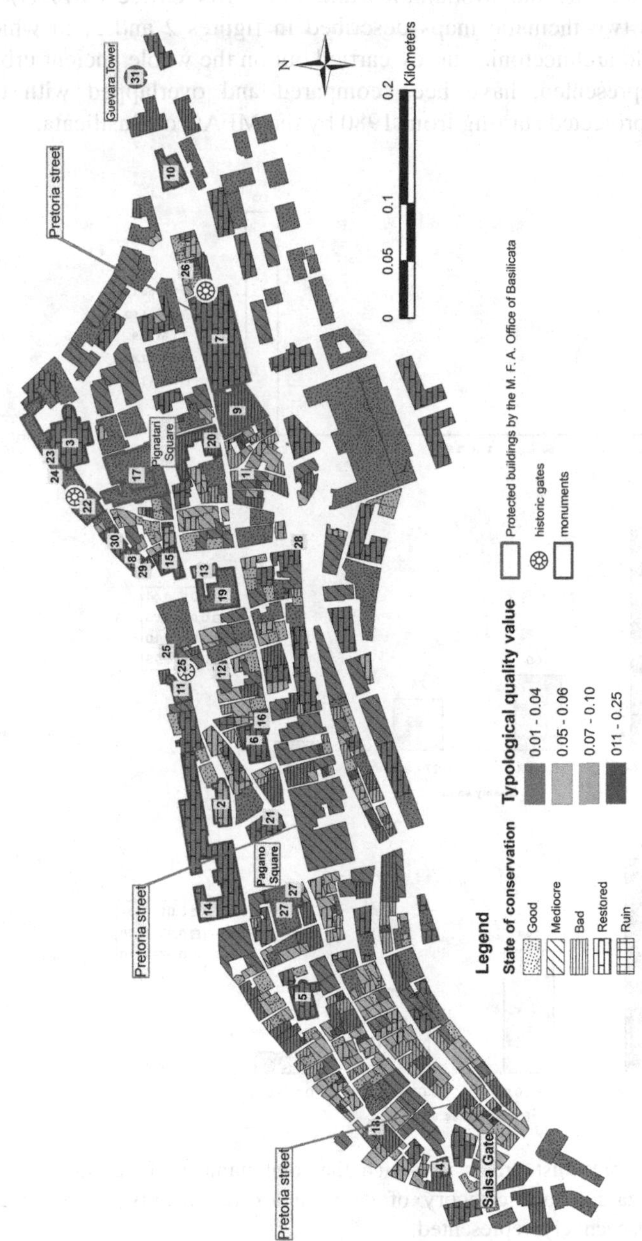

Fig. 3. The typological quality values compared with the state of conservation of each building

Afterwards, we compared the layer with the state of conservation in field surveyed and the buildings protected by the Monuments and Fine Arts Office (MFAO) of Basilicata (Fig. 5). The two thematic maps described in figures 2 and 3, in which results of census and field architectonic survey carried out on the whole ancient urban site of Potenza are represented, have been compared and overlapped with the buildings surveyed and protected starting from 1980 by the MFAO of Basilicata.

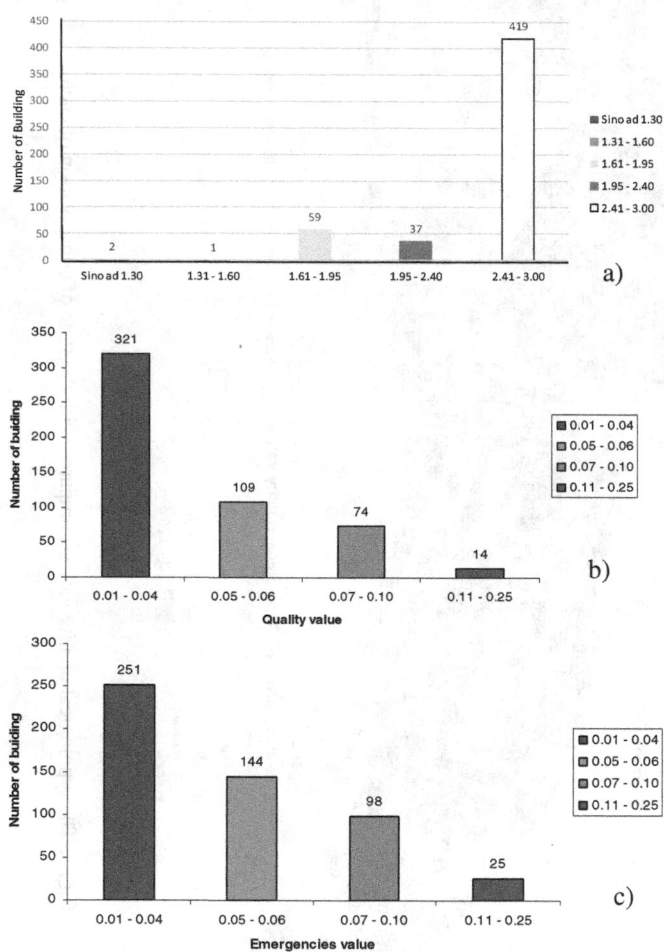

Fig. 4. The figure shows three histograms in which the total number of buildings of the historical centre of Potenza for each category of decay index (a), quality value (b) and emergency value (c) are, respectively, represented.

The degradation of the constructive materials or Decay Index (DIx), evaluated using the method proposed by Lazzari et al. [4], is mainly due to the deteriorating action of the weather agents and structural damage.

We collect typological characteristics of individual buildings through a field survey using a pre-designed form; this forms the basis for a GIS-based analysis to detect priorities of intervention for structural recovery and management of surveyed buildings in relation to seismic risk. The vulnerability of the historical buildings has been evaluated through a decay index (Dlx) calculation assigning to each considered element a weight value depending on its state of conservation, distinguishing between three main classes of damage:

1) structural damage, such as cracks, collapses of walls, strains and collapses of the roofs caused by seismic events, mediocre building materials and techniques, geological instability factors (SD);
2) decay of structures, architectural surfaces, frames and decorations due to moisture (DM);
3) decay of architectural surfaces, frames and decorations due to the lack of maintenance (DLM).

The assigned weight, varying from bad (1) to good (3) condition, has been multiplied for an amplifying factor, equal to 3 for DM and DLM classes and equal to 6 for SD class that represents a heavier condition for conservation of a building. The *DIx* has been obtained dividing the weighted total score (Wt) by sum of coefficients (Σ_{coef}) determined in function of number of floors (p), as shown in table 2 and figure 5:

$$DI_x = \frac{Wt}{\sum coef} \tag{3}$$

where, $\sum_{coef} = 22 + 3p$ (4)

Table 2. Example of decay index (DIx) calculation applied to the building 9, Biscotti Palace, located in Pretoria street

BISCOTTI PALACE							
	Score	1	2	3			
coverange	(multiplying 3)		X		=		6
façade	(multiplying 3)	X			=		3
doors and windows	(multiplying 1)	X			=		1
common areas	(multiplying 3)		X		=		6
base P. T.	(multiplying 3)		X		=		6
base others levels	(multiplying 3p)		X		=	1	8
humidity walling	(multiplying 3)	X			=		3
crack	(multiplying 6)	X			=		6
	TOTAL SCORE WEIGHTED					4	9
					DECAY INDEX	**1, 7**	**4**

Instructions				Estimate decay index		
Humidity	1 serius	2 average	3 soft	Until 1.30	EXTREME	
Crack	1 alarming	2 serius	3 restricted	1.31-1.60	VERY HIGH	
Others levels	1 bad	2 moderate	3 good	1.61-1.95	**HIGH**	
p= number levels				1.96-2.40	SENSITIVE	
				2.41-3.00	RESTRICTED	
decay index i s obtained by dividing the total score weighted and total of coefficient (22+3p)						

This index represents a quantitative expression of the conservation of the historical building, in which either structural or not structural features have been considered.

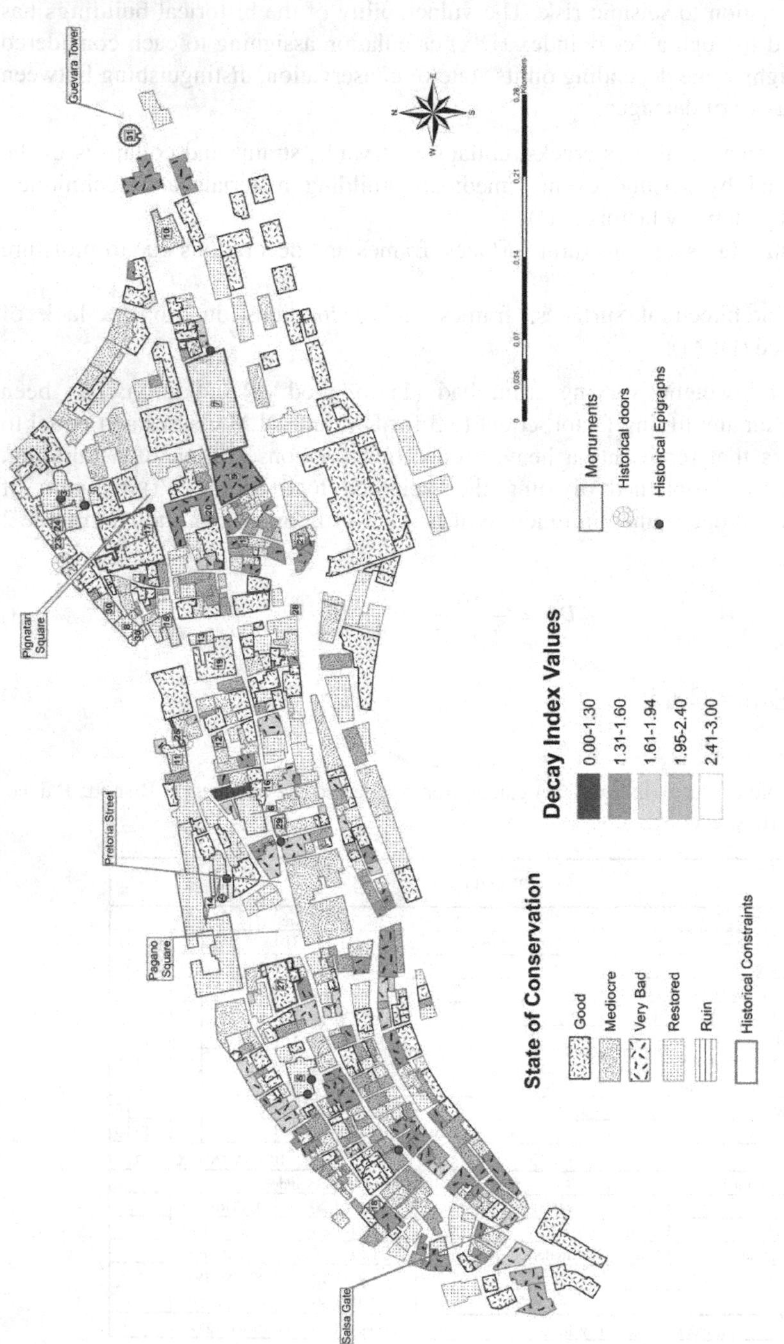

Fig. 5. The decay index values compared with the state of conservation of each building.

2.2 Assessment of Vulnerability

In this study the physical vulnerability of cultural heritage objects has been established using spatial multi-criteria evaluation based on their state of conservation which determines the level of exposure to climate weathering. Vulnerability is analyzed in this context as the potential for physical impact on the cultural heritage assets with basically the same structural type, hence having similar damage performance.

In order to assess the structural vulnerability of the buildings we considered the 1980's Irpinian earthquake macroseismic damage scenario reconstructed by Danese et al. [11, 12]. A remarkable search of unpublished data has been implemented in a geodatabase, such as a topographic map (1:4,000), a geological and geomorphological map, borehole logs, geotechnical parameters, geophysical data, historical macroseismic data at building scale and historical photographs. Starting from historical macroseismic data, a damage scenario has been reconstructed (Fig. 6) considering five damage levels ($D_{1\text{-}5}$, low damage - total building collapse) according to the European Macroseismic Scale [13]

The total vulnerability results from two components linked to decay index-state of conservation and macroseismic damage level surveyed after the 1980's Irpinian earthquake. The figure 7 shows as the 19% of buildings is characterized by a bad/ruin state of conservation and the 28% suffered macroseismic damage between D_3 and D_5 levels. These statistical data demonstrate a base-vulnerability due to the constructive typology (masonry) to which it is added the material degradation due to climatic and anthropic factors (rainfall, thermal oscillation, abandonment, pollution).

The results obtained through GIS analysis allowed us to extrapolate various statistics relating to the entire historic centre of Potenza. In particular, the 33% of buildings appears to be in a state of conservation mediocre, followed buildings in good state (27%) and restored (21%). Only two buildings appear to be ruins, while the remaining 19% is in poor condition, a clear sign of a state of disrepair after the earthquake.

The calculation of the decay index showed that the 81% of buildings obtained a value falling in a range between 2.83-3.00. Only 12% of the buildings has a high DIx and 7% appears to be high.

The overlapping of informative levels coming of typological quality map, typological emergencies map, state of conservation map and buildings protected from Monuments and Fine Arts Office of Basilicata permitted to evidence that several buildings, such as Pignatari Palace, still not subjected to protection need of urgent structural interventions in order to avoid their definitive loss (Tab. 3).

Table 3. Synthesis of typological constructive elements, decay index (DIx) and state of conservation surveyed for historical and monumental buildings of Potenza

Buildings	Name	Protected buildings by the M. F. A. Office of Basilicata	N° Typologies	N° Typological Elements	State of conservation	Lateral Surface (m^2)	Decay index	Decay evaluation	Emergencies value	Quality value
1	Beato Bonaventura Church	no	6	8	Restructured	91.6	2.88	Low	0.087	0.065
2	San Francesco Church	no	8	40	Restructured	565.2	3.00	Low	0.071	0.014
3	San Gerardo Church	no	8	47	Restructured	1227.4	3.00	Low	0.038	0.007
4	Santa Lucia Church	no	5	13	Restructured	275.0	3.00	Low	0.047	0.018
5	San Michele Church	no	7	49	Restructured	677.2	3.00	Low	0.072	0.010
6	SS Trinità Church	no	7	38	Restructured	658.9	3.00	Low	0.058	0.011
7	San Luca Monastery	yes	7	157	Restructured	2133.7	3.00	Low	0.074	0.003
8	Historic walls	no	6	38	Restructured	583.0	3.00	Low	0.065	0.010
9	Biscotti Palace	yes	5	22	Bad	1387.1	1.74	High	0.016	0.004
10	Bonifacio Palace	no	5	49	Mediocre	1338.2	2.80	Low	0.037	0.004
11	Castellucci Palace	no	6	62	Mediocre	1279.5	2.82	Low	0.048	0.005
12	Corrado Palace	no	8	51	Restructured	591.1	2.91	Low	0.086	0.014
13	Fascio Palace	no	7	138	Restructured	559.6	3.00	Low	0.247	0.013
14	Government building	no	7	258	Restructured	5835.6	3.00	Low	0.044	0.001
15	Municipal building	no	4	88	Restructured	1159.3	3.00	Low	0.076	0.003
16	Giuliani Palace	no	5	15	Good	588.1	2.89	Low	0.026	0.009
17	Loffredo Palace	yes	6	130	Good	2956.5	3.00	Low	0.044	0.002
18	Marsico Palace	no	5	14	Good	293.3	2.89	Low	0.048	0.017
19	Navarra Palace	no	9	121	Good	2017.5	3.00	Low	0.060	0.004
20	Pignatari Palace	no	6	37	Bad	2417.3	1.88	High	0.015	0.002
21	Riviello Palace	no	6	24	Restructured	401.8	2.90	Low	0.060	0.015
22	Scafarelli Palace	no	7	51	Good	1129.8	3.00	Low	0.045	0.006
23	Vescovile Palace	no	6	100	Good	2261.9	3.00	Low	0.044	0.003
24	Towers	yes	7	9	Good	182.6	3.00	Low	0.049	0.038
25	Stabile theatre	yes	9	80	Good	2409.5	3.00	Low	0.033	0.004
26	San Gerardo Temple	no	3	4	Good	51.9	3.00	Low	0.077	0.058
27	Aragonese tower	no	6	8	Restructured	953.1	3.00	Low	0.017	0.012
28	Aragonese tower	no	6	12	Restructured	484.3	3.00	Low	0.013	0.060
29	Guevara Tower	yes	4	8	Restructured	558.5	3.00	Low	0.014	0.007

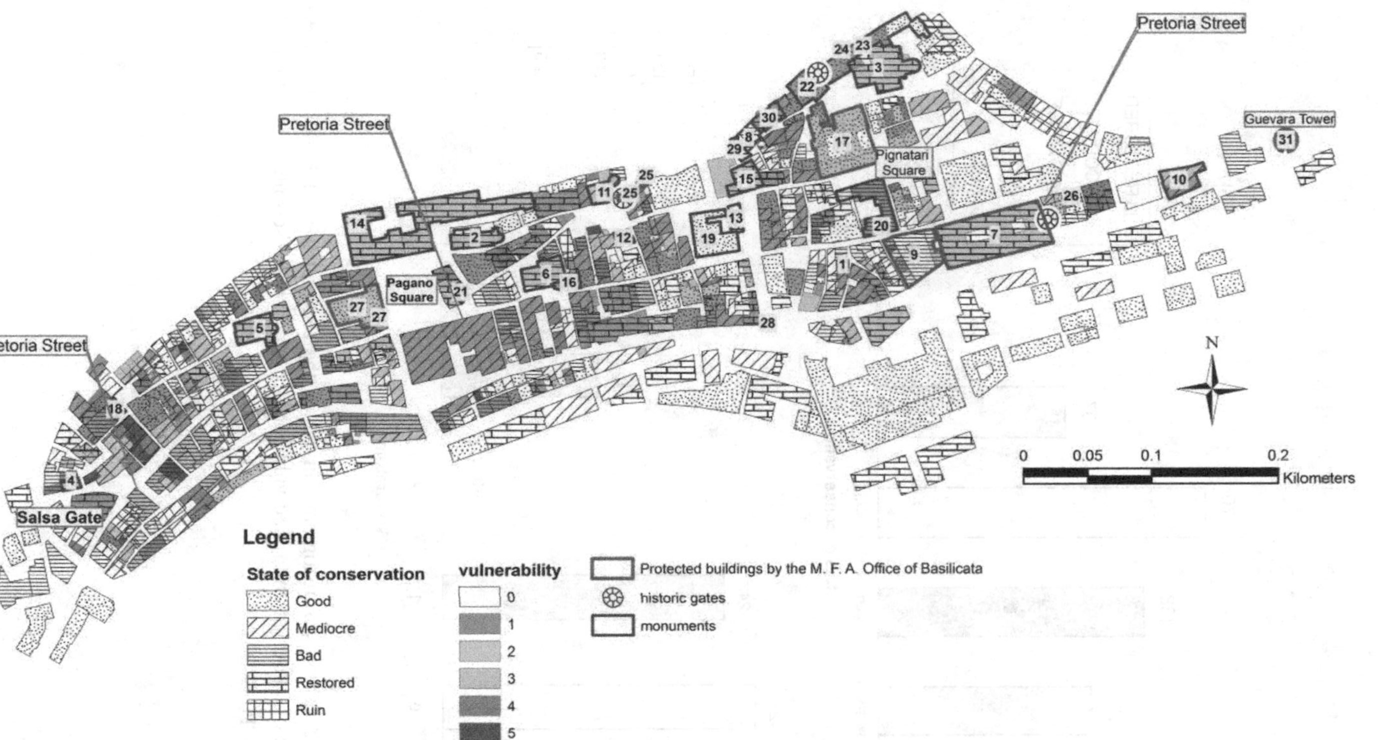

Fig. 6. - Vulnerability map obtained from macroseismic damage levels of 1980 Irpinian earthquake compared with the of conservation

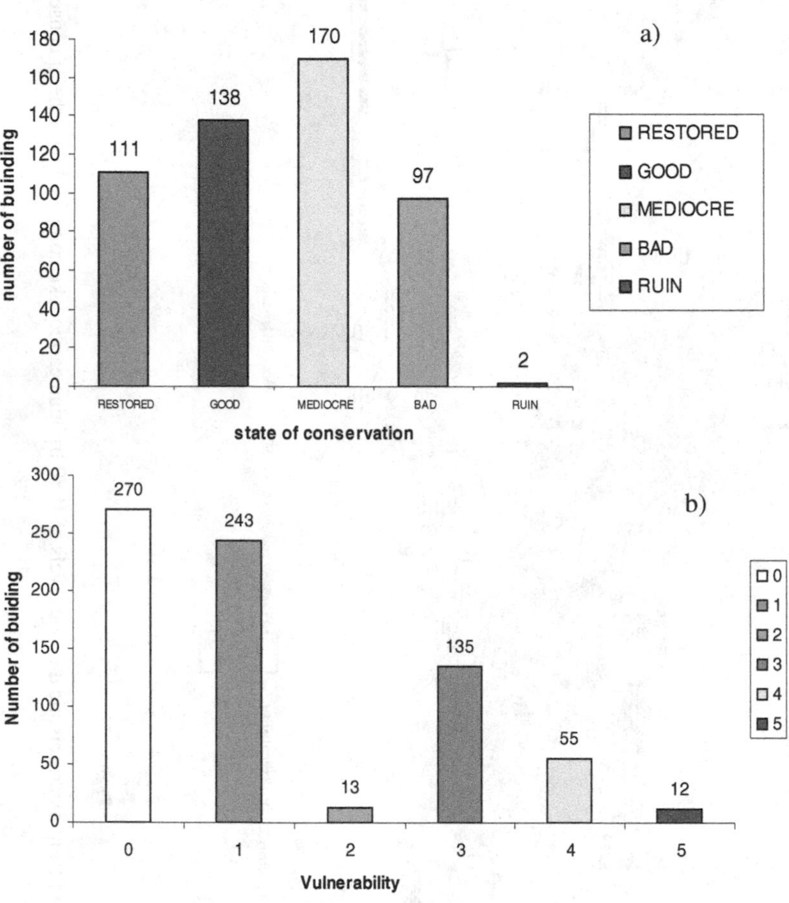

Fig. 7. The figure shows two histograms, in which the total number of buildings for each categories of state of conservation (a) and vulnerability due to macroseismic damage level (b), are, respectively, represented

3 Results an Final Remarks

To complete the spatial analysis for the identification of critical areas of the historical center of Potenza it was necessary to transform the vector files to raster on which to apply functions to map algebra. To this end we assigned weights of 1 to 3 equivalent to a specific level (low, medium, high or ruin, bad/mediocre, good/ restructured), as shown in Table 4 a, b and c.

This procedure has been used for the quality typological, the amount typological and the state of conservation, while for the decay index values were used, in line with the pattern applied in the method (Table 2), inserting negative values (from - 2 to 0), inversely proportional to the others (Table 4d).

Table 4. Class values applied to each thematic raster

QUALITY VALUE	
0.01-0.03	LOW (1)
0.04-0.06	MEDIUM (2)
0.07-0.09	HIGH (3)

A)

STATE OF CONSERVATION	
RUIN	1
BAD-MEDIOCRE	2
GOOD- RESTRUCTURED	3

B)

EMERGENCIES VALUE	
0.01-0.03	LOW (1)
0.04-0.06	MEDIUM (2)
0.07-0.25	HIGH (3)

C)

DECAY INDEX	
EXTREME-VERY HIGH	-2
HIGH	-1
SENSITIVE-RESTRICTED	0

D)

Using the raster calculator all values of each raster were added creating a final map (Fig. 8), following compared with the vulnerability raster map (Fig.9). Starting from the observation of figure 8 it is possible to note that the area of the Potenza where there are the main problems is the central and eastern part of the historical center from Salsa Gate until to Pagano Square.

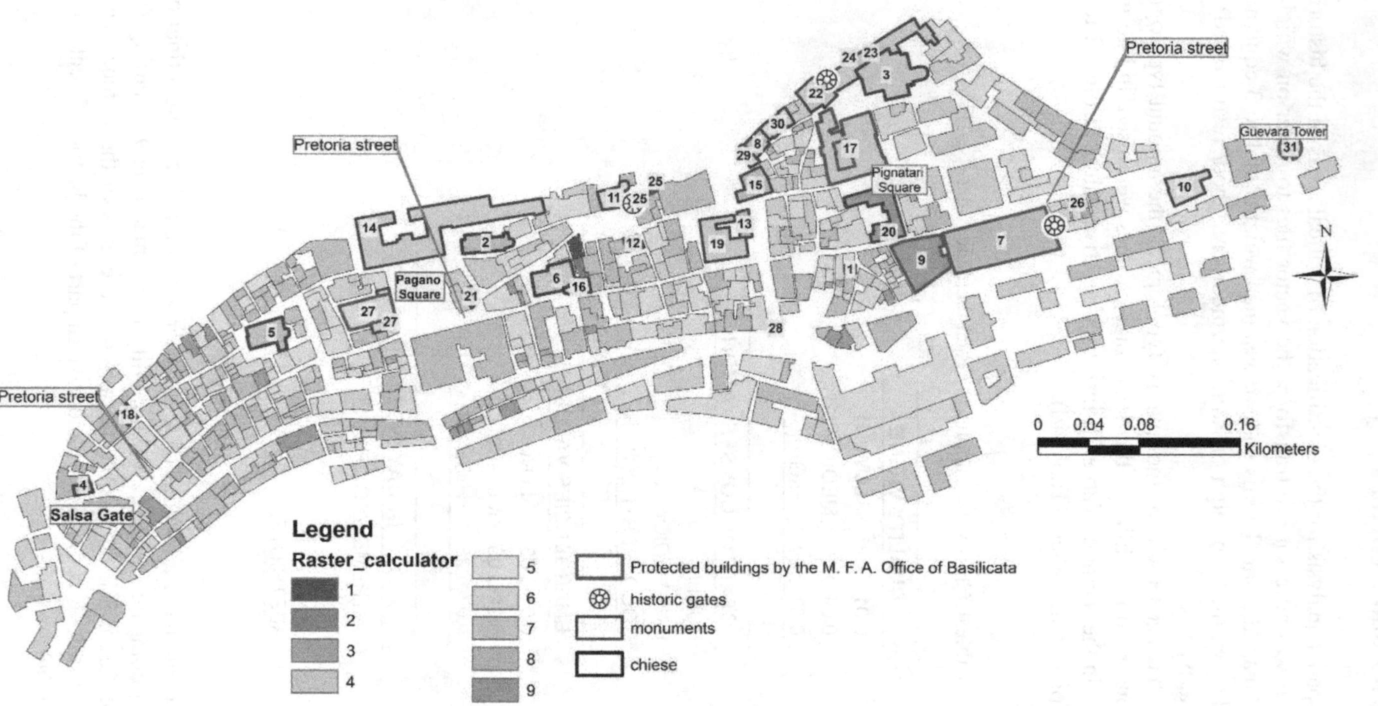

Fig. 8. Final raster map to be used for determining priorities for action

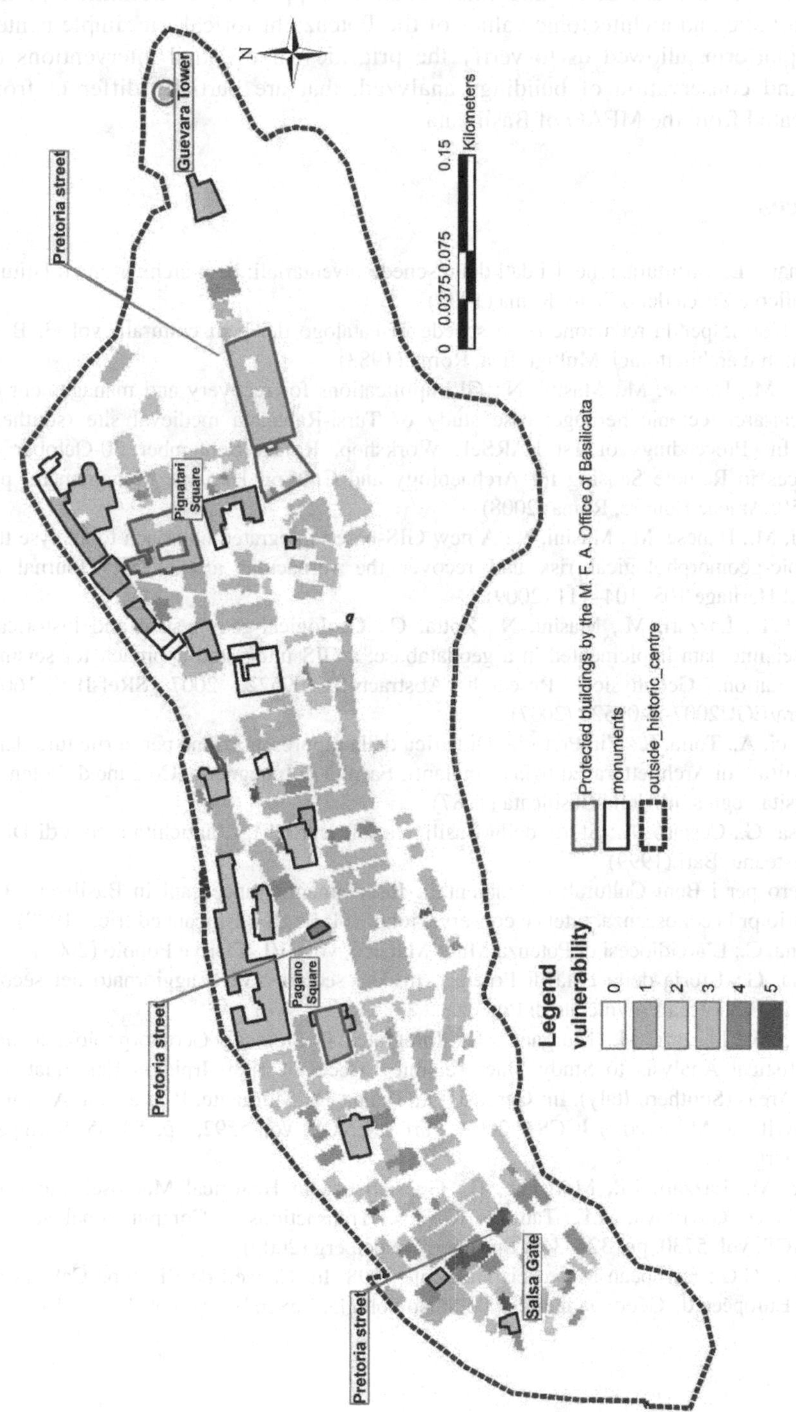

Fig. 9. Structural vulnerability raster map

In conclusion, the use of an integrated cognitive approach for definition of the historical-artistic and architectonic values of the Potenza historical site, implemented in a GIS platform, allowed us to verify the priorities of eventual interventions of recovery and conservation of buildings analyzed, that are partially different from those indicated from the MFAO of Basilicata.

References

1. Cavagnaro, L.: Strutturazione dei dati delle schede inventariali: beni architettonici, Istituto Poligrafico e Zecca dello Stato, Roma (1993)
2. ICCD, Norme per la redazione delle schede di catalogo dei beni culturali, vol. 3, Beni ambientali e architettonici, Multigrafica, Roma (1983)
3. Lazzari, M., Danese, M., Masini, N.: GIS applications for recovery and management of historical-architectonic heritage: case study of Tursi-Rabatana medieval site (southern Italy). In: Proceedings of 1st EARSeL Workshop, Rome, September 30-October 4. Advances in Remote Sensing for Archaeology and Cultural Heritage Management, pp. 347–350. Aracne Editore, Roma (2008)
4. Lazzari, M., Danese, M., Masini, N.: A new GIS-based integrated approach to analyse the anthropic-geomorphological risk and recover the vernacular architecture. Journal of Cultural Heritage 10S, 104–111 (2009)
5. Gizzi, F.T., Lazzari, M., Masini, N., Zotta, C.: Geological-geophysical and historical-macroseismic data implemented in a geodatabase: a GIS integrated approach for seismic microzonation. Geophysical Research Abstracts 9, 09522, 2007 SRef-ID: 1607-7962/gra/EGU2007-A-09522 (2007)
6. De Nucci, A., Tolla, E.: Via Pretoria. Didattica della rappresentazione per la rilettura della città, Istituto di Architettura Edilizia e Impianti, Facoltà di Ingegneria, Comune di Potenza, Università degli studi della Basilicata (1987)
7. De Rosa, G., Cestaro, A.: Storia della Basilicata 1 (a cura di), L'antichità a cura di Dinu Adamesteanu, Bari (1999)
8. Ministero per i Beni Culturali e Ambientali, Insediamenti francescani in Basilicata. Un repertorio pr la conoscenza, tutela e conservazione, vol. I e II. Basilicata editrice (1988)
9. Palestina, C.: L'arcidiocesi di Potenza Muro Marsico, vol. II/1. Clero e Popolo (2001)
10. Rendina, G.: Istoria della città di Potenza, ms del secolo XVIII, aggiornato nel secolo XVIII, in Biblioteca Provinciale di Potenza
11. Danese, M., Lazzari, M., Murgante, B.: Integrated Geological, Geomorphological and Geostatistical Analysis to Study Macroseismic Effects of 1980 Irpinian Earthquake in Urban Areas (Southern Italy). In: Gervasi, O., Taniar, D., Murgante, B., Laganà, A., Mun, Y., Gavrilova, M.L. (eds.) ICCSA 2009, Part I. LNCS, vol. 5592, pp. 50–65. Springer, Heidelberg (2009)
12. Danese, M., Lazzari, M., Murgante, B.: Geostatistics in Historical Macroseismic Data Analysis. In: Gavrilova, M.L., Tan, C.J.K. (eds.) Transactions on Computational Science VI. LNCS, vol. 5730, pp. 324–341. Springer, Heidelberg (2009)
13. Grünthal, G.G.: European Macroseismic Scale 1998. In: Conseil de l'Europe Cahiers du Centre Européen de Géodynamique et de Séisomologie, Luxembourg, vol. 15 (1998)

A Methodological Approach to Integrate Ontology and Configurational Analysis

Antonia Cataldo[1], Valerio Di Pinto[2], and Antonio M. Rinaldi[3]

[1] Ministero dell'Istruzione, dell'Universitá e della Ricerca
via Ponte della Maddalena, 55 - 80142 Napoli
[2] Dipartimento di Ingegneria Civile, Edile e Ambientale - 80125 via Claudio, 21 - Napoli, Italy
Universitá di Napoli Federico II
[3] Dipartimento di Ingegneria Elettrica e delle Tecnologie dell'Informazione
80125 via Claudio, 21 - Napoli, Italy
IKNOS-LAB Intelligent and Knowledge Systems - LUPT
80134 via Toledo, 402, - Napoli, Italy
Universitá di Napoli Federico II
{cataldo,valerio.dipinto,antoniomaria.rinaldi}@unina.it

Abstract. The problems related to the management of information increase everyday in several contexts. This process needs of more and more effective and efficient Knowledge Management techniques. Such techniques lead to a suitable organization (and therefore to a more detailed analysis) of the phenomena, which in turn suggests how to choose the actions to perform. Information represents a strategic resource for the subjects in charge of territorial management: it is in fact raw material, job tool and final product. In a complex research scenario such as the territorial science, some fundamental elements, useful to analyze and resolve several problems, can be represented using a framework based on an ontological approach pushed on the reduction of the space as a network. The science of complex networks is indeed leading the quest for a renewed approach to territorial phenomena in many cases. Amongst the different territorial planning fields, we chose as case study the issue of urban planning and design. This choice is motivated by the cultural and scientific innovations, which led in recent years to a completely new interpretation of a city as a system of elements each other connected. The paper aim is to give a first conceptual framework to integrate ontological aspects of urban elements with topological features of the city; these latter accounted as property of the city-network. The topological information revealed by the configurational analysis of the city-network will be improved with a formal solution of conceptual misunderstanding and semantic ambiguity using a suitable ontology-based model.

1 Introduction

Cities, as complex entities, seem to challenge descriptions, so that they can hardly be defined and treated in disciplinarily terms. Planners and urban designers have always tried to use simplified concepts and notions, thus emphasizing hierarchies, regular geometries and the separation of parts from wholes. In this broad framework, assessing and modelling complexity in cities is gaining momentum, leading the quest for a new

B. Murgante et al. (Eds.): ICCSA 2014, Part II, LNCS 8580, pp. 693–708, 2014.

"urban science". One approach that is increasingly engaging scholars and practitioners rethinks the way we can look at the city and its problems and potential; another approach that starts from space and urban life at the same time, hence grasping the two dimensions of a place: cognitive and geographic. The city is considered as uniform fact. For this purpose we need to imagine cities as complex systems consisting of many variables that interact with each other, assuming the space as a primary element in such dynamics rather than the mere and inactive background of social and economic phenomena. Since the configurational approach was first pioneered in the early 80' by Bill Hillier, under the notion of Space Syntax [23], the application to cities and open spaces of the science of complex networks mainly developed in structural sociology and physics has flourished, providing new knowledge for urban regeneration and sustainable urbanism. The key idea of the theory is that the city is a configurational entity, or rather that the urban environment is expressed through the reciprocal spatial relationships between all of its elements. As a science-based and human-focused approach, Space Syntax has brought to light key underlying structures in the city, which have a direct bearing on sustainability in that they seem to show that the spatial form of the unplanned city is already a reflection of the relations between environmental, economic and socio-cultural forces. Do not lack ideas for the discipline evolution over the last decade. We could trace a general tendency for the opening of the urban network science to the use of geo-statistics and the recourse to data-base already structured and shareable. In this sense, is gaining momentum a new methodological approach, known as Multiple Centrality Assessment, developed by Sergio Porta at the University of Strathclyde. Configurational approach, under any notion, provides a concise and effective overview on how a city "operates", but it is unable to render a formal representation of the meaning of urban elements in terms of knowledge. During recent years, several approaches have been proposed to represent knowledge. Some of them are based on ontologies to delete or at least smooth conceptual or terminological mess and to have a common view of same information. The ontological aspects of information are intrinsically independent from information representation, so information itself may be isolated, recovered, organized and integrated with respect to its contents. Using an ontological approach will be possible to formally describe the various elements recognized in the urban environment. In addition, we will define the different types of conceptual relationships between these elements that will come together in the topology.

Our research project will aim to allow it, basing its logic on a formal representation of the urban space semantic, implemented on a network model, assumed as the syntax of the same space. In our knowledge representation (KR) every urban feature, formally defined and represented, has a position on the urban network, characterized by its configurational value: a location aware formal representation of the city. We call this knowledge structure as Configurational Ontology.

The paper is organized as follows: in section 2 we present a theoretical background about the main aspects of our research; section 3 gives a description of the whole proposed framework to define and implement our Configurational Ontology; a preliminary case study example is presented and discussed in section 4. Eventually, conclusions are in section 5.

2 Theoretical Background

In this section we give a comprehensive overview of existing literature in our domain of interest to understand the various dimensions of our work. In fact, it is necessary to take into account both models for knowledge representation and management and approaches for the configurational analysis.

2.1 Ontologies and KR Languages

Our approach starts from the *modeling view* of knowledge acquisition [9], where the modeling activity must establish a correspondence between a knowledge base and two separate subsystems: the agent's behavior (i.e., the problem-solving expertise) and its own environment (the problem domain) (see also [16,32,18]). This vision is in contrast with the *transfer view*, wherein a knowledge base is a repository of knowledge extracted from one expert's mind. Using the modeling view approach, knowledge is much more related to the classical notion of truth as correspondence to the real world, and it is less dependent on the particular way an intelligent agent pursues its goals. Although knowledge representation is a basic step in the whole process of knowledge engineering, a part of the AI research community seems to have been much more interested in the nature of reasoning than in the nature of "real world" representation. This tendency has been especially evident among the disciples of the called logicist approach: in their well-known textbook on AI, Genesereth and Nilsson [17] explicitly state the "essential ontological promiscuity of AI" and devote just a couple of pages to the issue of conceptual modeling. They admit, however, that it is still a serious open problem. The issues of representation are also addressed in the same way in [31]. The dichotomy between reasoning and representation is comparable with the philosophical distinction between epistemology and ontology, and this distinction is important to better understand our aim and approach. Epistemology can be defined as "the field of philosophy which deals with the nature and sources of knowledge" [26]. The usual logicistic interpretation is that knowledge consists of propositions whose formal structure is the source of new knowledge. The inferential aspect seems to be essential to epistemology (at least in the sense that this term assumes in AI): the study of the "nature" of knowledge is limited to its superficial meaning (i.e., the form), since it is mainly motivated by the study of the inference process. Ontology, on the other hand, can be seen as the study of the organization and the nature of the world independent of the form of our knowledge about it. A formal definition of ontology is proposed in [19] according to which *"an ontology is a formal and explicit specification of a shared conceptualization"*; *conceptualization* refers to an abstract model of a specific reality in which the component concepts are identified; *explicit* means that the type of the used concepts and the constraints on them are well defined; *formal* refers to the ontology propriety of being "machine-readable"; *shared* refers to the fact that an ontology captures the consensual knowledge, accepted by a group of persons. We also consider other definitions of ontology: in [25] *"an ontology defines the basic terms and relations comprising the vocabulary of a topic area, as well as the rules for combining terms and relations to define extensions to the vocabulary"*. This definition indicates the way to proceed in order to construct an ontology: i) identification of the basic terms and their relations; ii) agreeing on the rules to arrange them; iii) definition of terms and relations between concepts. From this perspective, an ontology includes not only the terms that are explicitly defined in it, but also those one that can be derived using defined rules and properties. Thus an ontology can be seen as a set of "terms" and "relations" among them, denoting the concepts that are used in a specific domain. Previous approaches to the need for "tools" to represent knowledge, both for inferring and organizing it. From this point of view, one of the most important advances in the knowledge representation applications is derived from proposing [24], studying [5,6] and developing [7,14,4] languages based on the specification of objects (concepts) and the relationships among them. The main features of these languages are the following:

object-orientedness all information about a specific concept is stored in the concept itself (in contrast, for example, to rule-based systems);

generalization/specialization these properties are basic aspects of the human cognition process [24]; these languages have mechanisms to cluster concepts into hierarchies where higher-level concepts represent more general attributes than the lower-level ones, which inherit the general concept attributes but are more specific, presenting additional features of their own;

reasoning the capability to infer the existence of information not explicitly declared by the existence of a given statement;

classification given an abstract description of a concept, there are mechanisms to determine whether a concept can have this description. This feature is a special form of reasoning.

Object orientation and generalization/specialization help human users in understanding the represented knowledge; reasoning and classification guide an automatic system in building a knowledge representation, as the system knows what it is going to represent. Our approach arises from the above considerations and is also suggested by the work of Guarino [20]. When a KR formalism is constrained in such a way that its intended models are made explicit, it can be classified as belonging to the ontological level [20] introduced in the distinctions proposed in [6], where KR languages are classified according to the kinds of primitives offered to the user. At the (first order) *logical level*, the basic primitives are predicates and functions, which are given formal semantics in terms of relations among objects of a domain. No particular assumption is made, however, regarding the nature of such relations, which are completely general and content-independent. The *epistemological level* was introduced by Brachman in order to fill the gap between the logical level, where primitives are extremely general, and the conceptual level, where they acquire a specific intended meaning that must be taken as a whole, without any consideration of its internal structure. At the *ontological level*, the ontological commitments associated with the language primitives are specified explicitly. Such a specification can be made in two ways: either by suitably restricting the semantics of the primitives or by introducing meaning postulates expressed in the language itself. In both cases, the goal is to restrict the number of possible interpretations, characterizing the meaning of the basic ontological categories used to describe the domain: the ontological level is therefore the level of meaning. At the *conceptual level*, primitives have a definite cognitive interpretation, corresponding to language-independent concepts such as elementary actions or thematic roles. The skeleton of the domain structure is already given, independently of an explicit account of the underlying ontological assumptions. Finally, primitives at the *linguistic level* refer directly to lexical categories. In the last few years, a huge number of methodologies and tools to create, manage, represent, and match ontologies have been proposed and implemented in several contexts, often for ad hoc purposes (e.g., to map well-known knowledge bases or represent specific knowledge domains); a complete discussion of these is outside the scope of our work, but useful books, surveys and ad hoc models are available in the literature (e.g., [12,10,34,13,30]).

2.2 Configurational Approach to the City

A formal model represents a basic joint element between the human activity and the environmental system, where human abilities and skills represent the continuous research of a global eco-systemic balance. Once understood that the process of understanding the urban elements is priority and propaedeutic for the definition of actions for urban management and services implementation we need to individuate those homogeneous territorial contexts, which contain highly related and characterizing factors. To this aim we must try to resolve conceptual misunderstanding and semantic ambiguity and, on the other hand, we need a precise and accurate description of our knowledge, i.e. of the terminologies representing our concepts, which influence goals and entail actions to be implemented.

A city could be well interpreted as a complex network, constituted by nodes and arches standing for urban space elements and their mutual relations. As noted by Michael Batty in 2008[3], this simplifying model of the city is not rooted primarily in Euclidean space but deals as much with topologies, such as social networks. This suggests ways in which our longstanding physical approach to cities can be consistently linked to phenomena that only obliquely manifest themselves in physical terms. The science of complex network is therefore the leading approach in this field. As pointed out by Porta et al. in 2005[27], its entire theoretical background was pioneered in the first '80, by the seminal work of Bill Hillier and Julienne Hanson under the notion of Space Syntax[23]. It is founded on the idea that is possible to relate the urban space's collective cognition and use to the topological properties of its representation as a urban network, obtained dividing the space into a set of convex sub-spaces [22]. Space syntax methods were initially developed in order to allow architectural space to be represented and its pattern properties quantified so that comparisons could be made between differently designed buildings or urban areas. The aim of the research was to develop an understanding of the way that spatial design and social function were related. It was soon found that a primary effect of spatial *configuration* on social function resulted from the way that space patterns determined pedestrian movement patterns and so co-presence between people in space. These findings show that between 50 and 80% of the variance in pedestrian flows from location to location in an environment can be explained in terms of variations in configurational properties of those locations in the network. Since co-presence is a prerequisite for both communication and transaction in socio-economic life, space syntax research suggests that spatial configuration itself plays a critical role in determining this. Hillier's approach to urban space is so hardly pivoted on the idea that the city is a configurational entity. Configuration, simply defined as simultaneous existing relations, is about the composition of the built form from the parts that are in a unique relationship with each other. The analytical core of the space syntax methodology is the concept of closeness centrality in social networks [15]: the index of *integration*. It is stated to be *"so fundamental that it is probably in itself the key to most aspects of human spatial organization"* [22]. The integration of one street has been defined as the *"shortest journey routes between each link (or a space) and all of the other in the network (defining shortest in terms of fewest changes in direction)"* [22]. This measure of centrality, as better explained in the following sections, is carried out on an urban network of dual type, which simplifiyng means reducing streets into nodes[27]. In the process of building the dual graph what gets lost is something very relevant, beside its somehow questionable importance for the human cognitive experience of spaces, for any human sensorial experience of space: *distance*. One street will be represented in the dual graph as one point. The dual approach characterize Space Syntax to be a purely topolgical method to analize city as a network. In recent years Space Syntax started being used in simulation mode, in order to support experimentation and inform architectural and urban design. Usually, different options of a project are tested with space syntax and subsequently compared. This provides valuable feedback to designers and has already resulted in a significant portfolio of analysed projects, including major schemes by the architects Richard Rogers and Norman Foster [29]. However, despite its growing success and the fascinating questions on the use of space that it has raised, some of Space Syntax findings remain controversial in the academic community. Notably, is the dual approach to make the major concerns arising. In a dual graph, in fact, it is possible to find one node (a street) with a conceptually unlimited number of edges (intersections between streets), a number which heavily depends on the actual length of the street itself. Thus, the longer the street in reality, the more central it is likely to be in the dual graph, which counters the experiential concept of accessibility that is conversely related to how close is the destination to all origins, like in transportation models. Another consequence of the dual reduction of streets into points is that it makes impossible to account for the variations which so often characterize one single street, variations that may easily become very significant for lengthy streets that cross wide urban areas. Primarily to face these problems, many efforts have been making to identify a primal approach (returning an urban network where the streets are edges and their intersections are nodes) which do not dissipate the assets of the results achieved by Space Syntax in topological terms. The most notable landing in this frame is the model known as Multiple Centrality Assessment (MCA). It has been introduced by Porta and others in a series of studies published since 2006[28]. This new model is based on the idea of using central road lines between nodes as edges and their junctions as nodes of the city network. This makes it possible to introduce the geometry as a local properties of edges, hence offering deeper analysis outputs while still preserving the topological structure of the real street network. In addiction of the use of primal graphs and the use of metric distance, MCA introducts un urban network analysis the common use of many different indices of centrality. Compared to dual approaches (Space Syntax), MCA are multifold: it is not based on any generalization criterion (uninterrupted linearity of axial lines) therefore is more legible and feasible; it is fit to access the road-centreline-between-nodes world standard datasets, including those for traffic engineering or geo-mapping; it is more comprehensive and probably more realistic, in that it couples the topological informations with the metric consistency of the system; it gives a set of multifaceted pictures of reality, rather than just one (closeness-based centrality measures)[33]. On the other hand, the large number of lines and the complexity of the measures of centrality make slower MCA computation than that of Space Syntax[27]. Ultimately, the two approaches represent two opportunities to study in the city, absolutely complementary to urban phenomena. If, in fact, the best match between the urban network and real space, as well as the native presence of the spatial consistency as a net attribute, makes MCA a very powerful tool for the study of sub-urban scale phenomena, Space Syntax is preferable to the study at the global city scale, due to the best balance between computational costs and data reliability.

Our paper aim is to reach out the development of an integrated framework, based on topological relationships between geographical elements as well on their typology, used to represent knowledge in urban areas. Cities have been traditionally investigated and represented from sectorial viewpoints, each time endorsing functional aspects rather than anthropological, that are shape-based. The formers are well explained by the set of relationships between urban spaces, or rather by the topology they express in terms of the key-concept of spatial configuration: meaning relations which take account of other relations in a complex. The shape of the city , or rather the typology of the city, has been conversely ever proposed as the ending of personal and experiential tale, and so unsuitable to a modelling approach.

At this stage of our research we want integrate this two elements, trying to disprove the idea that they are two sides of the same coin, in behalf of considering them as two complementary informative layers. With this aim we will put together the modelling properties of city-typology in terms of formal representation of knowledge (ontology) and the modelling properties of city-topology in terms of spatial configuration (urban network analysis).

3 The Proposed Framework

In this section we describe the proposed framework to implement our Configurational Ontology. We first present the used formal model to implement our ontology then we discuss about the approach and the techniques we used for the analysis of the urban environment. Eventually, a formal definition of Configurational Ontologyis proposed.

3.1 Ontology Representation

In this section we define a formal model for KR using a conceptualization as much as possible close to the way in which the concepts are organized and expressed in human language [8]. In reference to urban studies, it could improve the reliability of city modeling tools, furnishing them a robust support to contemplate the human common interpretation of the space. Specifically in reference to urban network configurational analysis, ontologies promise to be the natural complementary element to enhance the analysis of city-user cognition of built environment.

The used ontology based model is composed by a triple $< S, P, C >$ where:

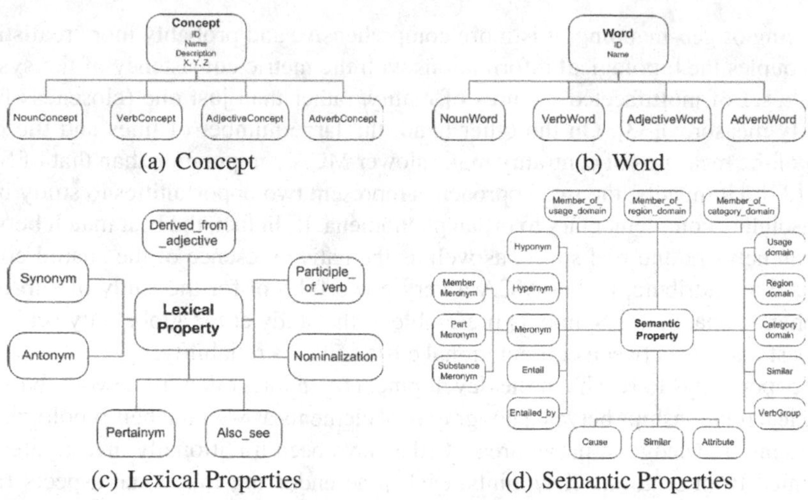

(a) Concept (b) Word

(c) Lexical Properties (d) Semantic Properties

Fig. 1. Model components

S is a set of objects;
P is the set of properties used to link the objects in S;
C is a set of constraints on P.

In this context we consider *words* as objects; the properties are *linguistic relations* and the constraints are *validity roles* applied on linguistic properties with respect to the considered term category. In our approach the knowledge is represented by an ontology implemented w.r.t. a semantic network [2]. A semantic network can be seen as a graph where the nodes are concepts and the arcs are relations among concepts. A concept is a set of words which represent an abstract idea. In recent years, several languages have been proposed to represent ontologies and, even if those languages have several differences, they share some common aspects based on the specification of objects (concepts) and the relationships among them. These languages have a different expressive power and, starting from some considerations from a previous authors' work [1], it is our opinion that OWL [11] is the best language for the purpose of the proposed approach.Therefore we describe the semantic network implementing the ontology in OWL using the defined model. In particular we use the DL version of OWL because it has enough effectiveness to describe the ontology. The DL version allows the declaration of disjoint classes which are used, for example, to assert that a word belong to a syntactic category. Moreover it allows the declaration of union classes used to specify domains and property ranges to relate concepts and words belonging to different lexical categories.

We formally describe the ontology schema and the corresponding semantic network representation using OWL. Every node, both concept and word, is an OWL individual. The connecting edges in the semantic network are represented as *ObjectProperties*. This properties have some constraints that depend on the syntactic category or on the kind of property (semantic or lexical). For example the hyponymy property can relate only nouns to nouns or verbs to verbs; on the other hand a semantic property links concepts to concepts and a syntactic one relates word forms to word forms. Concept and word attributes are considered with *DatatypeProperties*, which relate individuals with a predefined data type. Each word is related to the represented concept by the ObjectProperty *hasConcept* while a concept is related to words that represent it using the ObjectProperty *hasWord*. These are the only properties able to relate words with concepts and vice versa; all the other properties relate words to words and concepts to concepts. Concepts, words and properties are arranged in a class hierarchy, resulting from the syntactic category for concepts and words and from the semantic or lexical type for the properties. In figure 1 the hierarchies used to represent the objects of interest in our model are shown. Figures 1(a) and 1(b) show that the two main classes are Concept, in which all the objects have defined as individuals and Word which represent all the terms in the semantic network. These classes are not supposed to have common elements therefore we have defined them as disjoint. The class Word define the logical model of the word forms used to express a concept. On the other hand, the class Concept represents the word meaning related to a word form. We can see that the subclasses have been derived from the related categories. There are some union classes useful to define properties domain and codomain. We define some attributes for Concept and Word respectively. In particular Concept has: *Name* that represents the concept name; *Description* that gives a short description of concept; *X, Y, Z* that localize a concept in a 3D space. On the other hand Word has *Name* as attribute that is the word name. Moreover for all elements we define an *ID* within the WordNet offset number or a user defined ID. The semantic and lexical properties are arranged in a hierarchy (see figure 1(c) and 1(d)). In table 1(a) some of the considered properties and their domain and range of definition are shown. The use of domain and codomain reduces the property range application; however the model so far described does not have a perfect behavior in some cases. For example the model does not know that even if the hyponymy property is defined on the sets of nouns and verbs, if it is applied on the set of nouns it has as range the set of nouns, otherwise if it is applied to the set of verbs it has as range the set of verbs. In table 1(c) there are some of defined constraints and we specify on which classes they have been applied with respect to the considered properties; the table shows the matching range too. Sometimes the existence of a property between two or more individuals entails the existence of other properties. For example, being the concept dog a hyponym of animal, we can assert that animal is a hypernymy of dog. We represent in OWL this characteristics by means of property features. The table 1(b) shows several of those properties and their features.

3.2 Urban Networks and Spatial Cognition

What is the city? *"The city is two things: a collection of buildings linked by urban space, and a complex system of human activity linked by interaction"* [35]. They could be called as the physical and the social layer. The key goal for urban thinkers and practitioners should be to connect one to other. But they are commonly inclined to put the layers on different levels, in effect seeing the "other" layer through the foregrounded one[35]. Is the city in any case two things or one? There are good reasons to account it as a one thing: *unique*. But how can a physical process relate to a social process?

Table 1. Model features

(a) Properties

Property	Domain	Range
hasWord	Concept	Word
hasConcept	Word	Concept
hypernym	NounsAnd VerbsConcept	NounsAnd VerbsConcept
holonym	NounConcept	NounConcept
entailment	VerbWord	VerbWord

(b) Property features

Property	Features
hasWord	*inverse* of hasConcept
hasConcept	*inverse* of hasWord
hyponym	*inverse* of hypernym; *transitivity*
hypernym	*inverse* of hyponym; *transitivity*
cause	*transitivity*
verbGroup	*symmetry* and *transitivity*

(c) Model constraints

Costraint	Class	Property	Constraint range
AllValuesFrom	NounConcept	hyponym	NounConcept
AllValuesFrom	VerbConcept	hyponym	VerbConcept
AllValuesFrom	NounConcept	attribute	AdjectiveConcept
AllValuesFrom	AdjectiveConcept	attribute	NounConcept
AllValuesFrom	NounWord	synonym	NounWord
AllValuesFrom	AdjectiveWord	synonym	AdjectiveWord
AllValuesFrom	AdverbWord	synonym	AdverbWord
AllValuesFrom	VerbWord	also_see	VerbWord

This places theoretical obstacles. Urban network analysis was pioneered in late 1970s by Bill Hillier as an attempt to address this kind of question, under the notion of **Space Syntax**. Its genesis was in the 1960s urban revolution, as a way to dominate the contradiction between the impressive architectures that were flourishing and the un-urban nature of their spaces. Hillier's apporach began from the observation that space is the common ground of the physical and social layers in the city. Under this premise, the main theoretical contribution that Space Syntax gave to the existing collection of ideas in urban space analysis, is the introduction of the concept of *configuration*.

In syntax terms, spatial configuration means relations between spaces which take into account other relations. Space syntax theorises that certain configurational measures of centrality in graphs express a potential to embody or transmit social ideas, and then pushes this potential to spatial structures, by gegrafically linking the graph to the space of the system under examination[23]. These measures are not by chance borrowed from structural centrality in social networks. The idea of point centrality as applied to interaction among people was introduced in the late 1940s as a way to account for the individual role in human communication. Many years of research have highlighted that centrality was related to group efficiency in problem-solving, perception of leadership and the personal satisfaction of participants[15].

There are three main distinct intuitive conceptions of point centrality, and many measures for each of them. The simplest and perhaps the most intuitively conception is that centrality is function of the *degree* of a point on the net, that is the number of connections a node has to other nodes. It is generally known as *Degree Centrality*, and could be measured as the sum of the direct links of a point in a net. Another simple idea is that the independence of a point is determined by its *closeness* to all other points in the graph. The *Closeness Centrality* of a point could be measured by inverting the sum of the topological geodesic distances from that point to all other points in the graph. Another view of point centrality is based upon the frequency with which a point falls between pairs of other points on the topological geodesic paths connecting them. A point that falls on the communication paths between other points exhibits a potential for control their communication. It is this potential for control that defines the centrality of these points. The *Betweenness Centrality* of a point could be measured as the probability that the point falls on a randomly selected geodesic connecting any pairs of other points. All these point measures of centrality could be easily extended to the whole graph as a property of the network. Space Syntax is traditionally pivoted on closeness centrality, in terms of *Integration Index*. Theoretically, the integration index shows the cognitive complexity of reaching a street, and is often argued to predict the pedestrian use of a street: the easier it is to reach a street, the more popular it should be. The last evolutions of the discipline have been developing betweenness-based measures also.

But how does Space Syntax effectively works in operational terms, or rather, how Space Syntax build the city-graph and perform its measures? This is one of the most controversial elements of the discipline. From the very first steps, Space Syntax has been characterized by an approach to the city as a network very unusual. Consistent with the social interpretation that it render on the city, Space Syntax perform the network making process by the reduction of the city to a specifical set of elementary spaces of copresence: the minimal set of maximum size convex spaces in which the urban public space is divisible. On this map, known as the *convex map*, we can draw the minimal set of lines crossing the convex spaces that oversee them all, getting the *axial map*. It is properly the inversion of the urban graph: axial lines represent the nodes of the graph, while their intersections are the edges. On this graph we can perform the centrality measures, both on individual points (axial lines) and on the entire graph (urban network). To associate the meaning of centrality measures to the real urban space, it is therefore necessary to take the opposite road. This method is known as *dual* or *indirect* approach to the urban network analysis. Although it may seem very expensive, it is actually a process very carefull of computational resuorces. This is due to the case of having a highly simplified network, as well as a system where the geometry is proscribed. In the Space Syntax urban graph, in fact, there is no concept of geometric consistency of the space, even as local properties of nodes (which represent streets) in favor of a purely topological approach.

In figure2 the key-aspects of Space Syntax operational methodology are shown. The urban settlement (figure 2(a)) is split into its main components: the subset of closed spaces (figure 2(b)) and the subset of fully accesible public open spaces (figure 2(c)). Following the methodology above described, Axial Map (figure 2(d)) and urban graph (figure 2(e)) could be created. In figure 2(d) the axial lines are numbered to highligth their correspondence to graph nodes, also numbered(figure 2(e)). The relationship between axial lines and real urban space is shown in figure 2(f).

The study of urban networks is the most advanced approach to the analysis and understanding of the city as a place or rather as a social-based phenomenon. In this sense, the urban network analysis allows us to understand the contribution that urban structure provides to the arising of urban phenomena. It also allows us to infer how urban space is perceived by city-users as well as the result of this perception. In this last sense, the science of complex networks in urban areas, it is the natural substrate for the development of human-oriented systems and applications to city fruition.

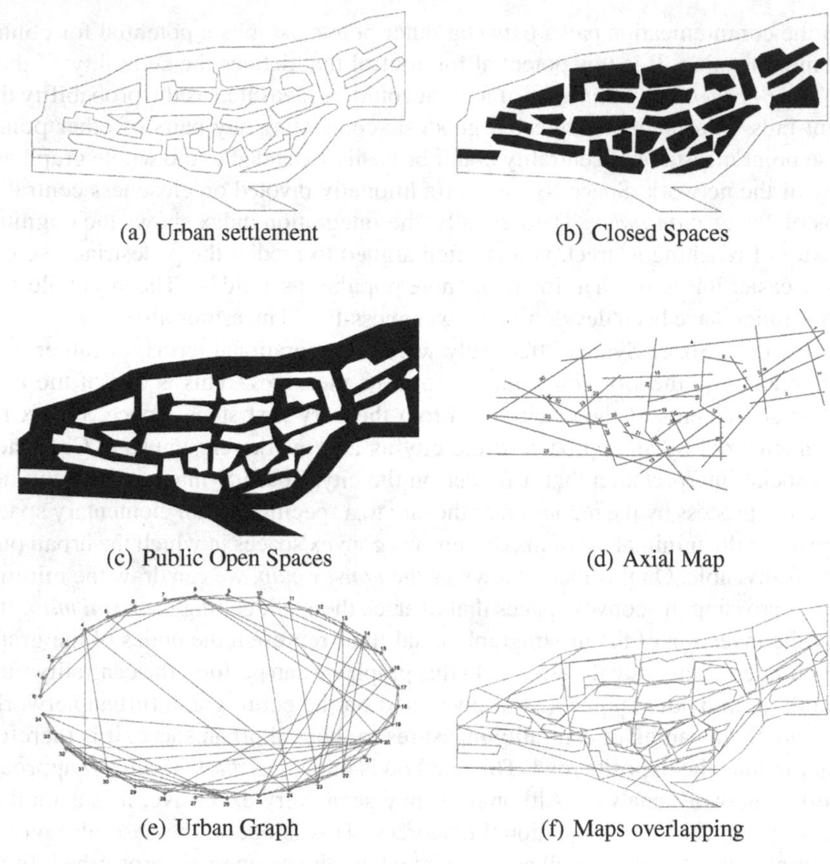

Fig. 2. Space Syntax operational approach

3.3 Configurational Ontology

We are now in the position of define our Configurational Ontology. Considering the configurational indexes as previously described, and leaning them to a specific geographic location (i.e. a location on the network), the same will take on a local significance. In such a case, we give to the term "local" the meaning of "in relation to a specific context". Otherwise, in configurational analysis, the same term "local" refers to a method of index calculation. Specifically, it means to reduce the analysis radius to a limited number of topological steps or to a predetermined metric neighborhood, in order to explore the city at different scale levels[21].

On this assumption, considering the existence of a local urban element, formally described by the implemented ontological model,we can define the Configurational Ontology as:

$$CO = O(C^d, C^c, C^b)$$

Where O is the ontology as represented in our model, and C^x are the following network point centrality measures:

$$C^d = \sum_{i=1}^{n} a(p_i, p_k)[15]; \quad C^c = \frac{n-1}{\sum_{i=1}^{n} d(p_i, p_k)}[15]; \quad C^b = \sum_{i=1}^{n}\sum_{j>i}^{n}(\frac{g_{ij}(p_k)}{g_{ij}})[15]$$

where:

$$a(p_i, p_k) = \begin{cases} 1, & \text{if } p_i \text{ and} p_k \text{ are connected by a line} \\ 0, & \text{otherwise} \end{cases}$$

$d(p_i, p_k) =$ the number of edges in the geodesic linking p_i andp_k
$g_{ij} =$ the number of geodesics linking p_i andp_k
$g_{ij}(p_k) =$ the number of geodesics linking p_i andp_k that contain p_k

The above configurational indexes C^x are the topological attributes of the novel Configurational Ontology.

4 Case Study Example

In this section we show a case study about the creation of a Configurational Ontology. The first step of our framework is a pre-consensual phase during which we try to solve conceptual misunderstandings and semantic ambiguities and also to generate a precise and accurate description of our knowledge. In this context, the definition of a common and shared "language" is the basis in a formalization of knowledge on the topic of urban knowledge. Starting from the assumptions mentioned above, building a *glossary of terms* can be seen as a first step toward urban element interpretation, or rather as a first attempt to formalize a shared knowledge. The sharing process is one of the harder issues both in our context and, generally speaking, in the process of ontology formalization. Therefore, the choice of signifiers and meanings related to the urban element arises from the analysis of shared international documents and legal tools that directly or indirectly affect the city itself. In this way, we define all the elements that characterize a city, the relations between them, and their connotative and denotative meanings. Using the proposed methodology, all these components with their specializations (i.e., signifiers and meanings) have been arranged in domain glossary. In this first ontology implementation we use WordNet a well know general informative resource widely used in the knowledge engineering research community. All information in WordNet is organized using linguistic properties. The basic unit in WordNet is the synset, a logic set of words related by the synonymy property. Each synset is a concept in WordNet. All the synsets are related to the others by pointers that represent linguistic properties. Two kinds of relations are represented by pointers: lexical and semantic. Lexical relations hold between word forms; semantic relations hold between word meanings. Examples of those relations are hypernymy/hyponymy, antinomy, entailment, and meronymy/holonymy.

A sketch of the glossary is shown in table 2. This glossary shows a description about several concepts related to the *church* concept.

Table 2. A part of the domain ontology glossary

Terms	Gloss
building, edifice	a structure that has a roof and walls and stands more or less permanently in one place
house of worship	any building where congregations gather for prayer
church, church building	a place for public (especially Christian) worship;
cathedral	any large and important church
mosque, masjid, musjid	(Islam) a Muslim place of worship
apse, apsis	a domed or vaulted recess or projection on a building especially the east end of a church; usually contains the altar
vestry, sacristy	a room in a church where sacred vessels and vestments are kept or meetings are held

Using the proposed model we arrange all the terms of the glossary in an ontology implemented by means of a semantic network. The figure 3 shows the semantic networks and the related OWL description.

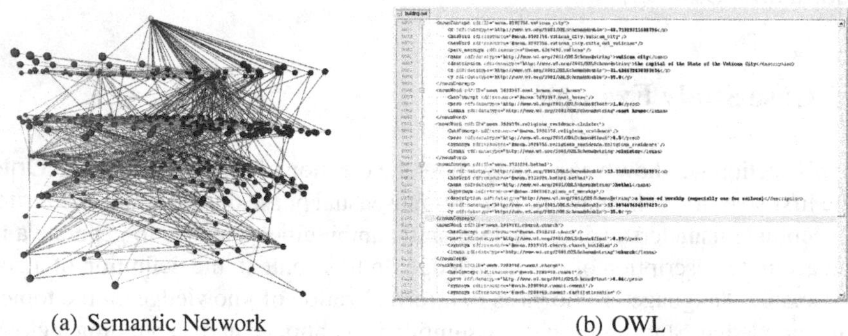

(a) Semantic Network (b) OWL

Fig. 3. Ontology Representations

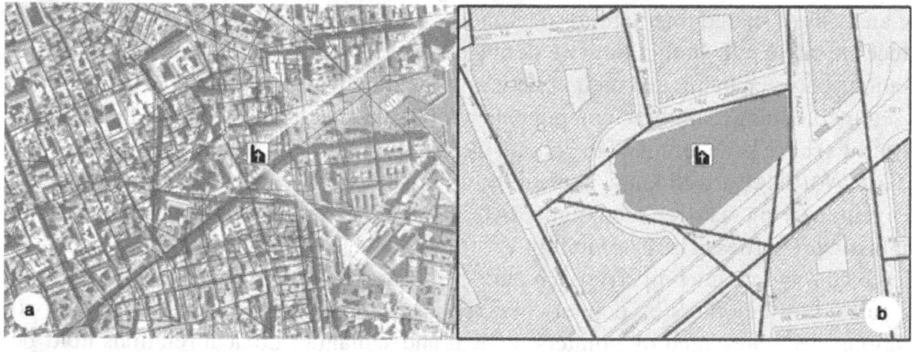

Fig. 4. A church in Space Syntax: San Pietro *ad Aram* on an orthoimage (a) , and a zoomed aerial photogrammetry (b), of Naples (IT), as background of the Axial Map (blue lines)

The figure 4 shows how Space Syntax models the urban neighborhood of the Church of San Pietro *ad Aram* in Naples, Italy. The axial map of the city is overlapped to an orthoimage (figure 4(a)) and then to an aerial photogrammetry at a narrower scale (figure 4(b)).

In the assumption of the existence of one of the urban element formally represented in the ontology we have implemented (i.e. Church), by the use of the equation $CO = O(C^d, C^c, C^b)$, as well as the topological values of its indexes (C^x), the specific Configurational Ontology is created.

5 Conclusions

In this paper a novel knowledge structure to represent urban environment has been proposed. Using an ontology based model and a configurational analysis tool we give an integrated representation of urban elements both from a conceptual point of view and from a topological one. Combining these information we formally describe knowledge about a city by means of a Configurational Ontology. A complete use case has been discussed to show a real implementation of our framework. Current research effort is regarding the use of other configurational indexes to take into account other topological information of urban networks, leading to more argumentative indications about the city. Moreover, we are implementing Configurational Ontology in a recommendation system to help city-users in the seeking of useful information about specific urban domains.

References

1. Albanese, M., Maresca, P., Picariello, A., Rinaldi, A.M.: Towards a multimedia ontology system: an approach using tao_xml. In: Proceedings of the 11th International Conference on Distributed Multimedia Systems (DMS 2005), pp. 52–57 (2005)
2. Albanese, M., Picariello, A., Rinaldi, A.M.: A semantic search engine for web information retrieval: an approach based on dynamic semantic networks. In: ACM SIGIR Semantic Web and Information Retrieval Workshop, SWIR 2004 (2004)
3. Batty, M.: The size, scale, and shape of cities. Science 319(5864), 769–771 (2008)
4. Bobrow, D.G., Winograd, T.A.: An overview of krl, a knowledge representation language. Technical report, Stanford, CA, USA (1976)
5. Brachman, R.J.: What's in a concept: Structural foundations for semantic nets. International Journal of Man-Machine Studies 9(2), 127–152 (1977)
6. Brachman, R.J.: On the epistemological status of semantic networks. In: Findler, N.V. (ed.) Associative Networks: Representation and Use of Knowledge by Computers, pp. 3–50. Academic Press, Orlando (1979)
7. Brachman, R.J., Schmolze, J.: An overview of the Kl-ONE knowledge representation system. Cognitive Science 9(2), 171–216 (1985)
8. Cataldo, A., Rinaldi, A.M.: An ontological approach to represent knowledge in territorial planning science. Computers, Environment and Urban Systems 34(2), 117–132 (2010)
9. Clancey, W.: The Knowledge Level Reinterpreted: Modelling Socio-Technical Systems. International Journal of Intelligent Systems 8, 33–49 (1993)
10. Cristani, M., Cuel, R.: A survey on ontology creation methodologies. International Journal on Semantic Web & Information Systems 1(2), 49–69 (2005)

11. Dean, M., Schreiber, G.: OWL Web Ontology Language Reference. Technical Report W3C (February 2004), http://www.w3.org/TR/2004/REC-owl-ref-20040210/
12. Denny, M.: Ontology tools survey. XML.com (2004) (revisited)
13. Euzenat, J., Shvaiko, P.: Ontology Matching. Springer, Heidelberg (2007)
14. Fox, M.S., Wright, J.M., Adam, D.: Experiences with srl: an analysis of frame-based knowledge representations. In: Proceedings from the First International Workshop on Expert Database Systems, Redwood City, CA, USA, pp. 161–172. Benjamin-Cummings Publishing Co., Inc. (1986)
15. Freeman, L.: Centrality in social networks conceptual clarification. Social Networks (1), 215–239 (1978)
16. Gaines, B.: Modeling as Framework for Knowledge Acquisition Methodologies and Tools. International Journal of Intelligent Systems 8, 155–168 (1993)
17. Genesereth, M.R., Nilsson, N.J.: Logical Foundation of Artificial Intelligence. Morgan Kaufmann, Los Altos (1987)
18. Gruber, T.R.: Toward principles for the design of ontologies used for knowledge sharing. International Journal of Human-Computer Studies, 907–928 (1993)
19. Gruber, T.R.: A translation approach to portable ontology specifications. Knowl. Acquis. 5(2), 199–220 (1993)
20. Guarino, N.: The ontological level. In: Casati, R., Smith, B.B., White, G. (eds.) Philosophy and the Cognitive Sciences. Holder-Pichler-Tempsky, Vienna (1994)
21. Hillier, B.: Spatial sustainability in cities. organic patterns and sustainable forms. In: Proceedings of the 7th International Space Syntax Symposium, Stockholm, Sweden, pp. K01:1–K01:20. KTH (2009)
22. Hillier, B.: Space is the machine. Cambridge University Press, United Kingdom (1996)
23. Hillier, B., Hanson, J.: The social logic of space. Cambridge University Press, United Kingdom (1984)
24. Minsky, M.: A framework for representing knowledge. Technical report, Cambridge, MA, USA (1974)
25. Neches, R., Fikes, R., Finin, T., Gruber, T., Patil, R., Senator, T., Swartout, W.R.: Enabling technology for knowledge sharing. AI Mag. 12(3), 36–56 (1991)
26. Nutter, J.T.: Epistemology. In: Shapiro, S. (ed.) Encyclopedia of Artificial Intelligence. John WyleyS (1998)
27. Porta, S., Crucitti, P., Latora, V.: The urban network analysis of urban streets: A primal approach. Environment and Planning B: Planning and Design 33(5), 705–725 (2005)
28. Porta, S., Crucitti, P., Latora, V.: The network analysis of urban streets: A dual approach. Physica A: Statistical Mechanics and its Applications 369(2), 853–866 (2006)
29. Ratti, C.: Urban texture and space syntax: some inconsistencies. Environment and Planning B: Planning and Design 31(4), 487–499 (2004)
30. Rinaldi, A.M.: A content-based approach for document representation and retrieval. In: Proceedings of the Eighth ACM Symposium on Document Engineering, DocEng 2008, pp. 106–109. ACM, New York (2008)
31. Russell, S., Norvig, P.: Artificial Intelligence: A Modern Approach, 2nd edn. Prentice-Hall (2003)
32. Schreiber, G., Wielinga, B., Breuker, J.: KADS: A Principled Approach to Knowledge-Based System Development. Academic Press, London (1993)
33. Strano, E., Viana, M., da Fontoura Costa, L., Cardillo, A., Porta, S., Latora, V.: Urban street networks, a comparative analysis of ten european cities. Environment and Planning B: Planning and Design 40(6), 1071–1086 (2013)
34. van Assem, M., Gangemi, A., Schreiber, G.: RDF/OWL Representation of WordNet. W3C Working Draft (2006)
35. Vaughan, L.: The spatial syntax of urban segregation. Progress in Planning (67), 205–294 (2007)

If Appleseed Had an Open Portal: Making Sense of Data, SEIS and Integrated Systems for the Maltese Islands

Saviour Formosa

Department of Criminology, Faculty for Social Wellbeing, University of Malta, Humanities B, Msida MSD 2080
saviour.formosa@um.edu.mt

Abstract. Much sought and realistically distant, an open data system can serve as the Holy Grail for many a policy-maker and decision taker as well as the operational entities involved in the field. The steady seeding of data-related legislative tools has aided the setting up of exploratory and active systems that serve the concept of data-information-knowledge-action to academia, the general public and the implementing agencies. Legislation, inclusive of Data Protection, Freedom of Information, Public Sector Information, Aarhus, INSPIRE, SEIS and the still embryonic SENSE, have all managed to create a new reality that may be too complex for some still caught in a jurassic analogue stage where data hoarding might still be prevalent and little effort is made to jump to the post-modern reality. Efforts to push the process through various domains such as census, environment protection, spatial development and crime have helped the Maltese Islands to create a scenario that is ripe for a national data infrastructure, inter-entity data exchange, open data structuring, and free dissemination services. This process enhances the knowledge-base and reduces redundancy, whilst creating new challenges on how to make sense of all the data being made available, particularly in the interpretation or misinterpretation of the outputs. The paper reviews Malta's process to go through the birth pains of SEIS as an open data construct, through to the dissemination of various spatial datasets and the first open portals pertaining to the various regulatory directives.

Keywords: open data, Aarhus, SEIS, INSPIRE, Malta, data interoperability, geoportal, LIDAR, spatial data, integration.

1 The Long and Winding Road

1.1 Johnny Appleseed's Legacy

Access to data posited many a dilemma for systems integration and dissemination. The transitional process from data to information to knowledge to action has been tackled from different perspectives, ranging from policymaking, through impact assessments to decision making exercises and recently to the integration of disparate datasets within integrated systems and eventually ported to the web for dissemination purposes. Each sector can be taken as a research topic in isolation, however the main

B. Murgante et al. (Eds.): ICCSA 2014, Part II, LNCS 8580, pp. 709–722, 2014.

fulcrum of the process revolves around the creation of a framework of policies and technologies that enable the exchange of spatial data across the different thematic and technological domains. This led to the establishment of a series of data-management processes aimed at setting-up and maintaining information resources structures through Spatial Data Infrastructures (SDIs) with early investigative work on conceptualization and international initiatives by Masser and Craglia [1] [2] [3] [4]. The drive was enhanced with inter-organisational studies by Nedovic-Budic and Pinto [5] as well as the work of the individual persons who pushed the initiative [6] and the eventual creation of an established SDI framework [7] that was also taken up at international level by the Global Spatial Data Infrastructure Association [8] and at national level [9]. Not exclusively anchored to the generic data management disciplines, this process nonetheless finds broad scope in this field particularly due to availability of specialised tools employed in environmental monitoring and reporting.

This paper reviews the mythical Appleseed one-core-at-a-time process employed in the implementation of a shared environmental information system (SEIS) [10] for the Maltese Islands which emulated the seeding with sequential implementation measures. Through the implementation of an ERDF [11] project entitled "Developing National Environmental Monitoring Infrastructure and Capacity", Malta embarked on a process that points towards the implementation of an open data structure, with a main output being the delivery of a SEIS geoportal. The final output, based on a specific target to create a SEIS, based on the environmental themes of air, water, noise, radiation, soil and marine [12] [13] resulted in a comprehensive innovative system that serves as an initial launching pad for open data [14]. The steps taken outline a description of the basic data definitions, the legislative mechanisms, the international-reporting requirements, the tools available and projects that tackle the means to reach an open data construct.

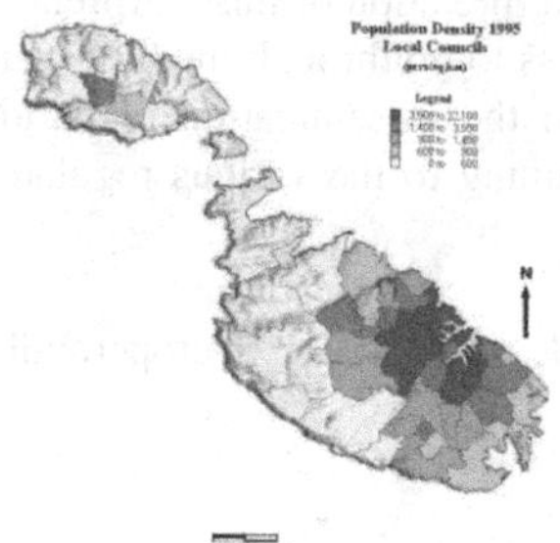

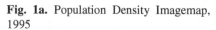

Fig. 1a. Population Density Imagemap, 1995

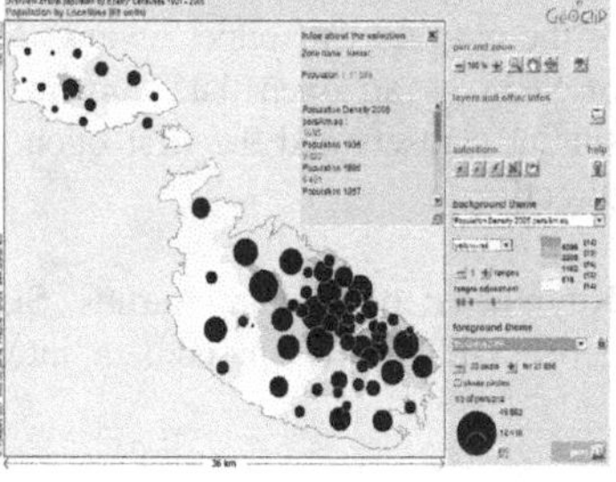

Fig. 1b. Census Interactive Map, 2005

Initial attempts to provide for an open portal [15] [16] [17], albeit limited by technology and/or lack of regulatory tools served to investigate user access and usage issues, with some basic Imagemap/GIS-client [18] (Figure 1a) and Interactive GIS [19] (Figure 1b) deliverables. The activities were based on the transposition of international directives, inclusive of the Data Protection Directive [20], the Public

Sector Information Directive [21], the Aarhus Directive [22], the INSPIRE Directive [23], as well as the national initiative pertaining to the Freedom of Information Act [24].

1.2 Fertile Fields

A review of the main agencies involved in the spatial data exercise show that there exist a wide range of disparate datasets that are either non-conformant with INSPIRE or do not fall under the legislative tools mentioned above. The agencies involved include the main IT agency MITA (responsible for INSPIRE), MEPA (responsible for development planning and the environment, landuse/land cover, GMES, Copernicus, GEO, GEOSS, Aarhus and EEA-related reporting) and other entities.

A number of government departments and entities make use of the base map owned by MEPA although their GIS operating architecture differs from one to another, inclusive of the Agriculture Department, Fisheries Department, Land Registry, utilities such as Enemalta Corporation, the Water Services Corporation, cable and telephony companies (private), the Malta Resources Authority and the Transport Malta. Other structures include defense and civil protection, as well as a number of other departments, corporations and authorities that own data in various structures and which still need to conform to internationally recognized data standards. The INSPIRE Directive provides the framework for this structure but an NSDI (National Spatial Data Infrastructure) would ensure the integrative processing required for such a system. The Maltese Islands have embarked and delivered on the pilot domains emanating from the ERDF project [11] and developed a SEIS [25] in order to ensure conformity for monitoring and reporting.

1.3 Trying to Make SENSE

SEIS is not the first or only open-data conveyor for data, with such precursors including ENPI-SEIS (environmental protection project focusing on networking and open access through free tools) and a parallel project entitled ICT-ENSURE aimed at the management and dissemination of environmental information within a single informational infrastructure entitled SISE. Other initiatives include SEIS-BASIS (database structure on environmental monitoring programmes), NESIS (EEA-related state of play at national level), TESS (decision-making functional system), HUMBOLDT (aimed at the implementation of a European Spatial Data Infrastructure (ESDI), LENVIS (management systems for environment and health) and ORCHESTRA (risk-based management system for disaster reduction).

Spatially-targeted activities have also been implemented or are in such a phase, amongst which one can find the Copernicus Land Monitoring Services, the EEA's systems inclusive of the CLC runs (land-cover analysis), NATURA 2000, LUCAS and CCDA projects (protection zones), PLAN4ALL (landuse planning), GEO (earth observation), GMES (monitoring and environmental security), GEOSS, (system of systems), and GENESIS (synergic exercise between INSPIRE and the previous initiatives listed above that ensure integration of information in line with the single

system envisaged in ICT-ENSURE). A lacunae identification initiative (GIGAS) was aimed to look at the gaps between these systems and also to point entities towards the requirements of a systems-approach data gathering structure (SANY) where sensors on the ground can gather information in real-time in a cohesive whole [12] [13].
SEIS established itself as a mainstay for such projects through its location-based services that bring together spatial, social and physical domains within a place-based structure. Such is made possible through its WMS, WFS, WCS and other services. The parallel SEIS-development, entitled SENSE aims to enable the sharing of European and national state of the environment and that allows for cross-country selection and support to SOE Online, the latter targeted to create a forum for the state of the environment.

1.4 The Maltese Initiative

Introducing a state to high-end information systems that encompass total national coverage is no mean task, even for such a small state as Malta with its 316 square kilometer area. Introducing a new paradigm in data creation and dissemination targeting spatial analysis points to a whole new reality [26] [27]. The Maltese Islands, through access to the European Regional Development Fund, managed to create a process aimed at environmental research that included innovative tool creation and scans that will help analysts to monitor the environment and related offences committed on the environment. In a process initiated by the author in 2006 and concluded in 2014, the SEIS-based activity resulted in the creation of fundamental datasets that also bring Maltese terrestrial and bathymetric baseline information to the public domain [25]. These activities have been carried out as part of a €4.6 million project, entitled Developing National Environmental Monitoring Infrastructure and Capacity, which also entailed the monitoring of air, water, soil, radiation, noise and marine themes [11]. This project was co-financed by the European Regional Development Fund, which provides 85% of the project's funding and the Government of Malta, which finances the rest under Operational Programme 1 - Cohesion Policy 2007-2013 - Investing in Competitiveness for a Better Quality of Life. Involving international experts from a number of countries and expert input from JRC, EEA, EC and other entities, whilst at a national scale, implementing partners included MEPA as project leader, the University of Malta, MRA, NSO and the Environmental Health Directorate.

SEIS in the Maltese Islands was based on a three-pronged approach; the alignment of its environmental structures to the varied legislative tools, the creation of integrated systems and the design of a reporting infrastructure. The main remit was to ensure such through the take-up of the SEIS initiative, where in 2008, the EU Commission published a Communication (COM (2008) 46 Final) "Towards a Shared Environmental Information System", which sets out an approach to modernise and simplify the collection, exchange and use of the data and information required for the design and implementation of environmental policy, according to which the current, mostly centralised systems for reporting are progressively replaced by systems based on access, sharing and interoperability. The overall aim was to maintain and improve the quality and availability of information required for environmental policy, in line with better regulation, while keeping the associated administrative burdens to a minimum.

Malta took part in the development of the SEIS, both at EU and national levels. The development of the national component of this system is particularly important for Malta, because it would streamline and simplify reporting processes to the EU – an essential consideration for a relatively small national administration, which nonetheless has the same reporting requirements as much larger countries. But the benefits of the SEIS for Malta are not limited to improved reporting procedures to the EU. At the national level, the system would simplify, reduce costs, and increase effectiveness at all stages of environmental data cycle. This, in turn, would translate into more and better quality information being available for a variety of purposes at a considerably lesser cost than is the case at present.

The project aimed to develop the Malta component of the Shared Environmental Information System at a time when MEPA's geoportal (mapserver) was not deemed to be a comprehensive environmental information system. This project was tasked with an analysis of the current systems in place to process environmental monitoring data and data flows required, the design of the SEIS for Malta, and the development of such a web-based environmental information system. The project had to result in the creation of a web-based environmental information system, on the basis of existing platforms, as well as on the basis of any other additional platforms and components that may be required, to achieve full interoperability and functionality of the Maltese component of the SEIS, in line with the applicable guidelines and best practices in this field. The deliveries had to include a web-based GIS dedicated to environmental monitoring data incorporating MEPA's aerial orthophotos and basemaps available at the start of the project in 2010, as well as newly acquired satellite imagery, oblique aerial imagery, LiDAR terrain datasets and bathymetric data acquired through the ERDF project, of which the SEIS component was a part. Moreover, the SEIS had to be developed using an ArcGIS Server platform, based on system migration from an ArcInfo database to an ArcGIS geodatabase structure. This issue was set out due to the perceived need to fit such a system within the organisation's requirements at the time, which in turn could have also resulted in its main limitation, due to cross-system incompatibility as against a full-open structure. Also, the tendering process as such, limited the possibilities for alternate and innovative developments. One main issue concerned the need for the SEIS to be a modular and scalable system which is flexible to meet the varying demands of usage and applications over time.

The deliverables [11] were structured in a phased approach that sought to actuate:

- A review report on all requirements and parameters for the development and operation of the Maltese component of the SEIS and a proposal for the design and development of the SEIS;
- A report on the proposed ArcGIS geodatabase design for the SEIS based on an ArcGIS server architecture;

- A prototype and pilot of the SEIS implemented and tested;
- A final version of the customised SEIS with a dedicated geoportal implemented and put into operation following feedback on the previous phase.

2 They Came Before

The ERDF project SEIS component was one in a series of initiatives that set the stage for this encompassing system. With initiatives such as the Census web-mapping project [28], the National Protection Inventory [29] and the SEIS-precursor Ambjent project [16], the process entailed the move from an image-mapping system (Figure 2a) to an early interactive prototype system. Such was based on the creation of spatial entities and attribute designations that were integrated with digitised card material (Figure 2b), integrated with pseudo-3d graphical interfaces (Figure 2c) and eventually to dynamic query systems (Figure 2d).

The dissemination technologies available at the time were used as surrogates towards this advanced system with data integration proposed through accessibility made possible by Image-Maps and map-server options. The resultant information system was envisaged to deliver a layered approach where users could access data that is available in an immersive clickable scenario through direct linking to spatial entities (points, lines, areas).

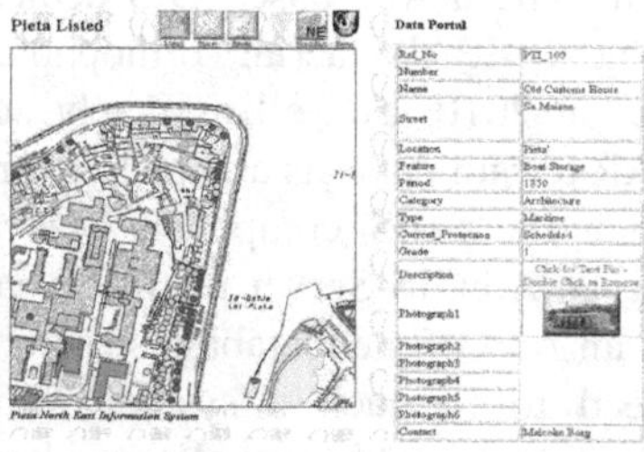

Fig. 2a. Imagemap 1996

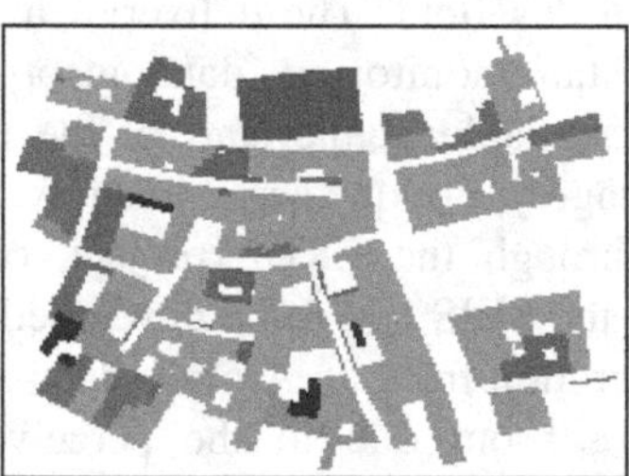

Fig. 2b. GIS layer

Fig. 2c. 3D extrusion

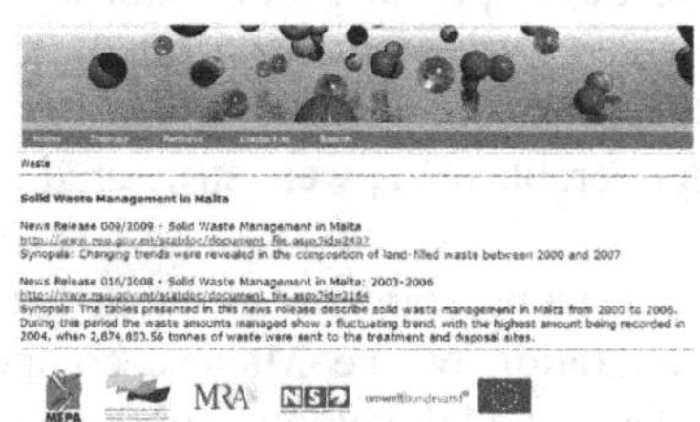

Fig. 2d. Query Interface

In addition, the system would incorporate links to multimedia, imagery, walkthroughs, thematic data and access to a dynamic array of live information systems. The case was the same for the subsequent Census mapping exercises, the MEPA mapserver (Figure 3a), the Plan4All geoserver (Figure 3b), amongst others, however few had yet to envisage a system as proposed by SEIS, which was only made possible through the foundation laying of the implementation rules laid out by INSPIRE, Aarhus and the SEIS initiative.

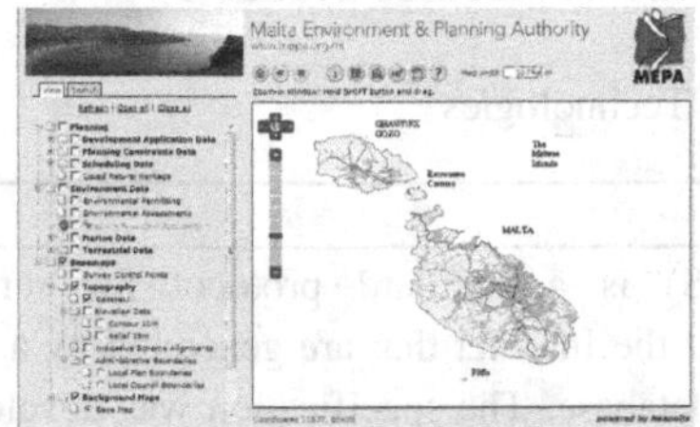

Fig. 3a. MEPA mapserver

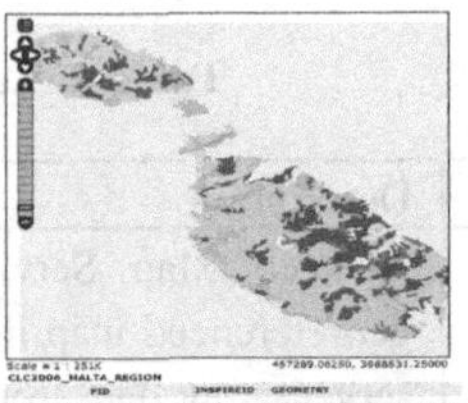

Fig. 3b. Plan4All geoserver

3 Implementing SEIS

3.1 A 6-Stepped Approach

With Malta being one of the first countries to initiate SEIS implementation, the aim was one to deliver information interoperability. It aimed to upgrade the methods employed to gather data, to streamline the reporting processes, to introduce implementation rules and to create a spatial data infrastructure as well as launch a visualisation and dissemination tool. The main aim was to develop data management and ingestion systems, allow for online data editing options, allow for data export and metadata viewing. In addition the system had to ensure that the country did not need to reinvent the wheel every time an information packet is requested from the EU and international conventions but would develop a search tool for the metadata and in turn enable automated reporting processes as required, particularly the European Environment Agency priority dataflows, always in line with INSPIRE, Aarhus and other legislative tools.

A 6-Step methodology was adopted [12] [13] to ensure the base setting for the SEIS:

1. Analysis of the target Data Model (INSPIRE Data Specifications and EEA reporting schemas)
2. Analysis of the Source Data (MEPA)
3. Conceptual design of the geodatabase
    a. to include all the INSPIRE elements for which a correspondence with the source data has-been found

   b. to include all the additional elements not existing in the INSPIRE data model but present in the source data
   c. to include the INSPIRE elements not existing in the source data
   d. to include all the elements existing in the EEA reporting schemas
4. Preparation and filling-in of the matching table (MT)
5. Creation of the geodatabase structure, using different tools, according to the theme concerned
6. Import of the geodatabase in SQL Server (provided also an ESRI geodatabase for each theme, as an additional resource available).

Table 1. Standards & Technologies

Standard	Description
OGC WMS	A Web Map Service (WMS) is a standard protocol for serving georeferenced map images over the Internet that are generated by a map server using data from a GIS database. The specification was developed and first published by the Open Geospatial Consortium in 1999.
OGC WMS - T	A WMS server can provide support to temporal requests. This is done by providing a TIME parameter with a time value in the request. WMS specifies that the basic format used for TIME requests is based on the ISO 8601:1988(E) "extended" format.
OGC WFS	The Open Geospatial Consortium Web Feature Service Interface Standard (WFS) provides an interface allowing requests for geographical features across the web using platform-independent calls.
OGC WCS	The Open Geospatial Consortium Web Coverage Service Interface Standard (WCS) provides an interface allowing requests for geographical coverages across the web using platform-independent calls.
ANSI SQL	The geodatabase will follow the ANSI/ISO SQL specifications
INSPIRE	INSPIRE is "an EU initiative to establish an infrastructure for spatial information in Europe that will help to make spatial or geographical information more accessible and interoperable for a wide range of purposes supporting sustainable development".
Z39.50	Z39.50 is a client–server protocol for searching and retrieving information from remote computer databases. It is covered by ANSI/NISO standard Z39.50, and ISO standard 23950. The standard's maintenance agency is the Library of Congress. Z39.50 is widely used in library environments and is often incorporated into integrated library systems and personal bibliographic reference software. Interlibrary catalogue searches for interlibrary loan are often implemented with Z39.50 queries.
CSW	The OGC Catalog Service defines common interfaces to discover, browse, and query metadata about data, services, and other potential resources. Web Catalog Service includes several profiles including Catalog Service - Web.

Source: Bonozountas, M., and Karampourniotis, I., (2013), p. 15

The resultant system had to deliver a SEIS portal that conformed to international standards, conditions and technologies set out by the same legislative and working documents described earlier. Table 1 describes the standards and technologies identified for the SEIS-Malta portal [13].

3.2 The Resultant Interface

Learning from the outcomes of the precursor exercises, particularly the Plan4All project, which had indicated that it was sometimes difficult to bring together the different datasets across the different thematic social and physical fields and required stringent rules for inter-operability, the SEIS project learned from the need to 'listen' to the outcomes of conceptual models that served as a veritable exercise in comprehensiveness due to their holistic and detailed approach. The main issues in the Maltese context deal with the fact that the conceptual models reflect their name: they are concepts that require tweaking and need to consider different levels of conformity: local-national (NUTS 2,3,4,5 as compared to NUTS 1) and national-super-national (NUTS 1 as compared to EU). The CLC1990-2000-2006 runs proved that this can be done if one uses a harmonisation of the top-down (model) and bottom-up approach (users-data creation), whilst remaining loyal to the legislative requirements.

Based on a GeoNetwork Open Source (GNOS) approach, the Malta SEIS webportal delivered various services that went beyond the precursors of ambjent.org.mt and the census webmaps and also beyond the development of the Plan4All geoserver. The new seismalta.org.mt portal was resultant of the ERDF project.

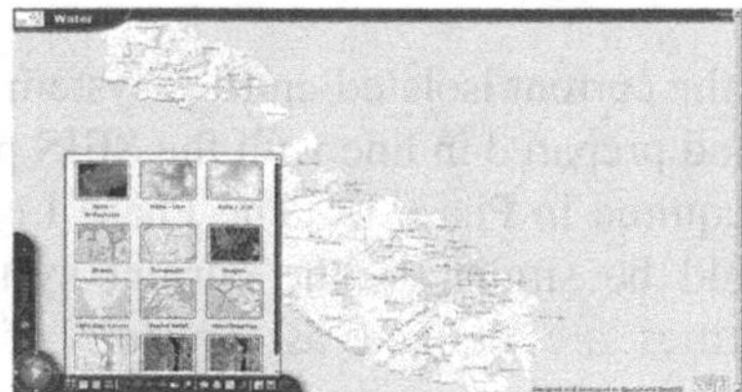

Fig. 4a. Basemapping

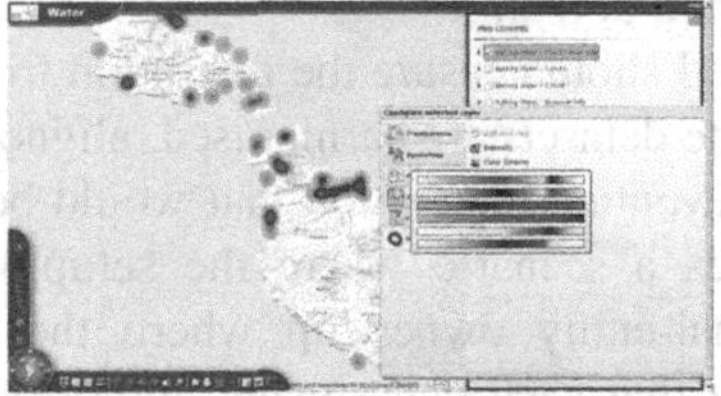

Fig. 4b. Heatmaps

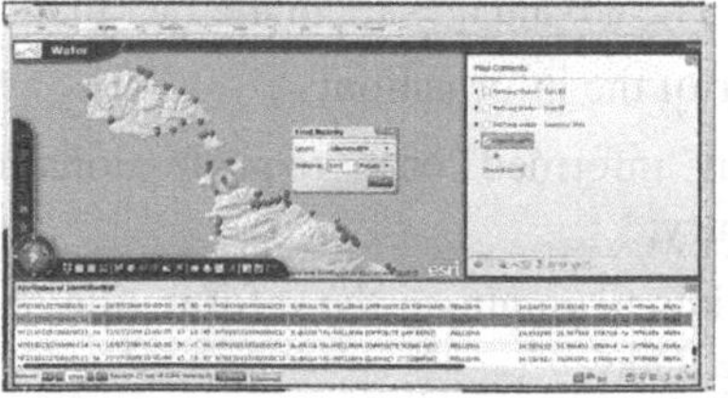

Fig. 4c. Data outputs

Fig. 4d. 3D viewer

SEISmalta.org.mt offers a veritable plethora of services, made possible through the ERDF geoportal, which requires the ArcGIS Silverlight API, employing visual tools, though the latter has served many a criticism by users who do not wish to install the tool or use other platforms, something that needs to be rectified over the updated versions of the portal, especially if it needs to be full open. The services cover: Download Services (Figure 4a), Ingestion Service, a Data Quality Service, a Sensor Observation Service for real-time data input, as well as a Feedback Service to the geoportal administering agency. In addition, the SEIS output includes a reporting Notification Service, a Registry Service for new databases, a Reporting Service, which service queries the database and prepares reporting outputs for EC and national requirements' consumption. Figure 4 depicts the Seismalta portal's basemaps (Figure 4a), thematic heatmaps (Figure 4b), data portal (Figure 4c) and a 3D topographic viewer (Figure 4d).

3.3 Post-SEIS

Post-SEIS project conclusion, the entities are in the process of establishing a wider national inter-agency approach, where the main tenet is based on the underlying strategy for data management built on a 'gather-once / use-many' approach. Such ensures that data is gathered once but used by all without incurring further costs, access and implementation bottlenecks, whilst at the same time employing one tool for dissemination services through an enhancement of the SEIS portal. This proposal looks at the setting up of an organisation through a two-phased approach where an entity is tasked with implementing short-term targets, such as the creation of a SEIS-base-data structure for all entities and in the long-term tasked with the integration of all these systems into one entity with dedicated thematic expertise across the diverse GI-enabled domains.

Phase I should ensure the migration from the current isolated-entities system to one where the datasets are harmonised, aligned and prepared in line with the SEIS process for the eventual integration that would be required in Phase II. The Phase I concept envisages a scenario where the setup would be similar to the current system of individual-entity ownership where the entities are defined as "owners of data" meaning that each department, authority, corporation or organisation is responsible for collecting, maintaining and managing data relevant to the running of its activities and operations. This data will be shared with other entities in an open mode and free disseminated through the SEIS-based tools. The advantages lie in the fact that:

- the data is maintained by the owner of the information;
- updating of the system is done in an "informed" or more professional manner rather than straight forward data entry;
- the organisation itself and its officials maintaining the information are made responsible and accountable for the data;
- this system also allows the other entities to create their value-added data on to the same datasets which the 'guardian' entity can then decide to implement as part of that dataset.

It is imperative that each dataset has to comply with legislative implementation rules, even for those that do not fall under the diverse Directives. It is vital that the data inputted in the system, once the necessary data collection exercise is carried out, will be almost completely error free. Thus, it is of the utmost importance, that the project is set in the right perspective and that there are clear guidelines and standards to which all participants within the system would have to abide by. This at a time when another MED project entitled HOMER [30] is specifically focusing on the Open Data theme, as based on the Public Sector Information Directive and its update. In addition, new issues are cropping up with the emergent EU level e-Reporting systems ensconced within the Structured Implementation and Information Framework (SIIF) concept [31]. Such a situation calls for an interesting development that places the Islands at an advantage for takeup through the enhancement of the SEIS geoportal into a wider thematic construct, going beyond environmental domains into highly integrated societal systems.

3.4 Tasting the Apples

The project's trust serves its main purpose only if its functionality is translated into tangible outcomes and usability. The SEIS output has been both augured and criticised by users in terms of ease of use and requirements to install additional tools due to browser constraints. The main users were professionals in the field and students who regularly reviewed the site outputs for their research studies. Non-governmental organisations welcomed the initiative though highlighted the issue that now that data was being made accessible, it was difficult to interpret without expert input. The latter issue is interesting since it posits a state of affairs that users may not be willing to take up new technologies beyond their wow-factor, that data and especially open-data can be overwhelming due to its large volumes, that users find themselves lost in receipt of data even when supplied with lineages and all metadata. Interestingly, when challenged with the fact that data is shared by all and that it is available in real or quasi-real time, users showed both disbelief and worry, ironically due to the fact that they must now criticise themes based on scientific facts as against opinion or second-hand comments from reports; NGOs and experts alike now have the tools and the data to reach informed opinions on their say and offer data-backed feedback for social-change initiatives. This process also serves to increase the number of researchers who were previously holding back due to access issues.

Having sowed the initial SEIS portal in the Maltese Islands, it is time to compare and contrast the system with the new developments in SEIS coming out from the EEA, the Austrian Umweltbundesamt and similar initiatives.

4 Conclusion

In conclusion, the Malta SEIS geoportal depicts an integrated system based on a geodatabase that includes those INSPIRE elements where source data has been found in conjunction with other elements that were not required by INSPIRE but were available within the source dataset. The basic requirements emanating from the ERDF project requirements were satisfied, whilst additional services as yet not possible due to space and bandwidth restrictions have been provided through alternative measures such as physical pickup of the data that measures at 600Gb and counting.

In reviewing the process to establish a framework for the development of the system in Malta, the project established various factors, mainly on the potential uses of such a system, the need for user consideration and feedback as well as the need to ensure that there is conformity to the regulations that guide such developments.

With limitations imposed by the same procedural process as outlined in this paper, such a project would have overcome benefitted from more 'openness' on systems choice, creation of various parallel tools for comparative analysis of the outputs and a critical approach to similar systems under development in other countries. The latter, though entering the scene late in the day, would have co-benefitted from the successes and pitfalls of the Malta SEIS.

However, the SDI concept and its SEIS initiate is a phenomenon that will not go away, as the cores have been planted and the roots established.

References

1. Masser, I.: All shapes and sizes: the first generation of national spatial data infrastructures. International Journal of Geographical Information Science 13(1), 67–84 (1999)
2. Masser, I.: GIS Worlds - Creating Spatial Data Infrastructures. Urisa Journal, 1–3 (2005)
3. Masser, I.: Some Priorities for SDI Related Research. In: Proceeding of the FIG Working Week 2005 and GSDI-8: From Pharaohs to Geoinformatics, Cairo, Egypt, April 16-21 (2005)
4. Craglia, M., Masser, I.: Geographic information and the enlargement of the European Union: Four national case studies. URISA Journal 14(2), 43–52 (2002)
5. Nedovic-Budic, Z., Pinto, J.K.: Information sharing in an interorganizational GIS environment. Environment and Planning B: Planning and Design 27(3), 455–474 (2000)
6. Craig, W.J.: White Knights of Spatial Data Infrastructure: The Role and Motivation of Key Individuals. Urisa Journal 16(2), 5–13 (2005)
7. Janssen, K., Dumortier, J.: Legal Framework for a European Union Spatial Data Infrastructure: Uncrossing the Wires. In: Research and Theory in Advancing Spatial Data Infrastructure Concepts. ESRI Press, Redlands California (2007)
8. Douglas, G., Nebert, D. (eds.): The SDI Cookbook, Version 2.0 (Technical Working Group Chair, Global Spatial Data Infrastructure (GSDI) (2004), `http://www.gsdi.org/` (accessed on April 14, 2014)
9. Farrugia, A.: Implications of EU Accession on Environmental Spatial Data: a Malta Case Study, unpublished MSc GIS Science dissertation. Manchester Metropolitan University, Manchester (2006)
10. European Commission: Shared Environment Information System, `http://ec.europa.eu/environment/seis/` (accessed on March 10, 2014)
11. Malta Environment & Planning Authority: Developing National Environmental Monitoring Infrastructure and Capacity, MEPA, Floriana, Malta (2009)
12. Bonozountas, M., Karampourniotis, I.: MALTA-SEIS: Deliverable D2.1Report of Analysis and Detailed Proposal for SEIS, CT3067/2010 – 02, Malta (2012)

13. Bonozountas, M., Karampourniotis, I.: MALTA-SEIS, Report of Analysis and Detailed Proposal for SEIS, CT3067/2010 – 02 D2_31-01-2, Marousi, Greece (2013)
14. MEPA: Service Tender for the Design of the Shared Environmental Information System (SEIS) and development of a web-based GIS interface, ERDF 156: Developing national environmental monitoring infrastructure and capacity 3067/2010, Floriana (2010)
15. MEPA: MEPA mapserver, `http://www.mepa.org.mt/Planning/index.htm?MapServer.htm&1` (accessed on February 29, 2014)
16. MEPA: Further Institution Building in the Environment Sector, `http://www.ambjent.org.mt` (accessed on March 15, 2014)
17. Formosa, S., Magri, V., Neuschmid, J., Schrenk, M.: Sharing integrated spatial and thematic data: the CRISOLA case for Malta and the European project Plan4all process. Future Internet 2011 3(4), 344–361 (2011)
18. NSO: Census of Population and Housing (1995), `http://www.mepa.org.mt/Census/index.htm` (accessed on February 29, 2014)
19. NSO: Census of Population and Housing (2005), `http://www.mepa.org.mt/index.htm?links.htm&1` (accessed on February 29, 2014)
20. Official Journal of the European Union: "Directive 95/46/EC of the European Parliament and of the Council of 24 October 1995 on the protection of individuals with regard to the processing of personal data and the free movement of data," L 281, 23/11/1995, (24 October 1995)
21. Official Journal of the European Union: Directive 2003/98/EC of the European Parliament and of the Council of 17 November 2003 on the re-use of public sector information, L 345, 31/12/2003 (November 17, 2003)
22. Official Journal of the European Union: Directive 2003/4/EC of the European Parliament and of the Council of 28 January 2003 on Public access to environmental information and repealing Council Directive 90/313/EEC (Aarhus), L 041, 14/02/2003, pp. 0026–0032 (January 28, 2003)
23. Official Journal of the European Union: Directive 2007/2/EC of the European Parliament and of the Council of 14 March 2007 establishing an Infrastructure for Spatial Information in the European Community (INSPIRE), L108, vol. 50 (April 25, 2007)
24. Government of Malta: Freedom of Information Act, CAP496, Valletta, Malta (September 1, 2012)
25. MEPA: Malta SEIS, `http://www.seismalta.org.mt` (accessed on February 29, 2014)
26. Formosa, S., Sciberras, E., Formosa Pace, J.: Taking the Leap: From Disparate Data to a Fully Interactive SEIS for the Maltese Islands. In: Murgante, B., Gervasi, O., Misra, S., Nedjah, N., Rocha, A.M.A.C., Taniar, D., Apduhan, B.O. (eds.) ICCSA 2012, Part II. LNCS, vol. 7334, pp. 609–623. Springer, Heidelberg (2012)
27. Martirano, G., Bonazountas, M., Formosa, S., Nolle, M., Sciberras, E., Vinci, F.: INSPIRE EF Data Specifications to develop the SEIS-Malta Geodatabase for the Air Quality data management. In: INSPIRE Conference, Istanbul, Turkey (June 2012)
28. Formosa, S.: Coming of Age: Investigating the Conception of a Census Web-Mapping Service for the Maltese Islands. Unpublished MSc thesis Geographical Information Systems. University of Huddersfield, United Kingdom (2000), `http://www.tcnseurope.org/census/1995/index.htm` (accessed on February 29, 2012)

29. Borg, M., Formosa, S.: The Malta NPI project: Developing a fully-accessible information system. In: Wise, S., Craglia, M. (eds.) GIS and Evidence-Based Policy Making, Innovations 10. Taylor & Francis, Boca Raton (2008)
30. HOMER: HOMER MED project, http://www.homerproject.eu (accessed on March 15, 2014)
31. European Commission: Communication on Implementing EU environment legislation – Questions & Answers,
http://europa.eu/rapid/press-release_MEMO-12-159_en.htm
(accessed on March 15, 2014)

Involving Citizens in Public Space Regeneration: The Experience of "Garden in Motion"

Sara Lorusso[1], Michele Scioscia[2], Gerardo Sassano[2], Antonio Graziadei[2], Pasquale Passannante[3], Sara Bellarosa[3], Francesco Scaringi[3], and Beniamino Murgante[5]

[1] Giardino in Movimento Association
[2] WOP architettura e paesaggio
[3] NUR Association
[4] Basilicata 1799 Association
[5] School of Engineering, University of Basilicata

{lorusso.sara,sbellarosa}@gmail.com, {gsassano,mscioscia,agraziadei}@woplab.it, passannantep@yahoo.it, f.scaringi@me.com, beniamino.murgante@unibas.it

Abstract. The paper illustrates a Placemaking process developed in Potenza Municipality (Southern Italy), based on an interpretation of the theories by the French landscape architect Gilles Clément. A laboratory has been organized in a residual area of the city, famous for an architectural monument, the bridge designed by Sergio Musmeci. The Internet allows a continuous online storytelling of work, creating citizens engagement on projects or choices and producing creativity and knowledge circulation. In this perspective "Garden in Motion" initiative produced new important processes for the community life, just like in Gilles Clément's "Garden in motion", where the processes of nature are favoured and spontaneous plants put in condition to grow and move freely.

Keywords: Citizens Participation, Placemaking, Garden in Motion, Smart communities, Urban Space Regeneration.

1 Parks, Public Participation, Placemaking

In recent years protests frequently occurred in cities all over the world. Generally, symbols of such protests are parks, but behind parks, people claim an improvement of urban quality, more services, a greater involvement in decisions and more generally a better quality of life and a wider welfare.

Last year the international public opinion supported the protests in Turkey to save Gezi Park, one of the last small green spaces in Beyoğlu (Istanbul), which was threatened by a shopping centre project. During the latter days, new protests occurred in Turkey, which led to social networking ban. This suggests that even 2013 protests were directed to a request of democracy increase and Gezi Park was just a symbol, a way to gather more attention.

B. Murgante et al. (Eds.): ICCSA 2014, Part II, LNCS 8580, pp. 723–737, 2014.

Jacobs [5] observed that quality of life was closely linked to the ability of a community to self determine its conditions of everyday life. The greater is self-organization ability, the wider are the possibilities of producing social capital.

In order to support this kind of activities, Davidoff (1965) argues that the role of a planner should not only be limited to analyse social problems and try to propose possible solutions, but he should be a sort of "advocate" of categories that do not have enough power and financial resources, able to mediate between the plurality of community interests, in order to pursue the general interest.

The situation highlighted by Davidoff [3] fifty years ago is still current today; the representativeness of several groups of citizens continues to be a problem accentuated by increasing gap between citizens and institutions. This is manly due to the typical decision maker's behaviour, principally concentrated in relationships with self-referring groups with the only purpose to protect their interests rather than to listen to the community.

Great part of urban renewal programmes can reach, in the best cases, the medium level of Arnstein Ladder [1]; consequently citizens do not believe in public participation involvement or prefer other more bottom-up forms, such as Placemaking [10].

This approach has been applied in Potenza, a small municipality located in southern Italy, following some initiatives pursued by several local associations. An abandoned area, important for its architectural symbols, has been chosen in order to build a participatory design process and to give back to the city a shared space with no public expense.

This activity led to a working group composed by architects, sociologists, engineers, journalists and philosophers, called “Garden in motion” [2]. "Garden in motion" is also a bottom-up urban renewal project of an abandoned space in Potenza municipality, famous for an architectural monument, the bridge designed by Sergio Musmeci. This area, for a lot of time used as an uncontrolled parking, has been transformed into a garden, with children playground and spaces for walking, reading and cultural events.

On July 2013, the above mentioned group organized a free workshop involving young architects, engineers and agronomists. The area below Musmeci Bridge has been cleaned and analysed; existing plants have been catalogued. Participants studied possibilities of space use, also interviewing local residents and listening to old tales: Who lived there? Was there a community? Who does frequent the area now?

At the end of the research, workshop participants proposed some designs in order to decorate the area. Proposals have been published (online/offline) and shared with citizens. The final project has been developed with the voluntary participation of many citizens.

2 A Bottom-Up Approach in City Design: The Dichotomy between Formal and Informal Spaces

Participation can be defined as a dichotomy between formal and informal spaces such as the collective contribution to urban design, through activism phenomena of that trigger informal spaces, starting from critics to functionalism patterns, which ultimately would lead to estrangement and liabilities between citizens.

In order to better specify what kind of functionalism is referenced in the project for the city it is useful to start from critics by Lefebvre [7] opposed to this sociological theory.

Lefebvre considers functionalism as a coincidence between rationality and functionality that would bring function to create reality. He states that "the new cities showed merits and deficiencies, more evident than merits, of functionalism, when it wants to create the framework and the conditions of daily life."

A space between the formal place of the city, designed and planned without any participation form of the future users will be a place with almost exclusively unique ways of interaction. People living that space are unable to have unusual behaviours, compared to the prescribed forms, without being accused of madness.

Considering a kind of anachronistic functionalism, still contemplated today, the city designed and governed without participatory paths appears to be a space of silence.

The silence, defined as lack of response to the aims proposed in the city project, is the nourishment for the continuous replication of formal places.

Apathy that characterizes citizens could be generated by situations described in the International Situationist [6]: “It is important to redevelop the area around them, to build for them, without distracting them from worries transmitted through the eyes and ears”.

The city expands leaving on one side empty places without function and on the other side multiplying functions and places consumption, transforming its inhabitants in space consumers.

On one hand there are formal places that require feasible behaviours and actions only if recognized in dominating economic and social system, on the other hand at the edge of the city are located the informal places that allow intents sharing.

Away from the project, the "manufacturer" inhabitant in empty spaces of the city claims his right of movement and action, which up to nowadays has been denied.

In several cases, the “design machine” does not give meaning and function to informal spaces if they are not attractors of economic interests.

All this brings us to an issue dealt with by Lefebvre: the Right to the City [8].

The re-appropriation of informal spaces by single or informal groups of citizens through reuse or urban regeneration actions are an expression of the rights to the city as an active proposition in countering individuals separation and specialization of places imposed by formal city.

Controlling behaviours and actions that people practicing in the interstices of free cities, like simple deviance from the common sense, is reductive because they are not the result of individual strategies of re-appropriation, implemented according to place characteristics, but acts more or less aware of collective design of places and their potentiality.

Harvey [4] argues that the “right to the city" is more than an individual freedom to access to resources offered by town, but it is the right to change ourselves by changing the city.

While individuals and individuality are not recognizable in informal spaces, it is possible to outline political and organization aspects that collectivity uses to govern the informal space.

3 The Experience of "Garden in Motion"

The "Garden in motion" design and implementation come from several experiences developed in past years in this area. The attention was focused on Musmeci Bridge, an extremely important engineering and architectural artifact, crossing the Basento river, full of symbolic values and historical memory of the city, completely abandoned today. Under the bridge there is an dismissed industrial area with green spaces, which requires a reconversion. The location of the industrial area on Basento river dates back to post World War II, when some mechanical and steel industries have been located in this zone taking into consideration the proximity of the area with the railroad which connected the city of Potenza with Tyrrhenian and Ionian coasts. The importance of this area is crucial because, due to the worldwide economic crisis, a lot of activities have been abandoned. Consequently the area could be converted to leisure, sports and cultural activities.

Fig. 1. The underside of Musmeci Bridge

The motivation is due to the position. This area has a good level of accessibility, it is one of the few flat zones of the city, it is close to Rossellino park, it is close to a Roman bridge (Saint Vito bridge), it is very close to a zone where an important participatory process of reconversion to park of an old pig breeding [9] is on-going and the municipality has also a project of a fluvial park around the Basento river. For this reason, the most important cultural associations of the city, since several years, organize many cultural events in the area to attract media attention to the zone and to remind its importance to the whole city population.

"Garden in Motion" is an initiative testing an innovative approach to design and suggests a different way to enjoy a monument, to live an urban fragment, to take care of a collective space. Based on an interpretation of the theories by the French landscape architect Gilles Clément, especially related to the concepts of third landscape and Garden in Motion, it was decided to conduct a laboratory located in a residual area of the city of Potenza, in proximity to the bridge crossing Basento river, designed by the Italian architect Sergio Musmeci.

This bridge was built at the end of the sixties of the last century and it is one of the few elements of architectural interest in Potenza city. The bridge was built entirely with reinforced concrete and its forms are the result of a complex engineering research conducted by Musmeci, who pursued the maximum correspondence between shape and structure. The result is a work characterized by a complex and unexpected spatiality, perceived through the pedestrian path under his deck, which offers picturesque views and interactions with the river landscape and the city. Because of these values the bridge was recognized in 2003, among the first Italian works of contemporary architecture under the protection by the Ministry of Heritage and Cultural Activities, becoming a monument like Coliseum and Santa Maria del Fiore.

Fig. 2. An external view of Musmeci Bridge

The importance of the relationship between the monument and the environmental context in which it is placed had already been affirmed by Venice Charter in 1964 and placed in the centre of the action of protection. Later on, the declaration ICOMOS of Xi'An [13] reiterated the importance of the contribution that the context provides to the value of the monument.

The relationship with natural environment, past and present social practices, uses and activities and other forms of intangible heritage that create form and space have been included in that concept.

The same declaration also paid attention to the importance of documentation understanding environment interpretation in inclusive and multidisciplinary ways.

Focusing the attention on the bridge as a monument, Garden in Motion represented an important moment of awareness of its monumental value and an opportunity to build new communitarian values.

The context has been studied and analysed in all its aspects: results of these analyses were the basis of the design workshop and interventions implementation.

Great emphasis was placed on community education and public awareness to achieve conservation objectives and to improve means of protection and management.

A new awareness of the importance of the relationship between the monument and the context has been the basis of subsequent activities of collective use of the area.

In this perspective, the "Garden in Motion" initiative has given (and continues to give) its contribution in the activation of some processes of great importance for the life of a community, just like in Gilles Clément's Garden in motion, where the processes of nature are favoured and spontaneous plants put in condition to grow and move freely. The interest in the monument and the participation to social dynamics are like seeds of wild plants sown in a field.

In this sense, and according to the etymological interpretation of the term "monument", the "Garden in Motion" experience activated human and social energies fundamental in a community life that not only transformed the physical environment of the bridge, but they also created a number of perception practices and a place usage.

The experience was composed of two different and integrated phases.

The objectives of the first phase "Designing the garden under the bridge" (June 2013) can be classified into three main activities: plans, strategies and communication.

The aims of the second phase "Making the garden under the bridge" (July 2013) were the realization, in self-construction, of interventions designed during the first phase.

3.1 The Workshop

The basic idea was to induce to a garden design taking into account informal uses of Potenza, which citizens already put into practice in that place.

In addition, there was the will to *second*, using the term adopted by Clement, informality of place, through a proposal of self-built furniture, connections with formal city and construction of a privileged point of view of Musmeci bridge.

"The aesthetics of natural disorder", as defined by Clement have been reproduced in the first phase of the workshop through analysis of needs. Workshop participants were invited to explore needs expressed by citizens mapping through senses, signs of presence and action of man. Each participant revised the information collected in a project which expresses a function for the garden, a formal strategy of connection to the city. Considering again Clement's theory, participants no more played the role of architect sculptors but that of garden makers with the only purpose to accompany with their own projects transformations already present in that place.

Starting from the garden of spontaneous plants, recognized and labelled with qr-code, workshop participants imagined to activate a process of re-appropriation of a degraded and abandoned public space.

Three main issues have been considered: 1) the limit, defined as visibility from outside, permeability, border protection, access; 2) pathways, walk or stand in the garden; 3) pairs of connections, garden/bridge, garden/river.

Participants to workshop designed seven projects, but only four have been realized because of the typology of materials.

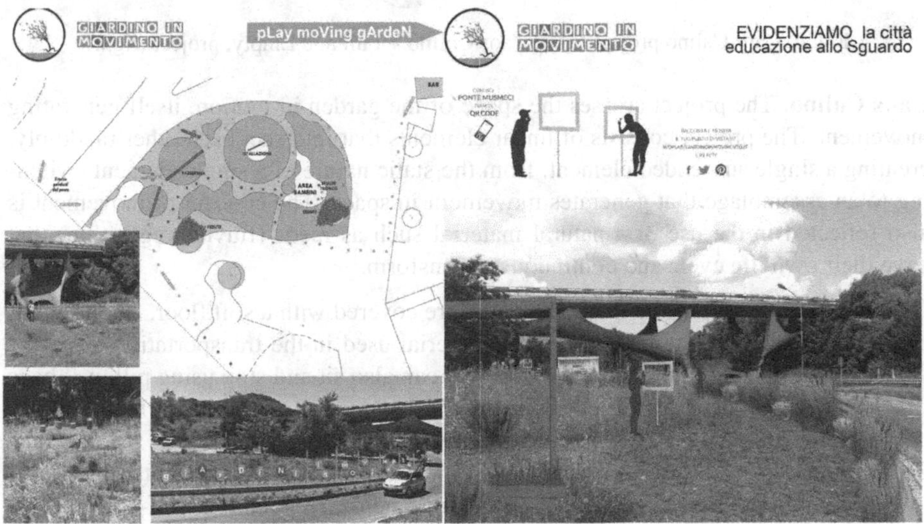

Fig. 3. pLay moVing gArdeN project (left), Highlighting the city, education to sight project (right)

pLay moVing gArdeN. The project is based on the idea of a game implying citizen participation. Using coloured paths, signals and playful elements, it is possible to enter into a relationship with the garden and its surroundings (bridge, road, river, bars, etc.). Borders do not separate but create a relationship.

Highlighting the City, Education to Sight. The project aim is to accompany garden visitors to a discovery travel that takes place through the view, freeing them from the liability where they live in everyday life. Visitors become explorers and adventurers. The frames suggest glimpses of the landscape in which the protagonist is Musmeci Bridge; not only plants, but also visitors themselves become actors in the garden, animating it looking at the city.

Fig. 4. Caos Calmo project (left), Caos Calmo + Full and Empty, project (right)

Caos Calmo. The project crosses the space of the garden in motion, itself generating movement. The project consists of linear elements that intersect each other randomly, creating a single suspended element: from the static nature of a single element switching to an assemblage that generates movement in space. The concept of movement is also reflected in the use of a natural material such as reeds (fluvial vegetation) that have their own life cycle and continuously transform.

Caos Calmo + Full and Empty. The paths are covered with a soft floor, a continuous carpet, consisting of jute bags (recycled material used in the transportation of coffee from different places in the world) where you can also sit and stop using pillows (bags filled with plant material). The texture of the bags, sewn together by hand, allows native plants to find a way to continue to grow and conquer new spaces.
Each project has also taken into account some fundamental requirements: the use of low-cost and low-impact materials (recycled and/or recyclable); self-construction realization; temporary installations, reversible and easily removed.
The realization of the projects has been strongly influenced by materials possible to find and by the number of people available in the days of work.

3.2 Engaging the Community in Realizing the Garden in Motion

Sirky [11], describing communities development patterns or interest groups, states that groups managing problems of collective resources assume the shared respect of cooperation rule.

In most cases a strong interest towards a problem\good\space makes the spontaneous action of involved people more effective, than previous policies adopted in good management by responsible Agencies\Local Authorities.

Today we are immersed in technology with continuous connections which improve ideas and experiences exchange.

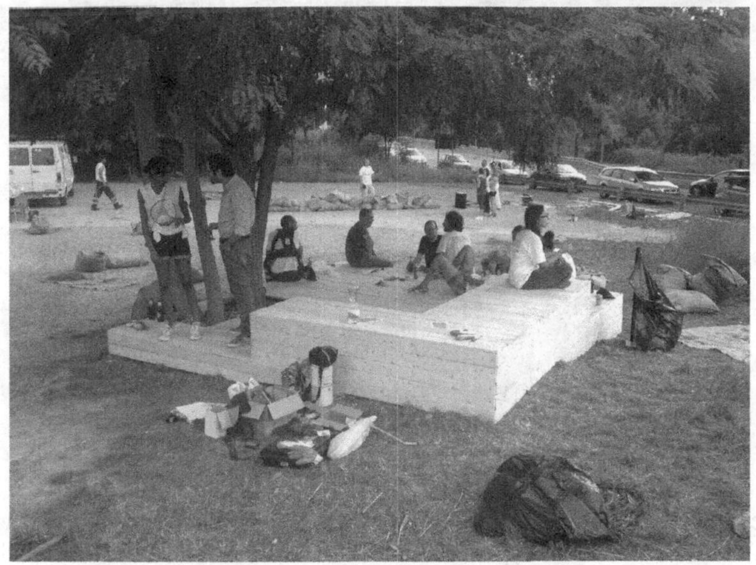

Fig. 5. Garden in Motion Building

The stronger is knowledge circulation, the greater is community growth. For this reason, citizens participation in urban space design and management has consequences, also in terms of innovation.

Digital approaches and technologies also allow to build a strong storytelling of community actions, able to go much farther than city limits. It is possible to build an internal engagement, to activate citizens on projects or choices and produce circulation creativity and knowledge. It is no longer to promote actions or projects, such as a marketing bottom-up participation.

The internet and the great possibilities offered by digital technologies allow us to tell the experience, making sure that the story becomes part of the experience itself. The context is typical of a provincial town, which grew haphazardly, with an increasing poverty, little investment in services by local authorities and lack of experience in citizens participatory processes.

The choice of the area to be regenerated is not random: it is an abandoned area, forgotten by the institutions, behind a monument of great architectural value, but almost unknown in the city, Musmeci bridge.

During a workshop, some practitioners - agronomists, sociologists, architects, engineers - put the area under observation and designed a number of proposals with zero impact.

The first phase, analysis of the space below the bridge, collaborative design of some furniture and design of communication campaign, had a very strong impact on citizens.

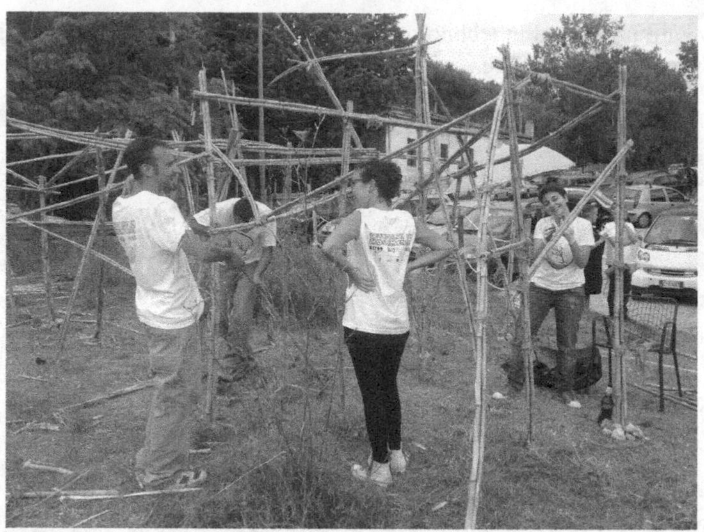

Fig. 6. Garden in Motion Building

The continuous online storytelling of work (blogs, social networks, wiki approaches) added news about hyperlocal context to mainstream. But it was mainly the online space to explain the story of what was happening in the city. Also, this space has been built in a participatory and spontaneous way using pictures, text and comments. The population was made curious with small actions based on urban games.

Fig. 7. The urban game created to promote the initiative in the city and to engage citizens

The participation to storytelling experiment design has become a way to participate to the entire project development and to the regeneration of the identified space.

Some plants, for example, have been disseminated in the city, photographed, geolocated, posted on the internet. Without receiving instructions, citizens began to track the traces of the game, joining to the story.

Meanwhile, the exposition of project proposals prepared by practitioners represented a further inclusion invitation for other citizens. Which project would you choose? Citizens have commented, selected, suggested, becoming both part of the "Garden in Motion community" and active nodes.

At the same time traders, private enterprises, administrators encouraged by the enthusiasm created around the initiative, helped providing materials, equipments, and even food.

Hundreds have joined to the second phase, mainly based on the construction of the furniture of the area under Musmeci bridge. This number of participants was almost unexpected for the reality of the city. That community was then able to supervise the area.

Fig. 8. Garden in Motion Building

The storytelling of the positive experience of "Garden in Motion" has activated new interest in associations, artists, single citizens that have filled the area of cultural events, according to a logic of cooperation, proposing new ideas.

Weinberg [12] highlights that network is always smarter than individuals, even when the latter are very skilled. It is in the network that ideas are mixed, taken together, combined, revised. It is also a matter of community connections.

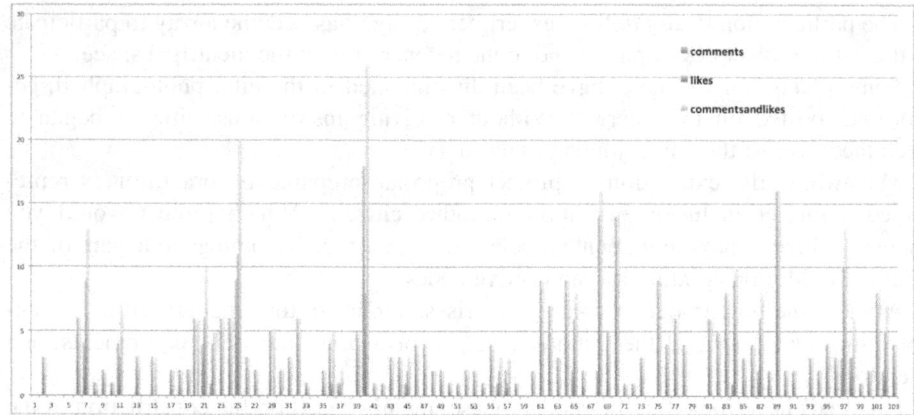

Fig. 9. Number of Likes and Comments on "Garden in motion" Facebook group

The analysis of activities on Facebook group is very interesting. The organization was mainly based in a like page and a group. The figure above shows the number of likes and comments of the group, with the highest peaks of activity in the period of the major cultural events organized in "Garden in motion".

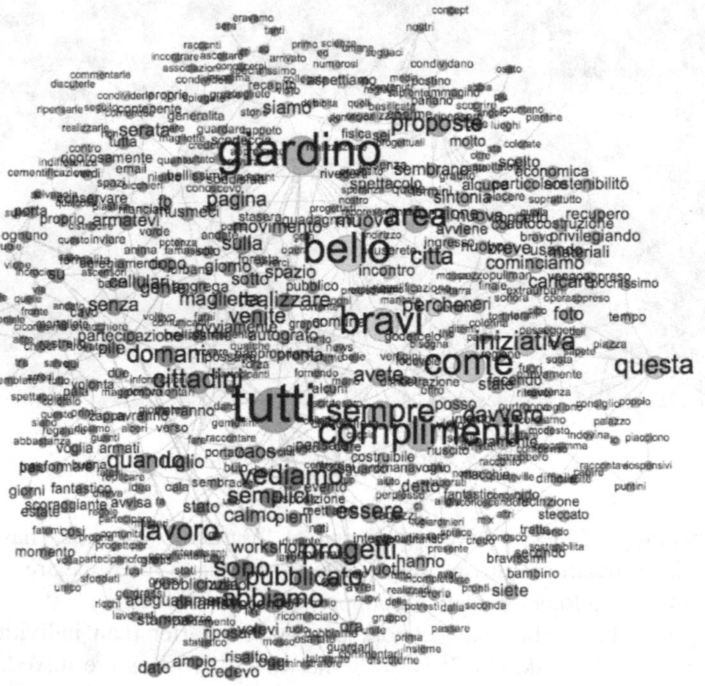

Fig. 10. Semantic analysis on "Garden in motion" Facebook group and like page

The above figure shows the semantic analysis developed merging the Facebook group and like page. The main word is obviously Garden, but also other words are important, such as beautiful, congratulations, city, projects, proposals, initiative, etc.

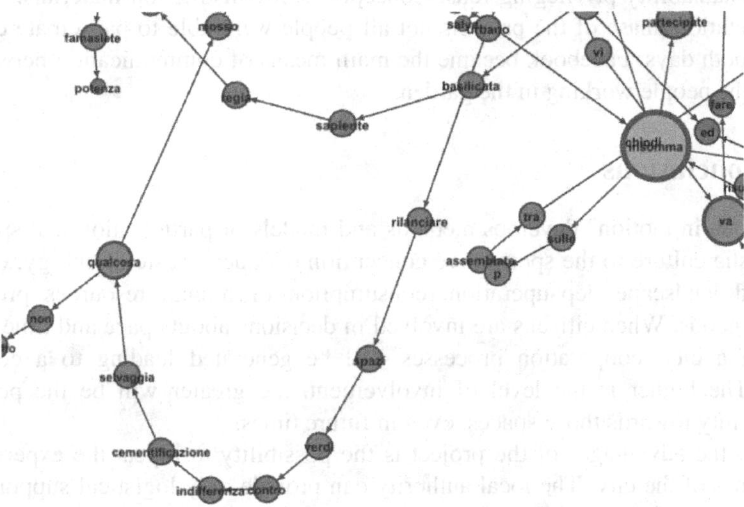

Fig. 11. Semantic network generated using word in "Garden in motion" Facebook group

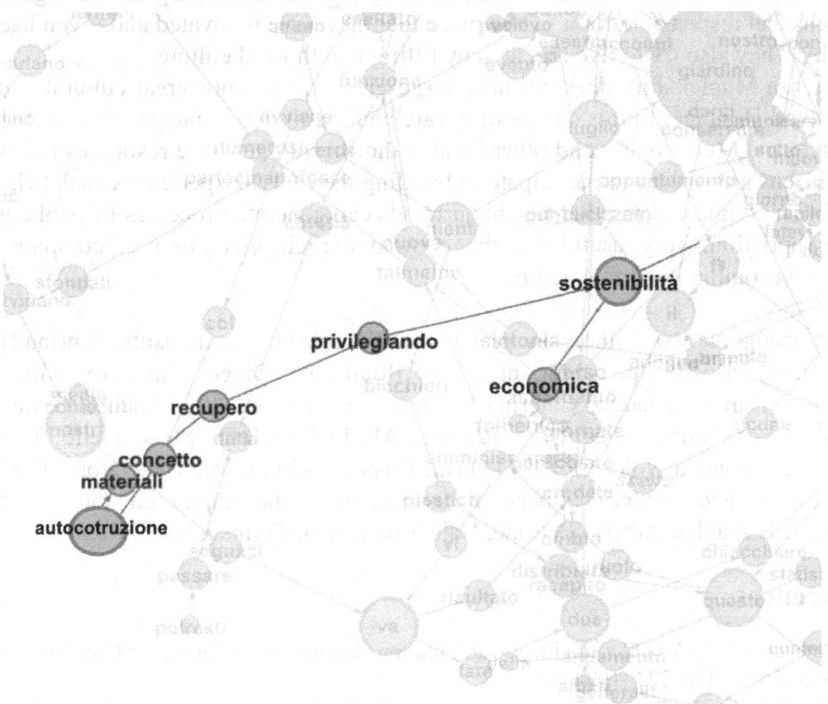

Fig. 12. Semantic network generated using word in "Garden in motion" Facebook group

Figure 11 and 12 show two semantic networks generated using word in "Garden in motion" Facebook group. In the first network, the sequence is revitalize-green-spaces-against-indifference-overbuilding-with no rules. The second network is economic-sustainability-privileging-reuse-concept- self-construction-materials. In the implementation phase of the projects not all people were able to participate continuously in both days: Facebook became the main means of communication between the city and the people working in the garden.

4 Conclusions

"The garden in motion" develops methods and models of participation that spread an antagonistic culture to the speculative conception of space, waste of energy, environment and landscape depauperation, consumption of human resources promoting common goods. When citizens are involved in decisions about space and time organization of a city, cooperation processes will be generated leading to a collective benefit. The higher is the level of involvement, the greater will be the perceived responsibility towards those spaces, even in future times.

One of the advantages of the project is the possibility to repeat the experience in other places of the city. The local authority can provide only logistical support, without investment of economic resources. The city participates to the design phase selecting proposals, building decorations, cleaning the area where, at the end of the project, neighbourhood events, cultural events, meetings with a social purpose are organized. Citizens feel responsible for a public place that they have renovated and given back to the city. They also try to live that area by filling it with local culture.

Potenza Municipality does not have large green spaces, nor great cultural and architectural Heritage. In this city poverty rate and social disadvantage have doubled in recent years. Municipality and other local authorities do not have resources to invest, but citizens are willing to participate in building their city. Experiences such "Garden in Motion" replicated in other neighbourhoods can generate processes of public good re-appropriation, very useful for the community: citizens observe, compare and choose, becoming important actors.

Acknowledgements. Authors are grateful to all workshop participants, Sabrina Bianchi, Hilary Bochicchio, Mariantonietta Bonifacio, Diego Calocero, Giuseppe D'Emilio, Danilo Eduardo, Caterina Fidanza, Antonio Galante, Marilina Giannotta, Marica Grano, Gianleo Gruosso, Rossana Machiella, Ketty Mancaniello, Daniela Mancini, Antonella Nolè, Carmen Peluso, Francesco Pitta, Mara Salvatore, Pierluigi Santopietro, and other members of associations who supported the initiative, Giuseppe Biscaglia, Alberto Petrone, Valentina Russo, Carmela Dacchille.

References

1. Arnstein, S.R.: A ladder of citizen participation. Journal of the American Planning Association 35(4), 216–224 (1994)
2. Clement, G.: Il giardino in movimento. Da La Vallée al giardino planetario. Quaderni Quodlibet (2011) (trad.it. E Borio)

3. Davidoff, P.: Advocacy and Pluralism in Planning. Journal of the American Institute of Planners 31(4), 331–338 (1965)
4. Harvey, D.: The right to the city. New Left Review 53 (September-October 2008)
5. Jacobs, J.: The Death and Life of Great American Cities. Random House, New York (1961)
6. L'internazionale situazionista 1958-69, vol. 6 "Commenti", Nautilius, Torino, p. 26 (1994)
7. Lefebvre, H.: Critica della vita quotidiana, pp. 229–238. Dedalo (1993)
8. Lefebvre, H.: Le droit à la Ville. Anthropos, Paris (1968)
9. Murgante, B.: Wiki-Planning: The Experience of Basento Park in Potenza (Italy). In: Borruso, G., Bertazzon, S., Favretto, A., Murgante, B., Torre, C. (eds.) Geographic Information Analysis for Sustainable Development and Economic Planning: New Technologies, pp. 345–359. Information Science Reference IGI Global, Hershey (2012), doi:10.4018/978-1-4666-1924-1.ch023
10. Schneekloth, L.H., Shibley, R.G.: Placemaking: the art and practice of building communities. Wiley, New York (1995)
11. Shirky, C.: Surplus Cognitivo. Creatività e generosità nell'era digitale, Codice Edizioni (2010)
12. Weinberg, D.: Rethinking Knowledge Now That the Facts Aren't the Facts, Experts Are Everywhere, and the Smartest Person in the Room Is the Room. Basic Books, New York (2012)
13. Xi'AnIcomos declaration, `http://www.international.icomos.org/charters/xian-declaration-sp.pdf` (last accessed March 28, 2014)

Smart City or Smurfs City

Beniamino Murgante[1] and Giuseppe Borruso[2]

[1] University of Basilicata, 10,Viale dell'Ateneo Lucano, 85100 Potenza, Italy
[2] University of Trieste, P. le Europa 1, 34127 Trieste, Italy
beniamino.murgante@unibas.it, giuseppe.borruso@econ.units.it

Abstract. Very often the concept of smart city is strongly related to the wide-spread of mobile applications, completely forgetting the essence of a city with its connected problems. The real challenge in future years will be the huge population migration from rural areas to cities. It is fundamental to manage this phenomenon with clever approaches in order to save money and environment. This paper develops some considerations on these aspects trying to lead the discussion in a correct direction.

Keywords: Smart city, Smart communities, Urban Planning, Open data, Citizens as sensors, Governance.

1 Introduction

Nowadays an approach combining in a narrow way the concept of Smart City to the sudden spread of electronic devices is very common. There is a deep conviction that the implementation of a Smart City is simply an exasperated use of applications for Smartphone or tablet.

Often the attention is exclusively focused on mobile applications forgetting that there is also a city. These approaches, despite having a certain degree of usefulness, when completely disconnected from the context, especially from the city in its essence, can produce a waste of resources. When complex computer systems are proposed it is crucial to ask "are they really useful to the city?".

This common belief evokes urban scenarios inspired by Ridley Scott's movie "Blade Runner". The idea of a city with many vendors should lead to a vision of cities similar to a Pioneer advertisement[1] of late '80s, very popular in Italy, where each person was "wearing" one or more televisions that constituted a barrier to the outside world determining a robot behaviour. This approach to smart cities will lead to a "flood" of electronic devices in our cities connected to improbable goals to be achieved.

If a city has a structural mobility problem it is quite impossible to solve it only by means of a smartphone. The term "smart" is today very popular and is adopted also in common language and in all kind of advertisement. In order to describe the adoption

[1] Pioneer advertisement Pioneer Blue velvet.mpg http://youtu.be/5rMI_aVYtR0

B. Murgante et al. (Eds.): ICCSA 2014, Part II, LNCS 8580, pp. 738–749, 2014.

of this term in everyday speaking it could be useful to adopt a parallelism with Smurfs cartoons.

Everything is smart such as in the Smurf world we have Smurf-Forest, Smurf-berry Smurf-strawberries, etc. It is very common in participation processes, smart participation, to find an interview to the major of a city or to the director of a journal called smart interview or to find the term smart questionnaire for paper form distributed to a sample of citizens.

Very often the concept of smart city is strongly related to the widespread of mobile applications, completely forgetting the essence of a city with its connected problems. In order to bring the smartness concept in a correct approach it is important to highlight the challenges that cities will face in the next years.

2 The Real Challenges of Cities

A study developed by "the Economist" [1] highlights that despite the United States and the European Union have a comparable total population, in the U.S. 164 million people live in 50 major metropolitan areas, while in Europe there are only 102 million metropolitan areas inhabitants. This difference leads to surprising consequences in terms of productivity and incomes. Gross Domestic Product of European metropolitan areas is 72% of GDP of the 50 largest American cities.

An article on "Wall Street Journal" [2] shows how major metropolitan areas of United States produce a higher GDP than the economies of entire nations.

Another article on "Washington Post" [3] emphasizes how in 31 American states one or two metropolitan areas account for the vast majority of the nation's economical production and in 15 other states, a large metropolitan area alone produces the most part of GDP. The seventeen major metropolitan areas generate 50% of United States Gross Domestic Product.

Urbanization is also different in terms of city size classes in the two areas. In Europe 67% of urban inhabitants live in medium size urban centres, smaller than 500,000 inhabitants; while just 9.6% are located in cities bigger than 5 millions inhabitants. In the U.S.A. one upon five urban inhabitants lives in major cities of more than 5 millions people.

From these statistics it is very easy to understand that, despite common opinions against big cities quality of life, in most cases living in large cities becomes a necessity. Glaeser [4] defines the city as the greatest invention of mankind. Using the advantages of agglomeration principle, a city emphasizes strengths of a society. Despite the evolution of modern and contemporary cities has led to disadvantages resulting from congestion, urban poverty and security, today living in an urban context, even of not high quality, involves more benefits than living in remote areas. Consequently, cities play a central role for humanity, offering the opportunity to learn from each other, face to face. Despite economic contexts and production patterns have been radically changed, a city always represents the most vital element of the economy of a nation. Generally, in every developed country, cities are the economic heart and the most densely populated places, very attractive for people who want to exchange knowledge. While in the past advantages were closely related to the reduction of transportation and distribution costs, today cities have got huge benefits in economic terms due to exchange of ideas, therefore there is the transition from an idea of a city founded on the concept of location to a city based on interaction [5].

In the next few years an increase in world population of 2.3 billion people will occur, with an average increase of population in urban areas by 30% [6].

These scenarios can be inserted in a larger picture, in which cities already host the majority of world urban population. Western and industrialized countries already host nearly 80% of urban population, while developing countries to-date set at 47%. Asia and Africa are expected to overcome 50% of urban people by 2020 and 2035, respectively. Urban population is forecasted to increase of 72 per cent by 2050, changing from 3.6 billion people in 2011 to 6.3 billion in 2050 [9].

Within 2020 urban population of China will reach 60% of the total and more than 100 million people will migrate to metropolitan areas or contribute to the creation of new urban centres.

This phenomenon is not only limited to countries where a rapid economic development is occurring, such as China [7] and India [8], but also in Europe, as highlighted by "World Urbanization Prospects" United Nations report [9], where in 2050 almost 90% of the population will live in urban areas.

Obviously, an "urban" lifestyle implies a lower level of sustainability, more energy consumption, more pollution, more waste production, etc.. In China 45 airports will be realized within the next five years, cities will produce 80% of carbon emissions, urban areas will consume 75% of energy and 50% of water supply losses will take place in cities.

Some alarming predictions highlighted at the Rio de Janeiro conference of 1992 are taking place. The planet's resources are used by 20% of the population, but with the economic growth of countries such as China, India, Russia and Brazil, with an elevated number of inhabitants could completely blow up the environmental balance of the planet. Therefore, clever approaches to save money and environment are needed. We cannot reproduce an urban development based on the same model that has governed the process of urbanization occurred since the Industrial Revolution until today. It is necessary to move from an approach based on pure physical growth of the city, to one founded on the ability to use in a correct and efficient way energy, water and other resources and to provide a good quality of life. In practice, Cities should become smarter in programming and planning management and use of existing resources.

3 From Location to Interaction

As far as the world is getting more and more urbanized, cities are becoming the most visible footprint of humans on the planet. However, defining what is urban and what is not, or finding a unique definition of city does not represent an easy task.

Many of the characters of cities can be found in other human made artificial landscapes but, taken alone, they are not sufficient to discriminate between cities and other environments.

From a geographical point of view different principles can be followed to define a city. A demographic point of view deals with population and its concentration in a given area. However, how many citizens are required for differentiating a city from a village or a minor settlement is not uniform in different countries or areas of the world.

Another principle is based on quantity, shape and concentration of buildings, but here also the concept varies from area to area, and a same concentration of building could define an industrial area as well as other kinds of settlements.

A third principle deals with the concentration of activities in cities, and particular with the fact that activities are various and different from agriculture, as well as serving an extra-urban demand.

Furthermore, cities can be defined as places where some activities and functions are located and concentrated. In such way a city is not only characterized from demographic and infrastructural points of view but rather by the functions played in that particular environment, holding a concentration of buildings, infrastructure and people.

Functions in cities are of particular type as they are generally different from agriculture and serving a wider range than that of the proper physical coverage of the city.

The consideration is also that, although more than 50 % of human population live in cities, actually they are quite rare on the Earth's surface and play their functions over a wider range that is served.

So functions played by a city are both dedicated to those people that live in the city in their day by day needs (i.e., schools, retail, etc.), or city serving, and those activities that are the essence of the city and make it special, like university, specialized medical doctors, etc., that work for both the strictly defined inhabitants of the city but in particular call people from outside the city to benefit by such activities, defined as city forming. A wider surrounding area of the city is therefore served, that implying a gravitation towards the city and an interaction, thought as flows, movements of people from outside the city towards the city itself.

So a city is not just based on steady, fixed elements as buildings, infrastructure and localized economic activities, but on movements, too. Typically commuting is identifying metropolitan areas defining the range of a city in terms of its (physical) attractiveness over a certain geographical distance.

A key element in doing that is the distance decay function, stating that the amount of interaction among people and places tend to decrease - with different slopes and speeds - as distance from the place increases.

Interaction and distance decay is applicable at different scales and in different contexts: in the already mentioned commuting case, the amount of people heading to a city for working activities from the surrounding settlements tends to decrease as the distance of neighbouring settlements decreases. Similarly in analyses made on telecommunication traffic, interaction decreases with distance.

In such sense, usually cities are seen as nodes in a network system, characterized by linear elements linking nodes and flows on such links.

However, in general the attention in geographical terms is dedicated to cities and their physical and functional features, at the urban scale, while at the extra-urban scale the focus is on how different cities are organized and linked in comparison with other ones, with whom they maintain a relationship in terms of commuting, political presence, economic environment, etc.

The same network metaphor can however be moved in the internal part of a city, identifying places where people gather and interact more, like squares, shopping malls and high streets, public offices, etc.

So geographers consider cities and their regions as systems of nodes, networks and flows, organized in a network or hierarchical system. However, the attention is generally focused on places and on the interaction between people and places. Recently, cities are more and more seen as complex systems, needing an even more integrated approach. In particular, then, the huge availability of data, often coming from users of portable devices and ICT social networks, provides suggestions and data sources to go more in depth on the issue, and moving the attention on interaction between people and happening in places.

Mike Batty [5] suggests that "to understand cities we must view them not simply as places in space but as systems of networks and flows".

According to Bettencourt [17], a city is a complex system characterized by a twofold soul: it " works like a star, attracting people and accelerating social interaction and social outputs in a way that is analogous to how stars compress matter and burn brighter and faster the bigger they are." Also, "Cities are massive social networks, made not so much of people but more precisely of their contacts and interactions. These social interactions happen, in turn, inside other networks – social, spatial, and infrastructural – which together allow people, things, and information to meet across urban space."

4 Smart City, Smart Cities, Smartness or Dumbness

One of the challenges lays in the definition of 'Smart City', or trying to understand the level of smartness that a city can have. Although a certain agreement on elements and indicators defining a Smart City is set, not such optimism can be directed towards their meaning and transformation into active practices.

Six axes represent the backbone of a Smart City, with smartness translated into economy, society, mobility, people, governance and environment. In all of them attention is given to the opportunity promised by modern ICT to boost such axes, optimizing and making cities more efficient. The philosophy behind the Smart City is strongly related to the Sustainable City, in which environmental, social and economic dimensions are considered as part of the development to be pursued, to allow present and future generations equity in living conditions. The difference lays mainly in the role played by technology, and ICT in particular, in allowing a more efficient management and organization of the different parts of life in cities.

However, how is that translated into the real world? Sustainability in urban contexts involves public participation. Possibly the Local Agenda 21 has been one of the first cases in which a bottom up approach was suggested into political action at local level, in such sense anticipating – and putting the basis for – to-date public participation in planning, also helped and speeded up by social networks and media. In Smart Cities public participation is central and, of course, is boosted by new technologies, as social networks and media and it must therefore rely on a consistent network and infrastructure, allowing data and information flowing and sharing. However, the bottom up approach is possible in the Smart City also by means of citizens and urban users building and realizing their own services and activities, therefore meeting needs they do know and feel, often better than the final decision and policy makers.

Nevertheless, and as a paradox, the Smart City concept is often translated into a 'techy' top down approach and consequent solution, with a single (set of) decision maker(s) preparing supposed valuable solutions for citizens. This is the case of new investments towards 'Smart Cities' in which hi-tech tools are proposed and realized as centralized systems to control several aspects related to energy efficiency, transport, house access, etc.. In such a big infrastructure, projects are implemented, coupling hardware network infrastructure and control systems, as well as more traditional, although generally technologically advanced, real estate investments.

Rio de Janeiro and Song-Do Smart City are among these examples. In the former case a control system was sold to Rio De Janeiro to monitor traffic in real time, in the second one a brand new 'Smart City' or 'Smart Suburb' was built from a blueprint in a greenfield area, off Seoul and close to the new South Corean international airport, having in mind energy efficiency and saving, quality of life, a planned environment for business, living, working, etc.. These examples are the offspring of a planned centralized system, often not so flexible to incorporate innovation: as an example, Song-Do was based on RFID technology and not ready to adapt to new communication tools as smartphones and tablets – whose role in locating sensors and devices helping us in automatizing activities was completely unconsidered or underestimated.

On the other hand, the bottom up approach is based on how citizens or city-users live and interact with the city, and develop their own applications and solutions for the different uses of a city. Similarly to what happened in the past, with new utilities and infrastructure both serving cities' expansion and also shaping it, technology is influencing how we live and set our relations with other people and places. As a trivial example, on one side new devices and tools induce us clustering close to free wi-fi hotspots; on the other hand people usually gathering in popular places induce authorities or private activities to set and reinforce wireless sensors.

Therefore a similarity with other physical infrastructures (roads, electricity cables, fresh water pipes) arises, but how we use now what flows on such infrastructure is quite different – and often unexpected – if compared with what we used to. So it risks or tends to be for the physical infrastructure of the smart city, or the hardware composing the digital layer superimposed over the city. And that suggests that the bottom up approach in a very 'open' way should be based on the setting of an infrastructure (and a set of rules) and should allow people to 'flow', interact and develop their own activities.

Another issue related to 'smartness' can be put onto the international differences the smart city concept has got. Asia, and particularly South East of Asia, are working on housing and on expanding cities and the issue is related to the urban model to be adopted for brand new cities or neighbourhoods, while other geographical areas, as

Europe and the US, hold older urban structures and heritage. A Smart City in an urbanizing world means building brand new settlements from scratch, often in greenfields and from a blueprint. On the other hand, a Smart City based on an existing urban fabric, stratified in years of history – as in Europe or even in some US cities – requires optimization and reuse. In the former case a Smart City appears as a 'new town', a planned city in which functions and activities are organized. Often this is also translated into new suburbs or mid-size cities to be realized, in such sense following a suburbanized scheme already seen in other contexts, with the difference that smartness is put primarily onto energy efficiency and technological devices. On a more traditional urban fabric, smartness is more related to the challenge of rethinking a city in a smarter way, therefore optimizing it particularly in terms of interaction existing between citizens or city-users and the 'hard', infrastructural component of the city, not just building brand new settlements or suburbs that, in a non sustainable way, would consume soil and space.

5 The Pillars of a Smart City

The risks today are in focusing on just the technological side of "smartness", maybe without a tight connection neither among techy initiatives, nor –what is even worse- with spatial and urban planning activities. We do not deny ICT is central in setting a technological infrastructure as the backbone of the growing flow of data and information. The role of infrastructure of both serving and boosting urban growth and expansion was already mentioned, having a heritage, since their shape and fabric remain in time and influence different periods and generations.

So a focused planning is needed, but not to be limited to the short term but to persist.

In such terms a true Smart City acts as an "enabling platform for the activities that citizens are able to develop, linking those inherited from the past to those that can be realized in the future, so it is not focused just on applications but on the possibility that citizens realize them" [10]. Doing so is possible by thinking about it in terms of [14] three main pillars:

1. connections - as networks and technological infrastructures;
2. data – open and public or public interest data to allow the development of innovative solutions and the interaction between users/citizens and the city;
3. sensors - these including citizens [11] [12] [13] able to actively participate in a bottom up way to city activities.

Such pillars need to be accompanied by an urban governance able to harmonize them and particularly to represent a set of minimum 'driving rules', regulating a smart city in a neutral way, without entering too much into details concerning contents and applications developed by citizens, urban users, private companies, etc..

In such sense, a correct approach to Smart Cities should in some way try to resolve problems typical of urban areas and not just those of niches of users. As an example, our urban areas are often profiled on a category of users: generally male, in his productive age, driving a car, therefore cutting out other important parts of urban population, as young and elderly people, as well as the female component. [15]. So, a purely 'techy' Smart approach risks to approach just those people actively using ICT (mainly mobile) technologies. Therefore the technological layer needs to be linked to the spatial context where it is applied, as cities are different from each other. One of the key elements in planning is verifying the compatibility and complementarity of a plan with other ones just ended or to be licensed in a short time, other than considering the possible overlapping with similar initiatives [16].

It is important to use the big impact of technologies on new forms of policy and planning. The six axes of smartness not only need to be connected to technology, but also to be connected to the added value that innovation can lead to programs and plans already issued.

6 City and Open Data

As mentioned above connections, sensors and Open-Data are the *smart cities* pillars adopting an approach based on the transition from the concept of *government* to the concept of *governance*. The essence is a background vision of the city able to transform the "impulse" resulting from the pillars activities to be performed into the individual application domains, the six *smart city* axes, *Economy*, *Governance*, *Living*, *People*, *Environment* and *Mobility*.

A lot of people everyday talk about Open Data, in the same way of smart cities, without getting more in the detail of the real meaning and the great opportunities that could arise from their correct use.

In most of the cases the concept of Open Data is based on uploading a file in portable document format (pdf) on a website allowing the download to everybody. When a public agency share a file in pdf format a monitoring authority should take action and if necessary sanctioning it, because a public employee spends his time to put constraints to data and in another government agency another public employee will waste much more time to use that data just because of these constraints. The PDF format was created to allow documents or drawings printouts, often in printing services, without using the software that produced these data, but simply employing a pdf file reader.

Tim Berners-Lee proposed an Open Data classification scheme associating the stars to the level of quality [18]. The lowest level is based on providing an open license, making the data available on a web site without defining the specific type of format (usually the files are in pdf format). The only purpose of this type of data is to inform, it is only possible to read or print them. The second level aim is to provide data preserving the original structure, allowing also to manipulate them. It is a small improvement even if data remain in a proprietary format. Three stars Open Data allow manipulation and management of data and adopt a non-proprietary format ensuring a better interoperability. The upper level maintains interoperability properties of data and improves availability on the network through the use of semantic web standards [19] (W3C (RDF, OWL, SKOS, SPARQL, ecc.). Five-stars Open data are Linked Open Data.

The limit of this classification is that spatial aspects are not considered at all.

In the introduction to the book "Geocomputation and urban planning" [22], the authors cite the famous paper by Franklin [21], who in 1992 quoted that 80% of all organisational information contain some references to geography. After the publication of this book, a lot of discussions started on social networks and blogs [20] on how was it possible that in 1992 80% of information contained a spatial component. This book was published in 2009 and up to date, after only few years, the situation is completely changed: each mobile phone has a GPS and Google OpenStreetMap transformed geographical information from a specialist interest to a mass phenomenon and probably 100% of data have a spatial relation. Consequently ignoring spatial aspects as an intrinsic component of data is a big mistake.

The spatial component has always been underestimated, sometimes intentionally sometimes for ignorance. In the first experiences of implementing master plans in spatial information system, data were deliberately shifted from the original coordinates in a lot of cases and the values of the translation were jealously guarded as access codes of a bank account. The main aim was to avoid overlapping of planning tools with other layers, allowing to discover the level of subjectivity of some decisions. In Italy, for instance, there is a great tradition in creating barriers to the immediate overlapping of information layers: cartographical maps and cadastral maps have always been produced at different scales to allow a certain subjectivity to technical bureaus of municipalities.

A comprehensive approach to open data should consider Open Geospatial Consortium (OGC) standards and INSPIRE directive.

Nowadays, data represent an unused big economic potential, because if they were available to everybody, collective imagination could create new companies and produce additional business to existing companies. The great part of these possible business initiatives should be based on applications for smartphones and tablets, which in 100% of the cases require a spatial component.

Considering the classic application for parking, there is a great difference if the application allows only ticket purchase, or if it indicates also where the nearest free parking is located. Consequently, open data for this type of application should be distributed at least as Web Feature Service (WFS) OGC Standard.

It is crucial to radically change public authorities approach: very often the term service is synonymous of contract.

A municipality does not have to pursue a contract for parking application, but it has to make *Open Data* available in Web Feature Service OGC Standard, allowing to local start-ups to produce an application ot to re-use an application produced for other municipalities. The municipality receives a free service and the enterprise gains with advertisements and if someone does not like the advertisement can delete it by paying one euro. Local authorities save money and contribute to create or consolidate enterprises in the field of innovation.

To achieve this goal it is essential that authorities produce and distribute high quality data.

7 Smart Citizens or Devices?

What about portable and mobile devices when talking of smart cities? How smart are we in using smartphones, tablets and all the family of portable devices? How we work, navigate, spend our free time is mainly based on mobile devices, todate smartphones and tablets, which diffusion has widely overcome that of more traditional desktop and laptop pcs.

How we use such devices is however still very limited to some kinds of uses and applications and quite far to exploit their potential. Figures help us, reminding that accessing the web is the main use of such devices, together with social networks and media (facebook, instagram, twitter, google+ just to cite a few of them), or with downloading apps dedicated to information retrieval, sport, games, meteo, maps, etc. Professional uses have been on top of the ranking from the beginning, so emailing, live meeting and calendars were features differentiating smart phones from more traditional mobile phones. That made the initial fortune of a company like R.I.M. – *Research In Motion* which actually created the smartphone concept and the popular BlackBerry platform, now suffering – and actually losing – the competition of giant ICT players as Apple and Android. Such a competition is also a symptom of a blurring of personal and professional uses, creating a generation of users whose activities have no more a marked spatial and temporal separation. So a question arises: is the use we make of smart devices really smart? When we talk about Smart Cities and Communities is the use of such devices really helping us in reaching such targets?

Fig. 1. What is in your Smartphone?

Probably, we are far from reaching a really smart and complete use of such devices, similarly with what happened with standard pcs and the software running on them: spreadsheets or database management systems, as a matter of example, are generally designed for a wealth of uses the most of users would not probably rely on in their all life, and so probably will happen with smartphones and their apps. We are facing a very wide and extensive coverage of mobile devices that however appear as Formula 1 or NASCAR racing cars driven in a peak time urban traffic jam, queuing at crossroads.

As in Figure 1, Smartphones and portable devices in general can be viewed in different ways as tools to connect accounts to social media or to check emails and contacts, but capable of hosting several tools and applications actually enhancing our capacity to act as real mobile sensors [5] . Our smartness as citizens should therefore be that of using the potential of such devices to exploit our interaction with the city to monitor it and highlight both positive and negative aspects and help its better management. Private companies and public bodies already use data we in a more or less aware way share, as positional and movement data, which allow estimating traffic jams, public transport time, etc. Also, our preferences in checking-in and doing particular activities in certain places is already monitored and both allow private companies to target marketing campaigns and products, as well as would – and hopefully will – allow planners and scholars to better understand how cities shape themselves from a social – not only in the ICT way! - point of view.

8 Conclusions

There is a widespread belief that the realization of a Smart City is based on an extreme use of applications for smart phones and tablets. Very often the attention has been focused exclusively on device applications forgetting that there is a city.
Whenever automation through mobile applications is proposed it is important to consider its effects on the city. When someone proposes a complex technological system it is important to ask "is it really useful for the city?".

In a lot of cases programs, which declared objectives mainly related to urban aspects, have been purely transformed in programs, based on ICT improvement. It is evident that in these experiences, the program has lost sight of its main original goal during implementation. In the first lesson of strategic planning course, it is usually explained that when building, a correct program it is important, as a first step, to identify who are the beneficiaries. In most "technology driven" programs, often this principle is not taken into account or is forgotten during the implementation. Technologies can represent a fundamental support in improving the efficiency and the effectiveness of cities planning and management, but it is important to have more clearly in mind that technologies are the means and not the target.

References

1. Hilber, C., Cheshire, P.: Concrete gains. America's big cities are larger than Europe's. That has important economic consequences. The Economist (2012), `http://www.economist.com/node/21564536` (last access December 2013)

2. Dougherty, C.: U.S. Cities With Bigger Economies Than Entire Countries. Wall Street Journal (2012), http://blogs.wsj.com/economics/2012/07/20/u-s-cities-with-bigger-economies-than-entire-countries/ (last access December 2013)
3. Cillizza, C.: The case for big cities, in 1 map. Washington Post (2014), http://www.washingtonpost.com/blogs/the-fix/wp/2014/02/19/you-might-not-like-big-cities-but-you-need-them/ (last access December 2013)
4. Glaeser, E.: Triumph of the City: How Our Greatest Invention Makes Us Richer, Smarter, Greener, Healthier, and Happier. Penguin Books (2011)
5. Batty, B.: The New Science of Cities. The MIT Press (2013)
6. Manyika, J., Remes, J., Dobbs, R., Orellana, J., Schaer, F.: Urban America: US cities in the global economy. McKinsey Global Institute, McKinsey & Company (2012)
7. Woetzel, J., Mendonca, L., Devan, J., Negri, S., Hu, Y., Jordan, L., Li, X., Maasry, A., Tsen, G., Yu, F.: Preparing for China's urban billion. McKinsey Global Institute, McKinsey & Company (2009)
8. Sankhe, S., Vittal, I., Dobbs, R., Mohan, A., Gulati, A., Ablett, J., Gupta, S., Kim, A., Paul, S., Sanghvi, A., Sethy, G.: India's urban awakening: Building inclusive cities, sustaining economic growth. McKinsey Global Institute, McKinsey & Company (2010)
9. United Nations, Department of Economic and Social Affairs, Population Division. World Urbanization Prospects (2011), http://esa.un.org/unup/pdf/WUP2011_Final-Report.pdf (last access December 2013)
10. De Biase, L.: L'intelligenza delle Smart Cities (2012), http://blog.debiase.com/2012/04/intelligenza-delle-smart-city/
11. Goodchild, M.F.: Citizens as Voluntary Sensors: Spatial Data Infrastructure in the World of Web 2.0. International Journal of Spatial Data Infrastructures Research 2, 24–32 (2007)
12. Goodchild, M.F.: Citizens as sensors: the world of volunteered geography. GeoJournal 69(4), 211–221 (2007), doi:10.1007/s10708-007-9111-y
13. Goodchild, M.F.: NeoGeography and the nature of geographic expertise. Journal of Location Based Services 3, 82–96 (2009)
14. Murgante, B., Borruso, G.: Cities and Smartness: A Critical Analysis of Opportunities and Risks. In: Murgante, B., Misra, S., Carlini, M., Torre, C.M., Nguyen, H.-Q., Taniar, D., Apduhan, B.O., Gervasi, O. (eds.) ICCSA 2013, Part III. LNCS, vol. 7973, pp. 630–642. Springer, Heidelberg (2013)
15. Archibugi, F.: Introduzione alla pianificazione strategica in ambito pubblico. Alinea Editrice (2002)
16. Tonucci, F.: La città dei bambini. Un nuovo modo di pensare la città. Editori Laterza (2004)
17. Bettencourt, L.M.A.: The Origins of Scaling in Cities. Science, 1438–1441 (June 21, 2013)
18. Berners-Lee T. 5 ⋆ Open Data, http://5stardata.info/#addendum4
19. About W3C Standards, http://www.w3.org/standards/about.html
20. Ball, M.: Reference for 80% of Data Contains Geography Quote, Spatial Sustain: Promoting Spatial Design for a Sustainable Tomorrow (2009), http://www.sensysmag.com/spatialsustain/reference-for-80-of-data-contains-geography-quote.html
21. Franklin, C.: An Introduction to Geographic Information Systems: Linking Maps to databases. Database 15, 13–21 (1992)
22. Murgante, B., Borruso, G., Lapucci, A.: Geocomputation and Urban Planning. SCI, vol. 176. Springer, Heidelberg (2009)

Territorial Specialization in Attracting Local Development Funds: An Assessment Procedure Based on Open Data and Open Tools

Francesco Scorza and Giuseppe Las Casas

Laboratory of Urban and Regional Systems Engineering,
School of Engineering, University of Basilicata,
10, Viale dell'Ateneo Lucano, 85100, Potenza, Italy
francescoscorza@gmail.com, giuseppe.lascasas@unibas.it

Abstract. The Concentration issue by EU Cohesion Policy started new instances in territorial research according with the objective of improving effectiveness of public choices in local development. The paper, after a review of New Cohesion Policy, describes a process of territorial assessment of ROPs oriented to the analysis of territorial specialization in attracting funds. The process is based on Open Data overcoming the dependence from proprietary data formats and software towards interoperability.

Keywords: New Cohesion Policy, Concentration, Specialization, Impact Assessment, Open data, Regional Development, Open-Cohesion.

1 Introduction

The New Cohesion Policy, developed in the context of Europe 2020 agenda, opens to an integrated place-based approach for the improvement of territorial and social cohesion. Smart growth, sustainable growth and inclusive growth for EU 2020 represent overall goals to be achieved under the comprehensive approach defined by Barca [1] as 'place based approach'. As the authors already discussed [2] concerning the issue of territorial impact assessment of regional development policies, the relevant instance comes from knowledge management in regional programming practice. It means data availability, open access to datasets in "near real-time"[1], participation, knowledge sharing, key actors effective involvement in planning process.

The "concentration" issue coming from EU 2020 Cohesion Policy still reflects ambiguity in interpretation [3] and not structured implementation in Regional

[1] We refer to the effectiveness of a 'policy monitoring system' providing data concerning regional programs implementation according to the current status. Today, in the information explosion era it is more useful an on going datasets tuned with the actul implementation status of a program, instead of a final and checked dataset provided years after the closure of a program.

B. Murgante et al. (Eds.): ICCSA 2014, Part II, LNCS 8580, pp. 750–757, 2014.

Programs. From a "thematic concentration" to a "spatial concentration", several attempt are going to be developed in an uncertainty framework.

If a "thematic concentration" reflects more a traditional approach considering a panel of main objectives and goal, it could represents an affective procedure if a proper context analysis identified ex-ante specific needs and priorities coming from local specializations and local communities needs (in other words "place based"). A "spatial-concentration" should produce a map of cohesion programming based on clear and informed decisions expressing the awareness of 'where' to invest in order to maximize the effects of cohesion policies. There is not a ex-ante solution in order to ensure the achievement of regional development results but a balance between a thematic generalization of objectives and a concrete spatial awareness of development precondition should be investigated,

The contribution of 'open data' to the impact assessment of EU Operative Programs appears to be mature in concept but still week in accuracy of available data bases. We used for the research data from the project 'opencoesione' by Italian Ministry for Territorial Cohesion. The Italian Ministry engaged with this unstoppable process of collecting and sharing data for improving citizens commitment on public policies. It developed a web service distributing data on investments policies developed by National and Regional Operative Programs 2007/2013 matching together data from regional and national administrations. The results are analysed in the paragraph number four of the paper with the application of spatial analysis techniques for the evaluation of spatial effects.

In this paper, after a short framework review of New Cohesion Policy issues, we describe a process of territorial impact assessment of Regional Operative Programs investments oriented to the analysis of territorial specialization in attracting funds. The process in completely based on Open Data analysis through Open Tools (software and web services) in order to demonstrate that the integration of such resources overcomes the dependence from proprietary data formats and proprietary software towards interoperability and open information.

2 EU Cohesion and Strategic 'Place' Concentration

EU cohesion policies include different areas of intervention and generally are carried out in order to promote the principle of redistributing opportunities among European regions and territories. It is the largest area of expenditure for European Union and it is possible to affirm that policy analysis tends to overlook the evaluation stage of such complex strategies while a proper assessment practice .

The EU Cohesion Policy is actually interpreted as the main tool in order to achieve the Europe 2020 target addressing a wide range of EU economic, environmental and social objectives. It represents a driven tool toward a new concept of Europe with smart, sustainable and inclusive growth. It currently offers both examples of significant economic and environmental "win-wins" and of "tradeoffs" that fail to offer net added value.

The reform of cohesion approach can be highlighted in two main concept areas including a wide spread of arguments and objectives:

- Investment choices: "where to spend more, where to spend less"
- Investment better - via improved Cohesion Policy governance and tools

This complex policy framework is based on the key objective of "achieving greater economic and social cohesion in the European regions". Anyway some critical consequences could be derived from the point of view of convergence process for lagged regions: the imbalance between regional objectives and financial resources; the existence of serious difficulties for complying with earmarking; and the unknown effects of other policies on regional convergence [4].

The principle of concentration is widely stressed within New Cohesion Policy framework.

The concentration of EU efforts is contra-posed to the indiscriminate distribution of funding ('raining models'). This interpretation could intend that investment promoted by Regional Operative Programs (ROPs) should be focused on specific and circumscribed instances generating effective local development processes. We are in the case of redistributing investment effectiveness and positive outcomes on local communities, instead of realizing a well balanced €/citizen rate within a region.

This synthetic and not exhaustive remark allows us to highlight two relevant aspects of concentration principle: the concentration on objectives and the territorial concentration of investments. If the first appears to be not so far from traditional behavior of managing EU cohesion policies, the second level looks more at the 'place based' approach and expresses the importance of selecting territorial specification.

To the procedural concept through which decisions are taken binds, as a consequence, the concept of evaluation as a process closely linked to the project cycle [5]. It is reductive to solve the problems of evaluation by the application of a techniques set related to limited issues.

In the "evaluation cycle" [6] it is possible to identify three types of evaluation each connected to one or another phase of the project cycle.

Connected to the interpretation depending on evaluation approach to local development processes, appeared the thesis of "renationalization" of cohesion policies [7]. This approach reinforced the role of national administration in driving the implementation of ROPs at regional ad local level. While the negotiation of Member States at the top level of Cohesion Policy hierarchy was previously considered the primary role of National Authorities, now the importance is mainly focussed on the implementation phase. This idea fits more with the 'place based' approach in terms of local specific needs interpretation. In following section we describe the use of an open data service provided by Italian Ministry for Territorial Cohesion. The project 'opencoesione'[2] collected and distributed data on Operative Programs implementation in Italy for the programming period 2007/2013.

[2] http://www.opencoesione.gov.it/

3 Open Data for Effective Spatial Evaluation of Cohesion Policies

Today data availability is not the main problem in territorial investigation but new instances emerged in terms of data management, certification and standard exchange protocols [8]. Many people and organizations collect a wide range of different data in order to perform their own tasks.

The Open data, and in particular 'open government data', are an huge resource still largely untapped. The Government role is particularly important in this sense, not only for the quantity and the centrality of the data collected, but also because most of the government data are public by law, and therefore should be made open and available for anyone to use.

According to the Open Knowledge Foundation Italia [9] there are many circumstances in which we can expect that the open data have significant value.

We are interested in the Measuring the impact of public policies as the extraction of new knowledge by combining different sources of data and the identification of regularities that emerge from the analysis of large masses of data represent the core of the application we propose for the evaluation of ROPs impact at local scale.

In Italy there are many initiatives opening of information assets undertaken by public central and local administrations. The portal dati.gov.it, (available since 2011) is a milestone in the process of opening a new era for innovation and transparency in the public administration.

Actually we can affirm that the practice of open data has been extended, but a lot of work and efforts should still be pay in order to get affective services for data integration.

The project Open Cohesion provides an open data service concerning cohesion policies effects with a orientation toward planning processes.

The publication of the data in an accessible format and reusable on their corporate websites shows the willingness of the government to move in a systematic way towards a structure of transparency that encourages the active participation of citizens and the re-use of data. The service pursues the objective of improving Citizen Engagement on investments policies, and offers a data set with specific information concerning project funded by the current programming period 2007-2013 matching implementation data from regional and national administrations entitled of Ops management.

Open-Government and Open-Data represent the two faces of the same coin.

But, to ensure that data are really "open", they has to be provided in an open and non-proprietary format, without particular restrictions of licenses, reusable and integrable, easily searchable on the web through databases, catalogs and search engines, directly accessible via Internet protocols, network-accessible in network quickly, immediately and at any time, and transmitted directly interchangeable between all users on the network. The data must also be supported by metadata and should allow the export in order to use on-line and off-line, integration, manipulation and share.

We consider the classification of data "open" according to stars scoring by Tim Berners-Lee [10]. Concerning territorial pourpose five stars open linked data includes the adoption of OCG standards

Open Coesione appears to be a result of Open Governement Approach and concerning the Tim Berners-Lee we can affirm that it provides only three stars open data with the opportunity to get spatial dimension information through external elaboration. In this direction, previous researches demonstrated the added value of providing open spatial information concerning development programs. We refer to the experience developed by PIT Marmo Platano Melandro (Basilicata – IT) during the EU programming period 2000-2006 with a web gis service for the spatialization of development policies [11].

While in that experience the main effort was in territorial data production we have to affirm that today it is possible to develop accurate spatial analysis concerning the distribution of EU funded investments with public open data. In the next section we describe the adopted methodology.

4 Toward the Elaboration of Investments Spatial Dimension

The innovative element in the assessment approach we proposed depends on the punctual territorialisation of projects and interventions.

This approach changes the perspective of the assessment because it reinforces the selection criteria on the basis of the request for territorial "specialization" in programming New Cohesion Policy.

Our approach is based on open dataset distributed according to interoperable formats, managed by open-sources software and application, with a strong relationship with web-based services.

The issue of territorial impact assessment of development policies is a domain in which different approaches produce different results that often represent solutions for a specific purpose, serving a specific process of socio-economic and territorial planning without a framework methodology validated under a scientific or technical point of view.

The research aims to provide answers to the demand for territorial specializations analysis oriented to the construction of policy choices to be developed within the EU's 2014-2020 operational planning tools. The proposed approach, based on information concerning the implementation of the instruments 2007-2013, develops a interpretation model that allows a progressive monitoring of on-going processes. A territorial monitoring system that allows at a detailed scale punctual information.

The territorial context of the implementation is the Agri Valley. An inland area of the Basilicata Region in which coexists structural problems for the socio-economic development [12].

5 Open Source and Web-Based Tools: An Innovative Procedure

In relation to the structure of the information sources we used, it has been implemented a procedure 'ad hoc' that exclusively refers to open-source tools. In this section we cite the operational steps and tools used (see next figure)

The figure shows the total of 551 projects localized in the study context. The proposed procedure allows to achieve high accuracy in punctual localization of interventions and projects.

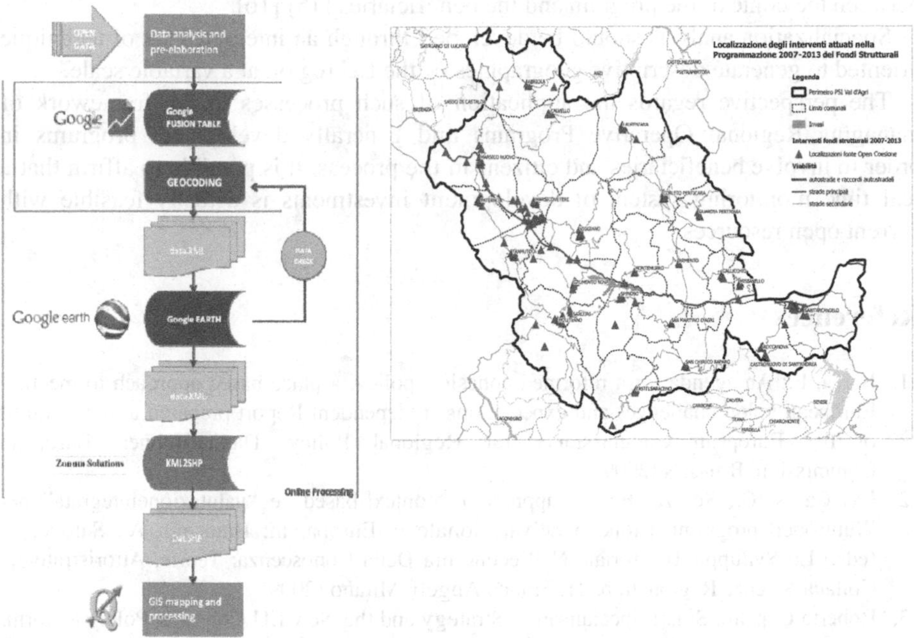

Fig. 1. Territorialisation procedure flow chart [13] and investments point pattern

6 Conclusions and Perspectives

Place based approach will bring to innovations in EU cohesion management. Where outcomes indicators measure the implementation of cohesion operative program [14] other efforts should be addressed to the identification of local specialization. It could generate not a fix picture of a context, places and communities evolve continuously especially as a reaction to the huge changes brought by economic crisis. Main issues connected with the instances of the New Cohesion Policies are: The need of a clear identification of the combined place-specific characteristics in each region; a clear identification of the appropriate territorial context in order to implement effectively "smart specializations". Open data phenomena represent an useful process that already driven the research from data production to exploitation of the informative value of several data sources available for everybody. But data and data analysis technique cannot bring to useful information. Regional science has the task to produce effective 'places' interpretation in order to support public decision in incoming generation of EU ROPs. We are in the case in which it is relevant to use numerous data sources and indicators assuming a variable rate of approximation in the accuracy of the datasets.

The information management and exchange implies problem in interoperability between sources, procedures and technologies. In the field of Regional development the ontological approach provided alternative interpretation models of the interaction between the context, the program and the beneficiaries [15] [16].

Specialization analysis should be developed through an integrated set of technique oriented to generate descriptive geographies of the EU region at a variable scale.

The perspective regards the application of such processes in the framework of managing Regional Operative Programs and generally development programs in order to involve beneficiaries and citizens in the process. It is possible to affirm that a real time monitoring system of development investments is actually feasible with current open resources.

References

1. Barca, F.: An agenda for a reformed cohesion policy: a place-based approach to meeting European union challenges and expectations. Independent Report prepared at the Request of the European Commissioner for Regional Policy, DanutaHübner. European Commission, Brussels (2009)
2. Las Casas, G., Scorza, F.: Un approccio "context-based" e "valutazioneintegrata" per ilfuturodellaprogrammazioneoperativaregionale in Europa. In: Bramanti, A., Salone, C. (eds.) Lo Sviluppo Territoriale Nell'economia Della Conoscenza: Teorie, Attoristrategie. Collana Scienze Regionali, p. 41. Franco Angeli, Milano (2009)
3. Roberta Capello. Smart Specialisation Strategy and the New EU Cohesion Policy Reform: Introductory Remarks. Scienze Regionali 2014, Fascicolo: 1, pp. 5–13 (2014), doi:10.3280/SCRE2014-001001
4. Las Casas, G., Scorza, F.: Un approccio "context-based" e "valutazioneintegrata" per ilfuturodellaprogrammazioneoperativaregionale in Europa. In: Bramanti, A., Salone, C. (eds.) Lo Sviluppo Territoriale Nell'economia Della Conoscenza: Teorie, Attoristrategie. Collana Scienze Regionali, vol. 41. Franco Angeli, Milano (2009)
5. Institute for European Environmental Policy (IEEP) (2011) Cohesion Policy and Sustainable Development, Executive Summary Retrived at (Maj 2013), http://ec.europa.eu/regional_policy/sources/docgener/studies/pdf/sustainable_development/sd_executive_summary.pdf
6. Hoerner, J., Stephenson, P.: Theoretical Perspectives on Approaches to Policy Evaluation in the EU: The Case of Cohesion Policy. Public Administration 90, 699–715 (2012), doi:10.1111/j.1467-9299.2011.02013.x
7. Las Casas, G.B.: Processo di piano edesigenze informative. In: Clemente, F. (ed.) Pianificazione del Territorio e Sistema Informativo. Ed. F. Angeli, Milano (1984)
8. Lombardo, S. (ed.): La valutazione del processo di piano. Contributiallateoria e al metodo, CollanaScienzeregionali. Franco Angeli, Milano (1995)

9. Mancha-Navarro, T., Garrido-Yserte, R.: Regional policy in the European Union: The cohesion-competitiveness dilemma. Regional Sci Policy & Practice 1, 47–66 (2008), doi:10.1111/j.1757-7802.2008.00005.x
10. Scorza, F.: Improving EU cohesion policy: The spatial distribution analysis of regional development investments funded by EU structural funds 2007/2013 in italy. In: Murgante, B., Misra, S., Carlini, M., Torre, C.M., Nguyen, H.-Q., Taniar, D., Apduhan, B.O., Gervasi, O. (eds.) ICCSA 2013, Part III. LNCS, vol. 7973, pp. 582–593. Springer, Heidelberg (2013)
11. Open Knowledge Foundation Italia, Open data handbook (May 2013), http://opendatahandbook.org/ (retrived)
12. Tim Burner Lee (2009), http://www.w3.org/
13. Murgante, B., Tilio, L., Lanza, V., Scorza, F.: Using participative GIS and e-tools for involving citizens of MarmoPlatano – Melandro area in European programming activities, special issue on E-Participation in Southern Europe and the Balkans. Journal of Balkans and Near Eastern Studies 13(1), 97–115 (2011) ISSN:1944-8953, doi:10.1080/19448953.2011.550809
14. ENI in Basilicata, Local Report. ENI (2012)
15. Barca, F., McCann, P.: Methodological note: outcome indicators and targets-towards a performance oriented EU cohesion policy and examples of such indicators are contained in the two complementary notes on outcome indicators for EU2020 entitled meeting climate change and energy objectives and improving the conditions for innovation. Research and Development (2011), http://ec.europa.eu/regional_policy/sources/docgener/evaluation/performance_en.htm (accessed October 1, 2011)
16. Scorza, F., Casas, G.B.L., Murgante, B.: That's ReDO: Ontologies and Regional Development Planning. In: Murgante, B., Gervasi, O., Misra, S., Nedjah, N., Rocha, A.M.A.C., Taniar, D., Apduhan, B.O. (eds.) ICCSA 2012, Part II. LNCS, vol. 7334, pp. 640–652. Springer, Heidelberg (2012)
17. Las Casas, G., Scorza, F.: Redo: applicazioniontologiche per la valutazionenellaprogrammazioneregionale, Italian. Journal of Regional Science - Scienze Regionali 10(2), 133–140 (2011), doi:10.3280/SCRE2011-002007
18. Scorza, F., Casas, G.L., Murgante, B.: Overcoming Interoperability Weaknesses in e-Government Processes: Organizing and Sharing Knowledge in Regional Development Programs Using Ontologies. In: Lytras, M.D., Ordonez de Pablos, P., Ziderman, A., Roulstone, A., Maurer, H., Imber, J.B. (eds.) WSKS 2010. CCIS, vol. 112, pp. 243–253. Springer, Heidelberg (2010)

Using Spatiotemporal Analysis in Urban Sprawl Assessment and Prediction

Federico Amato, Piergiuseppe Pontrandolfi, and Beniamino Murgante

University of Basilicata, 10,Viale dell'Ateneo Lucano, 85100 Potenza, Italy
fdrc.amato@gmail.com,
{piergiuseppe.pontrandolfi,beniamino.murgante}@unibas.it

Abstract. The importance of soil resource protection is now universally recognized, but despite a lot of debates and principles enunciation, in the last decades the soil was consumed at a rate of 8 m^2 per second. In this paper a simulation model has been proposed based on two methods: Joint information uncertainty and Weights of Evidence in order to analyse and predict new built-up areas. The proposed model has been applied to Pisticci Municipality in Basilicata region (Southern Italy). This area is a significant example, because of high landscape values and, at the same time, of a lot of developing pressure due to touristic activities along the coastal zone.

Keywords: Urban planning, Soil Consumption, Urban sprawl, Built-up areas, Sustainability.

1 Urban Sprawl: Phenomenon Definition and Analysis

Since more than a decade, the European Union recognizes soil as a common good and considers it as a finite resource with an inestimable value. Consequently, soil is strongly linked to agricultural and zoo-technical production, and therefore it intensely influences human nutrition.

The current gradual and steady world population growth is mainly due to the introduction of intensive farming technologies, which produced the maximum exploitation of soil potential. The cultivation of a single plant species, monoculture, on one hand ensures a greater amount of food production, whilst, on the other hand has a high environmental cost, using a high extent petroleum derivatives to move machinery, to fertilize and protect crops from pests. Monoculture is one of the main effects of green revolution. It started following the end of Second World War, and, together with the processes of market liberalization and globalization, it has led to a growth of food production by 140%, 200% and 280% in Africa, Latin America and Asia, respectively. Industrialization of agriculture has produced two particularly effective consequences. The first one is related to occupation. The use of machinery for agricultural production has greatly reduced labour supply in the sector: in Europe 44% of the land is cultivated, but the primary sector contributes only to 5.5% of employment.

B. Murgante et al. (Eds.): ICCSA 2014, Part II, LNCS 8580, pp. 758–773, 2014.

The second effect is the abandonment of countrysides and the connected-urbanization process of rural population. In Italy, from 1951 to 1991 employees in agriculture passed from 44% to 9% of the population. This demographic phenomenon has developed in parallel with the expansion of cities and settlements.

For this reason, today the main threat to agri-food sector is represented by the expansion of urban areas, because it occurs mainly with great disadvantage of rural areas. This point has great importance, because having less fertile land available means to weaken the potential productive of a country and to amplify the dependence by food and forage, increasing transport and, consequently, pollution.

Soil, however, is not only the place of agriculture. It is also the location of exchanges of energy and matter with other environmental compartments.

It is an open system, and the main biogeochemical cycles pass through soil. Soil is actively involved in the hydrological cycle, because it intercepts most of the water derived from precipitation. Sliding surface is strongly influenced by soil use. Paved or built areas limit or eliminate soil absorption capacity by rainfall. In land use planning, it is important to consider water surface drainage, because it affects erosion and flooding. Carbon cycle is the second biogeochemical cycle where soil plays a crucial role. The greatest amount of carbon is not concentrated in the atmosphere, but in forests and soils. It is estimated that the amount of carbon dioxide contained in forest biomass and in forest humus is 1.5-4 times higher than carbon dioxide contained in the atmosphere, respectively [1]. Given the enormous amount of carbon stored in the soil, land use change can lead to a significant increase in CO_2 emissions in the atmosphere.

The inestimable soil value is exemplarily summarized in the definition provided by the European Union [2], which considers it as "the upper layer of the earth's crust, formed by mineral particles, organic matter, water, air and living organisms".

2 A Quantitative Model to Assess and Predict Land Use Changes

In last decades a lot of models have been proposed to analyse land use changes [4, 5, 9, 11, 12, 13, 14, 15, 16, 17]. The aim of this study is to propose a model which, on one side, is able to measure variations occurred in land use, and, therefore, to determine soil consumption, and, on the other side, is capable to predict future changes.

The simulation is based on two methods. *Joint information uncertainty* [7] has been adopted in estimating spatial relationships between variables.

Subsequently, *Weights of Evidence* [7] method has been adopted to determine a probability map. More in particular, a raster map has been produced where the value of each pixel is included in a scale ranging between 0 and 100, where the value indicates the conditional probability that a certain land use change occurs.

The model adopts three land use maps in raster format at different dates T0, T1 e T2 and the spatial variables, also in raster format, that geographically identify factors, such as constraints or planning regulations which contribute to influence soil transformations. The model has been applied two times: the first one in order to determine a calibration, the second in order to produce a simulation.

Model calibrate aim is to adequately define the relationship between spatial variables in order to produce simulation results coherent with actual land use changes. Maps at dates T0 and T1 are used as input in order to create a simulation at time T2. At this point it is possible to compare the simulation T2 with actual land use at T2.

When the two maps are appreciably similar, the model will be properly calibrated, and it will be able to produce simulations which well describe actual land use changes. The simulation adopts the already calibrated model with maps at T1 and T2 dates in order to produce a land use at T3 time.

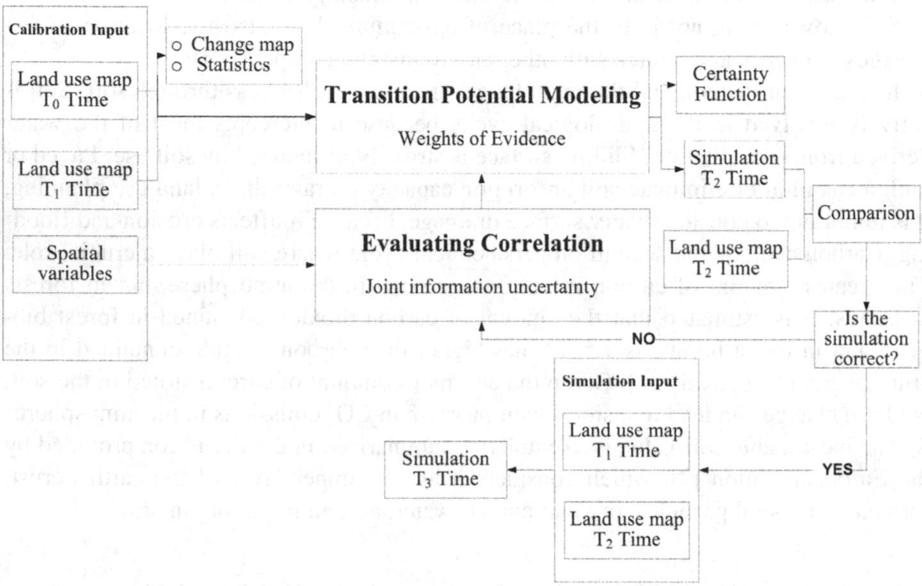

Fig. 1. The scheme shows the sequence of the different phases of model calibration and simulation

Evaluating variables correlation, we suppose to detect two politomic variables, able to assume a number of values greater than two, in a sample of n subjects.

	y_1	y_2	..	y_j	..	y_p	
x_1	n_{11}	n_{12}	..	n_{1j}	..	n_{1p}	$n_{1\cdot}$
x_2	n_{21}	n_{22}	..	n_{2j}	..	n_{2p}	$n_{2\cdot}$
:	:	:		:		:	
x_i	n_{i1}	n_{i2}	..	n_{ij}	..	n_{ip}	$n_{i\cdot}$
:	:	:		:		:	
x_k	n_{k1}	n_{k2}	..	n_{kj}	..	n_{kp}	$n_{k\cdot}$
:	:	:		:		:	
x_m	n_{m1}	n_{m2}	..	n_{mj}	..	n_{mp}	$n_{m\cdot}$
	$n_{\cdot 1}$	$n_{\cdot 2}$	..	$n_{\cdot j}$	..	$n_{\cdot p}$	n

Fig. 2. Contingency table

A double-entry table containing non-negative numbers can represent data. This is called a contingency table with $(m-1)(p-1)$ degrees of freedom. It is a table with the joint distribution of two variables. More particularly, rows in the table identify n values of X variable, while columns identify p values of Y variable.

The generic cell n_{ij} represents the absolute joint frequency, i.e. the number of statistical units in the sample that joint a feature associated with mode x_i of the i-th row and a feature associated with mode y_j in the j-th column. The total of each row represents the frequency with which the mode of the first variable X was observed how.

Joint information uncertainty is part of entropy measures [6]. These are calculated by the analysis of contingency table, and can be used to measure the association between two variables. Assuming to transform n_{ij} values in proportion to the area, dividing each element by the total n:

$$p_{ij} = \frac{n_{ij}}{n}$$

and also defining:

$$p_i = \frac{n_i}{n}; \quad p_j = \frac{n_j}{n}$$

entropy measures can be defined using the value of p_{ij} as probability estimation. This value is dimensionless, and can be then compared to measurements with Pearson chi-square, with the advantage of not having dependencies on measurement units.

Considering map A and map B and considering the contingency table as known, A and B entropies are defined as:

$$H(A) = -\sum_{j=1}^{m} p_j \ln p_j$$

$$H(B) = -\sum_{i=1}^{m} p_i \ln p_i$$

where ln indicates the natural logarithm. The joint entropy of two maps combination will be defined as

$$H(A,B) = -\sum_{i=1}^{m}\sum_{j=1}^{m} p_{ij} \ln p_{ij}$$

The joint information uncertainty is used as strength measure of the association, and it is defined as

$$U(A,B) = 2\left[\frac{H(A) + H(B) - H(A,B)}{H(A) + H(B)}\right]$$

always with a value between 0 and 1.

When the two maps, and therefore the variables, are completely independent of each other, we will have $H(A,B) = H(A) + H(B)$ e $U(A,B) = 0$, and in the case of perfect dependence we will have $H(A,B) = H(A) = H(B) = 1$ and $U(A,B) = 1$.

Once determined the correlation between variables, the conditional probability is quantified by Weights of Evidence method (WoE), which is based on Bayes' theorem, which states that given a predisposing factor B, divided into thematic classes, and an event s, the conditional probability that the event s occurs in the presence of the i-th class B_i is equal to:

$$P(s|B_i) = \frac{\{P(B_i|s) \times P(s)\}}{P(B_i)}$$

where:

- $P(B_i|s)$ is the conditional probability of B_i given the event s;
- $P(s)$ is the a priori probability that event s, also called source, occurs in the study area (AS);
- $P(B_i)$ is the probability of finding class B_i in the study area (AS).

The conditional probability that the event s will not occur in correspondence to the class B_i is instead defined by the relation:

$$P(s|B_i\wedge) = \frac{\{P(B_i\wedge|s) \times P(s)\}}{P(B_i\wedge)}$$

where:

- $P(s|B_i\wedge)$ is the conditional probability of not having the class B_i assigned to event s;
- $P(s)$ is the a priori probability that event s occurs in the study area (AS);
- $P(B_i\wedge)$ is the a priori probability that class B_i is not present in the study area (AS).

In operative terms, it is possible to calculate:

- $P(s) = (\text{area s})/(\text{area AS})$
- $P(B_i) = (\text{area } B_i)/(\text{area AS})$
- $P(B_i\wedge) = (\text{area } B_i\wedge)/(\text{area AS})$
- $P(s|B_i) = (\text{area s\_}B_i/\text{area}B_i)/P(B_i)$
- $P(B_i|s) = (\text{area s\_}B_i/\text{area}B_i)/P(s)$
- $P(s|B_i\wedge) = (\text{area s\_}B_i\wedge/\text{area}B_i\wedge)/P(B_i\wedge)$
- $P(B_i\wedge|s) = (\text{area s\_}B_i\wedge/\text{area}B_i\wedge)/P(s)$

From a purely mathematical point of view, it may be useful to express the probability in terms of Odds. This is the ratio of probability occurrence of an event and probability that this event does not happen. In this case the two previous equations can be rewritten in the form:

$$O(s|B_i) = \frac{O(B_i|s)}{O(B_i)} O(s)$$

$$O(s|B_i\hat{}) = \frac{O(B_i\hat{}|s)}{O(B_i\hat{})} O(s).$$

From these we can calculate odds natural logarithm obtaining:

$$\ln O(s|B_i) = W^{+}_{B_i} + \ln O(s)$$

$$\ln O(s|B_i\hat{}) = W^{-}_{B_i} + \ln O(s)$$

where W^+ is the positive weighted value that should be assigned when i-th class of factor B is present, while W^- is the negative weighted value assigned when i-th class of factor B is absent. The two weighted values can be obtained according to the expressions:

$$W^{+}_{B_i} = \ln \frac{P(B_i|s)}{P(B_i|s\hat{})}$$

$$W^{-}_{B_i} = \ln \frac{P(B_i\hat{}|s)}{P(B_i\hat{}|s\hat{})}.$$

W^+can be seen as the ratio between the probability of finding a thematic class in which the event occurred and the probability of finding the same class in the absence of the considered event; while W^- can be seen as the ratio of the probability of not finding a thematic class in which the event occurred and the probability of not finding the same class in an area where the considered event is not present.

It is possible to observe that the greater the weighted value of the considered class in predicting the event under investigation will be, the greater the value W^+ will be, and similarly the lower the weighted value of the considered class in predicting the event under investigation will be, the greater the value W^- will be.

The difference between positive weighted-value and negative weighted-value is defined contrast:

$$C = W^+ - W^-$$

and is an effective measure of the correlation between the analyzed thematic class and the considered events represented on a cartographic base.

A thematic class that produces null contrast values is not significant for analysis purposes, while a contrast positive value indicates a positive correlation of the thematic class compared to the examined events, and negative values suggest that the spatial distribution of thematic classes is independent in relation to the considered phenomena. In other words, the more the occurrence of an event within a thematic class is greater than its random distribution, the more W^+ will assume positive values and W^- negative values. Therefore, in case there is a positive spatial correlation, the number of events within the thematic class is higher than those randomly obtained, and vice versa for a negative spatial correlation in the considered thematic class the number of events is less than those due to a random distribution.

The Weights of Evidence method is particularly useful in land use changes prediction, because it effectively interprets relationships between these variations and the adopted spatial variables.

3 Applying the Model to a Real Case Study

The proposed model has been applied to a real case study. The study area is Pisticci Municipality in Basilicata region (Southern Italy). Basilicata is a region with low population density (575 505 inhabitants distributed on 9,992 km^2, corresponding to 57.5 inhabitants/km^2) that, partly because of bad policy, partly due to a lack of large private investors, in past decades has not been able to exploit its territory with high landscape values to develop tourism. Pisticci municipality is a significant example.

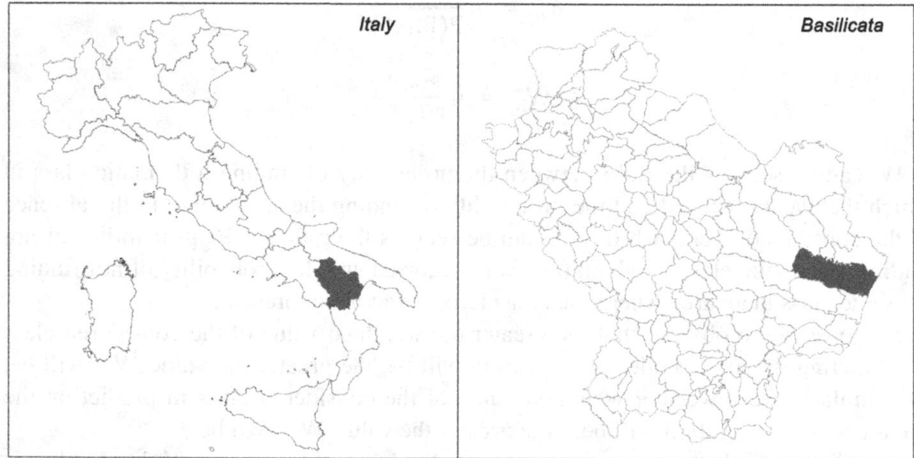

Fig. 3. Left: location of Basilicata region (highlighted in red) within Italian peninsula. Right: Pisticci Municipality (highlighted in red) within Basilicata Region.

Third center of Basilicata for inhabitants number, Pisticci is located in an area between Cavone and Basento rivers. Two important highways, S.S. Basentana and S.S. Jonica, also cross the territory. Ridges and uplands characterize the northern part of the municipality, while the area close to Ionian Sea has mostly a flat conformation.

The coastal area, characterized by strong tourism pressure, constitutes one of the most relevant areas to be analysed in the field of land use changes. The municipal area is therefore characterized by great diversity, due to the varied morphology, and has led in earlier times to locate the settlement along the main ridge. Only later, in the late twentieth century, building development has moved downstream, and in recent times has involved the coast. In addition to the main centre, other important settlements, as Marconia and Tinchi hamlets, have been developed in the valley, while small rural villages and tourist sites characterize the coast. In the area closest to the historical settlement of Pisticci, large areas used for natural pasture are measured. These areas are the steepest in the whole municipality and are characterized by arid and clay soil, mostly not irrigated. These characteristics define the Calanchi (geological formations determined by erosion on clay soils), which contribute to uniquely characterize the landscape of the area. Moving towards Tinchi and Marconia, it is immediately noticeable how the original vocation of these settlements was agriculture, and still today the soil surrounding the extra-urban area of these two centres is characterized by the presence of crops, irrigated or not. Data are measured through surveys carried out on the map at 1:25,000 scale of 1952 by Italian Military Geographic Institute (IGM), on IGM maps at 1:50,000 of 1972, and on ortho-photos from 1982, 1989, 1997, 2000 and 2006 available in Web Map Service (WMS) format on Italian National Geoportal (www.pcn.minambiente.it/). The increase of built-up areas at the various dates firstly coincides with the development of the two settlements of Tinchi and Marconia, then with housing development of the coastal zone.

Table 1. Quantity of built-up areas [ha] measured at different times

	1956	**1972**	**1982**	**1989**	**1997**	**2000**	**2006**
Unbuilt [ha]	23088	23077	23058	23018	23010	23007	23005
Built [ha]	28	39	58	98	106	109	111

In a less intense but steadily process over time, an increase of buildings number within extra-urban area is observed, especially in correspondence of main transport infrastructures.

Table 2. Input data used for model implementation

Category	**Variable**
Built-up areas	Built-up areas in 1956 Built-up areas in 1982 Built-up areas in 1989 Built-up areas in 2006
Geomorphological system	Slope Sun exposure
Environmental system	Maritime constraint Hydrogeological restrictions Forestry constraint Areas subjected to environmental regeneration Landscape plan (mouth of Basento river) Not transformable areas
Settlement system	Urban Implementation Plans and Shores Plan Building density
Relational system	Distance from the main road

The case of Pisticci is a significant case in the context of Basilicata Region, because it is one of the most populated municipalities, and it is in part supported by agricultural economy, like great part of municipalities included in this coastal zone, in part by tourism. Consequently soils, especially in the coastal area, are constantly suspended between agricultural and urban uses. Moreover, the presence, in addition to the historical settlement, of other two centres of significant dimension, means that there are three urban contexts, and therefore three components capable to change the positional rent in surrounding soils. The application of the model is limited for the lack of basic spatial information, which is typical in great part of Italy. Unable to use a time series data on the actual land use, we decided to use only data on built-up areas. The application is therefore designed to measure sprawl occurred in the area and to estimate the dimension that this phenomenon can have in the future. Obviously in this way it is not possible to measure total soil consumption, but still it is possible to identify the extent of the main component of the phenomenon. Table 2 shows input data used for model implementation.

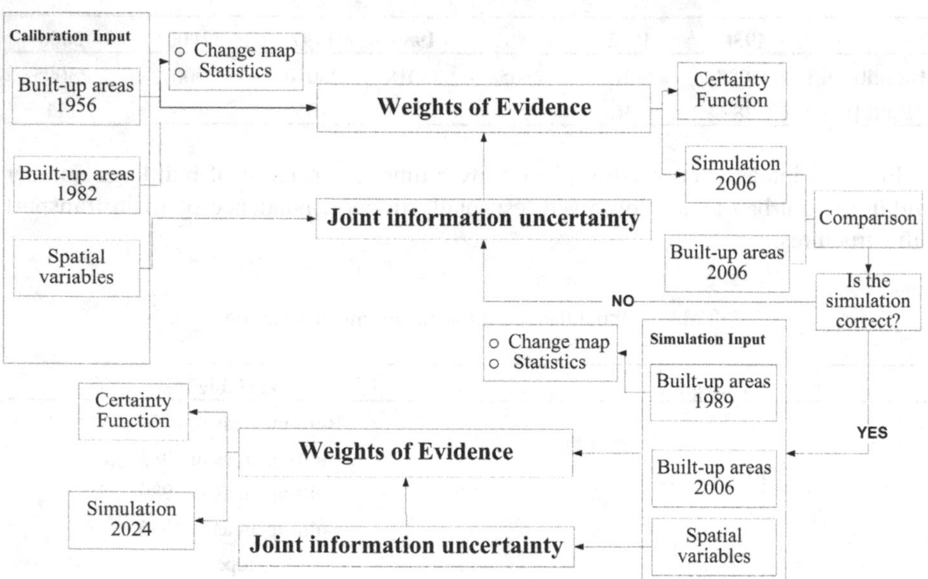

Fig. 4. The scheme shows the sequence of the different phases for the calibration and simulation of the model concerning the application to Pisticci municipality

In model execution, therefore, during the calibration phase the map at T_0 time of built-up areas in 1956 and the map at T_1 time of built-up areas in 1982 have been used. In this way, the simulation has been performed using Weights of Evidence, getting built-up areas map in 2006 as output. This simulation has been compared with the reference map at T_2 time, i.e. actual built-up areas map in 2006 and when the error has been considered insignificant the simulation phase has been implemented. In the second phase built up areas in 1989 and 2006 have been adopted as input data, running again the WoE with the same contrast parameters previously calculated, in order to obtain a simulation of built-up areas in 2024.

4 Results

Simulation results should be analysed under two different profiles: quantitative and positional. Initial assessments should refer to the quantitative aspects, purely comparing numerical terms of the amount of built-up areas provided by the simulation with those measured in 2012, without considering at this moment the geographical distribution of these quantities.

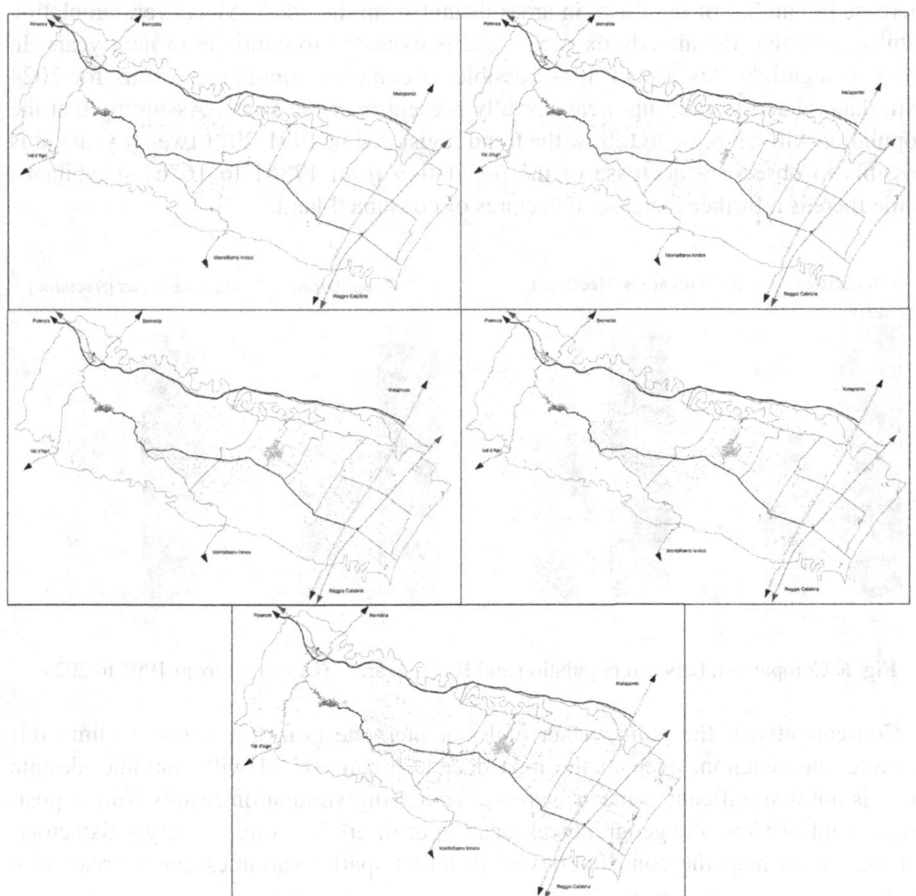

Fig. 5. From top to bottom, from right to left: the images show built-up areas within Pisticci Municipality in red, in 1956, 1982, 1989, 2006, respectively and in simulation for 2024

This comparison can be significant when comprehensively analysing the current situation of Pisticci Municipality in terms of demography and building development.

Observing the comparison between population and built-up areas (figure 6), it is very evident that large building production generated in twenty years (between 1991 and 2011) is far from being supported by a real demand increase of resident population, which is in slight decrease. Observing the comparison between households and number of dwellings, it is clear that the growth in number of available dwellings has occurred at a rate much higher than the growth in number of households. All comparisons carried out are exclusively based on resident population data, not taking into account how the economy of Pisticci municipality in the last twenty years, had a development in tourism field, which consequently produced an increase of housing demand. The demographic phenomena related to tourism, however, can not justify the high soil consumption measured, and especially can not justify the increase in number of buildings in areas distant from the coast. Moreover, simulation results show that the already existing trend is expected to continue in next years. In order to highlight this aspect, it is possible to compare simulation results for 2024 with data related to built-up areas actually present over the years. Assuming that the population will continue to follow the trend registered in 1991-2011 twenty years, it is possible to observe a decrease of the population from 17361 to 16761 inhabitants, while there is a further increase of hectares of consumed land.

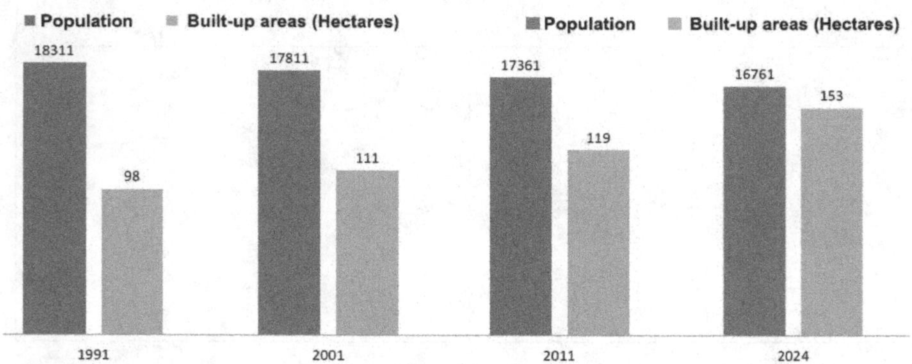

Fig. 6. Comparison between population and Built-up areas (Hectares) from 1991 to 2024

Consequently, if the policymaker will not intervene in a drastic way to limit this housing phenomenon, even in the next decade urban sprawl will continue, despite there is not a significant demand increase. Analysing simulation results from a positional point of view, the geographical location of future buildings is very satisfactory. On the output map the contributions of different spatial variables can be read with clarity.

Table 3. Quantity of built-up areas [ha] measured at different times

	1956	1982	1989	2006	2012	56-12	12-24	2024
Pisticci Municipality	28	58	98	111	125	97	28	153
Tinchi	1.62	6.05	9.06	9.73	10.43	8.81	5.72	16.15
Marconia	1.19	8.91	26.39	28.10	30.59	29.4	7.10	37.69
Coastal area	0.14	0.85	3.95	7.94	15.68	15.54	7.01	22.69

The new buildings are never included on constrained areas or on soils where it is not possible to build. Instead, they are very dense in correspondence of settlements, thickening built-up areas located along the road that connects the historic part of Pisticci with the centres of Tinchi and Marconia. Data shown in Table 3 include also edifications occurred within urban areas according to planning documents. However, although this rate is due to planned changes and therefore not strictly included in sprawl phenomenon, these changes have also produced soil consumption. In the centre of Tinchi, the amount of new built-up areas realized since post-World War II up to 1982 is very high (4,43 hectares). In the same period the amount of built-up areas throughout the whole municipal area was 30 hectares. In other words, 14.76% of edifications realized between 1956 and 1982 were concentrated in Tinchi. The percentage is even higher considering that the centre of Marconia has been developed in the same period, determining great part of the 30 hectares consumed. Most of the 30 hectares not built in the centre of Marconia, therefore, was built in Tinchi.

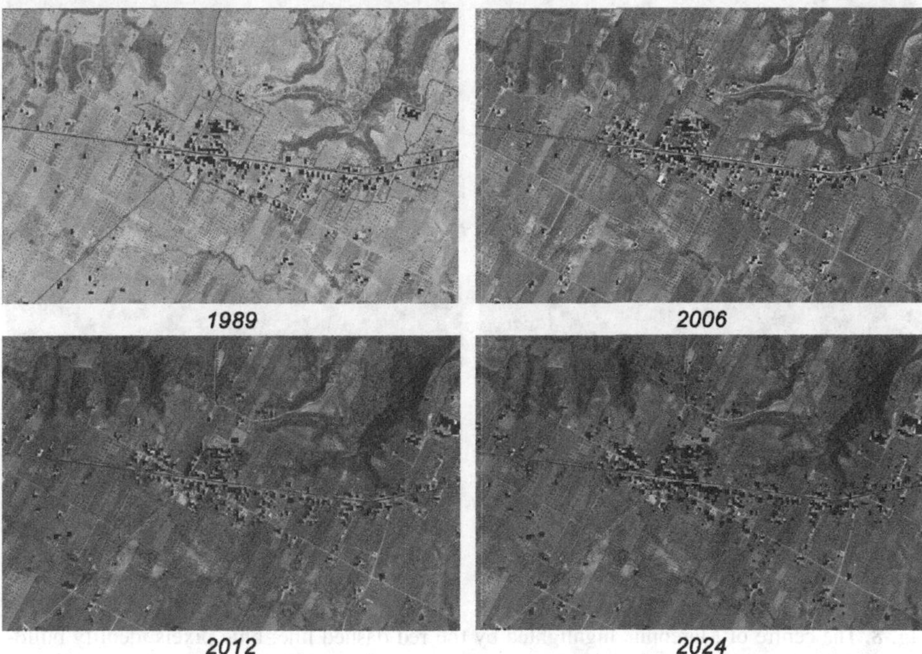

Fig. 7. The centre of Tinchi, highlighted by the red dashed line; blue pixels identify buildings at different dates

The comparison between built-up areas measured in the period 1991-2012 with resident population data provided by ISTAT (Italian Institute for Statistics) shows that the rapid growth of built-up areas is accompanied, instead, by a slight fluctuation of resident population. In the period 1991-2001 a slight increase occurred, while in the next decade a small reduction took place with 26 residents less than 1991 in 2011. The simulation considers that in 2024 Tinchi will have total built-up areas equal to 16.15 hectares. The simulation map (figure 7) shows as such an increase was homogeneously distributed between the centre of Tinchi and its surrounding areas. In both situations, however, the simulation provides a concentration of new buildings in proximity of already built areas, defining a densification trend towards the existing settlements. The development of Marconia from 1956 has been continuous. In 1956 the settlement was mainly agricultural, with 1.19 hectares built, which became 8.91 in 1982. The largest increase, however, occurred in the eighties when 17.48 hectares were built areas. In only seven years, therefore, built-up areas grew with the impressive speed of 2.49 hectares per year. Building has also continued during the nineties, and still continues nowadays, although at slower rates. In 2006 there were 1.71 hectares more than in 1989, and in 2012 2.49 ha more than 2006. It is possible to make two considerations. The first one is that rate of construction in the late nineties and the early years of the current decade was less than that calculated in the previous decade.

Fig. 8. The centre of Marconia, highlighted by the red dashed line; blue pixels identify buildings at different dates

The second is that, in contrast with the national trend, from 2006 up to 2012 there has been a new acceleration in building. The case of Marconia is significant because for the first time an increase of built-up areas is accompanied by an increase of resident population. This is mainly due to migratory movements within the municipality, from the historical settlement of Pisticci to Marconia. Analysing the decade 2001-2011, it is possible to note that although the population is still growing, the increase is less than in the previous decade, while built-up areas increased with a rate greater than the period 1991-2001. The last situation analysed is the part closest to the coast.

More specifically, what occurred to soils located at a maximum distance of 3 km from the shore line will be evaluated. Building activities in this strip are mainly due to the desire to create an environment suitable for the reception of tourists during the summer season. Therefore, it is not possible in this case to compare sprawl with data on resident population, because such data do not consider people who live only for short periods in the municipality. Land use changes in the coastal strip of Pisticci municipality are quite alarming. In 1989, there were 3.95 hectares of built-up areas, which became 7.94 in 2006. In 2012, the datum is almost doubled to 15.68 hectares of built-up areas. These increases are mostly due to the realization and completion of touristic villages. The increase of buildings scattered along the main roads in the areas should be also considered [3].

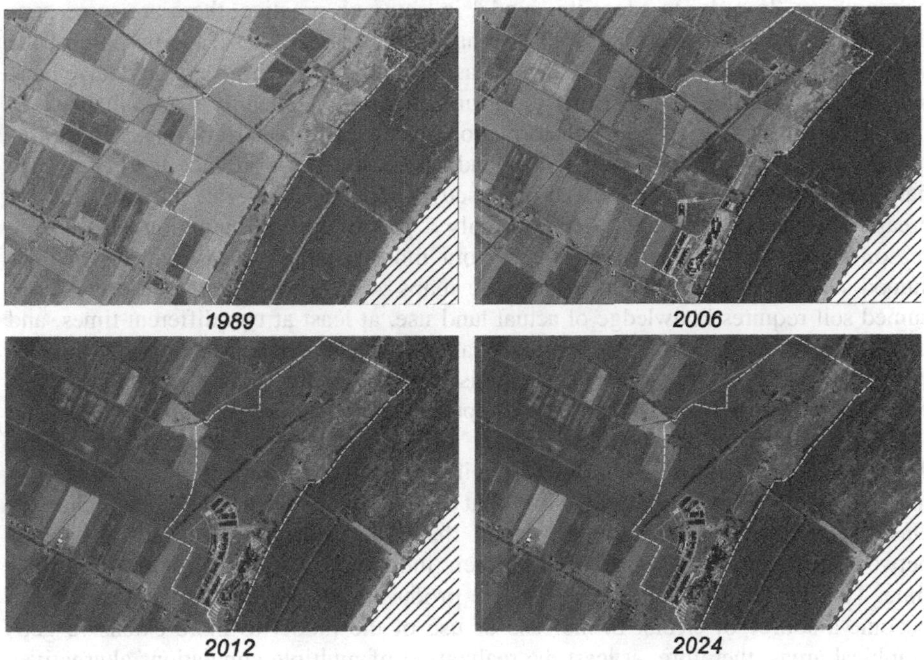

Fig. 9. An example of touristic village, highlighted by the yellow dashed line; blue pixels identify buildings at different dates

Also in this case, the simulation accounts for the considerable increase in building speed in the period 2006-2012. The prediction of additional 7.01 hectares of new built-up areas from 2012 until 2024 can represent a threat. The increase in built-up land in coastal areas can be compared with what occurred along the Adriatic Sea coast of Italy since the early seventies. In this period the development of the touristic sector produced a huge soil consumption along the Adriatic coast from North to South indifferently [8]. As already mentioned, Basilicata region is not historically part of the main Italian touristic destinations. This trend changed in recent years, as highlighted in the case of Pisticci, and together with the development of tourism an excessive building up activity is taking place. The hope is that after what happened along the Adriatic coast and following the alarming results of many researches, including the present one, may encourage policy-makers to pursue sustainable tourism development.

5 Conclusions

The objective of this study was to verify the adequacy of the spatio-temporal model in land use changes measurement and prediction. This would be an important analytical tool available to planners and policy makers in order to address their own choices. The experience is partially successful: in the previous sections we have seen how the simulation carried out in Pisticci Municipality produced a realistic trend of sprawl phenomenon. In order to adopt this model in common practice, there are some gaps that need to be improved. First of all, we have to recognize that if the purpose of the study was to achieve a land use change measure, then the experience would have been substantially a failure. In fact, the term land use refers not only to the increase in built-up areas, but to all soil transformation that lead to waterproofing, and consequently to the loss of ecosystem values and of the ability to produce food or to take part in biogeochemical cycles. The proposed experience, however, measures the increase of built-up areas, and, therefore, only a part of soil consumption causes. The reason why it was not possible to carry out more complete measurements of consumed soil is the lack of detailed land use maps at different dates. Measure of consumed soil requires knowledge of actual land use, at least at two different times, and unfortunately, these data are not currently available. But the real intention of this paper was to verify the reliability of the statistical models for spatial-temporal analysis of land use changes. In this sense, the experience was a success. Certainly, we can identify a number of aspects that should be subject of further consideration in future. The presented model, for example, determines, through a complex system of weighing of spatial variables, the probability that a soil, it is better to say a pixel, changes class in land use classification. However, the basic assumption is that the balance between different variables, as well as their individual weights stays constant over time. This, especially when predictions are produced at very huge time intervals, can become a source of errors. In the case of use of the model to more extensive geographical areas, therefore, at least the realization of multiple simulations alternatives is desirable, in order to produce different scenarios according to the different weights that different variables may assume over time. Moreover, this model is not able to compute the economic aspects that instead are fundamental in the analysis of housing market phenomena. In fact, the decrease of housing development occurred since 2005-2006 and still prosecuting, is due exclusively to the changed international and national economic conditions, and not to designed regulations to limit land use changes. In the specific case of Pisticci municipality, the inability to take into account this variable did not affect the final result, because comparing measured data of built-up areas in 2006 with those measured in 2012, an increase in built-up areas has occurred. Consequently, there was no analogy with the observed dynamics at international and national level. However, this has basically represented a chance circumstance, and future experiences should include aspects related to real estate market trends in prediction models, too.

References

1. Zampogno, L., Cattaneo, T. (eds.): Suolo bene commune – dalla convenzione europea del paesaggio al governo sostenibile del territorio. Legambiente Lombardia Onlus (2012)
2. COM/2006/0231 a firma della Commissione al parlamento Europeo, al Consiglio, al Comitato Economico e Sociale ed al Comitato delle Regioni in data 22/09/2006 "Strategia tematica per la protezione del suolo"
3. Agostinacchio, M., Ciampa, D., Diomedi, M., Olita, S.: The management of air pollution from vehiculartraffic by implementing forecasting models. In: Sustainability, Eco-Efficiency and Conservation in Transportation Infrastructure Asset Management - Proceedings of the 3rd International Conference on Tranportation Infrastructure, ICTI 2014, pp. 549–560 (2014)
4. Romano, B., Zullo, F.: Models of Urban Land Use in Europe: Assessment tools and criticalities. International Journal of Agricultural and Environmental Information Systems (IJAEIS) 4(3), 80–97 (2013), doi:10.4018/ijaeis.2013070105
5. Koomen, E.: Modelling Land-Use Change: Progress and Applications, Springer (2007)
6. Bonham-Carter, G.F.: Geographic information system for geoscientist: modelling with GIS, pp. 243–247. Pergamon Press (1994)
7. Smith, E.P., Lipkovich, I., Ye, K.: Weight of evidence: quantitative estimation of probability of impact. Dept. of Statistic, Virginia Tech (2002)
8. Romano, B., Zullo, F.: The urban transformation of Italy's Adriatical coastal strip: Fifty years of unsustainability. Land Use Policy 38 (2014)
9. Perchinunno, P., Rotondo, F., Torre, C.M.: The Evidence of Links between Landscape and Economy in a Rural Park. International Journal of Agricultural and Environmental Information Systems 3(2), 72–85 (2012), doi:10.4018/jaeis.2012070105
10. Murgante, B., Danese, M.: Urban versus rural: the decrease of agricultural areas and the development of urban zones analyzed with spatial statistics. Int. J. Agric. Environ. Inform. Syst. 2(2), 16–28 (2011), http://dx.doi.org/10.4018/jaeis.2011070102
11. Modica, G., Vizzari, M., Pollino, M., Fichera, C.R., Zoccali, P., Di Fazio, S.: Spatio-temporal analysis of the urban-rural gradient structure: an application in a Mediterranean mountainous landscape (Serra San Bruno, Italy). Earth Syst. Dyn. 3(2), 263–279 (2012), http://dx.doi.org/10.5194/esd-3-263-2012
12. Martellozzo, F.: Forecasting high correlation transition of agricul-tural landscapes into urban areas: diachronic case study in North East-ern Italy. Int. J. Agric. Environ. Inform. Syst (IJAEIS) 3(2), 22–34 (2012), http://dx.doi.org/10.4018/jaeis.2012070102
13. Evaluation of Urban Sprawl from space using open source technologies
14. NolèG., M.B., Calamita, G., Lanorte, A., Lasaponara, R.: Evaluation of Urban Sprawl from space using open source technologies. Ecological Informatics (2014), doi:http://dx.doi.org/10.1016/j.ecoinf.2014.05.005
15. Nolè, G., Lasaponara, R., Lanorte, A., Murgante, B.: Quantifying Urban Sprawl with Spatial Autocorrelation Techniques using Multi-Temporal Satellite Data. Int. J. Agric. Environ. Inform. Syst. 5(2) (2014)
16. Clarke, K.C., Gaydos, L.: Loose-coupling a cellular automaton model and GIS: Long-term urban growth prediction for San Francisco and Washington/Baltimore. Int. J. of Geographic Inf. Sci. 12, 699–714 (1998)
17. Cerreta, M., Poli, G.: A Complex Values Map of Marginal Urban Landscapes: An Experiment in Naples (Italy). International Journal of Agricultural and Environmental Information Systems 4(3), 41–62 (2013), doi:10.4018/ijaeis.2013070103

A New Design Method for Managing Spatial Vagueness in Classical Relational Spatial OLAP Architectures

Elodie Edoh-Alove[1,2], Sandro Bimonte[1], and Yvan Bédard[2]

[1] Irstea, TSCF, 9 avenue Blaise Pascal CS20085, 63178 Aubière, France
{elodie.edoh-alove,sandro.bimonte}@irstea.fr
[2] Department of Geomatics Sciences and Centre for Research in Geomatics, Laval University, Quebec city, Quebec, Canada
yvan.bedard@scg.ulaval.ca

Abstract. Spatial Data Warehouses (SDW) and Spatial OLAP (SOLAP) systems are well-known Business Intelligence technologies that aim to support multidimensional and online analysis of huge volumes of datasets with spatial reference. Spatial vagueness is one of the most neglected imperfections of spatial data. Although several works propose new ad-hoc models for handling spatial vagueness in information systems, the implementation of those models in Spatial DBMS and SDW is still in an embryonic state. Thus, in this paper, we present a new design method for SOLAP datacubes that allows handling vague spatial data analysis issues. This method relies on a risk management method applied to the potential risks of data misinterpretation and decision-makers' tolerance levels to those risks. We also present a system implementing our method.

Keywords: Spatial vagueness, Spatial Data Warehouse, Spatial OLAP, Design method, Risks management method, Spatial Datacubes.

1 Introduction

Spatial OLAP (SOLAP) can be defined as" a visual platform built especially to support rapid and easy spatiotemporal analysis and exploration of data following a multidimensional approach comprised of aggregation levels" [1]. The analysis results are available in interactive cartographic, tabular and diagram displays. The explored data are stored in a Spatial Data Warehouse (SDW) as datacube, that is the multidimensional model implementation [2]. Usual end-users of those technologies are first and foremost decision-makers who are rarely fully aware of the issues related to spatial data uncertainty [13]. And yet, spatial data are primarily "false but useful" models about the reality of interest [12] since they define geographic phenomena by means of crisp boundaries even if it is not always possible to define exactly where objects begin and end (spatial vagueness). With such a choice of representation, a clear gap is created between majority of real world phenomena and their formal representation in spatial databases [10]. The SOLAP end-users are then exposed to erroneous analysis due to the uncertainty issues on data sources.

B. Murgante et al. (Eds.): ICCSA 2014, Part II, LNCS 8580, pp. 774–786, 2014.

To the best of our knowledge, only some recent researches introduce spatial vagueness in the spatio-multidimensional model [6] and SOLAP operators in order to reduce the analysis errors. In the meantime [7] have proposed an approach based on Fuzzy Set theory and Tessellation to deal with spatial vagueness in SOLAP datacubes, approach experimented in coastal erosion.

Usually, existing SOLAP systems (e.g. Map4Decision, GeWOlap, etc.) [4] are based on a relational architecture where a Spatial DBMS is used to store and manage (spatial) data, a spatial OLAP server, such as GeoMondrian, to implement spatial data cubes and SOLAP operators, and an OLAP client enhanced with interactive cartographic displays to visualize and trigger SOLAP queries. In this paper, motivated by the need of using existing SOLAP systems, with vector levels of datacubes and simple polygons to represent the spatial data, we adopt a risk management approach that involves end-users into the design of the SOLAP datacube, to better deal with the uncertainty. We thus present a new design method for classic SOLAP datacubes that allows handling vague spatial data analysis issues by involving end-users' tolerance levels to the risks of misinterpreting the uncertain data and analysis results. Our method is based on an interactive and iterative design approach allowing the definition and validation of SOLAP datacubes prototypes with the help of end-users. Basically, the method takes into account risks of misinterpretation, which are defined in teamwork by datacube end-users and spatial data experts, and users' tolerance levels (to those risks). It proposes to the end-users a set of SOLAP datacube prototypes iteratively tailored to their tolerance levels, i.e. it makes necessary adjustments depending on their reactions when using the prototypes.

Moreover, we present in this paper the new tool elaborated to support our method.

The paper is structured the following way: An agricultural case study describing the motivation of our research is introduced in Section 2; Section 3 presents our method, and its implementation is described in Section 4; Section 5 presents related work.

2 Motivations

In this section we present an agri-environmental SOLAP data cube (Fig. 1), adapted from [21], which will be used in this paper to illustrate our contributions. Sewage sludge produced by wastewater treatment plant can be used as crop fertilizer in agriculture. Farmers spread sewage sludge on cultivated plot to fertilize soil. Sewage sludge can contain different elements such as trace metals Plumb, Cadmium, Chromium, … Some of those trace metals are essential to the functioning of the biological process (e.g. Copper, Chromium, Nickel). However, at high concentrations, trace metals provided by sludge can become toxic for different forms of life. So, from an environmental point of view, it is very important to carefully monitor the activity of sewage sludge spreading.

Here, we consider a SOLAP datacube to analyze and explore data related to the sludge spreading activities in agriculture. The intended SOLAP data cube initial Platform Independent Model (PIM in the MDA method [8]) has been elaborated using the

ICSOLAP UML Profile presented in [24] (Fig. 1). The SOLAP data cube has a temporal dimension (*Time*) a thematic dimension representing products spread (*Products*) and a spatial dimension (*Location*). The *Location* dimension has a hierarchy *LocationH* composed of spread zones level and watershed level. A spread zone is a zone on which sewage sludge can be present. The spread zone geometry is vague (see fig. 2); this geometry is composed of:

(1) a certain part (drawn in green) and
(2) an uncertain part which is the space between the certain part and the red boundary.

The green zone is the limit of an agricultural plot on which sludge has been spread. The geometries of agricultural plots come from the French geographic information system. Because of imprecision related to spreading activity (e.g., the imprecision of the tractor equipment and potentially strong winds), it is possible that in, some cases, sludge has been spread outside the green zone. It is the reason why a larger limit has been defined (the red boundary); we consider that it is not realistic that sludge has been spread outside the red limit during the spreading of the plot.

The measure *ProductFlow* (Fig. 1) in this SOLAP datacube is the flow of trace metals (quantity of trace metals in grams divided by spread surface in square meters) provided by each sludge spreading. The flow depends on the developed surface area. It is aggregated along the hierarchies using the average function. An example of factual data is shown on Table 1.

Table 1. Sample of the fact table "SludgeFact"

Elements	Month	Spread Zones	ProductFlow (g/m2) for certain part	ProductFlow (g/m2) for uncertain part
Zinc	January-2000	Plot1	4.09	3.09
Zinc	May-2000	Plot1	5.41	4.09

Using a SOLAP datacube implementing this model, decision-makers can monitor *the average flow of products for each plot per year* (Q1). This value is also useful because there is a regulation on the amount of products contained in the sludge spread. Moreover, *aggregating this measure at the watershed level allows evaluating the concentration of products at the watershed level* (Q2).

We note that in January of the year 2000, the flow of Zinc for Plot1 is 4.09g/m2 when considering the allowed surface (green zone), while it is 3.09g/m2 when considering the surface of the red polygon. This means that a risk of over evaluation of the measure *ProductFlow* exists and it is related to the vagueness of the geometry of spread zones (members of level *Spread Zones*).

We define this risk of misinterpretation as Risk-Geometry, a risk related exclusively to the vague geometries of members of the spatial dimension.

We now analyze the aggregation on the Watershed level. For a given watershed, we consider in the aggregation, the spread zones region that are included into that watershed. When a spread zone is not fully contained into a watershed, we apply a classical split method to aggregate the measures [9]. In other term, we weight the measure value on the surface of the spread zone included into the surface of the watershed. In our case study if we consider only the green zone for the Plot1, then in 2000, the flow of Zinc in the Watershed 1 is 0,75g/m2. However, if we consider the maximal surface (red polygons) then the flow of Zinc applied in the watershed is 1,65g/m2. Since such a difference happens for every Plot along the watershed, for several spreads, the aggregated value for the watershed will be faulty. There is a risk to under evaluate the flow of products in the watersheds if the spread zone is considered only in its spreading suitable limits.

We define this risk as Risk-Aggregation, a risk associated to the aggregation of measures associated to members with vague geometries.

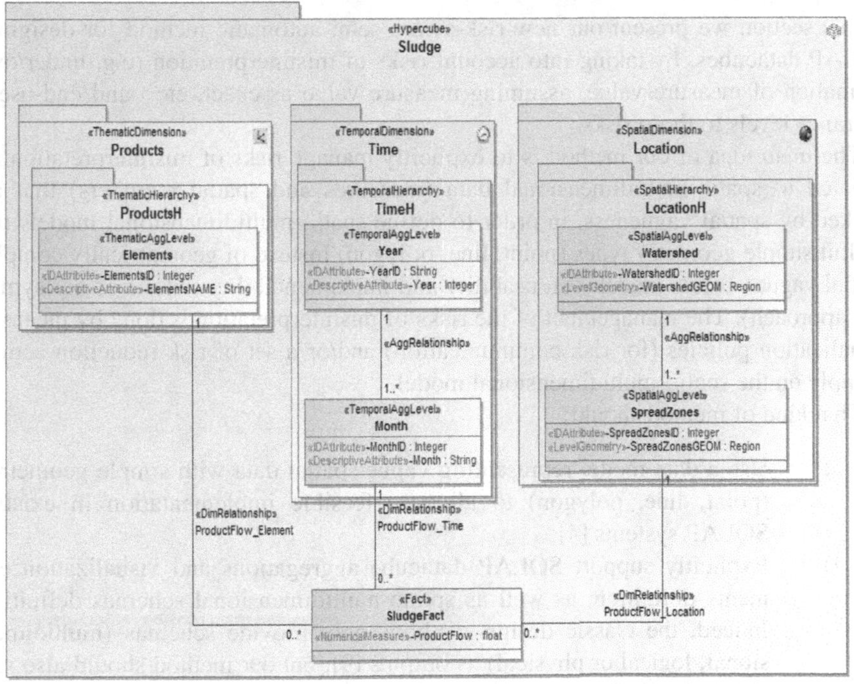

Fig. 1. Sludge spreading classic SOLAP datacube model

A watershed represents a region inside of which all surface waters converge to rivers, lakes or another waterbody.

Fig. 2. Cartographic representation of spread zones spatial vagueness

3 Risk-Aware Design of SOLAP applications: RADSOLAP

In this section we present our new risk-aware semi-automatic method for designing SOLAP datacubes, by taking into account risks of misinterpretation (e.g. under/over estimation of measure value, assuming measure value as exact, etc.) and end-users' tolerance levels to those risks.

The main idea of our method is to explicitly manage risks of misinterpretation associated to spatio-multidimensional data (measures and spatial members) that are marked by spatial vagueness, in order to define spatio-multidimensional models that exploit simple geometry types (point, line, polygon) instead of geometrically complex spatial vague objects models (cf. usability and implementability criteria of our symbiotic approach). The management of the risks of misinterpretation is done by means of visualization policies (for risk communication) and/or a set of risk reduction actions to apply on the spatio-multidimensional model.

That kind of method should:

I. Use a data model representing vague spatial data with simple geometries (point, line, polygon) to allow a feasible implementation in existing SOLAP systems [4];
II. Explicitly support SOLAP datacube aggregations and visualization elements definition, as well as spatio-multidimensional schemas definition. Indeed, the classic design methods only provide schemas (multidimensional, logical or physical) as outputs [9], but our method should also specifies the different pertinent and authorized aggregation operations, as well as visualization parameters to communicate the risks if needed;

III. Allow the possibility of an implementation according to the rapid prototyping paradigm [3]: the method must offer the possibility to go back to some of the key steps of the design in order to revise the choices made and refine the datacube modeling. Since our design method will define visualization parameters and change the spatio-multidimensional model, the SOLAP application end-users need to "play" with prototypes to validate the resulting datacubes, before SOLAP experts move to the real implementation.

The steps of our RADSOLAP method are the following (they are reflected on Fig. 3):

1. Users informally define the needed SOLAP system functional requirements, identifying vagueness on spatial data sources.
2. SOLAP experts create a multidimensional schema (PIM) starting from the users' analysis needs defined in step 1. Vague geometric members and measures are then identified on this initial schema. For example in our case study, spread zones are identified as being vague.
3. Decision-makers informally identify and assess risks associated to multidimensional elements marked with spatial vagueness. In our case study for example, the Risk Geometry of over evaluating the product flow brought on spread zones is identified and a tolerance level of 0 is expressed.
4. SOLAP experts extend the previously defined spatio- multidimensional schema by adding information on risks and associated tolerance levels to the model.
5. SOLAP experts choose a set of risks management actions according to the defined tolerance levels, between possible actions associated to given tolerance levels. Those actions are then applied on the schema in an iterative process that calls for risks reassessment by the users. For example, for the Risk Geometry on spread zones, the tolerance level being 0, SOLAP experts can choose the action "delete level" in a list of possible actions; this action will delete the whole *SpreadZones* level in the spatio-multidimensional model (Figure 4).
6. Decision-makers manually feed sample realistic data into the prototype.
7. Decision-makers access and explore these sample data using simple pivot tables of the SOLAP client, so as to validate the prototype.

   – If the prototype is validated but visualization policies are not, then return to step 5 to test other actions.
   – If the prototype is not validated, return to step 1 if tolerance levels cannot be changed otherwise return to step 3.

   For example in our case study users can go back to step 1 and choose to consider the spread zones in their maximal extent.

8. Once the prototype is validated and declared true to analysis requirements, data are collected, ETL is designed, and the prototype is engineered by the SOLAP experts.

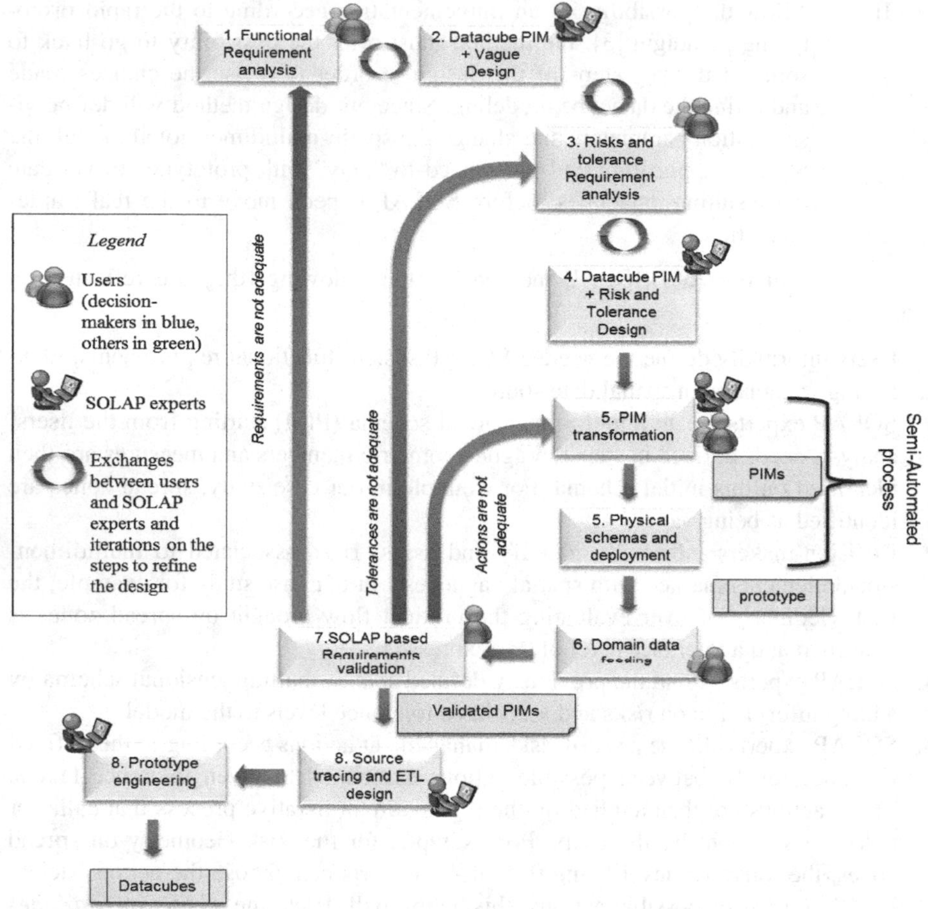

Fig. 3. RADSOLAP method

4 RADTool: A Tool to support the RADSOLAP Method

The global architecture of the RADTool is based on the ProtOLAP system proposed in [3] (Figure 5). ProtOLAP is based on a classical Relational SOLAP architecture as described in this section. ProtOLAP is a framework for rapid prototyping of datacube. The main idea of ProtOLAP is to allow datacube designers to automatically and incrementally implement datacube schemas and feed them with sample data, and provide end-users with real OLAP clients to allow them to test the produced datacube, in order to ultimately validate the spatio-multidimensional schema. We think that such method is highly beneficial in our approach. Indeed, as stated before, several SOLAP datacubes can be proposed to the decision-makers, each one with different schemas, data and visualization policies. It could be difficult for the end-users to have an idea of the impact of all actions associated to their tolerance levels on the resulting SOLAP datacubes, their exploration and visualization, without really "playing" with them.

For example, if one of the strategies applied is to remove a level, they may not be really sure it still fits the analyses needs until they perform the SOLAP datacube exploration. Moreover, ETL procedures are usually complex and time and resources consuming. A rapid prototyping will help them see what they can expect with the choices made, and then decide which SOLAP datacube better fits their use before moving the whole project to the costly ETL process phase.

The ProtOLAP architecture is based on a Relational OLAP platform and is composed of four tiers:

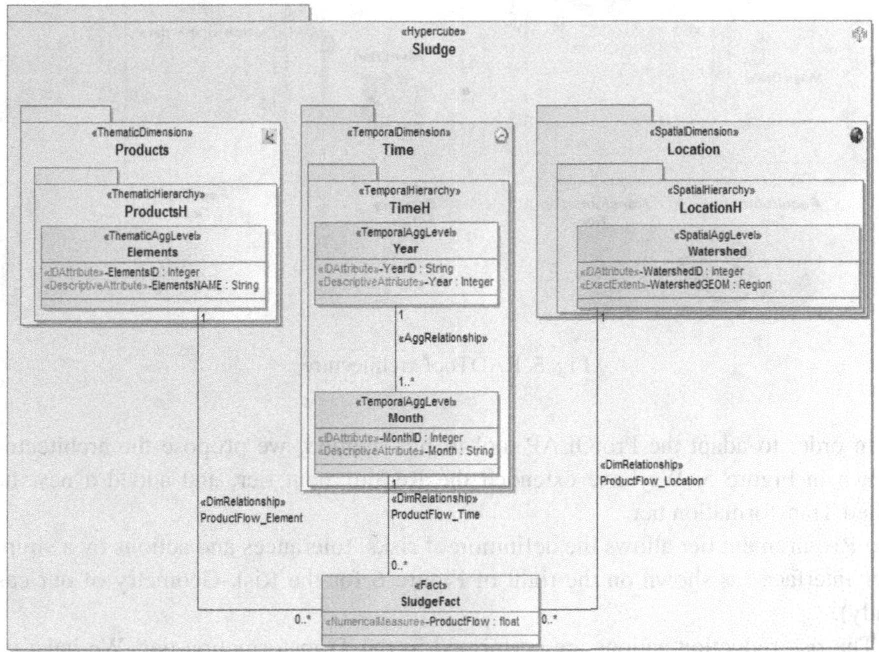

Fig. 4. Final SOLAP datacube PIM

- The Requirement tier, where SOLAP experts draw a UML-based PIM, using the ICSOLAP UML profile [5] and the MagicDraw CASE tool;
- The Deployment tier, that includes the Oracle Relational DBMS, the Mondrian SOLAP server, and a tool that creates relational schema for Oracle and metadata for Mondrian starting from the PIM;
- The Feeding tier, that automatically generates a visual interface through which users can feed the datacube stored in deployment tier with application domain data;
- The Analysis tier, that allows decision-makers to query data stored in the deployment tier using the JRubik OLAP client.

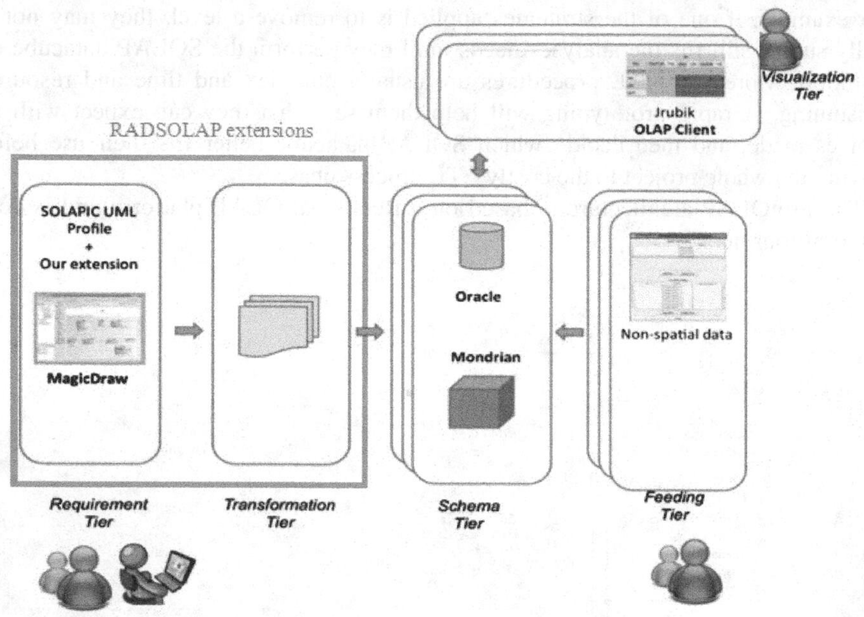

Fig. 5. RADTool architecture

In order to adapt the ProtOLAP tool to our method, we propose the architecture shown in Figure 5. We have extended the Requirement tier, and added a new tier called Transformation tier.

The Requirement tier allows the definition of risks, tolerances and actions by a simple user interface (as shown on the right of Figure 6 for the Risk-Geometry of our case study).

The risk reduction actions are performed in the Transformation tier. We have implemented datacube PIM transformation functions such as `DeleteLevel`, `ChangeAggregator` and `CommunicateRisk`. The `DeleteLevel` function for example removes a specific level from the initial PIM.

In the Schema tier, our solution automatically translates the new schema into corresponding Spatial DBMS and OLAP server schemata. That allows users to feed the datacube with some real sample data at the Feeding tier.

Finally in the Analysis tier, the OLAP client is delivered to decision-makers to validate: (i) the spatio-multidimensional schemas and aggregation functions, (ii) the choice of tolerance levels and actions.

An example of the final datacube prototype implemented in our case study where *SpreadZones* level is not deleted by actions, but a visual policy is used, is shown on Figure 7. We can note that the Risk-Geometry is communicated using a red color.

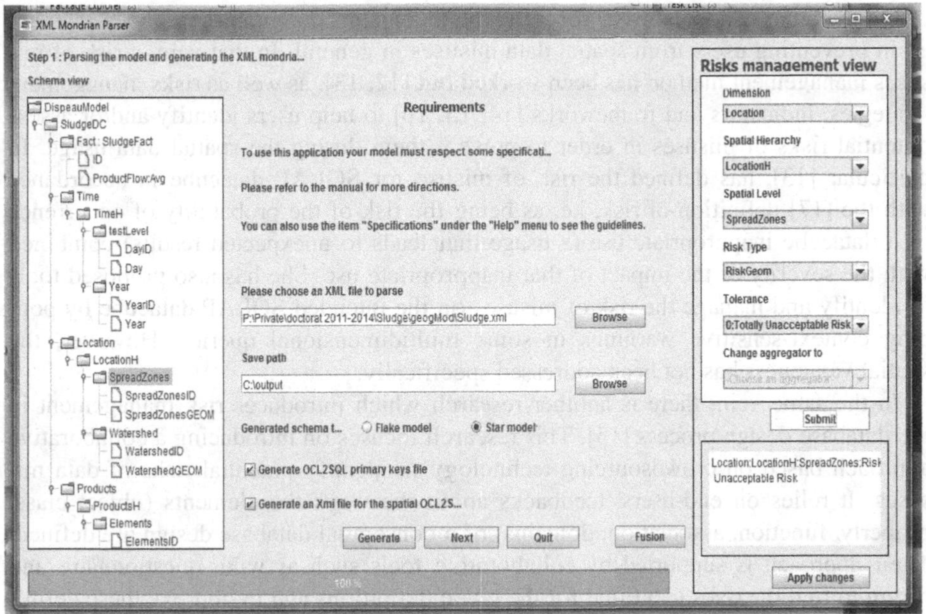

Fig. 6. RADSOLAP system: requirement tier for risks, tolerance and action setting

	Location							
Time	– All Location.LocationHs	+ Adour-Garonne	– Rhone-Mediterranee-Corse	Plot 1	Plot 2	Plot 3	Plot 4	Plot 8
–All Time.TimeHs	4,88	4,64	4,94	5,34	4,12	5,63	5,30	4,66
+2000	4,98	5,27	4,93	5,14	4,23	5,63	5,30	4,01
+2001	4,71	4,01	4,94	5,53	4,00			5,30

Fig. 7. RADSOLAP method: analysis tier

5 Related Work

The practical integration of vague objects models (used to represent vague spatial data) in SOLAP systems is in an embryonic state and it is only very recently that [7] have proposed an algorithmic approach based on Fuzzy Set theory to deal with fuzzy boundaries of coastal erosion risk areas, among others. Also, [6] have extended the multidimensional model in order to exploit vague objects exact models [11]. They introduced the term "vague" in the multidimensional concepts, redefining spatial attributes, measures, dimensions and hierarchies. There is no implementation tools proposed with their new definitions.

Thus, even though integrating vague objects in SOLAP systems is a good approach to reduce the uncertainty related to spatial vagueness, there is still much to do in order to design, implement and exploit SOLAP datacubes with vague objects. As a matter of fact, existing tools (SOLAP server and client) and Spatial DBMS [10] are not designed to manage vague objects.

On the other hand, also very recently, the Geomatics community has been interested in preventing users from spatial data misuses in general. In that vein, a risk of misuses management method has been worked out [12, 13], as well as risks management strategies, indicators and frameworks [14, 15, 16] to help users identify and/or assess potential risks of misuses in order to prevent them during the spatial data usage. In particular [13], has defined the risk of misuse for SOLAP datacube in accordance with the [17] definition of risk, i.e. as being the risk of the probability of occurrence of a datacube inappropriate use (a usage that leads to unexpected results) combined with the severity of the impact of that inappropriate use. She has also proposed tools to identify and manage the risk of misuses on the intended SOLAP datacube by popping context-sensitive warnings in some multidimensional queries. However, the spatial vagueness has not been addressed specifically.

In the same vein, there is another research which introduces risk management in the database design process [15]. This research focuses on introducing a collaborative approach based on crowdsourcing technology to identify potential risks of data misuses. It relies on end-users' feedbacks about the ways the elements (object class, property, function, association, domains) of a conceptual database design are defined. Their approach is supported by collaborative tools such as wiki, questionnaire and forum to find the risks identified for the given definitions and to improve these definitions if decided so. If not, other risk management strategies are adopted. It is the database design team who selects the best risk management strategy for each risk identified, not the crowd of end-users. This may lead to modifications to their database design. Their work is generic for any type of spatial database and in this regard, it can be seen as complementary to the work presented in this paper.

Note that [7], [15] and our own researches come from the same research group and were thought to be complementary in the way the quality issue is addressed. Other works by [13], [14] and [12, 16] also come from our research group and aim at different but complementary objectives.

Regarding the data cube design methodologies, the classical design and implementation process adopted for (spatial) data cubes in the literature is composed of these main steps: functional requirements analysis, conceptual design, logical schema elaboration and physical schema elaboration for the design phase, and ETL (Extract Transform and Loading) process followed by the data cube deployment for the implementation phase. Note that to design and implement SOLAP data cubes, designers commonly follow methods such as the MDA one [18] and/or agile prototyping based methods [3].

With that said, different researchers in the OLAP field, have specifically put the focus on the extraction of the multidimensional schema from either the users' needs (expressed in SQL queries or ontologies – user-driven approach), the available data sources (databases relational or logical models – data-driven approach) or from both users and data sources (hybrid approach). It led to the proposition of multidimensional modeling methods where requirements, conceptual and/or logical data cube schema are derived in an automatic or semi-automatic way according to the type of approach implemented.

The analysis of the design methodologies shows that (spatial) data uncertainty issues do not explicitly influence the resulting data cube multidimensional elements (i.e. facts, dimensions, hierarchies, measures and aggregations) definition. Instead, uncertainty and data cube quality are principally addressed during the ETL process and/or reporting phase [19, 20].

6 Conclusion and Future Work

In this paper we have presented our new design method for conventional SOLAP applications that allows handling vague spatial data analysis issues by means of SOLAP datacubes risks of misinterpretation and decision-makers' tolerance levels to those risks. The method allows decision-makers, organization geospatial data and systems users as well as SOLAP experts to collaborate in designing SOLAP datacubes: (1) that decision-makers can easily explore and analyze; (2) that can be implemented in the existing SOLAP systems and (3) that handle the spatial vagueness on the sources. We have also introduced the RADTool, a technical tool that supports this risk-aware design method. The RADTool allows the design of SOLAP datacubes PIM by exploiting the information on the risks and the end-users tolerance levels. It also helps generate automatically SOLAP datacubes prototypes that end-users can visualize and explore in order to validate the whole design.

At the moment we are working on proving the forcefulness of our method on the real French National sludge spread monitoring database [21].

References

1. Bédard, Y.: Spatial OLAP. In: Forum annuelsur la R-D, Géomatique VI: Un monde accessible (1997)
2. Salehi, M., Bédard, Y., Rivest, S.: A Formal Conceptual Model and Definitional Framework for Spatial Datacubes. Geomatica 64(3), 313–326 (2010)
3. Bimonte, S., Edoh-Alove, E., Nazih, H., Kang, M., Rizzi, S.: ProtOLAP: Rapid OLAP Prototyping with On-Demand Data Supply. In: DOLAP 2012, San Fransisco, CA, USA (2013)
4. Bimonte, S.A.: Web-Based Tool for Spatio-Multidimensional Analysis of Geographic and Complex Data. IJAEIS 1(2), 42–67 (2010)
5. Bimonte, S., Boulil, K., Pinet, F., Kang, M.: Design of Complex Spatio-multidimensional Models with the ICSOLAP UML Profile - An Implementation in MagicDraw. In: ICEIS (1), pp. 310–315 (2013)
6. Siqueira, T.L.L., de Aguiar Ciferri, C.D., Times, V.C., Ciferri, R.R.: Towards Vague Geographic Data Warehouses. In: Xiao, N., Kwan, M.-P., Goodchild, M.F., Shekhar, S. (eds.) GIScience 2012. LNCS, vol. 7478, pp. 173–186. Springer, Heidelberg (2012)
7. Jadidi, A., Mostafavi, M., Bédard, Y., Long, B.: Towards an Integrated Spatial Decision Support System to Improve Coastal Erosion Risk Assessment: Modeling and Representation of Risk Zones. In: FIG Working Week 2012, Rome, Italy, pp. 6–10 (2012)
8. OMG, MDA Guide, Version 1.0.1, Object Management Group (2003)

9. Malinowski, E., Zimányi, E.: Advanced Data Warehouse Design: From Conventional to Spatial and Temporal Applications, Data-Centric Systems and Applications. Springer, Heidelberg (2008) ISBN 978-3-540-74404-7
10. Pauly, A., Schneider, M.: VASA: An algebra for vague spatial data in databases. Information Systems 35(1), 111–138 (2010)
11. Bejaoui, L.: Qualitative topological relationships for objects with possibly vague shapes: implications on the specification of topological integrity constraints in transactional spatial databases and in spatial data warehouses. Université Blaise Pascal (2009)
12. Gervais, M., Bedard, Y., Levesque, M.-A., Bernier, E., Devillers, R.: Data Quality Issues and Geographic Knowledge Discovery. In: Geographic Data Mining and Knowledge Discovery, pp. 99–115 (2009)
13. Lévesque, M.-A.: Formal Approach for a better identification and management of risks of inappropriate use of geodecisional data. In: Geomatics, Laval University (2008)
14. Roy, T.: Nouvelle méthode pour mieux informer les utilisateurs de portails Web sur les usages inappropriés de données géospatiales. In: Geomatics Department, Laval University: Quebec City, Quebec, Canada, p. 145 (2013)
15. Grira, J., Bédard, Y., Roche, S.: Revisiting the Concept of Risk Analysis within the Context of Geospatial Database Design: A Collaborative Framework. World Academy of Science, Engineering and Technology 75 (2013)
16. Gervais, M., Bédard, Y., Rivest, S., Larrivée, S., Roy, T.: Enquête canadienne sur la qualité des données géospatiales et la gestion du risque. Centre for Research in Geomatics, Laval University, Quebec City, Canada (2012)
17. ISO, ISO 9000: Quality Management Systems:Fundamentals and Vocabulary (2000)
18. Glorio, O., Trujillo, J.: An MDA Approach for the Development of Spatial Data Warehouses. In: Song, I.-Y., Eder, J., Nguyen, T.M. (eds.) DaWaK 2008. LNCS, vol. 5182, pp. 23–32. Springer, Heidelberg (2008)
19. Dyreson, C.E., Pedersen, T.B., Jensen, C.S.: Incomplete information in multidimensional databases. In: Multidimensional Databases, pp. 282–309. IGI Publishing (2003)
20. Pedersen, T.B., Jensen, C.S., Dyreson, C.E.: Supporting imprecision in multidimensional databases using granularities. IEE (1999)
21. Soulignac, V., Barnabe, F., Rat, D., David, F.: SIGEMO: un système d'information pour la gestion des épandages de matières organiques - Du cahier de charges à l'outil opérationnel. Ingénieries (47), 37–42 (2006)

Growing Sustainable Behaviors in Local Communities through Smart Monitoring Systems for Energy Efficiency: RENERGY Outcomes

Francesco Scorza[1], Alessandro Attolico[2], Vincenzo Moretti[2], Rosalia Smaldone[2], Domenico Donofrio[2], and Giuseppe Laguardia[2]

[1] Laboratory of Urban and Regional Systems Engineering, School of Engineering, University of Basilicata, 10, Viale dell'Ateneo Lucano, 85100, Potenza, Italy
francescoscorza@gmail.com

[2] Province of Potenza, Planning and Civil Protection Office, P.zza M. Pagano, 85100, Potenza, Italy
<name>.<surname>@provinciapotenza.it

Abstract. EU 2020 agenda started new planning processes at territorial level connected to the challenge od sustainability goals defined at European level. Territorial administration entitled of territorial planning started to consider as a key aspect the energy planning with several implications. The paper started from the experience developed in the framework of RENERGY transnational cooperation project by the Province of Potenza and remarks the role of ICT applications for community involvement as a success factor to obtain the ambitious EU 20-20-20 targets. Conclusions regard the implementation dimension designed by the Province of Potenza in order to realize a pilot application producing open data in the field of energy performance of public and/or private interventions.

Keywords: Energy Efficiency, Covenant of Majors, EU 2020, New Cohesion Policy, Regional Planning, Smart Monitoring Systems.

1 Introduction

Sustainability in energy planning at territorial scale represents a new challenge speeded out in the framework of EU 2020 agenda. It means additional tasks for territorial administrations engaged with traditional planning issues and managing the territorial dynamics in a framework of strong community claims concerning development perspectives and strategic visions.

How to manage this issue effectively? This is, in other words, the new challenge regarding sustainability in territorial planning. The complexity of a "context based" [1] approach (i.e. Place based approach [2]) managing together territorial development, environmental preservation, sustainable energy development, and socio economic growth appear more as a chimera than as a strategy. Especially in the "time of the crisis" which addressed on EU policy and regulations a share of negligence.

B. Murgante et al. (Eds.): ICCSA 2014, Part II, LNCS 8580, pp. 787–793, 2014.

Several documents regard the long term EU policy vision is concerning sustainable development [3] [4] [5] and, more recently in the Europe 2020 strategy [6] the instances of knowledge and innovation, competitive economy, social and territorial cohesion were oriented to enrich the umbrella objective “to promote a smart, sustainable and inclusive EU growth”.

The territorial planning faced the new strategic vision of sustainable Europe in several operative way. One of the most effective tools are represented by the “Covenant of Mayors” [7]. One of the results of such policy is the number of SEAP (sustainable energy action plan). A new operative planning tool addressing energy sustainability at local level (municipal level).

The experience of the Province of Potenza appears to be relevant for the role played in this framework. In facts it had the role of coordinating the participation of local Municipalities to the EU policy, promoting the commitment of local politicians and decision makers to reinforce local perspectives according to EU 2020 objectives.

As the Province represents an intermediate public body in the Italian low, it supported this process including sustainable energy instances within the Provincial Territorial Master Plan recently approved as requirements of conditionality (in other words a precondition) for further development in EU 2014-2020 programming period.

Concerning energy sustainability emerged three main components according to the Province of Potenza experience. Such analysis reflects the structure and the outcomes of an International Cooperation Project RENERGY developed by the Province as lead partner during last months.

We considered the following domains as intervention areas but also as matters of investigation in order to propose and exchange effective solution in the field of sustainable energy planning:

- 'Policy Making';
- 'Community Involvement'
- 'Market Uptake'.

During the RENERGY Project activities some general instances were identified: the need for effective tools to support decisions ('Policy Making') aimed at raising awareness of local communities to energy efficiency ('Community Involvement') and with positive impacts in the field of Market.

This paper, after discussing the general framework of sustainable energy planning according to RENERGY Project visions and methodology highlight the potential contribution of ICT and SMAT technological applications as tools supporting Sustainable energy planning at municipal and territorial level. Such deepening allow us to identify the opportunity of such technology in supporting the growth of sustainable behaviour in citizens and communities toward the issues of energy sustainability optimizing local policies and EU 2020 target achievement.

Conclusions regard the implementation dimension designed by the Province of Potenza in order to realize a pilot application in this domain.

2 RENERGY Experience: The Framework for Implementing Innovations in Energy Planning

RENERGY is mainly a good practice exchange project involving several partners around Europe. The project developed several outcomes according to the three main areas: 'Policy Making'; 'Community Involvement'; 'Market Uptake'. The final goal of the project is the Local Implementation Plan implementing selected good practices at partner level.

Considering the EU policy framework the reference policy of the project is represented by the 'Covenant of Mayors', a framework initiative launched by the European Commission to support the implementation of the 20-20-20 targets at a local scale. It is based on a voluntary commitment of local and regional authorities to engage the overall objectives of increasing energy efficiency (EE) and renewable energy (RES) [7].

As RENRGY activities also demonstrated, one of the main action to carry out in order to do this is the involvement of local communities promoting participation and education activities to increase people trust in energy efficiency and to encourage investments and opportunities of cooperation among citizens, enterprises and public administrations.

A contribution could come by the implementation of ICT tools in order to give people the opportunity to exchange data, knowledge ad to compare results of public and private investments according to a 2.0 approach and collaborative data production on the web.

Such issue was considered by the Province of Potenza in developing a innovative and experimental project in the framework of RENERGY in order to test SMART Monitoring System in public building with a web interface allowing citizen to get information about energy consumption and energy indicators in real time.

We believe that the development of a a pilot application designed to evaluate the effectiveness of innovative ICT applications for real-time energy monitoring could contribute as a decisions support system (with positive remarks in 'Policy Making'), it will be also aimed at raising awareness of local communities to energy efficiency (with positive remarks in 'Community Involvement') and with positive impacts in the field of 'Market Uptake'.

Issues connected with each pillars of RENERGY Approach are:

POLICY MAKING

Considering direct experiences of project partners in Covenant of Major applications at territorial level, it appeared as a critical points in EE and RES policy development the lack of punctual data. Such information are absolutely necessary in order to define, under an operative point of view, the decision making process concerning policy and project implementation and public investments effective management. An open approach in monitoring EE and RES applications, based on open software and open web application, promoted under the public responsibility and connected to effective community involvement actions, could generate a bottom up process involving also citizens, private sectors etc. in producing open data concerning Energy performances.

COMMUNITY INVOLVEMENT

The main dimension of people perception of the benefits coming from EE and RES applications lies only on the economic benefits derived from saving on energy costs. But how to generate a common and shared consciousness regarding the benefits of ‘sustainable behaviours’ in energy consumption?

If a student in a school or a citizen in a public office has the opportunity to get information (through an ICT user-friendly real-time interface) concerning energy consumption and can monitor the results of a responsible behaviour (i.e. turning off the light on a sunny day) in terms of energy efficiency, we expect that he goes to replicate such behaviour elsewhere with global positive results.

So the PA intends to develop a prototype of Smart Monitoring System – based on open source framework and adopting ‘open-data’ approach, applied on public buildings, promoted through open testing acts in order to start this positive experience in citizens awareness.

MARKET UPTAKE

Also considering what ‘open data’ explosion era is generating in terms of exploitation of open datasets for providing services in different fields and generating also business opportunities the PA assumes that the EE and RES private sector operators could strongly benefit from such data availability as they have an additional tools in order to identify EE and RES demand through open data. In particular we consider two areas of market interactions:

1. Public demand: an open monitoring system on EE of public building could allow the private sector to produce effective solution and proposals in order to start public-private partnerships, ESCO agreements etc. in order to operatively develop intervention project.
2. Private demand: people showing proper energy performances in private housing and enterprises showing own energy consumption could generate the match between supply and demand for energy efficiency services, and also the results will be shared on the web in order to promote competition among operators and services providers

Such general implication fits with the overall objective of this pilot application: “to increase the awareness of operators and local communities interested in management/use of public buildings concerning energy consumption through ‘real-time data’ allowing experiment and evaluation of energy saving and energy consumption rationalization”.

An innovative aspect is to include a collaborative approach in monitoring energy systems. In fact it is usual today that each energy producer/consumer, also private people, hold a monitoring system based on sensors network, also very accurate and sophisticated, but not integrated in a wider public platform. The aim of the Pilot Application is to develop such platform according with the public responsibility to support Covenant of Mayors agreements and to give citizens the opportunity to integrate own systems in order to collaborate in producing data on EE and RES and other information. Such approach could produce a positive impact in community involvement processes.

3 Spatial Distribution of Migrants: The Italian Case

This SMART Monitoring System intend to reinforce the operative dimension of Covenant of Major at territorial level. The Province of Potenza will play a role of ICT service provider for other territorial public administrations and private people in order to allow people see the results of such complex action promoting sustainability in a user friendly ICT web interface.

The general objective of the RENERGY project includes the dimension of 'local/regional sustainable energy policies effectiveness' as a precondition for investments in Green Economy and contributes for local economy, jobs and quality of life.

On the operative point of view the availability of a comprehensive monitoring system appears to be a precondition for policy making, community involvement and market uptake as data/information availability still represent a structural barrier for sustainable energy effective development.

The Pilot Application aims to ensure up to date technological infrastructure (at pilot project level) oriented to adopt open source, open data and low budget sensoring system technologies following the way of SMART Communities policies and applications widely spread in EU.

Through this ICT 2.0 Infrastructure citizens and other territorial bodies will support and reinforce ' efficient behaviour ' for citizens; will reinforce 'close cooperation between local communities, energy producers/suppliers and public authorities' with web based platform; will stimulate ' local energy business sector ' providing data streaming supporting project development and private opportunities for investment in a specific area of interest: public building EE.

The Pilot Application, even on a short time frame, will increase more the competences of Project Partners focussing the new ICT applications and opportunities for a better management of EE and RES process.

The platform will represent a virtual place where local administrations, researchers, citizens and SME can interact (2.0 approach) and exchange opportunities. This relevant aspect is deeply coherent with RENERGY purpose: to develop 'synergy from the very beginning between users, producers, businesses'

This effort represents also an operative contribution to the improvement of open-government framework. In fact it is a tool for generating collaboration and commitment on sustainable development policies also producing open-data (in real time) on effects of such policies. In fact Open-Government and Open-Data represent the two faces of the same coin.

If we consider Tim Berners-Lee [8] classification of open data. This Pilot Application will produce the highest level of rating. In fact it will generate five-star data contains also link to other data provided in other contexts ("open linked data") also including geographic dimension according with territorial planning needs

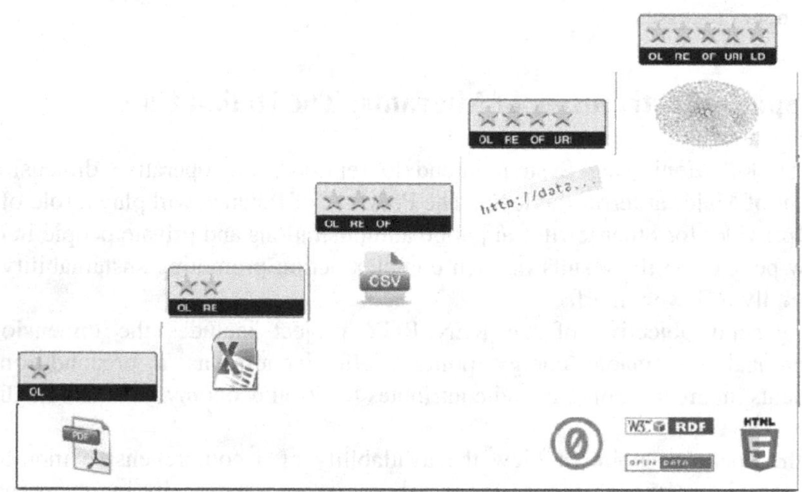

Fig. 1. Five Stars OPEN DATA from: http://5stardata.info/

4 Conclusions

This work presents a perspective for the improvement of sustainable energy policies and applications through innovative approaches integrating ICT tool and SMART procedures oriented to support community involvement processes.

This is a relevant change in traditional policy making approach in energy saving. In fact we experiences top-down applications, many times not coordinated in a strategic vision, oriented to face some punctual and specific interventions. It is the case of public building renovation or RES application on specific plants or purpose.

The Covenant of Mayor, through the operative tools od SEAP improved the local administration approach towards a planning approach which requested a detailed analysis of the current situation and a strategy to achieve the sustainability goals thought a list of possible interventions.

In this framework the active involvement of local communities rests in the background, while a bottom up approach based on people commitment could bring more effective results especially in the short term.

We look at the people commitment in sustainable use of energy as a key element for EU 20-20-20 strategy.

The Province of Potenza integrated such issue in the Provincial Master Plan and improved operatively such approach through the Pilot Application described in this paper from the strategic point of view.

It is possible to affirm that conscious local communities could promote specialization in planning and managing the process of growing sustainable model of energy production/use according whit specific context needs [9].

References

[1] Las Casas, G., Scorza, F.: Un approccio "context-based" e "valutazione integrata" per il futuro della programmazione operativa regionale in Europa. In: Bramanti, A., Salone, C. (eds.) Lo sviluppo territoriale nell'economia della conoscenza: teorie, attori strategie. Collana Scienze Regionali, vol. 41. FrancoAngeli, Milano (2009)

[2] Barca, F.: An agenda for a reformed cohesion policy: a place-based approach to meeting European union challenges and expectations. Independent report prepared at the request of the European Commissioner for Regional Policy, Danuta Hübner, European Commission, Brussels (2009)

[3] COM, 264 final, European Commision (2001)

[4] COM, 658 final, European Commision (2005)

[5] COM, 772 final, European Commision (2008)

[6] COM, 639 final, European Commision (2010)

[7] Covenant of Mayors, European Commission (2013)

[8] Tim Burner Lee (2009), http://www.w3.org/

[9] Scorza, F.: Improving EU cohesion policy: the spatial distribution analysis of regional development investments funded by EU structural funds 2007/2013 in Italy. In: Murgante, B., Misra, S., Carlini, M., Torre, C.M., Nguyen, H.-Q., Taniar, D., Apduhan, B.O., Gervasi, O. (eds.) ICCSA 2013, Part III. LNCS, vol. 7973, pp. 582–593. Springer, Heidelberg (2013)

Author Index

Agryzkov, Taras 330
Aliano, Giovanna Alessia 677
Almeida, Regina 253
Alvelos, Filipe 211, 237, 290
Alves, Cláudio 180
Amato, Federico 758
Antunes, Carlos Henggeler 92
Aparicio, Guillermo 115
Attolico, Alessandro 787

Balun, Marek 394
Barreto, João Pedro 92
Battino, Silvia 629
Bédard, Yvan 774
Bellarosa, Sara 723
Bencardino, Massimiliano 579
Bimonte, Sandro 774
Bollini, Letizia 642, 652
Borruso, Giuseppe 629, 738
Bourchtein, Andrei 437
Bourchtein, Ludmila 437
Brásio, Ana S.R. 169
Brega, José Remo Ferreira 342

Campagna, Michele 598
Casado, Leocadio G. 104, 115
Cataldo, Antonia 693
Cho, Gi-hwan 315
Cho, Youngsong 381
Claassen, G.D.H. 47
Clautiaux, François 180
Costa, M. Fernanda P. 140, 227
Cutini, Valerio 502

da Fonseca, Fernando Pereira 611
da Silva, Antônio Nélson Rodrigues 611
de Carvalho, José Manuel Valério 180, 211
De Mare, Gianluigi 547, 563
de Moura, Carlos A. 352
de Paiva Oliveira, Alcione 532
De Palma, Rinaldo 652
de Sousa Câmara, Jean Henrique 532
de Souza, Wagner Dias 532
Dias, Diego Roberto Colombo 342
Dias, Joana M. 1, 17, 60, 278
Di Pinto, Valerio 693
do Carmo Lopes, Maria 1, 17, 278
Donato, Carlo 629
Donofrio, Domenico 787
Dourado, Marco 195

Edoh-Alove, Elodie 774

Farah, Jihad 469
Fernandes, Edite M.G.P. 126, 140
Fernandes, Florbela P. 140
Fernandes, Natércia C.P. 169
Ferreira, Brígida 1, 17, 278
Filho, Jarbas Nunes Vidal 532
Fiore, Pierfrancesco 563
Florentino, Pablo Vieira 486
Formosa, Saviour 709
Francisco, Rogério B. 227

García, Inmaculada 104, 115
Gideon 315
Gnecco, Bruno Barberi 342
Graziadei, Antonio 723
Greco, Ilaria 579
Greene, Eugene 368
G-Tóth, Boglárka 115
Guimarães, Marcelo de Paiva 342
Guitouni, Adel 267

Hendrix, Eligius M.T. 47, 104, 115
Henriques, Roberto 404
Herrera, Juan F.R. 104

Karmacharya, Ashish 453
Kim, Deok-Soo 381
Kim, Donguk 381
Kim, Jae-Kwan 381
Klimentova, Xenia 237
Krichen, Saoussen 267

Laguardia, Giuseppe 787
Las Casas, Giuseppe 750
Lazzari, Maurizio 663, 677
Lecci, Agostino 663

Lecci, Nicola 663
Lee, Yuan-Shin 381
Lisboa-Filho, Jugurta 532
Lopes, Henrique 290
Lopes, Manuel 290
Lorusso, Sara 723

Marques, Maria do Céu 60
Masri, Hela 267
Mazzeo, Giuseppe 520
Meireles, José 195
Monteiro, Vasco 404
Moretti, Vincenzo 787
Mukhopadhyay, Asish 368
Müller, Hartmut 453
Murgante, Beniamino 723, 738, 758

Naoumova, Natalia 437
Nesticò, Antonio 547, 563
Netek, Rostislav 394
Nota, Rossella 652

Oliver, José L. 330
Oulamara, Ammar 76

Painho, Marco 404
Papa, João Paulo 342
Papa, Rocco 520
Passannante, Pasquale 723
Patriziano, Maria Serena 677
Pedroso, João Pedro 32
Peixoto, Aruquia 352
Pereira, Ana I. 154
Pereira, Gilberto Corso 486
Pietra, Riccardo 652
Pipolo, Ornella 563
Pontrandolfi, Piergiuseppe 758

Ramadas, Gisela C.V. 126
Ramos, Rui A.R. 611
Raposo, Carolina 92
Rietz, Jürgen 180
Rinaldi, Antonio M. 693
Riva Sanseverino, Eleonora 420
Riva Sanseverino, Raffaella 420
Rocha, Ana Maria A.C. 126, 195, 227, 278
Rocha, Humberto 1, 17, 278
Rocha, Maria Célia Furtado 486
Romanenko, Andrey 169
Rufino, José 154

Sanwar Hosen, A.S.M. 315
Sassano, Gerardo 723
Sassi, Ons 76
Scaccianoce, Gianluca 420
Scaringi, Francesco 723
Scioscia, Michele 723
Scorza, Francesco 750, 787
Silva, Elsa 211
Skala, Vaclav 301
Smaldone, Rosalia 787
Smolik, Michal 301

Teixeira, Ana Paula 253
Tortosa, Leandro 330
Trevelin, Luis Carlos 342

Vaccaro, Valentina 420
Vaz, Eric 404
Viana, Ana 237
Vicent, José F. 330

Würriehausen, Falk 453

Printed in the United States
By Bookmasters